AQA A Level Maths

Year 1 / AS Level

2018 UPDATED Large Data Set Car Data

BRIDGING EDITION

Series Editor

David Baker

Authors

Katie Wood, Brian Jefferson, David Bowles, Eddie Mullan, Garry Wiseman, John Rayneau, Mike Heylings, Paul Williams, Rob Wagner

OXFORD UNIVERSITY PRESS

OXFORD
UNIVERSITY PRESS

Great Clarendon Street, Oxford, OX2 6DP, United Kingdom

Oxford University Press is a department of the University of Oxford.

It furthers the University's objective of excellence in research, scholarship, and education by publishing worldwide. Oxford is a registered trade mark of Oxford University Press in the UK and in certain other countries.

First published in 2018

British Library Cataloguing in Publication Data
Data available

978-0-19-843642-3

10 9 8 7 6 5 4 3

Paper used in the production of this book is a natural, recyclable product made from wood grown in sustainable forests.
The manufacturing process conforms to the environmental regulations of the country of origin.

Printed and bound by CPI Group (UK) Ltd, Croydon, CR0 4YY

Acknowledgements

Series Editor: David Baker

Authors
Katie Wood, Rob Wagner, David Bowles, Brian Jefferson, Eddie Mullan, Garry Wiseman, John Rayneau, Mike Heylings, Paul Williams

Editorial team
Dom Holdsworth, Ian Knowles, Matteo Orsini Jones, Felicity Ounsted, Sarah Dutton

With thanks also to Phil Gallagher, Anna Cox, Katherine Bird, Keith Gallick, Linnet Bruce and Amy Ekins-Coward for their contribution.

Although we have made every effort to trace and contact all copyright holders before publication, this has not been possible in all cases. If notified, the publisher will rectify any errors or omissions at the earliest opportunity.

p5, **p180**, **p261**, **p297** iStock; **p68**, **p106**, **p109**, **p134**, **p137**, **p180**, **p210**, **p219**, **p232**, **p258**, **p288**, **p334**, **p339**, **p352**, **p355**, **p372** Shutterstock; **p71** Eric Krouse/ Dreamstime; **p210** George Bernard/Science Photo Library; **p232** Hulton Archive/Stringer/Getty Images; **p235** Ian Keirle/Dreamstime; **p372** Science Photo Library

Message from AQA

This student book has been approved by AQA for use with our qualification. This means that we have checked that it broadly covers the specification and we are satisfied with the overall quality. We have not, however, reviewed the MyMaths and InvisiPen links, and have therefore not approved this content.

We approve books because we know how important it is for teachers and students to have the right resources to support their teaching and learning. However, the publisher is ultimately responsible for the editorial control and quality of this book.

Please note that mark allocations given in assessment questions are to be used as guidelines only: AQA have not reviewed or approved these marks. Please also note that when teaching the AQA A Level Maths course, you must refer to AQA's specification as your definitive source of information. While the book has been written to match the specification, it cannot provide complete coverage of every aspect of the course.

Full details of our approval process can be found on our website: www.aqa.org.uk

Contents

About this book 1

Chapter 1: Algebra 1
Introduction 5
Bridging Unit 1
Topic A: Indices and surds 6
Topic B: Solving linear equations and rearranging formulae 10
Topic C: Factorising quadratics and simple cubics 13
Topic D: Completing the square 16
Topic E: The quadratic formula 17
Topic F: Line graphs 20
Topic G: Circles 27
1.1 Argument and proof 32
1.2 Index laws 36
1.3 Surds 40
1.4 Quadratic functions 44
1.5 Simultaneous equations 50
1.6 Lines and circles 54
1.7 Inequalities 60
Summary and review 66
Exploration 68
Assessment 69

Chapter 2: Polynomials and the binomial theorem
Introduction 71
Bridging Unit 2
Topic A: Expanding brackets 72
Topic B: Algebraic division 73
Topic C: Cubic, quartic and reciprocal graphs 77
2.1 Expanding and factorising 84
2.2 The binomial theorem 88
2.3 Algebraic division 94
2.4 Curve sketching 98
Summary and review 104
Exploration 106
Assessment 107

Chapter 3: Trigonometry
Introduction 109
Bridging Unit 3
Topic A: Trigonometry 1 110
Topic B: Trigonometry 2 118
3.1 Sine, cosine and tangent 122
3.2 The sine and cosine rules 128
Summary and review 132
Exploration 134
Assessment 135

Chapter 4: Differentiation and integration
Introduction 137
Bridging Unit 4
Topic A: Coordinate geometry 138
4.1 Differentiation from first principles 142
4.2 Differentiating ax^n and Leibniz notation 146
4.3 Rates of change 150
4.4 Tangents and normals 156
4.5 Turning points 162
4.6 Integration 168
4.7 Area under a curve 172
Summary and review 176
Exploration 180
Assessment 181

Chapter 5: Exponentials and logarithms
Introduction 185
Bridging Unit 5
Topic A: Exponentials and logarithms 186
5.1 The laws of logarithms 190
5.2 Exponential functions 194
5.3 Exponential processes 200
5.4 Curve fitting 204
Summary and review 208
Exploration 210
Assessment 211

Assessment, chapters 1–5: Pure 213

Chapter 6: Vectors
Introduction 219
6.1 Definitions and properties 220
6.2 Components of a vector 224
Summary and review 230
Exploration 232
Assessment 233

Chapter 7: Units and kinematics
Introduction 235
Bridging Unit 7
Topic A: Kinematics 236

7.1 Standard units and basic dimensions 240
7.2 Motion in a straight line - definitions and graphs 242
7.3 Equations of motion for constant acceleration 248
7.4 Motion with variable acceleration 252
Summary and review 256
Exploration 258
Assessment 259

Chapter 8: Forces and Newton's laws
Introduction 261
Bridging Unit 8
Topic A: Forces and Newton's laws 262
8.1 Forces 1 266
8.2 Dynamics 1 272
8.3 Motion under gravity 276
8.4 Systems of forces 280
Summary and review 286
Exploration 288
Assessment 289

Assessment, chapters 6–8: Mechanics 293

Chapter 9: Collecting, representing and interpreting data
Introduction 297
Bridging Unit 9
Topic A: Averages 298
Topic B: Measures of spread 302
Topic C: Histograms 306
9.1 Sampling 308
9.2 Central tendency and spread 312
9.3 Single-variable data 320
9.4 Bivariate data 328
Summary and review 332
Exploration 334
Assessment 335

Chapter 10: Probability and discrete random variables
Introduction 339
10.1 Probability 340
10.2 Binomial distribution 346
Summary and review 350
Exploration 352
Assessment 353

Chapter 11: Hypothesis testing 1
Introduction 355
Bridging Unit 11
Topic A: Hypothesis testing 356
11.1 Formulating a test 360
11.2 The critical region 366
Summary and review 370
Exploration 372
Assessment 373

Assessment, chapters 9–11: Statistics 375

Mathematical formulae 379

Mathematical notation 382

Answers 386

Index 447

About this book

This book has been specifically created for those studying the AQA 2017 Mathematics AS and A Levels. It's been written by a team of experienced authors and teachers, and it's packed with questions, explanation and extra features to help you get the most out of your course.

To give purpose and context to what you're about to learn, every chapter starts with an exciting, and often unexpected, example of how the maths is used in real life.

On the chapter **Introduction page**, the **Orientation box** explains what you should already know, what you will learn in this chapter, and what this leads to.

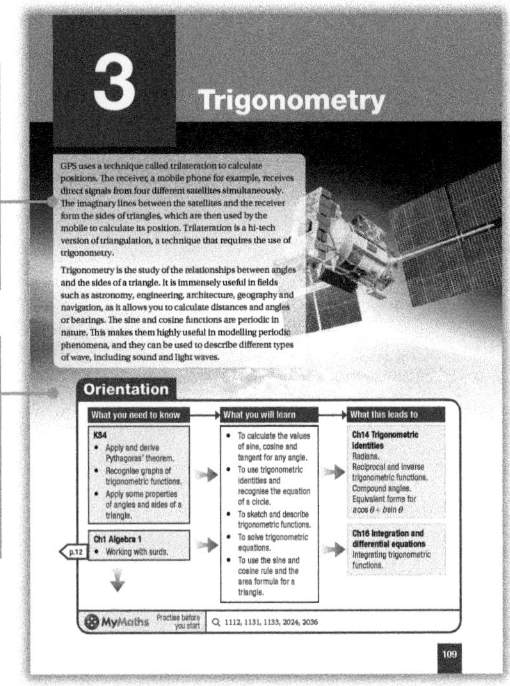

3 Trigonometry

GPS uses a technique called trilateration to calculate positions. The receiver, a mobile phone for example, receives direct signals from four different satellites simultaneously. The imaginary lines between the satellites and the receiver form the sides of triangles, which are then used by the mobile to calculate its position. Trilateration is a hi-tech version of triangulation, a technique that requires the use of trigonometry.

Trigonometry is the study of the relationships between angles and the sides of a triangle. It is immensely useful in fields such as astronomy, engineering, architecture, geography and navigation, as it allows you to calculate distances and angles or bearings. The sine and cosine functions are periodic in nature. This makes them highly useful in modelling periodic phenomena, and they can be used to describe different types of wave, including sound and light waves.

Orientation

What you need to know	What you will learn	What this leads to
KS4 • Apply and derive Pythagoras' theorem. • Recognise graphs of trigonometric functions. • Apply some properties of angles and sides of a triangle. Ch1 Algebra 1 (p.12) • Working with surds.	• To calculate the values of sine, cosine and tangent for any angle. • To use trigonometric identities and recognise the equation of a circle. • To sketch and describe trigonometric functions. • To solve trigonometric equations. • To use the sine and cosine rule and the area formula for a triangle.	Ch14 Trigonometric identities Radians. Reciprocal and inverse trigonometric functions. Compound angles. Equivalent forms for $a\cos\theta + b\sin\theta$ Ch16 Integration and differential equations Integrating trigonometric functions.

MyMaths Practise before you start 1112, 1131, 1133, 2024, 2036

109

The Bridging Units

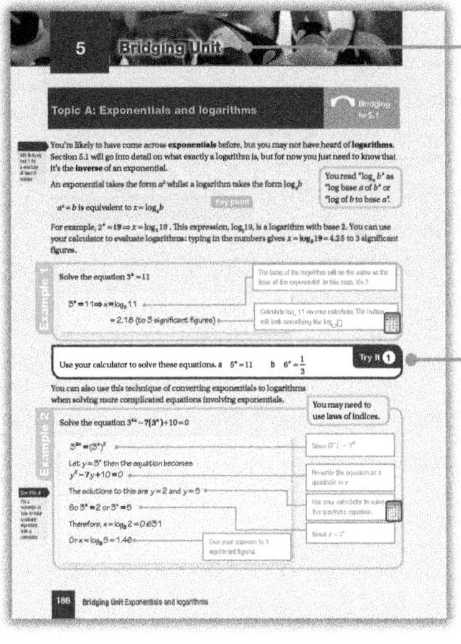

5 Bridging Unit

Topic A: Exponentials and logarithms — Bridging to 5.1

You're likely to have come across **exponentials** before, but you may not have heard of **logarithms**. Section 5.1 will go into detail on what exactly a logarithm is, but for now you just need to know that it's the **inverse** of an exponential.

An exponential takes the form a^x whilst a logarithm takes the form $\log_a b$

You read "$\log_a b$" as "log base a of b" or "log of b to base a".

$a^x = b$ is equivalent to $x = \log_a b$

For example, $2^x = 19 \Rightarrow x = \log_2 19$. This expression, $\log_2 19$, is a logarithm with base 2. You can use your calculator to evaluate logarithms: typing in the numbers gives $x = \log_2 19 = 4.25$ to 3 significant figures.

Example 1

Solve the equation $3^x = 11$

$3^x = 11 \Rightarrow x = \log_3 11$

$= 2.18$ (to 3 significant figures)

Try it 1

Use your calculator to solve these equations. a $5^x = 11$ b $6^x = \frac{1}{3}$

You can also use this technique of converting exponentials to logarithms when solving more complicated equations involving exponentials.

You may need to use laws of indices.

Example 2

Solve the equation $3^{2x} - 7(3^x) + 10 = 0$

$3^{2x} = (3^x)^2$

Let $y = 3^x$ then the equation becomes

$y^2 - 7y + 10 = 0$

The solutions to this are $y = 2$ and $y = 5$

So $3^x = 2$ or $3^x = 5$

Therefore, $x = \log_3 2 = 0.631$

Or $x = \log_3 5 = 1.46$

186 Bridging Unit Exponentials and logarithms

Bridging sections at the beginning of chapters 1, 2, 3, 4, 5, 7, 8, 9 and 11 help you to recall the GCSE skills you will need in the chapter, and start to use the skills you need at A Level.

Bridging sections pair **worked examples** with **Try it** questions, designed to help you practice the skills used in the example.

Bridging exercises provide practice in **Fluency and skills** at a level of challenge between GCSE and A level.

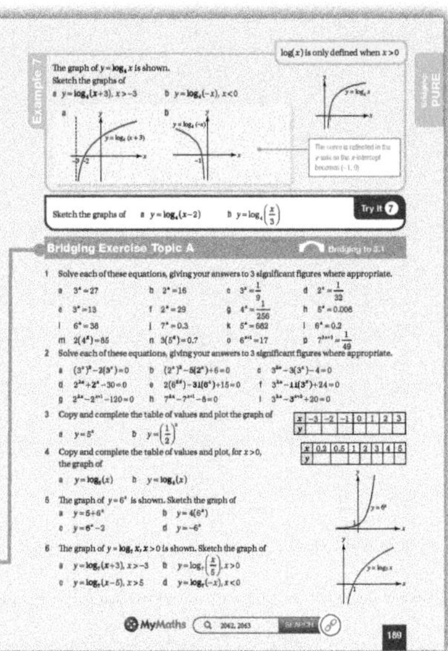

Example 7

The graph of $y = \log_4 x$ is shown.

Sketch the graphs of

a $y = \log_4(x+3)$, $x > -3$ b $y = \log_4(-x)$, $x < 0$

$\log(x)$ is only defined when $x > 0$

Try it 7

Sketch the graphs of a $y = \log_4(x-2)$ b $y = \log_4\left(\frac{x}{3}\right)$

Bridging Exercise Topic A — Bridging to 5.1

1 Solve each of these equations, giving your answers to 3 significant figures where appropriate.

a $3^x = 27$ b $2^x = 16$ c $3^x = \frac{1}{9}$ d $2^x = \frac{1}{32}$

e $3^x = 13$ f $2^x = 29$ g $4^x = \frac{1}{256}$ h $5^x = 0.008$

i $6^x = 36$ j $7^x = 0.3$ k $5^x = 662$ l $6^x = 0.2$

m $2(4^x) = 85$ n $3(5^x) = 0.7$ o $6^{x+1} = 17$ p $7^{3x+1} = \frac{1}{49}$

2 Solve each of these equations, giving your answers to 3 significant figures where appropriate.

3 Copy and complete the table of values and plot the graph of

a $y = 5^x$ b $y = \left(\frac{1}{2}\right)^x$

x	−3	−2	−1	0	1	2	3
y							

4 Copy and complete the table of values and plot, for $x > 0$, the graph of

x	0.2	0.5	1	2	3	4	5
y							

5 The graph of $y = 6^x$ is shown. Sketch the graph of

6 The graph of $y = \log_7 x$, $x > 0$ is shown. Sketch the graph of

MyMaths 2062, 2063

189

Every section in the main chapters starts by covering the basic **Fluency and skills**, then builds on these techniques by looking at **Reasoning and problem-solving**.

Strategy boxes help build problem-solving techniques.

7.1 Standard units and basic dimensions

Fluency and skills

All quantities in mechanics are defined in terms of three **fundamental quantities** or **dimensions**: mass, length and time. Quantities, or dimensions, are measured in units.

Some SI (Système International d'Unités) base units you'll have come across before are kilogram / kg (mass), metre / m (length), and second / s (time).

Kinematics is the study of motion. In kinematics, you will meet distance, displacement, speed, velocity and time. These are derived quantities that you can describe in terms of the fundamental quantities (mass, length and time).

Vector	Scalar	Fundamental quantities	SI Units
Displacement	Distance	length	metres (m)
Velocity	Speed	$\frac{\text{length}}{\text{time}}$	metres per second (m s^{-1} or m/s)
Acceleration		$\left(\frac{\text{velocity}}{\text{time}}\right) = \frac{\text{length}}{(\text{time})^2}$	metres per second squared (m s^{-2} or m/s^2)

Mechanics also involves the derived quantities force and weight.

Force = mass × acceleration. The SI unit is the newton (N). Key point

Weight is the force of gravity on an object. An object with mass m kg has weight mg N, where g is the acceleration due to gravity. On Earth, this is 9.81 m s^{-2} to 3 sf. If you were on the Moon, your mass would be the same but your weight would be less. In common speech, you might use mass and weight to mean the same thing, but make sure you don't do this in maths.

Correct formulae are dimensionally consistent. If, for example, $a = b + c$ and a is a velocity, then b and c also have the dimensions of a velocity. You must also use the same units throughout, and so you may need to convert some units before carrying out calculations for a formula to work.

Example 1

Express a speed of 15 km h^{-1} in m s^{-1}

$15\,\text{km h}^{-1} = 15\,000\,\text{m h}^{-1}$ — 1 km = 1000 m

$15\,000\,\text{m h}^{-1} = \frac{15\,000}{3600}\,\text{m s}^{-1} = 4\frac{1}{6}\,\text{m s}^{-1}$ — 1 h = 60 × 60 = 3600 s

Exercise 7.1A Fluency and skills

1 State the quantity described by these units.
 a Newtons, N b Kilograms, kg
 c Metres per second, m s^{-1}
 d Metres per second squared, m s^{-2}

2 Convert
 a 8.5 km to m b 2.3 m to mm
 c 482 cm to m d 1650 m to km
 e 72 km h^{-1} to m s^{-1} f 14 m s^{-1} to km h^{-1}
 g 25 cm s^{-1} to km h^{-1} h 2.4 m^2 to cm^2
 i 1.4 kg to g j 1.6 tonnes to kg

3 A car travels 70 km in 35 minutes. Evaluate its speed in
 a km h^{-1} b m s^{-1}

4 Work out the distance, in km, travelled in a quarter of an hour by a car that has a constant speed of 20 m s^{-1}

5 A particle has an acceleration of 200 km h^{-2}. Express this in m s^{-2}

6 The force F (in N) on an object is related to its mass m (in kg) and its acceleration a (in m s^{-2}) by $F = ma$. Work out, in kg, the mass of an object if a force of 0.25 kN (kilonewtons) accelerates it at 20 km min^{-2}

MECH

Reasoning and problem-solving

Strategy

When answering a question involving units

1. Convert units if they're inconsistent and perform any necessary calculations.
2. Check that dimensions have been conserved and that your final answer is in the correct units.

Example 2

u and v are velocities, a is acceleration and s is displacement. Use the formula $v^2 = u^2 + 2as$ to work out s if $u = 24\,\text{km h}^{-1}$, $v = 32\,\text{km h}^{-1}$ and $a = 0.005\,\text{m s}^{-2}$. Give your answer in kilometres.

$u = \frac{24 \times 1000}{3600} = 6.67\,\text{m s}^{-1}$ and $v = \frac{32 \times 1000}{3600} = 8.89\,\text{m s}^{-1}$ — (1) Convert velocities to m s^{-1}

$8.89^2 = 6.67^2 + 2 \times 0.005s$

$s = \frac{8.89^2 - 6.67^2}{0.01} = 3456.8\,\text{m}$ — (2) Substitute values and solve for s

$s = 3.46$ km (to 3 sf) — (3) Convert your answer to km.

Exercise 7.1B Reasoning and problem-solving

1 A runner travels 3900 m at 8 km h^{-1}. Find, in minutes, the time she takes.

2 In the formula $s = ut + \frac{1}{2}at^2$, u is velocity, a is acceleration, t is time and s is displacement. Find the value of s if $u = 4\,\text{km h}^{-1}$, $a = 0.01\,\text{m s}^{-2}$ and $t = 40$ minutes. Give your answer in km.

3 A station platform is 180 m long. A train of length 120 m passes it at 30 km h^{-1}. How long will it take for the train to pass completely?

Challenge

4 A liquid of density 1.2 g cm^{-3} is flowing at 3 km h^{-1} through a cylindrical pipe of radius 5 cm. Given that density = $\frac{\text{mass}}{\text{volume}}$, and that for a cylinder with radius r and height h its volume is given by $\pi r^2 h$, calculate the mass, in kg, of the liquid emerging from the pipe in 30 seconds.

MyMaths 2183 SEARCH

176 Units and kinematics Standard units and basic dimensions 177

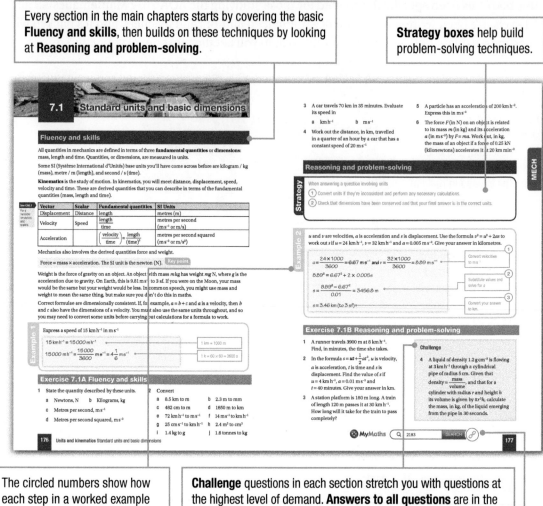

The circled numbers show how each step in a worked example is linked to the strategy box.

Challenge questions in each section stretch you with questions at the highest level of demand. **Answers to all questions** are in the back of this book, and **full solutions are available free** online.

Links to **MyMaths** provide a quick route to **extra support and practice**. Just log in and key the code into the search bar.

At the end of each chapter, a **What Next** box provides links to further support based on how well you've understood the content.

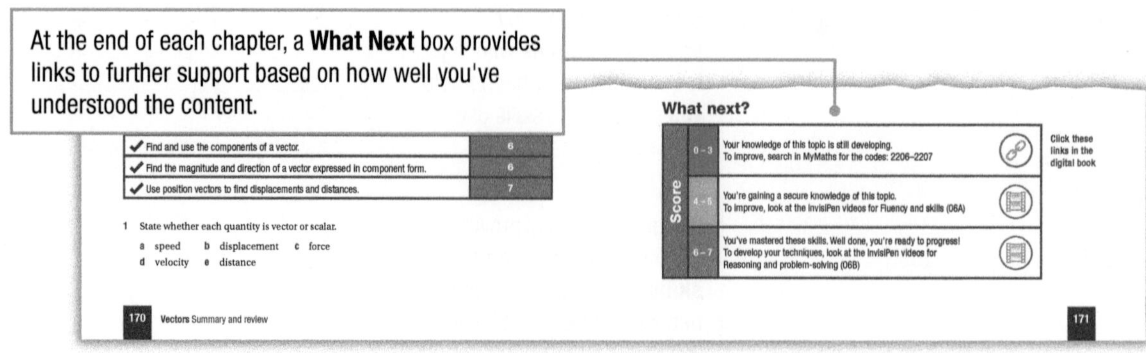

✔ Find and use the components of a vector. 6
✔ Find the magnitude and direction of a vector expressed in component form. 6
✔ Use position vectors to find displacements and distances. 7

1 State whether each quantity is vector or scalar.
 a speed b displacement c force
 d velocity e distance

What next?

Score	
0–3	Your knowledge of this topic is still developing. To improve, search in MyMaths for the codes: 2206–2207
4–5	You're gaining a secure knowledge of this topic. To improve, look at the InvisiPen videos for Fluency and skills (06A)
6–7	You've mastered these skills. Well done, you're ready to progress! To develop your techniques, look at the InvisiPen videos for Reasoning and problem-solving (06B)

Click these links in the digital book

170 Vectors Summary and review 171

Links to **ICT resources** on Kerboodle show how technology can be used to help understand the maths involved.

ICT Resource online

To investigate gradients of chords for a graph, click this link in the digital book.

Support for when and how to use **calculators** is available throughout this book, with links to further demonstrations in the **digital book**. Unless otherwise stated this book assumes that your calculator can do the required minimum according to specification guidelines. That is, it can perform an iterative function and it can compute summary statistics and access probabilities and value from standard statistical distributions.

Try it on your calculator

You can use a calculator to evaluate the gradient of the tangent to a curve at a given point.

d/dx(5X² - 2X, 3)

28

Activity

Find out how to calculate the gradient of the tangent to the curve $y = 5x^2 - 2x$ where $x = 3$ on *your* calculator.

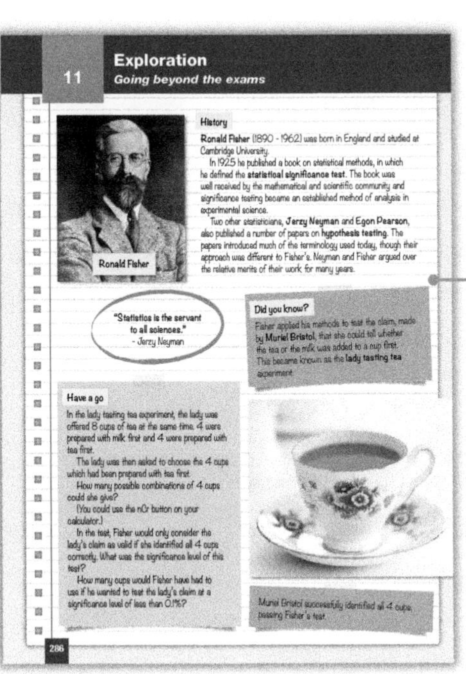

11 Exploration
Going beyond the exams

History

Ronald Fisher (1890 - 1962) was born in England and studied at Cambridge University.

In 1925 he published a book on statistical methods, in which he defined the **statistical significance test**. The book was well received by the mathematical and scientific community and significance testing became an established method of analysis in experimental science.

Two other statisticians, **Jerzy Neyman** and **Egon Pearson**, also published a number of papers on **hypothesis testing**. The papers introduced much of the terminology used today, though their approach was different to Fisher's. Neyman and Fisher argued over the relative merits of their work for many years.

Ronald Fisher

"Statistics is the servant to all sciences."
- Jerzy Neyman

Did you know?

Fisher applied his methods to test the claim, made by **Muriel Bristol**, that she could tell whether the tea or the milk was added to a cup first. This became known as the **lady tasting tea** experiment.

Have a go

In the lady tasting tea experiment, the lady was offered 8 cups of tea at the same time. 4 were prepared with milk first and 4 were prepared with tea first.

The lady was then asked to choose the 4 cups which had been prepared with tea first.

How many possible combinations of 4 cups could she give?

(You could use the nCr button on your calculator.)

In the test, Fisher would only consider the lady's claim as valid if she identified all 4 cups correctly. What was the significance level of this test?

How many cups would Fisher have had to use if he wanted to test the lady's claim at a significance level of less than 0.1%?

Muriel Bristol successfully identified all 4 cups, passing Fisher's test.

286

At the end of every chapter, an **Exploration page** gives you an opportunity to explore the subject beyond the specification.

Assessment sections at the end of each chapter test everything covered within that chapter. Further **synoptic assessments** covering Pure, Mechanics and Statistics can be found at the end of chapters 5, 8 and 11*

Dedicated questions throughout the statistics chapters will familiarise you with the **Large data set.** These questions have been updated to support AQA's 2018 data set (car data).

9 Assessment

1 The estimated mean of the data in the table is 11

x	$0 \le x < 4$	$4 \le x < 8$	$8 \le x < 12$	$12 \le x < 16$	$16 \le x < 20$
Frequency	5	2	13	a	8

Calculate the value of the missing frequency, a. Select the correct answer.

A 0.857 **B** 4 **C** 12 **D** 0.182 **[1 mark]**

2 The maximum temperature, t, was recorded every day one year in July.

You are given $\sum t = 686$ and $\sum t^2 = 15598$

6 The large data set contains data on 20 BMW, petrol fuelled, convertible cars registered in London. The table shows a copy of the measured values of hydrocarbon emissions, in g km^{-1}, for the cars.

0.040	0.048	0.035	0.086	0.120	0.120	0.125	0.035	0.097	0.062
0.040	0.035	0.125	0.048	0.038	0.034	0.020	0.172	0.079	0.037

a Calculate **i** the mean, $\bar{x}$ **ii** the sample standard deviation, s [2]

b An outlier is defined as any value greater than $\bar{x} + 2s$ or less than $\bar{x} - 2s$
Identify any outliers in the data set. [2]

c Find the median of the data set. [1]

d Explain whether the mean, median or mode is a better representative of a typical value for this data set. [2]

*Please note that mark allocations given in assessment questions are to be used as guidelines only: AQA have not reviewed or approved these marks.

1 Algebra 1

Supply and demand is a well-known example of how maths helps us model real situations that occur in the world. Economists and business analysts use simultaneous equations to model how changes in price will affect both the supply of, and demand for, a particular product. This allows them to forecast the optimum price–the price at which both supply and demand are optimised.

Algebra is a branch of maths that includes simultaneous equations, along with many other topics such as inequalities, surds and polynomial functions. Algebra is used to model real world occurrences in fields such as economics, engineering and the sciences, and so an understanding of algebra is important in a wide range of different situations.

Orientation

What you need to know	What you will learn	What this leads to
KS4 • Understand and use algebraic notation and vocabulary. • Simplify and manipulate algebraic expressions. • Rearrange formulae to change the subject. • Solve linear equations.	• To use direct proof, proof by exhaustion and counter examples. • To use and manipulate index laws. • To manipulate surds and rationalise a denominator. • To solve quadratic equations and sketch quadratic curves. • To understand and use coordinate geometry. • To understand and solve simultaneous equations. • To understand and solve inequalities.	**Ch12 Algebra 2** Functions. Parametric equations. Algebraic and partial fractions. **Ch17 Numerical Methods** Simple and iterative root finding. Newton-Raphson root finding.

MyMaths Practise before you start 1170, 1171, 1928, 1929

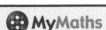

1 Bridging unit

Topic A: Indices and surds

Bridging to Ch1.2, 1.3

You can apply the rules of indices and surds to simplify algebraic expressions. The following expressions can be simplified in **index form**:

Key point

$x^a \times x^b = x^{a+b}$ $\quad$ $x^a \div x^b = x^{a-b}$ $\quad$ $(x^a)^b = x^{ab}$

Example 1

Simplify these expressions. **a** $2x^3 \times 3x^5$ **b** $12x^7 \div 4x^3$ **c** $(3x^5)^3$

a $2x^3 \times 3x^5 = 6x^{3+5}$

$= 6x^8$

Multiply the coefficients together and use $x^a \times x^b = x^{a+b}$

b $12x^7 \div 4x^6 = \dfrac{12x^7}{4x^6}$

$= 3x$

Since $\frac{12}{4} = 3$ and $x^a \div x^b = x^{a-b}$ so $\frac{x^7}{x^6} = x^1$ which we just write as x

c $(3x^5)^3 = 3^3(x^5)^3$

Both the 3 and the x^5 must be raised to the power 3

$= 27x^{15}$

Since $(x^a)^b = x^{ab}$

Try It 1

Simplify these expressions.

a $5x^3 \times 2x^7$ **b** $18x^9 \div 3x^2$ **c** $(2x^6)^4$ **d** $\left(\dfrac{x^3}{3}\right)^2$

Roots can also be expressed using indices, such that the square root of x is written as $\sqrt{x} = x^{\frac{1}{2}}$

In general:

Key point

The nth root of x is written $\sqrt[n]{x} = x^{\frac{1}{n}}$, and this can be raised to a power to give $\sqrt[n]{x^m} = x^{\frac{m}{n}}$

Key point

A power of -1 indicates a reciprocal, so $x^{-1} = \dfrac{1}{x}$ and, in general, $x^{-n} = \dfrac{1}{x^n}$

Example 2

Evaluate each of these without using a calculator.

a $25^{0.5}$ **b** 6^{-2} **c** $8^{\frac{2}{3}}$

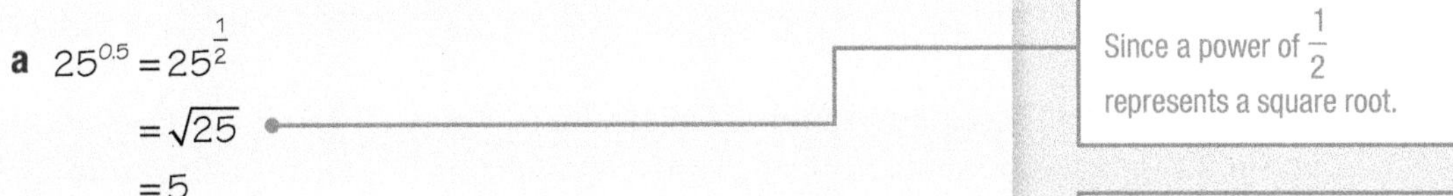

a $25^{0.5} = 25^{\frac{1}{2}}$

$= \sqrt{25}$ — Since a power of $\frac{1}{2}$ represents a square root.

$= 5$

b $6^{-2} = (6^2)^{-1}$

$= \frac{1}{6^2}$ — Since a power of −1 represents a reciprocal.

$= \frac{1}{36}$

c $8^{\frac{2}{3}} = \left(8^{\frac{1}{3}}\right)^2$ — Always calculate a root before a power.

$= 2^2$ — Since the cube root of 8 is 2

$= 4$

Try It 2

Evaluate each of these without a calculator.

a $36^{\frac{1}{2}}$ **b** $27^{\frac{2}{3}}$ **c** $64^{-0.5}$ **d** $\left(\frac{1}{2}\right)^4$

Example 3

Write these expressions in simplified index form.

a $\sqrt[3]{x}$ **b** $\frac{2}{x^3}$ **c** $\frac{2x}{\sqrt{x}}$

a $\sqrt[3]{x} = x^{\frac{1}{3}}$

b $\frac{2}{x^3} = 2x^{-3}$

c $\frac{2x}{\sqrt{x}} = \frac{2x}{x^{\frac{1}{2}}}$ — Since $\sqrt{x} = x^{\frac{1}{2}}$

$= 2x^{1-\frac{1}{2}}$ — Subtract the powers, remembering that $x = x^1$

$= 2x^{\frac{1}{2}}$

Try It 3

Write these expressions in simplified index form.

a $\sqrt[5]{x^2}$ **b** $\frac{3}{\sqrt{x}}$ **c** $\frac{3x^2}{\sqrt{x}}$ **d** $\frac{\sqrt{x}}{3x}$

A **surd** is an **irrational** number involving a root, for example $\sqrt{2}$ or $\sqrt[3]{7}$.
You can multiply and divide surds using the rules:

Key point

$$\sqrt{a}\times\sqrt{b}=\sqrt{ab} \text{ and } \frac{\sqrt{a}}{\sqrt{b}}=\sqrt{\frac{a}{b}}$$

An irrational number is a real number that cannot be written as a fraction $\frac{a}{b}$, where a and b are integers with $b\neq 0$

You can simplify surds by finding square-number factors, for example $\sqrt{12}=\sqrt{4}\sqrt{3}=2\sqrt{3}$. It may also be possible to simplify expressions involving surds by collecting like terms or by **rationalising the denominator**. Rationalising the denominator means rearranging the expression to remove any roots from the denominator.

Key point

To rationalise the denominator, multiply both the numerator and denominator by a suitable expression:

$\frac{1}{\sqrt{a}}\times\frac{\sqrt{a}}{\sqrt{a}}=\frac{\sqrt{a}}{a}$ (multiply numerator and denominator by $\sqrt{a}$)

$\frac{1}{a+\sqrt{b}}\times\frac{a-\sqrt{b}}{a-\sqrt{b}}=\frac{a-\sqrt{b}}{a^2-b}$ (multiply numerator and denominator by $a-\sqrt{b}$)

$\frac{1}{a-\sqrt{b}}\times\frac{a+\sqrt{b}}{a+\sqrt{b}}=\frac{a+\sqrt{b}}{a^2-b}$ (multiply numerator and denominator by $a+\sqrt{b}$)

Example 4

Simplify these expressions without using a calculator.

a $\sqrt{18}+5\sqrt{2}$ **b** $\frac{6}{\sqrt{3}}$ **c** $\frac{2}{1-\sqrt{5}}$

a $\sqrt{18}=\sqrt{9}\sqrt{2}$
$=3\sqrt{2}$ — 9 is a square-number factor of 18 so you can simplify $\sqrt{18}$

Therefore $\sqrt{18}+5\sqrt{2}=3\sqrt{2}+5\sqrt{2}$
$=8\sqrt{2}$ — Collect like terms.

b $\frac{6}{\sqrt{3}}=\frac{6\sqrt{3}}{\sqrt{3}\sqrt{3}}$ — Rationalise the denominator by multiplying numerator and denominator by $\sqrt{3}$
$=\frac{6\sqrt{3}}{3}$
$=2\sqrt{3}$ — Since $6\div 3=2$

c $\frac{2}{1-\sqrt{5}}=\frac{2(1+\sqrt{5})}{(1-\sqrt{5})(1+\sqrt{5})}$ — Rationalise the denominator by multiplying numerator and denominator by $1+\sqrt{5}$
$=\frac{2(1+\sqrt{5})}{-4}$ — $(1-\sqrt{5})(1+\sqrt{5})=1-\sqrt{5}+\sqrt{5}-5$
$=1-5=-4$
$=-\frac{1}{2}(1+\sqrt{5})$

Try It 4

Simplify these expressions without using a calculator.

a $3\sqrt{28}-\sqrt{7}$ **b** $\frac{4}{\sqrt{3}}$ **c** $\frac{3}{1+\sqrt{2}}$ **d** $\frac{\sqrt{5}}{\sqrt{5}-2}$

Bridging Exercise Topic A

 Bridging to Ch1.2, 1.3

1 Evaluate each of these without using a calculator.

a $49^{\frac{1}{2}}$ b $27^{\frac{1}{3}}$ c 5^{-1} d $64^{-\frac{1}{3}}$

e $9^{\frac{3}{2}}$ f $16^{\frac{3}{4}}$ g $125^{-\frac{2}{3}}$ h $\left(\frac{1}{2}\right)^3$

i $\left(\frac{1}{9}\right)^{-2}$ j $\left(\frac{4}{9}\right)^{\frac{1}{2}}$ k $\left(\frac{9}{16}\right)^{-0.5}$ l $\left(\frac{27}{8}\right)^{-\frac{2}{3}}$

2 Simplify these expressions fully without using a calculator.

a $\sqrt{8}$ b $\sqrt{75}$ c $2\sqrt{24}$ d $3\sqrt{48}$

e $\sqrt{20}+\sqrt{5}$ f $\sqrt{27}-\sqrt{12}$ g $5\sqrt{32}-3\sqrt{8}$ h $\sqrt{50}+3\sqrt{125}$

i $\sqrt{68}+3\sqrt{17}$ j $3\sqrt{72}-\sqrt{32}$ k $4\sqrt{18}-2\sqrt{3}$ l $6\sqrt{5}+\sqrt{50}$

3 Simplify these expressions fully without using a calculator.

a $\frac{1}{\sqrt{7}}$ b $\frac{2}{\sqrt{8}}$ c $\frac{12}{\sqrt{3}}$ d $\frac{\sqrt{8}}{\sqrt{12}}$

e $\frac{1}{1+\sqrt{3}}$ f $\frac{2}{1+\sqrt{2}}$ g $\frac{8}{1-\sqrt{5}}$ h $\frac{2}{\sqrt{5}-1}$

i $\frac{\sqrt{2}}{2+\sqrt{3}}$ j $\frac{2\sqrt{3}}{\sqrt{6}-2}$ k $\frac{1+\sqrt{2}}{1-\sqrt{2}}$ l $\frac{3+\sqrt{5}}{\sqrt{5}-3}$

4 Expand the brackets and fully simplify each expression.

a $(1+\sqrt{2})(3+\sqrt{2})$ b $(1+\sqrt{2})(3-\sqrt{2})$ c $(1-\sqrt{2})(3+\sqrt{2})$ d $(1-\sqrt{2})(3-\sqrt{2})$

e $(\sqrt{3}+2)(4+\sqrt{3})$ f $(\sqrt{3}+2)(4-\sqrt{3})$ g $(\sqrt{3}-2)(4+\sqrt{3})$ h $(\sqrt{3}-2)(4-\sqrt{3})$

i $(\sqrt{6}+1)(\sqrt{2}+3)$ j $(\sqrt{6}+1)(\sqrt{2}-3)$ k $(\sqrt{6}-1)(\sqrt{2}+3)$ l $(\sqrt{6}-1)(\sqrt{2}-3)$

5 Write each of these expressions in simplified index form.

a $x^3\times x^7$ b $7x^5\times 3x^6$ c $5x^4\times 8x^7$ d $x^8\div x^2$

e $8x^7\div 2x^9$ f $3x^8\div 12x^7$ g $(x^5)^7$ h $(x^2)^{-5}$

i $(3x^2)^4$ j $(6x^5)^2$ k $\sqrt{x^3}$ l $\sqrt[4]{x^5}$

m $\frac{5\sqrt{x}}{x}$ n $2x\sqrt{x}$ o $\frac{x^2}{3\sqrt{x}}$ p $x^3(x^5-1)$

q $x^3(\sqrt{x}+2)$ r $\frac{x+2}{x^3}$ s $\frac{\sqrt{x}+3}{x}$ t $\frac{(3-x^3)}{\sqrt{x}}$

u $(\sqrt{x}+3)^2$ v $\frac{3+\sqrt{x}}{x^2}$ w $\frac{1-x}{2\sqrt{x}}$ x $\frac{\sqrt{x}+2}{3x^3}$

MyMaths 1033, 1045, 1064, 1065, 1951 SEARCH

Topic B: Solving linear equations and rearranging formulae

This topic recaps the **balance** method to solve problems involving linear equations, and both the **elimination** and **substitution** methods to solve linear simultaneous equations.

You can solve linear equations and inequalities using the **balance** method where the same operation is applied to both sides.

Example 1

Solve the equation $7x-5=3x-2$

$4x-5=-2$ — Subtract $3x$ from both sides of the equation.

$4x=3$ — Add 5 to both sides of the equation.

$x=\frac{3}{4}$ — Divide both sides of the equation by 4

Try It 1

Solve the equation $3x+8=5x-6$

Example 2

Solve the inequality $5(x-2)\leq 2x+1$

$5x-10\leq 2x+1$ — First expand the brackets.

$3x-10\leq 1$ — Subtract $2x$ from both sides.

$3x\leq 11$ — Add 10 to both sides.

$x\leq\frac{11}{3}$ — Divide both sides by 3

Try It 2

Solve the inequality $7x-4>x+8$

When solving inequalities, remember that multiplying or dividing by a negative number will reverse the inequality sign. For example, $5>3$ but $-5<-3$

Equations and formulae can be rearranged using the same method as for solving equations.

Example 3

Rearrange $Ax-3=\frac{x+B}{2}$ to make x the subject.

$2Ax-6=x+B$ — Multiply both sides by 2

$2Ax-6-x=B$ — Subtract x from both sides.

$2Ax-x=B+6$ — Add 6 to both sides.

$x(2A-1)=B+6$ — Factorise the side involving x

$x=\frac{B+6}{2A-1}$ — Divide both sides by $(2A-1)$ to make x the subject.

Bridging: PURE

Try It 3

Rearrange $3(x+A)=Bx+1$ to make x the subject.

You can solve linear simultaneous equations using the **elimination** method, as shown in Example 4. The solutions to simultaneous equations give the point of intersection between the lines represented by the two equations.

Example 4

Solve the simultaneous equations $5x-4y=17$, $3x+8y=5$

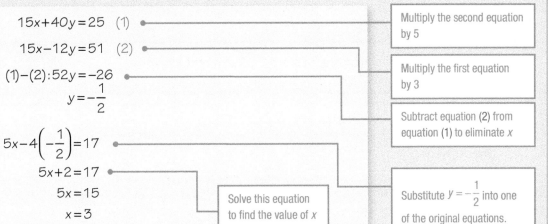

$15x+40y=25 \quad (1)$ — Multiply the second equation by 5

$15x-12y=51 \quad (2)$ — Multiply the first equation by 3

$(1)-(2): 52y=-26$ — Subtract equation (2) from equation (1) to eliminate x

$y=-\frac{1}{2}$

$5x-4\left(-\frac{1}{2}\right)=17$ — Substitute $y=-\frac{1}{2}$ into one of the original equations.

$5x+2=17$ — Solve this equation to find the value of x

$5x=15$

$x=3$

Try It 4

Solve the simultaneous equations $2x+5y=1, 3x-2y=-27$

Calculator

Try it on your calculator

You can use a calculator to solve linear simultaneous equations.

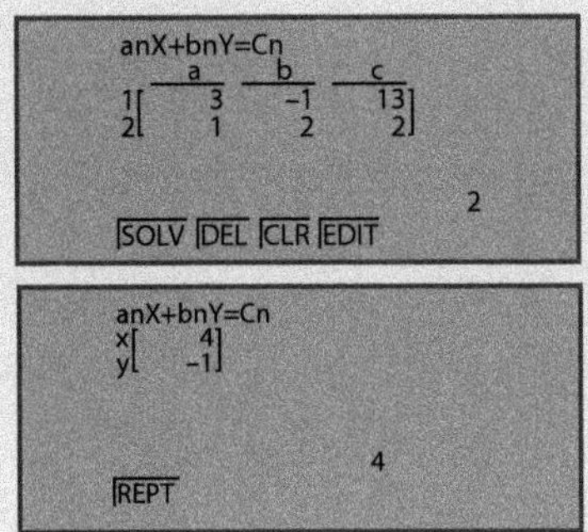

Activity

Find out how to solve the simultaneous equations $3x-y=13$ and $x+2y=2$ on *your* calculator.

The example shows you that the lines $5x-4y=17$ and $3x+8y=5$ intersect at the point $\left(3,-\frac{1}{2}\right)$

If you are given the equation of two lines where y is the subject then the easiest way to solve these simultaneously is to use the **substitution** method as shown in the next example.

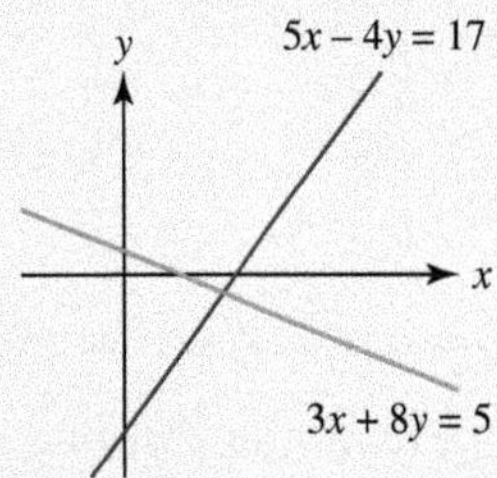

Example 5

Find the point of intersection between the lines with equations $y=2x+5$ and $y=7-3x$

$2x+5=7-3x$ — Substitute $2x+5$ for y in the second equation.

$5x+5=7$

$5x=2$

$x=0.4$ — Solve to find the value of x

$y=2(0.4)+5$

$=5.8$ — Substitute $x=0.4$ into either of the original equations to find the y-coordinate.

So the lines intersect at the point (0.4, 5.8)

Find the point of intersection between the lines $y=3x+4$ and $y=6x-2$ **Try It 5**

Calculator

Try it on your calculator

You can use a graphics calculator to find the point of intersection of two lines.

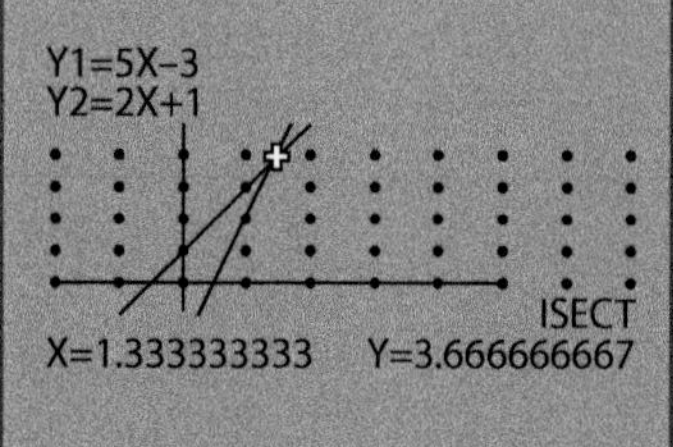

Activity

Find the point of intersection of the lines $y=5x-3$ and $y=2x+1$ on *your* graphics calculator.

Bridging Exercise Topic B

Bridging to Ch1.4

1 Solve each of these linear equations.

a $3(2x+9)=7$ b $7-3x=12$ c $\frac{x+4}{5}=7$ d $2x+7=5x-6$

e $8x-3=2(3x+1)$ f $\frac{2x+9}{12}=x-1$ g $2(3x-7)=4x$ h $7-2x=3(4-5x)$

2 Solve each of these linear inequalities.

a $\frac{x}{2}+7\geq 5$ b $3-4x<15$ c $5(x-1)>12+x$ d $\frac{x+1}{3}>2$

e $8x-1\leq 2x-5$ f $3(x+1)\geq\frac{x-3}{2}$ g $3(2x-5)<1-x$ h $x-(3+2x)\geq 2(x+1)$

3 Rearrange each of these formulae to make x the subject.

a $2x+5=3A-1$ b $x+u=vx+3$ c $\frac{3x-1}{k}=2x$ d $5(x-3m)=2nx-4$

e $(1-3x)^2=t$ f $\frac{1}{x}=\frac{1}{p}+\frac{1}{q}$ g $\frac{1}{x^2+k}-6=4$ h $\sqrt{x+A}=2B$

4 Use algebra to solve each of these pairs of simultaneous equations.

a $5x+12y=-6,\ x+5y=4$ b $7x+5y=14,\ 3x+4y=19$ c $2x-5y=4,\ 3x-8y=5$

d $3x-2y=2,\ 8x+3y=4.5$ e $5x-2y=11,\ -2x+3y=22$ f $8x+5y=-0.5,\ -6x+4y=-3.5$

5 Use algebra to find the point of intersection between each pair of lines.

a $y=8-3x,\ y=2-5x$ b $y=7x-4,\ y=3x-2$ c $y=2x+3,\ y=5-x$

d $y+5=3x,\ y=-5x+7$ e $y=\frac{1}{2}x+3,\ y=5-2x$ f $y=3(x+2),\ y=7-2x$

Topic C: Factorising quadratics and simple cubics

Expressions such as $5x^2+x$, $2x^2+4$ and x^2+2x-1 are called **quadratics** and can sometimes be factorised into two linear factors. There are three types of quadratics to consider:

1. Quadratics of the form ax^2+bx have a common factor of x so can be factorised using a single bracket and removing the highest common factor of the two terms, e.g. $6x^2+8x=2x(3x+4)$
2. Quadratics of the form x^2+bx+c will sometimes factorise into two sets of brackets. You need to find two constants with a product of c and a sum of b, e.g. $x^2-3x+2=(x-2)(x-1)$ since $-2\times-1=2$ and $-2+-1=-3$
3. Quadratics of the form ax^2-c will factorise if a and c are square numbers. This is called the **difference of two squares**, e.g. $4x^2-9=(2x+9)(2x-9)$

Example 1

Factorise each of these quadratics.

a $9x^2+15x$ **b** $x^2+3x-10$ **c** x^2-16

a $9x^2+15x=3x(3x+5)$ — The highest common factor of $9x^2$ and $15x$ is $3x$

b $x^2+3x-10=(x+5)(x-2)$ — You need to find two constants with a product of -10 and a sum of 3: $5\times-2=-10$ and $5+-2=3$ so the constants are -2 and 5

c $x^2-16=(x+4)(x-4)$ — x^2 and 16 are both square numbers.

Try It 1

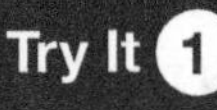

Factorise each of these quadratics.

a $14x^2-7x$ **b** x^2-5x+4 **c** x^2-25

When factorising quadratics of the form ax^2+bx+c with $a\neq1$, first split the bx term into two terms where the coefficients multiply to give the same value as $a\times c$

Example 2

Factorise each of these quadratics.

a $3x^2+11x+6$ **b** $2x^2-9x+10$

a $3x^2+11x+6=3x^2+9x+2x+6$ — Split $11x$ into $9x+2x$ since $9\times2=18$ and $3\times6=18$

$=3x(x+3)+2(x+3)$ — Factorise the first pair of terms and the second pair of terms.

$=(3x+2)(x+3)$

b $2x^2-9x+10=2x^2-4x-5x+10$ — Split $9x$ into $-4x-5x$ since $-4\times-5=20$ and $2\times10=20$

$=2x(x-2)-5(x-2)$ — Factorise the first pair of terms and the second pair of terms.

$=(2x-5)(x-2)$

Try It 2

Factorise each of these quadratics.

a $5x^2+21x+4$ **b** $6x^2+7x-3$ **c** $8x^2-22x+5$

You can use the factors of ax^2+bx+c to find the roots of the **quadratic equation** $ax^2+bx+c=0$

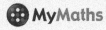

MyMaths 1156, 1157, 1181, 1950, 1959, 1960 SEARCH

Example 3

Use factorisation to find the roots of these quadratic equations.

a $4x^2+12x=0$ **b** $5x^2=21x-4$

a $4x^2+12x=4x(x+3)$ — Factorise the quadratic.

$4x(x+3)=0 \Rightarrow 4x=0$ or $x+3=0$ — One of the factors must be equal to zero.

If $4x=0$ then $x=0$ and if $x+3=0$ then $x=-3$ — Solve to find the roots.

b $5x^2-21x+4=0$ — Rearrange so you have a quadratic expression equal to zero.

$5x^2-21x+4=5x^2-20x-x+4$ — Write $-21x=-x-20x$ since $-20\times-1=20$ and $5\times4=20$

$=5x(x-4)-(x-4)$

$=(5x-1)(x-4)$ — Factorise the quadratic.

$(5x-1)(x-4)=0 \Rightarrow 5x-1=0$ or $x-4=0$ — The product is zero so one of the factors must be equal to zero.

If $5x-1=0$ then $x=\frac{1}{5}$ and if $x-4=0$ then $x=4$ — Solve to find the roots.

Try It 3

Find the roots of these quadratic equations.

a $6x^2-12x=0$ **b** $4x^2=23x-15$

A quadratic function has a **parabola** shaped curve.

When you sketch the graph of a quadratic function you must include the coordinates of the points where the curve crosses the x and y axes.

See Ch2.4
For more information on curve sketching.

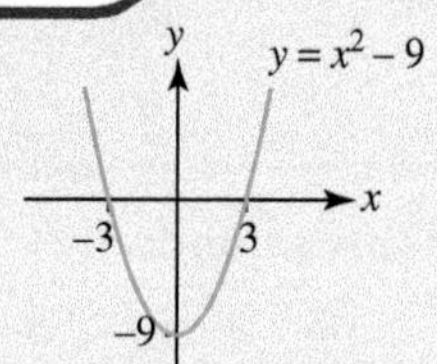

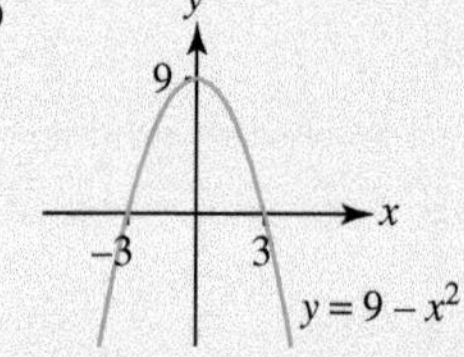

Example 4

Sketch these quadratic functions.

a $y=x^2+x-6$ **b** $y=-x^2+4x$

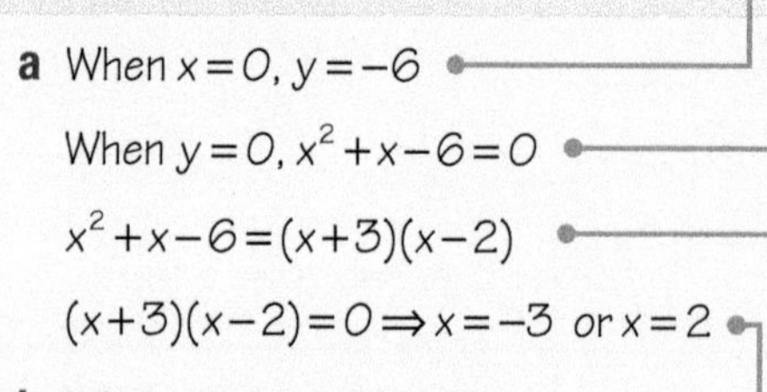

a When $x=0$, $y=-6$ — Find the y-intercept by letting $x=0$

When $y=0$, $x^2+x-6=0$ — Find the x-intercept by letting $y=0$

$x^2+x-6=(x+3)(x-2)$ — Factorise to find the roots.

$(x+3)(x-2)=0 \Rightarrow x=-3$ or $x=2$

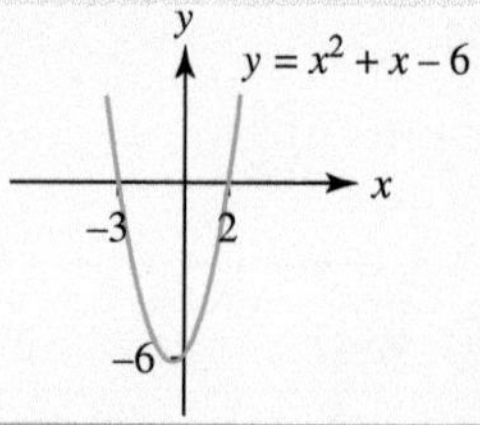

Sketch the parabola and label the y-intercept of −6 and the x-intercepts of −3 and 2

Graph: $y=x^2+x-6$, with intercepts −3, 2 and −6.

b When $x=0$, $y=0$ — Find the y-intercept by letting $x=0$

When $y=0$, $-x^2+4x=0$ — Find the x-intercept by letting $y=0$

$-x^2+4x=-x(x-4)$

$-x(x-4)=0 \Rightarrow x=0$ or $x=4$ — Factorise to find the roots.

Graph: $y=-x^2+4x$, with intercepts 0 and 4.

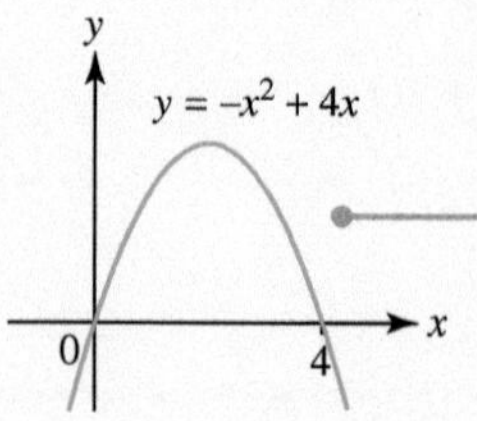

Sketch the parabola, it will be this way up since the x^2 term in the quadratic is negative. Label the x and y intercepts.

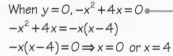

Sketch these quadratic functions.

Try It 4

a $y=x^2-25$ b $y=x^2+10x+25$ c $y=5x-x^2$

Bridging Exercise Topic C

Bridging to Ch1.4

1 Fully factorise each of these quadratics.

a $3x^2+5x$ b $8x^2-4x$ c $17x^2+34x$ d $18x^2-24x$

2 Factorise each of these quadratics.

a x^2+5x+6 b $x^2-7x+10$ c x^2-5x-6 d $x^2+3x-28$

e x^2-x-72 f $x^2+2x-48$ g $x^2-12x+11$ h $x^2-5x-24$

3 Factorise each of these quadratics.

a x^2-100 b x^2-81 c $4x^2-9$ d $64-9x^2$

4 Factorise each of these quadratics.

a $3x^2+7x+2$ b $6x^2+17x+12$ c $4x^2-13x+3$ d $2x^2-7x-15$

e $2x^2+3x-5$ f $7x^2+25x-12$ g $8x^2-22x+15$ h $12x^2+17x-5$

5 Fully factorise each of these quadratics.

a $16x^2-25$ b $4x^2-16x$ c $x^2+13x+12$ d $3x^2+16x-35$

e x^2+x-12 f $100-9x^2$ g $2x^2-14x$ h $20x^2-3x-2$

6 Use factorisation to find the roots of these quadratic equations.

a $21x^2-7x=0$ b $x^2-36=0$ c $17x^2+34x=0$ d $6x^2+13x+5=0$

e $4x^2-49=0$ f $x^2=7x+18$ g $x^2-7x+6=0$ h $21x^2=2-x$

i $17x=5x^2+6$ j $16x^2+24x+9=0$ k $9x^2+4=12x$ l $40x^2+x=6$

7 Sketch each of these quadratic functions, labelling where they cross the x and y axes.

a $y=x(x-3)$ b $y=-x(3x+2)$ c $y=x(3-x)$ d $y=(x+2)(x-2)$

e $y=(x+4)^2$ f $y=-(2x+5)^2$ g $y=(x-5)(x+2)$ h $y=(x+1)(5-x)$

8 Sketch each of these quadratic functions, labelling where they cross the x and y axes.

a $y=x^2+6x$ b $y=3x^2-12x$ c $y=x^2-121$ d $y=x^2-3x-10$

e $y=-x^2+3x$ f $y=15x-10x^2$ g $y=49-x^2$ h $y=-x^2+2x+3$

i $y=x^2-4x+4$ j $y=-x^2+14x-49$ k $y=3x^2+4x+1$ l $y=-2x^2+11x-12$

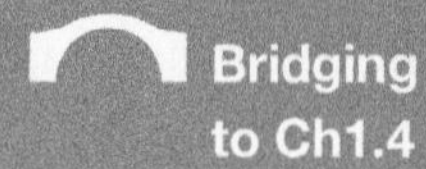

Topic D: Completing the square

Some quadratics are **perfect squares** such as $x^2-8x+16$ which can be written $(x-4)^2$. For other quadratics you can **complete the square**. This means write the quadratic in the form $(x+q)^2+r$

Key point

The completed square form of x^2+bx+c is $\left(x+\frac{b}{2}\right)^2-\left(\frac{b}{2}\right)^2+c$

If you have an expression of the form ax^2+bx+c then first factor out the a, as shown in Example 1

Example 1

Write each of these quadratics in the form $p(x+q)^2+r$ where p, q and r are constants to be found.

a x^2+6x+7 **b** $-2x^2+12x$

a $x^2+6x+7=\left(x+\frac{6}{2}\right)^2-\left(\frac{6}{2}\right)^2+7$ — The constant term in the bracket will be half of the coefficient of x

$=(x+3)^2-9+7=(x+3)^2-2$

b $-2x^2+12x=-2\left[x^2-6x\right]$ — First factor out the coefficient of x^2 then complete the square for the expression in the square brackets.

$=-2\left[(x-3)^2-9\right]=-2(x-3)^2+18$

Try It 1

Write each of these quadratics in the form $p(x+q)^2+r$

a x^2+22x **b** $2x^2-8x-6$ **c** $-x^2+10x$

Key point

The turning point on the curve with equation $y=p(x+q)^2+r$ has coordinates $(-q, r)$, this will be a minimum if p is positive and a maximum if p is negative.

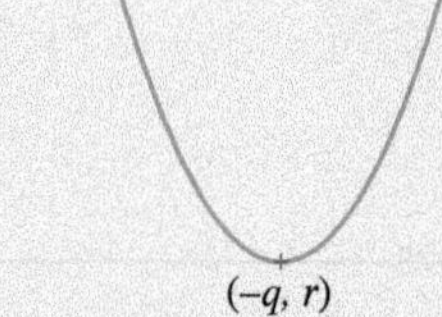

Example 2

Find the coordinates of the turning point of the curve with equation $y=-x^2+5x-2$

$-x^2+5x-2=-\left[x^2-5x+2\right]$ — First factor out the −1 then complete the square for the expression in the square brackets.

$=-\left[\left(x-\frac{5}{2}\right)^2-\frac{25}{4}+2\right]$

$=-\left[\left(x-\frac{5}{2}\right)^2-\frac{17}{4}\right]=-\left(x-\frac{5}{2}\right)^2+\frac{17}{4}$

So the maximum point is at $\left(\frac{5}{2},\frac{17}{4}\right)$ — The curve is at its highest point when the bracket is equal to zero: $x-\frac{5}{2}=0\Rightarrow x=\frac{5}{2}$

Try It 2

Find the coordinates of the turning point of each of these curves and state whether they are a maximum or a minimum.

a $y=x^2-3x+1$ **b** $y=-x^2-7x-12$ **c** $y=2x^2+4x-1$

Bridging Exercise Topic D

Bridging to Ch 1.4

1 Write each of these quadratic expressions in the form $p(x+q)^2+r$

a x^2+8x b x^2-18x c x^2+6x+3 d $x^2+12x-5$

e $x^2-7x+10$ f x^2+5x+9 g $2x^2+8x+4$ h $3x^2+18x-6$

i $2x^2-10x+3$ j $-x^2+12x-1$ k $-x^2+9x-3$ l $-2x^2+5x-1$

2 Use completing the square to find the turning point of each of these curves and state whether it is a maximum or a minimum.

a $y=x^2+14x$ b $y=x^2-18x+3$ c $y=x^2-9x$ d $y=-x^2+4x$

e $y=x^2+11x+30$ f $y=-x^2+6x-7$ g $y=2x^2+16x-5$ h $y=-3x^2+15x-2$

Topic E: The quadratic formula

Bridging to Ch1.4

You can solve a quadratic equation using the **quadratic formula**. The quadratic formula can also be used to quickly determine how many roots a quadratic equation has.

Key point

The quadratic formula for $ax^2+bx+c=0$ is $x=\dfrac{-b\pm\sqrt{b^2-4ac}}{2a}$

Example 1

Solve the equation $3x^2-5x-7=0$ using the quadratic formula.

$a=3, b=-5, c=-7$

$$x=\frac{-(-5)\pm\sqrt{(-5)^2-4\times3\times(-7)}}{2\times3}$$

$$=\frac{5\pm\sqrt{109}}{6}$$

$=2.57$ or -0.91 (to 2 dp)

Substitute into the formula, taking care with negatives.

Use your calculator to give answer as a decimal:

$\frac{5+\sqrt{109}}{6}=2.57$ and

$\frac{5-\sqrt{109}}{6}=-0.91$

You can also use the equation solver on your calculator to solve quadratic equations.

See Ch1.4
For a calculator method for solving quadratic equations.

Try It 1

Use the quadratic formula to solve the quadratic equation $7x^2-4x-6=0$

1185

Inside the square root of the quadratic formula you have the expression $b^2 - 4ac$. This expression is called the **discriminant**. You can use the discriminant to determine how many roots the equation has.

Key point

1. If $b^2 - 4ac < 0$ then the equation has no real roots.
2. If $b^2 - 4ac > 0$ then the equation has two real roots.
3. If $b^2 - 4ac = 0$ then the equation has one real root.

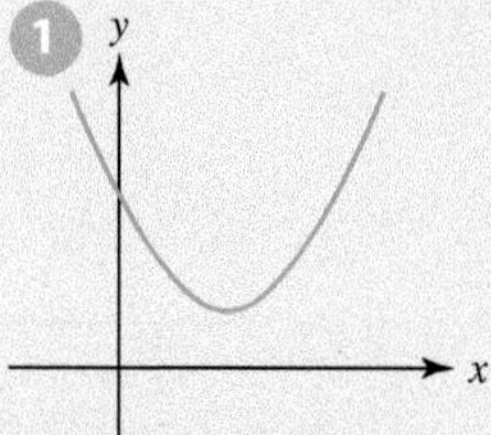

The curve does not cross the x-axis so the discriminant is negative.

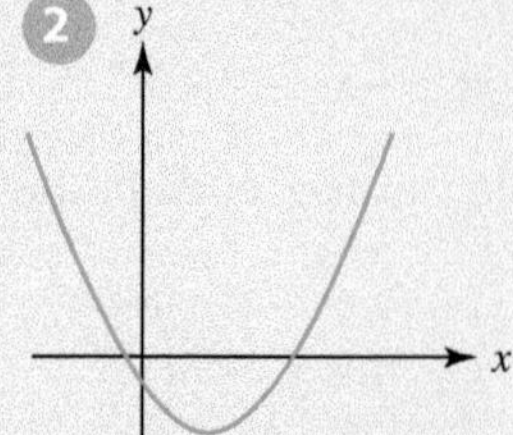

The curve crosses the x-axis twice so the discriminant is positive.

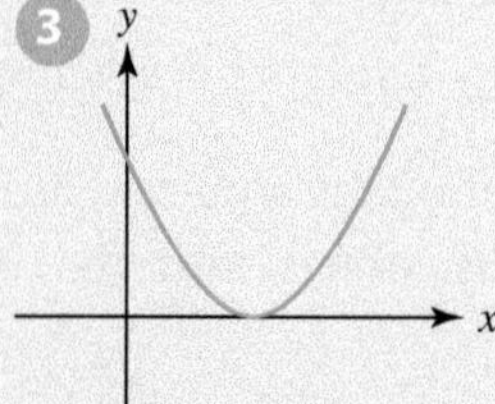

The curve touches the x-axis once so the discriminant equals zero.

Example 2

Given that the quadratic equation $x^2 + 3x + k + 1 = 0$ has exactly one solution, find the value of k

$a = 1, b = 3, c = k + 1$

So $b^2 - 4ac = 3^2 - 4 \times 1 \times (k+1)$ — Find the discriminant.

$= 5 - 4k$

$5 - 4k = 0 \Rightarrow k = \frac{5}{4}$ — The equation has exactly one solution so the discriminant is zero.

Try It 2

Given that the quadratic equation $kx^2 - x + 5 = 0$ has exactly one solution, find the value of k

Example 3

Given that the quadratic equation $5x^2 + 3x - k = 0$ has real solutions, find the range of possible values of k

$a = 5, b = 3, c = -k$

So $b^2 - 4ac = 5^2 - 4 \times 5 \times (-k)$ — Find the discriminant.

$= 25 + 20k$

$25 + 20k \geq 0 \Rightarrow k \geq -\frac{5}{4}$ — The equation has real solutions so the discriminant is greater than or equal to zero.

Try It 3

Given that the quadratic equation $x^2 + 3x - k = 0$ has real solutions, find the range of possible values of k

Example 4

Given that the quadratic equation $-x^2+7x+3-k=0$ has no real solutions, find the range of possible values of k

$a=-1, b=7, c=3-k$

So $b^2-4ac=7^2-4\times(-1)\times(3-k)$ — Find the discriminant.

$=61-4k$

$61-4k<0 \Rightarrow k>\frac{61}{4}$ — The equation has no solutions so the discriminant is negative.

Try It 4

Given that the quadratic equation $kx^2-7x+1=0$ has no real solutions, find the range of possible values of k

Bridging Exercise Topic E

Bridging to Ch1.4

1 Use the quadratic formula to solve each of these equations.

a $7x^2+3x-8=0$ **b** $-x^2+4x-2=0$ **c** $x^2-12x+4=0$

2 Work out how many real solutions each of these quadratic equations has.

a $x^2-5x+7=0$ **b** $7-2x-3x^2=0$ **c** $4x^2-28x+49=0$

3 Choose a possible equation from the box for each of the graphs.

a

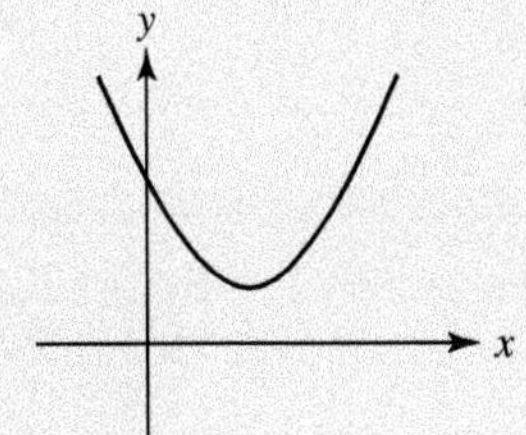

b

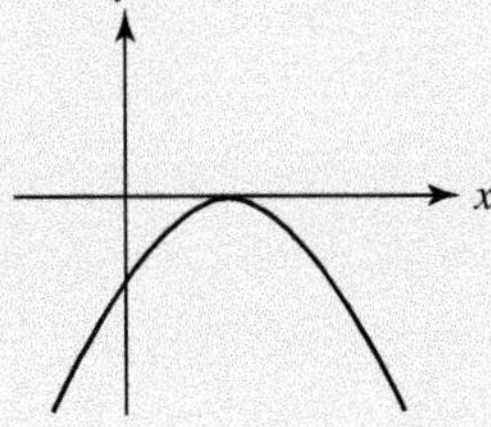

c

d

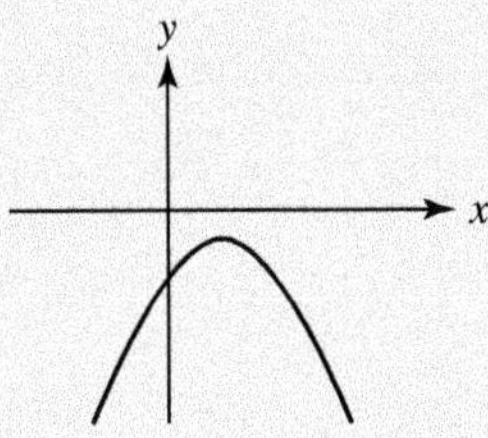

$y=-4x^2+12x-9$

$y=-x^2+2x-4$

$y=7x^2-5x+4$

$y=-x^2+x+6$

$y=6x^2-x-15$

4 Find the value of k in each equation given that they each have exactly one solution.

a $3x^2+2x-k=0$ **b** $kx^2-x+4=0$ **c** $2x^2+5x+k-5=0$

5 Find the range of possible values of k for each equation given that they all have real solutions.

a $x^2+3x-3k=0$ **b** $kx^2-7x+4=0$ **c** $-x^2+6x-k-2=0$

6 Find the range of possible values of k for each equation given that they all have no real solutions.

a $5x^2-x+2k=0$ **b** $-kx^2+4x+5=0$ **c** $6x^2-5x+3-2k=0$

MyMaths 1160 SEARCH

Topic F: Line graphs

This topic recaps how you can calculate key properties of straight line graphs when given two points on the line, in particular: the gradient, the length of a line segment, the midpoint of a line segment, the equation of the perpendicular bisector of a line segment, and the equation of the line.

The gradient of a line is a measure of how steep it is.

Key point

The gradient, m, of a line between two points (x_1, y_1) and (x_2, y_2) is given by $m = \frac{y_2 - y_1}{x_2 - x_1}$

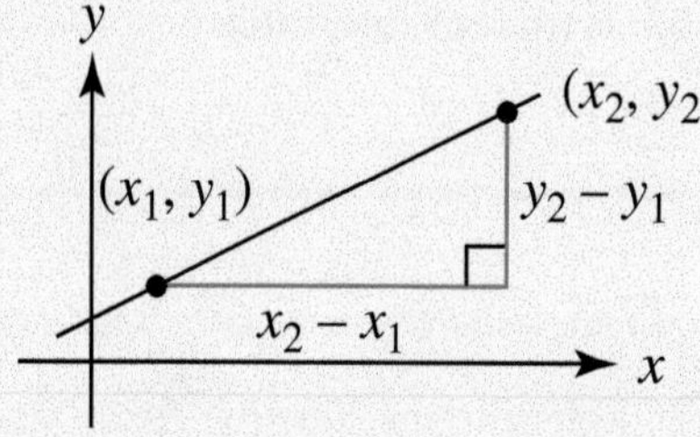

Example 1

Calculate the gradient of the line through the points $A(1, -6)$ and $B(-5, 2)$

$m = \frac{2-(-6)}{(-5)-1}$ — Use $m = \frac{y_2 - y_1}{x_2 - x_1}$ with $x_1 = 1$, $x_2 = -5$ and $y_1 = -6$, $y_2 = 2$

$= \frac{8}{-6}$

$= -\frac{4}{3}$ — The line has a negative gradient so slopes down from left to right.

Try It 1

Find the gradient of the line through each pair of points.

a (1, 7) and (4, 8) **b** (8,–2) and (4, 6) **c** (–8, 7) and (–4,–7)

You also can find the length of a line segment, d, between two points using Pythagoras' theorem.

Key point

The length of the line segment, d, between two points (x_1, y_1) and (x_2, y_2) is $d = \sqrt{(x_2 - x_1)^2 + (y_2 - y_1)^2}$

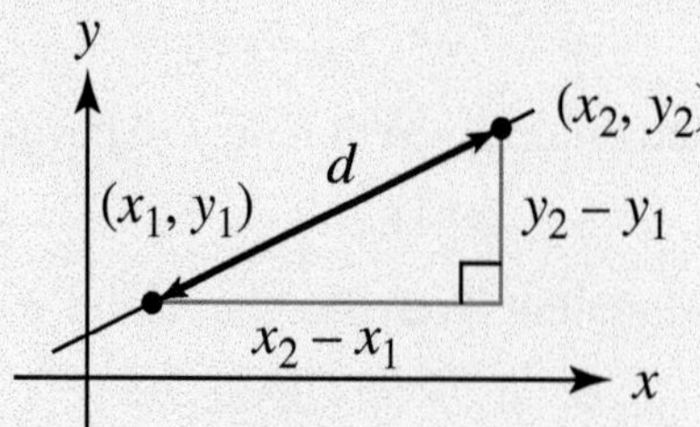

Example 2

Calculate the exact distance between the point (5, 1) and (6, –4)

$d=\sqrt{(6-5)^2+(-4-1)^2}$

$=\sqrt{1^2+(-5)^2}$

$=\sqrt{26}$

Use $d=\sqrt{(x_2-x_1)^2+(y_2-y_1)^2}$ with $x_1=5$, $x_2=6$ and $y_1=1$, $y_2=-4$

Leave answer as a surd since this is exact.

Try It 2

Calculate the exact distance between each pair of points.

a (5, 2) and (7, 4) **b** (6,–4) and (–3, –1) **c** $(\sqrt{2}, 4)$ and $(4\sqrt{2}, -5)$

The midpoint of a line segment is half-way between the points at either end.

Key point

The midpoint of the line segment from (x_1, y_1) to (x_2, y_2) is $\left(\frac{x_1+x_2}{2}, \frac{y_1+y_2}{2}\right)$

Example 3

The points *A* and *B* have coordinates (–4, –9) and (6, –2) respectively. Find the midpoint of *AB*

$\text{Midpoint}=\left(\frac{(-4)+6}{2}, \frac{(-9)+(-2)}{2}\right)$

$=\left(\frac{2}{2}, \frac{-11}{2}\right)$

$=(1, -5.5)$

Use $\left(\frac{x_1+x_2}{2}, \frac{y_1+y_2}{2}\right)$ with $x_1=-4$, $x_2=6$ and $y_1=-9$, $y_2=-2$

Try It 3

Calculate the midpoint of the line segment between each pair of points.

a (1, 9) and (2, 5) **b** (–2, 3) and (–5, –7) **c** (6.4, –9.3) and (–2.6, –3.7)

You need to be able to work with equations of straight lines.

Key point

The equation of a straight line is $y=mx+c$ where m is the gradient and c is the y-intercept.

MyMaths 1153, 1312, 1314 SEARCH

Example 4

Work out the gradient and the y-intercept of each of these lines.

a $y=\frac{1}{2}x+4$ **b** $y+x=5$ **c** $-2x+3y+7=0$

a Gradient $=\frac{1}{2}$ and y-intercept $=4$

Since $y=mx+c$ where m is the gradient and c is the y-intercept.

b $y=5-x$

Rearrange to make y the subject.

So gradient $=-1$ and y-intercept $=5$

c $3y=-7+2x$

Rearrange to make y the subject.

$y=-\frac{7}{3}+\frac{2}{3}x$

So gradient $=\frac{2}{3}$ and y-intercept $=-\frac{7}{3}$

Try It 4

Work out the gradient and the y-intercept of each line.

a $y=8-2x$ **b** $2y+x=3$ **c** $6x-9y-4=0$

You can write the gradient of a line in terms of a known point on the line (x_1, y_1), the general point (x, y), and the gradient, m.

$m=\frac{y-y_1}{x-x_1}$ or alternatively $y-y_1=m(x-x_1)$

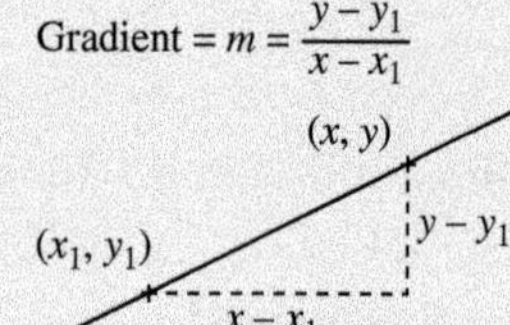

Key point

The equation of the line with gradient m through the point (x_1, y_1) is $y-y_1=m(x-x_1)$

If you have the coordinates of two points on a line then you can find the equation of the line. First use $m=\frac{y_2-y_1}{x_2-x_1}$ to find the gradient of the line then substitute into $y-y_1=m(x-x_1)$. Sometimes you will then need to rearrange the equation into a specific form.

Example 5

Find the equation of the line through the points (3, 7) and (4, –2) in the form $y=mx+c$

$m=\frac{(-2)-7}{4-3}$

$=-9$

First use $m=\frac{y_2-y_1}{x_2-x_1}$ to find the gradient.

So the equation is $y-7=-9(x-3)$

Use $y-y_1=m(x-x_1)$ with $(x_1,y_1)=(3, 7)$, or you could use the point (4, –2) instead.

$y-7=-9x+27$

$y=-9x+34$

Expand the brackets and rearrange to the correct form.

Find the equation of the line through each pair of points.

Try It 5

a (3, 7) and (2, 9) **b** (5, –1) and (7, 5) **c** (–3, –4) and (7, 2)

Lines with the same gradient are **parallel**. For example, $y=5x+2$ is parallel to $y=5x-7$, because the gradients are the same.

Example 6

The line l_1 has equation $2x+6y=5$. The line l_2 is parallel to l_1 and passes through the point (1, –5). Find the equation of l_2 in the form $ax+by+c=0$ where a, b and c are integers.

$l_1: 2x+6y=5 \Rightarrow 6y=5-2x$

$\Rightarrow y=\frac{5}{6}-\frac{2}{6}x$ — Rearrange to make y the subject so you can see what the gradient is.

The gradient of l_1 is $-\frac{2}{6}$ which simplifies to $-\frac{1}{3}$

Therefore the gradient of l_2 is $-\frac{1}{3}$ — Since l_1 and l_2 are parallel.

So the equation of l_2 is $y-(-5)=-\frac{1}{3}(x-1)$ — Use $y-y_1=m(x-x_1)$ to write the equation of l_2

$\Rightarrow y+5=-\frac{1}{3}(x-1)$

$\Rightarrow -3y-15=x-1$ — Multiply both sides by –3 so that all coefficients are integers.

$\Rightarrow x+3y+14=0$ — Rearrange to the correct form.

Try It 6

The line l_1 has equation $3x-2y=8$. A second line, l_2 is parallel to l_1 and passes through the point (3, –2). Find the equation of l_2 in the form $ax+by+c=0$ where a, b and c are integers.

Lines that meet at a right angle are **perpendicular**. The gradients of two perpendicular lines multiply to give –1. For example, a line with gradient 5 is perpendicular to a line with gradient $-\frac{1}{5}$ since $5\times\left(-\frac{1}{5}\right)=-1$

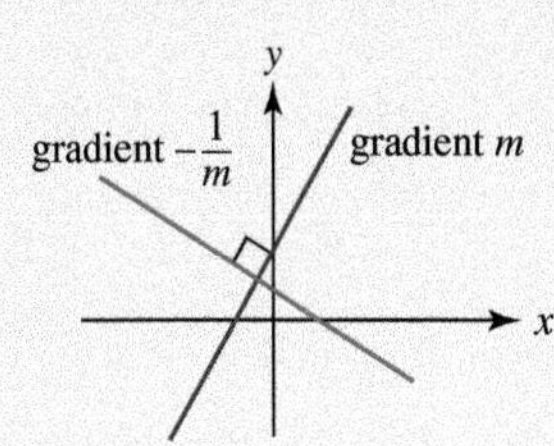

Key point

If the gradient of a line is m then the gradient of a perpendicular line is $-\frac{1}{m}$ since $m\times\left(-\frac{1}{m}\right)=-1$

MyMaths 1153, 1312, 1314 SEARCH

Example 7

Decide whether or not each line is parallel or perpendicular to the line $y=4x-1$

a $2x+8y=5$ **b** $20x+5y=2$ **c** $16x-4y=5$

First note that the gradient of $y=4x-1$ is 4

a $2x+8y=5 \Rightarrow 8y=5-2x$ — Rearrange to make y the subject.

$\Rightarrow y=\frac{5}{8}-\frac{1}{4}x$ — The gradient is $-\frac{1}{4}$

$4\times\left(-\frac{1}{4}\right)=-1$ so this line is perpendicular to $y=4x-1$ — Since the product of the gradients is -1

b $20x+5y=2 \Rightarrow 5y=2-20x$

$\Rightarrow y=\frac{2}{5}-4x$ — Rearrange to make y the subject.

The gradient is -4 so this line is neither parallel nor perpendicular to $y=4x-1$

c $16x-4y=5 \Rightarrow 4y=16x-5$

$\Rightarrow y=4x-\frac{5}{4}$

The gradient is 4 so this line is parallel to $y=4x-1$

Try It 7

Decide whether or not each line is parallel or perpendicular to the line $y=4-3x$

a $3x+6y=2$ **b** $5x-15y=7$ **c** $18x+6y+5=0$

Example 8

The line l_1 has equation $7x+4y=8$. The line l_2 is perpendicular to l_1 and passes through the point (7, 3). Find the equation of l_2 in the form $ax+by+c=0$ where a, b and c are integers.

$l_1: 7x+4y=8 \Rightarrow 4y=-7x+8$

$\Rightarrow y=-\frac{7}{4}x+2$ — Rearrange to make y the subject so you can see what the gradient is.

So the gradient of l_1 is $-\frac{7}{4}$ and the gradient of l_2 is $\frac{4}{7}$ — Since $\left(-\frac{7}{4}\right)\times\frac{4}{7}=-1$

So the equation of l_2 is $y-3=\frac{4}{7}(x-7)$ — Use $y-y_1=m(x-x_1)$ to write the equation of l_2

$\Rightarrow 7y-21=4(x-7)$ — Multiply both sides by 7 so that all coefficients are integers.

$\Rightarrow 7y-21=4x-28$

$\Rightarrow 4x-7y-7=0$ — Rearrange to the correct form.

Try It 8

The line l_1 has equation $4x+6y=3$. A second line, l_2 is perpendicular to l_1 and passes through the point $(-1, 5)$. Find the equation of l_2 in the form $ax+by+c=0$ where a, b and c are integers.

The **perpendicular bisector** of a line segment passes through its midpoint at a right angle.

Example 9

Find the equation of the perpendicular bisector of the line segment joining (3, –4) and (9, –6)

Midpoint is $\left(\frac{3+9}{2}, \frac{-4+(-6)}{2}\right) = (6, -5)$ — Use $\left(\frac{x_1+x_2}{2}, \frac{y_1+y_2}{2}\right)$

Gradient of line segment is $\frac{-6-(-4)}{9-3} = -\frac{2}{6} = -\frac{1}{3}$ — Use $m = \frac{y_2-y_1}{x_2-x_1}$

So the perpendicular bisector has gradient $m = 3$ — Since they are perpendicular and $3 \times \left(-\frac{1}{3}\right) = -1$

The equation of the perpendicular bisector is $y-(-5) = 3(x-6)$ — Use $y - y_1 = m(x - x_1)$

or $y = 3x - 23$

Try It 9

Find the equation of the perpendicular bisector of the line segment joining (2, –3) and (–12, 5)

Bridging Exercise Topic F

Bridging to Ch1.6

1 Find the gradient of the line through each pair of points.

a (3, 7) and (2, 8) b (5, 2) and (–4, –6) c (1.3, 4.7) and (2.6, –3.1)

d $\left(\frac{1}{2}, \frac{1}{3}\right)$ and $\left(\frac{3}{4}, \frac{2}{3}\right)$ e $(\sqrt{3}, 2)$ and $(2\sqrt{3}, 5)$ f $(3a, a)$ and $(a, 5a)$

2 Calculate the exact distance between each pair of points.

a (8, 4) and (1, 3) b (– 3, 9) and (12, –7) c (5.9 ,6.2) and (–8.1, 3.8)

d $\left(\frac{1}{5}, -\frac{1}{5}\right)$ and $\left(\frac{3}{5}, -\frac{4}{5}\right)$ e $(5, -3\sqrt{2})$ and $(2, \sqrt{2})$ f $(k, -3k)$ and $(2k, -6k)$

3 Find the coordinates of the midpoint of each pair of points.

a (3, 9) and (1, 7) b (2, –4) and (–3, –9) c (2.1, 3.5) and (6.3, –3.7)

d $\left(\frac{2}{3}, -\frac{1}{2}\right)$ and $\left(-\frac{5}{3}, -\frac{3}{2}\right)$ e $(6\sqrt{5}, 2\sqrt{5})$ and $(-\sqrt{5}, \sqrt{5})$ f $(m, 2n)$ and $(3m, -2n)$

4 Work out the gradient and the y-intercept of these lines.

a $y=7x-4$ b $y+2x=3$ c $x-y=4$ d $3x+2y=7$

e $5x-2y=9$ f $5y-3x=0$ g $x+6y+3=0$ h $3(y-2)=4(x-1)$

5 Find the equation of the line through each pair of points.

a (2, 5) and (0, 6) b (1, –3) and (2, –5) c (4, 4) and (7, –7)

d (8, –2) and (4, –3) e (–3, –7) and (5, 9) f $(\sqrt{2}, -\sqrt{2})$ and $(3\sqrt{2}, 4\sqrt{2})$

6 Which of these lines is either parallel or perpendicular to the line with equation $y=6x+5$?

a $2x+12y+3=0$ b $18x+3y=2$ c $3x-\frac{1}{2}y+5=0$

MyMaths 1153, 1312, 1314 SEARCH

7 Which of these lines is either parallel or perpendicular to the line with equation $y=\frac{2}{3}x-4$?

a $24x+16y+3=0$ b $6x+9y+2=0$ c $2x-3y=7$

8 Which of these lines is either parallel or perpendicular to the line with equation $6x+12y=1$?

a $2y=5-x$ b $9x=18y+4$ c $10x-5y+3=0$

In questions **9–13**, give your answers in the form $ax+by+c=0$ where a, b and c are integers.

9 The line l_1 has equation $y=5x+1$

a Find the equation of the line l_2 which is parallel to l_1 and passes through the point (3, –3)

b Find the equation of the line l_2 which is perpendicular to l_1 and passes through the point (–4, 1)

10 The line l_1 has equation $y=3+\frac{1}{2}x$

a Find the equation of the line l_2 which is parallel to l_1 and passes through the point (–1, 5)

b Find the equation of the line l_2 which is perpendicular to l_1 and passes through the point (6, 2)

11 The line l_1 has equation $3x+y=9$

a Find the equation of the line l_2 which is parallel to l_1 and passes through the point (8, –2)

b Find the equation of the line l_2 which is perpendicular to l_1 and passes through the point (–1, –1)

12 The line l_1 has equation $6x+5y+2=0$

a Find the equation of the line l_2 which is parallel to l_1 and passes through the point (4, 0)

b Find the equation of the line l_2 which is perpendicular to l_1 and passes through the point (12, 3)

13 The line l_1 has equation $6x-2y=1$

a Find the equation of the line l_2 which is parallel to l_1 and passes through the point $\left(\frac{1}{2}, 1\right)$

b Find the equation of the line l_2 which is perpendicular to l_1 and passes through the point $\left(-1, -\frac{1}{2}\right)$

14 Find the equation of the perpendicular bisector of the line segment joining each pair of points.

a (5, –7) and (–3, 5) b (–5, –9) and (5, 5) c (–6, 2) and (4, 12)

d (2, –7) and (–1, 2) e (–13, –5) and (15, –12)

15 Find the point of intersection between these pairs of lines.

a $y=5x-4$ and $y=3-2x$ b $y=8x$ and $y=3x-10$

c $y=7x-5$ and $y=-\frac{1}{2}x+5$ d $y=\frac{1}{4}x+7$ and $y=5x-\frac{5}{2}$

16 Find the point of intersection between these pairs of lines.

a $2x+3y=1$ and $3x-y=7$ b $3x-2y=4$ and $x+y=8$

c $5x-7y=3$ and $2x+8y=3$ d $-8x+5y=1$ and $3x+18y+7=0$

Topic G: Circles

Bridging to Ch1.6

You can use the centre and radius of a circle to define its equation, and to define the equation of a tangent to circle at a given point. You can also find points of intersection between a circle and a line or chord.

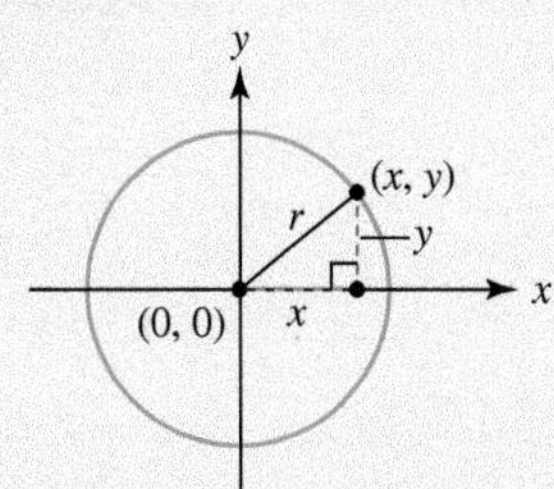

Using Pythagoras' theorem, a circle of radius r, with centre at the origin, has equation $x^2+y^2=r^2$

Following a similar method, you can write down the equation of a circle with centre (a, b) and radius r, using a general point (x, y) on the circle, as shown in the diagram.

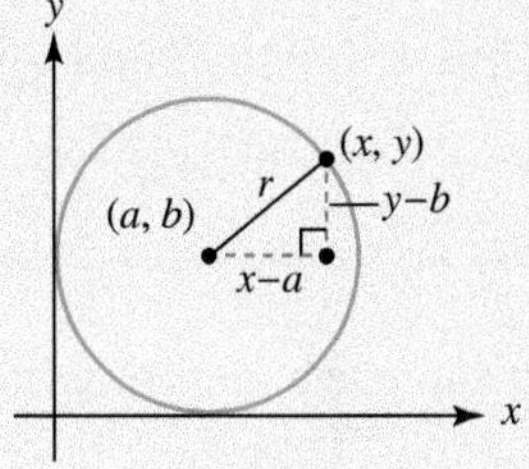

The horizontal distance between the centre (a, b) and the point on the circle (x, y) is the difference between the x-coordinates. The vertical distance between the centre (a, b) and the point on the circle (x, y) is the difference between the y-coordinates.

Using Pythagoras' theorem: $r^2=(x-a)^2+(y-b)^2$

Key point

A circle of radius r and centre (a, b) has equation $(x-a)^2+(y-b)^2=r^2$

Example 1

a Find the centre and radius of the circle with equation $(x-5)^2+(y+1)^2=9$

b Write the equation of a circle with centre (–3, 7) and radius 4

a The centre is at $(5,-1)$ — Equation is $(x-5)^2+(y-(-1))^2=9$ so $a=5$ and $b=-1$

The radius is $\sqrt{9}=3$ — Remember to find the positive square root.

b $a=-3, b=7$ and $r=4$

So equation is $(x+3)^2+(y-7)^2=16$ — Remember to square the radius.

Try It 1

a Find the centre and radius of the circle with equation $(x+2)^2+(y-8)^2=25$

b Write the equation of a circle with centre (7, –9) and radius 8

If you have the equation of a circle in expanded form then you can complete the square, as shown in Topic D, to write it in the form $(x-a)^2+(y-b)^2=r^2$ which will enable you to state the centre and radius.

MyMaths 1152, 1953 SEARCH

Example 2

Find the centre and radius of the circle with equation $x^2+y^2-8x+4y+2=0$

$x^2-8x+y^2+4y+2=0$

$(x-4)^2-16+(y+2)^2-4+2=0$

$(x-4)^2+(y+2)^2=18$

So the centre is $(4, -2)$ and the radius is $\sqrt{18}=3\sqrt{2}$

Group the terms involving x and the terms involving y

Complete the square for x^2-8x and y^2+4y

Try It 2

Find the centre and radius of the circles with these equations.

a $x^2+y^2-10y+16=0$ **b** $x^2+y^2+6x-12y=0$

You can use a diameter of a circle to find the equation of the circle.

Key point

If AB is the diameter of a circle then

- the centre of the circle is the midpoint of AB
- the radius of the circle is half the length of the diameter AB

Example 3

Find the equation of the circle with diameter AB where A is $(3, -8)$ and B is $(-5, 4)$

Centre is $\left(\frac{3+(-5)}{2}, \frac{(-8)+4}{2}\right)$
$=(-1, -2)$

Radius is $\frac{1}{2}\sqrt{(-5-3)^2+(4-(-8))^2}$
$=\frac{1}{2}\sqrt{(-8)^2+(12)^2}$
$=2\sqrt{13}$

So the equation of the circle is $(x+1)^2+(y+2)^2=52$

The centre is the midpoint of AB. Use $\left(\frac{x_1+x_2}{2}, \frac{y_1+y_2}{2}\right)$

The radius is half of the length of AB

Use $(x-a)^2+(y-b)^2=r^2$ and remember to square the radius: $(2\sqrt{13})^2=52$

Try It 3

Find the equation of the circle with diameter AB where A is $(4, 6)$ and B is $(2, -4)$

A **tangent** to a circle is a line which is perpendicular to a radius of the circle. Note that a tangent will intersect a circle exactly once.

You can use these facts to find the equation of a tangent to a circle.

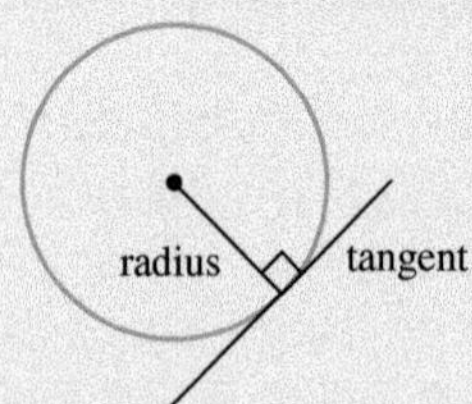

Example 4

A circle has equation $(x+3)^2+(y-7)^2=26$

a Show that the point (–4, 2) lies on the circle.

b Find the equation of the tangent to the circle that passes through the point (–4, 2)

a $(-4+3)^2+(2-7)^2=(-1)^2+(-5)^2$
$=1+25$
$=26$ so (–4, 2) lies on the circle.

Substitute $x=-4$, $y=2$ into the equation.

b Centre of circle is (–3, 7)

Gradient of radius is $\frac{2-7}{-4-(-3)}=\frac{-5}{-1}=5$

Use $m=\frac{y_2-y_1}{x_2-x_1}$

A tangent is perpendicular to a radius so gradient of tangent is $-\frac{1}{5}$

Since $\left(-\frac{1}{5}\right)\times 5=-1$

Therefore equation of tangent is $y-2=-\frac{1}{5}(x+4)$

Use $y-y_1=m(x-x_1)$ with $(x_1, y_1)=(-4, 2)$

Try It 4

A circle has equation $(x-1)^2+(y+4)^2=50$

a Show that the point (6, 1) lies on the circle.

b Find the equation of the tangent to the circle that passes through the point (6, 1)

You can find the point of intersection of a line and a circle by solving their equations simultaneously. You will need to use the **substitution** method of solving simultaneous equations.

If the line intersects the circle twice then it is a **chord**.

chord

Example 5

The line $x+3y=12$ and the circle $(x+3)^2+(y-7)^2=4$ intersect at the points A and B

a Find the coordinates of A and B

b Calculate the length of the chord AB

a $x=12-3y$

Rearrange the equation of the line to make either x or y the subject (whichever is easiest).

$(12-3y+3)^2+(y-7)^2=4$

Substitute for x (or y) in the equation of the circle.

$\Rightarrow(15-3y)^2+(y-7)^2=4$

$\Rightarrow 225-90y+9y^2+y^2-14y+49=4$

$\Rightarrow 10y^2-104y+270=0$

Simplify, then use the equation solver on your calculator.

$\Rightarrow y=5.4$ or $y=5$

$x=12-3(5.4)\Rightarrow x=-4.2$

Substitute the values of y into the rearranged equation of the line to find the values of x

$x=12-3(5)=-3$

So they intersect at $A(-4.2, 5.4)$ and $B(-3, 5)$

The line and the circle will intersect twice unless the line is a **tangent** to the circle.

(*Continued on next page*)

b Length of chord $AB = \sqrt{(-3-(-4.2))^2+(5-5.4)^2}$

$= \sqrt{1.2^2+(-0.4)^2}$

$= \frac{2}{5}\sqrt{10}$ (=1.26 to 3 significant figures)

Use $d = \sqrt{(x_2 - x_1)^2 + (y_2 - y_1)^2}$

You can find points of intersection using a graphics calculator.

Try It 5

The line $3x + y = 5$ intersects the circle $x^2 + (y-4)^2 = 17$ at the points A and B

a Find the coordinates of A and B

b Calculate the length of the chord AB

Example 6

Show that $x - y = 12$ is a tangent to the circle $(x-6)^2 + (y+2)^2 = 8$

$y = x - 12$ — Rearrange the equation of the line to make either x or y the subject.

$(x-6)^2 + (x-12+2)^2 = 8$ — Substitute for y (or x) in the equation of the circle.

$\Rightarrow (x-6)^2 + (x-10)^2 = 8$

$\Rightarrow x^2 - 12x + 36 + x^2 - 20x + 100 = 8$ — Expand the brackets.

$\Rightarrow 2x^2 - 32x + 128 = 0$ — Simplify.

$b^2 - 4ac = (-32)^2 - 4 \times 2 \times 128 = 0$ — If the discriminant is zero then there is exactly one solution.

So they meet once only.

Hence $x - y = 12$ is a tangent to $(x-6)^2 + (y+2)^2 = 8$

To show that a line is a tangent to a circle you can show that they only intersect once.

Try It 6

Show that $2x - y + 11 = 0$ is a tangent to the circle $(x-5)^2 + (y-1)^2 = 80$

Bridging Exercise Topic G

Bridging to Ch1.6

1 Write the equations of these circles.

a circle with radius 7 and centre (2, 5)

b circle with radius 4 and centre (–1, –3)

c circle with radius $\sqrt{2}$ and centre (–3, 0)

d circle with radius $\sqrt{5}$ and centre (4, –2)

2 Find the centre and the radius of the circles with these equations.

a $(x-5)^2+(y-3)^2=16$ b $(x+3)^2+(y-4)^2=36$ c $(x-9)^2+(y+2)^2=100$

d $(x+3)^2+(y+1)^2=80$ e $(x-\sqrt{2})^2+(y+2\sqrt{2})^2=32$ f $\left(x+\frac{1}{4}\right)^2+\left(y+\frac{1}{3}\right)^2=\frac{25}{4}$

3 Find the centre and the radius of the circles with these equations.

a $x^2+2x+y^2=24$ b $x^2+y^2+12y=13$ c $x^2+y^2-4x+3=0$

d $x^2+y^2+6x+8y+2=0$ e $x^2+y^2-8x-10y=3$ f $x^2+y^2+14x-2y=5$

g $x^2+y^2+5x-4y+3=0$ h $x^2+y^2-3x-9y=2$ i $x^2+y^2-x+7y+12=0$

4 Find the equation of the circle with diameter AB where the coordinates of A and B are

a (3, 5) and (1, 7) b (4, –1) and (2, –5) c (1, –3) and (–9, –6)

d (–3, –7) and (8, –16) e $(\sqrt{2}, 4)$ and $(-\sqrt{2}, 6)$ f $(4\sqrt{3}, -\sqrt{3})$ and $(-2\sqrt{3}, -5\sqrt{3})$

5 Determine whether each of these points lies on the circle with equation $(x-3)^2+(y+2)^2=5$

a (5, 3) b (1, –1) c (4, 3) d (2, 0)

6 Determine which of these circles the point (–3, 2) lies on.

a $(x-5)^2+y^2=68$ b $(x+2)^2+(y+1)^2=8$ c $(x-6)^2+(y-2)^2=81$

7 A circle has equation $(x-1)^2+(y+1)^2=10$. Find the equation of the tangent to the circle through the point (2, –4). Write your answer in the form $ax+by+c=0$ where a, b and c are integers.

8 A circle has equation $(x+3)^2+(y+7)^2=34$. Find the equation of the tangent to the circle through the point (0, –2). Write your answer in the form $ax+by+c=0$ where a, b and c are integers.

9 A circle has equation $x^2+(y-8)^2=153$. Find the equation of the tangent to the circle through the point (3, –4). Write your answer in the form $y=mx+c$

10 A circle has equation $(x+4)^2+y^2=20.5$. Find the equation of the tangent to the circle through the point (0.5, –0.5). Write your answer in the form $y=mx+c$

11 Find the points of intersection, A and B, between these pairs of lines and circles.

a $x+y=5$, $x^2+y^2=53$ b $y+1=0$, $(x-1)^2+(y+2)^2=17$

c $2x-y+7=0$, $(x-2)^2+(y+1)^2=36$ d $y=2x+1$, $(x+4)^2+(y+6)^2=10$

12 The line $3x-9y=6$ intersects the circle $(x+7)^2+(y+3)^2=10$ at the points A and B

a Find the coordinates of A and B b Calculate the length of the chord AB

13 The line $2x+4y=10$ intersects the circle $(x+5)^2+(y-2)^2=20$ at the points A and B

a Find the coordinates of A and B b Calculate the length of the chord AB

14 Show that the line $y=x-3$ is a tangent to the circle $(x-3)^2+(y+2)^2=2$

15 Show that the line $4x+y=34$ is a tangent to the circle $(x+1)^2+(y-4)^2=68$

16 Show that the line $x+3y=25$ is a tangent to the circle $x^2+(y-5)^2=10$

17 Show that the line $y=2x+3$ does not intersect the circle $(x-1)^2+(y+4)^2=1$

18 Show that the line $3x+4y+2=0$ does not intersect the circle $(x+3)^2+(y-6)^2=9$

MyMaths 1152, 1953 SEARCH

1.1 Argument and proof

Fluency and skills

A **proof** is a logical argument for a mathematical statement. It shows that something *must* be either true or false.

The most simple method of proving something is called **direct proof**. It's sometimes also called deductive proof. In direct proof, you rely on statements that are already established, or statements that can be assumed to be true, to show by deduction that another statement is true (or untrue).

Statements that can be assumed to be true are sometimes known as **axioms**.

Examples of statements that can be assumed to be true include 'you can draw a straight line segment joining any two points,' and 'you can write all even numbers in the form $2n$ and all odd numbers in the form $2n - 1$'

Key point

To use direct proof you

- Assume that a statement, P, is true.
- Use P to show that another statement, Q, must be true.

Example 1

Use direct proof to prove that the square of any integer is one more than the product of the two integers either side of it.

Let the integer be n
The two numbers on each side of n are $n-1$ and $n+1$ — *Assume that this statement is true.*
The product of these two numbers is $(n-1)(n+1)$

$(n-1)(n+1) = n^2 - 1$ — *Expand the brackets to get the square number n^2*
So $n^2 = (n-1)(n+1) + 1$ — *Rearrange to show the required result.*

So the square of any integer is one more than the product of the two integers either side of it.

Another method of proof is called **proof by exhaustion**. In this method, you list all the possible cases and test each one to see if the result you want to prove is true. All cases must be true for proof by exhaustion to work, since a single counter example would disprove the result.

Key point

To use proof by exhaustion you

- List a set of cases that exhausts all possibilities.
- Show the statement is true in each and every case.

Example 2

Prove, by exhaustion, that p^2+1 is not divisible by 3, where p is an integer and $6 \le p \le 10$

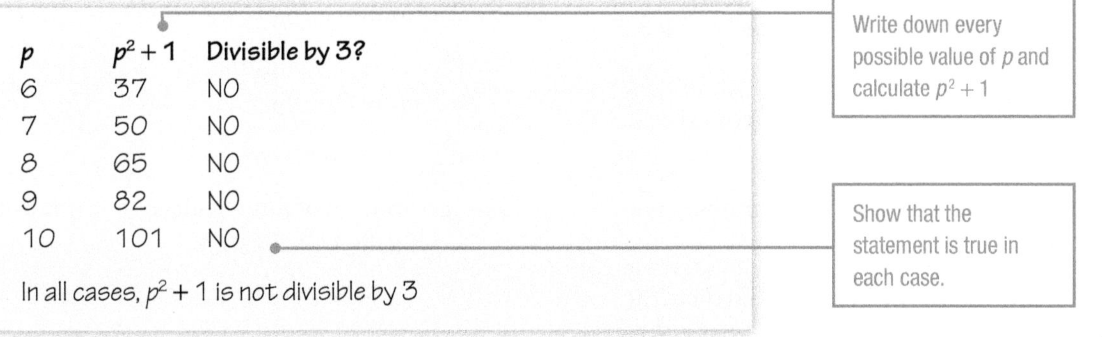

p	p^2+1	Divisible by 3?
6	37	NO
7	50	NO
8	65	NO
9	82	NO
10	101	NO

In all cases, p^2+1 is not divisible by 3

Write down every possible value of p and calculate p^2+1

Show that the statement is true in each case.

You can also *disprove* a statement with **disproof by counter example**, in which you need to find just one example that does not fit the statement.

Example 3

Prove, by counter example, that the statement 'n^2+n+1 is prime for all integers n' is false.

Let $n=4$

$4^2+4+1=21=3\times7$

21 has factors 1, 3, 7 and 21, so is not prime.

This disproves the statement 'n^2+n+1 is prime for all integers n'

Show that the statement is false for one value of n

Exercise 1.1A Fluency and skills

Use direct proof to answer these questions.

1 Prove that the number 1 is *not* a prime number.

2 Prove that the sum of two odd numbers is always even.

3 Prove that the product of two consecutive odd numbers is one less than a multiple of four.

4 Prove that the mean of three consecutive integers is equal to the middle number.

5 a Prove that the sum of the squares of two consecutive integers is odd.

b Prove that the sum of the squares of two consecutive even numbers is always a multiple of four.

6 Show that the sum of four consecutive positive integers has both even factors and odd factors greater than one.

7 Prove that the square of the sum of any two positive numbers is greater than the sum of the squares of the numbers.

8 Prove that the perimeter of an isosceles right-angled triangle is always greater than three times the length of one of the equal sides.

9 a and b are two numbers such that $a=b-2$ and the sum and product of a and b are equal. Prove that neither a nor b is an integer.

10 If $(5y)^2$ is even for an integer y, prove that y must be even.

Use proof by exhaustion in the following questions.

11 Prove that there is exactly one square number and exactly one cube number between 20 and 30

12 Prove that, if a month has more than five letters in its name, a four letter word can be made using those letters.

MyMaths 2252, 2253 SEARCH

13 Prove that, for an integer x, $(x+1)^3 \geq 3^x$ for $0 \leq x \leq 4$

14 Prove that no square numbers can have a last digit 2, 3, 7 or 8

Give counter examples to disprove these statements.

15 The product of two prime numbers is always odd.

16 When you throw two six-sided dice, the total score shown is always greater than six.

17 When you subtract one number from another, the answer is always less than the first number.

18 Five times any number is always greater than that number.

19 If $a > b$, then $a^b > b^a$

20 The product of three consecutive integers is always divisible by four.

Reasoning and problem-solving

Strategy

To prove or disprove a statement

1. Decide which method of proof to use.
2. Follow the steps of your chosen method.
3. Write a clear conclusion that proves/disproves the statement.

Example 4

Prove that the sum of the interior angles in any convex quadrilateral is 360°

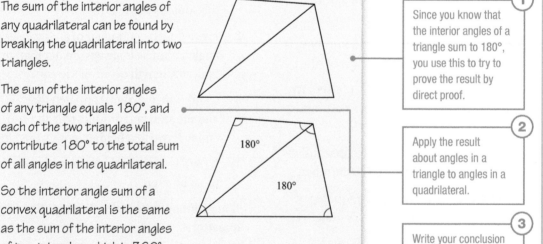

The sum of the interior angles of any quadrilateral can be found by breaking the quadrilateral into two triangles.

The sum of the interior angles of any triangle equals 180°, and each of the two triangles will contribute 180° to the total sum of all angles in the quadrilateral.

180°

180°

So the interior angle sum of a convex quadrilateral is the same as the sum of the interior angles of two triangles, which is 360°

1. Since you know that the interior angles of a triangle sum to 180°, you use this to try to prove the result by direct proof.

2. Apply the result about angles in a triangle to angles in a quadrilateral.

3. Write your conclusion clearly.

Example 5

Jane says that there are exactly three prime numbers between the numbers 15 and 21 (inclusive).

Is she correct? Use a suitable method of proof to justify your answer.

NUMBER	PRIME?
15	NO → $15 = 1 \times 15, 3 \times 5$
16	NO → $16 = 1 \times 16, 2 \times 8, 4 \times 4$
17	YES → $17 = 1 \times 17$
18	NO → $18 = 1 \times 18, 2 \times 9, 3 \times 6$
19	YES → $19 = 1 \times 19$
20	NO → $20 = 1 \times 20, 2 \times 10, 4 \times 5$
21	NO → $21 = 1 \times 21, 3 \times 7$

There are exactly two prime numbers between 15 and 21, so Jane is wrong (she said there were three).

(1)(2) Use proof by exhaustion to check all the numbers within the range of values.

(3) Write a clear conclusion that proves or disproves the statement.

PURE

Exercise 1.1B Reasoning and problem-solving

1 P is a prime number and Q is an odd number.

  a Sue says PQ is even, Liz says that PQ is odd and Graham says PQ could be either. Who is right? Use a suitable method of proof to justify your answer.

  b Sue now says that $P(Q+1)$ is always even. Is she correct? Use a suitable method of proof to justify your answer.

2 Use a suitable method of proof to show whether the statement '*Any odd number between 90 and 100 is either a prime number or the product of only two prime numbers*' is true or false.

3 Use a suitable method of proof to prove that the value of $9^n - 1$ is divisible by 8 for $1 \le n \le 6$

4 Is it true that 'all triangles are obtuse'? Use a suitable method of proof to justify your answer.

5 Prove that the sum of the interior angles of a convex hexagon is 720°

6 Martin says that 'All quadrilaterals with equal sides are squares.' Use a suitable method of proof to show if his statement is true or false.

7 Prove that the sum of the interior angles of a convex n-sided polygon is $180(n-2)°$

8 Use a suitable method of proof to prove or disprove the statement 'If $m^2 = n^2$ then $m = n$.'

9 The hypotenuse of a right-angled triangle is $(2s + a)$ cm and one other side is $(2s - a)$ cm. Use a suitable method of proof to show that the square of the remaining side is a multiple of eight.

10 Use a suitable method of proof to show that, for $1 \le n \le 5$,

$$\frac{1}{1\times2}+\frac{1}{2\times3}+\frac{1}{3\times4}+\ldots+\frac{1}{n\times(n+1)}=\frac{n}{n+1}$$

Challenge

11 A teacher tells her class that any number is divisible by three if the sum of its digits is divisible by three. Use a suitable method to prove this result for two-digit numbers.

MyMaths 2252, 2253 SEARCH

1.2 Index laws

Fluency and skills

The algebraic term $3x^5$ is written in **index form**. The 3 is called the **coefficient**. The x part of the term is called the **base**. The 5 is called the **power**, or **index**, or **exponent**. $3x^5$ means $3 \times x \times x \times x \times x \times x$

Indices follow some general rules.

Key point

Rule 1: Any number raised to the power zero is 1 $\quad x^0 = 1$

Rule 2: Negative powers may be written as reciprocals. $\quad x^{-n} = \frac{1}{x^n}$

Rule 3: Any base raised to the power of a unit fraction is a root. $\quad x^{\frac{1}{n}} = \sqrt[n]{x}$

Rule 1 has an exception when $x = 0$, as 0^0 is undefined.

$x^{\frac{1}{2}} = \sqrt{x}$ and $x^{\frac{1}{3}} = \sqrt[3]{x}$

You don't normally write the '2' in a square root.

You can combine terms in index form by following this simple set of rules called the **index laws**.

To use the index laws, the bases of all the terms must be the same.

Key point

Law 1: To multiply terms you add the indices. $\quad x^a \times x^b = x^{a+b}$

Law 2: To divide terms you subtract the indices. $\quad x^a \div x^b = x^{a-b}$

Law 3: To raise one term to another power you multiply the indices. $\quad \left(x^a\right)^b = x^{a \times b}$

By combining the third general rule and the third index law you can see that $\sqrt[b]{\left(x^a\right)} = x^{\frac{a}{b}} = \left(\sqrt[b]{x}\right)^a$

So $\sqrt[3]{\left(125^4\right)} = \left(\sqrt[3]{125}\right)^4 = 5^4 = 625$

Example 1

Simplify these expressions, leaving your answers in index form.

a $2m^4n^2 \times 3m^3n^9$ $\quad$ **b** $4d^{\frac{5}{3}} \div 2d^{\frac{1}{3}}$

a $2 \times 3 \times m^{4+3} \times n^{2+9} = 6m^7n^{11}$ — Use $x^a \times x^b = x^{a+b}$

b $4d^{\frac{5}{3}-\frac{1}{3}} \div 2 = 2d^{\frac{4}{3}}$ — Use $x^a \div x^b = x^{a-b}$

Exercise 1.2A Fluency and skills

Simplify the expressions in questions **1** to **44**. Show your working.

1 4^3

2 $(-3)^5$

3 $7^8 \div 7^4$

4 $c^7 \times c^4$

5 $(-p^3)^4$

6 $-(p^3)^4$

7 $(2c^{-3})^6$

8 $d^7 \times d^3 \times d^4$

9 $2e^3 \times 5e^4 \times 7e^2$

10 $4f^2 \times -3f^4 \times 9f^6$

11 $24g^{12} \div 6g^6$

12 $-44k^{44} \div 11k^{-11}$

13 $12f^2 \times 4f^4 \div 6f^3$

14 $(12e^{13} \div 6e^4) \div 3e^7$

15 $3a \times 5b$

16 $5w \times 4x \times (-6x)$

17 $2d \times 3e \times 4f^2$

18 $3h^6 \times (-3h^8)$

19 $5r^5s^6 \times r^3s^4$

20 $5r^5s^6 \div r^3s^4$

21 $(g^2h^3) \times (-g^7h^5) \times (ghi^4)$

22 $(g^2h^3) \times (-g^7h^5) \div (ghi^4)$

23 $(-20z^9y^6) \div (-4z^4y)$

24 $\sqrt{36u^{36}}$

25 $(36u^{36})^{\frac{1}{2}}$

26 $\sqrt[3]{125t^{27}}$

27 $\sqrt[3]{-125t^{27}c^{12}}$

28 $(5)^{-1}$

29 $\left(\frac{1}{5}\right)^{-1}$

30 $6u^0$

31 $-50f^0$

32 $7y^0 - 4z^0$

33 4^{-2}

34 2^{-10}

35 $(3w)^{-2}$

36 $(3w^{-2})^{-2}$

37 $64^{\frac{3}{2}}$

38 $1024^{\frac{1}{5}}$

39 $1024^{\frac{4}{5}}$

40 $16^{\frac{3}{4}}$

41 $16^{\frac{-3}{4}}$

42 $\left(\frac{36}{49}\right)^{\frac{1}{2}}$

43 $\left(\frac{36}{49}\right)^{\frac{-1}{2}}$

44 $\left(\frac{36}{49}\right)^{\frac{3}{2}}$

45 If $5^n = 625$, find the value of n

46 If $3^m = 243$, find the value of m

47 If $6^{2t+1} = 216$, find the value of t

48 If $(2^{2b})(2^{-6b}) = 256$, find the value of b

Reasoning and problem-solving

Strategy

To solve problems involving indices

1. Use the information in the question to write an expression or equation involving indices.
2. Apply the laws of indices correctly.
3. Simplify expressions as much as possible.
4. Give your answer in an appropriate format that is relevant to the question.

Example 2

A swimming pool has a volume of $16s^2$ cubic metres.

a How long does it take to fill, from empty, if water is pumped in at a rate of $4s^{-3}$ cubic metres per minute?

b If it takes 128 minutes to fill the swimming pool, calculate the value of s

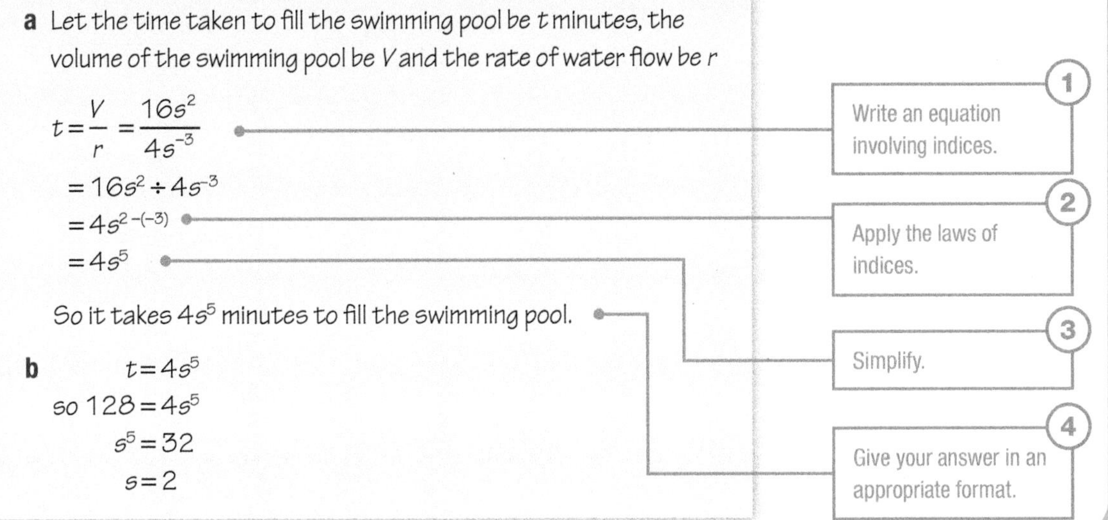

a Let the time taken to fill the swimming pool be t minutes, the volume of the swimming pool be V and the rate of water flow be r

$t = \frac{V}{r} = \frac{16s^2}{4s^{-3}}$

$= 16s^2 \div 4s^{-3}$

$= 4s^{2-(-3)}$

$= 4s^5$

So it takes $4s^5$ minutes to fill the swimming pool.

b $t = 4s^5$

so $128 = 4s^5$

$s^5 = 32$

$s = 2$

Example 3

A rectangular flower bed has sides of length x and $8x$

Around it are 6 further flower beds, each with an area equal to the cube root of the larger flower bed.

Calculate the total area covered by the 6 smaller flower beds, giving your answer in index form.

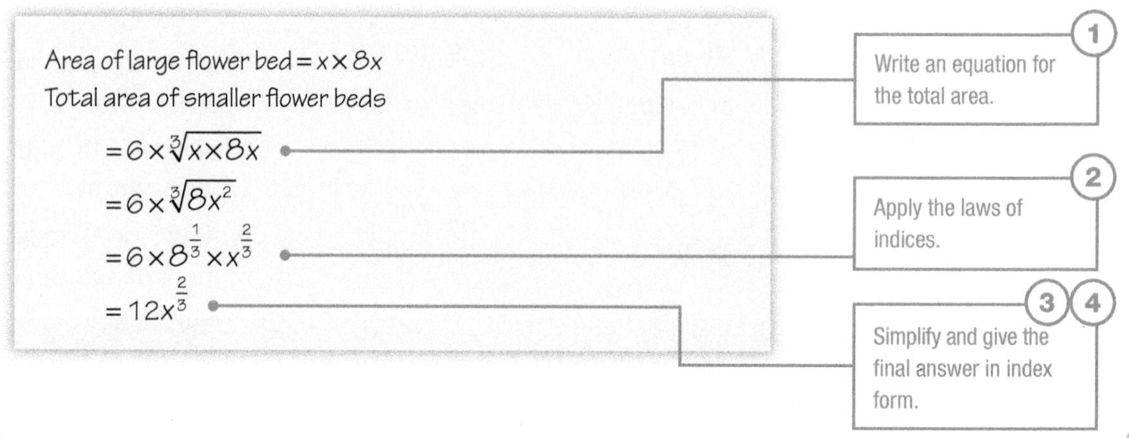

Area of large flower bed $= x \times 8x$

Total area of smaller flower beds

$= 6 \times \sqrt[3]{x \times 8x}$ — (1) Write an equation for the total area.

$= 6 \times \sqrt[3]{8x^2}$

$= 6 \times 8^{\frac{1}{3}} \times x^{\frac{2}{3}}$ — (2) Apply the laws of indices.

$= 12x^{\frac{2}{3}}$ — (3)(4) Simplify and give the final answer in index form.

Example 4

The brain mass (kg) of a species of animal is approximately one hundredth of the cube root of the square of its total body mass (kg).

a Write a formula relating brain mass, B, to total body mass, m, using index form.

b Use your formula to calculate

i The approximate brain mass of an animal of mass 3.375 kg,

ii The approximate total body mass of an animal with brain mass 202.5 g.

a $B = \dfrac{\sqrt[3]{m^2}}{100}$ — (1) Write a formula for the brain mass.

$= \dfrac{m^{\frac{2}{3}}}{100}$ — (2) Apply the index laws correctly.

b i $B = \dfrac{3.375^{\frac{2}{3}}}{100}$

$= 0.0225$ kg or 22.5 g — (4) Give your answer in an appropriate form.

ii $0.2025 = \dfrac{m^{\frac{2}{3}}}{100}$ — Convert the brain mass into kilograms and substitute into the formula.

$m^{\frac{2}{3}} = 20.25$

$m = 20.25^{\frac{3}{2}}$ — Use inverse operations to find m

$= 91.125$ kg

Exercise 1.2B Reasoning and problem-solving

1 a Work out the area of a square of side $2s^2$ inches.

b Work out the side length of a square of area $25p^4q^6$ cm^2.

2 a Work out the circumference and area of a circle of radius $3w^5$ ft.

b The volume of a sphere is $\frac{4}{3}\pi\times\text{radius}^3$ and the surface area is $4\pi\times\text{radius}^2$.

Work out the surface area and volume of a sphere of radius $3w^4$ ft.

3 What term multiplies with $4c^2d^3$, $5de^2$ and $3c^2e^3$ to give 360?

4 Work out the volume of a cuboid with dimensions $4p^2q^3$, $3pq^2$ and $\sqrt{9p^4q^0}$
Give your answer in index form.

5 a The area of a rectangle is $8y^5z^7$ and its length is $4y^2z^3$. Work out its width.

b Explain why the area of a rectangle of sides $\sqrt[3]{8m^{-3}n^6}$ and $\sqrt{16m^2n^{-4}}$ is independent of m and n. What is the area?

6 A cyclist travels $4b^2c^{\frac{1}{2}}$ miles in $3b^2c$ hours. What is her average speed?

7 a The volume of a cylinder is $8\pi c^2d$ cm^3. The radius of the cylinder is $2cd^{-1}$ cm. What is its height?

b Explain why the volume of a cylinder of radius $3s^2t^{-1}$ and height $(5st)^2$ is independent of t. What is the volume?

8 A disc of radius $3v^2z^{-2}$ cm is removed from a disc of radius $4v^2z^{-2}$ cm. What is the remaining area?

9 a Work out the hypotenuse of a right-angled triangle with perpendicular sides of length $5n^{\frac{1}{2}}$ and $12n^{\frac{1}{2}}$

b Work out the area of the right-angled triangle described in part **a**

10 a To work out the voltage, V volts, in a circuit with current i amps and resistance r ohms, you multiply the current and resistance together.

Work out the voltage in a circuit of resistance $3m^4n^{-4}$ ohms carrying a current of $6m^{-2}n^{-3}$ amps.

b The power, W watts, in a circuit with current i amps and resistance r ohms, is found by multiplying the resistance by the square of the current. Work out the power when the current and resistance are the same as in the circuit in part **a**

11 The kinetic energy of a body is given by $\frac{1}{2}\times\text{mass}\times\text{velocity}^2$
Work out the kinetic energy when mass $= m$ and velocity $= 9x^{\frac{3}{4}}c^{\frac{3}{4}}$

12 You can find your Body Mass Index (BMI) by dividing your mass (kg) by the square of your height (m). If your mass is $3gt$ kg and your height is $4gt^{-1}$ m, what is your BMI?

13 In an electrical circuit, the total resistance of two resistors, t_1 and t_2, connected in parallel, is found by dividing the product of their resistances by the sum of their resistances. Work out the total resistance when $t_1 = 3rs^2$ ohms and $t_2 = 5rs^2$ ohms.

Challenge

14 In a triangle with sides of length a, b and c the semi-perimeter, s, is half the sum of the three sides. The area of the triangle can be found by subtracting each of the sides from the semi-perimeter in turn (to give three values), multiplying these expressions and the semi-perimeter altogether and then square rooting the answer.

a Write a formula for the area of the triangle involving s, a, b and c

b Use your formula to work out the area of a triangle with sides, $12xy$, $5xy$ and $13xy$

c You could have found the area of this triangle in a much easier way. Explain why.

1.3 Surds

Fluency and skills

Key point

A **rational number** is one that you can write exactly in the form

$$\frac{p}{q}$$

where p and q are integers, $q \neq 0$

Numbers that you cannot write exactly in this form are **irrational numbers**. If you express them as decimals, they have an infinite number of non-repeating decimal places.

Some roots of numbers are irrational, for example, $\sqrt{3} = 1.732...$ and $\sqrt[3]{10} = 2.15443...$ are irrational numbers.

Key point

Irrational numbers involving roots, $\sqrt[n]{\ }$ or $\sqrt{\ }$, are called **surds**.

You can use the following laws to simplify surds

$$\sqrt{a} \times \sqrt{b} = \sqrt{ab}$$

$$\frac{\sqrt{a}}{\sqrt{b}} = \sqrt{\frac{a}{b}}$$

You usually write surds in their simplest form, with the smallest possible number written inside the root sign.

You can simplify surds by looking at their factors.

You should look for factors that are square numbers.

For example $\sqrt{80} = \sqrt{16} \times \sqrt{5} = 4\sqrt{5}$

If $\sqrt{a}$ and $\sqrt{b}$ cannot be simplified, then you cannot simplify $\sqrt{a} + \sqrt{b}$ or $\sqrt{a} - \sqrt{b}$ for $a \neq b$

Example 1

Simplify these expressions. Show your working.

a $\sqrt{7} \times \sqrt{7}$ **b** $\sqrt{5} \times \sqrt{20}$ **c** $\sqrt{\frac{1}{9}} \times \sqrt{9}$

d $\sqrt{80} - \sqrt{20}$ **e** $\sqrt{63} + \sqrt{112}$ **f** $\sqrt{\frac{4}{3}} + \sqrt{\frac{25}{3}}$

a $\sqrt{7} \times \sqrt{7} = 7$ **b** $\sqrt{5} \times \sqrt{20} = \sqrt{100} = 10$ **c** $\sqrt{\frac{1}{9}} \times \sqrt{9} = \frac{1}{3} \times 3 = 1$

d $4\sqrt{5} - 2\sqrt{5} = 2\sqrt{5}$ **e** $3\sqrt{7} + 4\sqrt{7} = 7\sqrt{7}$ **f** $2\sqrt{\frac{1}{3}} + 5\sqrt{\frac{1}{3}} = 7\sqrt{\frac{1}{3}}$

Calculations are often more difficult if surds appear in the denominator. You can simplify such expressions by removing any surds from the denominator. To do this, you multiply the numerator and denominator by the same value to find an **equivalent fraction** with surds in the numerator only. This is easier to simplify.

This process is called **rationalising the denominator**.

Key point

If the fraction is in the form

$\frac{k}{\sqrt{a}}$, multiply numerator and denominator by $\sqrt{a}$

$\frac{k}{a \pm \sqrt{b}}$, multiply numerator and denominator by $a \mp \sqrt{b}$

$\frac{k}{\sqrt{a} \pm \sqrt{b}}$, multiply numerator and denominator by $\sqrt{a} \mp \sqrt{b}$

Example 2

Rationalise these expressions. Show your working.

a $\frac{4}{\sqrt{5}}$ **b** $\frac{6+\sqrt{7}}{9-\sqrt{7}}$

a $\frac{4}{\sqrt{5}} = \frac{4}{\sqrt{5}} \times \frac{\sqrt{5}}{\sqrt{5}} = \frac{4 \times \sqrt{5}}{\sqrt{5} \times \sqrt{5}} = \frac{4\sqrt{5}}{5}$ — Multiply top and bottom by $\sqrt{5}$

b $\frac{6+\sqrt{7}}{9-\sqrt{7}} = \frac{6+\sqrt{7}}{9-\sqrt{7}} \times \frac{9+\sqrt{7}}{9+\sqrt{7}}$ — Multiply top and bottom by $9+\sqrt{7}$

$= \frac{54 + 9\sqrt{7} + 6\sqrt{7} + 7}{(9-\sqrt{7})(9+\sqrt{7})}$

$= \frac{61 + 15\sqrt{7}}{81 - 9\sqrt{7} + 9\sqrt{7} - \sqrt{7}\sqrt{7}}$

$= \frac{61 + 15\sqrt{7}}{81 - 9\sqrt{7} + 9\sqrt{7} - 7} = \frac{61+15\sqrt{7}}{74}$

Exercise 1.3A Fluency and skills

Complete this exercise without a calculator.

1 Classify these numbers as rational or irrational.

a $1+\sqrt{25}$ **b** π^2

c $4-\sqrt{3}$ **d** $\sqrt{21}$

e $\sqrt{169}$ **f** $(\sqrt{8})^2$

g $(\sqrt{17})^3$

2 For each of these expressions, show that they can be written in the form $a\sqrt{b}$ where a and b are integers.

a $\sqrt{4} \times \sqrt{21}$ **b** $\sqrt{8} \times \sqrt{7}$

c $\sqrt{75}$ **d** $\sqrt{27}$

e $\frac{\sqrt{800}}{10}$ **f** $(\sqrt{8})^3$

g $(\sqrt{17})^3$ **h** $2\sqrt{3} \times 3\sqrt{2}$

i $5\sqrt{6} \times 7\sqrt{18}$ **j** $4\sqrt{24} \times 6\sqrt{30}$

3 Show that these expressions can be expressed as positive integers.

a $\frac{\sqrt{128}}{\sqrt{2}}$ **b** $\frac{\sqrt{125}}{\sqrt{5}}$

4 Show that these expressions can be written in the form $\frac{a}{b}$, where a and b are positive integers.

a $\frac{\sqrt{32}}{\sqrt{200}}$ **b** $\frac{\sqrt{50}}{\sqrt{72}}$

5 Show that these expressions can be written in the form $a\sqrt{b}$, where a and b are integers.

a $\sqrt{54}$ b $\sqrt{432}$

c $\sqrt{1280}$ d $\sqrt{3388}$

e $\sqrt{2}\times\sqrt{20}$ f $\sqrt{2}\times\sqrt{126}$

g $\sqrt{20}+\sqrt{5}$ h $\sqrt{18}-\sqrt{2}$

i $\sqrt{150}-\sqrt{24}$ j $\sqrt{75}+\sqrt{12}$

k $\sqrt{27}-\sqrt{3}$ l $\sqrt{5}+\sqrt{45}$

m $\sqrt{363}-\sqrt{48}$ n $\sqrt{72}-\sqrt{288}+\sqrt{200}$

6 Show these expressions can be written in the form $a+b\sqrt{c}$, where a, b and c are integers.

a $\left(3\sqrt{6}+\sqrt{5}\right)^2$

b $\left(\sqrt{2}+3\right)\left(4+\sqrt{2}\right)$

c $\left(\sqrt{2}-3\right)\left(4-\sqrt{2}\right)$

d $\left(3\sqrt{5}+4\right)\left(2\sqrt{5}-6\right)$

e $\left(5\sqrt{3}+3\sqrt{2}\right)\left(4\sqrt{27}-5\sqrt{8}\right)$

7 Rationalise the denominators in these expressions and leave your answers in their simplest form. Show your working.

a $\dfrac{1}{\sqrt{13}}$ b $\dfrac{8}{\sqrt{6}}$

c $\dfrac{\sqrt{11}}{2\sqrt{5}}$ d $\dfrac{3}{\sqrt{2}-1}$

e $\dfrac{3\sqrt{7}\times5\sqrt{4}}{6\sqrt{7}}$ f $\dfrac{13\sqrt{15}-2\sqrt{10}}{4\sqrt{75}}$

g $\dfrac{5}{8-\sqrt{5}}$ h $\dfrac{2\sqrt{2}}{4+\sqrt{2}}$

i $\dfrac{\sqrt{6}-\sqrt{5}}{\sqrt{6}+\sqrt{5}}$ j $\dfrac{3\sqrt{11}-4\sqrt{7}}{\sqrt{11}-\sqrt{7}}$

8 Rationalise the denominators and simplify these expressions. a, b and c are integers.

a $\dfrac{a+\sqrt{b}}{\sqrt{b}}$ b $\dfrac{a+\sqrt{b}}{a-\sqrt{b}}$

c $\dfrac{\sqrt{a}+b\sqrt{c}}{b\sqrt{c}}$ d $\dfrac{\sqrt{a}-b\sqrt{c}}{\sqrt{a}+\sqrt{b}}$

Reasoning and problem-solving

Strategy

To solve problems involving surds

1. Use the information given to form an expression involving surds.
2. If possible, simplify surds. They should have the lowest possible number under the root sign.
3. Rationalise any denominator containing surds.

Example 3

The sides of a parallelogram are $\sqrt{27}$ m and $2\sqrt{12}$ m, and it has a perpendicular height of $\dfrac{10}{\sqrt{3}}$ m.

a Work out the perimeter of the parallelogram.

b Work out the area of the parallelogram.

Give your answers in their simplest form.

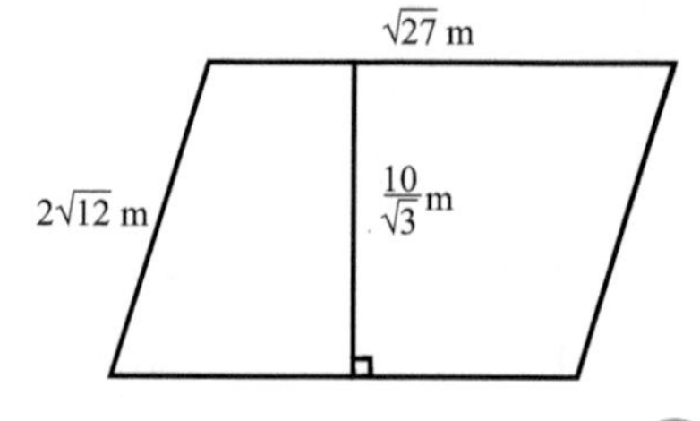

a Perimeter

$2(\sqrt{27}+2\sqrt{12})=2(3\sqrt{3}+4\sqrt{3})=14\sqrt{3}$ m

b Area = base × perpendicular height

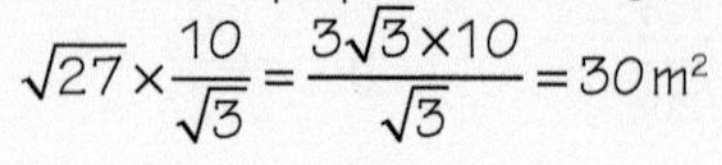

$\sqrt{27}\times\dfrac{10}{\sqrt{3}}=\dfrac{3\sqrt{3}\times10}{\sqrt{3}}=30\text{ m}^2$

1 Form an expression involving surds.

2 Simplify.

2 3 Simplify and rationalise the denominator.

Exercise 1.3B Reasoning and problem-solving

1 A rectangle has sides of length $2\sqrt{3}$ cm and $3\sqrt{2}$ cm. What is its area? Show your working.

2 a A circle has radius $9\sqrt{3}$ cm. Show that its area is $243\pi\text{ cm}^2$.

b A circle has area $245\pi\text{ m}^2$. What is its diameter? Show your working.

3 a A car travels $18\sqrt{35}$ m in $6\sqrt{7}$ s. Work out its speed, showing your working.

b A runner travels for 5 s at $\frac{8}{\sqrt{5}}\text{ m s}^{-1}$.

Work out how far she ran in simplified form. Show your working.

4 A cube has sides of length $(2+\sqrt{7})$ m. Work out its volume in simplified surd form.

5 Rectangle A has sides of length $3\sqrt{3}$ m and $\sqrt{5}$ m. Rectangle B has sides of length $\sqrt{5}$ m and $\sqrt{7}$ m. How many times larger is rectangle A than rectangle B? Give your answer in its simplest surd form, showing your working.

6 A right-angled triangle has perpendicular sides of $2\sqrt{3}$ cm and $3\sqrt{7}$ cm. Calculate the length of the hypotenuse. Show your working and give your answer in simplified form.

7 Base camp is $5\sqrt{5}$ miles due east and $5\sqrt{7}$ miles due north of a walker. What is the exact distance from the walker to the camp? Show your working.

8 The arc of a bridge forms part of a larger circle with radius $\frac{12}{\sqrt{3}}$ m. If the arc of the bridge subtends an angle of 45°, show that the length of the bridge is $\sqrt{3}\,\pi$ m.

9 The equation of a parabola is $y^2 = 4ax$

Find y when $a = 6-\sqrt{6}$ and $x = \frac{6+\sqrt{6}}{10}$

Show your working and give your answer in simplified form.

10 Show that the ratio of the volumes of two cubes of sides $6\sqrt{8}$ cm and $4\sqrt{2}$ cm is 27

11 The top speeds, in m s^{-1}, of two scooters are given as $\frac{12}{\sqrt{a}}$ and $\frac{17\sqrt{a}}{3a}$, where a is the volume of petrol in the tank. Find the difference in top speed between the two scooters if they both contain the same volume of petrol. Give your answer in surd form, showing your working.

12 The area of an ellipse with semi-diameters a and b is given by the formula πab

Work out the area of an ellipse where

$a = \frac{5}{4+\sqrt{3}}$ m and $b = \frac{8}{4-\sqrt{3}}$ m

Show your working.

13 The force required to accelerate a particle can be calculated using $F = ma$, where F is the force, m is the mass of the particle and a is the acceleration. Showing your working, find F when $m = 8\sqrt{6}$ and $a = \frac{5}{2+\sqrt{6}}$

14 An equilateral triangle with side length $5\sqrt{6}$ inches has one vertex at the origin and one side along the positive x-axis.

The centre is on the vertical line of symmetry, $\frac{1}{3}$ of the way from the x-axis to the vertex.
Work out the distance from the origin to the centre of mass of the triangle. Show your working.

Challenge

15 The Indian Mathematician Brahmagupta (598 – 670) developed a formula to calculate the area of a cyclic quadrilateral.

If the sides of the quadrilateral are a, b, c and d, and $s = \frac{a+b+c+d}{2}$, the area is $A = \sqrt{(s-a)(s-b)(s-c)(s-d)}$

A quadrilateral with side lengths $5+5\sqrt{2}$, $3+3\sqrt{2}$, $6+4\sqrt{2}$ and $2+4\sqrt{2}$ cm is inscribed in a circle.

Prove that $A = 2(1+\sqrt{2})\sqrt{15(11+8\sqrt{2})}$

1.4 Quadratic functions

Fluency and skills

A **quadratic function** can be written in the form $ax^2 + bx + c$, where a, b and c are constants and $a \neq 0$

A **quadratic equation** can be written in the general form $ax^2 + bx + c = 0$

Curves of quadratic functions, $y = ax^2 + bx + c$, have the same general shape. The curve crosses the y-axis when $x = 0$, and the curve crosses the x-axis at any **roots** (or solutions) of the equation $ax^2 + bx + c = 0$

Quadratic curves are symmetrical about their **vertex** (the turning point). For $a > 0$, this vertex is always a **minimum** point, and for $a < 0$ this vertex is always a **maximum** point.

When $a > 0$, a quadratic graph looks like this.

When $a < 0$, a quadratic graph looks like this.

Example 1

The quadratic equation $3x^2 - 20x - 7 = 0$ has solutions $x = -\frac{1}{3}$ and $x = 7$
Sketch the curve of $f(x) = 3x^2 - 20x - 7$, showing where it crosses the axes.

When $x = 0$, $y = 3(0)^2 - 20(0) - 7 = -7$ — Find the y-intercept.

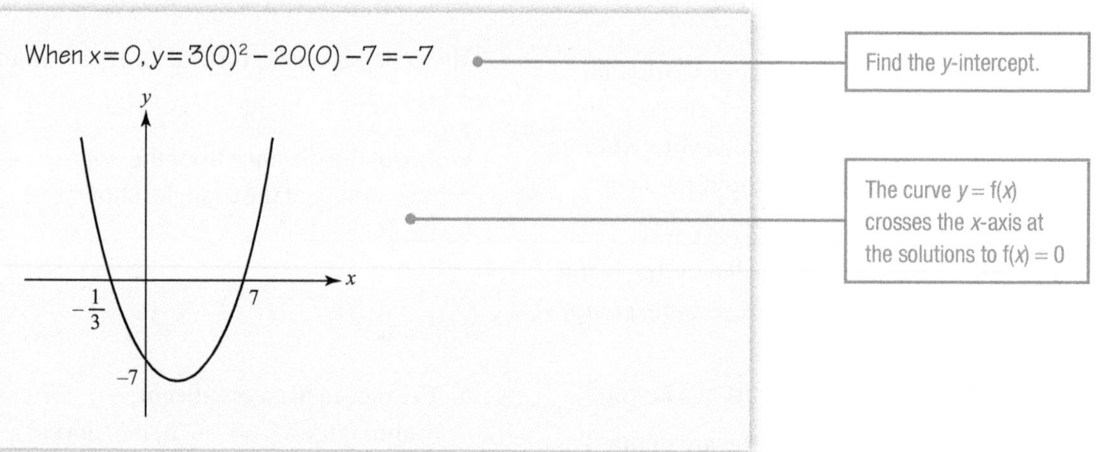

The curve $y = f(x)$ crosses the x-axis at the solutions to $f(x) = 0$

Calculator

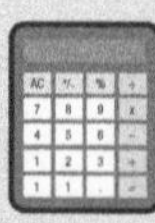

Try it on your calculator

You can sketch a curve on a graphics calculator.

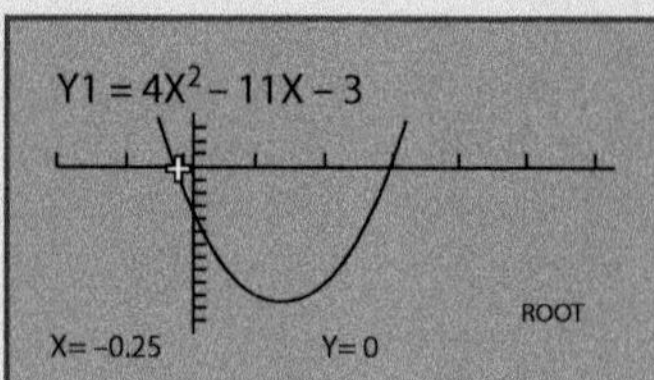

Activity
Find out how to sketch the curve $y = 4x^2 - 11x - 3$ on *your* graphics calculator.

You can solve some quadratics in the form $ax^2+bx+c=0$ by **factorisation**. To factorise a quadratic, try to write it in the form $(mx+p)(nx+q)=0$

Key point

A quadratic equation that can be written in the form $(mx+p)(nx+q)=0$ has solutions $x=-\frac{p}{m}$ or $x=-\frac{q}{n}$

Example 2

Find the solutions of the quadratic equation $6x^2+17x+7=0$ by factorisation.

$6x^2+17x+7=0$

$6x^2+3x+14x+7=0$ — Split the x term so that the two coefficients multiply to give ac. $6\times7=42$ and $3\times14=42$

$3x(2x+1)+7(2x+1)=0$ — Factorise the first pair of terms, then the second pair of terms. Take out a factor which is common to both pairs.

$(2x+1)(3x+7)=0$ — Factorise the full expression.

$x=-\frac{7}{3}$ or $x=-\frac{1}{2}$

Sometimes a quadratic will not factorise easily. In these cases you may need to **complete the square**.

Key point

Any quadratic expression can be written in the following way. This is called completing the square.

$$ax^2+bx+c\equiv a\left(x+\frac{b}{2a}\right)^2+q$$

You'll need to find the value of q yourself. It will be equal to $c-\frac{b^2}{4a}$

When $a=1$ and $q=0$, the expression is known as a **perfect square**. For example, $x^2+6x+9=(x+3)^2$

Perfect squares have only one root, so a graph of the quadratic function touches the x-axis only once, at its vertex.

Example 3

By completing the square, find all the solutions of $4-3x^2-6x=0$

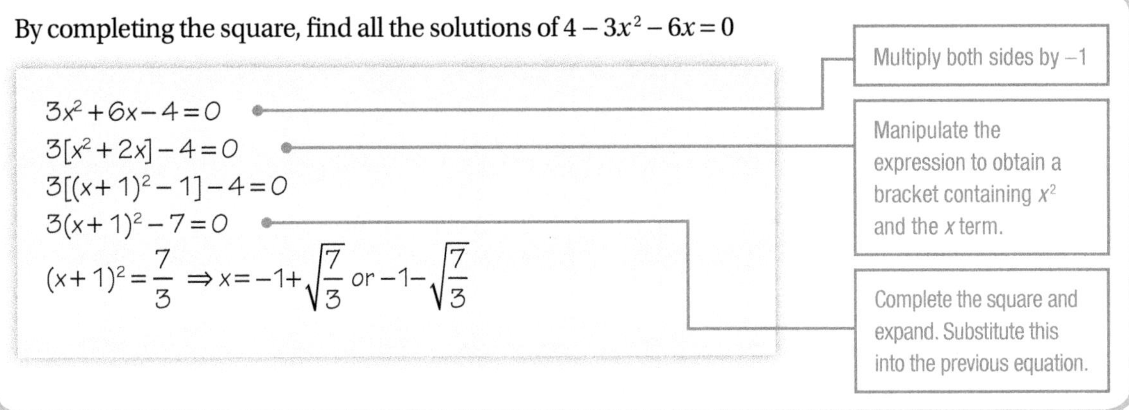

$3x^2+6x-4=0$ — Multiply both sides by -1

$3[x^2+2x]-4=0$ — Manipulate the expression to obtain a bracket containing x^2 and the x term.

$3[(x+1)^2-1]-4=0$

$3(x+1)^2-7=0$ — Complete the square and expand. Substitute this into the previous equation.

$(x+1)^2=\frac{7}{3}\Rightarrow x=-1+\sqrt{\frac{7}{3}}$ or $-1-\sqrt{\frac{7}{3}}$

Completing the square is a useful tool for determining the maximum or minimum point of a quadratic function.

Example 4

Complete the square to determine the minimum point of the graph of $f(x) = 2x^2 + 12x + 16$

$f(x) = 2x^2 + 12x + 16 = 2(x^2 + 6x) + 16 = 2[(x+3)^2 - 9] + 16$
$= 2(x+3)^2 - 2$

At minimum point $x = -3$

$f(-3) = 2(0)^2 - 2 = -2 \Rightarrow$ minimum point $(-3, -2)$

$(x+3)^2 \geq 0$, so the minimum point is when $(x+3)^2 = 0$

By writing the equation, $ax^2 + bx + c = 0$, $a \neq 0$, in completed square form you can derive the **quadratic formula** for solving equations.

$$ax^2 + bx + c = 0$$

$$a\left[x^2 + \frac{b}{a}x\right] + c = 0$$

$$a\left[\left(x + \frac{b}{2a}\right)^2 - \frac{b^2}{4a^2}\right] + c = 0$$

$$\left(x + \frac{b}{2a}\right)^2 - \frac{b^2}{4a^2} = \frac{-c}{a}$$

$$\left(x + \frac{b}{2a}\right)^2 = \frac{b^2}{4a^2} - \frac{c}{a}$$

$$\left(x + \frac{b}{2a}\right)^2 = \frac{b^2 - 4ac}{4a^2}$$

$$x + \frac{b}{2a} = \frac{\pm\sqrt{b^2 - 4ac}}{2a}$$

$$x = \frac{-b \pm \sqrt{b^2 - 4ac}}{2a}$$

Key point

For constants a, b and c, $a > 0$, the solutions to the equation $ax^2 + bx + c = 0$ are

$$x = \frac{-b \pm \sqrt{b^2 - 4ac}}{2a}$$

The expression inside the square root is called the **discriminant**, Δ

Key point

$\Delta = b^2 - 4ac$

If the discriminant, Δ, is positive, it has two square roots. If Δ is 0, it has one square root. If Δ is negative, it has no real square roots. The value of Δ tells you whether a quadratic equation $ax^2 + bx + c = 0$ has two, one or no real solutions. This result is useful for curve sketching.

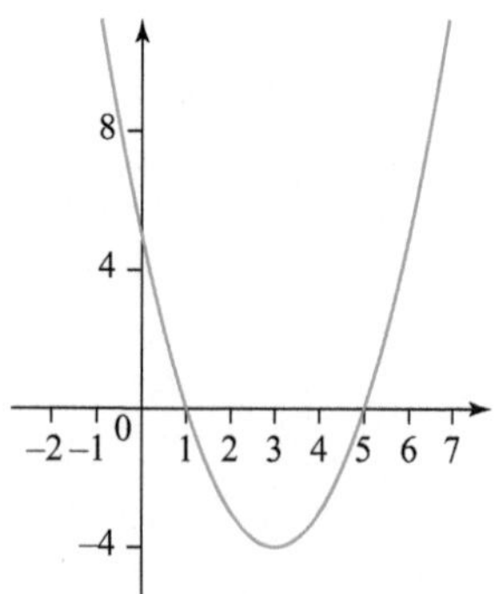

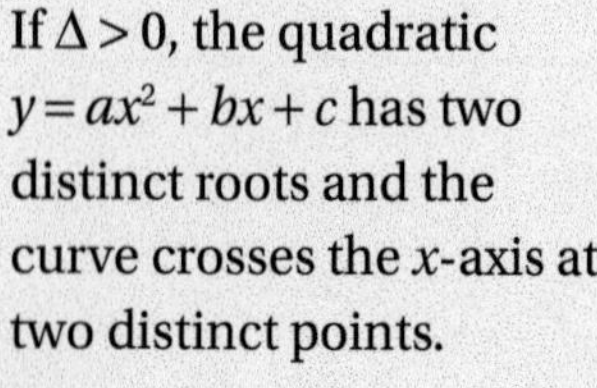

If $\Delta > 0$, the quadratic $y = ax^2 + bx + c$ has two distinct roots and the curve crosses the x-axis at two distinct points.

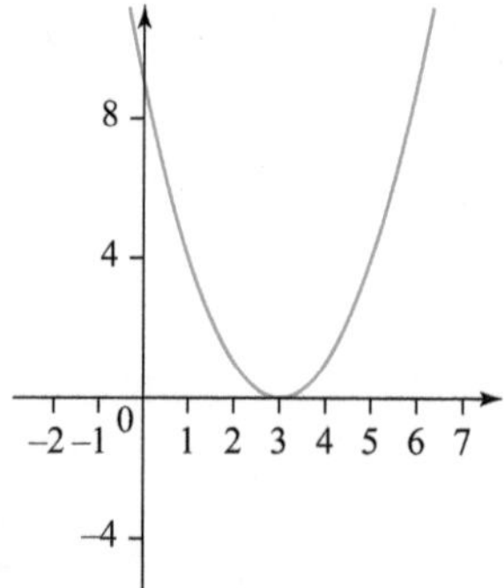

If $\Delta = 0$, the quadratic $y = ax^2 + bx + c$ has one repeated root and the x-axis is a tangent to the curve at this point.

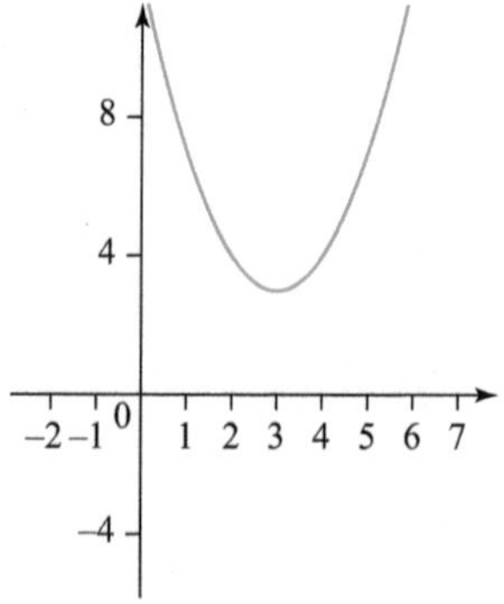

If $\Delta < 0$, the quadratic $y = ax^2 + bx + c$ has no (real) roots and the curve does not cross the x-axis at any point.

Example 5

Use the discriminant $\Delta = b^2 - 4ac$ to determine how many roots each of these quadratic equations have.

a $x^2 + 2x + 1 = 0$ **b** $x^2 + 2x - 8 = 0$ **c** $x^2 + 6x + 10 = 0$

a $\Delta = b^2 - 4ac = 2^2 - 4 \times 1 \times 1 = 0$

So $x^2 + 2x + 1 = 0$ has one repeated root.

b $\Delta = b^2 - 4ac = 2^2 - 4 \times 1 \times -8 = 36 > 0$

So $x^2 + 2x - 8 = 0$ has two distinct roots.

c $\Delta = b^2 - 4ac = 6^2 - 4 \times 1 \times 10 = -4 < 0$

So $x^2 + 6x + 10 = 0$ has no real roots.

Calculator

Try it on your calculator

You can solve a quadratic equation on a calculator.

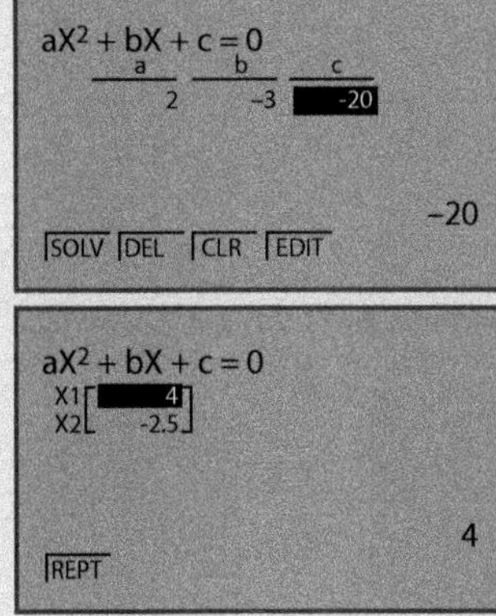

Activity

Find out how to solve $2x^2 - 3x - 20 = 0$ on *your* calculator.

Exercise 1.4A Fluency and skills

1 Solve these quadratic equations by factorisation.

a $x^2 - 18 = 0$ **b** $2x^2 - 6 = 0$

c $4x^2 + 5x = 0$ **d** $x^2 + 2\sqrt{3}x + 3 = 0$

e $2x^2 + 5x - 3 = 0$ **f** $3x^2 - 23x + 14 = 0$

g $16x^2 - 24x + 9 = 0$ **h** $18 + x - 4x^2 = 0$

2 For each quadratic function

i Factorise the equation,

ii Use your answer to part **i** to sketch a graph of the function.

a $f(x) = x^2 + 3x + 2$ **b** $f(x) = x^2 + 6x - 7$

c $f(x) = -x^2 - x + 2$ **d** $f(x) = -x^2 - 7x - 12$

e $f(x) = 2x^2 - x - 1$ **f** $f(x) = -3x^2 + 11x + 20$

3 Solve these quadratic equations using the formula. Write your answers both exactly (in surd form) and also, where appropriate, correct to 2 decimal places.

a $3x^2 + 9x + 5 = 0$ **b** $4x^2 + 5x - 1 = 0$

c $x^2 + 12x + 5 = 0$ **d** $28 - 2x - x^2 = 0$

e $x^2 + 15x - 35 = 0$ **f** $34 + 3x - x^2 = 0$

g $4x^2 - 36x + 81 = 0$ **h** $3x^2 - 23x + 21 = 0$

i $5x^2 + 16x + 9 = 0$ **j** $10x^2 - x - 1 = 0$

4 For each quadratic function, complete the square and thus determine the coordinates of the minimum or maximum point of the curve.

a $f(x) = x^2 - 14x + 49$ **b** $f(x) = x^2 + 2x - 5$

c $f(x) = -x^2 - 6x - 5$ **d** $f(x) = -x^2 + 4x + 3$

e $f(x) = 9x^2 - 6x - 5$ **f** $f(x) = -2x^2 - 28x - 35$

MyMaths 2014–2017, 2025, 2026, 2257 SEARCH

5 Complete the square to work out the exact solutions to these quadratic equations.

a $x^2-2x=0$ b $3-4x-x^2=0$
c $x^2-14x+33=0$ d $x^2+8x+10=0$
e $x^2-6x+9=0$ f $x^2+10x+24=0$
g $x^2+22x+118=0$ h $x^2-16x+54=0$
i $4x^2-12x+2=0$ j $9x^2+12x-2=0$
k $x^2+11x+3=0$ l $9x^2-30x-32=0$

6 Solve these quadratic equations using your calculator.

a $2x^2-6x=0$ b $x^2+2x-15=0$
c $x^2-5x-6=0$ d $8+2x-x^2=0$
e $2x^2-x-15=0$ f $6+8x-8x^2=0$

7 By evaluating the discriminant, identify the number of real roots of these equations.

a $x^2+2x-5=0$ b $13+3x-x^2=0$
c $x^2+5x+5=0$ d $-3+2x-x^2=0$
e $4x^2+12x+9=0$ f $-35+2x-x^2=0$
g $9x^2-66x+121=0$
h $-100-100x-100x^2=0$

Reasoning and problem-solving

Strategy

To solve a problem involving a quadratic curve

1. Factorise the equation or complete the square and solve as necessary.
2. Sketch the curve using appropriate axes and scale.
3. Mark any relevant points in the context of the question.

Example 6

The motion of a body, which has an initial velocity u and acceleration a, is given by the formula $s=ut+\frac{1}{2}at^2$, where s is the displacement after a time t

a By completing the square and showing all intermediate steps, sketch the graph of s against t when $u=8$ and $a=-4$

b What is happening to the body at the turning point of the graph?

a $s=8t-2t^2$
$=-2(t^2-4t)$
$=-2[(t-2)^2-4]=-2(t-2)^2+8$

So there is a turning point at (2, 8)

Where the curve crosses the x-axis,

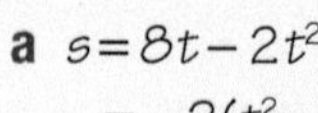

$2t(4-t)=0$
$t=0$ or 4

Therefore the curve cuts the x-axis at (0, 0) and (4, 0)

b The body is reversing direction. At this point it has zero velocity.

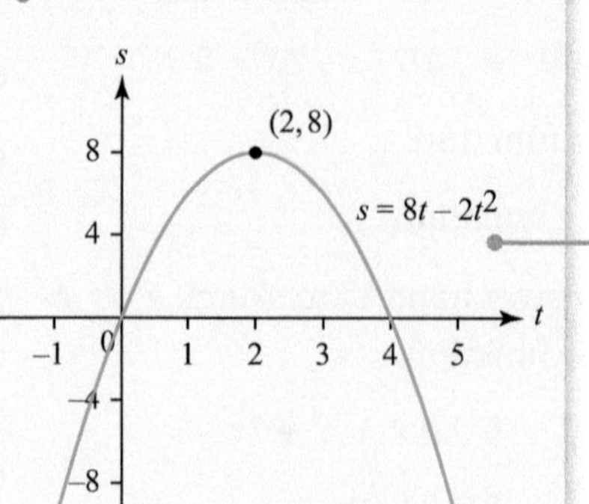

① Factorise and complete the square.

②③ Sketch the curve and mark on it any relevant points.

Exercise 1.4B Reasoning and problem-solving

1 A designer is lining the base and sides of a rectangular drawer, dimensions $2x$ cm by $3x$ cm by 5 cm high, with paper.

The total area of paper is $4070\ \text{cm}^2$.

a Write and solve an equation to find x

b Hence work out the volume of the drawer in litres.

2 Sam and his mother Jane were both born on January 1st.

In 2002, Sam was x years old and Jane was $2x^2 + 11x$ years old. In 2007, Jane was five times as old as Sam.

a Form and solve a quadratic equation in x

b Hence work out Jane's age when Sam was born.

3 A photo is to be pasted onto a square of white card with side length x cm. The photo is $\frac{3}{4}x$ cm long and its width is 20 cm less than the width of the card. The area of the remaining card surrounding the photo is $990\ \text{cm}^2$. Work out the dimensions of the card and photo.

4 A piece of wire is bent into a rectangular shape with area $85\ \text{in}^2$.

The total perimeter is 60 inches and the rectangle is x^2 in long.

a Form an expression for the area of the rectangle.

b By substituting z for x^2, form a quadratic equation in z

c Hence work out all possible values of x

5 A man stands on the edge of a cliff and throws a stone out over the sea. The height, h m, above the sea that the stone reaches after t seconds is given by the formula $h = 50 + 25t - 5t^2$

a Complete the square.

b Sketch the graph of $h = 50 + 25t - 5t^2$

c Use your graph to estimate

i The maximum height of the stone above the sea and the time at which it reaches this height,

ii The time when the stone passes the top of the cliff on the way down,

iii The time when the stone hits the sea.

6 A firm making glasses makes a profit of y thousand pounds from x thousand glasses according to the equation $y = -x^2 + 5x - 2$

a Sketch the curve.

b Use your graph to estimate

i The value of x to give maximum profit,

ii The value of x not to make a loss,

iii The range of values of x which gives a profit of more than £3250

7 The mean braking distance, d yards, for a car is given by the formula $d = \frac{v^2}{50} + \frac{v}{3}$, where v is the speed of the car in miles per hour.

a Sketch this graph for $0 \le v \le 80$

b Use your sketch to estimate the safe braking distance for a car driving at

i 15 mph ii 45 mph iii 75 mph

c A driver just stops in time in a distance of 50 yards.
How fast was the car travelling when the brakes were applied?

Challenge

8 Use a suitable substitution to solve $2(k^6 - 11k^3) = 160$ for k. Give your answers in exact form.

MyMaths 2014–2017, 2025, 2026, 2257 SEARCH

1.5 Simultaneous equations

Fluency and skills

The graphs of $3x - y = 7$ and $5x + y = 9$ are shown.

Only one pair of values for (x, y) satisfies both equations. This corresponds to the point of intersection of the two graphs. In this example it is $x = 2$ and $y = -1$

When you solve two equations together like this, they are called **simultaneous equations.**

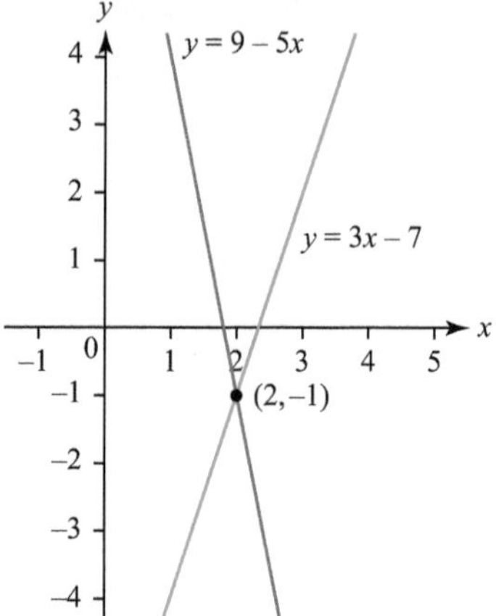

Key point

You can solve a pair of linear simultaneous equations

1. Graphically.
2. By eliminating one of the **variables.**
3. By substituting an expression for one of the variables from one equation into the other.

Example 1

Solve the simultaneous equations $2a - 5b = -34$ and $3a + 4b = -5$ by elimination.

$2a - 5b = -34$ (1) and $3a + 4b = -5$ (2) — Label the equations 1 and 2

$8a - 20b = -136$ and $15a + 20b = -25$ — Multiply equation (1) by four and equation (2) by five.

$23a = -161$ — Add equations to eliminate the b terms.

$a = -7$

$-14 - 5b = -34$ — To find b, substitute $a = -7$ into equation (1)

$-14 + 34 = 5b \Rightarrow b = 4$

$a = -7$ and $b = 4$ — You can check your answers by solving the simultaneous equations on your calculator.

Example 2

Solve the simultaneous equations $2c - 3d = 8$ and $4c + 5d = 5$ by substitution.

$2c - 3d = 8$ (1)

$4c + 5d = 5$ (2) — Label the equations 1 and 2

$c = \frac{8 + 3d}{2}$ — Rearrange equation (1) to make c the subject.

so $4\left(\frac{8 + 3d}{2}\right) + 5d = 5$ — Substitute for c in equation (2) and solve to find d

$2(8 + 3d) + 5d = 5 \Rightarrow d = -1$

$2c - 3(-1) = 8 \Rightarrow c = \frac{5}{2}$ — To find c, substitute $d = -1$ into equation (1)

So the solution is $c = \frac{5}{2}$ and $d = -1$

A straight line can intersect a quadratic curve at either two, one or zero points.

You should equate expressions for the curve and the line to find the points of intersection. You will then obtain a quadratic equation in the form $ax^2 + bx + c = 0$
You can use the discriminant, $\Delta = b^2 - 4ac$, to show if this equation has two, one or no solutions.

The diagram shows the graphs of $y = x^2 + 4x$, $y = -4x - 16$, $y = x$ and $y = x - 3$

$y = -4x - 16$ touches $y = x^2 + 4x$ at the point $(-4, 0)$
$y = x$ intersects $y = x^2 + 4x$ at the points $(0, 0)$ and $(-3, -3)$
$y = x - 3$ does not intersect $y = x^2 + 4x$ at any point.

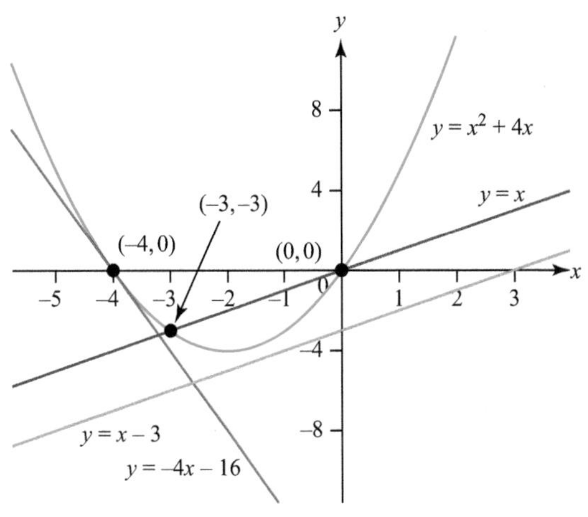

ICT Resource online

To investigate simultaneous equations, click this link in the digital book.

Example 3

Solve the simultaneous equations $y = x^2 + 4x$ and $y + 4x + 16 = 0$. Interpret your answers graphically.

$y = -4x - 16$ — Make y the subject.

$x^2 + 4x = -4x - 16$ — Substitute $y = -4x - 16$ into the quadratic.

$x^2 + 8x + 16 = 0 \Rightarrow (x+4)(x+4) = 0 \Rightarrow$ so $x = -4$ — Solve the resulting quadratic equation.

When $x = -4$, $y = -4 \times -4 - 16 = 0$

The solution is $x = -4$, $y = 0$

$\therefore$ the straight line $y + 4x + 16 = 0$ touches the curve $y = x^2 + 4x$ at a single point $(-4, 0)$, and is therefore a tangent to the curve. — Derive the nature of the roots of the quadratic equation.

Exercise 1.5A Fluency and skills

Solve the simultaneous equations from 1 to 19. You must show your working.

1 $x + y = 7$
$2x + y = 11$

2 $a + b = 7$
$2a - b = 11$

3 $2a + 3b = 8$
$a + 2b = 5$

4 $4x + 2y = 2$
$5x - 2y = 7$

5 $4c + 2d = 2$
$5c + 2d = 7$

6 $2e - f = 13$
$e - f = 5$

7 $2e - f = 13$
$e + f = 5$

8 $7g + 4h = 12$
$-5g + 4h = 12$

9 $7x + 4y = 12$
$-5x - 4y = 12$

10 $3m - 4n = -15$
$-3m - n = 0$

11 $-3m - 4n = -15$
$-3m - n = 0$

12 $4a - 21 = b$
$2b = 13 - 3a$

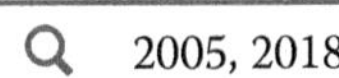

13 $5c+2d=9$
$3c=d-10$

14 $3c-4d=29$
$4c=13+3d$

15 $2e=13-3f$
$6f=e-4$

16 $x+y=3$
$x^2+y=3$

17 $g+h=1$
$g^2-h=5$

18 $3g+2h=13$
$h+2g^2=20$

19 $10m=7n+17$
$m=n^2$

20 Find the point(s) of intersection of the graphs $y^2=5x$ and $y=x$. Show your working.

21 Find the point(s) of intersection of the graphs $y^2=6x+7$ and $y=x+2$. Show your working.

22 Solve the simultaneous equations $x+y^2=2$ and $2=3x+y$, showing your working. Find the points of intersection.

23 Solve the simultaneous equations $y^2=-1-5x$, $y=2x+1$. Find the points of intersection, showing your working.

24 A curve has equation $xy=20$

A straight line has equation $y=8+x$

Solve the two equations simultaneously and show that the points of intersection are $(2, 10)$ and $(-10, -2)$

Reasoning and problem-solving

Strategy

To solve a simultaneous equations problem

1. Use the information in the question to create the equations.
2. Use either elimination or substitution to solve your equations.
3. Check your solution and interpret it in the context of the question.

Example 4

A rectangle has sides of length $(x+y)$ m and $2y$ m. The rectangle has a perimeter of 64 m and an area of 240 m^2. Calculate the possible values of x and y. Show your working.

The perimeter is $2[x+y+2y]=2x+6y$
and the area is $2y(x+y)=2xy+2y^2$

$\therefore 2x+6y=64$ and $2xy+2y^2=240$ — **1** Create the equations.

$x+3y=32$ or $x=32-3y$ (1)
$2xy+2y^2=240$ or $xy+y^2=120$ (2)

$(32-3y)y+y^2=120$ — Substitute $x=32-3y$ from equation (1) into equation (2)

$32y-2y^2=120$
$2y^2-32y+120=0$
$y^2-16y+60=0$
$(y-6)(y-10)=0$
$y=6$ or 10

$x=32-3\times6=14$ or $32-3\times10=2$ — **2** Substitute the values for y into equation (1) to obtain values for x

Checking $14\times6+6^2=84+36=120$ ✓
or $2\times10+10^2=20+100=120$ ✓

$\therefore$ the two possible values of x and y are
$x=14, y=6$ or $x=2, y=10$ — **3** Check your solution in equation (2) and interpret it in context.

Exercise 1.5B Reasoning and problem-solving

1 In a recent local election, the winning candidate had an overall majority of 257 votes over her only opponent. There were 1619 votes cast altogether.

Form a pair of simultaneous linear equations.

How many votes did each candidate poll?

2 A fisherman is buying bait. He can either buy 6 maggots and 4 worms for £1.14 or 4 maggots and 7 worms for £1.28

How much do maggots and worms cost individually? Show your working.

3 The straight line $y = mx + c$ passes through the points $(3, -10)$ and $(-2, 5)$

Find the values of m and c

4 This triangle is equilateral. Find the values of m and n

$8m$ $4n$ $2m + 5n - 6$

5 This triangle is isosceles. It has a perimeter of 150 cm. Find the values of p and q

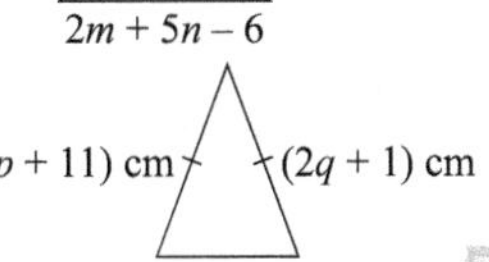

6 The ages of Florence and Zebedee are in the ratio 2 : 3
In 4 years' time, their ages will be in the ratio 3 : 4
Use simultaneous equations to calculate how old they are now. Show your working.

7 **a** Try to solve the simultaneous linear equations $y - 2x = 3$ and $4x = 2y - 6$

How many solutions are there? Explain your answer.

b Try to solve the simultaneous linear equations $y - 2x = 3$ and $4x = 2y - 8$

How many solutions are there? Explain your answer.

8 The equations of three straight lines and a parabola are $y + 2x + 4 = 0$, $y + 11x - 27 = 0$, $x - y + 3 = 0$ and $y = 2x^2 - 19x + 35$. One of the lines intersects the curve at two points, one 'misses' the curve and one is a tangent to the curve. Investigate the nature of the relationship between each of these lines and the curve, and calculate any real points of intersection.

9 Prove that the line $y = 2x - 9$ does not intersect the parabola $y = x^2 - x - 6$

10 The sums of the first n terms of two sequences of numbers are given by formulae $S_1 = 2n + 14$ and $2S_2 = n(n + 1)$. For which values of n does $S_1 = S_2$? Explain your results carefully.

11 A farmer has 600 m of fencing. He wants to use it to make a rectangular pen of area 16 875 m^2

Calculate the possible dimensions of this pen.

12 The equation $x^2 + y^2 = 25$ represents a circle of radius 5 units. Prove that the line $3x + 4y = 25$ is a tangent to this circle and find the coordinates of the point where the tangent touches the circle.

13 An ellipse has the equation $4y^2 + 9x^2 = 36$

Show that the line $y = 2x + 1$ intersects this ellipse at the points $\left(\frac{-8 \pm 12\sqrt{6}}{25}, \frac{9 \pm 24\sqrt{6}}{25}\right)$

Challenge

14 Two particles, A and B, move along a straight line. At a time, t, the position of A from a fixed point, O, on the line is given by the formula $x = 2 + 8t - t^2$ and that of B by $x = 65 - 8t$

a How far from O is each particle initially?

b Explain how you know that B is initially moving towards O

c Explain how you know A moves away from O and then moves back towards O

d What is the maximum distance of A from O?

e Calculate the first time when both particles are at the same distance from O

f In which directions are A and B moving at the time you calculated in part **e**?

MyMaths 2005, 2018 SEARCH

1.6 Lines and circles

Fluency and skills

The equation of a straight line can be written in the form $y = mx + c$

where m is the gradient and c is the y-intercept.

Key point

A straight line can also be written in the form
$y - y_1 = m(x - x_1)$
where (x_1, y_1) is a point on the line and m is the gradient.

You can rearrange the general equation of a straight line to get a formula for the gradient.

Key point

The gradient of a straight line through two points (x_1, y_1) and (x_2, y_2) is $m = \frac{y_2 - y_1}{x_2 - x_1}$

You can use Pythagoras' theorem to find the distance between two points.

Key point

The distance between two points (x_1, y_1) and (x_2, y_2) is given by the formula $\sqrt{(x_1 - x_2)^2 + (y_1 - y_2)^2}$

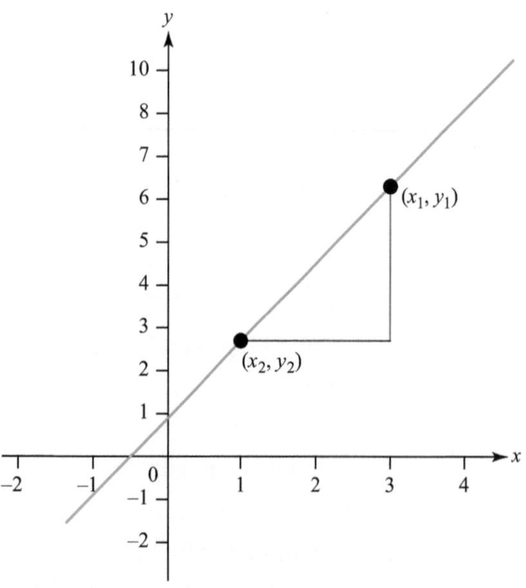

Key point

The coordinates of the midpoint of the line joining (x_1, y_1) and (x_2, y_2) are given by the formula

$$\left(\frac{x_1 + x_2}{2}, \frac{y_1 + y_2}{2}\right)$$

You can use the gradients of two lines to decide if they are **parallel** or **perpendicular**.

Key point

Two lines are described by the equations $y_1 = m_1x + c_1$ and $y_2 = m_2x + c_2$

If $m_1 = m_2$, the two lines are parallel.

If $m_1 \times m_2 = -1$, the two lines are perpendicular.

Example 1

A straight line segment joins the points $(-2, -3)$ and $(4, 9)$

a Work out the midpoint of the line segment.

b Work out the equation of the perpendicular bisector of the line segment. Give your answer in the form $ay + bx + c = 0$ where a, b and c are integers.

a $\left(\frac{-2+4}{2}, \frac{-3+9}{2}\right) = (1, 3)$

b Gradient of line segment $= \frac{9--3}{4--2} = 2$ — Use gradient $= \frac{y_2 - y_1}{x_2 - x_1}$

Gradient of perpendicular bisector $= -\frac{1}{2}$ — Use $m_1 \times m_2 = -1$

$y - 3 = -\frac{1}{2}(x - 1)$ — Use $y - y_1 = m(x - x_1)$

$2y - 6 = -x + 1$

$2y + x - 7 = 0$ — Multiply through by 2 and rearrange to the required form.

On a graph, the equation of any circle with centre (a, b) and radius r has the same general form.

The diagram shows a circle, centre $(1, 4)$ and radius 5, with a general point (x, y) shown on the circumference.

The vertical distance of the point (x, y) from the centre is $y - 4$ and the horizontal distance of the point (x, y) from the centre is $x - 1$.

Using Pythagoras' theorem for the right-angled triangle shown, you get $(x-1)^2 + (y-4)^2 = 5^2$

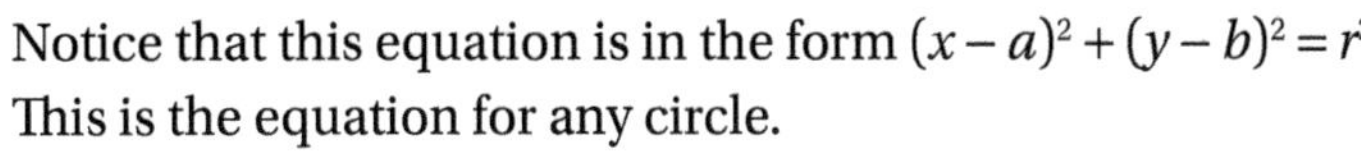

Notice that this equation is in the form $(x-a)^2 + (y-b)^2 = r^2$
This is the equation for any circle.

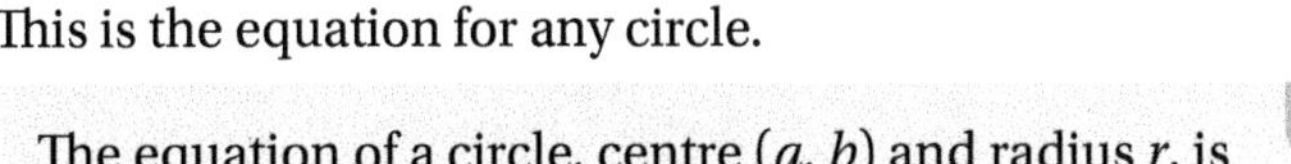

Key point

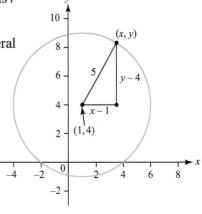

The equation of a circle, centre (a, b) and radius r, is

$$(x-a)^2 + (y-b)^2 = r^2$$

Key point

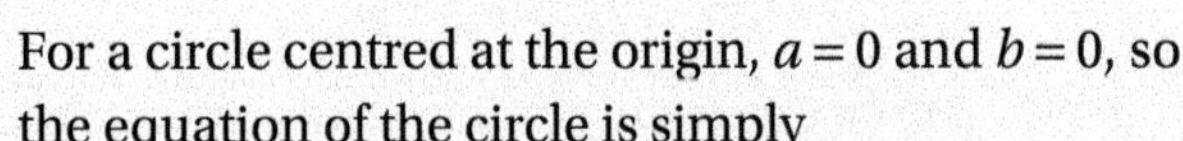

For a circle centred at the origin, $a = 0$ and $b = 0$, so the equation of the circle is simply

$$x^2 + y^2 = r^2$$

MyMaths 2001–2004, 2020, 2021 SEARCH

Example 2

Work out the equation of the circle with centre $(-4, 9)$, radius $\sqrt{8}$

Write your answer without brackets.

$(x+4)^2+(y-9)^2=8$ — Use $(x-a)^2+(y-b)^2=r^2$

$x^2+8x+16+y^2-18y+81-8=0$

$x^2+y^2+8x-18y+89=0$

Example 3

Work out the centre and radius of the circle $4x^2-4x+4y^2+3y-6=0$

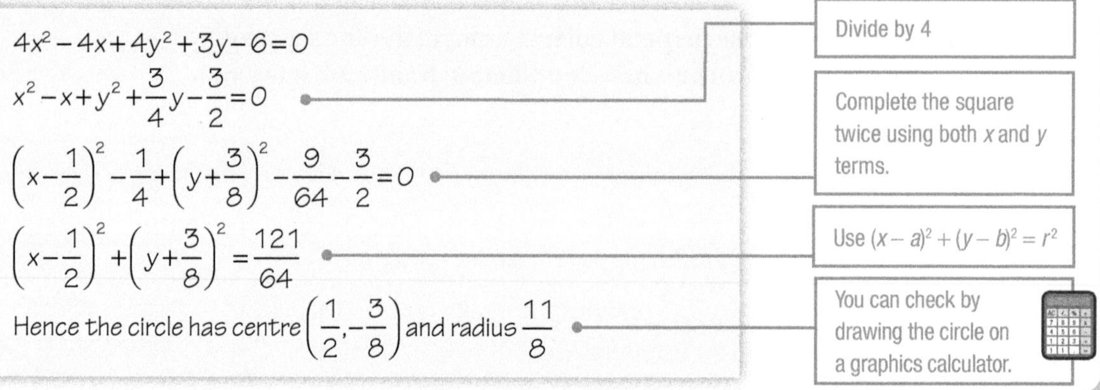

$4x^2-4x+4y^2+3y-6=0$
Divide by 4
$x^2-x+y^2+\frac{3}{4}y-\frac{3}{2}=0$
Complete the square twice using both x and y terms.
$\left(x-\frac{1}{2}\right)^2-\frac{1}{4}+\left(y+\frac{3}{8}\right)^2-\frac{9}{64}-\frac{3}{2}=0$
Use $(x-a)^2+(y-b)^2=r^2$
$\left(x-\frac{1}{2}\right)^2+\left(y+\frac{3}{8}\right)^2=\frac{121}{64}$
Hence the circle has centre $\left(\frac{1}{2},-\frac{3}{8}\right)$ and radius $\frac{11}{8}$
You can check by drawing the circle on a graphics calculator.

When you're working with equations of circles, it's useful to remember some facts about the lines and angles in a circle. You should have come across these before in your studies.

Key point

- If a triangle passes through the centre of the circle, and all three corners touch the circumference of the circle, then the triangle is right-angled.
- The perpendicular line from the centre of the circle to a chord bisects the chord (Point A in the diagram).
- Any tangent to a circle is perpendicular to the radius at the point of contact (B).

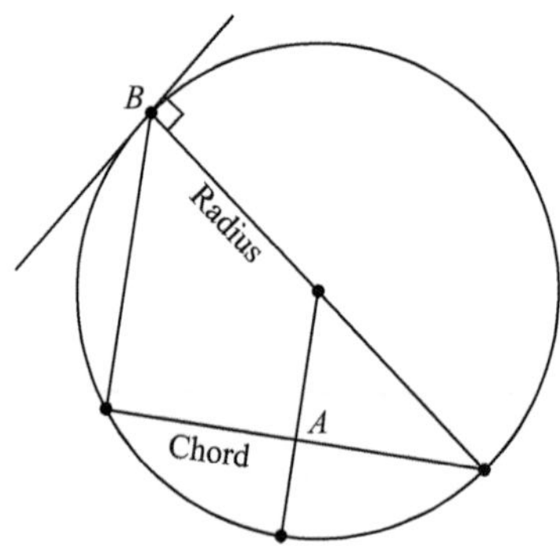

Exercise 1.6A Fluency and skills

1 a Write down the equation of the straight line with gradient $-\frac{2}{3}$ that passes through the point $(-4, 7)$. Give your answer in the form $ax+by+c=0$, where a, b and c are integers.

b Does the point $(13, 3)$ lie on the line described in part **a**?

2 Find the gradient and y-intercept of the line $4x-3y=8$

3 a Show that the lines $2x-3y=4$ and $6x+4y=7$ are perpendicular.

b Show that the lines $2x-3y=4$ and $8x-12y=7$ are parallel.

4 Write down the gradient and y-intercept of the line $\frac{2}{3}x+\frac{3}{4}y+\frac{7}{8}=0$

5 Calculate the gradient of the straight-line segment joining the points $(-5,-6)$ and $(4,-1)$

Hence write down the equation of the line.

6 Write, in both the form $y=mx+c$ and the form $ax+by+c=0$, the equation of the line with gradient -3 passing through $(-8,-1)$

7 Work out the midpoint and length of the line segment joining each of these pairs of points.

a $(2,2)$ and $(6,10)$

b $(-3,-4)$ and $(2,-3)$

c $(0,0)$ and $(\sqrt{5},2\sqrt{3})$

8 Which of these lines are parallel or perpendicular to each other?

$2x+3y=4$ $\quad 4x-5y=6$ $\quad y=4x+8$

$10x-8y=5$ $\quad 10x+8y=5$ $\quad 3y-12x=7$

$6x+9y=12$

9 a Write down the equation of the straight line through the point $(5,-4)$ which is parallel to the line $2x+3y-6=0$

b Write down the equation of the straight line through the point $(-2,-3)$ which is perpendicular to the line $3x+6y+5=0$

10 Write down the equations of each of these circles.

Expand your answers into the form $ax^2+bx+cy^2+dy+e=0$

a Centre $(1,8)$; radius 5

b Centre $(6,-7)$; radius 3

c Centre $(\sqrt{5},\sqrt{2})$; radius $\sqrt{11}$

11 Work out the centre and radius of each of these circles.

a $x^2+18x+y^2-14y+30=0$

b $x^2+12x+y^2+10y-25=0$

c $x^2-2\sqrt{3}x+y^2+2\sqrt{7}y-1=0$

12 Prove that the points $A(-10,-12)$, $B(6,18)$ and $C(-2,-14)$ lie on a semicircle.

13 Write the equation of the tangent to the circle with centre $(4,-3)$ at the point $P(-2,-1)$

14 $(-3,9)$ is the midpoint of a chord within a circle with centre $(7,-1)$ and radius 18

a Calculate the equation of the circle.

b Calculate the length of the chord.

c Complete the square to find the exact coordinates of the ends of the chord.

15 Write down the equations of each of the circles with diameters from

a $(0,0)$ to $(0,20)$

b $(2,6)$ to $(6,2)$

c $(4,-2)$ to $(-3,16)$

d $(-4,-5)$ to $\left(-\sqrt{2},\sqrt{5}\right)$

16 The circle with equation $x^2+y^2=25$ crosses the line $y=7-x$ at two points. Solve these simultaneous equations and find the points of intersection. Show your working.

17 a Write down the equation of the straight line with gradient $-\frac{1}{2}$ that passes through the point $(1,1)$

b Write down the equation of the circle with radius 3 and centre $(2,2)$

c The line in part **a** crosses the circle in part **b** at two points. Solve these simultaneous equations and find the coordinates of these two points. Show your working.

18 a Write down the equation of the straight line that passes through the points $(3,5)$ and $(-1,-3)$

b Write down the equation of the circle with centre $(1,0)$ and radius $\frac{17}{2}$

c The line in part **a** crosses the circle in part **b** at two points. Solve these simultaneous equations and find the coordinates of these two points. Show your working.

Reasoning and problem-solving

Strategy

To solve a problem involving a straight line or a circle

1. Choose the appropriate formulae.
2. Apply any relevant rules and theorems. Draw a sketch if it helps.
3. Show your working and give your answer in the correct form.

Example 4

a A diagonal of a rhombus has equation $2x-3y+8=0$ and midpoint $(-3, 7)$

Work out the equation of the other diagonal.

b One vertex of the rhombus on the original diagonal is $(14, 12)$

Work out the coordinates of the opposite vertex.

a $y=\frac{2}{3}x+\frac{8}{3}$

Gradient of this diagonal is $\frac{2}{3}$

$\therefore$ gradient of the other diagonal is $-\frac{3}{2}$

Hence the equation of other diagonal is

$y-7=-\frac{3}{2}(x+3)$, so

$2y+3x-5=0$

b The vertex of the diagonal is $(14, 12)$ and the midpoint is $(-3, 7)$

so $(-3,7)=\left(\frac{x_1+14}{2},\frac{y_1+12}{2}\right)$

so $x_1=-20$ and $y_1=2$

$\therefore$ the other vertex is $(-20, 2)$

(1) Write the equation of the diagonal in the form $y=mx+c$

(2) Diagonals of a rhombus are perpendicular so $m_1 \times m_2=-1$

(1) Use $y-y_1=m(x-x_1)$

(1) Use midpoint $=\left(\frac{x_1+x_2}{2},\frac{y_1+y_2}{2}\right)$

Example 5

$A(-7, 1)$, $B(11, 13)$ and $C(19, 1)$ are three points on a circle. Prove that AC is a diameter.

Gradient of $AB=\left(\frac{13-1}{11+7}\right)=\frac{2}{3}$

Gradient of $BC=\left(\frac{1-13}{19-11}\right)=-\frac{3}{2}$

$\frac{2}{3}\times-\frac{3}{2}=-1$ so AB is perpendicular to BC

Therefore ABC is 90° and, since the angle in a semicircle is a right angle, ABC is a semicircle and thus AC is a diameter.

(1) Use gradient $=\frac{y_2-y_1}{x_2-x_1}$

(2) Use $m_1 \times m_2=-1$ to prove that the lines are perpendicular.

(2)(3) Use angle in a semicircle theorem and answer the question.

Exercise 1.6B Reasoning and problem-solving

1 A quadrilateral has vertices $P(-15,-1)$, $Q(-3,4)$, $R(12,12)$ and $S(0,7)$. Write the equation of each side and identify the nature of the quadrilateral.

2 A giant kite is constructed using bamboo for the edges and diagonals and card for the sail. When mapped on a diagram, the ends of the long diagonal are at $P(2,-2)$ and $R(-14,-14)$ and the diagonals intersect at M. The short diagonal QS divides RP in the ratio 3 : 1 and $MP = MQ = MS$

Calculate

- **a** The coordinates of M,
- **b** The coordinates of Q and S,
- **c** The equations of the diagonals,
- **d** The equations of the sides of the kite,
- **e** The area of card needed to make the kite,
- **f** The total length of bamboo required for the structure.

3 On a map, three villages are situated at points $A(2,-5)$, $B(10,1)$ and $C(9,-6)$, and all lie on the circumference of a circle.

- **a** Find the equations of the perpendicular bisectors of AB and AC
- **b** Hence work out the centre and equation of the circle and show that the triangle formed by the villages is right-angled.

4 The equation of a circle, centre C, is $x^2 + y^2 - 4x - 12y + 15 = 0$

- **a** Prove the circle does not intersect the x-axis.
- **b** P is the point $(8, 1)$. Find the length CP and determine whether P lies inside or outside the circle.
- **c** Write the set of values of k for which $3y - 4x = k$ is a tangent to the circle.

5 A circle, centre C, has equation $x^2 + y^2 - 20x + 10y + 25 = 0$ and meets the y-axis at Q. The tangent at $P(16, 3)$ meets the y-axis at R. Work out the area of the triangle PQR

6 A park contains a circular lawn with a radius of 50 m. If the park is mapped on a set of axes, with the y-axis due north, this lawn is centred on the point $(-40, -80)$

A straight, underground water pipe runs through the park and its position is represented by the line $y + 2x = -60$

The town council wants to install a drinking fountain in the park. The fountain must be directly above the underground pipe, it must lie on the outer edge of the lawn, and it must be as close to the east side of the park as possible.

Determine the coordinates of the only possible location for the new drinking fountain.

Challenge

7 One diagonal of a rhombus has equation $2y - x = 20$. The two corners that form the other diagonal in the rhombus touch the edges of a circle with equation $x^2 + y^2 - 8x - 24y + 144 = 0$

- **a** Find the radius of the circle.
- **b** Find the equation of the other diagonal of the rhombus.

8 **a** Show that if (a, c) and (b, d) are the ends of a diameter of a circle, the equation of the circle is

$$(x-a)(x-b) + (y-c)(y-d) = 0$$

b The line segment with endpoints $(-3, 12)$ and $(13, 0)$ is the diameter of a circle. Work out the equation of the circle. Give your answer without brackets.

1.7 Inequalities

Fluency and skills

See p.382
For a list of mathematical notation.

You can express **inequalities** using the symbols < (less than), > (greater than), ≤ (less than or equal to) and ≥ (greater than or equal to).

Key point

You can represent inequalities on a number line.

For example

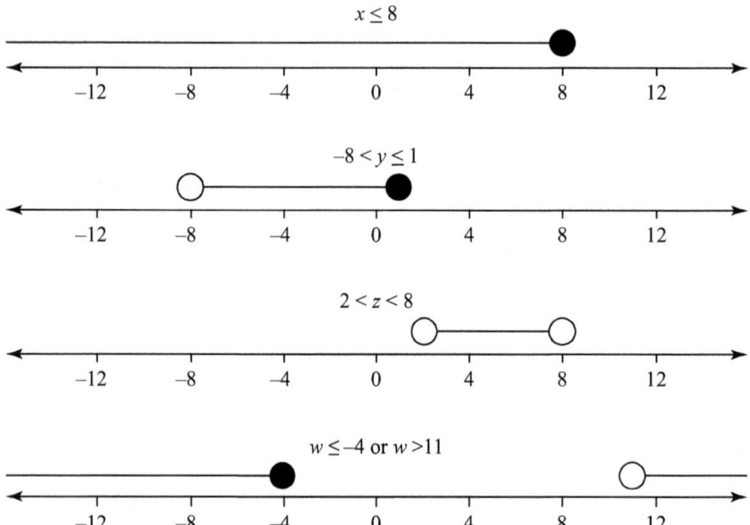

On a number line, you use a dot, ●, when representing ≤ or ≥, and you use an empty circle, ○, when representing < or >

You can also use set notation to represent inequalities.

For example, the last inequality could be represented in any of the following ways.

- $w \in \{w: w \leq -4 \text{ or } w > 11\}$
  w is an element of the set of values that are less than or equal to −4 or greater than 11
- $w \in \{w: w \leq -4\} \cup \{w: w > 11\}$
  w is an element of the union of two sets. This means w is in one set or the other.
- $w \in (-\infty, -4] \cup (11, \infty)$
  w is in the union of two intervals. Square brackets indicate the end value is included in the interval, round brackets indicate that the end value is not included in the interval.

To solve **linear inequalities** you follow the same rules for solving linear equations, but with one exception.

Key point

When you multiply or divide an inequality by a negative number, you reverse the inequality sign.

PURE

Example 1

Solve the inequality $4(3z+12) \leq 5(4z-8)$

$4(3z+12) \leq 5(4z-8)$

$12z+48 \leq 20z-40$ — Expand the brackets.

$12z-20z \leq -40-48$ — Subtract $20z$ and 48 from each side.

$-8z \leq -88$

$z \geq 11$ — Divide by −8. Remember to reverse the inequality sign.

Example 2

Shade each of these regions on a graph.

a $y-2x<6$

b $y+3x \leq 8$; $y-2x<4$; $y>0$

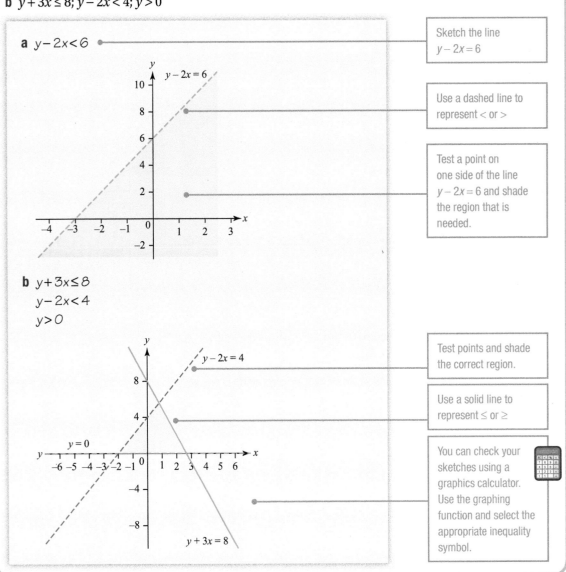

a $y-2x<6$ — Sketch the line $y-2x=6$

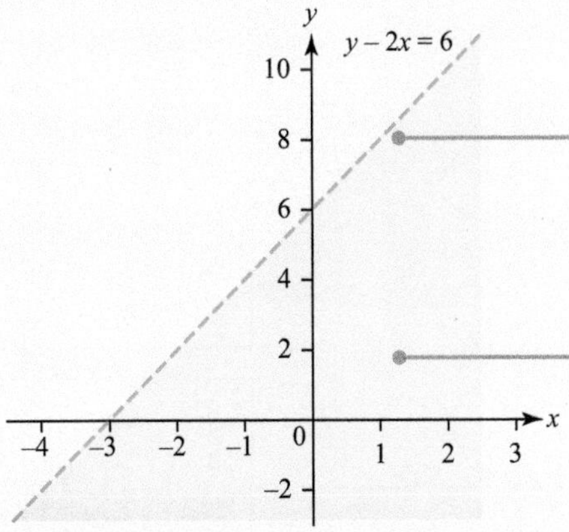

Use a dashed line to represent < or >

Test a point on one side of the line $y-2x=6$ and shade the region that is needed.

b $y+3x \leq 8$

$y-2x<4$

$y>0$

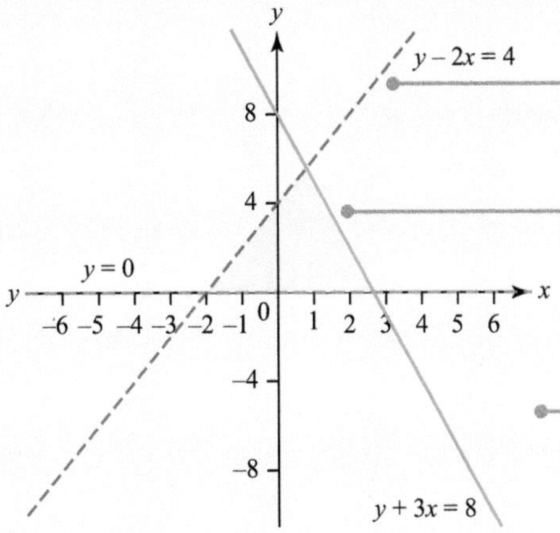

Test points and shade the correct region.

Use a solid line to represent ≤ or ≥

You can check your sketches using a graphics calculator. Use the graphing function and select the appropriate inequality symbol.

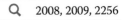

A **quadratic inequality** looks similar to a quadratic equation except it has an inequality sign instead of the '=' sign.

You can solve quadratic inequalities by starting the same way you would to solve quadratic equations. The answer, however, will be a range of values rather than up to two specific values.

Example 3

Solve the equation $x^2 + 4x - 5 = 0$ and sketch the graph $y = x^2 + 4x - 5$

Use your sketch to solve the inequality $x^2 + 4x - 5 \geq 0$

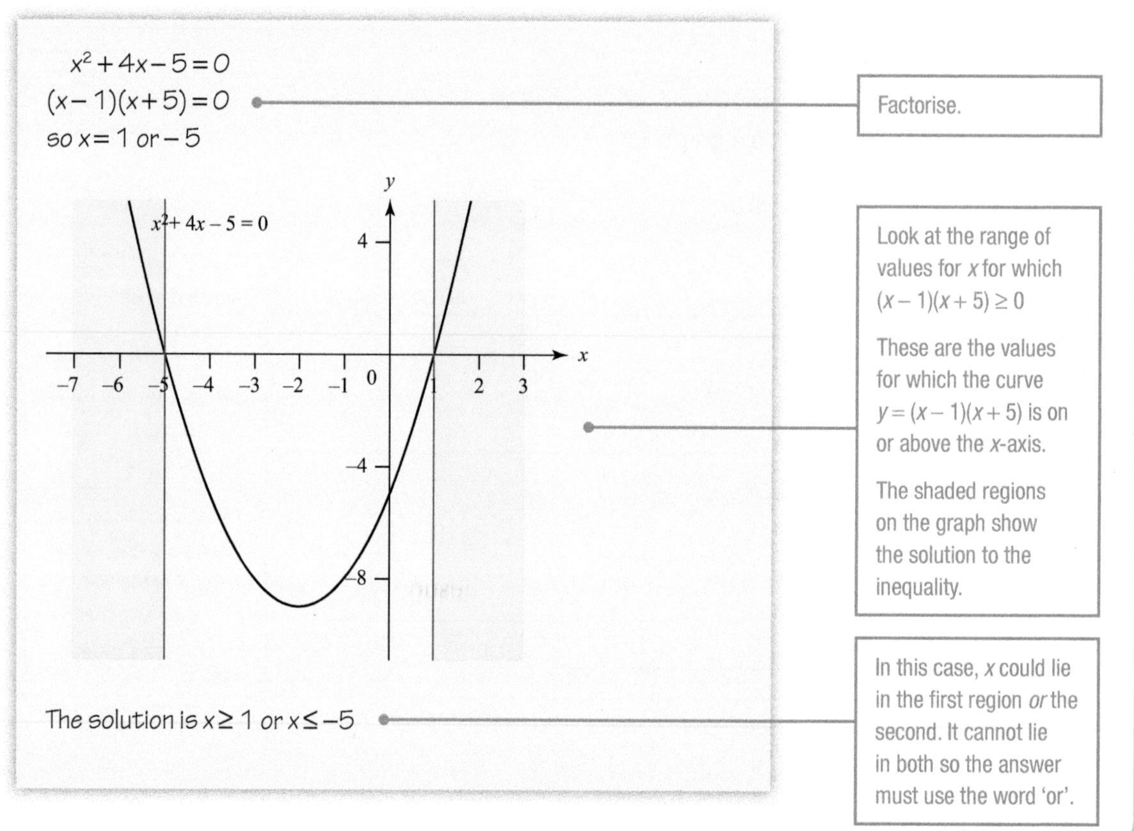

x² + 4x − 5 = 0
(x − 1)(x + 5) = 0
so x = 1 or −5
Factorise.
y
x
x² + 4x − 5 = 0
4
−4
−8
−7 −6 −5 −4 −3 −2 −1 0 1 2 3
Look at the range of values for x for which (x − 1)(x + 5) ≥ 0
These are the values for which the curve y = (x − 1)(x + 5) is on or above the x-axis.
The shaded regions on the graph show the solution to the inequality.
The solution is x ≥ 1 or x ≤ −5
In this case, x could lie in the first region or the second. It cannot lie in both so the answer must use the word 'or'.

You could also have solved the question in Example 3 using the factorised form, by considering signs.

The product of the two brackets is positive if they are both positive or both negative.

$x - 1 \geq 0$ *and* $x + 5 \geq 0$ only if $x \geq 1$

$x - 1 \leq 0$ *and* $x + 5 \leq 0$ only if $x \leq -5$

The solution is $x \geq 1$ or $x \leq -5$

Exercise 1.7A Fluency and skills

1 Show the following inequalities on a number line.

a $s \geq 14$

b $3 < u \leq 9$

c $v < 5$ and $v > 14$

d $r > 7$ and $r \leq 12$

2 Draw graphs to show these inequalities. You can check your sketches using a graphics calculator.

a $x > -4$

b $y \geq 5$

c $y + x < 6$

d $2y - 3x < 5$

e $3y + 4x \leq 8$

f $2y > 10 - 4x$

g $y < x + 4;\ y + x + 1 > 0;\ x \leq 5$

h $y \geq 2;\ x + y < 7;\ y - 2x - 4 \leq 0$

3 Find the values of x for which

a $2x - 9 > -6$

b $15 - 2x \geq 8x + 34$

c $2(4x - 1) + 6 < 15 - 3x$

d $3(x - 3) + 6(5 - 4x) \leq 54$

e $4(2x + 1) - 7(3x + 2) > 5(4 - 2x) - 6(3 - x)$

f $4\left(3x - \frac{1}{2}\right) + 2(8 - 3x) < 6\left(x + \frac{3}{2}\right) - 2\left(x - \frac{5}{2}\right)$

4 For each part **a** to **h**, sketch a suitable quadratic graph and use your sketch to solve the given inequality.

a $x^2 + x - 6 > 0$

b $x^2 + 11x + 28 < 0$

c $x^2 - 11x + 24 \leq 0$

d $x^2 - 2x - 24 \geq 0$

e $2x^2 - 3x - 2 > 0$

f $3x^2 + 19x - 14 < 0$

g $-3 + 13x - 4x^2 \leq 0$

h $6x^2 + 16x + 8 \geq 0$

5 Complete the square or use the quadratic formula to solve these inequalities to 2 dp.

Sketch graphs to help you with these questions.

a $x^2 + 2x - 7 > 0$

b $x^2 + 7x + 8 < 0$

c $x^2 - 12x + 18 \leq 0$

d $x^2 - 3x - 21 \geq 0$

e $3x^2 - 5x - 7 > 0$

f $4x^2 + 17x - 4 < 0$

g $5x^2 - 17x + 12 \leq 0$

h $6x^2 - 16x - 7 \geq 0$

6 These inequalities define a region of the x-y plane. In each case

i Write equations which define the boundaries of the region,

ii Use algebra to find the points where the boundaries intersect,

iii Draw a graph and shade the appropriate region.

a $y < 2x + 3;\ y > x^2$

b $x + y \leq 4;\ y > x^2 - 5x + 4$

c $y - 4x \leq 17;\ y \leq 4x^2 - 4x - 15;\ x \leq 4$

d $y - 2x - 20 < 0;\ y + 4x - 6 < 0;$
$y > x^2 - 5x - 24$

MyMaths 2008, 2009, 2256 SEARCH

Reasoning and problem-solving

Strategy

To solve a problem involving inequalities

1. Use the information in the question to write the inequalities.
2. Solve the inequalites and, if requested, show them on a suitable diagram.
3. Write a clear conclusion that answers the question.

Example 4

Alan travels a journey of 200 miles in his car. He is travelling in an area with a speed limit of 70 mph.

Write down and solve an inequality in t (hours) to represent the time his journey takes.

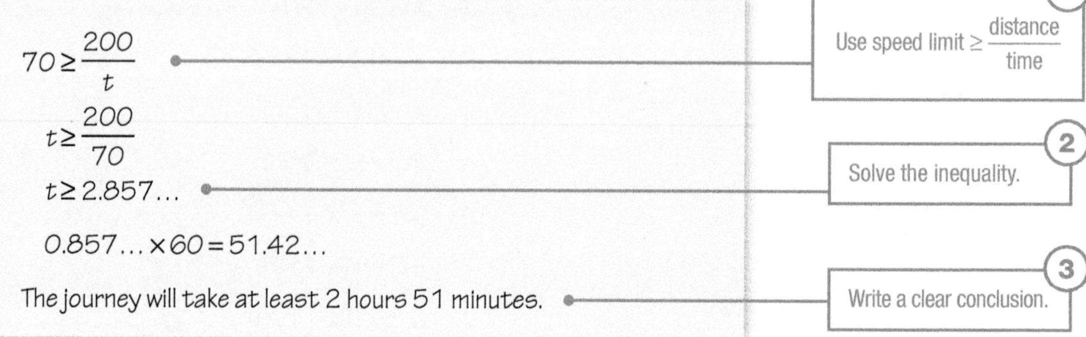

$70 \geq \frac{200}{t}$ — ① Use speed limit $\geq \frac{\text{distance}}{\text{time}}$

$t \geq \frac{200}{70}$

$t \geq 2.857...$ — ② Solve the inequality.

$0.857... \times 60 = 51.42...$

The journey will take at least 2 hours 51 minutes. — ③ Write a clear conclusion.

Example 5

An illustration in a book is a rectangle $(x - 7)$ cm wide and $(x + 1)$ cm long.

It must have an area less than 65 cm^2

Work out the range of possible values of x. Justify your answer.

$(x - 7)(x + 1) < 65$ — ① Write the inequality.

$x^2 - 6x - 72 < 0$

$(x - 12)(x + 6) < 0$

$-6 < x < 12$ — ② Solve the inequality.

However, since any side of a rectangle must be positive, it follows that $x - 7 > 0$, so $x > 7$

So the solution is $7 < x < 12$ — ③ x is both greater than 7 and smaller than 12 so you combine the two inequalities.

Exercise 1.7B Reasoning and problem-solving

1 a Children in a nursery range from six months old to 4 years and six months old inclusive.

Represent this information on a number line.

b The range of temperatures outside the Met Office over a 24 hour period ranged from −4°C to 16°C.

Represent this information on a number line.

2 On a youth athletics club trip there must be at least one trainer for every six athletes and the trip is not viable unless at least eight athletes travel. Due to illness there are fewer than six trainers available to travel. Represent this information as a shaded area on a graph.

3 In an exam, students take a written paper (marked out of 100) and a practical paper (marked out of 25).

The total mark, T, awarded is gained by adding together twice the written mark and three times the practical mark. To pass the exam, T must be at least 200. A student scores w marks in the written paper and p in the practical.

a Write an inequality in w and p

b Solve this inequality for

i $w=74$

ii $p=9$

c Can a student pass if she misses the practical exam?

4 The length of a rectangle, $(5m+7)$ cm, is greater than its width, $(2m+16)$ cm

What values can m take?

5 For Amanda's 18th birthday party, 110 family and friends have been invited and at most 10% will not be able to come. Food has been prepared for 105 people.

Write down inequalities for the number of people, n, who come to the party and have enough to eat.

Solve them and find all possible solutions.

6 A bag contains green and red discs. There are r red discs and three more green than red. The total number of discs is not more than twenty. Write appropriate inequalities and find all solutions.

7 A girl is five years older than her brother. The product of their ages is greater than 50. What ages could the sister be?

8 The length of a rectangle, $(5b-1)$ cm, is greater than its width, $(2b+9)$ cm. The area is less than 456 cm^2. Find the possible values of b

9 The sum, S, of the first n positive integers is given by the formula $2S=n(n+1)$. What are the possible values of n for values of S between 21 and 820?

10 The ages of two children sum to 10 and the product of their ages is greater than 16. Find all possible values of the children's ages.

Challenge

11 A firm makes crystal decanters.

The profit, £P, earned on x thousand decanters is given by the formula $P=-20x^2+1200x-2500$

a Solve the equation $-20x^2+1200x-2500=0$ giving your answer to two decimal places.

b Sketch the graph of $y=-20x^2+1200x-2500$

c Use your graph to estimate

i The values of x where the firm makes a loss,

ii The range of values of x for which the profit is at least £10 000. Check this algebraically.

1 Summary and review

Chapter summary

- To use direct proof, assume P is true and then use P to show that Q must be true.
- To use proof by exhaustion, show that the cases are exhaustive and then prove each case.
- To use proof by counter example, give an example that disproves the statement.
- $x^a \times x^b = x^{a+b}$, $x^a \div x^b = x^{a-b}$, $(x^a)^b = x^{ab}$
- $x^0 = 1$, $x^{-n} = \frac{1}{x^n}$, $x^{\frac{1}{n}} = \sqrt[n]{x}$, $x^{\frac{p}{r}} = \sqrt[r]{(x^p)}$ or $(\sqrt[r]{x})^p$
- You can write any rational number exactly in the form $\frac{p}{q}$, where p and q are integers.
- $\sqrt{A} \times \sqrt{B} = \sqrt{AB}$; $\frac{\sqrt{A}}{\sqrt{B}} = \sqrt{\frac{A}{B}}$
- You rationalise a fraction in the form $\frac{k}{\sqrt{a}}$ by multiplying top and bottom by $\sqrt{a}$
- You rationalise a fraction in the form $\frac{k}{a \pm \sqrt{b}}$ by multiplying top and bottom by $a \mp \sqrt{b}$
- You rationalise a fraction in the form $\frac{k}{\sqrt{a} \pm \sqrt{b}}$ by multiplying top and bottom by $\sqrt{a} \mp \sqrt{b}$
- Any function of x in the form $ax^2 + bx + c$ where $a \neq 0$ is called a quadratic function and $ax^2 + bx + c = 0$ is called a quadratic equation.
- You can solve a quadratic equation $ax^2 + bx + c = 0$ using a calculator, by factorisation, by completing the square, by using the quadratic formula $x = \frac{-b \pm \sqrt{b^2 - 4ac}}{2a}$, and graphically.
- If the discriminant $\Delta = b^2 - 4ac > 0$, the quadratic has two different roots. If $\Delta = b^2 - 4ac = 0$, the quadratic has one repeated root. If $\Delta = b^2 - 4ac < 0$, the quadratic has no real roots.
- You can use gradients of two straight lines to decide if they are parallel, perpendicular, or neither.
- The equation of a circle, centre (a, b) and radius r, is $(x - a)^2 + (y - b)^2 = r^2$
- If you multiply or divide an inequality by a negative number you reverse the inequality sign.

Check and review

You should now be able to...	Try Questions
✔ Use direct proof, proof by exhaustion and counter examples to prove results.	1, 2, 3
✔ Use and manipulate the index laws for all powers.	4
✔ Manipulate surds and rationalise a denominator.	5, 6
✔ Solve quadratic equation°s using a variety of methods.	7, 8
✔ Understand and use the coordinate geometry of the straight line and of the circle.	9, 10
✔ Understand and solve simultaneous equations involving only linear or a mix of linear and non-linear equations.	11, 12
✔ Solve linear and quadratic inequalities algebraically and graphically.	13, 14

1 Prove that the product of two odd numbers must be odd.

2 Prove that there is at least one prime number between the numbers 40 and 48

3 Is it true that for every number n, $\frac{1}{n} < n$? Give a reason for your answer.

4 Simplify

a $(-s^4)^3$ b $\sqrt{64c^{64}}$ c 3^{-4} d $(k^2)^{\frac{-3}{4}}$

5 a Express $\sqrt{275}$ in its simplest form.

b Rationalise the denominator of $\frac{3-\sqrt{a}}{\sqrt{a}+1}$

6 What is the length of the hypotenuse of a right-angled triangle with sides containing the right angle of length $3\sqrt{3}$ and $3\sqrt{5}$ cm?

7 a Solve the equation $2c^2+9c-5=0$ by factorisation.

b Solve, to 2 dp, the equation $5x^2+9x-28=0$

8 Sketch the quadratic curve $y=x^2-4x-1$

9 a Write down the equations of these lines.

i Gradient -6 passing through $(6,-7)$

ii Gradient $\frac{2}{3}$ passing through $(-3,4)$

b A square joins the points $(-2,1)$, $(2,4)$, $(5,0)$ and $(1,-3)$. Write the equations of its diagonals. Hence prove that they are perpendicular.

10 a Write the equations of the circles. The centre and radius are given for each.

i $(3,6)$; 8 ii $(-3,9)$; 4 iii $(-2,-7)$; 11

b Write the equation of the tangent, at point $P(-9,19)$, to the circle with centre $(-4,7)$ and radius 13

c A circle with centre $C(5,6)$ and radius 10 has $M(8,5)$ as the midpoint of a chord. Work out the coordinates of the ends of the chord.

11 Solve these equations simultaneously.

a $2x-5y=11$; $4x+3y=9$

b $2x-3y=5$; $x^2-y^2+5=0$

12 The line $y=3x+4$ intersects the curve $xy=84$ at two points. Work out their coordinates.

13 a Writing your answers in set notation, solve these inequalities.

i $12-3x \geq 7x+2$

ii $2(x-7)+5(6-3x) \leq 10$

b Solve these inequalities, giving your answers to 2dp.

i $x^2-14x+16 \leq 0$

ii $5x^2-13x-11 \geq 0$

14 Shade the regions represented by these inequalities.

a $2y-2x-7 \leq 0$; $x+y-7<0$; $y>0$

b $x+y \leq 8$; $y>(x-2)^2-4$

What next?

Score			
	0 – 7	Your knowledge of this topic is still developing. To improve, search in MyMaths for the codes: 2001–2005, 2008, 2009, 2014–2018, 2020, 2021, 2025, 2026, 2033–2037, 2252, 2253, 2255–2257	
	8–10	You're gaining a secure knowledge of this topic. To improve, look at the InvisiPen videos for Fluency and skills (01A)	
	11–14	You've mastered these skills. Well done, you're ready to progress! To develop your techniques, look at the InvisiPen videos for Reasoning and problem-solving (01B)	

Click these links in the digital book

History

Pierre de Fermat was a lawyer in 17th century France who studied mathematics as a hobby. He often wrote comments in the margins of the maths books that he read and, on one occasion, wrote about a problem set over a thousand years ago by Greek mathematician **Diophantus**.

The problem was to find solutions to the equation $x^n + y^n = z^n$ where x, y, z and n are all positive integers. Fermat wrote that he had discovered 'the most remarkable proof' that the equation has no solutions if $n \geq 3$, but that the margin was too small to contain it.

Despite many attempts, a copy of Fermat's proof was never found. No one else was able to prove or disprove it for over 350 years.

Have a go

Find the flaw in the following 'proof'.

$x = 1$	
$x^2 = 1$	Square both sides
$x^2 - 1 = 0$	Subtract 1 from both sides
$(x+1)(x-1) = 0$	Factorise
$x + 1 = 0$	Divide both sides by $(x - 1)$
$2 = 0$	Substitute $x = 1$

Research

Who eventually proved the result known as **Fermat's last Theorem**?

How long did it take them to complete the proof?

Investigation

How does Fermat's last theorem relate to **Pythagoras' Theorem**?

How can you use these diagrams to prove Pythagoras' Theorem?

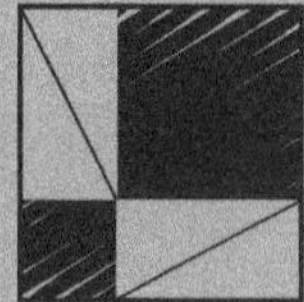

What are Pythagorean triples? How many are there?

"We cannot solve our problems with the same level of thinking that created them."
- Einstein

1 Assessment

1 Simplify $(3+2\sqrt{2})(5-4\sqrt{2})$. Choose the correct answer.

A $31-2\sqrt{2}$ **B** $-1+2\sqrt{2}$ **C** $-1-2\sqrt{2}$ **D** $15-10\sqrt{2}$ **[1 mark]**

2 What is the equation of the straight line that is perpendicular to $3x+2y=5$ and that passes through the point (4, 5)? Choose the correct answer.

A $3y-2x=7$ **B** $3y+2x=7$ **C** $2y-3x=7$ **D** $2y+3x=7$ **[1]**

3 **a** Simplify these expressions.

i $2^m \times 2^n$ **ii** $\frac{5^{m+1}}{5^{2n}}$ **iii** $(3^m)^2 \times \sqrt{(3^m)}$ **[5]**

b Given $\frac{16^p \times 8^q}{4^{p+q}} = 2^n$, write down the an expression for n in terms of p and q **[3]**

4 **a** Simplify these surds. You must show your working.

i $\frac{12}{\sqrt{3}}$ **ii** $\frac{3-\sqrt{7}}{1+3\sqrt{7}}$ **[6]**

b A rectangle *ABCD* has an area of 8 cm² and length $(3-\sqrt{5})$ cm.

Work out its width, giving your answer as a surd in simplified form. Show your working. **[3]**

5 **a** Express $x^2+6x+13$ in the form $(x+a)^2+b$ **[2]**

b Hence sketch the curve $y=x^2+6x+13$ and label the vertex, and the point where the curve cuts the y-axis. **[3]**

6 Solve these simultaneous equations.

$2x+y=3$ $3x^2+2xy+7=0$ **[8]**

7 Prove that the equation $x=1+\frac{2x-5}{x+4}$ has no real solutions. **[4]**

8 **a** Solve these inequalities.

i $3x-5<11-x$ **ii** $x^2-6x+5\leq 0$ **[5]**

b Show on a number line the set of values of x that satisfy both $3x-5<11-x$ and $x^2-6x+5\leq 0$ **[2]**

9 *PQR* is a right-angled triangle.
Write an exact expression for x, show your working. **[6]**

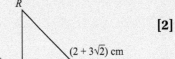

R
(2 + 3√2) cm
(3 + 2√2) cm
P
x cm
Q

10 The equation of a circle is $x^2+y^2-10x+2y-23=0$

a Work out **i** Its centre, **ii** Its radius. **[5]**

b The line $y=x+2$ meets the circle at the points P and Q. Work out, in exact form, the coordinates of P and Q **[5]**

11 The quadratic equation $(k+1)x^2-4kx+9=0$ has distinct real roots.
What range of values can k take? **[6]**

12 Prove that $\frac{a+b}{2}\geq\sqrt{ab}$ for all positive numbers a and b **[4]**

13 a Factorise the expression $2u^2 - 17u + 8$ **[2]**

b Hence solve the equation $2^{2x+1} - 17 \times 2^x + 8 = 0$ **[3]**

14 The straight line $y = mx + 2$ meets the circle $x^2 + y^2 + 4x - 6y + 10 = 0$

a Prove that the x-values of the points of intersection satisfy the equation
$(m^2 + 1)x^2 + 2(2 - m)x + 2 = 0$ **[4]**

b The straight line $y = mx + 2$ is a tangent to the circle $x^2 + y^2 + 4x - 6y + 10 = 0$
What are the possible values of m? Give your answers in exact form. **[5]**

15 a Given $9^2 = 3^n$, write down the value of n **[1]**

b Solve these simultaneous equations.

$3^{x+y} = 9^2 \qquad 4^{x-2y} = 8^4$ **[4]**

16 Decide which of these statements are true and which are false.
For those that are true, prove that they are true.
For those that are false, give a counter-example to show that they are false.

a If $a > b$ then $a^2 > b^2$ **[2]**

b $n^2 + n$ is an even number for all positive integers n **[3]**

c If a and b are real numbers then $b^2 \geq 4a(b - a)$ **[3]**

d $2^n - 1$ is prime for all positive integers n **[2]**

17 The diagram shows the parabola $y = 2x^2 - 8x + 9$ and the straight line $y = 4x - 5$

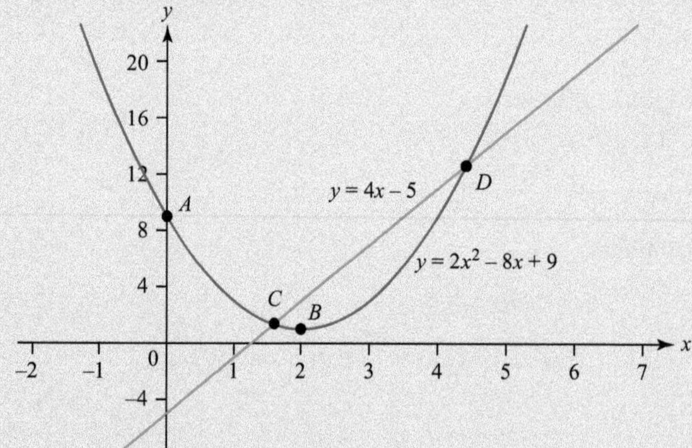

Work out the coordinates of the following points.

a A, the y-intercept of the parabola. **[1]**

b B, the vertex of the parabola. **[2]**

c C and D, the points of intersection of the line and parabola. **[4]**

18 Prove that the circle $x^2 + y^2 + 6x - 4y - 2 = 0$
lies completely inside the circle $x^2 + y^2 - 2x - 10y - 55 = 0$ **[9]**

2 Polynomials and the binomial theorem

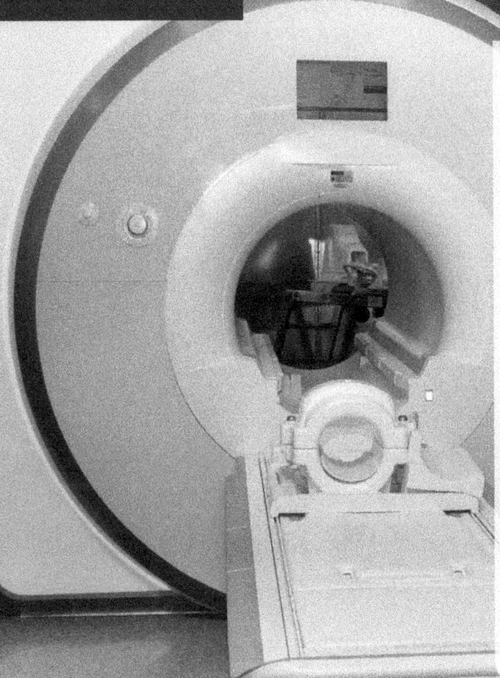

MRI scanners are powerful machines that produce detailed images of parts of the body. They use a combination of magnetic fields and radio waves to make the hydrogen nuclei in your body, which is mostly water (H_2O), resonate and spin in different directions. This creates small energy differences between nearby nuclei, which are detected by the machine. These differences split into 'multiplets,' whose relative strengths can be predicted using Pascal's triangle.

When repeatedly multiplying out brackets containing a pair of terms, $(x+y)^n$, the binomial theorem provides a shortcut to the final expression, and Pascal's triangle provides the coefficients. The binomial theorem is a basic result that has many applications in areas of mathematical modelling, such as the medical imaging example described.

Orientation

What you need to know

KS4

- To simplify and manipulate algebraic expressions, including collecting like terms and use of brackets.

What you will learn

- To manipulate, simplify and factorise polynomials.
- To understand and use the binomial theorem.
- To divide polynomials by algebraic expressions.
- To understand and use the factor theorem.
- To analyse a function and sketch its graph.

What this leads to

Ch4 Differentiation and integration
Using calculus for curve sketching.

Ch10 Probability and discrete random variables
The binomial probability distribution.

Ch13 Sequences
Binomial expansions.
Position-to-term and term-to-term rules.

Practise before you start

1150, 1155, 1156

2 Bridging Unit

Topic A: Expanding brackets

Bridging to Ch2.1

You know how to find the product of two binomials by multiplying every combination of terms together and simplifying. Take extra care when squaring a binomial, and remember that $(x+a)^2 = (x+a)(x+a) = x^2 + 2ax + a^2$ NOT $x^2 + a^2$

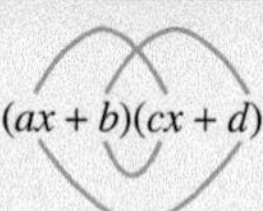

Example 1

Expand and simplify $(3x-5)^2$

$(3x-5)^2 = (3x-5)(3x-5)$ — Always write in this form until you are confident.

$= 9x^2 - 15x - 15x + 25$

$= 9x^2 - 30x + 25$ — Simplify the x-terms.

Try It 1

Expand and simplify **a** $(x-7)^2$ **b** $(5x+1)^2$

To find the product of three binomials, first expand any pair, then multiply by the third.

Example 2

Expand and simplify $(x+3)(x-2)(x+1)$

$(x+3)(x-2) = x^2 - 2x + 3x - 6$ — Expand the first two pairs.

$= x^2 + x - 6$ — Simplify $-2x + 3x$ to x

$(x^2 + x - 6)(x+1) = x^3 + x^2 + x^2 + x - 6x - 6$ — 3 terms × 2 terms = 6 terms

$= x^3 + 2x^2 - 5x - 6$ — Add the x^2-terms and simplify $x - 6x$ to $-5x$

Try It 2

Expand and simplify **a** $(x-3)(x-1)(x+1)$ **b** $(x+2)^2(x-4)$

Bridging Exercise Topic A

Bridging to Ch2.1

1 Expand and simplify each of these expressions.

a $(x-4)^2$ **b** $(x+6)^2$ **c** $(x-9)^2$ **d** $(x+5)^2$ **e** $(2x+1)^2$ **f** $(3x-2)^2$

g $(4x+3)^2$ **h** $(5x+2)^2$ **i** $(3-x)^2$ **j** $(7-2x)^2$ **k** $(8-3x)^2$ **l** $(10-9x)^2$

2 Expand and simplify each of these expressions.

a $(x+5)(x+2)(x+4)$ **b** $(x+2)(x+7)(x-1)$ **c** $(x-3)(x+8)(2-x)$

d $(x+6)(2x-5)(x-8)$ **e** $(3x+1)(2x-1)(x+5)$ **f** $(2x-3)(3x-4)(5-4x)$

g $(x+5)^2(x+9)$ **h** $(3-x)^2(x-8)$ **i** $(x+7)(x-9)^2$

j $(2x+3)^2(4-x)$ **k** $(3x+7)^2(x-8)$ **l** $(2x-11)^2(3-2x)$

Topic B: Algebraic division

Bridging to Ch2.3

When simplifying fractions, divide the numerator and denominator by their **highest common factor (HCF).** For example, $\frac{112}{140}$ can be simplified to its equivalent fraction $\frac{4}{5}$ by dividing numerator and denominator by their HCF: 28, which is the product of their common prime factors 7, 2 and 2.

Algebraic fractions can be simplified in the same way. You must first factorise the numerator and the denominator, then divide both the numerator and denominator by their highest common factor.

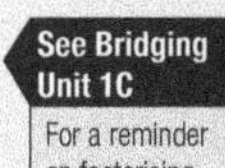
See Bridging Unit 1C
For a reminder on factorising.

For example, $\frac{x(x+1)(x-2)}{x^2(x+1)}=\frac{x-2}{x}$ since the common factor of $x(x+1)$ can be cancelled.

Example 1

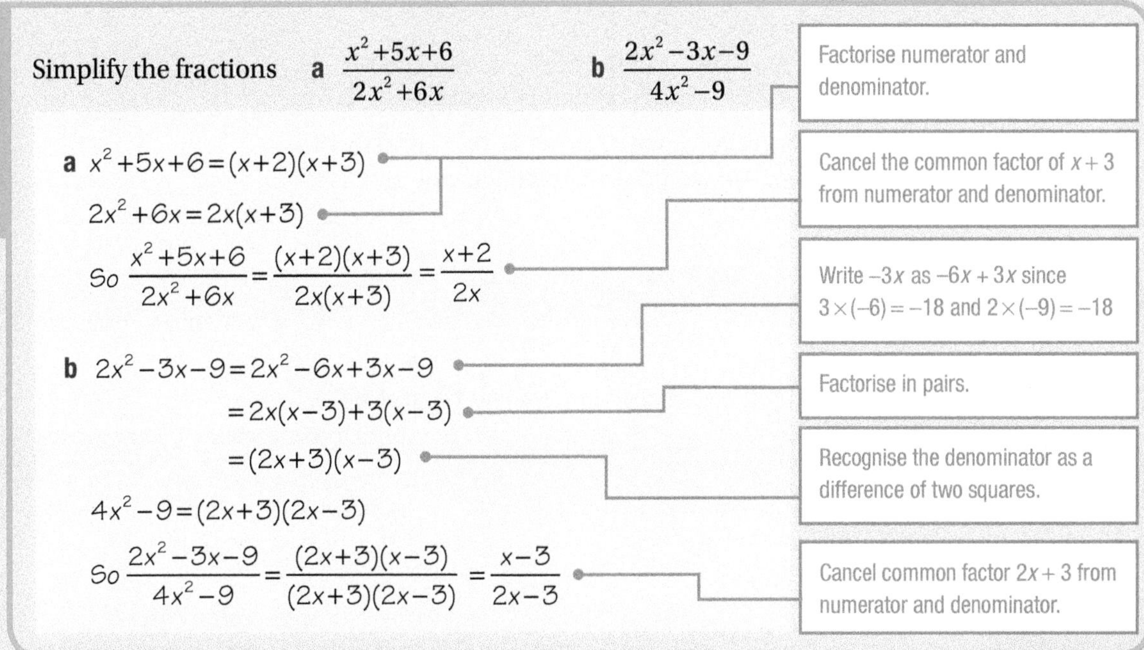

Simplify the fractions **a** $\frac{x^2+5x+6}{2x^2+6x}$ **b** $\frac{2x^2-3x-9}{4x^2-9}$

a $x^2+5x+6=(x+2)(x+3)$

$2x^2+6x=2x(x+3)$

So $\frac{x^2+5x+6}{2x^2+6x}=\frac{(x+2)(x+3)}{2x(x+3)}=\frac{x+2}{2x}$

b $2x^2-3x-9=2x^2-6x+3x-9$

$=2x(x-3)+3(x-3)$

$=(2x+3)(x-3)$

$4x^2-9=(2x+3)(2x-3)$

So $\frac{2x^2-3x-9}{4x^2-9}=\frac{(2x+3)(x-3)}{(2x+3)(2x-3)}=\frac{x-3}{2x-3}$

Factorise numerator and denominator.

Cancel the common factor of $x+3$ from numerator and denominator.

Write $-3x$ as $-6x+3x$ since $3\times(-6)=-18$ and $2\times(-9)=-18$

Factorise in pairs.

Recognise the denominator as a difference of two squares.

Cancel common factor $2x+3$ from numerator and denominator.

Try It 1

Simplify the fractions **a** $\frac{x(x+2)(x-1)}{x^2(x-1)}$ **b** $\frac{3x^2+2x-1}{x^2-1}$

If you can't simplify by cancelling common factors, you can use the method of **long division.** To use long division to divide the number 813 by 7

- See how many times 7 goes into 8, ignoring remainders: only once, so write 1 in the hundreds column.
- Then multiply 7 by this result and write it under 813: $7\times1=7$
- Subtract this number from the 8 above it to get a remainder of 1, and write that underneath again. Then copy down the 1 in the tens column.

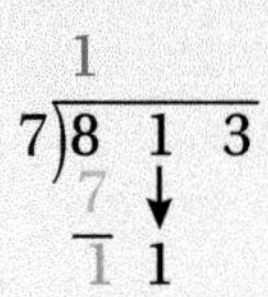

- Repeat this process with the 11: again, 7 goes into 11 once, so write a 1, this time in the tens column.
- $7\times1=7$, so write a 7 underneath, and $11-7=4$, so write a 4 underneath that. then copy down the 3

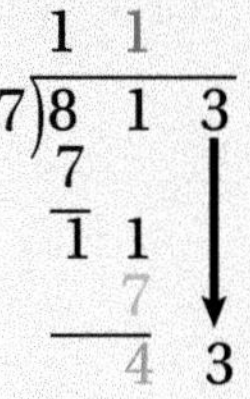

1151

- Finally, 7 goes into 43 six times, so write a 6 in the units column, and $7 \times 6 = 42$, so write 42 underneath.
- $43 - 42 = 1$, so write 1 under that.
- To complete the solution, divide this 1 by the original divisor (7) to get your remainder: $813 \div 7 = 116\frac{1}{7}$

$$\begin{array}{r} 116 \\ 7\overline{)813} \\ 7 \\ \hline 11 \\ 7 \\ \hline 43 \\ 42 \\ \hline 1 \end{array}$$

You can use long division in the same way to divide polynomials. For example, to work out $(6x^2 + x - 12) \div (2x + 3)$

- First divide $6x^2$ by $2x$ to give $3x$, and write this in the 'x-column'.
- Multiply $3x$ by $2x + 3$ to get $6x^2 + 9x$ and write this underneath
- Subtract this from $6x^2 + x$ to get $-8x$ and write that underneath again, then carry down the -12
- $2x$ goes into $-8x$ four times, so write -4 in the next column. Then multiply this by $2x + 3$ to get $-8x - 12$ and write that underneath.
- Subtracting these two rows gives 0, so the answer is $3x - 4$ exactly (if there had been a remainder, you would need to divide it by $2x + 3$)

$$\begin{array}{r} 3x \quad -4 \\ 2x+3\overline{)6x^2 \quad +x \quad -12} \\ 6x^2 \quad +9x \qquad\quad \\ \hline -8x \quad -12 \\ -8x \quad -12 \\ \hline 0 \end{array}$$

Example 2

Work out $(12x^2+7x+1)\div(3x-2)$

$$\begin{array}{r} 4x+5 \\ 3x-2\overline{)12x^2+7x+1} \\ 12x^2-8x \\ \hline 15x+1 \\ 15x-10 \\ \hline 11 \end{array}$$

$(12x^2+7x+1)\div(3x-2)=4x+5+\frac{11}{3x-2}$

First divide $12x^2$ by $3x$ to give $4x$ and write above the x-column.

Multiply the divisor by $4x$ and subtract. Copy down the 1

$15x \div 3x = 5$ so write this above the number column, then multiply by the divisor and subtract, giving a remainder of 11

Write the remainder over the divisor.

Try It 2

Use long division to work out $(5x^2+19x-1)\div(x+4)$

Another method of dividing polynomials involves **comparing the coefficients**. The **coefficient** is the number or constant that multiplies the variable. For example, for $2x^3-5x$, the coefficients of x^3 and x are 2 and -5. You can equate coefficients of the same variable from both sides of an identity.

Example 3

Given that $(ax+1)(x+3)^2 \equiv 3x^3+bx^2+cx+9$, find the values of the constants a, b and c

$(ax+1)(x+3)^2 \equiv (ax+1)(x^2+6x+9)$ — Expand the squared bracket.

$\equiv ax^3+6ax^2+9ax+x^2+6x+9$ — Multiply by the first bracket.

$\equiv ax^3+(6a+1)x^2+(9a+6)x+9$ — Collect like terms.

So $3x^3+bx^2+cx+9 \equiv ax^3+(6a+1)x^2+(9a+6)x+9$

(*Continued on the next page*)

Equate coefficients of x^3: $3 = a$

Equate coefficients of x^2: $b = 6a + 1 = 19$

Equate coefficients of x^3: $c = 9a + 6 = 33$

Therefore, $a = 3$, $b = 19$ and $c = 33$

The coefficients of x^3 on both sides of the identity must be equal so $a = 3$

Using $a = 3$

Try It 3

Given that $(x+a)(x-2)(x+1) \equiv x^3 - 5x^2 + bx + c$, find the values of the constants a, b and c

You can use the comparing coefficients technique and your knowledge of indices to divide algebraic expressions. As you saw in Example 2, dividing a quadratic polynomial by a linear polynomial gives a linear quotient and a constant remainder.

For example, for certain constants a, b and c

$$\frac{15x^2 - 14x + 5}{3x - 4} \equiv ax + b + \frac{c}{3x-4}$$

Dividend: $15x^2 - 14x + 5$; Divisor: $3x - 4$; Quotient: $ax + b$; Remainder: c

Use the general form for a linear expression.

This is an identity, true for all values of x

To find the values of a, b and c you multiply both sides of the identity by $3x-4$

$$15x^2 - 14x + 5 \equiv (ax+b)(3x-4) + c$$
$$\equiv 3ax^2 + (3b - 4a)x - 4b + c$$

Comparing coefficients of x^2 gives $3a = 15 \Rightarrow a = 5$

Comparing coefficients of x gives $-14 = 3b - 4a \Rightarrow 3b = 4a - 14 = 6 \Rightarrow b = 2$

Comparing constant terms gives $5 = -4b + c \Rightarrow c = 5 + 4b = 13$

Therefore, $(15x^2 - 14x + 5) \div (3x - 4) \equiv 5x + 2 + \frac{13}{3x-4}$

The same method can be used for higher-order polynomials, for example, dividing a cubic polynomial by a linear polynomial will give a quadratic quotient and a constant remainder.

Key point

Dividing a polynomial of order m by a polynomial of order n where $m \geq n$ will give

- A quotient of order $m - n$
- A remainder of order at most $n - 1$

For example, dividing two quadratics will give a constant quotient plus a linear remainder.

Example 4

Use the method of comparing coefficients to work out $(2x^3 - 7x^2 - 19x + 23) \div (x - 5)$

$$\frac{2x^3 - 7x^2 - 19x + 23}{x-5} \equiv ax^2 + bx + c + \frac{d}{x-5}$$

A cubic (order 3) divided by a linear expression (order 1) gives a quotient of order 3 – 1 = 2, i.e. a quadratic, and a remainder of order 1 – 1 = 0, i.e. a constant. Write the general expression for a quadratic and a constant remainder.

$$2x^3 - 7x^2 - 19x + 23 \equiv (ax^2 + bx + c)(x - 5) + d$$
$$\equiv ax^3 - 5ax^2 + bx^2 - 5bx + cx - 5c + d$$
$$\equiv ax^3 + (b - 5a)x^2 + (c - 5b)x - 5c + d$$

Multiply both sides by the divisor $x - 5$

(*Continued on the next page*)

$x^3: a=2$

$x^2: b-5a=-7 \Rightarrow b=5a-7=3$

$x: c-5b=-19 \Rightarrow c=5b-19=-4$

Constant: $-5c+d=23 \Rightarrow d=23+5c=3$

Therefore $(2x^3-7x^2-19x+23)\div(x-5)=2x^2+3x-4+\dfrac{3}{x-5}$

Compare coefficients of x^3, x^2, x and constant terms to find the values of *a*, *b*, *c* and *d*

Write out the solution.

Use the method of comparing coefficients to work out $(6x^2+11x-49)\div(3x-5)$

Bridging Exercise Topic B

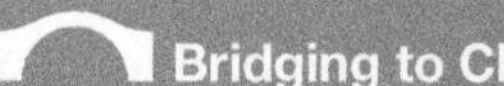

1 Simplify these fractions.

a $\dfrac{x(x-5)(x+2)}{x^3(x+2)}$ b $\dfrac{(x+3)^2}{x(x+3)}$ c $\dfrac{(x-4)}{2x(x-4)}$ d $\dfrac{x^2(x+5)}{x(x+5)^2}$

2 Simplify these fractions by first factorising the numerator and the denominator.

a $\dfrac{x^2-2x-8}{x^2+4x+4}$ b $\dfrac{x^2-10x+21}{x^2-x-6}$ c $\dfrac{x^2-3x-10}{x^2-10x+25}$ d $\dfrac{x^2+10x+24}{2x+8}$

e $\dfrac{x^2+6x}{x^2-36}$ f $\dfrac{3x^2+6x}{x^2-5x-14}$ g $\dfrac{5x^3+15x^2}{x^2+6x+9}$ h $\dfrac{x^2-64}{3x^2-24x}$

i $\dfrac{25-x^2}{45-4x-x^2}$ j $\dfrac{2x^2-x-28}{2x^3+7x^2}$ k $\dfrac{15x^2+7x-4}{10x^2+13x+4}$ l $\dfrac{x^3-100x}{6x^2+56x-40}$

m $\dfrac{12x^3+36x}{2x^2+6}$ n $\dfrac{42x^2-x-1}{36x^2-12x+1}$ o $\dfrac{9x^3-x}{24x^2-x-3}$ p $\dfrac{9x^2-34x-8}{2x^4-8x^3}$

3 Use long division to work out these expressions.

a $(2x^2-9x-16)\div(x-6)$ b $(6x^2+3x+2)\div(x-1)$ c $(5x^2+41x+41)\div(x+7)$

d $(6x^2+x-2)\div(2x-1)$ e $(15x^2+26x+5)\div(3x+4)$ f $(8x^2+6x-34)\div(4x+9)$

g $(3x^3+18x^2+9x+19)\div(x+5)$ h $(12x^3+4x^2+13x+4)\div(2x+1)$

4 Find the values of the constants *a*, *b* and *c* in each of these identities.

a $(x+3)(x-4)\equiv ax^2+bx+c$ b $(x-9)^2\equiv ax^2+bx+c$

c $(x+4)(x+2)(x-2)\equiv x^3+ax^2+bx+c$ d $(x+5)^2(x+8)\equiv x^3+ax^2+bx+c$

e $(ax+1)(x+3)(x-7)\equiv 5x^3+bx^2+cx-21$ f $(x+a)(x-1)(x+5)\equiv x^3+bx^2+cx+10$

g $9x^3+bx^2+cx-2\equiv(ax+1)^2(x-2)$ where $a>0$

h $(x+a)(x+b)(x-4)\equiv x^3-8x^2-5x+c$ where $b>0$

5 Use the method of comparing coefficients to work out each of these divisions.

a $(3x^2-19x-18)\div(x-7)$ b $(4x^2+27x+20)\div(x+6)$

c $(18x^2+3)\div(3x+1)$ d $(28x^2-120x-34)\div(2x-9)$

6 Use the method of comparing coefficients to simplify each of these fractions.

a $\dfrac{3x^3-2x^2-3x+6}{x-1}$ b $\dfrac{14x^3+27x^2+7x-4}{2x+3}$ c $\dfrac{3x^3+7x^2-11x+4}{x^2+3x-2}$ d $\dfrac{2x^3-x-1}{2x^2-4}$

Topic C: Cubic, quartic and reciprocal graphs

This topic recaps the main properties and shapes of cubic, quartic and reciprocal graphs. By calculating x and y intercepts you can sketch these graphs.

Key point

A sketch should always show:

- The shape of the curve.
- The position of the curve on the axes.
- Points of intersection with the x and y axes.

If the question calls for it, a sketch may also include:

- Maximum and minimum points.
- The location of a named point.

See Bridging Unit 1C
For a reminder of sketching quadratic graphs.

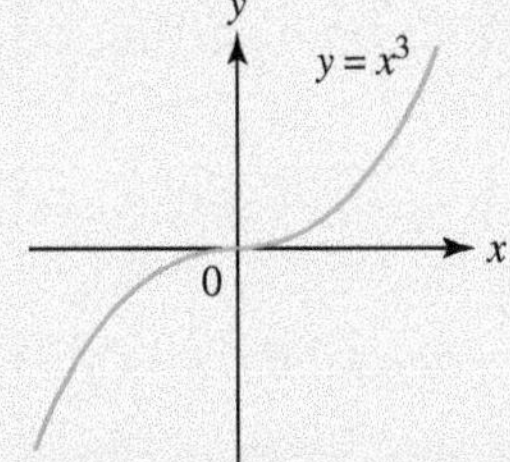

The graph of $y=x^3$ is shown. It crosses through the x-axis once only, at the point (0, 0), since the equation $x^3=0$ has just one real solution $x=0$

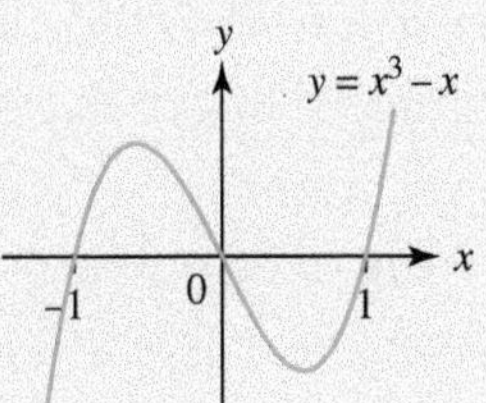

A cubic equation can have three real solutions, in which case it will cross the x-axis three times. For example, the equation $y=x^3-x$ crosses the x-axis at $x=-1$, $x=0$ and $x=1$ since the equation $x^3-x=0$ can be solved by factorising in this way.

$$\begin{aligned} x^3-x=0 &\Rightarrow x(x^2-1)=0 \\ &\Rightarrow x(x-1)(x+1)=0 \\ &\Rightarrow x=0, 1, -1 \end{aligned}$$

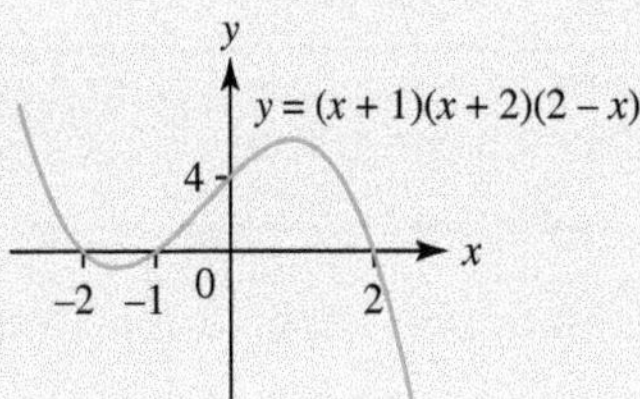

A cubic equation will look different if the coefficient of the x^3 term is negative. For example, the equation $y=(x+1)(x+2)(2-x)$ crosses the x-axis at $x=-1$, $x=-2$ and $x=2$, but if you expand $(x+1)(x+2)(2-x)$ you get $-x^3-x^2+4x+4$ so the coefficient of the x^3 term is negative.

You can also work out where a graph crosses the y-axis by substituting $x=0$ into the equation, in this case at $y=4$

Example 1

Sketch the graph with equation $y=x(x+4)(5-x)$

The equation $y=x(x+4)(5-x)$ crosses the x-axis at $x=0$, $x=-4$ and $x=5$

Since these are the solutions to $x(x+4)(5-x)=0$

Expanding $x(x+4)(5-x)$ gives $-x^3+x^2+20x$

So the coefficient of x^3 is negative.

You don't necessarily need to expand the whole expression to work out the coefficient of x^3

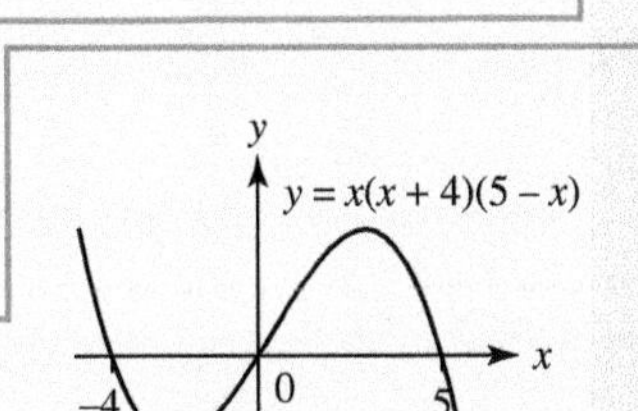

Sketch the graph. Remember to label the intercepts.

Sketch the graphs of **a** $y=(x+3)(x-4)(x-1)$ **b** $y=-x(x+1)(x-3)$ **Try It 1**

Sometimes a polynomial equation will have **repeated roots.** A repeated root occurs when the polynomial has a squared factor, so the same root is given twice. This means the x axis is a tangent to the curve at the root. For example, the quadratic equation $x^2+6x+9=0$ can be factorised and written as $(x+3)^2=0$, which means that there is a repeated root of $x=-3$. So the graph of $y=x^2+6x+9$ will just touch the x-axis at the point $x=-3$ as shown.

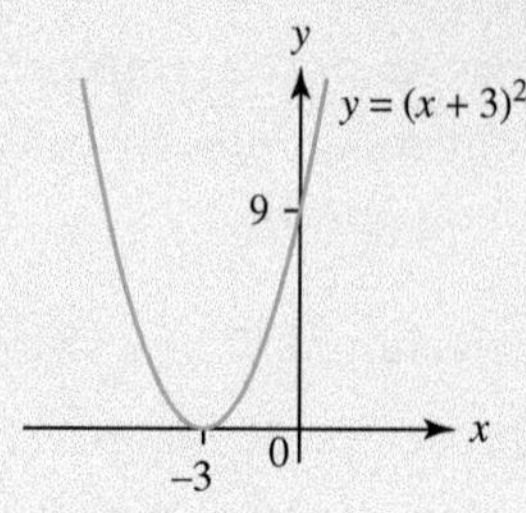

The same principle applies to cubic equations. For example the equation $y=x(x-3)^2$ crosses the x-axis at $x=0$ and touches the x-axis at $x=3$ because this is a repeated root.

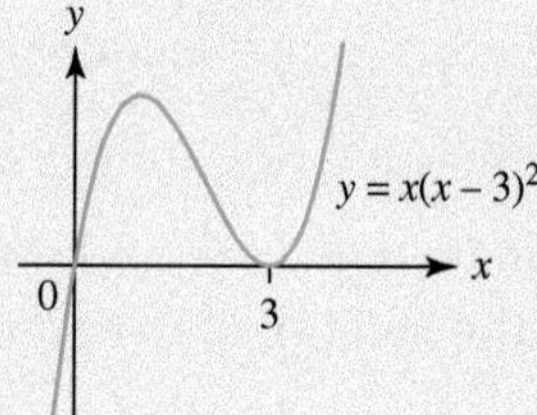

Example 2

Sketch the graph with equation $y=-(x+2)^2(x-5)$

$-(x+2)^2(x-5)=0$ has a root at $x=5$ and a repeated root at $x=-2$

Therefore the graph of $y=-(x+2)^2(x-5)$ will cross the x-axis at $x=5$ and touch at $x=-2$

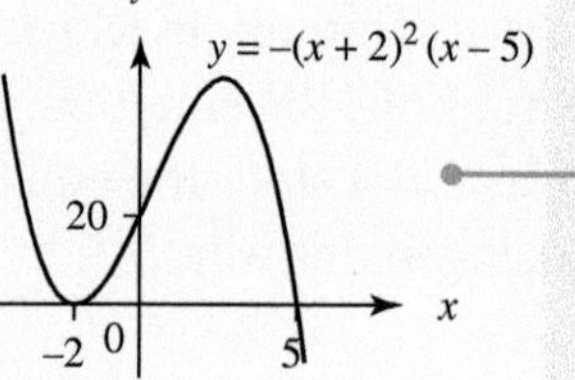

The coefficient of x^3 is negative so the graph is this way around.

When $x=0$, $y=-(2)^2(-5)=20$

Find the point of intersection with the y-axis.

Sketch the graphs of **a** $y=-x^2(x+7)$ **b** $y=(x-4)(x+3)^2$ **Try It 2**

You may need to factorise or use your calculator to solve the cubic and find the x-intercepts.

Example 3

Sketch the graph with equation $y=5x^3+2x^2$

$5x^3+2x^2=x^2(5x+2)$

$x^2(5x+2)=0$ has a root at $x=-\frac{2}{5}$ and a repeated root at $x=0$

Therefore the graph of $y=5x^3+2x^2$ will cross the x-axis at $x=-\frac{2}{5}$ and touch at $x=0$

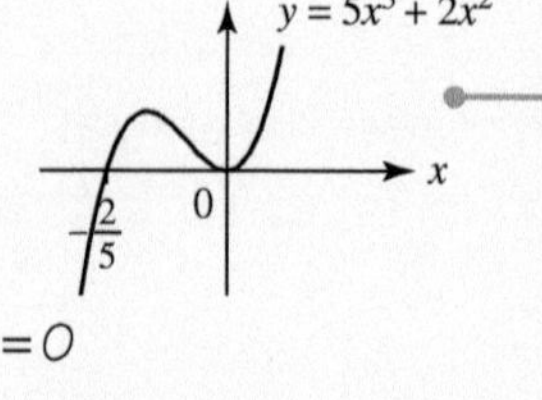

Sketch the graph and label the intercepts.

Sketch the graphs of **a** $y=2x^3+6x^2$ **b** $y=7x^2-14x^3$ **Try It 3**

A polynomial of order 4 is called a **quartic**. The graph of $y = x^4$ touches the x-axis just once at (0, 0) since the equation $x^4 = 0$ has one (repeated) real root of $x = 0$

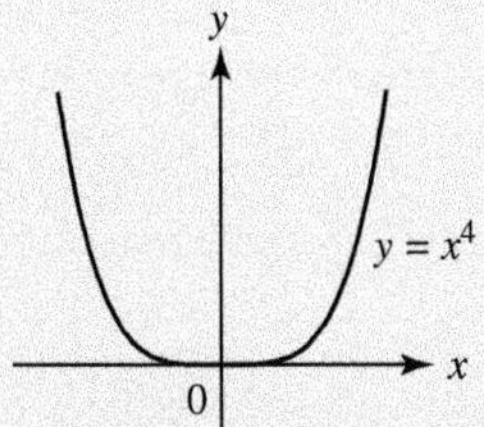

A quartic equation with four, distinct, real roots will cross the x-axis four times; for example $y = (x+3)(x+1)(x-2)(x-5)$ will cross the x-axis at $x = -3, -1, 2, 5$

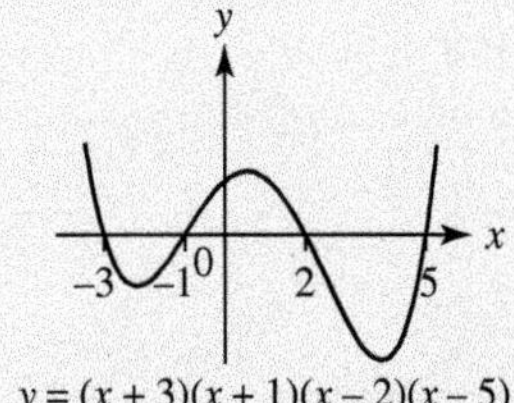

Example 4

Sketch the graph of $y = x(x+3)^2(5-x)$

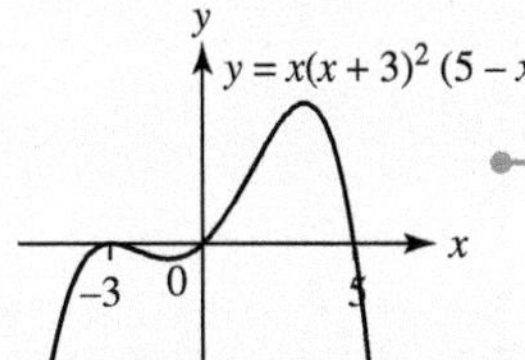

The coefficient of x^4 is negative so the graph is this way round.

There is a repeated root at $x = -3$ so the curve touches the x-axis at $(-3, 0)$ and crosses at $(0, 0)$ and $(5, 0)$

Try It 4

Sketch the graphs of **a** $y = 1 - x^4$ **b** $y = x^2(x-4)^2$

You can apply a transformation to the graph of $y = f(x)$ by considering the effect on the coordinates of points.

Key point

$y = f(x+a)$ subtracts a from the x-coordinates.

$y = f(x) + a$ adds a to the y-coordinates.

$y = f(ax)$ divides the x-coordinates by a

$y = af(x)$ multiplies the y-coordinates by a

Example 5

The graph of $y = f(x)$ is shown. Sketch the graphs of

a $y = f(x+2)$ **c** $y = f\left(\frac{x}{2}\right)$

c $y = 3f(x)$ **d** $y = -f(x)$

State the coordinates of the local maximum, point A, in each case.

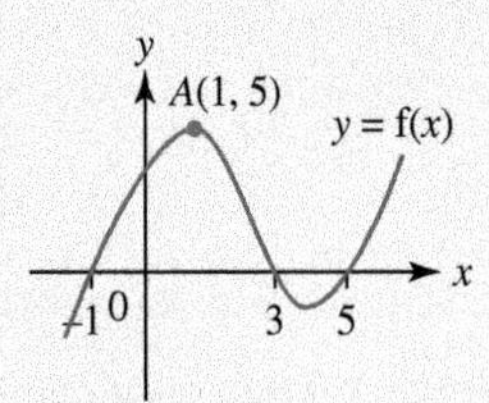

(*Continued on the next page*)

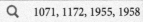

a

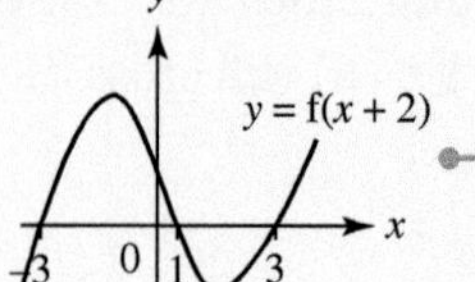

The maximum point moves to (−1, 5)

Shift the graph 2 units to the left by subtracting 2 from each x-coordinate.

b

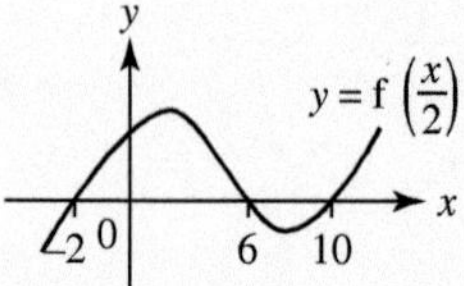

The maximum point moves to (2, 5)

Stretch the graph in the x-direction by multiplying each x-coordinate by 2

c

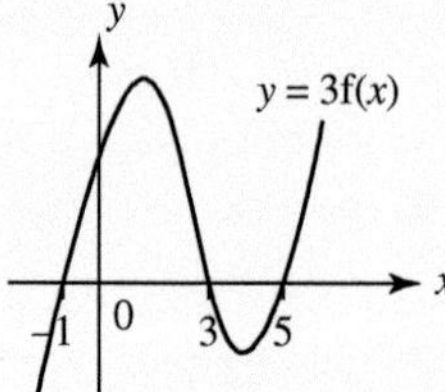

The maximum point moves to (1, 15)

Stretch the graph in the y-direction by multiplying each of the y-coordinates by 3

d

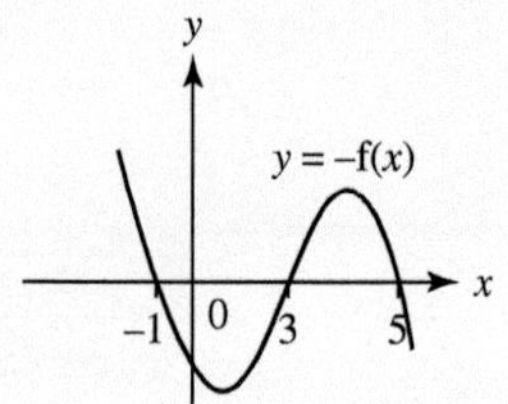

The maximum point is now a minimum point at (1, −5)

Reflect the graph in the x-axis by multiplying each of the y-coordinates by −1

Try It 5

The graph of $y=\mathrm{f}(x)$ is shown. Sketch the graphs of

a $y=\mathrm{f}(x-4)$ **b** $y=\mathrm{f}(-x)$

c $y=\mathrm{f}(2x)$ **d** $y=3\mathrm{f}(x)$

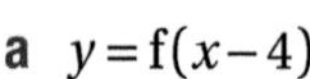

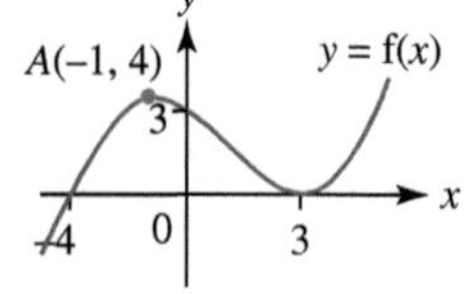

State the coordinates of the local maximum, point A, in each case.

The graph of $y=\frac{1}{x}$ is called a **reciprocal graph**.

It has asymptotes at $x=0$ and $y=0$ as shown.

The vertical asymptote, $x=0$, exists because $\frac{1}{0}$ is undefined. The horizontal asymptote, $y=0$, exists because there is no value of x for which $\frac{1}{x}=0$

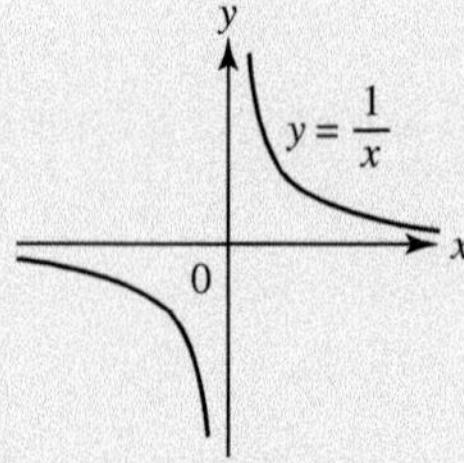

A line is an **asymptote** to a curve if, over an unbounded part of the curve, the curve tends towards the line but never reaches it.

Example 6

Sketch the graph of $y = -\frac{2}{x}$ and state the equations of the asymptotes.

Let $f(x) = \frac{1}{x}$, then $-2f(x) = -\frac{2}{x}$

Write the function in terms of $\frac{1}{x}$ so you can see how the basic function has been transformed.

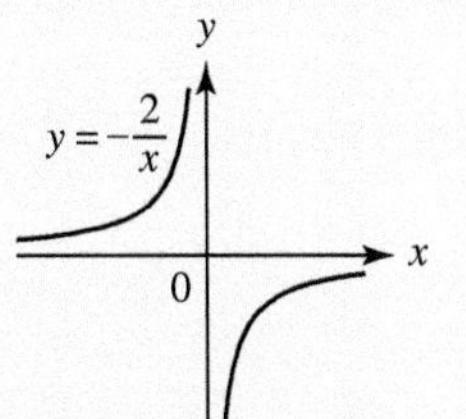

The graph is in the other two quadrants to $y = \frac{1}{x}$ due to the minus sign in the equation $y = -\frac{2}{x}$

The 2 indicates that the graph of $y = -\frac{1}{x}$ has been stretched vertically by a scale factor of 2

Asymptotes are $x = 0$ and $y = 0$

Try It 6

Sketch these graphs and state the equations of the asymptotes.

a $y = \frac{3}{x}$ **b** $y = -\frac{1}{x}$

When reciprocal graphs are translated horizontally the equation of the vertical asymptote will change. The graph of $y = \frac{1}{x-k}$ will have asymptotes at $x = k$ and $y = 0$

Example 7

Sketch the graph of $y = \frac{1}{x+3}$ and state the equations of the asymptotes.

Let $f(x) = \frac{1}{x}$, then $f(x+3) = \frac{1}{x+3}$

The basic function $\frac{1}{x}$ has been translated 3 to the left.

When $x = 0$, $y = \frac{1}{3}$ so this is the y-intercept.

Asymptotes are $x = -3$ and $y = 0$

Try It 7

Sketch the graph of $y = \frac{1}{x-5}$ and state the equations of the asymptotes.

When reciprocal graphs are translated vertically the equation of the horizontal asymptote will change. The graph of $y = k + \frac{1}{x}$ will have asymptotes at $x = 0$ and $y = k$

1071, 1172, 1955, 1958 SEARCH

Example 8

Sketch the graph of $y=2+\frac{1}{x}$ and state the equations of the asymptotes.

Let $f(x)=\frac{1}{x}$, then $2+f(x)=2+\frac{1}{x}$

The basic function $\frac{1}{x}$ has been translated 2 up.

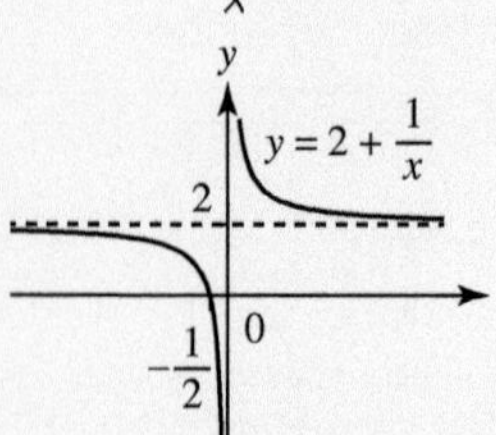

When $y=0$, $2+\frac{1}{x}=0 \Rightarrow \frac{1}{x}=-2$

$\Rightarrow x=-\frac{1}{2}$ so this is the x-intercept.

Asymptotes are $x=0$ and $y=2$

Try It 8

Sketch the graph of $y=-3+\frac{1}{x}$ and state the equations of the asymptotes.

Bridging Exercise Topic C

Bridging to Ch2.4

1 Sketch each of these cubic graphs.

a $y=-x^3$ b $y=(x+1)(x+2)(x+4)$ c $y=(x-2)(x+3)(x+5)$

d $y=x(x+1)(x-2)$ e $y=(5-x)(x+2)(x+6)$ f $y=-x(x+1)(x-7)$

g $y=x^2(x+3)$ h $y=(x-1)^2(x+4)$ i $y=-x(x+5)^2$

j $y=x^2(6-x)$ k $y=x(2x+1)(x-4)$ l $y=(x-5)(3x-1)^2$

2 Sketch each of these cubic graphs.

a $y=x^3+2x^2$ b $y=3x^3-12x$ c $y=6x^3+15x^2$

d $y=-x^3+7x^2$ e $y=8x^2-28x^3$ f $y=15x^3-10x^2$

g $y=x^3+3x^2-28x$ h $y=x^3-7x^2+10x$ i $y=-x^3-4x^2-3x$

j $y=-x^3+8x^2-15x$ k $y=x-4x^3$ l $y=-15x^3+x^2+2x$

3 Sketch each of these quartic graphs.

a $y=-x^4$ b $y=x(x+5)(x+1)(x-3)$

c $y=(x+4)(x+6)(x-2)(x-1)$ d $y=(x+2)^2(x-5)(4x-7)$

e $y=(3x-4)^2(x+6)(x-1)$ f $y=-(x+1)(2x+5)(x-7)(x-1)$

g $y=\frac{1}{3}(x+3)^2(3-x)^2$ h $y=(x+8)^2(1-x)(2x+1)$

4 The graph of $y = f(x)$ is shown. Sketch the graphs of

a $y = f(x+3)$ **b** $y = f(2x)$

c $y = 3f(x)$ **d** $y = -f(x)$

State the coordinates of the local maximum, point A, in each case.

5 The graph of $y = g(x)$ is shown. Sketch the graphs of

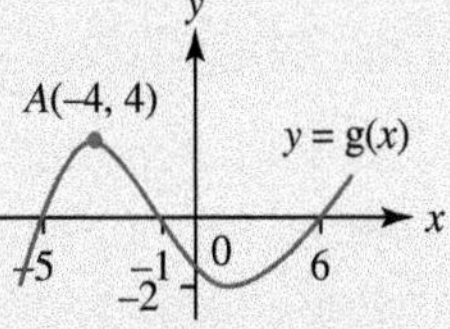

a $y = g(x-1)$ **b** $y = g(-x)$

c $y = g(x)+2$ **d** $y = 2g(x)$

State the coordinates of the maximum point A in each case.

6 The graph of $y = f(x)$ is shown. Sketch the graphs of

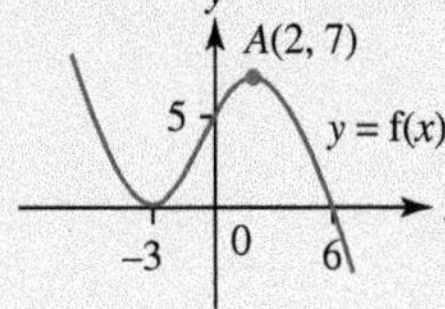

a 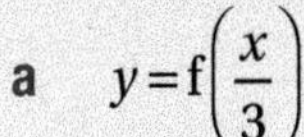 $y = f\left(\frac{x}{3}\right)$ **b** $y = f(x-3)$

c $y = 3f(x)$ **d** $y = -f(x)$

State the coordinates of the maximum point A in each case.

7 The graph of $y = g(x)$ is shown. Sketch the graphs of

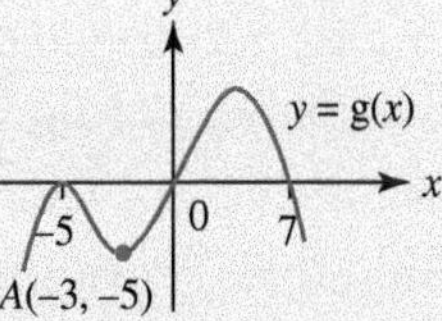

a $y = g(x-5)$ **b** $y = g(-x)$

c $y = 5+g(x)$ **d** $y = -2g(x)$

State the coordinates of the minimum point A in each case.

8 Sketch each of these reciprocal graphs.

a $y = -\frac{2}{x}$ **b** $y = \frac{1}{x+2}$ **c** $y = \frac{1}{x-9}$ **d** 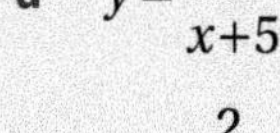 $y = \frac{2}{x+5}$

e $y = -\frac{1}{x+7}$ **f** $y = -\frac{3}{x-4}$ **g** $y = \frac{1}{10-x}$ **h** $y = \frac{2}{1-x}$

i $y = 1+\frac{1}{x}$ **j** $y = \frac{1}{x}+3$ **k** $y = -4+\frac{1}{x}$ **l** $y = 2-\frac{3}{x}$

9 Write down a possible equation for each of these curves.

a

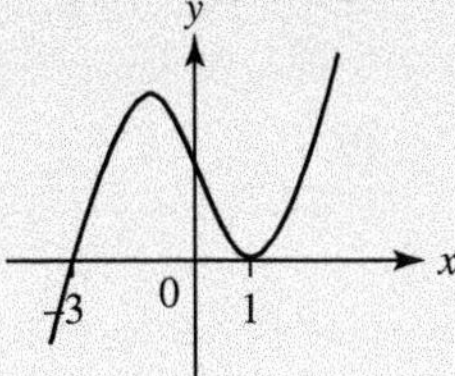

b

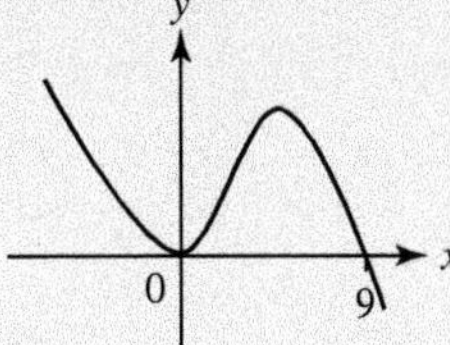

c

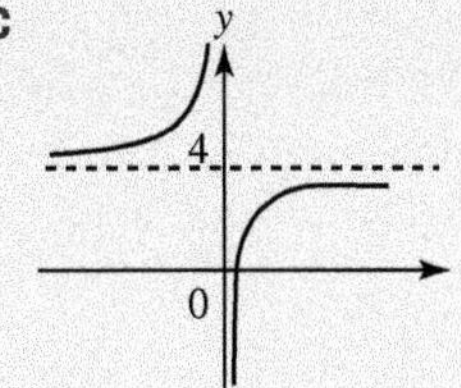

d

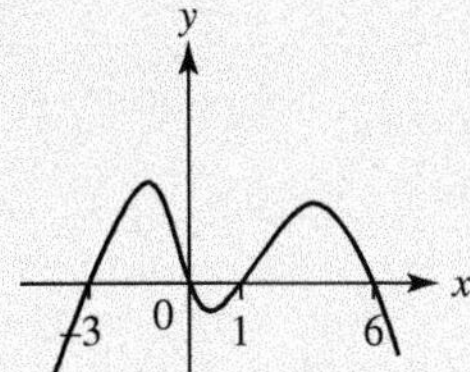

e

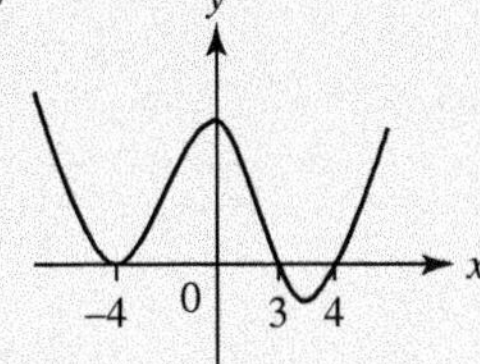

2.1 Expanding and factorising

Fluency and skills

A **polynomial** is an algebraic expression that can have constants, variables, **coefficients** and **powers** (also known as **exponents**), all combined using addition, subtraction, multiplication and division.

Key point

The highest power in a polynomial is called its **degree.**

All quadratics are of degree two.

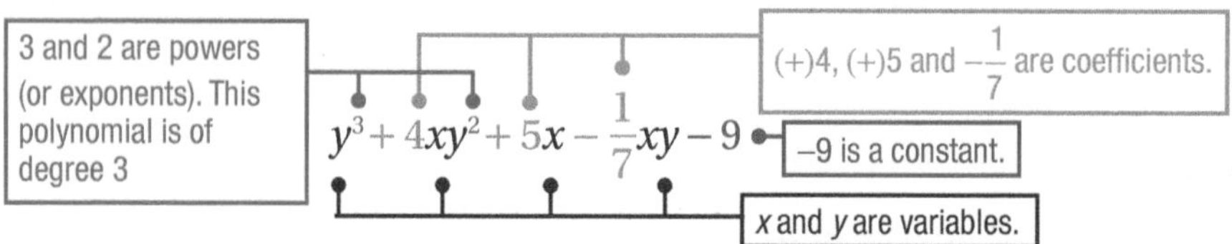

3 and 2 are powers (or exponents). This polynomial is of degree 3
(+)4, (+)5 and $-\frac{1}{7}$ are coefficients.
$y^3 + 4xy^2 + 5x - \frac{1}{7}xy - 9$
−9 is a constant.
x and y are variables.

You can simplify polynomials by collecting (adding or subtracting) **like terms**. You must *never* attempt to simplify a polynomial by dividing by a variable and the exponent of a variable can only be 0, 1, 2, 3, ...etc.

You can manipulate polynomials by **expanding**, **simplifying** and **factorising** them.

Example 1

Expand and simplify $(3x + 2y)^2 - (2x - 3y)^2$

$(3x+2y)^2 = (3x+2y)(3x+2y)$ — Expand the brackets before adding or subtracting polynomials.

$= 9x^2 + 12xy + 4y^2$ — Multiply each term in the first bracket by each term in the second bracket.

$(2x-3y)^2 = 4x^2 - 12xy + 9y^2$ so $-(2x-3y)^2 = -4x^2 + 12xy - 9y^2$

$9x^2 + 12xy + 4y^2$

$-4x^2 + 12xy - 9y^2$ — You may find it useful to write like terms vertically under each other.

$= 5x^2 + 24xy - 5y^2$ — Collect like terms to simplify the polynomial.

Key point

A statement that is true for all values of the variable(s) is called an **identity**.

You write an identity using the symbol $\equiv$

For example, $15x^3 + 8x^2 - 26x + 8 \equiv (3x^2 + 4x - 2)(5x - 4)$ is true for all values of x

It follows that $(3x^2 + 4x - 2)$ and $(5x - 4)$ are **factors** of $15x^3 + 8x^2 - 26x + 8$

Factorising is the opposite process to expanding brackets.

You can factorise polynomials by comparing coefficients.

Example 2

$(4x - 5)$ is a factor of the polynomial $12x^3 + 21x^2 - 61x + 20$

Factorise the polynomial completely.

$12x^3 + 21x^2 - 61x + 20 \equiv (4x - 5)(Ax^2 + Bx + C)$ — Use the fact that $4x - 5$ is a factor to write an identity.

$(4x - 5)(Ax^2 + Bx + C)$ — To expand, multiply each term in the first bracket by each term in the second bracket.

$\equiv 4Ax^3 + 4Bx^2 + 4Cx$ — To collect like terms write them under each other.

$\quad - 5Ax^2 - 5Bx - 5C$

$\equiv 4Ax^3 + (4B - 5A)x^2 + (4C - 5B)x - 5C$

This is identical to $12x^3 + 21x^2 - 61x + 20$

so the coefficients must all be the same.

so $4A = 12$ ① — Equate and compare coefficients for x^3, x^2, x and compare the constants.

$4B - 5A = 21$ ②

$4C - 5B = -61$ ③

$-5C = 20$ ④

$A = 3$ and $C = -4$ — Rearrange ① and ④

$4B - 5 \times 3 = 21$ — Substitute A = 3 into ②

$4B = 36$ so $B = 9$

$4(-4) - 5(9) = -61$ ✓ — Check by substituting the values into ③

So $12x^3 + 21x^2 - 61x + 20 \equiv (4x - 5)(3x^2 + 9x - 4)$ — State your answer clearly and check it by expanding the brackets. $(3x^2 + 9x - 4)$ cannot be factorised so this is the final answer.

Exercise 2.1A Fluency and skills

1 Write the degree of each of these expressions.

a $3 - 2x + x^2$ b $1 - 3x + 5x^4$ c $2x^2 - x + 1 - 4x^3$

2 Expand and simplify each of these expressions.

a $2x(3x + 8)$ b $2x(3x^2 + 8x - 9)$ c $(3y + 2)(4y - 7)$

d $3y(4y^2 + 8y - 7)$ e $(t - 5)^2$ f $(t + 3)(t - 5)^2$

3 Expand and simplify each of these expressions.

a $(x + 4)^2 + (x - 4)^2$ b $(5p + q)^2 - (5p - q)^2$

4 Factorise each of these expressions.

a $4m^3 + 6m^2$ b $16n^4 - 12n$ c $5p^4 - 2p^2 + 6p$ d $9y^2 - 15xy$

e $6x^2 - 3xy + 9x$ f $7yz - 21z^3$ g $4e(e - 2f) - 12ef$ h $p^2 - 100$

i $6q(3 - 2q) + 9q$ j $\frac{y}{5} - \frac{y^2}{15} + \frac{3y}{25}$ k $(d + 1)(d + 3) + (d + 1)(d - 5)$

l $w(2w + 3)(3w + 9) + w(2w - 11)(2w + 3)$

MyMaths 2006 SEARCH

5 Fully factorise these expressions.

a $4m^3 + 4m^2 - 15m$ b $7n^3 - 15n^2 + 2n$

6 Factorise this expression $3x(x+2)^2 + (x+2)(5x^2 + 2x - 6)$

7 Expand and simplify these expressions.

a $(5p+4q)^2 - (5p-4q)^2$ b $(x+y+z)^2 - (x-y-z)^2$

c $(x\sqrt{3}+4)^2 + (x\sqrt{3}-4)^2$ d $(x\sqrt{5}+4)^2 + (x\sqrt{3}-4)^2$

8 a (x^2+3x+9) is a factor of x^3+2x^2+6x-9. Work out the other factor.

b (x^2-2x+3) is a factor of $2x^3-11x^2+20x-21$. Work out the other factor.

c $(y^2+2y-15)$ is a factor of $2y^3+3y^2-32y+15$. Work out the other factor.

d $(z-2)$ is a factor of z^3+z^2-2z-8. Work out the other factor.

e $(2a+5)$ is a factor of $6a^3+7a^2-2a+45$. Work out the other factor.

f (x^2-4x+7) is a factor of $2x^3-5x^2+2x+21$. Factorise the polynomial fully.

g (k^2-3k+1) is a factor of $k^4+3k^3-24k^2+27k-7$. Work out the other factors.

Reasoning and problem-solving

Strategy

To factorise polynomials

1. Look for obvious common factors and factorise them out.
2. Write an identity and expand to compare coefficients.
3. Write your solution clearly and use suitable units where appropriate.

Example 3

The volume of a cylinder is $y^2 - 25y + 24$ ft^3

The base area is $(y-1)$ ft^2

Write an expression for its height.

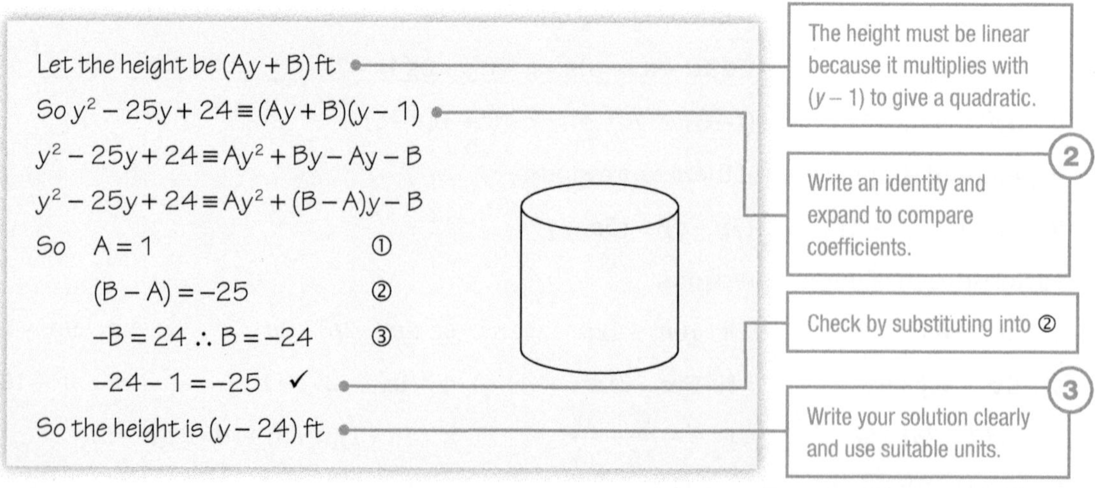

Let the height be $(Ay+B)$ ft
So $y^2 - 25y + 24 \equiv (Ay+B)(y-1)$
$y^2 - 25y + 24 \equiv Ay^2 + By - Ay - B$
$y^2 - 25y + 24 \equiv Ay^2 + (B-A)y - B$
So $A = 1$ ①
$(B-A) = -25$ ②
$-B = 24 \therefore B = -24$ ③
$-24 - 1 = -25$ ✓
So the height is $(y-24)$ ft
The height must be linear because it multiplies with $(y-1)$ to give a quadratic.
2 Write an identity and expand to compare coefficients.
Check by substituting into ②
3 Write your solution clearly and use suitable units.

Exercise 2.1B Reasoning and problem-solving

1 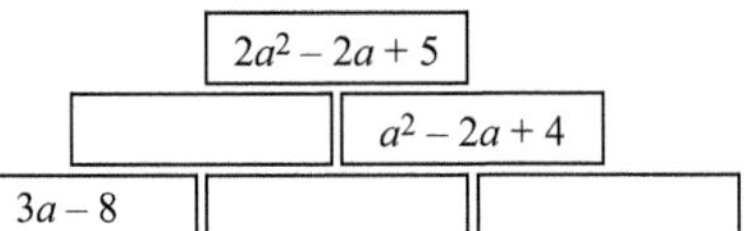

In this pyramid, each block is the sum of the two blocks vertically beneath.

Copy and complete the pyramid.

2 A square has side length $(4b - 7a)$ cm.

Write an expression for its area in expanded form.

3 A cuboid has sides of length $(c + 2)$, $(2c - 1)$ and $(3c - 7)$ cm. Write an expression for its volume in expanded form.

4 A square hole of side length $(a + 2)$ cm is cut from a larger square of side length $(2a + 5)$ cm. Without expanding any brackets, write the remaining part of the large square as a pair of factors.

5 A rectangle, sides $2a$ cm by a cm, has a square of side x cm cut from each corner. The sides are then folded up to make an open box. Work out the volume of this box.

6 A ball is thrown from ground level and its height, h ft, at time t s is given by the polynomial $h = 25t - 5t^2$

a When does the ball next return to ground level?

b What is the maximum height reached by the ball?

7 A body moves along a straight line from a point O where its position, x metres at time, t seconds is given by the equation $x = 3t^3 - 28t^2 + 32t$. Its velocity, v m s^{-1} and acceleration, a m s^{-2} at time t are given by the equations $v = 9t^2 - 56t + 32$ and $a = 18t - 56$

a Find the values of t when the body is at O, and find its velocity and acceleration at these times.

b Find the distance of the body from O and its velocity when its acceleration is zero.

c Find the value(s) of t when its velocity is zero, and find its acceleration at these times.

8 A rectangle has the dimensions shown. All lengths are given in centimetres.

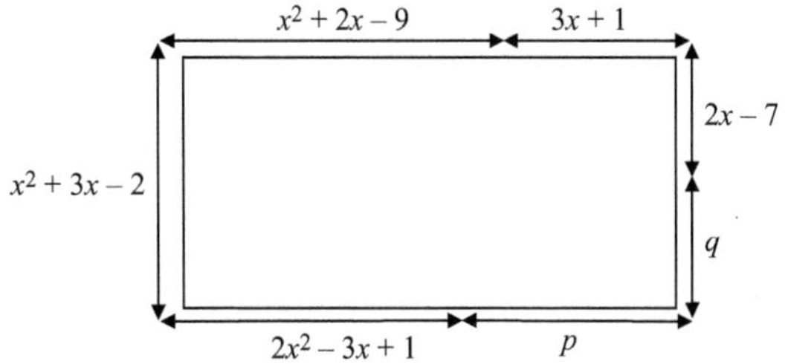

a Find p and q in terms of x

b Write expanded expressions for the rectangle's perimeter and area in terms of x

9 Kia has lost her calculator. Show how she can complete these calculations.

a $66.89^2 - 33.11^2$

b $(\sqrt{8})^3 - (\sqrt{2})^3$

10 A cuboid has volume $(2h^3 + 3h^2 - 23h - 12)$ cm^3. Its length is $(h + 4)$ cm and its width is $(h - 3)$ cm Work out the height of the cuboid.

11 The area of a trapezium is given by the polynomial $(2s^3 - 17s^2 + 41s - 30)$ cm^2. The perpendicular height is $(4s - 6)$ cm. Write an expression for the sum of the parallel sides.

12 The area of an ellipse is given by the formula πab, where a and b are half the lengths of the axes of symmetry. The area is $\pi(6t^3 - 5t^2 + 15t + 14)$ and $a = (3t + 2)$. Write an expression for b

Challenge

13 $V = \pi I[r^2 - (r - a)^2]$ is the volume of a circular pipe.

Find an expression for V in terms of p when $I = 4p + 5$, $r = 3p - 4$ and $a = p + 1$

PURE

2.2 The binomial theorem

Fluency and skills

You can expand $(1+x)^n$ where $n = 0, 1, 2, 3, \ldots$

EXPANSION	COEFFICIENTS
$(1+x)^0 \equiv 1$	1
$(1+x)^1 \equiv 1 + 1x$	1 1
$(1+x)^2 \equiv 1 + 2x + 1x^2$	1 2 1
$(1+x)^3 \equiv 1 + 3x + 3x^2 + 1x^3$	1 3 3 1
$(1+x)^4 \equiv 1 + 4x + 6x^2 + 4x^3 + 1x^4$	1 4 6 4 1
$(1+x)^5 \equiv 1 + 5x + 10x^2 + 10x^3 + 5x^4 + 1x^5$	1 5 10 10 5 1

Pascal's Triangle was published in 1654, but was known to the Chinese and the Persians in the 11th century.

The coefficients form a pattern known as **Pascal's triangle**.

Each coefficient in the triangle is the sum of the two coefficients above it.

Example 1

Use Pascal's triangle to write the expansion of $(1+2y)^6$ in ascending powers of y

The coefficients are 1, 6, 15, 20, 15, 6, 1

Write down the 6th row of Pascal's triangle.

$(1+(2y))^6$

$\equiv 1 + 6(2y) + 15(2y)^2 + 20(2y)^3 + 15(2y)^4 + 6(2y)^5 + (2y)^6$

$\equiv 1 + 12y + 60y^2 + 160y^3 + 240y^4 + 192y^5 + 64y^6$

Use the expansion of $(1+x)^n$, substituting $2y$ for x

Replacing 1 with a and x with b gives the **binomial expansion** $(a+b)^n$ where $n = 0, 1, 2, 3, \ldots$

A binomial expression has two terms.

As n increases you can see that again the coefficients form Pascal's triangle.

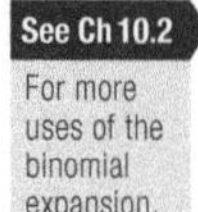

See Ch 10.2 For more uses of the binomial expansion.

$(a+b)^0 \equiv 1$

$(a+b)^1 \equiv 1a + 1b$

$(a+b)^2 \equiv 1a^2 + 2ab + 1b^2$

$(a+b)^3 \equiv 1a^3 + 3a^2b + 3ab^2 + 1b^3$

$(a+b)^4 \equiv 1a^4 + 4a^3b + 6a^2b^2 + 4ab^3 + 1b^4$

$(a+b)^5 \equiv 1a^5 + 5a^4b + 10a^3b^2 + 10a^2b^3 + 5ab^4 + 1b^5$

In each expansion, the power of a starts at n and decreases by 1 each term, so the powers are n, $n - 1, n - 2, \ldots, 0$

The power of b starts at 0 and increases by 1 each term, so the powers are $0, 1, 2, \ldots, n$

The sum of the powers of any individual term is always n

PURE

Example 2

Expand $(2 + 3t)^4$

$$(2+3t)^4$$

$$\equiv 2^4 + 4 \times 2^3 \times (3t) + 6 \times 2^2 \times (3t)^2 + 4 \times 2 \times (3t)^3 + (3t)^4$$

$$\equiv 16 + 96t + 216t^2 + 216t^3 + 81t^4$$

Use Pascal's triangle and the expansion of $(a + b)^4$ substituting 2 for a and $3t$ for b

It would be impractical to use Pascal's triangle every time you need to work out a coefficient—say, for example, you want to find the coefficient of x^6 in $(x + a)^{10}$

There is a general rule for finding this coefficient without needing to write out Pascal's triangle up to the tenth row.

Note that the first coefficient in each row is the 0th coefficient.

Key point

The rth coefficient in the nth row is ${}^nC_r \equiv \dfrac{n!}{(n-r)!r!}$

${}^nC_r$ is sometimes written as $\begin{pmatrix} n \\ r \end{pmatrix}$ or ${}_nC_r$

Key point

$n!$ stands for the product of all integers from 1 to n. You read it as n **factorial.**

For example, $6! = 6 \times 5 \times 4 \times 3 \times 2 \times 1 = 720$

Look for the factorial button on your calculator. It may be denoted $x!$

${}^nC_r$ is the choose function and you read it as 'n choose r'. It gives the number of possible ways of choosing r elements from a set of n elements when the order of choosing does not matter. For example, the number of combinations in which you can choose 2 balls from a bag of 5 balls is ${}^5C_2$

You use the choose function because there are several ways of getting certain powers from an expansion. For example, there are 3 ways of getting ab^2 from the expansion of $(a + b)^3$: a from either the first, second or third bracket and b from the other two brackets in each case. The term in ab^2 for the expansion of $(a + b)^3$ is therefore ${}^3C_1\, ab^2 = 3ab^2$

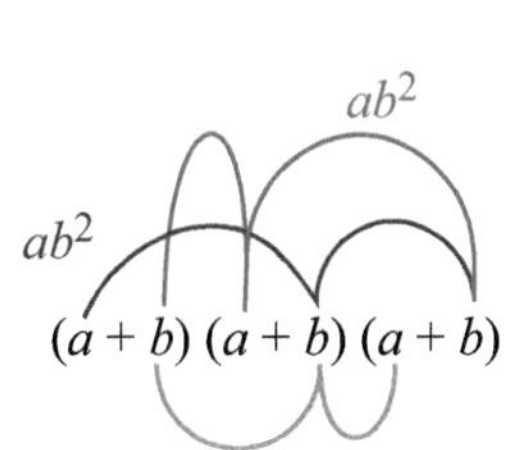

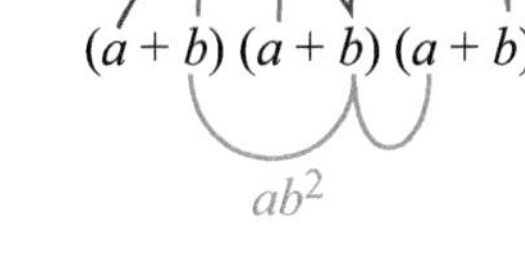

Example 3

A term in the expansion of $(y+2x)^9$ is given by ky^3x^6
Find the value of k

$^9C_6 \times y^3 \times (2x)^6 \equiv 84 \times y^3 \times 64x^6$

$\equiv 5376y^3x^6$

$k = 5376$

Use your calculator to find $^9C_6$ and work out 2^6

Simplify to find the value of k

The formula for the binomial expansion of $(a+b)^n$ is sometimes called the **binomial theorem**.

Key point

$$(a+b)^n \equiv a^n + {}^nC_1a^{n-1}b + {}^nC_2a^{n-2}b^2 + \ldots + {}^nC_ra^{n-r}b^r + \ldots + b^n$$

For the expansion of $(1+x)^n$ this gives

Key point

$$(1+x)^n \equiv 1+nx+\frac{n(n-1)}{2!}x^2+\frac{n(n-1)(n-2)}{3!}x^3+\ldots+x^n$$

Example 4

Write the term in z^4 in the expression $(2z-1)^{15}$. Simplify your answer.

Take $a = 2z$ and $b = -1$

$^{15}C_{11}(2z)^{15-11}(-1)^{11}$

$\equiv 1365 \times 16z^4 \times (-1)$

$= -21\,840z^4$

The powers add to 15 so the second power must be 11. Use the coefficient $^{15}C_{11}$

$(2z)^4 = 2^4z^4$

Exercise 2.2A Fluency and skills

1 Calculate the values of

a $5!$ **b** $7!$ **c** $11!$

2 Calculate the values of

a $^5C_2$ **b** $^9C_3$ **c** $^{11}C_7$ **d** $^{13}C_8$

3 Work out the values of

a $\binom{5}{3}$ **b** $\binom{10}{1}$ **c** $\binom{13}{5}$ **d** $\binom{20}{6}$

4 Use Pascal's triangle to find the expansions of each of these expressions.

a $(1+3x)^3$ **b** $\left(1-\frac{z}{2}\right)^5$

c $\left(1-\frac{m}{3}\right)^4$ **d** $\left(1+\frac{3x}{2}\right)^5$

5 Find the first four terms of these binomial expansions in ascending powers of x

a $(1+x)^8$ **b** $(1-3x)^7$

c $(1+2x)^9$ **d** $(2-3x)^6$

e $(x-2)^8$ **f** $(2x-1)^{10}$

6 Use Pascal's triangle to expand each of these expressions.

a $(2-4y)^3$ **b** $(3b+5)^4$ **c** $\left(4z-\frac{y}{3}\right)^5$

7 Find the first three terms of these binomial expansions in descending powers of x

a $(2+x)^6$ **b** $(1-2x)^8$

c $(3-x)^9$ **d** $(x+4)^7$

e $(2x+3)^{10}$ **f** $\left(\frac{x}{2}+4\right)^{11}$

8 Use the binomial theorem to expand each of these expressions.

a $(2+3t)^4$ **b** $(3-2p)^4$

c $(4p+3q)^5$ **d** $(3p-4q)^5$

e $(3z-2)^4$ **f** $\left(2z-\frac{1}{2}\right)^6$

g $\left(2+\frac{2x}{3}\right)^3$ **h** $\left(\frac{r}{3}+\frac{s}{4}\right)^8$

i $\left(\frac{x}{2}+\frac{y}{3}\right)^3$

9 Find the terms indicated in each of these expansions and simplify your answers.

a $(p+5)^5$ term in p^2

b $(4+y)^9$ term in y^5

c $(3+q)^{12}$ term in q^7

d $(4-3m)^5$ term in m^3

e $(2z-1)^{15}$ term in z^4

f $\left(z+\frac{3}{2}\right)^8$ term in z^6

g $(3x+4y)^5$ term in y

h $(2a-3b)^{10}$ terms in **i** a^5 and **ii** b^4

i $\left(4p+\frac{1}{4}\right)^3$ term in p^2

j $\left(4a-\frac{3b}{4}\right)^{11}$ terms in **i** a^5 and **ii** b^5

k $\left(\frac{a}{2}-\frac{2b}{3}\right)^{11}$ terms in **i** a^7 and **ii** b^5

10 Use the binomial theorem to expand each of these expressions.

a $(c^2+d^2)^4$ **b** $(v^2-w^2)^5$

c $(2s^2+5t^2)^3$ **d** $(2s^2-5t^2)^3$

e $\left(d+\frac{1}{d}\right)^3$ **f** $\left(2w+\frac{3}{w}\right)^4$

11 Use the binomial theorem to expand each of these brackets.

a $\left(x+\frac{2}{x}\right)^3$ **b** $(x^2-2)^4$

c $\left(x^2-\frac{1}{x}\right)^5$ **d** $\left(\frac{1}{x^2}+3x\right)^6$

12 Expand and simplify each of these expressions.

a $3x(2x-5)^5$ **b** $(2+x)^4(1+x)$

13 Expand and simplify each of these expressions.

a $(5-2x)^3+(3+2x)^4$

b $(1+3x)^5-(1-4x)^3$

14 Expand and fully simplify each of these expressions. Show your working.

a $(2+\sqrt{3})^4+(1-\sqrt{3})^4$

b $(1-\sqrt{5})^5-(2\sqrt{5}+3)^3$

15 Write down the first four terms of the expansion of each of these in ascending powers of x

a $(1+2x)^n$ **b** $(1-3x)^n$

where $n \in \mathbb{N}, n>3$

16 a Expand $(1+4x)^6$ in ascending powers of x up to and including the term in x^2

b Use your answer to part **a** to estimate the value of $(1.04)^6$

17 a Expand $(1-2x)^7$ in ascending powers of x up to and including the term in x^3

b Use your answer to part **a** to estimate the value of $(0.99)^7$

18 Use the binomial expansion to simplify each of these expressions. Give your final solutions in the form $a+b\sqrt{2}$

a $(1+\sqrt{2})^3$ **b** $(1-\sqrt{2})^5$

c $(3+2\sqrt{2})^4$ **d** $(\sqrt{2}-2)^6$

e $\left(1-\frac{1}{\sqrt{2}}\right)^3$ **f** $\left(\frac{\sqrt{2}}{3}+3\right)^4$

19 Use the binomial expansion to fully simplify each of these expressions.

Give your final answers in surd form.

a $(1+\sqrt{3})^4$ **b** $(1-\sqrt{5})^6$

c $(5-\sqrt{7})^5$ **d** $(2\sqrt{6}+5)^3$

e $(\sqrt{2}+\sqrt{6})^4$ **f** $(\sqrt{3}-\sqrt{2})^6$

Reasoning and problem-solving

Strategy

To construct a binomial expansion

1. Create an expression in the form $(1 + x)^n$ or $(a + b)^n$
2. Use Pascal's triangle or the binomial theorem to find the required terms of the binomial expansion.
3. Use your expansion to answer the question in context.

Example 5

A football squad consists of 13 players. Use the formula ${}^nC_r \equiv \dfrac{n!}{(n-r)!r!}$ to show that there are 78 possible combinations of choosing a team of 11 players from this squad.

$$^{13}C_{11} = \frac{13!}{(13-11)!11!}$$
$$= \frac{13\times12\times11\times10\times\ldots\times2\times1}{2!\times11\times10\times\ldots\times2\times1}$$
$$= \frac{13\times12}{2!}$$
$$= \frac{156}{2} = 78$$

2 Cancel the common factor 11!

Example 6

a Using the first *three* terms of the binomial expansion, estimate the value of 1.003^8

b By calculating the fourth term in the expansion show that the estimate from part **a** is accurate to 3 decimal places.

a $1.003^8 = (1 + 0.003)^8$

1 Rewrite in the form $(1 + x)^n$

First 3 terms

$$= 1 + nx + \frac{n(n-1)}{2!}x^2$$

2 Use the first 3 terms of the general expansion.

$$= 1 + 8(0.003) + 28(0.003)^2$$
$$= 1 + 0.024 + 0.000252$$
$$= 1.024252 \text{ (= 1.024 to 3 sf)}$$

3 Substitute values and simplify.

b $\dfrac{n(n-1)(n-2)}{3!}x^3 = 56(0.003)^3$

$= 0.000001512$

Adding this term will not affect the first three decimal places.

Exercise 2.2B Reasoning and problem-solving

PURE

1 How many possible ways are there to pick a 7's rugby team from a squad of 10 players?

2 How many possible ways are there to choose half of the people in a group of 20?

3 A cube has side length $(2s - 3w)$. Use the binomial expansion to find its volume.

4 Use Pascal's triangle to find the value of

a 1.05^6 correct to six decimal places,

b 1.96^3 correct to four decimal places.

5 Use the binomial theorem to work out the value of

a 1.015^5 correct to 4 decimal places,

b $\left(\frac{199}{100}\right)^{10}$ correct to five significant figures.

6 Use the binomial theorem to work out the value of $\left(\frac{13}{4}\right)^5$ correct to five decimal places.

7 Work out the exact value of the middle term in the expansion of $\left(\sqrt{3}+\sqrt{5}\right)^{10}$

8 a Find the coefficient of x^4 in the expansion of $(1 + x)(2x - 3)^5$

b Find the coefficient of x^3 in the expansion of $(x - 2)(3x + 5)^4$

9 Find, in the expansion of $\left(x^2 - \frac{1}{2x}\right)^6$, the coefficient of

a x^3 b x^6

10 Find, in the expansion of $\left(\frac{1}{t^2} + t^3\right)^{10}$, the coefficient of

a t^{10} b t^{-5}

11 The first three terms in the expansion of $(1+ax)^n$ are $1+35x+490x^2$. Given that n is a positive integer, find the value of

a n b a

12 Given that $(1+bx)^n \equiv 1-24x+252x^2+\ldots$ for a positive integer n find the value of

a n b b

13 In the expansion of $(1+2x)^n$, n a positive integer, the coefficient of x^2 is eight times the coefficient of x. Find the value of n

14 In the expansion of $\left(1+\frac{x}{2}\right)^n$, n a positive integer, the coefficients of x^4 and x^5 are equal. Calculate the value of n

15 Find an expression for

a $\binom{n}{n-1}$ b $\binom{n}{3}$

c $\binom{n}{n-2} - \binom{n+1}{n-1}$

Write your answers as polynomials in n with simplified coefficients.

16 Fully simplify these expressions.

a $\frac{n!}{(n+1)!}$ b $\frac{(n+3)!}{n(n+1)!}$

17 Find the constant term in the expansion of $(2+3x)^3\left(\frac{1}{x}-4\right)^4$

18 Find the coefficient of y^3 in the expansion of $(y+5)^3(2-y)^5$

Challenge

19 A test involves 6 questions.

For each question there is a 25% chance that a student will answer it correctly.

a How many ways are there of getting exactly two of the questions correct?

b What is the probability of getting the first two questions correct then the next four questions incorrect?

c What is the probability of getting exactly two questions correct?

d What is the probability of getting exactly half of the questions correct?

2.3 Algebraic division

Fluency and skills

In Section **2.1** you learned how to factorise a polynomial by writing the identity and comparing and evaluating constants.

You can also use the method of dividing the polynomial by a known factor. You can divide algebraically using the same method as 'long division' in arithmetic. It is an easier method than comparing coefficients when the polynomials are of degree 3 or higher.

ICT Resource online

To investigate algebraic division, click this link in the digital book.

Example 1

Use long division to divide $2x^4 + 7x^3 - 14x^2 - 3x + 15$ by $(x + 5)$

Give your answer in the form of a quotient and remainder.

$$
\begin{array}{r}
2x^3 - 3x^2 + x - 8 \\
(x+5)\overline{)\,2x^4 + 7x^3 - 14x^2 - 3x + 15} \\
\underline{2x^4 + 10x^3} \qquad\qquad\qquad\quad \\
-3x^3 - 14x^2 \qquad\qquad \\
\underline{-3x^3 - 15x^2} \qquad\qquad \\
x^2 - 3x \qquad \\
\underline{x^2 + 5x} \qquad \\
-8x + 15 \\
\underline{-8x - 40} \\
55
\end{array}
$$

So $(2x^4 + 7x^3 - 14x^2 - 3x + 15) \div (x + 5)$
$= (2x^3 - 3x^2 + x - 8)$ remainder 55

Divide the first term $2x^4$ by x. Write the answer, $2x^3$, on the top.

Write $(x + 5) \times 2x^3 \equiv 2x^4 + 10x^3$ on this line and subtract from the line above to give $-3x^3$

Write the $-3x^3$ and bring down the next term, $-14x^2$, to make $-3x^3 - 14x^2$ here.

Repeat this process until you get a quotient (and a remainder if there is one).

$(2x^3 - 3x^2 + x - 8)$ is the quotient. 55 is the remainder.

Example 2

Use long division to show that $(x - 2)$ is a factor of $f(x) = x^3 + 10x^2 + 11x - 70$

$$
\begin{array}{r}
x^2 + 12x + 35 \\
(x-2)\overline{)\,x^3 + 10x^2 + 11x - 70} \\
\underline{x^3 - 2x^2} \qquad\qquad\qquad \\
12x^2 + 11x \qquad \\
\underline{12x^2 - 24x} \qquad \\
35x - 70 \\
\underline{35x - 70} \\
0
\end{array}
$$

There is no remainder when $f(x)$ is divided by $x - 2$ so $x - 2$ is a factor of $f(x)$

Divide the first term x^3 by x. Write the answer, x^2, on top.

Multiply x^2 by $(x - 2)$
Write the answer, $x^3 - 2x^2$, underneath and subtract from the line above.

Write the answer, $12x^2$, and bring the next term down.

Repeat the process.

Example 1 shows that dividing f(x) by $(x - a)$ leaves you with a remainder, R

In general, for a polynomial f(x) of degree $n \geq 1$ and any constant a

$f(x) \equiv (x - a)\,g(x) + R$

Where g(x) is a polynomial of order $n - 1$ and R is a constant.

For the particular case when $x = a$, this gives

$f(a) = (a - a)\,g(a) + R$

$f(a) = R$

You can see from this that $f(a) = 0$ implies there is no remainder when f(x) is divided by $(x - a)$

> This calculation also demonstrates the **remainder theorem** which is beyond the scope of your A level course.

The **factor theorem** states that if $f(a) = 0$, $(x - a)$ is a factor of f(x) **Key point**

In Example 2 you saw that there is no remainder when $x^3 + 10x^2 + 11x - 70$ is divided by $(x - 2)$, which is equivalent to saying that $(x - 2)$ is a factor of $x^3 + 10x^2 + 11x - 70$

If you substitute $x = 2$ into the expression, the factor $(x - 2)$ is zero so the value of f(x) is zero.

You can check this by substitution, which gives $f(2) = 2^3 + 10(2)^2 + 11(2) - 70$

$= 8 + 40 + 22 - 70 = 0$

Example 3

Show that $(x + 3)$ is a factor of $2x^4 + 2x^3 - 9x^2 - 4x - 39$

$f(-3) = 2(-3)^4 + 2(-3)^3 - 9(-3)^2 - 4(-3) - 39$

$= 0$

$(x + 3)$ is a factor since $f(-3) = 0$

> $(x - a)$ is a factor if $f(a) = 0$, so to show $(x + 3)$ is a factor you need to show that $f(-3) = 0$

Example 4

Fully factorise the polynomial $2x^3 + 17x^2 - 13x - 168$

$f(x) = 2x^3 + 17x^2 - 13x - 168$

$f(1) = 2(1)^3 + 17(1)^2 - 13(1) - 168 = -162$

$f(1) \neq 0$ so $(x - 1)$ is not a factor

> Use trial and error with different values of a to find a case where $f(a) = 0$

$f(2) = 2(2)^3 + 17(2)^2 - 13(2) - 168 = -110$

$f(2) \neq 0$ so $(x - 2)$ is not a factor

$f(3) = 2(3)^3 + 17(3)^2 - 13(3) - 168 = 0$

$f(3) = 0$ so $(x - 3)$ is a factor

$$
\begin{array}{r}
2x^2 + 23x + 56 \\
(x-3)\overline{)\,2x^3 + 17x^2 - 13x - 168} \\
\underline{2x^3 - 6x^2} \qquad\qquad\quad \\
23x^2 - 13x \qquad\quad \\
\underline{23x^2 - 69x} \qquad\quad \\
56x - 168 \\
\underline{56x - 168} \\
0
\end{array}
$$

> Use long division to divide the polynomial by the factor to get a quadratic expression in x

So $2x^3 + 17x^2 - 13x - 168 \equiv (x - 3)(2x^2 + 23x + 56)$

$\equiv (x - 3)(2x + 7)(x + 8)$

> Use the result from the long division to express the polynomial in a partially factorised form.

> Factorise the quadratic to fully factorise the polynomial.

Exercise 2.3A Fluency and skills

1 Divide

a $x^2 - x - 90$ by $(x + 9)$

b $3x^2 - 19x - 14$ by $(x - 7)$

c $8x^2 + 14x - 15$ by $(2x + 5)$

2 Divide each polynomial by the given factor by comparing coefficients.

a $x^3 + 3x^2 - 11x + 7$ by $(x - 1)$

b $x^3 + 2x^2 - 4x - 3$ by $(x + 3)$

c $2x^3 + 9x^2 - 17x - 45$ by $(2x - 5)$

d $3x^3 - 14x^2 + 16x + 7$ by $(3x + 1)$

e $2x^4 - 17x^3 + 22x^2 + 65x - 9$ by $(2x - 9)$

3 Use long division to divide $4x^3 + 4x^2 - 8x + 5$ by $(x - 4)$

4 Use long division to show that $5x^3 + 11x^2 - 73x - 15$ is divisible by $(x - 3)$

5 Divide using long division

a $x^3 - 2x + 1$ by $(x - 1)$

b $x^3 - 10x^2 - 10x - 11$ by $(x - 11)$

c $6x^3 - 13x^2 - 19x + 12$ by $(3x + 4)$

d $6x^4 - 19x^3 + 23x^2 - 26x + 21$ by $(2x - 3)$

e $10x^4 + 33x^3 - 57x^2 + 5x + 1$ by $(5x - 1)$

6 Work out the values of **i** f(0) **ii** f(1) **iii** f(−1) **iv** f(2) **v** f(−2) when

a $f(x) = x^3 - 2x^2 + 10x$

b $f(x) = x^3 - 2x^2 - 2x - 2$

c $f(x) = x^3 - 3x^2 + x + 2$

d $f(x) = 2x^3 + x^2 - 5x + 2$

e $f(x) = x^3 - x^2 - 4x + 4$

7 a Show that $(x + 6)$ is a factor of $x^3 + 4x^2 - 9x + 18$

b Show that $(x - 8)$ is a factor of $2x^3 - 13x^2 - 20x - 32$

c Show that $(3x - 1)$ is a factor of $3x^3 + 11x^2 - 25x + 7$

d Show that $(5x + 2)$ is a factor of $10x^3 + 19x^2 - 39x - 18$

8 Fully factorise the polynomial $4x^3 + 27x^2 - 7x$

9 Fully factorise the polynomial $2x^3 + 9x^2 - 2x - 9$

10 a Factorise fully $x^3 + 3x^2 - 16x + 12$

b Factorise fully $x^3 - 6x^2 - 55x + 252$

c Factorise fully $6x^3 + 19x^2 + x - 6$

d Factorise fully $x^4 - 13x^2 - 48$

Reasoning and problem-solving

Strategy

To factorise a polynomial

1. Apply the factor theorem as necessary to find your first factor.
2. Divide the polynomial by the factor to get a quadratic quotient.
3. Factorise the quadratic quotient to fully factorise the polynomial.

Example 5

$(x + 1)$ is a factor of the polynomial $3x^3 + 8x^2 + ax - 28$. Fully factorise the polynomial.

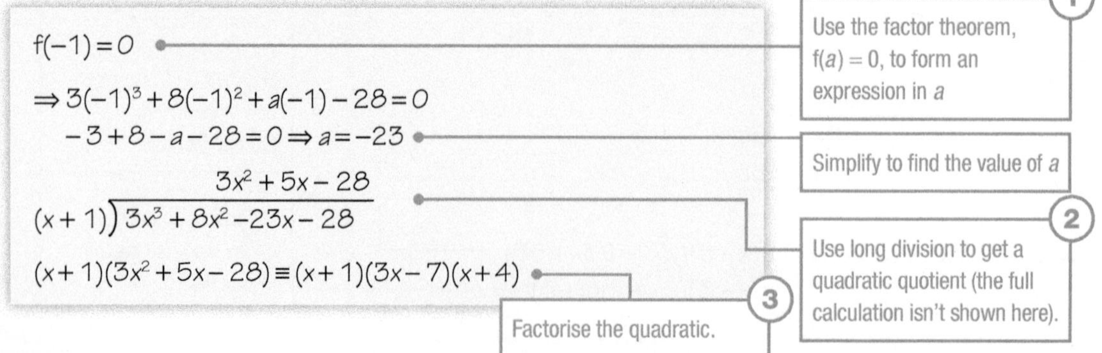

f(−1) = 0
⇒ 3(−1)³ + 8(−1)² + a(−1) − 28 = 0
−3 + 8 − a − 28 = 0 ⇒ a = −23
3x² + 5x − 28
(x + 1)) 3x³ + 8x² − 23x − 28
(x + 1)(3x² + 5x − 28) ≡ (x + 1)(3x − 7)(x + 4)
1
Use the factor theorem, f(a) = 0, to form an expression in a
Simplify to find the value of a
2
Use long division to get a quadratic quotient (the full calculation isn't shown here).
3
Factorise the quadratic.

Exercise 2.3B Reasoning and problem-solving

1 a $2x^4+px^3-6x^2+qx+6$ is divisible by $(x-1)$

Use this information to write an equation in p and q

b $2x^4+px^3-6x^2+qx+6$ is divisible by $(x+3)$

Use this information to write an equation in p and q

c Solve these equations simultaneously to find the values of p and q

2 a Work out the value of a when $2x^3+ax^2-4x+1$ is divisible by $(x-2)$

b Work out the value of b when $x^4+(b^2+1)x^3+bx^2+7x-15$ is divisible by both $(x+5)$ and $(x-1)$

c Work out the values of p and q when $2x^4+px^3-6x^2+qx+6$ is divisible by (x^2+2x-3)

3 $x^3-4x^2-31x+70$; $x^2+3x-10$ and $x^2-9x+14$ have one common factor. What is it?

4 What is the LCM of x^2+4x+3 and x^2+x-6?

5 What is the highest common factor of x^3+4x^2+x-6 and x^3+3x^2-x-3?

6 Find the LCM and HCF of $2x^2+x-21$ and $2x^2+15x+28$

7 Find the LCM and HCF of $x^3+7x^2-53x-315$ and $x^3+21x^2+143x+315$

8 $f(x)=x^3+9x^2+11x-21$ and $g(x)=x^3+2x^2-13x+10$

Find the common factor of $f(x)$ and $g(x)$ and show that it is also a factor of $f(x)-g(x)$

9 Find the values of a and b if $5x-4$ and $x+3$ are factors of $ax^2+33x+b$

10 a A circle's area is $\pi(4x^2-12x+9)$ m^2. Work out its radius.

b The volume of a square-based pyramid is $(2x^3-5x^2-24x+63)$ cm^3. The height is $(2x+7)$ cm. Work out the length of the side of the square base.

11 a The velocity of a moving body is $2t^3-19t^2+57t-54$ m s^{-1} at any time t. When is the body stationary?

b The acceleration of the same body is $6t^2-38t+57$ m s^{-2}. Work out

i The acceleration of the body when the velocities are zero,

ii The exact times when the acceleration is zero.

12 The volume of a cone is $\frac{\pi}{3}(3x^3-11x^2-15x+63)$ m^3

a Work out possible values of the radius and height of the cone in terms of x

b What is the range of possible values of x?

13 Part of a rollercoaster ride is modelled by the equation $h=t^3-12t^2+41t-30$ where h is the height above ground level in metres and t is the time in seconds. Work out

a At what times the ride is at ground level,

b When, between these times, the ride is above the ground level.

14 a A pyramid has a rectangular base. Its volume is given by $V=x^3+7x^2+14x+8$ cm^3 Work out the possible values for its dimensions.

b What is the range of possible values of x?

Challenge

15 A sphere, radius $(x+5)$ cm, has a concentric sphere, radius $(x-3)$ cm removed. Use the identity $A^3-B^3\equiv(A-B)(A^2+AB+B^2)$ to work out the volume of the shell. Give the volume in expanded form.

MyMaths 2042, 2043 SEARCH

2.4 Curve sketching

Fluency and skills

You can sketch the graph of a function without plotting a large number of points. A sketch should show the key features of a function.

- Its general shape including any symmetry,
- Its x- and y-intercepts.

Many sketches also show maximum points, minimum points and points of inflection.

You can use a graphics calculator to sketch curves.

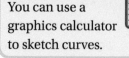

At a point of inflection the concavity of the curve changes: it bends in the other direction.

A **cubic** function can be written in the form $y = ax^3 + bx^2 + cx + d$, where a, b, c and d are constants and $a \neq 0$

Cubic curves take the form

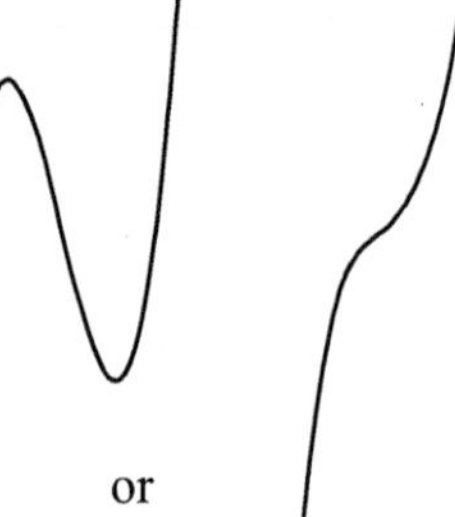

$y = ax^3 + bx^2 + cx + d,\ a > 0$

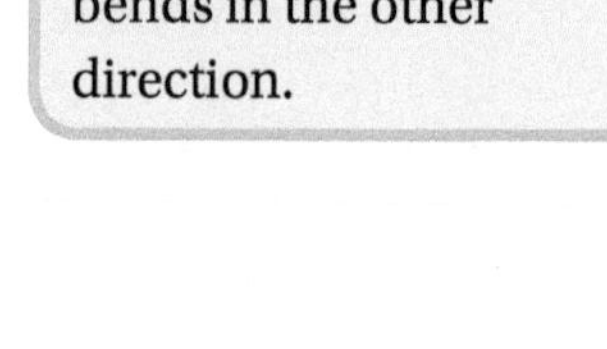

$y = ax^3 + bx^2 + cx + d,\ a < 0$

Reciprocal curves such as $y = \frac{1}{x}$ and $y = \frac{1}{x^2}$ exhibit interesting behaviour as they are undefined for certain values of x

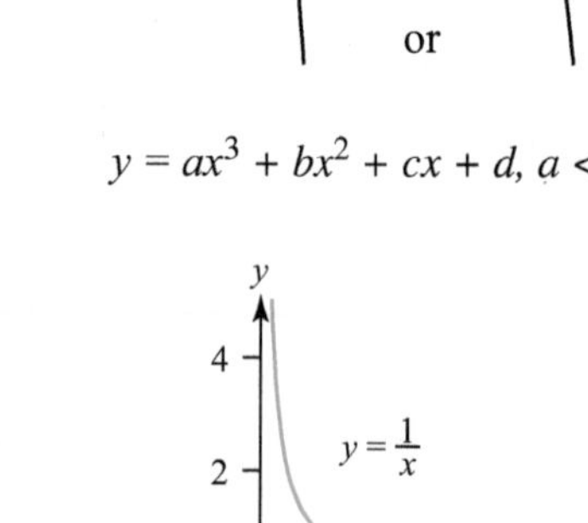

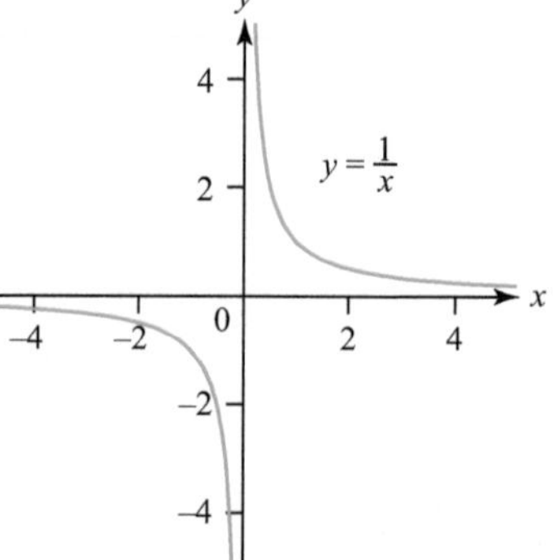

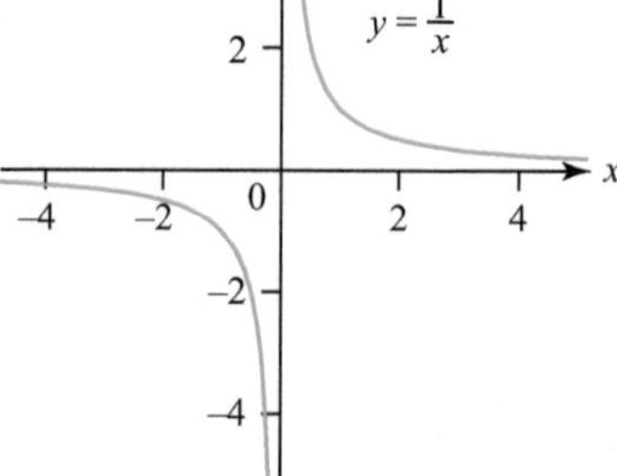

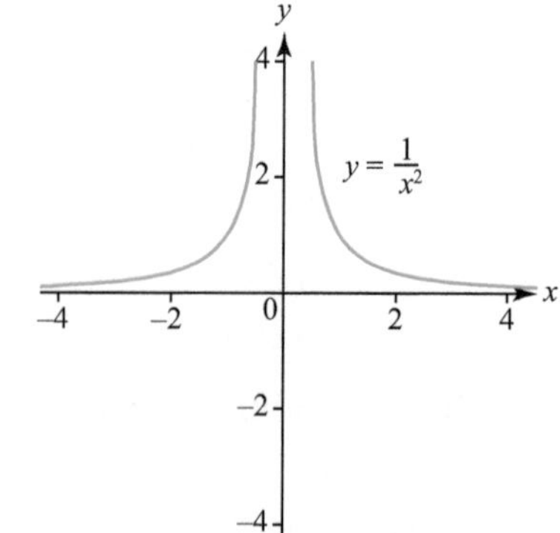

Both of these reciprocal functions are undefined for $x = 0$ as you cannot divide by 0. As x gets closer to 0, y approaches infinity, ∞, or negative infinity, $-\infty$

In both of these functions, as the magnitude of x gets bigger and bigger, y gets increasingly close to zero but never reaches zero.

The x- and y-axes are **asymptotes** to the curve in each case.

Key point

A line, l, is an asymptote to a curve, C, if, along some unbounded section of the curve, the distance between C and l approaches zero.

Example 1

a For a constant $a > 1$ sketch these curves on one set of axes.

i $f(x) = (a - x)(x + 1)(x + 2a)$ **ii** $g(x) = \dfrac{2}{x - a}$

b Show that there are no positive solutions to the equation $-(a - x)^2(x + 1)(x + 2a) - 2 = 0$

a i x-intercepts: $a - x = 0 \Rightarrow x = a$

$x + 1 = 0 \Rightarrow x = -1$

$x + 2a = 0 \Rightarrow x = -2a$

y-intercept: $x = 0 \Rightarrow y = a \times 1 \times 2a = 2a^2$

The coefficient of x^3 is

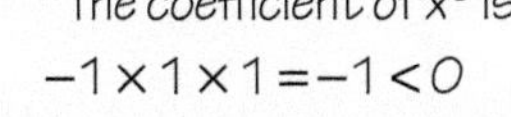

$-1 \times 1 \times 1 = -1 < 0$

a ii Undefined when $x - a = 0 \Rightarrow x = a$

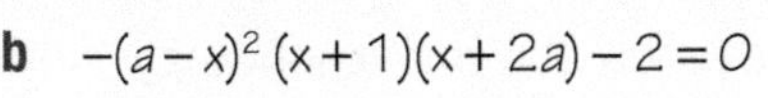

b $-(a - x)^2(x + 1)(x + 2a) - 2 = 0$

$(a - x)(x + 1)(x + 2a) = \dfrac{2}{-(a - x)}$

$(a - x)(x + 1)(x + 2a) = \dfrac{2}{(x - a)} \Rightarrow f(x) = g(x)$

The equation is satisfied at the points of intersection of f(x) and g(x). From the graph, the curves have two points of intersection and both have negative x-coordinates, so there are no positive solutions.

As the magnitude of x gets bigger and bigger the value of y gets closer to 0

Negative cubic shape, $a > 1$ and $-2a < -1$

You cannot divide by 0 so as x gets closer to a, y gets closer to ∞ or $-\infty$

Transformations can help you to see how different functions relate to one another.

You will work with four common transformations in this chapter.

y = af(x) is a vertical stretch of $y = f(x)$ with scale factor a

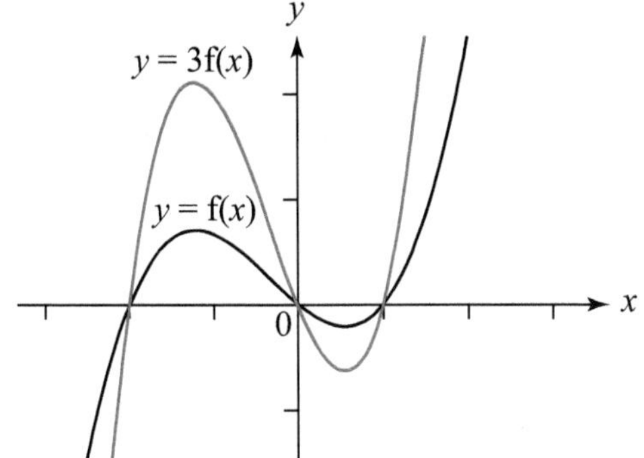

In the transformation $y = af(x)$, the x-values remain unchanged and each y-value is multiplied by a

Every point $(x, f(x))$ becomes $(x, af(x))$

Key point

If $a < 0$ the transformation $y = af(x)$ reflects the curve in the x-axis.

y = f(ax) is a horizontal stretch of $y = f(x)$ with scale factor $\dfrac{1}{a}$

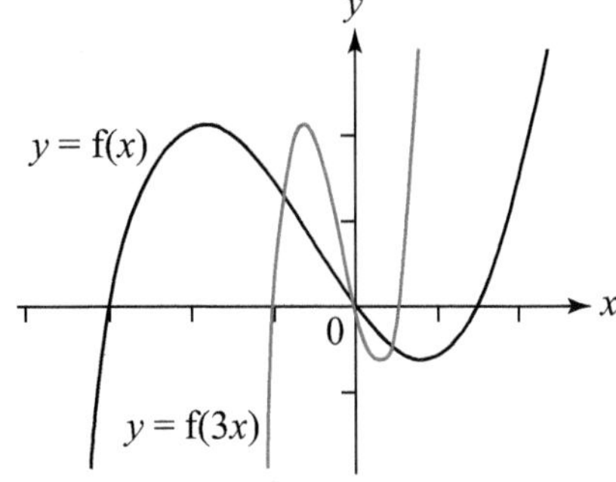

In the transformation $y = f(ax)$, each x-value is multiplied by a before the corresponding y-value is calculated.

Every point $(x, f(x))$ becomes $(x, f(ax))$

Key point

If $a < 0$ the transformation $y = f(ax)$ reflects the curve in the y-axis. If $-1 < a < 1$ the curve gets wider.

$y = f(x) + a$ is a translation of $y = f(x)$ by the vector $\begin{pmatrix} 0 \\ a \end{pmatrix}$

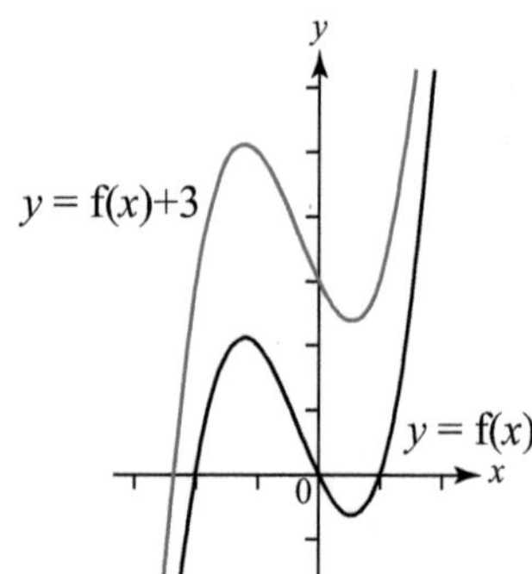

In the transformation $y = f(x) + a$, the x-values remain unchanged and each y-value is increased by a

Every point $(x, f(x))$ becomes $(x, f(x) + a)$

Key point

If $a < 0$ the transformation $y = f(x) + a$ translates the curve downwards.

$y = f(x + a)$ is a translation of $y = f(x)$ by the vector $\begin{pmatrix} -a \\ 0 \end{pmatrix}$

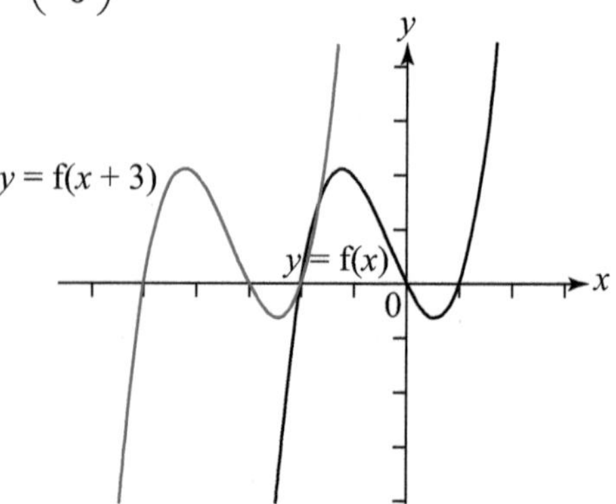

In the transformation $y = f(x + a)$, a is added to each x-value before the corresponding y-value is calculated.

Every point $(x, f(x))$ becomes $(x, f(x + a))$

Key point

If $a < 0$ the transformation $y = f(x + a)$ moves the curve to the right.

Example 2

The graph shows a sketch of the curve $y = f(x)$

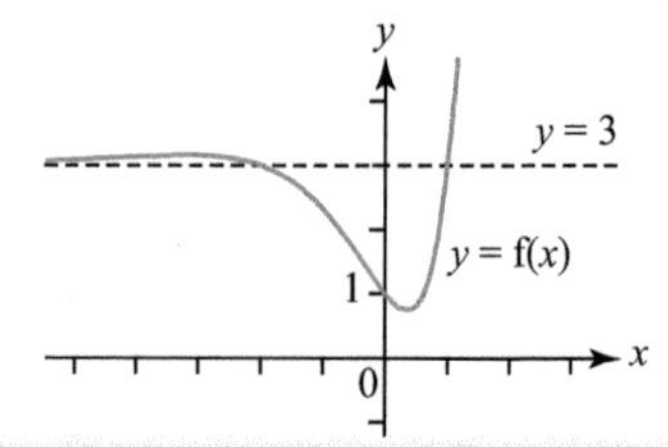

Sketch the curves

a $y = 2f(x)$ **b** $y = f(x) - 1$ **c** $y = f(x - 1)$ **d** $y = f(-x)$

The translated curve has asymptote $y = 2 \times 3 = 6$

a

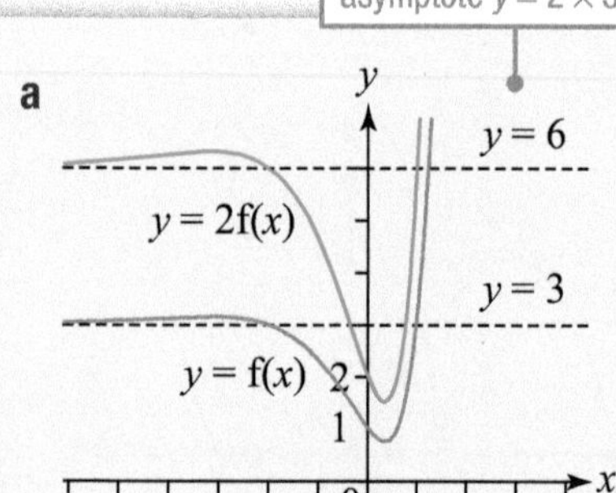

b

$y = f(x)$ $y = 3$

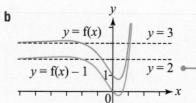

$y = f(x) - 1$ $y = 2$

The translated curve has asymptote $y = 3 - 1 = 2$

You don't have enough information to mark the y-intercept.

c

$y = 3$

$y = f(x)$ $y = f(x - 1)$

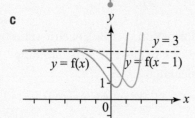

d

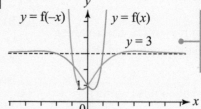

$y = f(-x)$ $y = f(x)$

$y = 3$

The curve is reflected in the y-axis.

Exercise 2.4A Fluency and skills

1 Evaluate all the x-intercepts for these graphs. Show your working.

a $y=x^2-x-6$ b $y=2x^2-9x-35$

c $y=x^3+8$ d $y=2x^3-54$

e $y=(x-3)^3$ f $y=(2x+5)^3-7$

2 Identify all the vertical and horizontal asymptotes for $y=\frac{3}{x-1}$. Show your working.

3 Evaluate all axes of symmetry in these graphs. Show your working.

a $y=x^2-8x-9$ b $y=(x+2)^4$

c $(y-3)^2=x+4$ d $y=(x-4)^2(x+3)^2$

Hence sketch the graph of each function.

4 Sketch the graphs of these functions.

a $y=x^3+3$ b $y=(x-3)^3$

c $y=-2x^3+3$ d $y=2(x+3)^3-1$

e $y=(2x+1)^3$ f $y=5+(3x-4)^3$

g $y=x^3-5x^2-14x$

h $y=(x+5)(x-6)(2x+1)$

i $y=\frac{-2}{x}$ j $y=\frac{4}{x+2}$

k $y=\frac{-5}{x-7}$

5 Sketch the graphs of these functions.

a $y=5x^3-2x^4$ b $y=5x^2+2x^3$

c $y=x^3-3x^2$ d $y=(1-x)(x+3)^2$

e $y=x^2(x-3)^2$ f $y=x^2(x+3)^2$

g $y=x^4-7x^3$ h $y=(x^2-4)(x^2-9)$

6 The graph of $y=f(x)$ is shown.

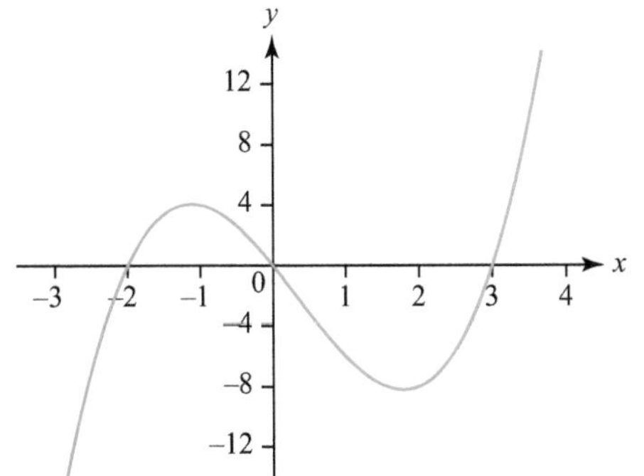

Sketch the graphs of

a $y=f(2x)$ b $y=f(x-2)$

c $y=f\left(\frac{x}{3}\right)$ d $y=f(x+3)$

e $y=f(-x)$ f $y=-f(x)$

7 The graph of $y=g(x)$ has a maximum point at $(-2, 5)$ and a minimum point at $(8, -4)$. State the coordinates of the maximum and minimum points of these transformed graphs.

a $y=g(4x)$ b $y=3g(x)$

c $y=g(x+7)$ d $y=g(x)+4$

e $y=\frac{1}{2}g(x)$ f $y=-g(x)$

g $y=g(-x)$ h $y=g\left(\frac{x}{2}\right)$

8 Describe each of the transformations in question **7**

9 $f(x)=x^3$. Write down the equation when the graph of $y=f(x)$ is

a Translated 3 units left,

b Translated 2 units up,

c Stretched vertically by scale factor 2,

d Stretched horizontally by scale factor 3

10 The graph of $y=f(x)$ is shown.

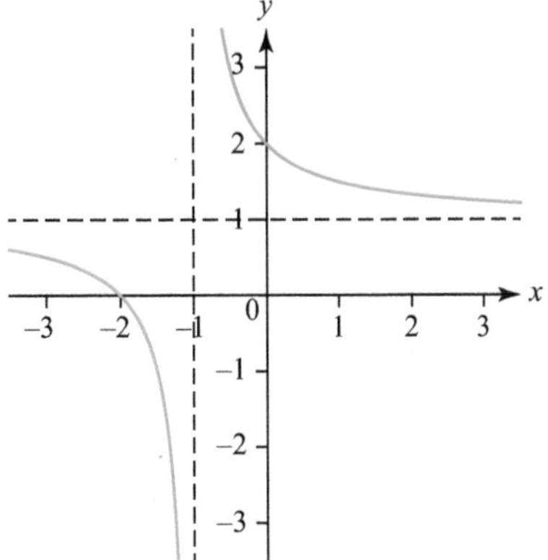

Sketch the graphs of

a $y=f(x+3)$ b $y=3f(x)$

c $y=f\left(\frac{x}{2}\right)$ d $y=f(x)+1$

Reasoning and problem-solving

Strategy

When sketching a graph

1. Define the function using any variables supplied in the question.
2. Identify the standard shape of the curve and identify any symmetry.
3. Identify any x-and y-intercepts and any asymptotes.
4. Apply any suitable transformations.
5. Show all relevant information on your sketch.

You can use graphs to show proportional relationships.

If y is proportional to x, you write $y \propto x$. This can be converted to an equation using a constant of proportionality, giving $y = kx$. The graph of y against x is a straight line through the origin with gradient k

If y is inversely proportional to x you write $y \propto \frac{1}{x}$ or $y = \frac{k}{x}$. The graph of $y = \frac{k}{x}$ is a vertical stretch, scale factor k, of the graph $y = \frac{1}{x}$

Example 3

A rectangle has a fixed area of 36 m^2. Its length, y m is inversely proportional to its width, x m.

a Write a formula for y in terms of x

b Without plotting exact points, sketch the graph of your function.

c Explain any asymptotes that the graph has.

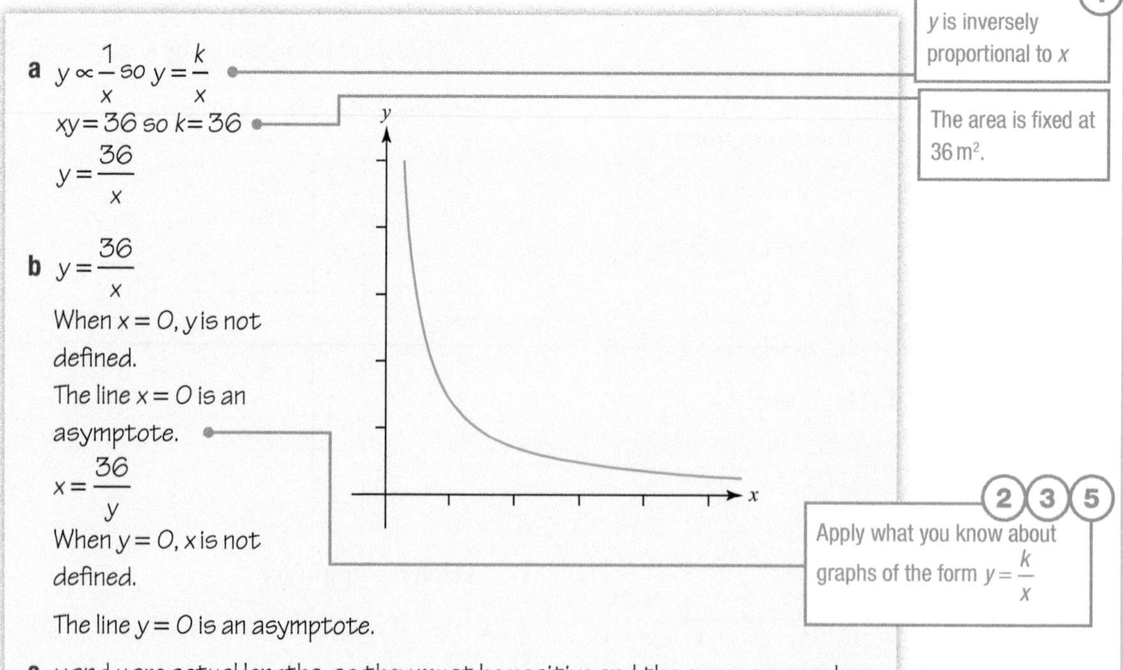

a $y \propto \frac{1}{x}$ so $y = \frac{k}{x}$
$xy = 36$ so $k = 36$
$y = \frac{36}{x}$
b $y = \frac{36}{x}$
When $x = 0$, y is not defined.
The line $x = 0$ is an asymptote.
$x = \frac{36}{y}$
When $y = 0$, x is not defined.
The line $y = 0$ is an asymptote.
c y and x are actual lengths, so they must be positive and the curve approaches the asymptotes as shown.
y
x
1
y is inversely proportional to x
The area is fixed at 36 m^2.
2 3 5
Apply what you know about graphs of the form $y = \frac{k}{x}$

Exercise 2.4B Reasoning and problem-solving

PURE

1 The radius, r, of a container is inversely proportional to its height, h

A container of radius 4 cm will have a height of 14 cm.

a Write an equation linking h and r

b Sketch a graph to illustrate this relationship.

2 The volume, v cm³ of water in a tank is proportional to the square-root of the time, t seconds. After 15 minutes the tank has 1800 cm³ of water in it.

a Write an equation linking v and t

b Sketch a graph to illustrate this relationship.

3 a Sketch the graphs of $y = \frac{1}{x+2}$ and $y = x^2(x-3)$ on the same axes.

b Use your answer to part **a** to explain how many solutions there are to the equation $x^2(x-3) = \frac{1}{x+2}$

4 The graph of $y = f(x)$ is shown.

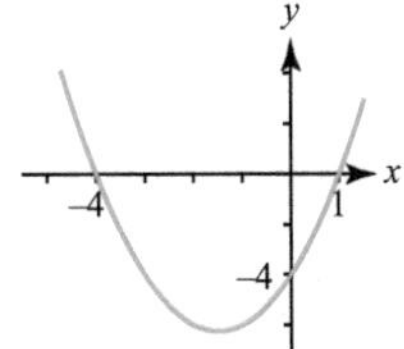

Give the equations for each of these transformations in terms of $f(x)$

a

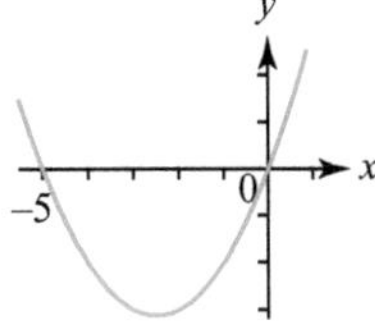

b

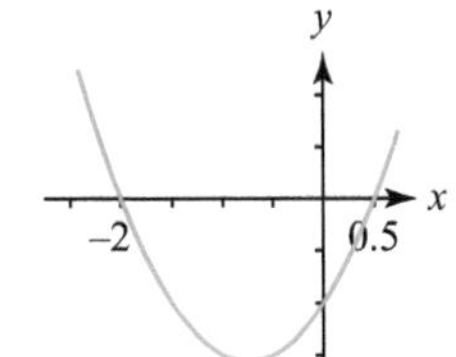

c

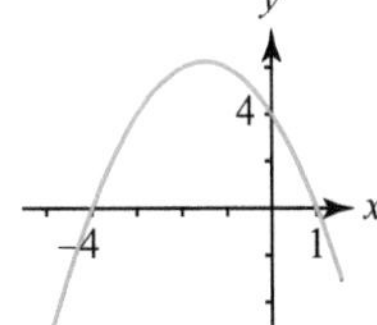

d

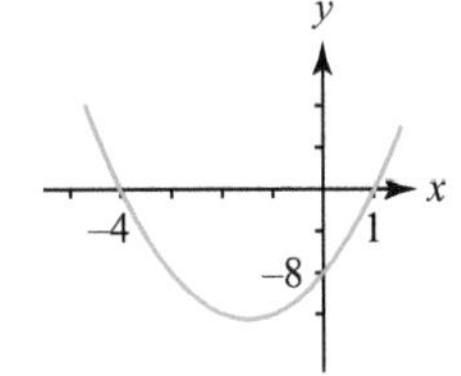

5 The graph of $y = x^3 + Ax^2 + Bx + C$ is shown.

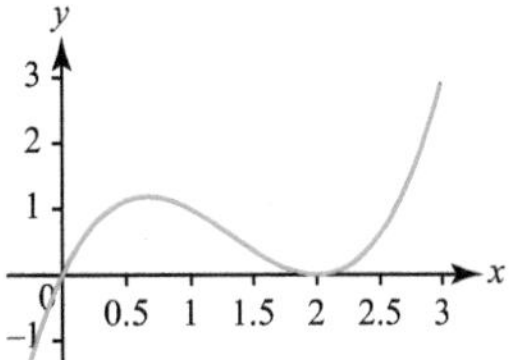

Find the values of the constants A, B and C

6 This is the graph of $y = f(x)$ where $f(x) = x^4 + Ax^3 + Bx^2 + Cx - 10$

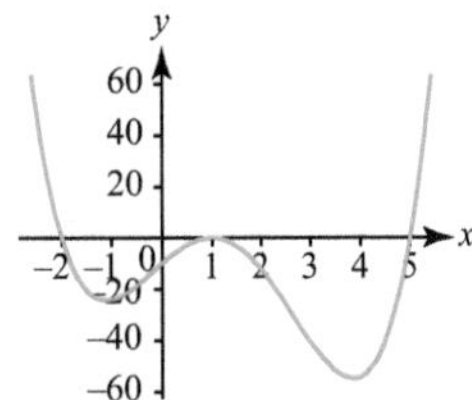

Find the values of the constants A, B, and C

7 The graph shown has the equation $y = A + \frac{B}{x+C}$

Find the values of A, B and C

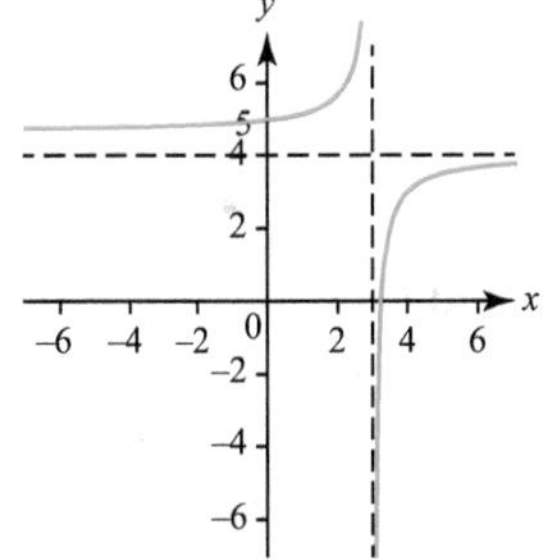

Challenge

8 For the graph of $y = ax^2 + bx + c$ where a, b and c are constants

a Explain the conditions for the graph to have a minimum point and the conditions for the graph to have a maximum point,

b Write down the coordinates of the maximum or minimum point,

c Write down the coordinates where the curve intersects the axes,

d Write down the equation of the line of symmetry of the curve.

MyMaths 2022–2024, 2027, 2258 SEARCH

2 Summary and review

Chapter summary

- The highest power in a polynomial expression is called its degree.
- When adding or subtracting polynomials, expand brackets before collecting like terms.
- Identities use the ≡ sign. Identities are true for all values of the variable(s).
- For $n = 0, 1, 2, 3, \ldots.$, the binomial expansions are

  $$(1+x)^n \equiv 1+nx+\frac{n(n-1)}{2!}x^2+\frac{n(n-1)(n-2)}{3!}x^3+\ldots+x^n$$

  and

  $$(a+b)^n \equiv a^n + {}^n\mathrm{C}_1 a^{n-1}b + {}^n\mathrm{C}_2 a^{n-2}b^2 + \ldots + {}^n\mathrm{C}_r a^{n-r}b^r + \ldots + b^n$$
- The coefficients of these expansions can be found from Pascal's triangle or from ${}^n\mathrm{C}_r \equiv \frac{n!}{(n-r)!r!}$
- You can divide algebraically using the same technique as for long division in arithmetic.
- The factor theorem states that if $\mathrm{f}(a) = 0$, then $(x - a)$ is a factor of $\mathrm{f}(x)$
- To sketch a graph you need to consider the symmetry, x- and y-intercepts, asymptotes, behaviour as x and/or y approaches $\pm\infty$, and any other obvious critical points. You can also use your knowledge of transformations.

Check and review

You should now be able to...	Try Questions
✔ Manipulate, simplify and factorise polynomials.	1–4, 17
✔ Understand, expand and use the binomial theorem.	7–11
✔ Divide polynomials by algebraic expressions.	6, 12, 14, 15
✔ Understand and use the factor theorem.	5, 13, 16
✔ Use a variety of techniques to analyse a function and sketch its graph.	18–23

1 Add together $2x^3 + 9x^5 + 11x^2 - 3x - 5x^4 - 12$ and $4x^2 - x^4 - 7x^5 + 3 + 12x - 5x^3$

2 Fully factorise $4n^3 + 4n^2 - 15n$

3 Expand and simplify these expressions.

a $(y-1)(y+3)(2y+5)$

b $(2z+1)(z-2)^2$

4 Factorise these expressions.

a $m(m+4)-(m+4)^2$

b $(d+1)^2 - 4(d+1)(d-1)$

5 The equation $2x^3 + ax^2 + bx + c = 0$ has roots -4, 3 and $\frac{7}{2}$. Find the values of a, b and c

6 Find the function that, when divided by $(x+3)$, gives a quotient of $(2x-3)$ and a remainder of -4

7 Use Pascal's triangle to write the expansion of $\left(1+\frac{m}{10}\right)^4$

Use your answer to evaluate the value of 1.1^4 to 4 decimal places.

8 Use the binomial theorem to expand $(2s^2 - 4t)^4$

9 Use Pascal's triangle to expand and simplify these expressions.

a $(1+\sqrt{3})^4$ **b** $(3-\sqrt{5})^5-(3+\sqrt{5})^5$

10 Find the constant term in the binomial expansion of $\left(w-\frac{3}{2w}\right)^{14}$

11 a $(2-ax)^9 \equiv 512+2304x+bx^2+cx^3+\ldots$ Find the values of a, b, and c

b Use your values of a, b, and c to find the first four terms in the expansion of $(1-x)(2-ax)^9$

12 Divide $2x^3-3x^2-26x+3$ by $(x+3)$

13 By successively evaluating f(1), f(−1), f(2), f(−2) and so on, find all the factors of x^3-4x^2+x+6

14 Divide $8x^3+14x^2-x+35$ by $(2x+5)$

15 Divide x^3-2x^2+3x+4 by $(x-2)$

16 Show that $(2x-3)$ is a factor of $4x^3-8x^2+x+3$

17 Factorise fully $2x^3+x^2-18x-9$

18 Susan attempted to transform the graph of $y=\mathrm{f}(x)$ into $y=\mathrm{f}(x-1)$

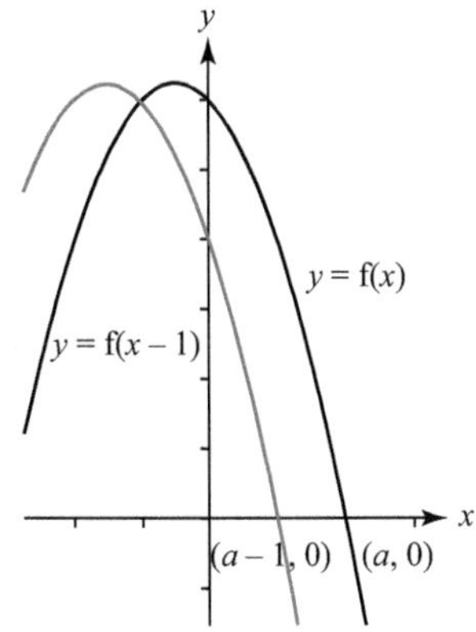

a Explain what mistake she has made.

b Sketch the graph of $y=\mathrm{f}(x-1)$

19 Sketch these curves on the same set of axes.

a $y=\frac{1}{x}$ **b** $y=\frac{4}{x}$ **c** $y=2+\frac{1}{x}$

20 Sketch the graph of $y=(x-6)^3$

21 A particle moves along a straight line from O, so that, at time t s, it is s m from O, given by the equation $s=t(2t-7)^2$

Sketch the graph and describe its motion fully.

22 A rectangular metal sheet, 16 in by 10 in, has squares of side x in removed from its corners.

The edges are turned up to form an open box.

a Show that the volume of this box is $V=160x-52x^2+4x^3$ in^3

b Sketch a graph to evaluate the value of x that gives the highest volume.

23 A particle moves along a straight line from O, so that, at time t seconds, it is s metres from O, given by the equation $s=t(t-4)^2$. Sketch the graph and describe its motion.

What next?

Score		
Score	0 – 11	Your knowledge of this topic is still developing. To improve, search in MyMaths for the codes: 2006, 2022–2024, 2027, 2041–2043, 2258
	12 – 17	You're gaining a secure knowledge of this topic. To improve, look at the InvisiPen videos for Fluency and skills (02A)
	18 – 23	You've mastered these skills. Well done, you're ready to progress! To develop your techniques, look at the InvisiPen videos for Reasoning and problem-solving (02B)

Click these links in the digital book

Exploration
Going beyond the exams

History

The binomial theorem is a formula for finding any power of a two-term bracket without having to multiply them all out. It has existed in various forms for centuries and special cases, for low powers, were known in Ancient Greece, India and Persia.

The triangular arrangement of the binomial coefficients is known as **Pascal's triangle**. It took its name from the 17th century mathematician **Blaise Pascal**, who studied its properties in great depth. Although the triangle is named after Pascal, it had been known about much earlier. A proof linking it to the binomial theorem was given by an Iranian mathematician Al-Karaji in the 11th century.

Around 1665, **Isaac Newton** developed the binomial theorem further by applying it to powers other than positive whole numbers. He showed that a general formula worked with any rational value, positive or negative.

Newton showed how the binomial theorem could be used to simplify the calculation of roots and also used it in a calculation of π, which he found to 16 decimal places.

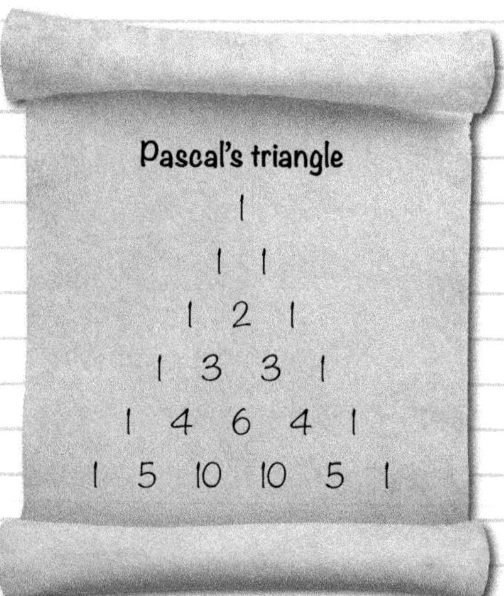

Pascal's triangle
1
1 1
1 2 1
1 3 3 1
1 4 6 4 1
1 5 10 10 5 1

Have a go

For small values of x, $(1 + x)^n \approx 1 + nx$

Use this result to estimate the value of

a) $(1.02)^4$
b) $(0.99)^5$
c) $(2.01)^5$

Find these values on a calculator and compare your results.

$(2.01)^5$ can be written in the form $2^5(1 + ...)^5$

"If I have seen further than others, it is by standing on the shoulders of giants."
- Isaac Newton

1 Factorise $2s^4 + 2s^3 - 12s^2$ Choose the correct answer.

A $(s+3)(s-2)$ **B** $s^2(s+3)(s-2)$ **C** $2(s^2+3)(s^2-2)$ **D** $2s^2(s+3)(s-2)$ **[1 mark]**

2 Simplify $\frac{3(x+1)-1}{2x^2+x-3} - \frac{1}{x-1}$ Choose the correct answer.

A $\frac{x+5}{(2x+3)(x-1)}$ **B** $\frac{1}{2x+3}$ **C** $\frac{3x+2}{x-1}$ **D** $\frac{1}{2x+1}$ **[1]**

3 **a** Simplify these expressions.

i $(2x-3)(6x+1)$ **ii** $(2a-3b)^2$ **iii** $(5x+2y)(x^2-3xy-y^2)$ **[6]**

b Given $\frac{ax^2+bx+c}{(3x+4)} \equiv (3x-4)$, evaluate the values of the constants a and c, and show that $b=0$ **[3]**

4 Write down the binomial expansion of $\left(1+\frac{1}{2}x\right)^8$ in ascending powers of x, up to and including the term in x^3. Simplify the terms as much as possible. **[6]**

5 **a** Factorise p^3-10p^2+25p **[2]**

b Deduce that $(2x+5)^3-10(2x+5)^2+25(2x+5) \equiv ax^2(2x+5)$, where a is a constant that should be stated. **[2]**

6 Show that $(x-3)$ is *not* a factor of $2x^3-5x^2+6x-7$ **[3]**

7 Show how algebra can be used to work out each of these without a calculator.

a 268^2-232^2 **[2]**

b $469 \times 548 + 469^2 - 469 \times 17$ **[2]**

c $\frac{65.1 \times 29.2 + 65.1 \times 35.9 - 91.7 \times 26.4 + 65.3 \times 26.4}{18.3^2 - 18.3 \times 5.4}$ **[5]**

8 Given that $(1+cx)^7 \equiv 1+21x+Ax^2+Bx^3+\ldots\ldots\ldots,$

a Work out **i** The value of c **ii** The value of A **iii** The value of B **[4]**

b Using your values of c, A and B, evaluate the coefficient of x^3 in the expansion of $(2+x)(1+cx)^7$ **[2]**

9 Express x^3-3x^2+5x+1 in the form $(x-2)(x^2+ax+b)+c$ **[3]**

10 **a** Write down the expansions of **i** $(x+y)^4$ **ii** $(x-y)^4$ **[4]**

b Show that $(\sqrt{5}+\sqrt{2})^4+(\sqrt{5}-\sqrt{2})^4=n$, where n is an integer to be found. **[4]**

11 Write down the term which is independent of x in the expansion of $\left(x^2+\frac{2}{x}\right)^9$ **[3]**

12 **a** Expand each of these in ascending powers of x up to and including the term in x^2

i $(1+2x)^6$ **ii** $(2-x)^6$ **[5]**

b Hence write down the first three terms in the binomial expansion of $(2+3x-2x^2)^6$ **[4]**

13 **a** Show that $(x-2)$ is a factor of $2x^3+x^2-7x-6$ **[2]**

b Show that the equation $2x^3+x^2-7x-6=0$ has the solutions 2, $-\frac{3}{2}$, and -1 **[5]**

14 Given that both $(x-1)$ and $(x+3)$ are factors of $ax^3+bx^2-16x+15$

a Evaluate the values of a and b **[6]**

b Fully factorise $ax^3+bx^2-16x+15$ **[3]**

c Sketch the graph of $y=ax^3+bx^2-16x+15$ **[3]**

d Solve the inequality $ax^3+bx^2-16x+15\geq 0$ **[2]**

15 a Expand $\left(x+\frac{1}{x}\right)^6$, simplifying the terms. **[7]**

b Hence write down the expansion of $\left(x-\frac{1}{x}\right)^6$ **[1]**

c Prove that the equation $\left(x+\frac{1}{x}\right)^6-\left(x-\frac{1}{x}\right)^6=64$ has precisely two real solutions. **[5]**

16 Prove these results

a ${}^{n+1}C_r \equiv {}^nC_r + {}^nC_{r-1}$ **[6]**

b ${}^{n+2}C_3 - {}^nC_3 \equiv n^2$ **[8]**

17 Here are five equations, labelled **i** - **v**, and five graphs, labelled **A** – **E**

i $y=\frac{1}{(x-2)^2}$ **ii** $y=1+\frac{1}{(x-1)}$ **iii** $y=-\frac{1}{x+1}$ **iv** $y=-\frac{1}{(x+1)^2}$ **v** $y=1+\frac{1}{x^2}$

A

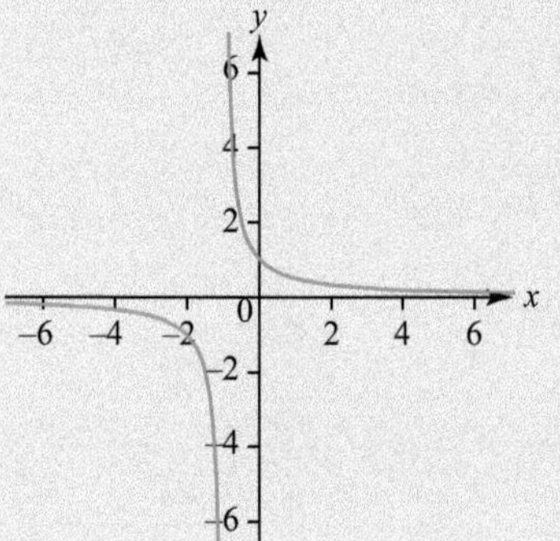

B

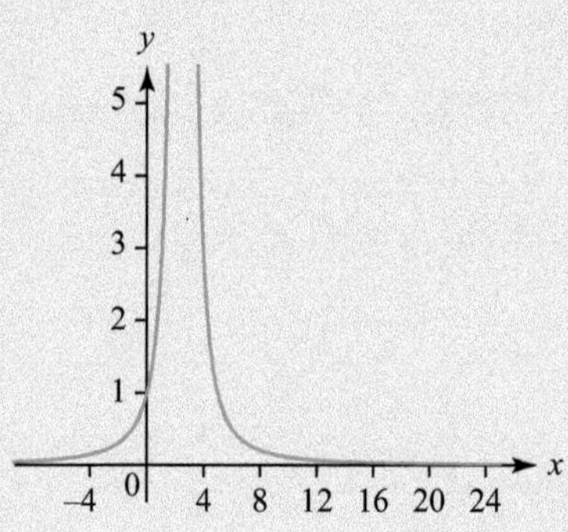

C

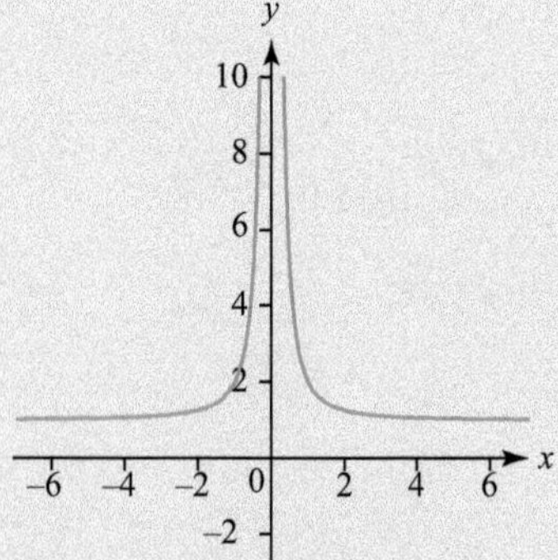

D

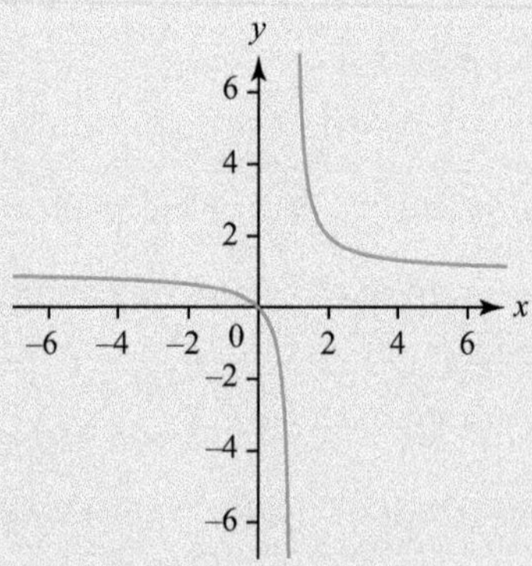

E

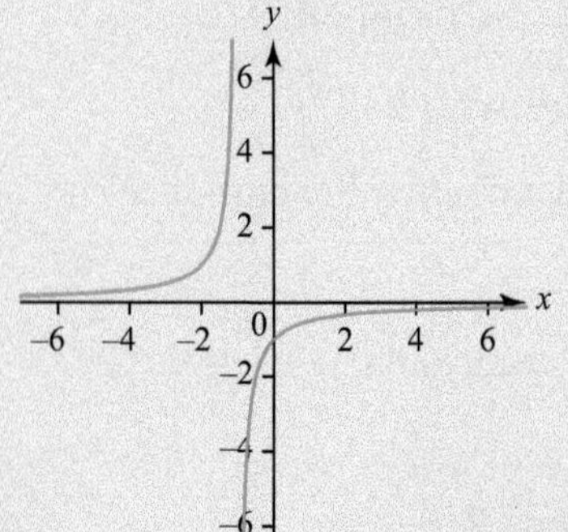

Four of the equations correspond to four of the graphs.

a Match the four equations to their graphs. **[4]**

b For the graph that has no equation, write down a possible equation. **[1]**

c For the equation that has no graph, sketch its graph. **[3]**

3 Trigonometry

GPS uses a technique called trilateration to calculate positions. The receiver, a mobile phone for example, receives direct signals from four different satellites simultaneously. The imaginary lines between the satellites and the receiver form the sides of triangles, which are then used by the mobile to calculate its position. Trilateration is a hi-tech version of triangulation, a technique that requires the use of trigonometry.

Trigonometry is the study of the relationships between angles and the sides of a triangle. It is immensely useful in fields such as astronomy, engineering, architecture, geography and navigation, as it allows you to calculate distances and angles or bearings. The sine and cosine functions are periodic in nature. This makes them highly useful in modelling periodic phenomena, and they can be used to describe different types of wave, including sound and light waves.

Orientation

What you need to know	What you will learn	What this leads to
KS4 • Apply and derive Pythagoras' theorem. • Recognise graphs of trigonometric functions. • Apply some properties of angles and sides of a triangle. **Ch1 Algebra 1** p.12 • Working with surds.	• To calculate the values of sine, cosine and tangent for any angle. • To use trigonometric identities and recognise the equation of a circle. • To sketch and describe trigonometric functions. • To solve trigonometric equations. • To use the sine and cosine rule and the area formula for a triangle.	**Ch14 Trigonometric identities** Radians. Reciprocal and inverse trigonometric functions. Compound angles. Equivalent forms for $a\cos\theta + b\sin\theta$ **Ch16 Integration and differential equations** Integrating trigonometric functions.

MyMaths Practise before you start: 1112, 1131, 1133, 2024, 2036

3 Bridging Unit

Topic A: Trigonometry 1

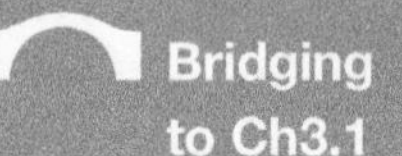

When dealing with right-angled triangles you can use the **sine, cosine and tangent ratios.**

Key point

For all right-angled triangles with angle θ:

$$\sin\theta = \frac{\text{opposite}}{\text{hypotenuse}},\ \cos\theta = \frac{\text{adjacent}}{\text{hypotenuse}},\ \tan\theta = \frac{\text{opposite}}{\text{adjacent}}$$

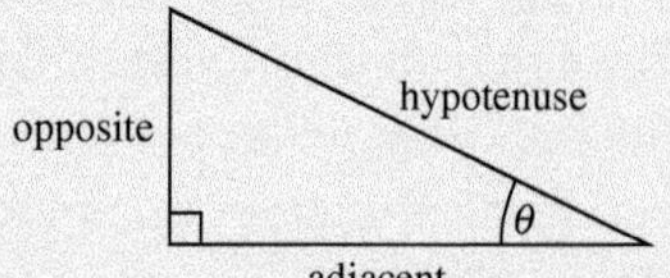

Example 1

Find the lengths of sides x and y in these right-angled triangles.

a 20°, 15 cm, x

b y, 75°, 6 cm

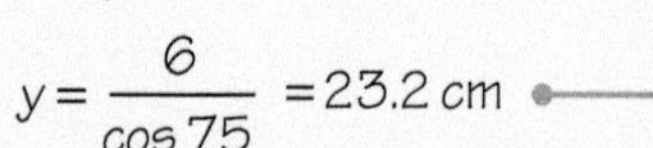

a $\sin 20 = \frac{x}{15}$ — x is the opposite side to the 20° angle and you know that the hypotenuse is 15 cm, so use $\sin\theta = \frac{\text{opposite}}{\text{hypotenuse}}$

$x = 15\sin 20 = 5.13$ cm — Rearrange to make x the subject.

b $\cos 75 = \frac{6}{y}$ — y is the hypotenuse and you know that the adjacent side to the 75° angle is 6 cm, so use $\cos\theta = \frac{\text{adjacent}}{\text{hypotenuse}}$

$y = \frac{6}{\cos 75} = 23.2$ cm — Rearrange to make y the subject.

Try It 1

Find the lengths of sides x and y in these right-angled triangles.

a

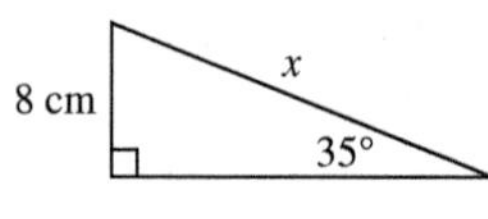

b

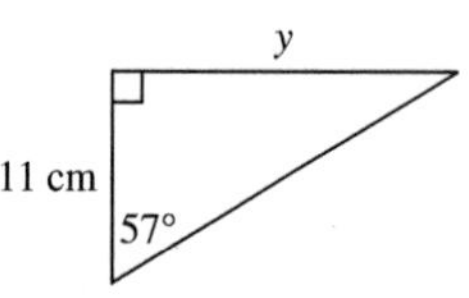

Example 2

Find the size of the angle marked x in the triangle.

9 cm, x, 5 cm

$\cos x = \frac{5}{9}$ — You know that the adjacent to the angle x is 5 cm and that the hypotenuse is 9 cm, so use $\cos\theta = \frac{\text{adjacent}}{\text{hypotenuse}}$

$x = \cos^{-1}\left(\frac{5}{9}\right) = 56.3°$ — Find the inverse function $\cos^{-1}$ on your calculator.

Try It 2

Find the size of the angle marked y in the triangle.

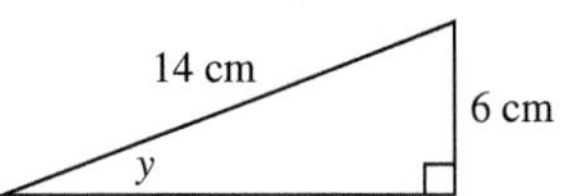

Sometimes you may need to use **Pythagoras' theorem** along with the trigonometric ratios to solve problems.

Key point

Pythagoras' theorem: $a^2 + b^2 = c^2$ for a right-angled triangle where c is the hypotenuse and a and b are the two shorter sides.

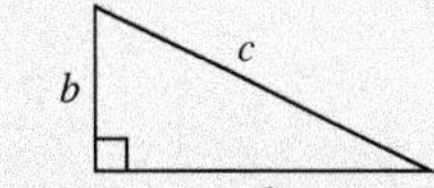

Example 3

Given that $\cos\theta = \frac{2}{3}$ for an acute angle θ, find the exact values of $\sin\theta$ and of $\tan\theta$

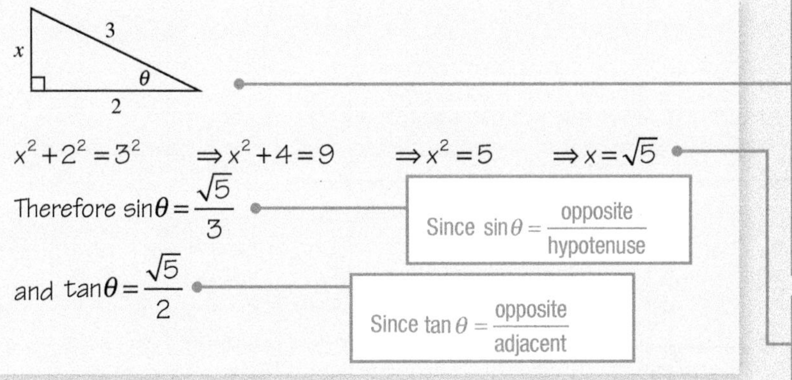

$x^2 + 2^2 = 3^2 \quad \Rightarrow x^2 + 4 = 9 \quad \Rightarrow x^2 = 5 \quad \Rightarrow x = \sqrt{5}$

Therefore $\sin\theta = \frac{\sqrt{5}}{3}$

and $\tan\theta = \frac{\sqrt{5}}{2}$

Since θ is an acute angle, you can solve this problem by drawing a right-angled triangle where $\cos\theta = \frac{2}{3}$

$\cos\theta = \frac{\text{adjacent}}{\text{hypotenuse}}$ so label the side adjacent to θ as 2 and the hypotenuse as 3

Use Pythagoras' theorem to find the missing length, x

Since $\sin\theta = \frac{\text{opposite}}{\text{hypotenuse}}$

Since $\tan\theta = \frac{\text{opposite}}{\text{adjacent}}$

Try It 3

Given that $\sin\theta = \frac{3}{4}$ for an acute angle θ, find the exact values of $\cos\theta$ and $\tan\theta$

You need to know what the graphs of trigonometric functions look like and where they cross the x and y axes.

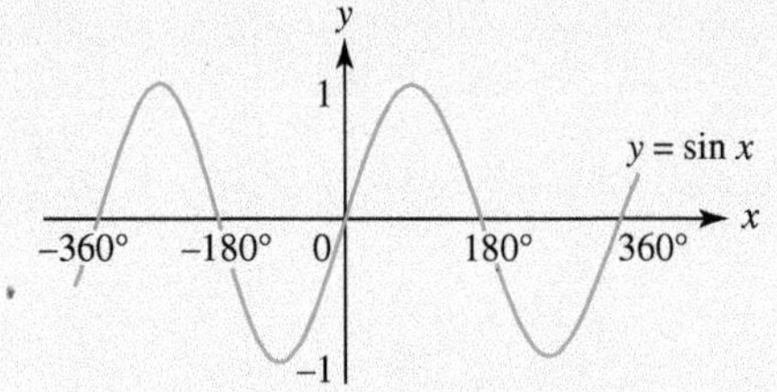

The period of $y = \sin x$ is 360° which means the curve repeats every 360°

The maximum value of $y = \sin x$ is 1 which occurs at $x = -270°, 90°, 450°$ etc.

The minimum value of $y = \sin x$ is –1 which occurs at $x = -90°, 270°, 630°$ etc.

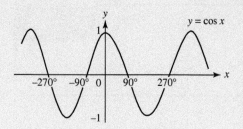

The period of $y = \cos x$ is 360° which means the curve repeats every 360°

The maximum value of $y = \cos x$ is 1 which occurs at 0°, ±360° etc.

The minimum value of $y = \cos x$ is –1 which occurs at = ±180°, ±540° etc.

The period of $y = \tan x$ is 180° which means the curve repeats every 180°

The graph of $y = \tan x$ has asymptotes at $x = \pm 90°$, $x = \pm 270$, …

See Ch 1.4

To see how to sketch a curve on a graphics calculator.

You can use the graphs to find all the solutions to a trigonometric equation in any range of values of x

Example 4

Find all the solutions to the equation $\sin x = 0.8$ in the range $-360° \le x \le 360°$

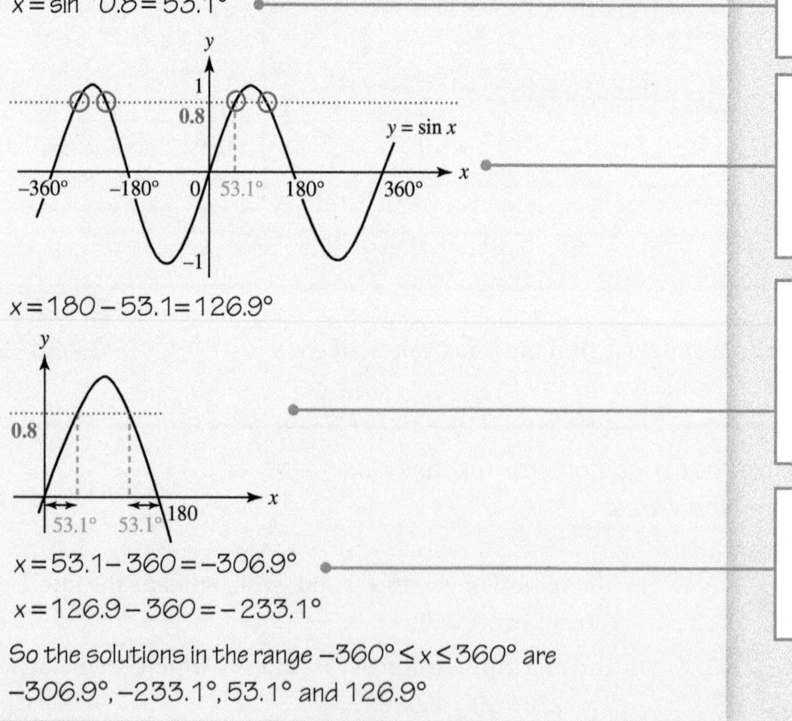

Find the first solution using your calculator.

Consider the graph of $y = \sin x$, notice that there are 4 solutions to $\sin x = 0.8$ in the range $-360° \le x \le 360°$

Because of the symmetry of the graph between 0 and 180°, the second solution can be found by subtracting the first solution from 180°

The graph repeats every 360° so further solutions can be found by adding or subtracting 360°

Try It 

Find all the solutions to the equation $\sin x = 0.2$ in the range $-360° \le x \le 360°$

Example 5

Find all the solutions to the equation $\cos x = -0.6$ in the range $0 \leq x \leq 720°$

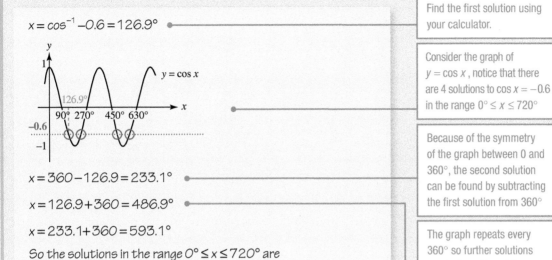

$x = \cos^{-1} -0.6 = 126.9°$ — Find the first solution using your calculator.

Consider the graph of $y = \cos x$, notice that there are 4 solutions to $\cos x = -0.6$ in the range $0° \leq x \leq 720°$

$x = 360 - 126.9 = 233.1°$ — Because of the symmetry of the graph between 0 and 360°, the second solution can be found by subtracting the first solution from 360°

$x = 126.9 + 360 = 486.9°$

$x = 233.1 + 360 = 593.1°$ — The graph repeats every 360° so further solutions can be found by adding or subtracting 360°

So the solutions in the range $0° \leq x \leq 720°$ are 126.9°, 233.1°, 486.9° and 593.1°

Try It 5

Find all the solutions to the equation $\cos x = 0.1$ in the range $-360° \leq x \leq 360°$

Example 6

Find all the solutions to the equation $\tan x = 0.5$ in the range $-180° \leq x \leq 360°$

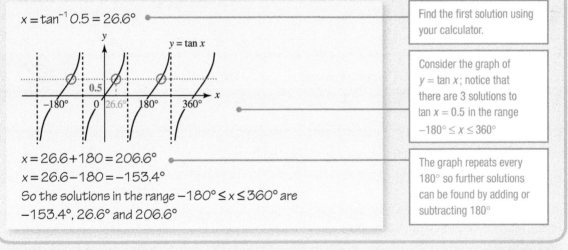

$x = \tan^{-1} 0.5 = 26.6°$ — Find the first solution using your calculator.

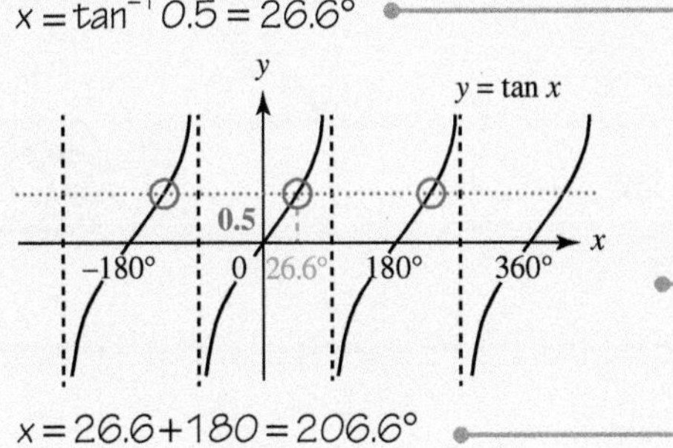

Consider the graph of $y = \tan x$; notice that there are 3 solutions to $\tan x = 0.5$ in the range $-180° \leq x \leq 360°$

$x = 26.6 + 180 = 206.6°$

$x = 26.6 - 180 = -153.4°$ — The graph repeats every 180° so further solutions can be found by adding or subtracting 180°

So the solutions in the range $-180° \leq x \leq 360°$ are −153.4°, 26.6° and 206.6°

Try It 6

Find all the solutions to the equation $\tan x = -0.3$ in the range $-360° \leq x \leq 360°$

MyMaths 1112, 1131, 1132, 1133, 1143 SEARCH

The trigonometric equation could require some rearranging first. For example, $2\sin\theta+1=3$ must first be rearranged so $\sin\theta$ is the subject.

Example 7

Find all the solutions to these equations in the range $0°\le\theta\le360°$

a $2\sin\theta=-\sqrt{2}$ **b** $5\cos\theta-1=3$

a $\sin\theta=-\dfrac{\sqrt{2}}{2}$ — Divide by 2

$\theta=\sin^{-1}\left(\dfrac{\sqrt{2}}{2}\right)=45°$ — Find the first solution using your calculator (if needed).

$\theta=180-45=135°$ — Because of the symmetry of the graph between 0 and 180°, the second solution can be found by subtracting the first solution from 180°

So the solutions in the range $0°\le\theta\le360°$ are 45° and 135°

b $5\cos\theta=4$ — Add 1 to both sides.

$\cos\theta=0.8$ — Divide both sides by 5

$\theta=\cos^{-1}0.8=36.9°$ — Find the first solution using your calculator.

$\theta=360-36.9=323.1°$ — Because of the symmetry of the graph between 0 and 360°, the second solution can be found by subtracting the first solution from 360°

So the solutions in the range $0°\le\theta\le360°$ are 36.9° and 323.1°

Try It 7

Find all the solutions to these equations in the range $-180°\le\theta\le180°$

a $\tan\theta+5=3$ **b** $8\cos\theta=-1$ **c** $7\sin\theta-3=-1$

The graph of $y=\sin 2x$ is a stretch of the graph $y=\sin x$ by scale factor $\frac{1}{2}$ in the x-direction as shown.

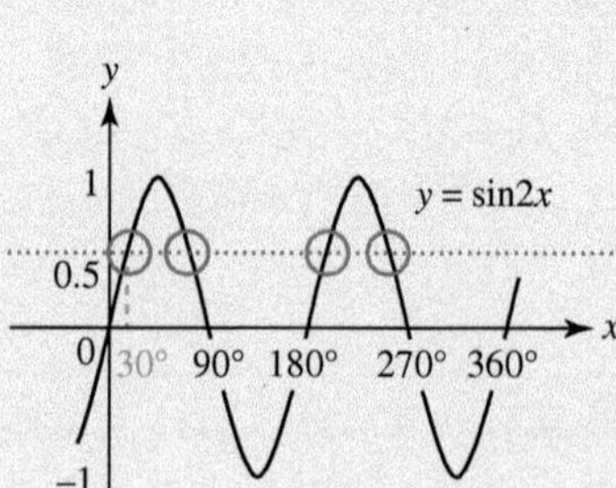

You can see on the graph that the equation $\sin 2x=0.5$ will have four solutions in the interval $0\le x\le360°$

To ensure you do not miss any of the required solutions, you should adjust the interval by multiplying each part by 2, giving $0\le2x\le720°$

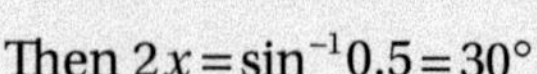

Then $2x=\sin^{-1}0.5=30°$

Also $2x=180-30=150°$, $2x=30+360=390°$ and $2x=150+360=510°$

So the four solutions for $2x$ in the interval $0\le2x\le720°$ are 30°, 150°, 390° and 510°. Divide each solution for $2x$ by 2 to give the values of x in the interval $0\le x\le360°$: $x=15°, 75°, 195°, 255°$

Example 8

Find all the solutions of the equation $\tan 3x = -1$ such that $-90 < x < 90°$

Adjust the interval: $-270° < 3x < 270°$ — Multiply each part by 3 since the equation involves $3x$

Then $3x = \tan^{-1}(-1) = -45°$

Also, $3x = -45 - 180 = -225°$ — Add or subtract multiples of 180° to find the other two solutions.

$3x = -45 + 180 = 135°$

So $x = -\frac{45}{3} = -15°$

$x = -\frac{225}{3} = -75°$

$x = \frac{135}{3} = 45°$ — Divide each solution in $-270° < 3x < 270°$ by 3 to find the solutions in $-90° < x < 90°$

Try It 8

Find all the solutions to $\cos 4x = -\frac{1}{2}$ in the interval $0 < x < 180°$

The curve $y = \cos(x - 30°)$ is a translation of the curve $y = \cos(x)$ by 30° in the positive x-direction as shown.

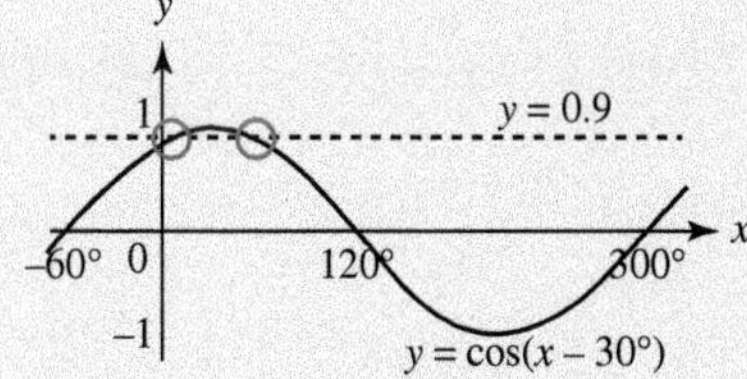

To solve the equation $\cos(x - 30°) = 0.9$ for x in the interval $0 \le x \le 360°$ you need to subtract 30° from each part of the expression, giving $-30° \le x - 30° \le 330°$

Then $x - 30 = \cos^{-1} 0.9 = 25.8°$

Also, $x - 30 = 360 - 25.8 = 334.2°$, but this is not in the interval $-30° \le x - 30° \le 330°$

To get the second solution, use the periodicity of the curve and subtract 360° to give the solution $x - 30 = -25.8°$

Then add 30 on to each value for $(x - 30)$ to give the final solutions $x = 25.8 + 30 = 55.8°$ and $x = -25.8 + 30 = 4.2°$

Example 9

Find all the solutions of the equation $\sin(x + 50°) = -0.8$ such that $-180° < x < 180°$

Adjust the interval: $-130° < x + 50° < 230°$ — Add 50° to each part.

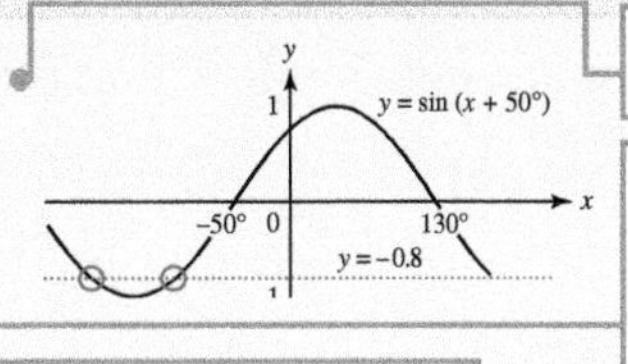

Then $x + 50 = \sin^{-1} 0.8 = 53.1°$

Also, $x + 50 = 180 + 53.1 = 233.1°$ — This is outside the interval $-130° < x + 50° < 230°$ so subtract 360° from 233.1 to give −126.9 which is in the interval.

Solution in interval is $x + 50 = -126.9°$

So $x = -53.1 - 50 = -103.1°$ — Subtract 50° from each solution in to find the solutions in $-180° < x < 180°$

$x = -126.9 - 50 = -176.9°$

Try It 9

Find all the solutions to $\tan(x + 100) = 0.3$ in the interval $0 < x < 360°$

Bridging Exercise Topic A

1 Find the side labelled x in each of these triangles.

a 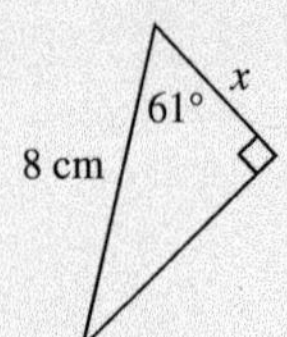

b 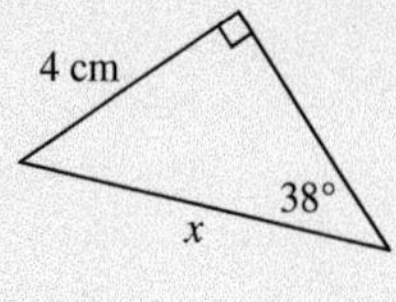

c 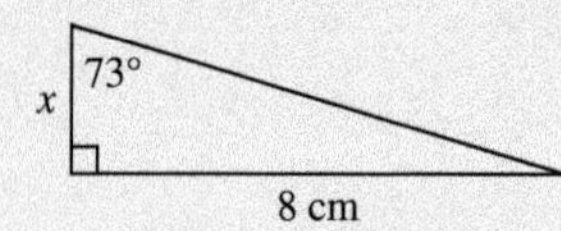

d 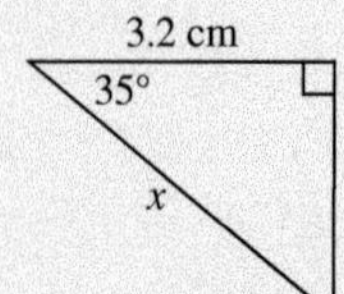

e 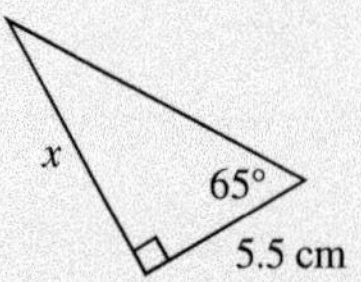

f 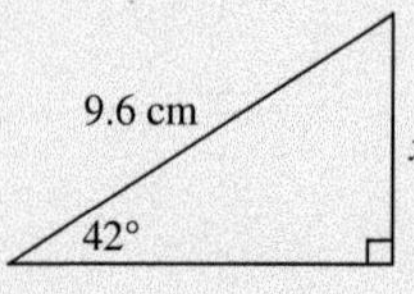

g 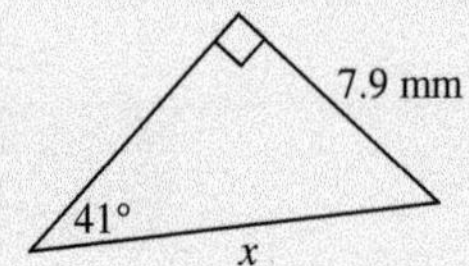

h 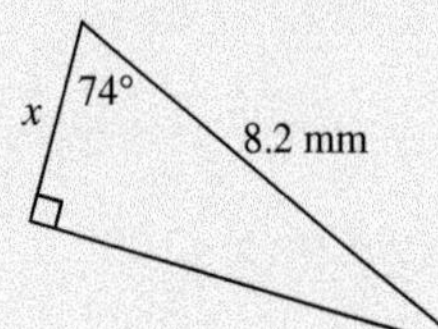

i 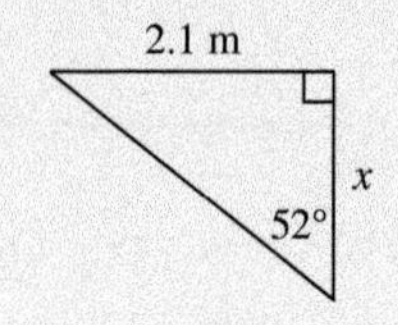

2 Find the angle labelled θ in each of these triangles.

a 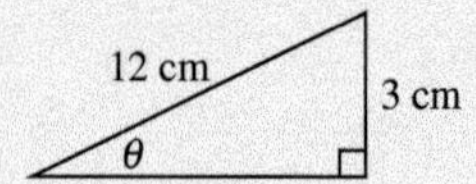

b 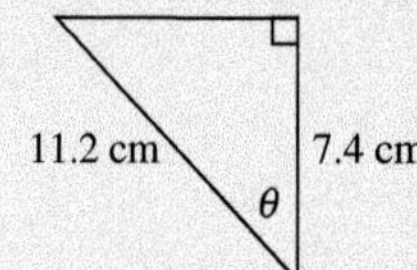

c 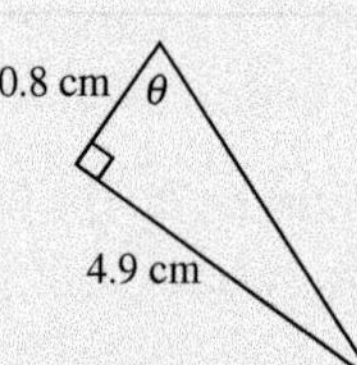

d 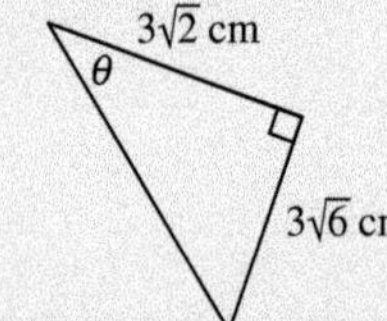

e 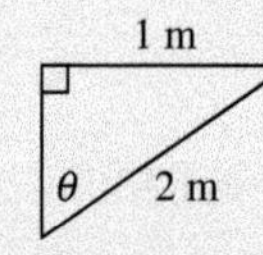

f 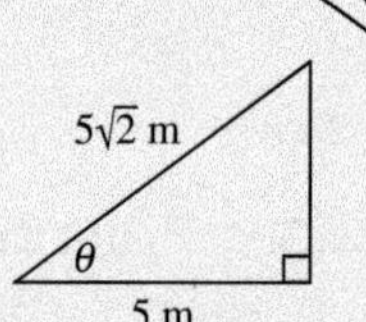

3 Find the missing side lengths in each of these triangles.

a 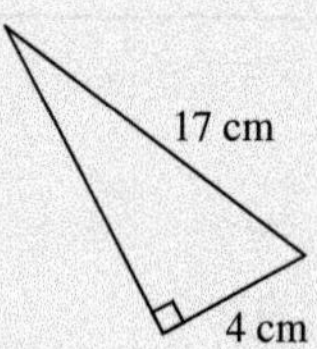

b 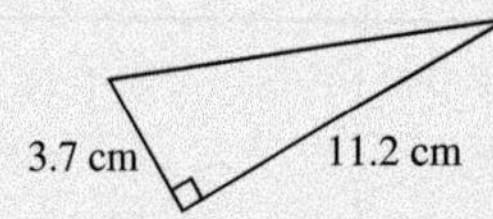

c 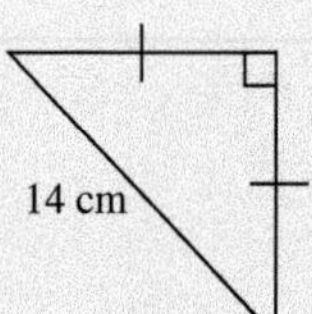

4 Given that $\cos\theta = \frac{3}{5}$ for an acute angle θ, find the values of $\sin\theta$ and $\tan\theta$

5 Given that $\tan\theta = \frac{1}{6}$ for an acute angle θ, find the exact values of $\sin\theta$ and $\cos\theta$

6 Given that $\sin\theta = \frac{1}{\sqrt{5}}$ for an acute angle x, find the exact values of $\cos\theta$ and $\tan\theta$

7 Solve each of these equations for $0° \le x \le 360°$

a $\sin x = 0.7$ **b** $\sin x = -0.5$ **c** $\sin x = 0.35$ **d** $\sin x = -0.27$

8 Solve each of these equations for $-360° \le x \le 360°$

a $\sin x = 0.5$ **b** $\sin x = -\frac{3}{4}$ **c** $\sin x = \frac{\sqrt{2}}{2}$ **d** $\sin x = -\frac{1}{3}$

9 Solve each of these equations for $0° \le x \le 360°$

a $\cos x = 0.3$ **b** $\cos x = 1$ **c** $\cos x = -0.9$ **d** $\cos x = -\frac{1}{4}$

10 Solve each of these equations for $-360° \le \theta \le 360°$

a $\cos\theta = \frac{1}{2}$ **b** $\cos\theta = -\frac{1}{3}$ **c** $\cos\theta = 0.15$ **d** $\cos\theta = -0.18$

11 Solve each of these equations for $-360° \le \theta \le 360°$

a $\tan\theta = 0.1$ **b** $\tan\theta = -1$ **c** $\tan\theta = -\frac{\sqrt{3}}{3}$ **d** $\tan\theta = 1.5$

12 Solve each of these equations for $0° \le x \le 360°$

a $3 + \sin x = 2$ **b** $5 + \tan x = 3$ **c** $\cos x - 1 = -0.5$ **d** $\tan x - 2 = 1$

13 Solve each of these equations for $-180° \le x \le 180°$

a $4\sin x = -3$ **b** $5\tan x = 3$ **c** $3\cos x = -2$ **d** $2\sin x = -\sqrt{2}$

14 Solve each of these equations for $0° \le \theta \le 360°$

a $3\cos\theta + 5 = 7$ **b** $2 + 5\sin\theta = 4$ **c** $5 - 3\tan\theta = 1$ **d** $6\cos\theta + 5 = 2$

15 Solve each of these equations for $-180° \le \theta \le 180°$

a $3\tan\theta - 4 = 7 + \tan\theta$ **b** $\sqrt{2}\cos\theta + 1 = \sqrt{2} + \cos\theta$

c $3\sin\theta + 2\sqrt{3} = \sqrt{3} + \sin\theta$ **d** $3 - 5\cos\theta = 2 + \cos\theta$

16 Solve each of these equations for $-360° \le x \le 360°$

a $2 - 5\cos x = \sin 30$ **b** $\cos 0 = 2 - \sin x$

c $3\tan 135 = 1 + \tan x$ **d** $\sin 60 + \cos x = 1$

17 Solve each of these equations for $-180° \le \theta \le 180°$

a $\tan 2\theta = \sqrt{3}$ **b** $\cos 2\theta = \frac{\sqrt{2}}{2}$ **c** $\sin 2\theta = -\frac{\sqrt{2}}{2}$ **d** $\cos 2\theta = 0.3$

18 Solve each of these equations for $0° \le \theta \le 90°$

a $\tan 4\theta = \frac{\sqrt{3}}{3}$ **b** $\sin 3\theta = 0.3$ **c** $\cos 4\theta = -0.6$ **d** $\tan 3\theta = -1$

19 Solve each of these equations for $-360° \le \theta \le 360°$

a $\sin\left(\frac{\theta}{3}\right) = 0.2$ **b** $\tan\left(\frac{\theta}{2}\right) = -\sqrt{3}$ **c** $\cos\left(\frac{\theta}{2}\right) = \frac{\sqrt{3}}{2}$ **d** $\tan\left(\frac{2\theta}{3}\right) = \frac{1}{3}$

20 Solve each of these equations for $-180° \le \theta \le 180°$

a $\tan(\theta + 45°) = \frac{1}{\sqrt{3}}$ **b** $\cos(\theta - 50°) = \frac{1}{4}$ **c** $\sin(\theta + 65°) = 0.7$ **d** $\cos(\theta - 140°) = -0.2$

21 Solve each of these equations for $0° \le \theta \le 360°$

a $\tan(\theta - 25°) = 0.8$ **b** $\sin(\theta + 10°) = -\frac{1}{5}$ **c** $\cos(\theta - 70°) = 0.15$ **d** $\tan(\theta + 200°) = -1$

MyMaths 1112, 1131, 1132, 1133, 1143 SEARCH

Topic B: Trigonometry 2

Bridging to Ch3.2

To solve problems involving non-right-angled triangles you use the **sine** rule and the **cosine** rule.

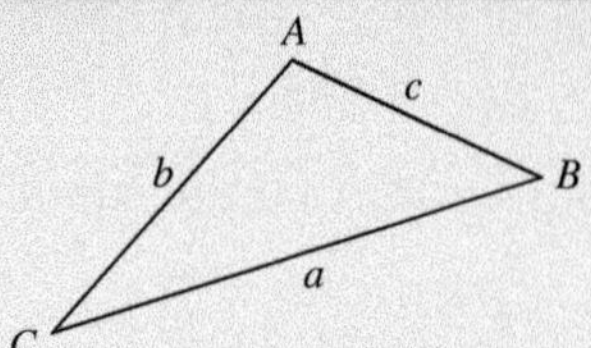

Key point

Sine rule: $\frac{\sin A}{a} = \frac{\sin B}{b} = \frac{\sin C}{c}$ to find an angle,

or write as $\frac{a}{\sin A} = \frac{b}{\sin B} = \frac{c}{\sin C}$ to find a side. It's important that sides and angles with the same letter are opposite.

Angles are written upper case and sides are lower case. Angle A is opposite side a and so on.

In order to use the sine rule you must have information about an opposite side and angle pair.

Example 1

Find the lengths of sides x and y in this triangle.

x
33°
9 cm
y
82°

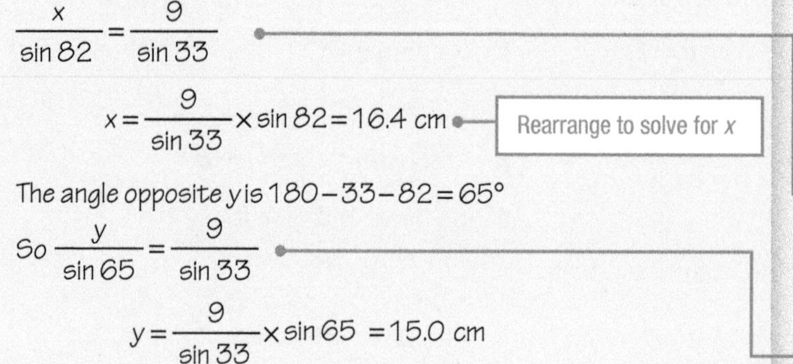

$$\frac{x}{\sin 82} = \frac{9}{\sin 33}$$

$$x = \frac{9}{\sin 33} \times \sin 82 = 16.4 \text{ cm}$$

Rearrange to solve for x

The angle opposite y is $180 - 33 - 82 = 65°$

So $\frac{y}{\sin 65} = \frac{9}{\sin 33}$

$$y = \frac{9}{\sin 33} \times \sin 65 = 15.0 \text{ cm}$$

The side of length 9 cm is opposite the 33° angle. The side x is opposite the 82° angle. So you can use the sine rule.

You could also use the other pair of x and 82°: $\frac{y}{\sin(65)} = \frac{16.4}{\sin(82)}$

Try It 1

Find the lengths of sides x and y in this triangle.

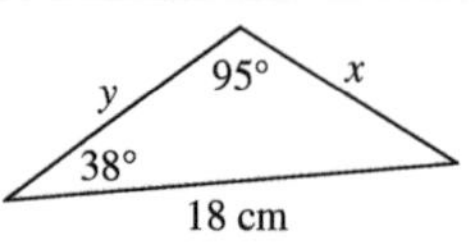

When finding an angle, remember that the equation $\sin\theta = k$ has two solutions in the range $0° < \theta < 180°$ when $0 < k < 1$. So you need to subtract the first solution you find from 180° to find a second solution then decide whether or not it is a possible solution for your triangle.

For example, if $A = 40°$, $a = 5$ and $b = 7$ then two different triangles could be formed:

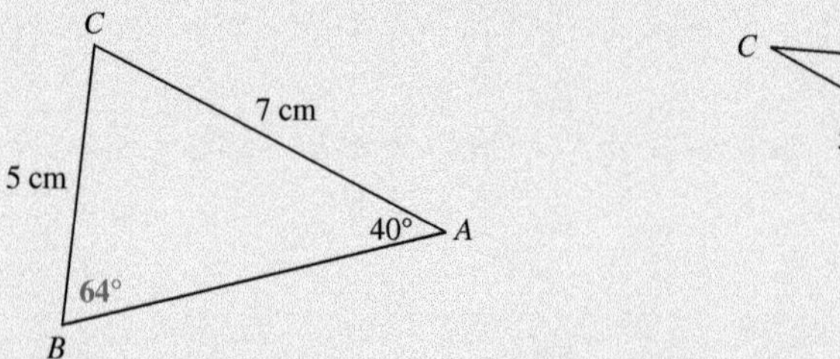

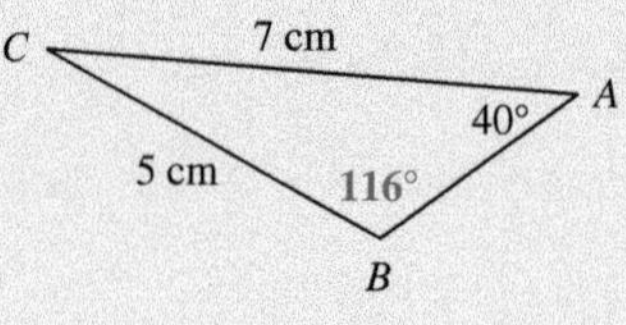

The sine rule gives the acute solution $B = 64°$, but $B = 180 - 64 = 116°$ is also a possible solution.

You need to check whether or not the obtuse solution will actually work. If it is too big then the angle sum of the triangle would be more than 180° which is not possible!

Example 2

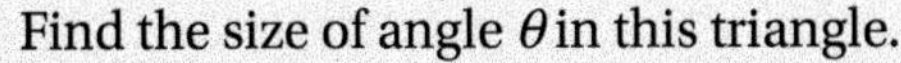

Find the size of angle θ in this triangle.

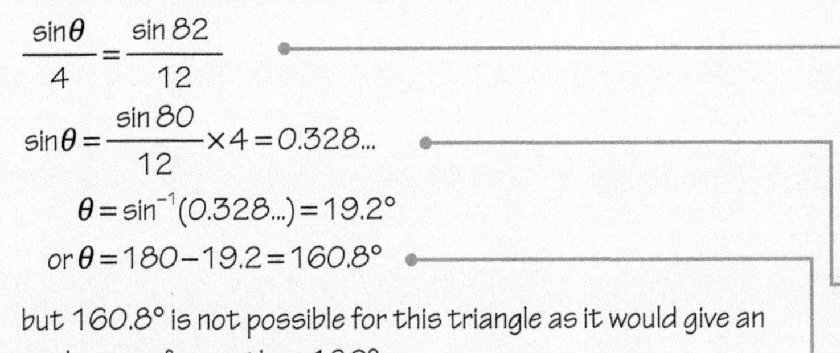

$$\frac{\sin\theta}{4} = \frac{\sin 82}{12}$$

$$\sin\theta = \frac{\sin 80}{12} \times 4 = 0.328...$$

$$\theta = \sin^{-1}(0.328...) = 19.2°$$

$$\text{or } \theta = 180 - 19.2 = 160.8°$$

but 160.8° is not possible for this triangle as it would give an angle sum of more than 180°

So, $\theta = 19.2°$

The side of length 12 cm is opposite the 80° angle so you can use the sine rule. The angle θ is opposite a side of length 4 cm.

Rearrange to solve for θ

Subtract from 180° to give other solution.

Try It 2

Find the size of angle θ in this triangle.

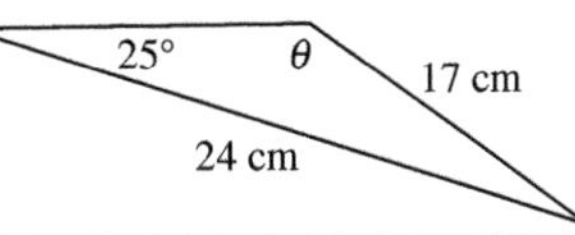

If you do not have information about an opposite side and angle pair then you will need to use the cosine rule.

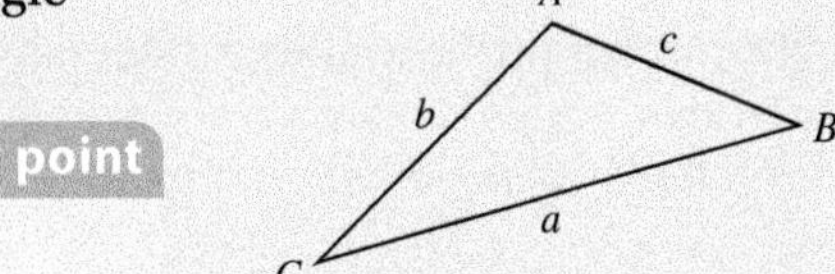

Key point

Cosine rule: $a^2 = b^2 + c^2 - 2bc\cos A$

(where angle A is opposite side a)

Example 3

Find the length of side x in this triangle.

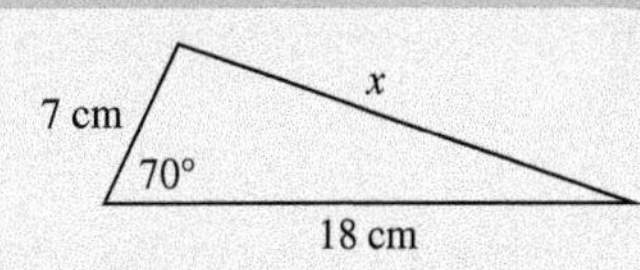

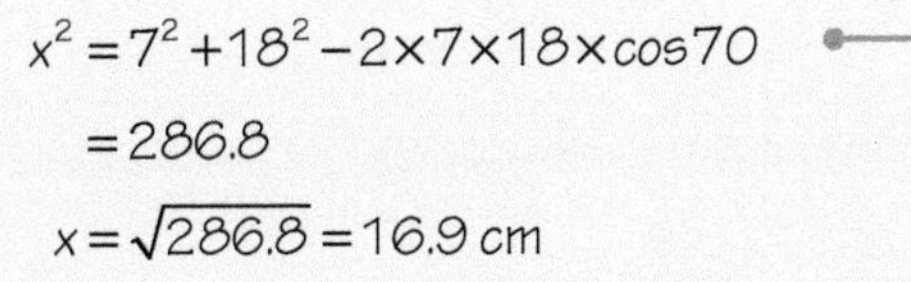

$$x^2 = 7^2 + 18^2 - 2 \times 7 \times 18 \times \cos 70$$

$$= 286.8$$

$$x = \sqrt{286.8} = 16.9 \text{ cm}$$

You do not have information about an opposite side and angle pair so you need to use the cosine rule.

x is the side you need to find so this is 'a' in the rule, which means the 70° angle is 'A' since it is opposite x. The other two sides are 'b' and 'c' in either order.

Try It 3

Find the length of side x in this triangle.

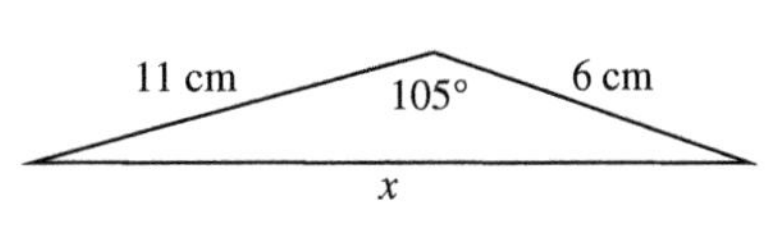

If you know the lengths of all three sides of a triangle then you can use the cosine rule to find one of the angles. You can either use $a^2 = b^2 + c^2 - 2bc\cos A$ and solve to find A, or you can use the rearranged version of the cosine rule:

Key point

Cosine rule: $\cos A = \dfrac{b^2 + c^2 - a^2}{2bc}$

(Remember that side a is opposite angle A)

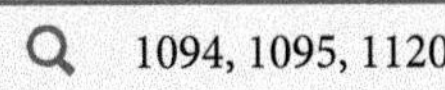

1094, 1095, 1120

Example 4

Find the size of angles x and y in this triangle.

$$\cos x = \frac{(12^2 + 9^2 - 7^2)}{2 \times 9 \times 12} = \frac{22}{27}$$

$$x = \cos^{-1}\left(\frac{22}{27}\right) = 35.4°$$

$$\cos y = \frac{9^2 + 7^2 - 12^2}{2 \times 9 \times 7} = -\frac{1}{9}$$

$$y = \cos^{-1}\left(-\frac{1}{9}\right) = 96.4°$$

You know the lengths of all three sides so use the cosine rule.

Make sure you use 12 as the side opposite to angle y

Try It 

Find the size of angles x and y in this triangle.

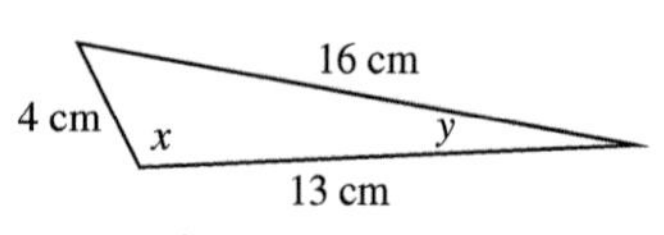

The equation $\cos\theta = k$ only has one solution in the range $0° \le \theta \le 180°$ so the cosine rule gives a unique solution.

You can use trigonometry to find the area of a triangle.

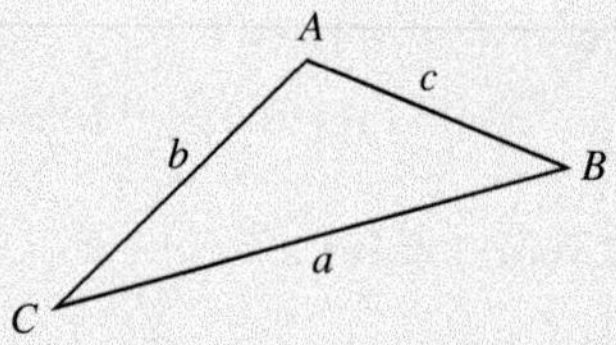

Key point

Area of triangle $= \frac{1}{2}ab\sin C$

where C is the angle between sides a and b

Example 5

Calculate the area of this triangle.

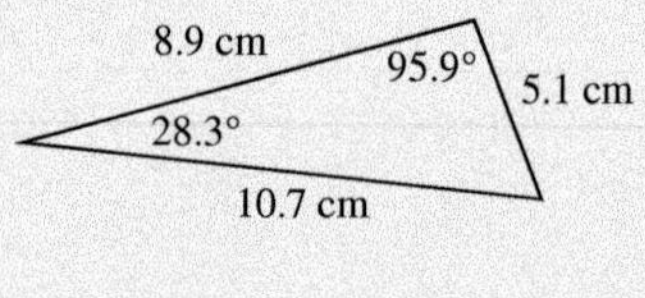

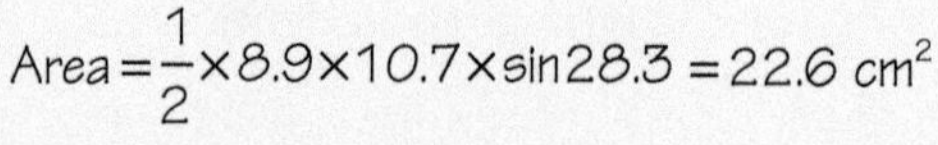

$$\text{Area} = \frac{1}{2} \times 8.9 \times 10.7 \times \sin 28.3 = 22.6 \text{ cm}^2$$

Or, using the 95.9° angle:

$$\text{Area} = \frac{1}{2} \times 8.9 \times 5.1 \times \sin 95.9 = 22.6 \text{ cm}^2$$

The 28.3° angle is between the 8.9 cm and 10.7 cm sides.

Try It 5

Calculate the area of this triangle

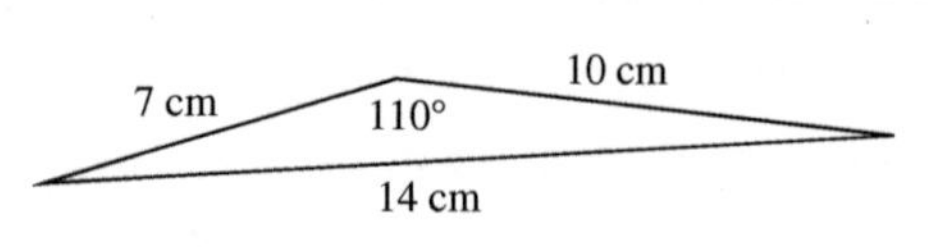

Bridging Exercise Topic B

Bridging to Ch3.2

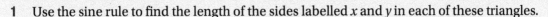

1 Use the sine rule to find the length of the sides labelled x and y in each of these triangles.

a 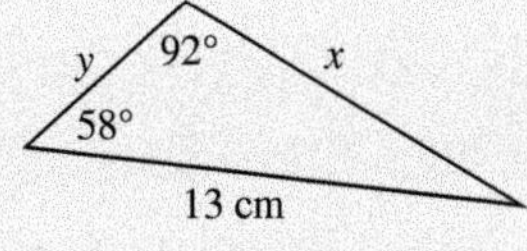

b 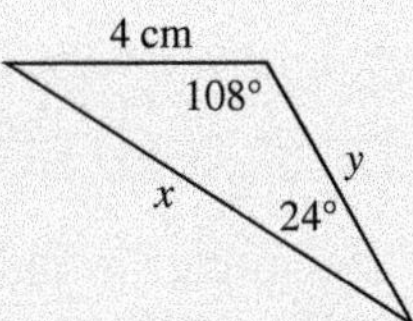

c 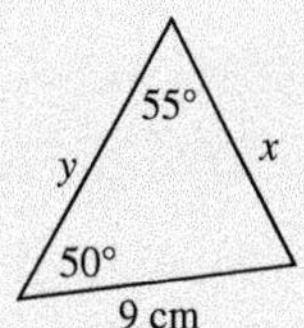

2 Use the sine rule to find the size of angle θ in each of these triangles. Explain whether the solution is unique in each case.

a 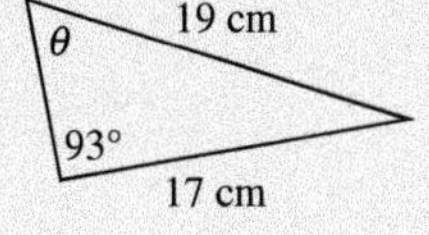

b 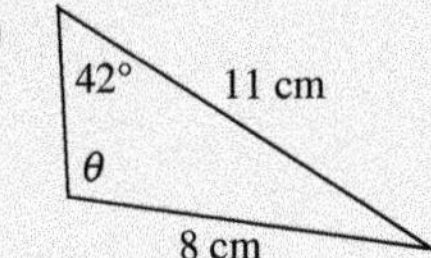

c 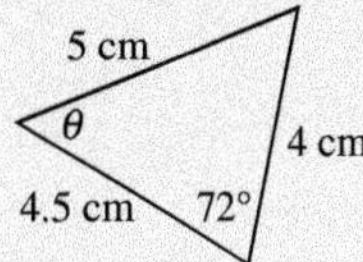

3 Use the cosine rule to find the length of the side x in each of these triangles.

a 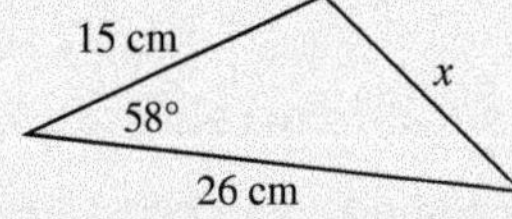

b 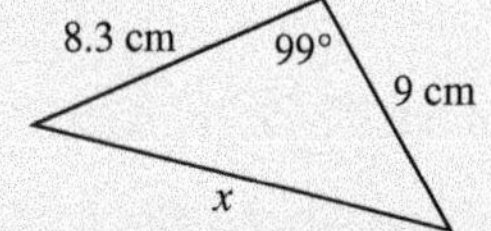

c 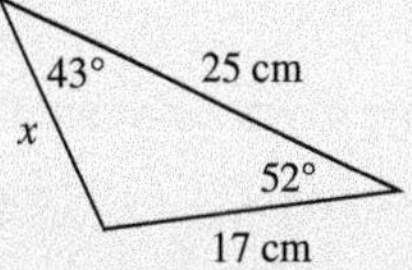

4 Use the cosine rule to find the size of the angles labelled x and y in each of these triangles.

a 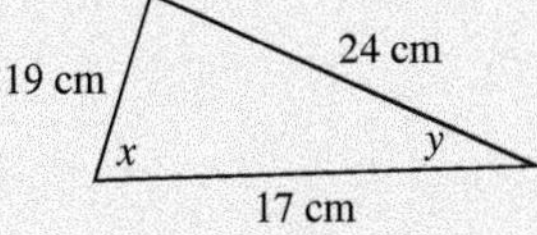

b 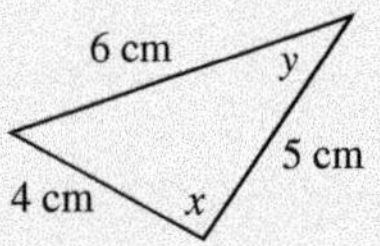

c 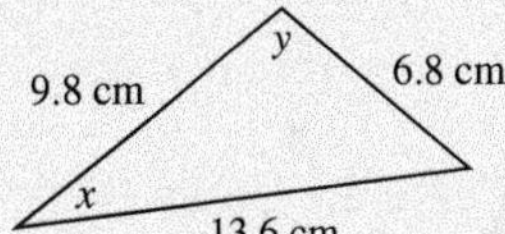

5 Find the length of the side labelled x in each of these triangles.

a 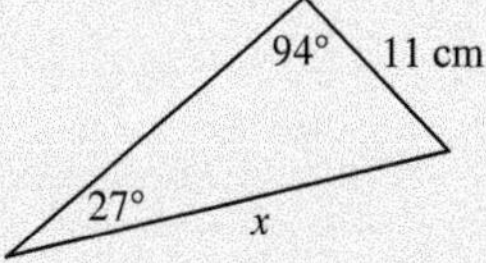

b 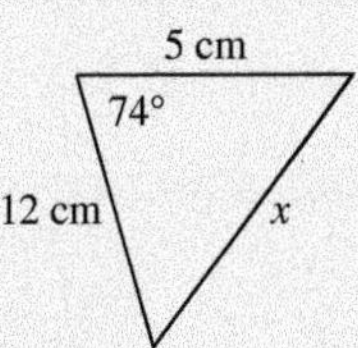

c 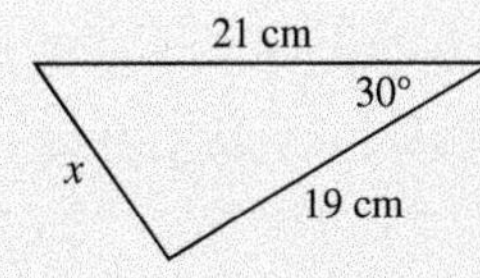

d 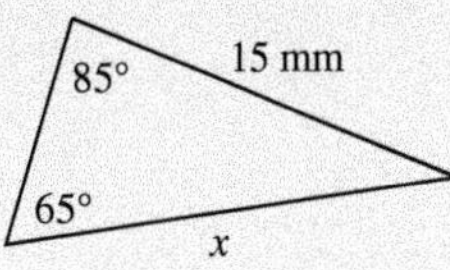

e 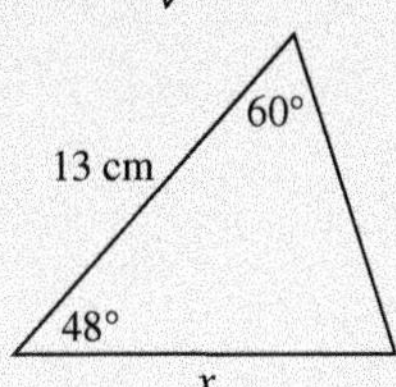

f 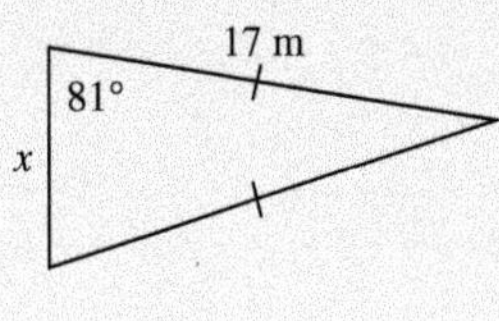

6 Find the size of the acute angle θ in each of these triangles.

a 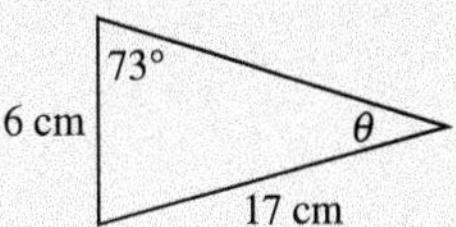

b 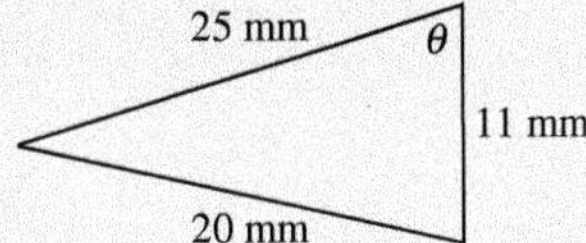

c 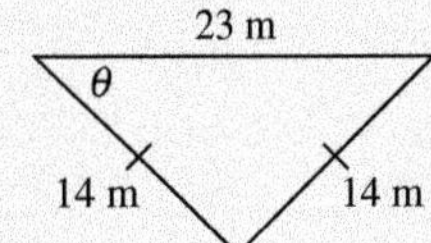

7 Triangle ABC is such that $AB = 5$ cm, $BC = 3$ cm and $AC = 7$ cm. Calculate the size of angle ABC

8 Triangle ABC is such that $AB = 24$ cm, $AC = 27$ cm, angle $ABC = 37°$ and angle $BCA = \theta$. Calculate θ

9 Find the area of each of the triangles in question **3**

MyMaths 1094, 1095, 1120 SEARCH

3.1 Sine, cosine and tangent

Fluency and skills

You can use trigonometry to find lengths and angles in right-angled triangles. This branch of mathematics is used in engineering, technology and many sciences.

Pythagoras' theorem for right-angled triangles is $a^2+b^2=c^2$

Dividing by c^2 gives $\frac{a^2}{c^2}+\frac{b^2}{c^2}=\frac{c^2}{c^2}$ or $\left(\frac{a}{c}\right)^2+\left(\frac{b}{c}\right)^2=1$

$\sin\theta=\frac{a}{c}$

$\cos\theta=\frac{b}{c}$

$\tan\theta=\frac{a}{b}$

See p.382 For a list of mathematical notation.

Key point

$\sin^2\theta+\cos^2\theta\equiv1$

$\tan\theta=\frac{a}{b}$

Dividing numerator and denominator of $\tan\theta$ by c gives a definition for $\tan\theta$ in terms of $\sin\theta$ and $\cos\theta$

Key point

$\tan\theta\equiv\frac{\frac{a}{c}}{\frac{b}{c}}\equiv\frac{\sin\theta}{\cos\theta}$

These two identities are true for all values of θ

Example 1

Calculate **a** $\sin\theta$ **b** $\tan\theta$ as surds, given that θ is acute and $\cos\theta=\frac{1}{\sqrt{3}}$

a $\sin^2\theta=1-\cos^2\theta=1-\left(\frac{1}{\sqrt{3}}\right)^2=\frac{2}{3}$

$\sin\theta=\sqrt{\frac{2}{3}}$ — θ is acute, so ignore $-\sqrt{\frac{2}{3}}$

b $\tan\theta=\frac{\sin\theta}{\cos\theta}$ — Use the identities.

$=\sqrt{\frac{2}{3}}\times\frac{\sqrt{3}}{1}=\sqrt{\frac{2\times3}{3\times1}}$ — Substitute values and simplify.

$\tan\theta=\sqrt{2}$

See Ch1.3 For a reminder of simplifying surds.

Example 2

Prove that $\tan\theta+\frac{1}{\tan\theta}\equiv\frac{1}{\sin\theta\cos\theta}$

$\tan\theta+\frac{1}{\tan\theta}\equiv\frac{\sin\theta}{\cos\theta}+\frac{\cos\theta}{\sin\theta}$ — Use $\tan\theta=\frac{\sin\theta}{\cos\theta}$

$\equiv\frac{\sin^2\theta+\cos^2\theta}{\sin\theta\cos\theta}$ — Add the two fractions.

$\equiv\frac{1}{\sin\theta\cos\theta}$ — Use $\sin^2\theta+\cos^2\theta=1$

You can use the unit circle to draw graphs of the trigonometric ratios. The point P moving around the circle, centre O, has coordinates $x = \cos\theta$ and $y = \sin\theta$

As $P(x, y)$ moves around the circle you can plot graphs of the values of y, x and $\frac{y}{x}$ for each value of θ

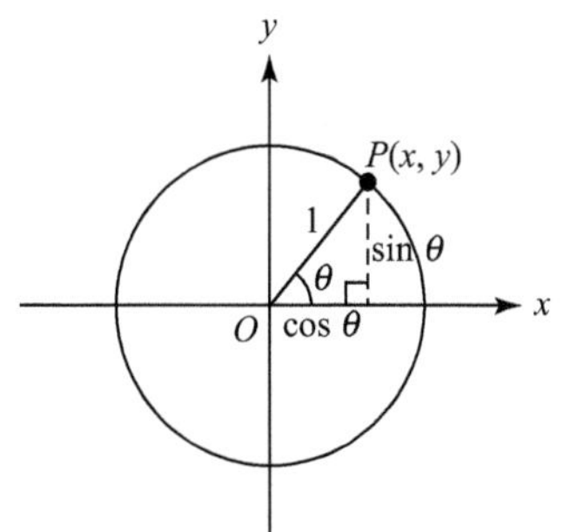

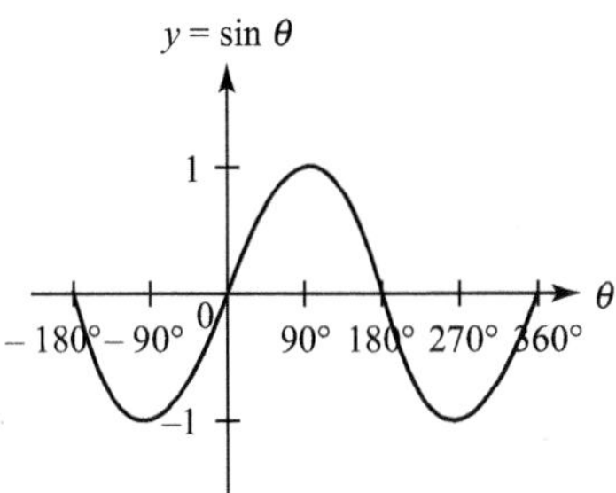

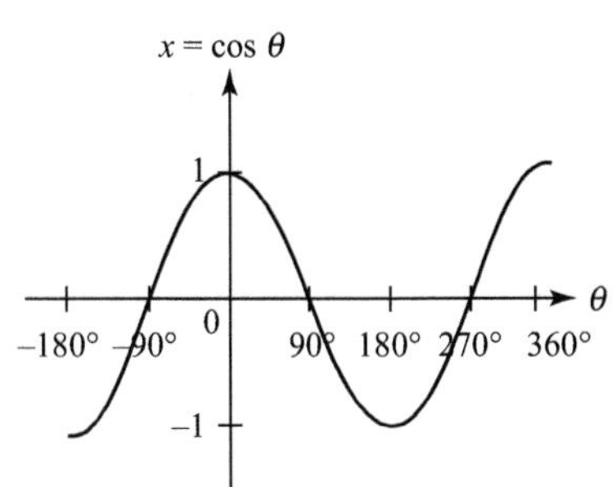

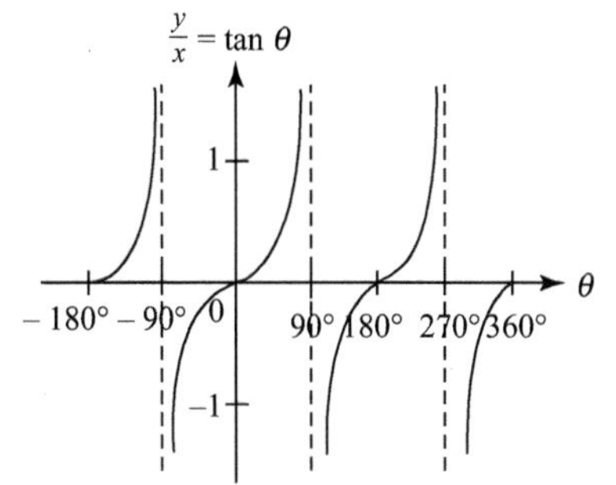

Extending the graphs for higher and lower values of θ shows they are all **periodic functions**, with a period of 360° for sine and cosine and 180° for tangent.

You can also see the symmetries from the graphs. For example, $y = \sin\theta$ has lines of symmetry at $\theta = -90°$, $\theta = 90°$, $\theta = 270°$, ... and it has rotational symmetry (order 2) about every point where the graph intersects the θ-axis.

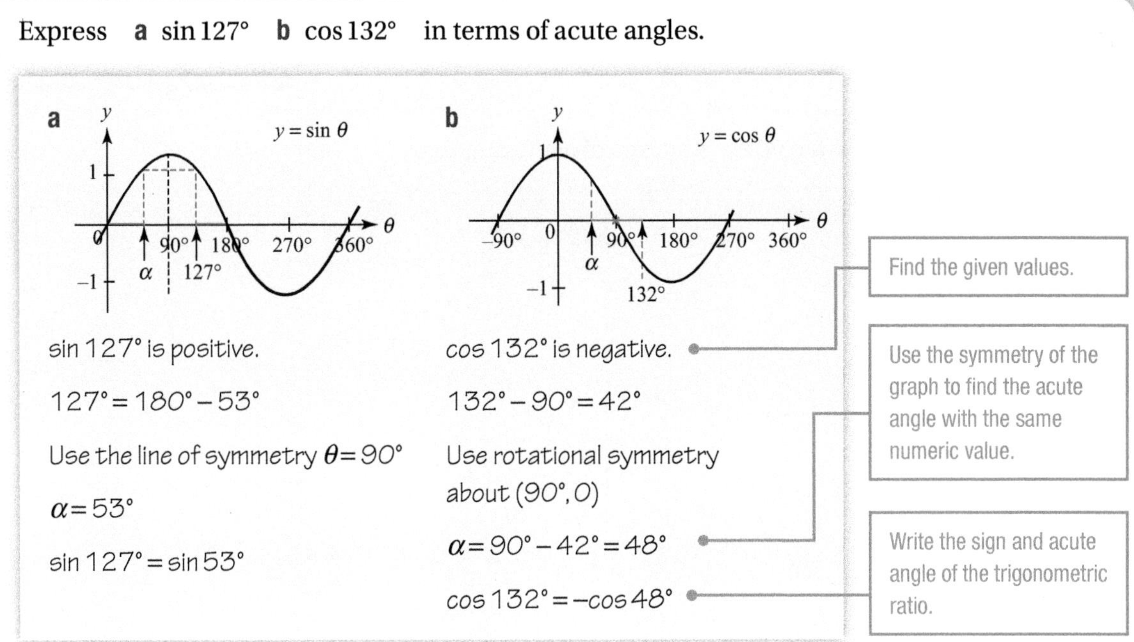

Example 3

Express **a** $\sin 127°$ **b** $\cos 132°$ in terms of acute angles.

a

$\sin 127°$ is positive.

$127° = 180° - 53°$

Use the line of symmetry $\theta = 90°$

$\alpha = 53°$

$\sin 127° = \sin 53°$

b

$\cos 132°$ is negative. — Find the given values.

$132° - 90° = 42°$

Use rotational symmetry about $(90°, 0)$

$\alpha = 90° - 42° = 48°$ — Use the symmetry of the graph to find the acute angle with the same numeric value.

$\cos 132° = -\cos 48°$ — Write the sign and acute angle of the trigonometric ratio.

The trigonometric graphs and equations can be transformed in the same way as quadratic and polynomial graphs.

See Ch2.4
For a reminder of transformations of graphs.

Another method for finding equivalent acute angles is to use a **quadrant diagram.**

Imagine a radius OP rotating about O through an angle θ from the positive x-axis. Whichever quadrant OP lies in, you can form an acute triangle with the x-axis as its base. Depending on the quadrant, the x and y coordinates of point P are positive or negative and so the sine, cosine and tangent of θ are also positive or negative.

You can use the word '**CAST**' as a mnemonic to help you remember where each ratio is positive.

CAST starts from the 4th quadrant and moves anticlockwise, as all the ratios are positive in the 1st quadrant.

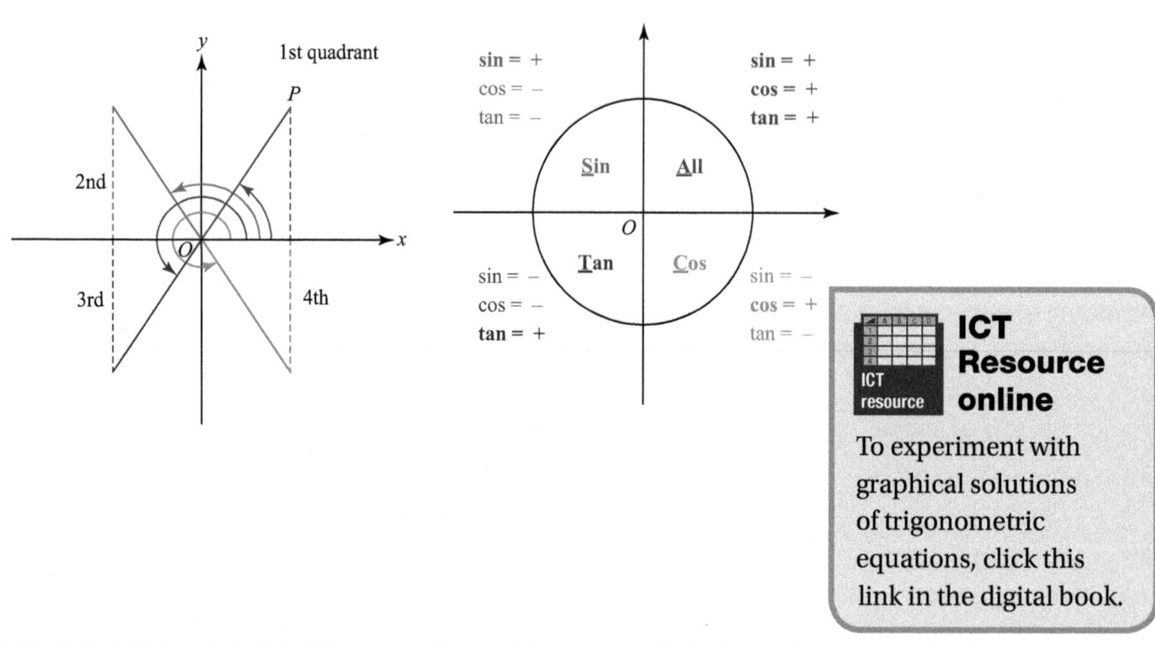

ICT Resource online

To experiment with graphical solutions of trigonometric equations, click this link in the digital book.

Example 4

Express **a** cos 132° **b** tan 683° in terms of acute angles.

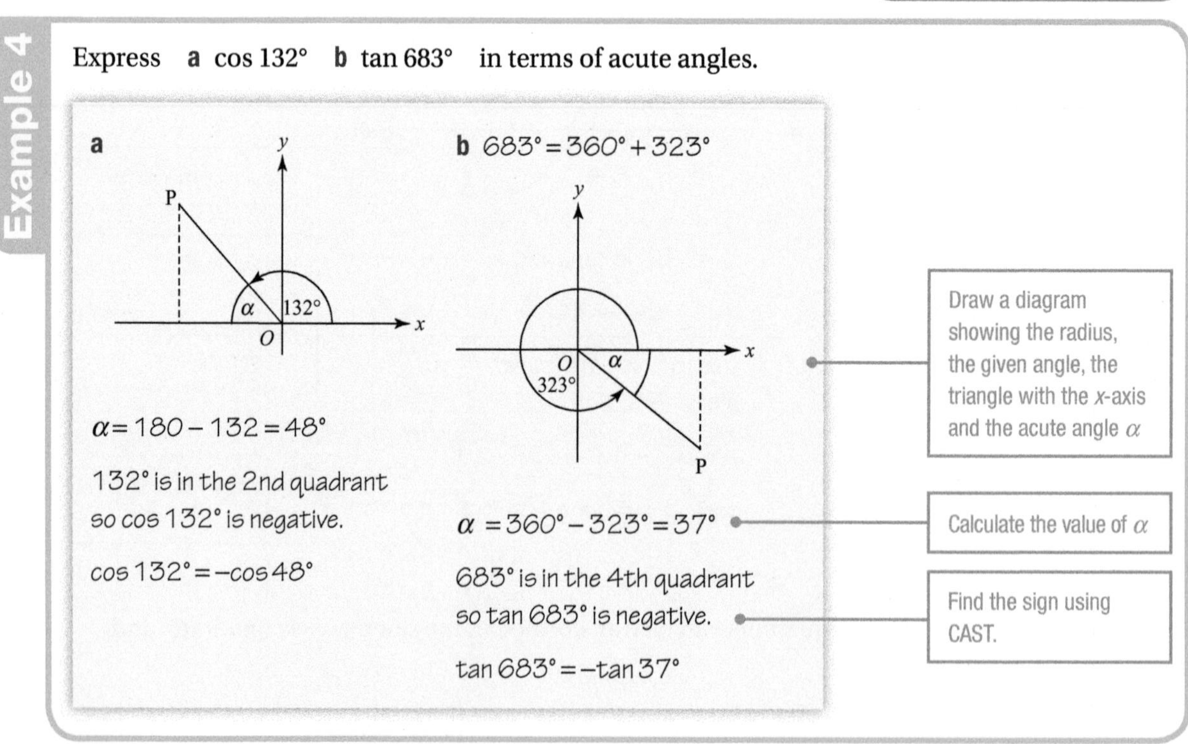

a

$\alpha = 180 - 132 = 48°$

132° is in the 2nd quadrant so cos 132° is negative.

$\cos 132° = -\cos 48°$

b $683° = 360° + 323°$

$\alpha = 360° - 323° = 37°$

683° is in the 4th quadrant so tan 683° is negative.

$\tan 683° = -\tan 37°$

Draw a diagram showing the radius, the given angle, the triangle with the x-axis and the acute angle α

Calculate the value of α

Find the sign using CAST.

Exercise 3.1A Fluency and skills

1 Use $\sin^2\theta+\cos^2\theta=1$ and $\tan\theta=\frac{\sin\theta}{\cos\theta}$ to calculate the value of $\sin\theta$ and $\tan\theta$, given that θ is acute and

a $\cos\theta=\frac{3}{5}$ b $\cos\theta=0.8$

c $\cos\theta=\frac{12}{13}$

2 Use the quadrant or graphical method to find these values in terms of acute angles.

a $\cos 190°$ b $\tan 160°$

c $\sin 340°$ d $\cos 158°$

e $\tan 215°$ f $\sin 285°$

Check your answers using the method that you didn't use the first time.

3 Copy and complete this table.

θ	–90°	0°	90°	180°	270°	360°
$\sin\theta$						
$\cos\theta$						
$\tan\theta$						

4 a Describe the line and rotational symmetries of the graphs of

i $y=\cos\theta$ ii $y=\tan\theta$

b Sketch these graphs for $-360°\le x\le 360°$

You can check your sketches on a calculator.

i $y=\sin 3x$ ii $y=\cos x-1$

iii $y=\tan\frac{1}{2}x$

5 Simplify

a $\frac{\sin\theta}{\sqrt{1-\sin^2\theta}}$ b $\sqrt{\frac{1-\cos^2\theta}{1-\sin^2\theta}}$

c $\frac{\sqrt{1-\sin^2\theta}}{\cos\theta}$ d $\tan\theta\cos\theta$

e $\frac{\sin\theta}{\tan\theta}$ f $\sin\theta\cos\theta\tan\theta$

6 Express, in terms of acute angles,

a $\sin 380°$ b $\tan 390°$

c $\cos 700°$ d $\tan(-42°)$

e $\cos(-158°)$ f $\sin(-203°)$

7 Solve these equations for $-180°\le\theta\le 180°$

a $\sin\theta=\cos\theta$ b $\sin\theta+\cos\theta=0$

8 Use the triangle to write these trigonometric ratios in surd form.

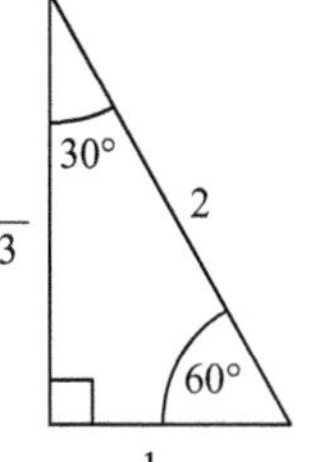

a $\sin 150°$ b $\cos 300°$

c $\tan 120°$ d $\sin 240°$

e $\cos(-60°)$ f $\tan(-150°)$

9 Use a calculator and give all the values of θ in the range –360° to 360° for which

a $\sin\theta=0.4$ b $\tan\theta=1.5$

c $\cos\theta=-0.5$

10 Use a calculator to find the smallest positive angle for which

a $\sin\theta$ and $\cos\theta$ are both positive and $\sin\theta=0.8$

b $\sin\theta$ and $\tan\theta$ are both negative and $\sin\theta=-0.6$

11 Solve these equations for $0°\le\theta\le 360°$

Show your working.

a $4\sin\theta=3$ b $3\tan\theta=4$

c $2\sin\theta+1=0$ d $3\cos\theta+2=0$

e $\tan\theta+3=0$ f $7+10\sin\theta=0$

g $4\cos\theta=-3$ h $4+9\tan\theta=0$

Reasoning and problem-solving

Strategy

To solve problems involving trigonometric ratios

1. Use trigonometric identities to simplify expressions.
2. Draw either a quadrant diagram or a trigonometric graph to show the information.
3. Use your knowledge of graphs, quadrant diagrams, symmetry and transformations to help you answer the question.

Example 5

Solve $5\cos 2\theta + 3 = 0$ for $0° \le \theta \le 180°$. Show your working.

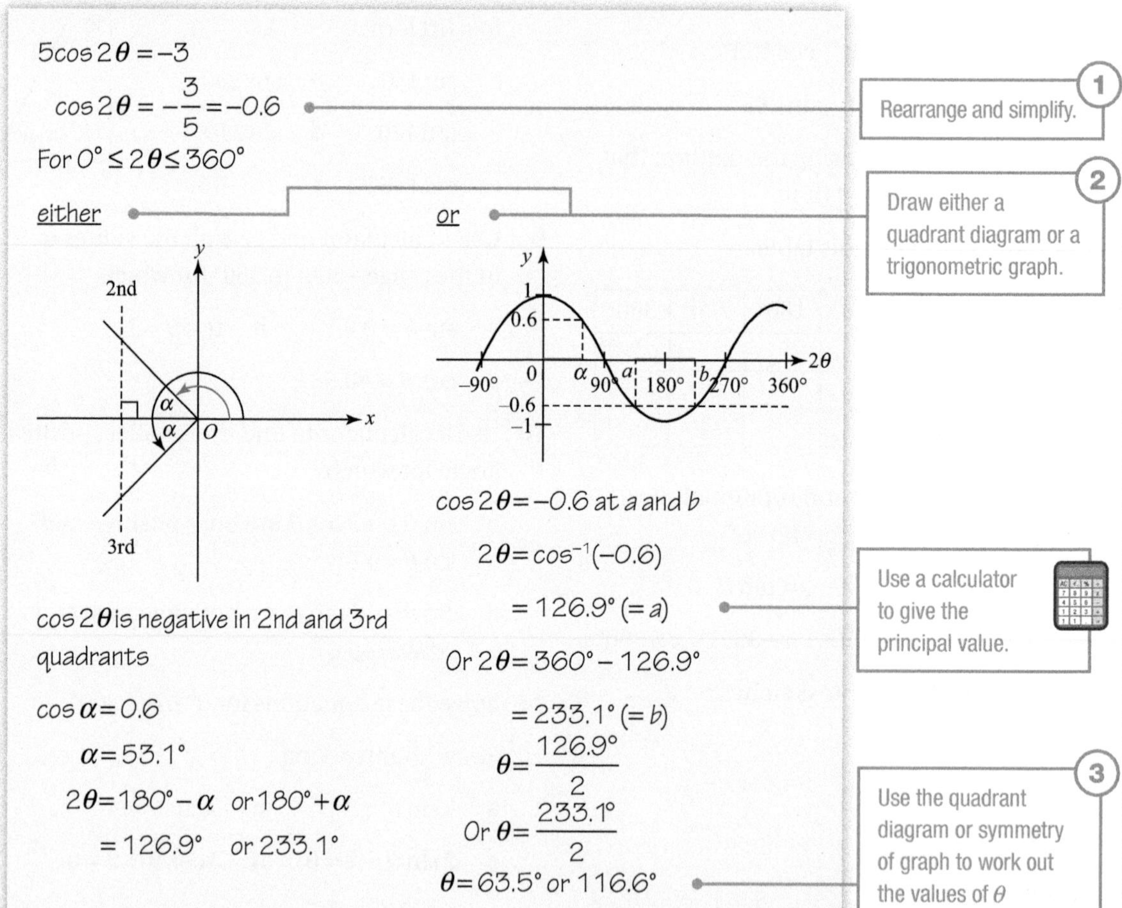

You can check your solution by solving the equation on a calculator.

Exercise 3.1B Reasoning and problem-solving

1 Solve these equations for $0° \le \theta \le 360°$

Show your working.

a $\sin(\theta + 30°) = \frac{1}{2}$ b $\cos(\theta - 30°) = \frac{1}{2}$

c $\tan(\theta + 20°) = 1$ d $2\sin(\theta + 30°) = -1$

2 Solve these equations for $0° \le \theta \le 180°$

a $2\sin 2\theta = 1$ b $3\tan 2\theta = 2$

c $5\cos 3\theta = 2$ d $5\sin 3\theta + 3 = 0$

e $\frac{1}{2}\tan 2\theta + 3 = 0$ f $4\sin\left(\frac{1}{2}\theta\right) = 3$

3 Give full descriptions of any two transformations, which map the graph of

a $y = \sin\theta$ onto $y = \cos\theta$

b $y = \tan\theta$ onto itself.

4

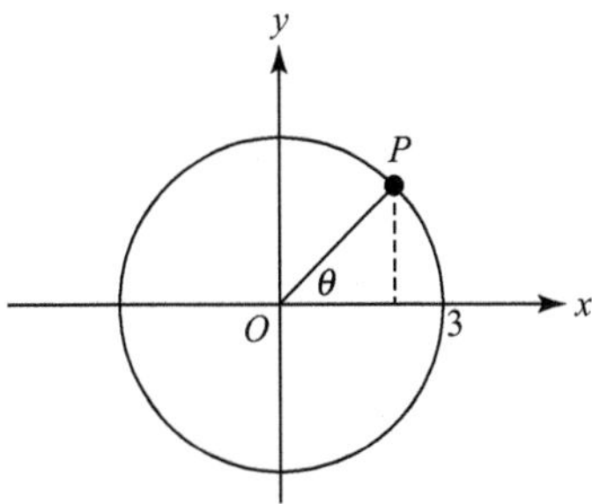

a Given a circle, centre O and radius 3, write the coordinates of point P

b Show these coordinates satisfy the equation of the circle $x^2 + y^2 = r^2$ and write the value of r

c Another circle has the equation $4x^2 + 4y^2 = 25$. What is its radius?

5 A circle, with centre at the origin, passes through the point $(6, 8)$. What is its equation in its simplest form?

6 Solve these equations for $-180° \leq \theta \leq 180°$

Show your working.

a $2\sin\theta = \cos\theta$

b $4\cos\theta = 5\sin\theta$

c $3\sin 2\theta - \cos 2\theta = 0$

d $3\sin 2\theta = 2\tan 2\theta$

e $3\sin\theta + \tan\theta = 0$

f $\sin\theta\cos\theta - \cos\theta = 0$

7 Solve these equations for $0° \leq \theta \leq 360°$

a $\sin^2\theta - 2\sin\theta + 1 = 0$

b $\tan^2\theta - \tan\theta - 2 = 0$

c $2\sin\theta + 2 = 3\cos^2\theta$

d $2\cos^2\theta + \sin^2\theta = 2$

e $2\cos\theta + 2 = 4\sin^2\theta$

f $5\sin\theta - 4\cos^2\theta = 2$

8 The graph of $y = a\sin b\theta$ has a maximum value of 5 and a period of 45°. Find the values of a and b. Show your working.

9 The depth of water, h metres, at point P on the seabed changes with the tide and is given by $h = 3 + 2\sin(30° \times t)$, where t is the time in hours after midnight.

a What is the greatest and least depth of water at P?

b What is the period of the oscillation of the tide?

c At what times do the high tides occur on this day?

10 Solve these equations for $0 \leq \theta \leq 360°$

a $7\cos\theta + 6\sin^2\theta - 8 = 0$

b $4\cos^2\theta + 5\sin\theta = 3$

11 a Draw an accurate graph of the function $y = 2\cos\theta + 3\sin\theta$ for $-180° \leq \theta \leq 180°$

b Solve the equation $2\cos\theta + 3\sin\theta = 0$ for this range of θ by using

i your graph,

ii an algebraic method.

12 Prove these identities.

a $\cos^4 x - \sin^4 x \equiv \cos^2 x - \sin^2 x$

b $\tan x + \dfrac{1}{\tan x} \equiv \dfrac{1}{\sin x \cos x}$

Challenge

13 Prove that $\dfrac{1 - \tan^2 x}{1 + \tan^2 x} \equiv 1 - 2\sin^2 x$ and

hence solve the equation $\dfrac{1 - \tan^2 x}{1 + \tan^2 x} = \dfrac{1}{2}$

for $0 \leq x \leq 360°$

3.2 The sine and cosine rules

Fluency and skills

You can use the sine and cosine rules to calculate lengths and angles in any triangle — not just right-angled triangles.

When you know a pair of opposite sides and angles, you can calculate other sides and angles using the **sine rule**.

Key point

The sine rule states that, for triangle *ABC*,

$$\frac{a}{\sin A}=\frac{b}{\sin B}=\frac{c}{\sin C} \quad \text{or} \quad \frac{\sin A}{a}=\frac{\sin B}{b}=\frac{\sin C}{c}$$

Use the left-hand version for sides and the right-hand one for angles.

Triangle *ABC* can be written ΔABC

Example 1

In ΔABC, angle $A=50°$, side $a=8$ cm and side $c=10$ cm. Calculate angles *B* and *C*, given that the triangle is acute.

$\frac{\sin C}{c}=\frac{\sin A}{a}$ — As *c* is known, use the sine rule to calculate angle *C* first.

$\frac{\sin C}{10}=\frac{\sin 50°}{8}$ — Substitute in the correct values.

$\sin C=\frac{\sin 50°}{8}\times 10=0.9575...$ — Rearrange to solve for sin *C*. Do not round answers during a calculation.

Sine is positive between 0° and 180°, so there are two possible values of *C*

The given triangle is acute, so $C=73.2°$

Angle $B=180°-50°-73.2°=56.8°$ — Use the angle sum of a triangle.

The data in Example 1 ($A=50°$, $a=8$ cm, $c=10$ cm) can also describe an obtuse triangle where $C'=180°-73.2°=106.8°$ and $B=180°-50°-106.8°=23.2°$. This is an example of the **ambiguous case**, where the initial data gives two possible triangles.

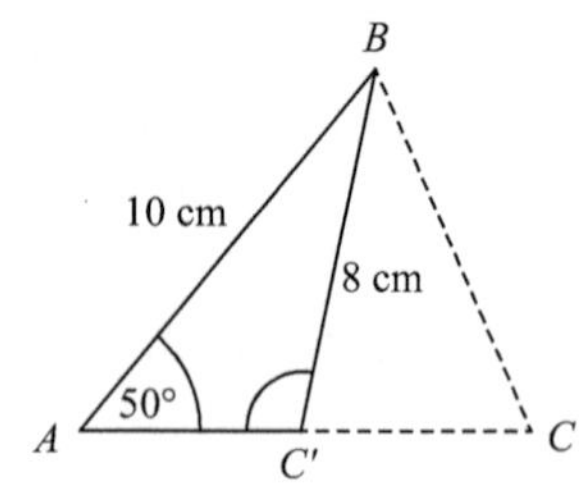

When you know two sides and the angle between them, you can use the **cosine rule** to calculate the third side. You can also use this rule to calculate angles when you know all three sides but no angles.

Key point

The cosine rule states that, for triangle *ABC*, $a^2=b^2+c^2-2bc\cos A$

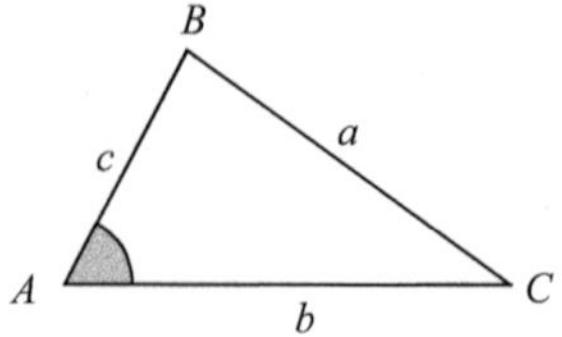

Alternatively, $b^2=a^2+c^2-2ac\cos B$ or $c^2=a^2+b^2-2ab\cos C$

Example 2

In ΔABC, $a = 4$ cm, $b = 9$ cm and $c = 6$ cm. Calculate angle B

$b^2 = c^2 + a^2 - 2ca\cos B$

You need to find angle B, so use the formula which has b^2 as the subject.

$9^2 = 6^2 + 4^2 - 2 \times 6 \times 4 \times \cos B$

$48\cos B = 36 + 16 - 81 = -29$

$\cos B = -\frac{29}{48}$

Rearrange to make $\cos B$ the subject.

Angle $B = 127.2°$

$\cos B$ is negative, so angle B is obtuse.

You can calculate the area of any triangle when you know any two sides and the angle between them.

Key point

Area of triangle $ABC = \frac{1}{2}ab\sin C$

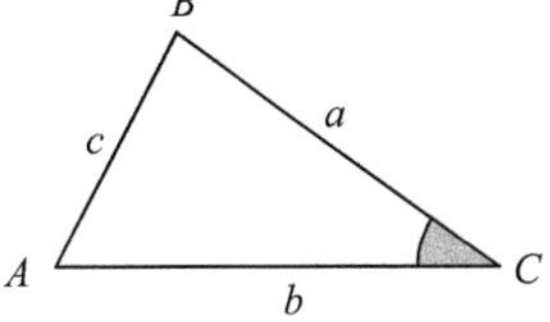

Exercise 3.2A Fluency and skills

1 Calculate the lengths BC and PR in these triangles.

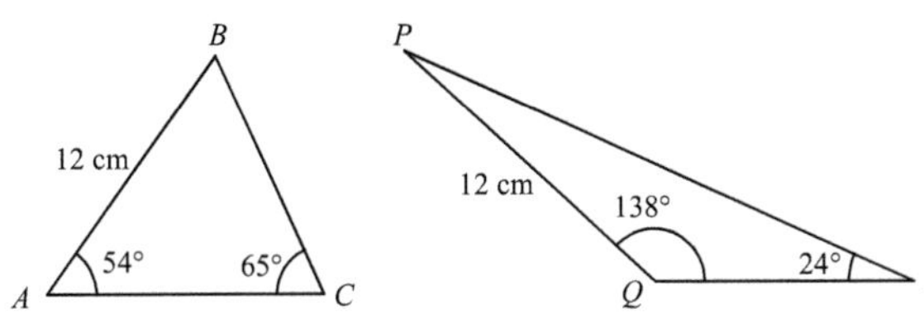

2 a Triangle ABC is acute with $AB = 9$ cm, $BC = 8$ cm and angle $A = 52°$. Calculate angle C

b Triangle EFG is obtuse with $EG = 11$ cm, $EF = 7$ cm and angle $G = 35°$. Calculate obtuse angle F

c Triangle HIJ is obtuse with $HI = 10$ cm, $IJ = 5$ cm and angle $H = 28°$. Calculate obtuse angle J

3 a Calculate the lengths BC and PR in these triangles.

b Calculate the areas of the triangles.

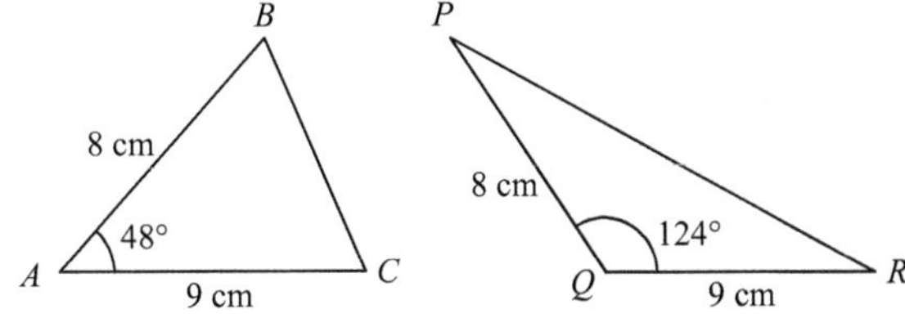

4 Calculate all the unknown sides and angles in these triangles. Give both solutions if the data is ambiguous.

a

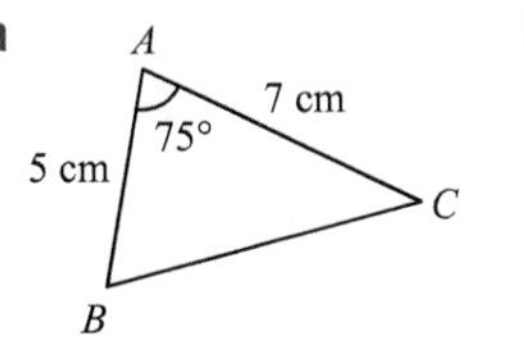

b

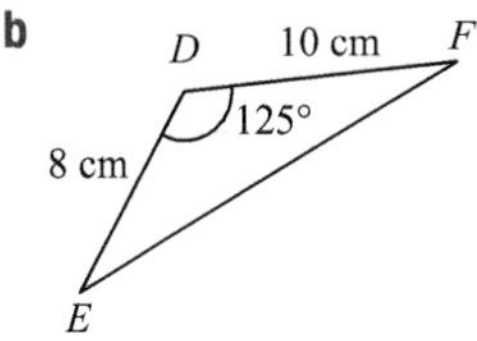

c

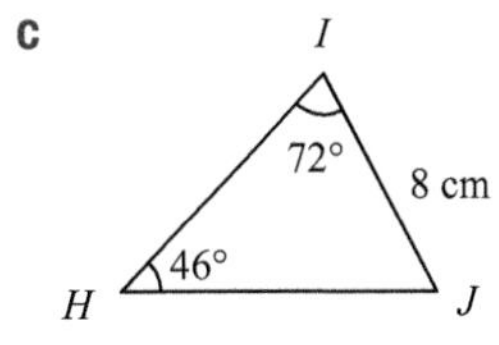

d

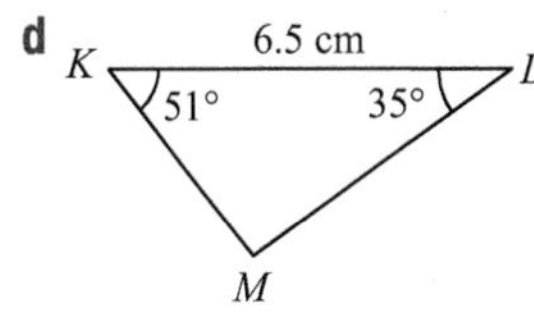

e

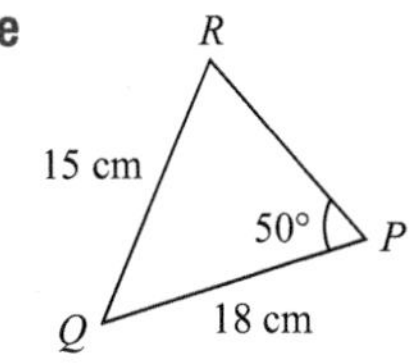

5 Calculate the unknown sides and angles in

a ΔABC where $a = 11.1$ cm, $b = 17.3$ cm and $c = 21.2$ cm

b ΔDEF where $d = 75.3$ cm, $e = 56.2$ cm and angle $F = 51°$

c ΔHIJ where with $h = 44.2$ cm, $i = 69.7$ cm and angle $J = 33°$

MyMaths 2045, 2046 SEARCH

6 Calculate the length x and the area of the triangle, giving your answers as surds.

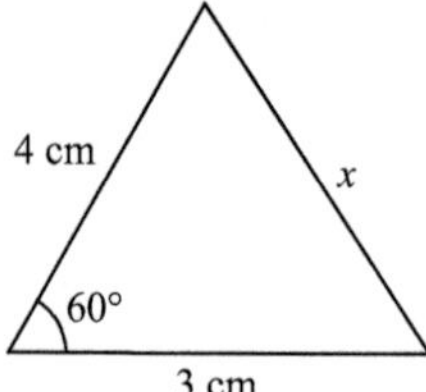

7 a In ΔABC, use the sine rule to show that there are two possible positions for vertex C (C_1 and C_2). Calculate the two possible sizes of angle C

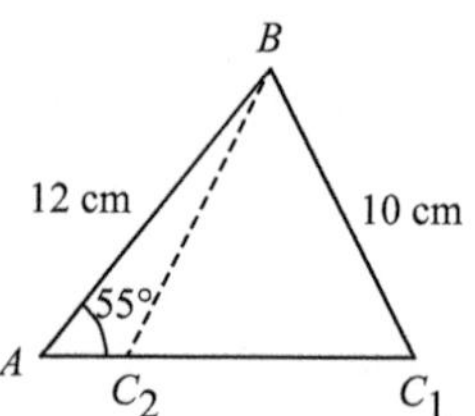

b In ΔXYZ, calculate the size of angle Z and explain why it is the only possible value.

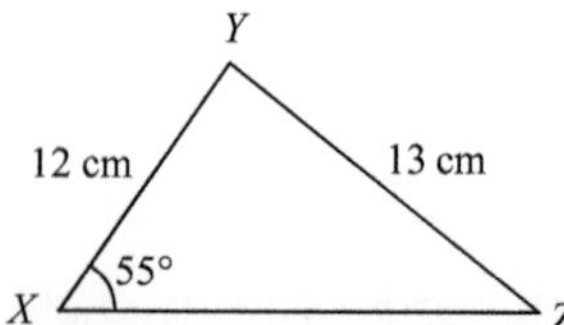

Reasoning and problem-solving

Strategy

To solve problems involving sine and cosine rules or the area formula

1. Draw a large diagram to show the information you have and what you need to work out.
2. Decide which rule or combination of rules you need to use.
3. Calculate missing values and add them to your diagram as you solve the problem.

Example 3

In ΔABC, angle $A = 49°$, angle $B = 76°$ and $c = 12$ cm. Calculate the unknown sides and angles, and calculate the area of the triangle.

Angle $C = 180° - 76° - 49° = 55°$

The sine rule gives $\frac{a}{\sin 49°} = \frac{12}{\sin 55°}$

$a = \frac{12 \sin 49°}{\sin 55°} = 11.055...$

$= 11.1$ cm (to 3 sf)

The sine rule gives

$\frac{b}{\sin 76°} = \frac{12}{\sin 55°}$

$b = \frac{12 \times \sin 76°}{\sin 55°}$

$= 14.2$ cm (to 3 sf)

The cosine rule gives

$b^2 = 12^2 + 11.055^2 - 2 \times 12 \times 11.055 \times \cos 76°$

$= 202.03$

$b = 14.213...$

$= 14.2$ cm (to 3 sf)

The two unknown sides are 11.1 cm and 14.2 cm.

The area of triangle $= \frac{1}{2} bc \sin A$

$= \frac{1}{2} \times 14.2 \times 12 \times \sin 49° = 64.3 \text{ cm}^2$

(1) Draw a diagram to show the information.

(2) Choose the sine rule because side c and angle C are now both known.

(3) Rearrange and calculate a

(2)(3) You can use either the sine rule or the cosine rule to calculate b

You can decide which rule is easier to use.

You could also use $\frac{1}{2} ac \sin B$ or $\frac{1}{2} ab \sin C$

Exercise 3.2B Reasoning and problem-solving

1 Calculate the area of ΔEFG, given that $e = 5$ cm, $f = 6$ cm and $g = 10$ cm.

2 **a** ΔABC has $AB = 5$ cm, $BC = 6$ cm and $AC = 7$ cm. Calculate the size of the smallest angle in the triangle.

b ΔEGF has $EF = 5$ cm, $FG = 7$ cm and $EG = 10$ cm. Calculate the size of the largest angle in the triangle.

3 ΔABC has $b = 4\sqrt{3}$ cm, $c = 12$ cm and angle $A = 30°$. Prove the triangle is isosceles.

4 A parallelogram has diagonals 10 cm and 16 cm long and an angle of 42° between them. Calculate the lengths of its sides.

5 **a** ΔDEF has sides $d = 3x$, $e = x + 2$ and $f = 2x + 1$. If angle $D = 60°$, show that the triangle is equilateral. Calculate its area as a surd.

b ΔPQR has an area of $\frac{3}{4}$ m². If $p = x + 1$, $q = 2x + 1$ and angle $R = 30°$, what is the value of x?

6 **a** Find two different expressions for the height h using ΔACP and ΔBCP

Hence, prove the sine rule. Also prove that the area of $\Delta ABC = \frac{1}{2}bc\sin A = \frac{1}{2}ca\sin B$

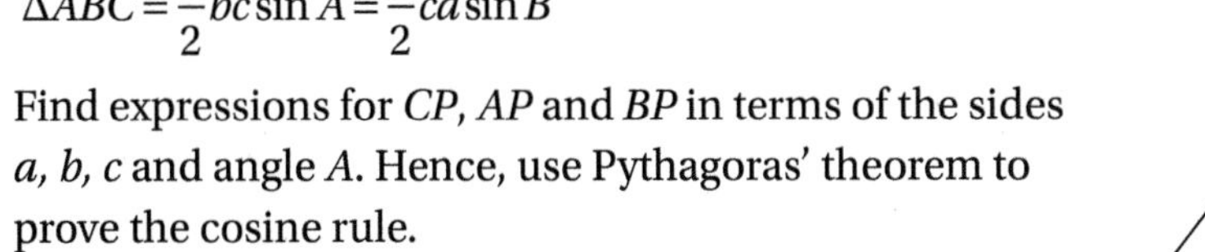

b Find expressions for CP, AP and BP in terms of the sides a, b, c and angle A. Hence, use Pythagoras' theorem to prove the cosine rule.

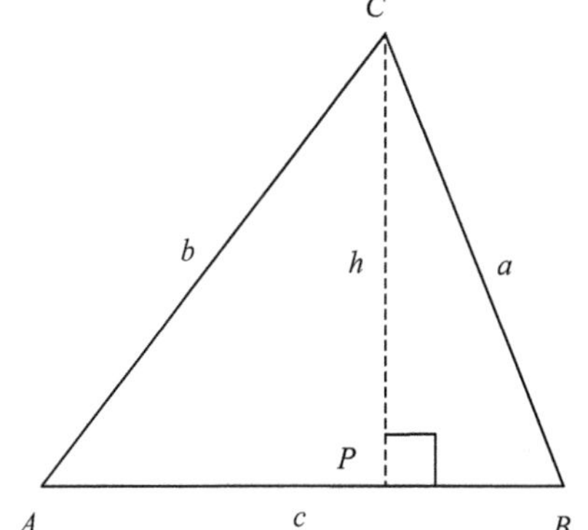

7 In ΔXYZ, $y = 2$ cm, $z = 2\sqrt{3}$ cm and angle $Y = 30°$. Prove that there are two triangles that satisfy this data and prove that one is isosceles and the other is right-angled.

8 The side opposite the smallest angle in a triangle is 8 cm long. If the angles are in the ratio 5 : 10 : 21, find the length of the other two sides.

9 Two circles, radii 7 cm and 9 cm, intersect with centres 11 cm apart. What is the length of their common chord?

10 The circumcircle of ΔABC has centre O and radius r, as shown in the diagram. Point P is the foot of the perpendicular from O to BC. Consider ΔBOP and prove that $2r = \frac{a}{\sin A}$. Hence, prove the sine rule.

11 A triangle has base angles of 22.5° and 112.5°. Prove that the height of the triangle is half the length of the base.

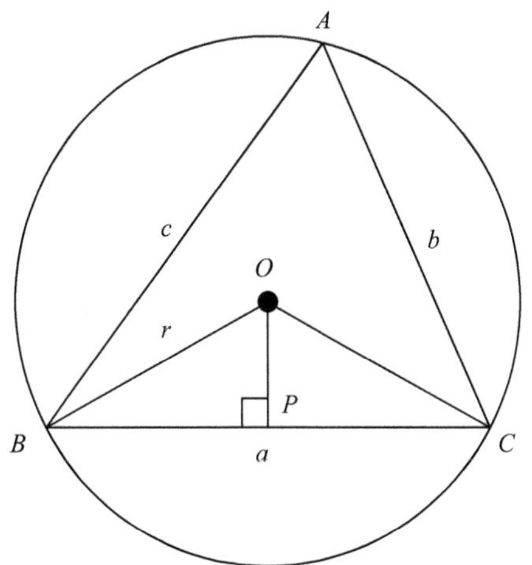

Challenge

12 In ΔXYZ, $x = n^2 - 1$, $y = n^2 - n + 1$ and $z = n^2 - 2n$
Prove that angle $Y = 60°$

3 Summary and review

Chapter summary

- Sine, cosine and tangent are periodic functions. Their graphs have line and rotational symmetry.
- The sine, cosine and tangent of any angle can be expressed in terms of an acute angle.
- The sign and size of the sine, cosine or tangent of any angle can also be found using a sketch graph of the function.
- The two identities $\sin^2\theta+\cos^2\theta\equiv 1$ and $\tan\theta\equiv\frac{\sin\theta}{\cos\theta}$ help you manipulate trigonometric expressions.
- The quadrant diagram can be used to find the sign and size of the sine, cosine or tangent of any angle. The mnemonic **CAST** helps you to remember in which quadrants the trigonometric ratios are positive.
- To solve a trigonometric equation, use identities to simplify it and then use a quadrant diagram or graph to find all possible angles.
- The sine and cosine rules are used to calculate unknown sides and angles in any triangle.
- The sine rule, $\frac{a}{\sin A}=\frac{b}{\sin B}=\frac{c}{\sin C}$, is used when you know a pair of opposite sides and angles.
- The cosine rule, $a^2=b^2+c^2-2bc\cos A$, is used when you know either two sides and the angle between them or all three sides.
- The area formula for a triangle, Area $=\frac{1}{2}ab\sin C$, uses two sides and the included angle.

Check and review

You should now be able to...	Try Questions
✔ Calculate the values of sine, cosine and tangent for angles of any size.	1, 3, 7
✔ Use the two identities $\sin^2\theta+\cos^2\theta\equiv 1$ and $\tan\theta\equiv\frac{\sin\theta}{\cos\theta}$, and recognise $x^2+y^2=r^2$ as the equation of a circle.	2, 6
✔ Sketch graphs of trigonometric functions and describe their main features.	4–5
✔ Solve various types of trigonometric equations.	8–12
✔ Use the sine and cosine rules and the area formula for a triangle.	13

1 Given that θ is acute, calculate the value of

a $\sin\theta$ and $\tan\theta$ when $\cos\theta=0.8$ b $\cos\theta$ and $\tan\theta$ when $\sin\theta=\frac{5}{13}$

2 Simplify a $\frac{\cos\theta\sqrt{1-\cos^2\theta}}{1-\sin^2\theta}$ b $\frac{\tan\theta(1-\sin^2\theta)}{\cos\theta}$

3 Express, in terms of acute angles,

a $\sin 190°$ b $\tan 260°$ c $\cos 140°$ d $\tan 318°$ e $\sin 371°$

f $\cos 480°$ g $\tan(-150°)$ h $\cos(-200°)$ i $\sin(-280°)$

4 Find the maximum value of y and the period for these functions, showing your working.

a $y = 4\sin x$ **b** $y = 5\sin 2x$ **c** $y = 6\cos 5x$

5 Sketch the graphs of $y = \sin x$ and $y = \cos x$ for $0° \le x \le 180°$ on the same axes.

a Use your graph to solve the equation $\sin x = \cos x$

b Solve the same equation algebraically to check your solutions.

6 **a** Show that the point $P(2\cos\theta, 2\sin\theta)$ lies on a circle and find its radius.

b Show that the point $Q(1, \sqrt{3})$ lies on the circle and write the value of θ at Q

7 Use the triangle to write these trig ratios in surd form.

a $\sin 135°$ **b** $\cos 225°$ **c** $\tan 315°$

d $\cos 405°$ **e** $\sin(-135°)$ **f** $\tan(-225°)$

8 Give all the values of θ in the range $-360°$ to $360°$ for which

a $\cos\theta = 0.7$ **b** $\tan\theta = 2.5$ **c** $\sin\theta = -0.5$

9 Solve these equations for $0° \le \theta \le 360°$, showing your working.

a $3\sin\theta = 2$ **b** $2\tan\theta = 7$ **c** $2\cos\theta + 1 = 0$ **d** $\cos\theta\tan\theta = -0.5$ **e** $\sin\theta\tan\theta = \frac{1}{4}$

10 Solve these equations for $-180° \le \theta \le 180°$. Show your working.

a $4\sin\theta = 3\cos\theta$ **b** $4\sin\theta = 3\tan\theta$ **c** $3\sin^2\theta = \tan\theta\cos\theta$

d $\sin(\theta - 20°) = \frac{\sqrt{3}}{2}$ **e** $\cos(\theta + 30°) = \frac{1}{2}$ **f** $\tan(\theta - 10°) = -1$

11 Solve these equations for $0° \le \theta \le 180°$, showing your working.

a $3\sin 2\theta = 1$ **b** $5\tan 2\theta - 2 = 0$ **c** $5\sin 3\theta - 1 = 0$

d $3\cos 3\theta - 2 = 0$ **e** $3\sin 2\theta - \cos 2\theta = 0$ **f** $2\sin\left(\frac{1}{2}\theta\right) - \cos\left(\frac{1}{2}\theta\right) = 0$

12 Solve these equations for $0° \le \theta \le 360°$

a $2\cos^2\theta + \sin\theta = 1$ **b** $\cos^2\theta + \cos\theta = \sin^2\theta$ **c** $6\sin^2\theta + 5\cos\theta = 5$

d $\tan^2\theta = 2 + \frac{1}{\cos\theta}$ **e** $1 + \sin\theta\cos^2\theta = \sin\theta$ **f** $4\sin^2\theta = 2 + \cos\theta$

13 Calculate the side BC, the angle E, and the area of each triangle.

a ΔABC where $AC = 8$ cm, Angle $A = 42°$ and Angle $B = 56°$

b ΔDEF where $DF = 6$ cm, $EF = 11$ cm and Angle $D = 124°$

What next?

Score		
	0 – 6	Your knowledge of this topic is still developing. To improve, search in MyMaths for the codes: 2045–2048, 2051–2053, 2257
	7 – 10	You're gaining a secure knowledge of this topic. To improve, look at the InvisiPen videos for Fluency and skills (03A)
	11 – 13	You've mastered these skills. Well done, you're ready to progress! To develop your techniques, look at the InvisiPen videos for Reasoning and problem-solving (03B)

Click these links in the digital book

3 Exploration

Going beyond the exams

History

Trigonometry, as we know it today, was largely developed between the 16th and 18th centuries. However, the foundations of trigonometry were laid as long ago as the 3rd century BC.

Hipparchus (190 BC – 120 BC), a Greek mathematician and astronomer regarded by many as the founder of trigonometry, constructed the first known trigonometric tables based on the lengths of chords in circles. Contributions to the early development of the theory were made by scholars from a number of countries, including Greece, Turkey, India, Egypt and China.

ICT

Use a spreadsheet to calculate and compare the values of $\sin(x + y)$ and $\sin x \cos y + \cos x \sin y$ for different values of x and y

x	y	$\sin(x + y)$	$\sin x \cos y + \cos x \sin y$

Use a 3D graph plotter to draw the graphs of $\sin(x + y)$ and $\sin x \cos y + \cos x \sin y$

What do you notice about the two graphs?

Have a go

The result $\sin(A + B) = \sin A \cos B + \cos A \sin B$ has been known in various forms since ancient times.

Prove that this result is the case, where A and B are acute angles, by considering the areas of the three triangles shown in the diagram.

Have a go

Adapt your proof from the other 'Have a go' for the case where one of the angles is acute and the other is obtuse.

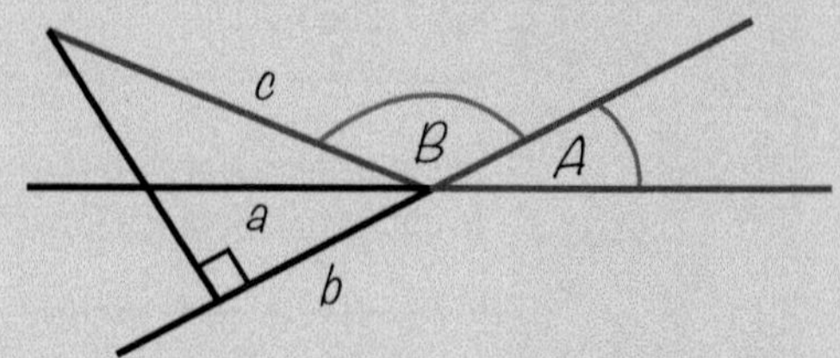

Assessment

1 a Solve $\cos x = 0.2$ for $0 \le x \le 360°$ Select the correct answer.

A $x = 78.5°$ B $x = 78.5°, 101.5°$ C $x = 78.5°, 281.5°$ D $x = 1.37°, 358.6°$ **[1 mark]**

b Solve $2\tan(x - 80°) = 3$ for $0 \le x \le 360°$. Select the correct answer.

A $x = 136.3, 316.3$ B $x = 160.5°, 340.5°$

C $x = 56.3°, 123.7°$ D $x = 56.3°, 236.3°$ **[1]**

2 For the curve with equation $y = \cos\frac{x}{2}$, select the coordinates of a

a maximum point,

A $(1, 0°)$ B $(0°, 1)$ and $(360°, 1)$ C $(1, 360°)$ D $(0°, 1)$ **[1]**

b minimum point.

A $(180°, -1)$ B $(-1, 180°)$ C $(360°, -1)$ D $(180°, 0)$ **[1]**

3 Sketch these graphs for $-180° \le x \le 180°$

a $y = \sin 2x$ **[3]**

b $y = \tan(x - 20°)$ **[4]**

On each diagram, show the coordinates where the curve crosses the x-axis and give the equations of any asymptotes.

4 $f(x) = \sin(x + 45°)$ for $0 \le x \le 360°$

a Sketch the graph of $y = f(x)$ and label the coordinates of intersection with the axes. **[3]**

b Write down the coordinates of the minimum and maximum points in this interval. **[3]**

c Solve the equation $\sin(x + 45°) = 0.3$ for x in the interval $0° \le x \le 360°$ **[4]**

5 The triangle DEF has $DE = 8$ m, $EF = 6$ m and $DF = 7$ m.

a Calculate the size of $\angle DEF$ **[3]**

b Calculate the area of the triangle. **[3]**

6 A triangle has side lengths 19 cm, 13 cm and x and angle 55° as shown.

Calculate the size of x **[3]**

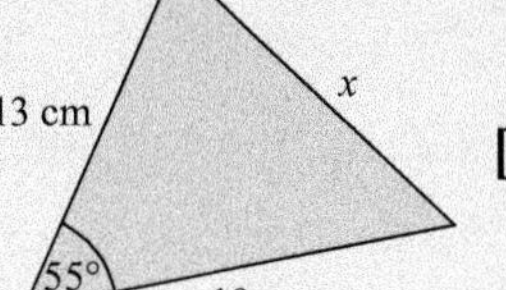

7 An equilateral triangle has area $\sqrt{3}$ square units.

Calculate the side lengths of the triangle. **[3]**

8 Solve the equations for x in the interval $-180° \le x \le 180°$

a $2\sin(x - 10) = -0.4$ **[4]**

b $\tan 3x = 0.7$ **[6]**

9 The curve C has equation $y = \tan(x - \alpha)$ with $-180° \le x \le 180°$ and $0° < \alpha < 45°$

a Sketch C and label the points of intersection with the x-axis. **[3]**

b Write down the equations of the asymptotes. **[2]**

c Solve the equation $\tan(x - \alpha) = \sqrt{3}$ for $-180° \le x \le 180°$ **[4]**

10 $f(x) = k\cos x$ where k is a positive constant.

a Sketch the graph of $y = f(x)$ for $0 \le x \le 360°$
Label the points of intersection with the coordinate axes. **[3]**

b Given that $k = 3$, solve the equation $k\cos x = \sin x$ for $0° \le x \le 360°$ **[4]**

11 $f(x) = \cos 2x$, $g(x) = 1 - \frac{x}{45}$

a Sketch $y = f(x)$ and $y = g(x)$ on the same axes for $0° \le x \le 360°$ **[5]**

b How many solutions are there to $f(x) = g(x)$? Justify your answer. **[1]**

12 A triangle ABC has $AB = 10$ cm, $BC = 16$ cm and $\angle BCA = 30°$

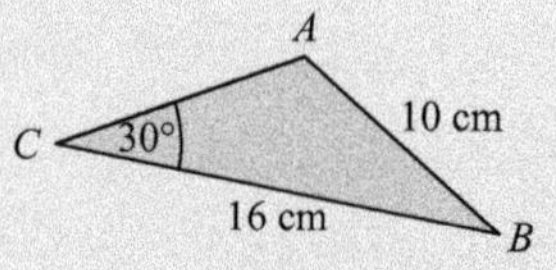

a Calculate the possible lengths of AC **[5]**

b What is the minimum possible area of the triangle? **[3]**

13 In the triangle CDE, $CD = 9$ cm, $CE = 14$ cm and $\angle CDE = 54°$

a Calculate the size of $\angle DEC$ **[3]**

b Explain why there is only one possible value of $\angle DEC$ **[2]**

14 In the triangle ABC, $BC = 9$ cm, $CA = 14$ cm and $\angle BCA = 46°$

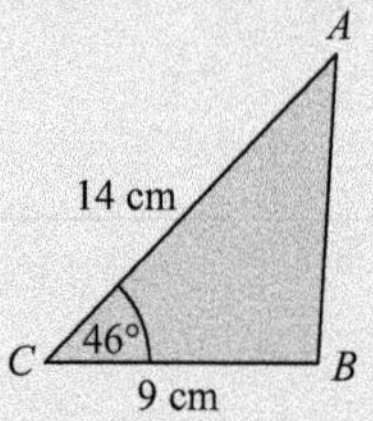

a Calculate the length of side AB **[3]**

b Calculate the size of the largest angle in the triangle. **[3]**

15 A triangle has side lengths 12 cm, 8 cm and 6 cm.
Calculate the size of the largest angle in the triangle. **[4]**

16 Solve the inequality $\sin x > \frac{\sqrt{2}}{2}$ for $0 \le x \le 360°$ **[4]**

17 a Show that $\cos\theta + \tan\theta\sin\theta \equiv \frac{1}{\cos\theta}$ **[3]**

b Hence solve $\cos\theta + \tan\theta\sin\theta = 2.5$ for $-180° \le \theta \le 180°$ **[3]**

18 Solve, for $0° \le x \le 360°$, the equations

a $\sin^2 x = 0.65$ **[5]**

b $\tan^2 x - 2\tan x - 3 = 0$ **[4]**

19 Solve, for $0 \le x \le 180°$, the equation $3\cos^2 2x = 2\sin 2x + 3$ **[8]**

20 $f(\theta) = 2\tan 3\theta + 5\sin 3\theta$
Find all the solutions to $f(\theta) = 0$ in the range $0° \le \theta \le 180°$ **[9]**

21 Given that $\sin x = \frac{1}{5}$ and x is acute, find the exact value of

a $\cos x$ **[2]** b $\tan x$ **[3]**

Give your answers in the form $a\sqrt{b}$ where a is rational and b is the smallest possible integer.

22 $2\cos^2\theta - k\sin\theta = 2 - k$

a Find the range of values of k for which the equation has no solutions. **[6]**

b Find the solutions in the range $0° \le \theta \le 360°$ when $k = 1$ **[4]**

23 Find an expression for $\tan\theta$ in terms of α, given that θ is acute and $\sin\theta = \alpha$ **[3]**

24 Solve the equation $\sin^4 x - 5\cos^2 x + 1 = 0$ for $0° \le x \le 360°$ **[8]**

25 The area of an isosceles triangle is 400 cm^2
Calculate the perimeter of the triangle given that one of the angles is 150° **[7]**

4 Differentiation and integration

Differentiation enables us to calculate rates of change. This is very useful for finding expressions for displacement, velocity and acceleration. For example, an expression for the displacement of an airplane in the sky informs us of the distance and direction of the airplane from its original position, at a given time. The first derivative of this expression gives the rate of change of displacement, that is, the velocity. The second derivative gives an expression for the rate of change of velocity, that is, the acceleration.

Differentiation and integration (which can be considered as reverse differentiation) belong to a branch of mathematics called calculus. Calculus is the study of change, and it's a powerful tool in modelling real-world situations. Calculus has many applications in a variety of fields including quantum mechanics, thermodynamics, engineering and economics, in modelling growth and movement.

Orientation

What you need to know	What you will learn	What this leads to
Ch1.6 Lines and circles (p.26) • The equation of a straight line where m is the gradient. • When two lines are perpendicular to each other: $m_1 \times m_2 = -1$	• To differentiate from first principles. • To differentiate terms of the form ax^n • To calculate rates of change. • To work out and interpret equations, tangents, normals, turning points and second derivatives. • To work out the integral of a function, calculate definite integrals and use these to calculate the area under a curve.	**Ch7 Units and kinematics** Velocity and acceleration as rates of change. Acceleration as a second derivative. Area under a velocity-time graph.
Ch2 Polynomials and the binomial theorem (p.43) • Binomial expansion formula. • Curve sketching.		**Ch15 Differentiation** Points of inflection. The product and quotient rules. The chain rule.
		Ch16 Integration Integration by parts, by substitution and using partial fractions.

MyMaths Practise before you start: 2002, 2004, 2022, 2041

4 Bridging Unit

Topic A: Coordinate geometry

Bridging to Ch4.2, 4.6

See Bridging Unit 1

For a reminder of gradients

You can use your knowledge of graphs to describe how the gradient of a polynomial changes with x, and you can use a **gradient function** to get information about the corresponding polynomial.

You can easily determine from a sketch whether part of a curve has a positive or negative gradient.

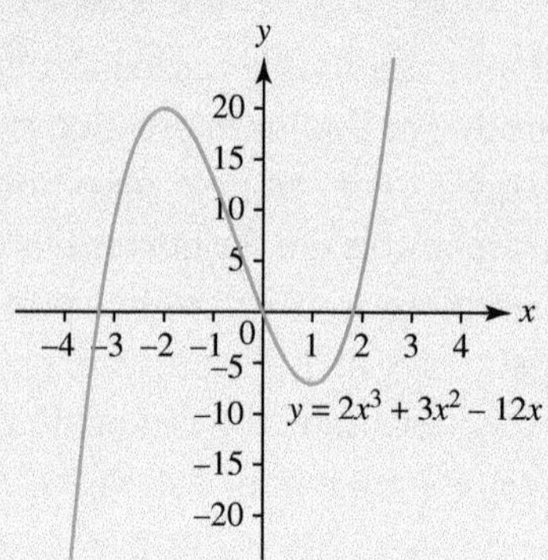

For example, this curve has a negative gradient when $-2<x<1$ and a positive gradient when $x<-2$ and $x>1$

When $x=-2$ and $x=1$ the gradient of the curve is zero, so the curve has a **maximum point** at $(-2, 20)$ and a **minimum point** at $(1, -7)$

You can estimate the gradient of the curve at a point by choosing two points (x_1, y_1) and (x_2, y_2) on the curve and using $\frac{y_2-y_1}{x_2-x_1}$

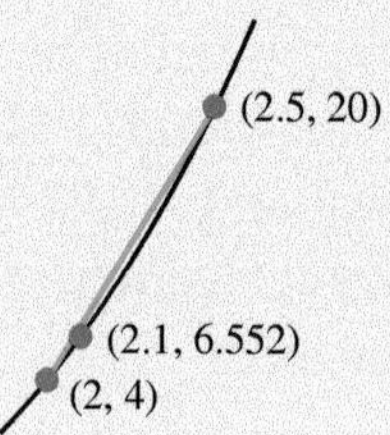

The two points should be close together since this will give a more accurate approximation. For example, to estimate the gradient at the point $(2, 4)$, take this point as (x_1, y_1)

If you chose $(2.5, 20)$ for (x_2, y_2) then this will give a gradient of

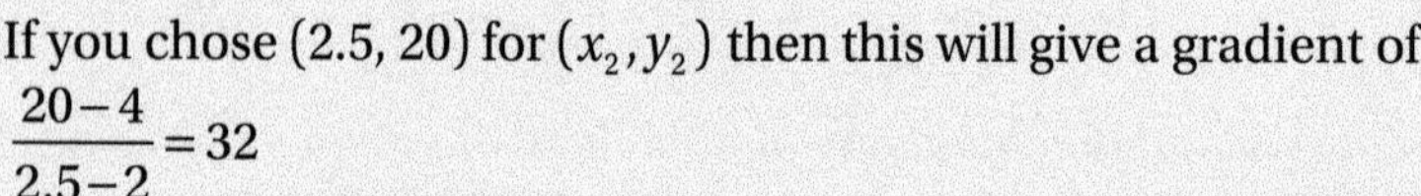

$$\frac{20-4}{2.5-2}=32$$

However, if you chose $(2.1, 6.552)$ for (x_2, y_2) then this will give a gradient of $\frac{6.552-4}{2.1-2}=25.52$ which is a more accurate estimate.

Example 1

For the graph shown,

a State the range of values of x for which the gradient is

i positive **ii** negative.

b State the coordinates of the point where the gradient is zero.

c Estimate the gradient of the curve at the point $(1, 0)$ using the chord

i from $(1, 0)$ to $(2, 7)$ **ii** from $(1, 0)$ to $(1.1, 0.61)$

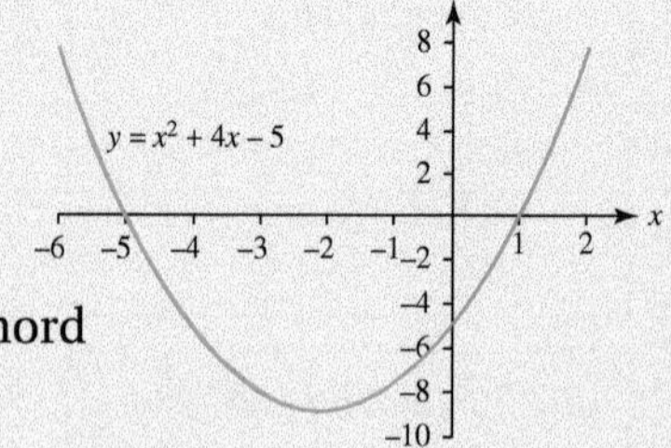

a i The curve has a positive gradient when $x>-2$

ii The curve has a negative gradient when $x<-2$

b The gradient of the curve is zero at the point $(-2, -9)$ — This is a minimum point.

c i Gradient of chord from $(1, 0)$ to $(2, 7)$ is $\frac{7-0}{2-1}=7$

ii Gradient of chord from $(1, 0)$ to $(1.1, 0.61)$ is $\frac{0.61-0}{1.1-1}=6.1$ — This is a more accurate approximation of the gradient of the curve at $(1, 0)$

Try It 1

For the graph shown,

a State the range of values of x for which the gradient is

i positive **ii** negative.

b State the coordinates of the point where the gradient is zero.

c Estimate the gradient of the curve at the point (5, 12) using the chord

i from (5, 12) to (6, 7) **ii** from (5, 12) to (5.5, 9.75)

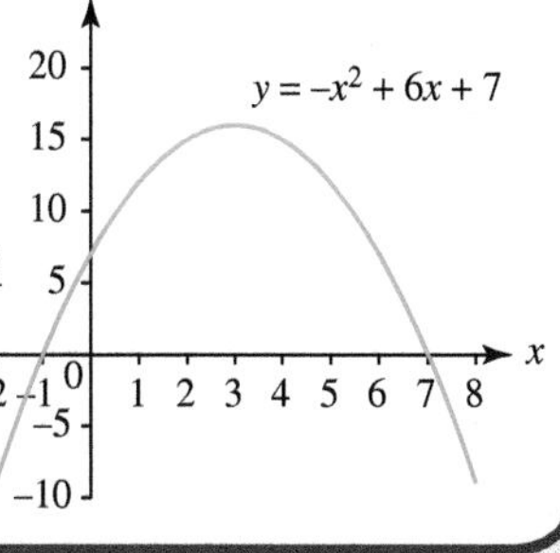

A curve $y=f(x)$ will have a gradient function called $f'(x)$. You can plot the graph of $y=f'(x)$ by estimating the gradient at several points on the curve $y = f(x)$.

Example 2

The graph of $y = f(x)$ is shown, where $f(x)=x^2+2x+3$

a State the gradient of the curve of $y=f(x)$ at the point (−1, 2)

b Use chords to estimate the gradient of the curve of $y=f(x)$ at the points (−2, 3), (0, 3) and (1, 6)

c Plot the graph of the gradient function $y=f'(x)$

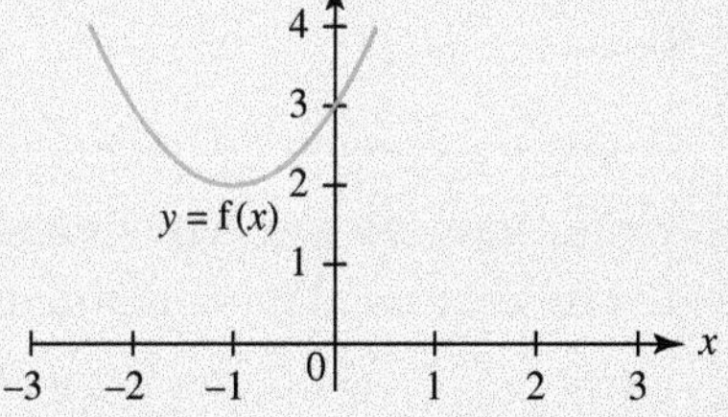

a (−1, 2) is a minimum point so the gradient is 0

b

x	Chord	Gradient
−2	(−2, 3) to (−1.9, 2.81)	$\frac{2.81-3}{-1.9--2}=-1.9$
0	(0, 3) to (0.1, 3.21)	$\frac{3.21-3}{0.1-0}=2.1$
1	(1, 6) to (1.1, 6.41)	$\frac{6.41-6}{1.1-1}=4.1$

Choose suitable chords to estimate the gradients.

c

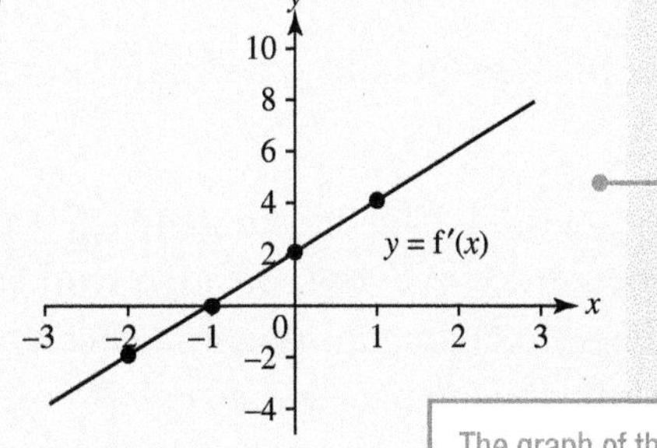

The graph of the gradient function is a straight line.

You can sense-check your answers by using the gradient function on your calculator.

Try It 2

$f(x)=5-x^2$

a State the gradient of the curve of $y=f(x)$ at the point (0, 5)

b Use chords to estimate the gradient of the curve of $y=f(x)$ at the points (−2, 1), (−1, 4) and (1, 4)

c Plot the graph of the gradient function $y=f'(x)$

MyMaths 2002, 2028 SEARCH

Example 3

$f(x)=\frac{1}{3}x^3+x$

You are given the value of the gradient at several points.

$f'(-3)=10, f'(-2)=5, f'(-1)=2, f'(1)=2, f'(2)=5, f'(3)=10$

Plot the gradient function $y=f'(x)$

The gradient function has a quadratic graph.

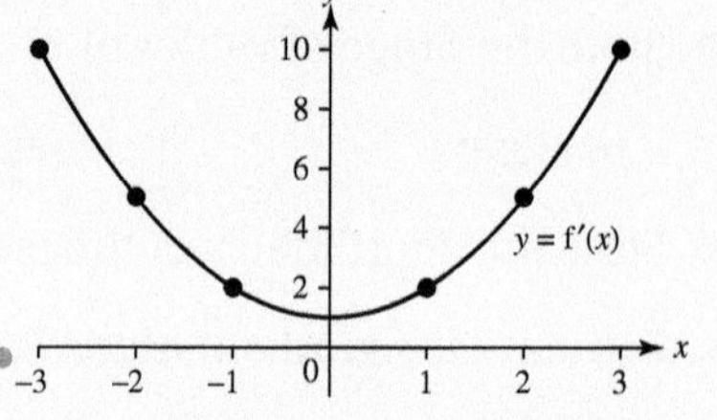

See Bridging Unit 2C For a reminder of curve sketching.

Try It 3

$f(x)=x^4+3$

The gradient of the curve $y=f(x)$ at various points is given in the table.

Use these points to plot the graph of the gradient function $y=f'(x)$

Point	Gradient
(–2, 19)	–32
(–1, 4)	–4
(0, 3)	0
(1, 4)	4
(2, 19)	32

You may have noticed a relationship between the order of the function and the order of its gradient function.

- The quadratic functions had linear gradient functions.
- The cubic function had a quadratic gradient function.
- The quartic function had a cubic gradient function.

Order of function	Order of gradient function
2	1
3	2
4	3

Key point

A polynomial function of order n will have a gradient function of order $n-1$

You can use the methods described to help you work out the gradient function of a curve using information about the curve. The converse, working out the equation of a curve using its gradient function, requires a bit more information. Each curve has only one, unique, gradient function but each gradient function might describe any one of an infinite set of curves. To know which curve the gradient function is describing, you need to know the coordinates of a point that lies on the curve.

For example, information about the gradient function might give you most of the equation of the curve: $y=x^2-3x+c$ If you also know it passes through the point (2, –1) then you can substitute for x and y and find the value of the constant c. In this case,

$-1=2^2-3(2)+c \Rightarrow c=1$ so $y=x^2-3x+1$

Example 4

Find the equation of the curve $y=x^3+3x^2-5x+c$ which passes through the point (–1, 5)

$5=(-1)^3+3(-1)^2-5(-1)+c$

Substitute $x=-1$ and $y=5$ into the equation.

$5=-1+3+5+c \quad \Rightarrow c=-2$

Therefore, the equation is $y=x^3+3x^2-5x-2$

Find the equation of the curve $y=x^3-2x^2+c$ which passes through (4, 16)

Try It 4

Bridging Exercise Topic A

Bridging to Ch4.2, 4.6

1 For the graph of $y=\frac{1}{4}x^2+\frac{1}{2}x-6$ shown,

a State the range of values of x for which the gradient is

i positive **ii** negative,

b Find the coordinates of the point where the gradient is zero,

c Estimate the gradient of the curve at the point (2, –4) by using the chord

i from (2, –4) to (3, –2.25) **ii** from (2, –4) to (2.2, –3.69)

2 For the graph of $y=\frac{1}{8}x^3-6x+10$ shown,

a State the range of values of x for which the gradient is

i positive **ii** negative,

b Find the coordinates of the points where the gradient is zero,

c Estimate the gradient of the curve at the point (0, 10) by using the chord

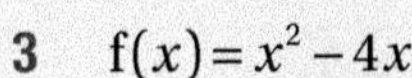

i from (0, 10) to (1, 4.125) **ii** from (0, 10) to $\left(\frac{1}{2}, \frac{449}{64}\right)$

3 $f(x)=x^2-4x$

a State the gradient of the curve of $y=f(x)$ at the point (2, –4)

b Estimate the gradient of the curve of $y=f(x)$ at the points (–1, 5), (0, 0) and (4, 0)

c Plot the graph of the gradient function $y=f'(x)$

4 State which of these graphs shows the gradient function of

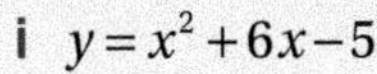

i $y=x^2+6x-5$ **ii** $y=x^4-4x^3+2$ **iii** $y=x^3-5$ **iv** $y=8+6x-x^2$

a

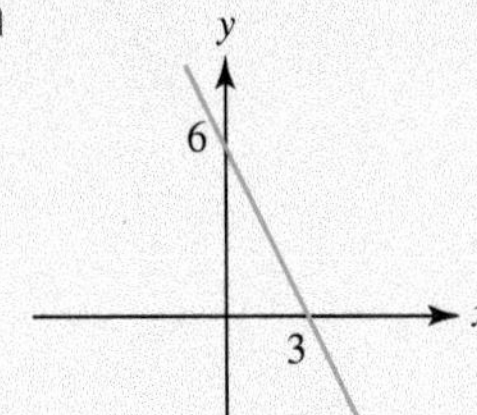

b

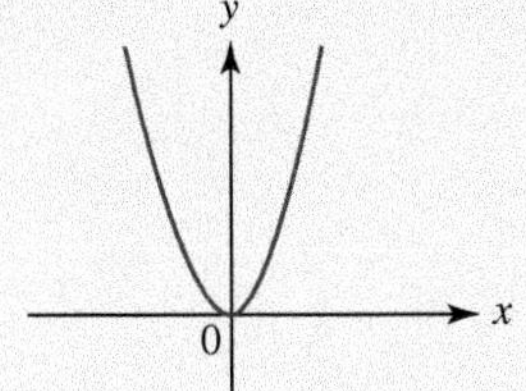

c

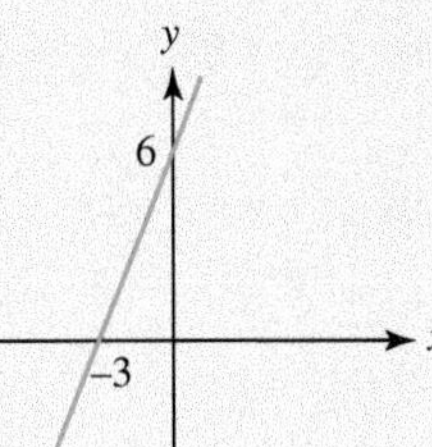

d

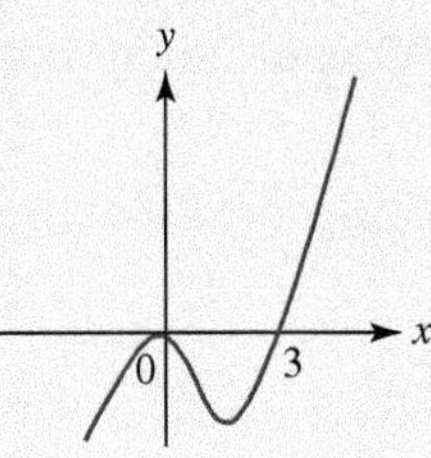

5 Find the equation of the curve $y=6x^2-8x+c$ which passes through (3, 12)

6 Find the equation of the curve $y=x^3+2x^2-3x+c$ which passes through (1, 3)

7 Find the equation of the curve $y=3x^3-7x^2+4x+c$ which passes through (–2, 4)

8 Given that $f(x)=x^3-5x+c$ and $f(2)=4$, find the value of c

9 Given that $f(x)=x^3+3x^2-2x+c$ and $f(-3)=8$, find the value of c

MyMaths 2002, 2028 SEARCH

4.1 Differentiation from first principles

Fluency and skills

When looking at the graph of a function, the gradient of its curve at any given point tells you the rate of change. Differentiation from first principles is a method of calculating the gradient and, therefore, the rate of change.

For example, you can work out the gradient of the function $y = x^2$ at the point $P(2, 4)$ using differentiation from first principles.

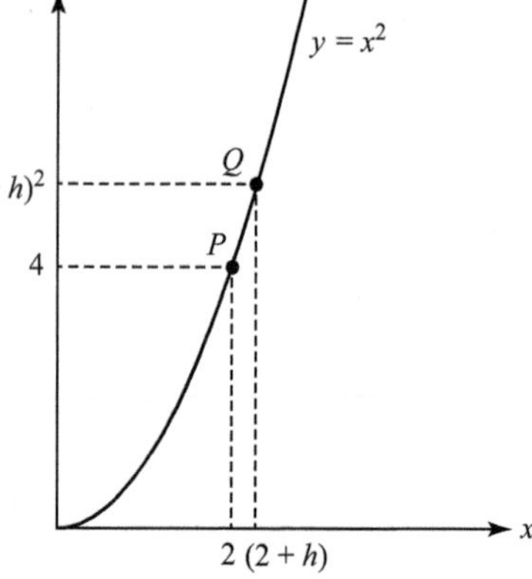

Define a point Q that lies on the curve, a tiny horizontal distance h from P, so that Q has coordinates $(2 + h, (2 + h)^2)$

PQ is the chord that connects the points.

The gradient of the chord PQ is given by

$$m_{PQ} = \frac{y_Q - y_P}{x_Q - x_P} = \frac{(2+h)^2 - 4}{(2+h) - 2}$$

$$= \frac{4 + 4h + h^2 - 4}{(2 + h) - 2} = \frac{4h + h^2}{h}$$

$$= 4 + h$$

As the distance between P and Q becomes very small, h very small and m_{PQ} approaches 4

Gradient at $P = 4$

ICT Resource online

To investigate gradients of chords for a graph, click this link in the digital book.

This method can be generalised for any function.

Consider the graph $y = f(x)$
Let the point P lie on the curve and have x-coordinate x
Its y-coordinate is then $f(x)$
Let the point Q also lie on the curve, h units to the right of P
Its coordinates are therefore $(x + h, f(x + h))$

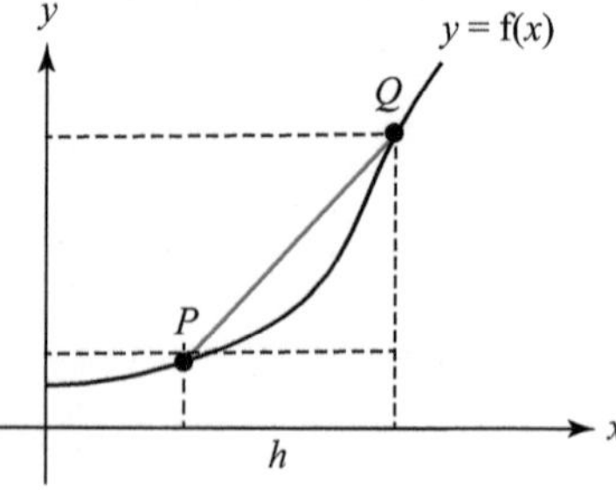

Key point

The **gradient** of PQ is given by

$$m_{PQ} = \frac{y_Q - y_P}{x_Q - x_P} = \frac{f(x+h) - f(x)}{(x+h) - x} = \frac{f(x+h) - f(x)}{h}$$

As h approaches 0, the point Q approaches P and the gradient of the chord PQ gets closer to the gradient of the curve at P

The gradient of the curve at P is defined as the **limiting value** of the gradient of PQ as h approaches 0. This limit is denoted by $f'(x)$ and is called the **derived function** or **derivative** of $f(x)$. See p.382 for a list of mathematical notation.

A limiting value, or **limit**, is a specific value that a function approaches or tends towards. "$\lim\limits_{h \to 0}$" followed by a function means the limit of the function as h tends to zero.

Key point

$$f'(x) = \lim_{h \to 0} \frac{f(x+h) - f(x)}{h}$$

Differentiation with this method is referred to as finding the derivative from **first principles**.

Example 1

Use differentiation from first principles to work out the derivative of $y = x^2$ and the gradient at the point $(3, 9)$

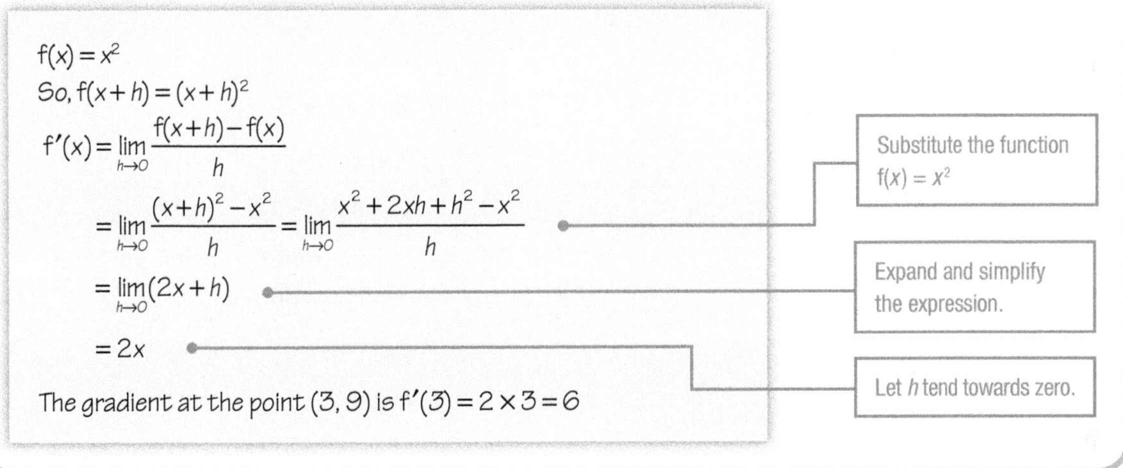

$f(x) = x^2$

So, $f(x+h) = (x+h)^2$

$f'(x) = \lim\limits_{h \to 0} \dfrac{f(x+h) - f(x)}{h}$

$= \lim\limits_{h \to 0} \dfrac{(x+h)^2 - x^2}{h} = \lim\limits_{h \to 0} \dfrac{x^2 + 2xh + h^2 - x^2}{h}$ — Substitute the function $f(x) = x^2$

$= \lim\limits_{h \to 0} (2x + h)$ — Expand and simplify the expression.

$= 2x$ — Let h tend towards zero.

The gradient at the point $(3, 9)$ is $f'(3) = 2 \times 3 = 6$

Derivatives give the rate of change and **constants** don't change. So a function multiplied by a constant will differentiate to give the derived function multiplied by the *same constant*.

If a function is itself a constant, then the derivative will be zero.

Key point

For a function $af(x)$, where a is a constant, the derived function is given by $af'(x)$

For a function $f(x) = a$, where a is a constant, the derived function $f'(x)$ is zero

Example 2

Differentiate **a** $3x^2$ **b** 7

a $f(x) = x^2, a = 3$ — $3x^2$ is in the form $af(x)$, where a is a constant.

So differentiating $3x^2$:

$3 \times 2x = 6x$ — From Example 1, $f'(x) = 2x$

b $f(x) = 7$ — $f(x) = a$, where a is a constant.

So $f'(x) = 0$

Exercise 4.1A Fluency and skills

1 Use the method shown in Example 1 to work out the gradient of these functions at the points given.

a $y=x^2$ at $x=3$ b $y=x^2$ at $x=4$

c $y=2x^2$ at $x=2$ d $y=5x^2$ at $x=1$

e $y=\frac{1}{2}x^2$ at $x=4$ f $y=\frac{3}{4}x^2$ at $x=10$

g $y=x^3$ at $x=1$ h $y=3x^3$ at $x=2$

2 Use the method of differentiation from first principles to work out the derivative and hence the gradient of the curve.

a $y=x^2$ at the point $(1, 1)$

b $y=3x^2$ at the point $(2, 12)$

c $y=x^3$ when $x=3$

d $y=4x-1$ at the point $(5, 19)$

e $y=\frac{1}{2}x^2$ at the point $(6, 18)$

f $y=x^2+1$ at the point $(2, 5)$

g $y=2x^3$ at the point $(1, 2)$

h $y=x^3+2$ at the point $(2, 10)$

3 Use differentiation from first principles to work out the gradient of the tangent to

a $y=2x^3$ at the point where $x=2$

b $y=3x^2+2$ at $(3, 29)$

c $y=\frac{x^2}{2}$ at the point $(4, 8)$

d $y=\frac{1}{2}x^2-4$ at the point $(2, -2)$

e $y=2x^2-1$ at the point $(-2, 7)$

f $y=x^2+x$ at the point where $x=1$

g $y=x^3+x$ at the point $(2, 10)$

4 Work out, from first principles, the derived function when

a $f(x)=2x^2$ b $f(x)=4x^2$

c $f(x)=6x^2$ d $f(x)=\frac{1}{2}x^2$

e $f(x)=x^2+1$ f $f(x)=2x-3$

g $f(x)=\frac{1}{3}x^3$ h $f(x)=2x^3+1$

5 a Work out, from first principles, the derived function where

i $f(x)=x+x^2$

ii $f(x)=x^2+x+1$

iii $f(x)=x^2+x-5$

iv $f(x)=x^2+2x+3$

v $f(x)=x^2-3x-1$

vi $f(x)=2x^2+5x-3$

b Work out, from first principles, the derived function of

i $f(x)=6$ ii $f(x)=0$

iii $f(x)=-2$ iv $f(x)=\pi$

c Work out, from first principles, the derived function of

i $f(x)=x$ ii $f(x)=-x$

iii $f(x)=2x+1$ iv $f(x)=4-3x$

6 Differentiate

a $y=2x^3-3$ b $y=3x^4-2x^2$

c $y=2x^5$ d $y=1-x^3$

e $y=x^2-2x^3$ f $y=x-x^2+x^3$

Reasoning and problem-solving

Strategy

To solve problems involving differentiation from first principles

1. Substitute your function into the formula $f'(x)=\lim_{h\to 0}\frac{f(x+h)-f(x)}{h}$
2. Expand and simplify the expression.
3. Let h tend towards 0 and write down the limit of the expression, $f'(x)$
4. Find the value of the gradient at a point (a, b) on the curve by evaluating $f'(a)$

Example 3

At which point on the curve $y = 3x^2$ is the gradient equal to 18?

$f(x) = 3x^2$ so, $f'(x) = \lim_{h \to 0} \frac{3(x+h)^2 - 3x^2}{h}$ — 1 Apply the algebra.

$= \lim_{h \to 0} \frac{3x^2 + 6xh + 3h^2 - 3x^2}{h} = \lim_{h \to 0}(6x + 3h)$

$f'(x) = 6x$ — 2 Find the derivative.

$f'(x) = 18$ So, $6x = 18$ $x = 3$ — 3 Use the derivative and the initial conditions to form an equation.

If $x = 3$ then $y = f(3) = 3 \times 3^2 = 27$

The gradient is 18 at the point (3, 27) — 4 Solve the equation you formed.

Exercise 4.1B Reasoning and problem-solving

1 At which point on the curve $y = 5x^2$ does the gradient take the value given? Show your working.

a 20

b 100

c 0.5

d Half of what it is at (2, 20)

e A third of what it is at (3, 45)

f Four times what it is at (1, 5)

2 At which **two** points on the curve $y = x^3$ does the gradient equal the value given? Show your working.

a 3 b 12 c 27

d 0.03 e $\frac{1}{3}$ f 1.47

3 For what value of x will these pairs of curves have the same gradient? Show your working.

a $y = x^2$ and $y = 2x$

b $y = 2x^2$ and $y = 12x$

c $y = 3x^2$ and $y = 15x$

d $y = ax^2$ and $y = bx$ where a and b are constants.

4 For what value(s) of x will these pairs of curves have the same gradient? Show your working.

a $y = x^3$ and $y = x^2 + 5x$

b $y = x^3$ and $y = 3x^2 + 9x$

c $y = x^3$ and $y = 2x^2 - x$

d $y = x^3$ and $y = 4x^2 - 4x$

e $y = x^2 + 3x + 1$ and $y = 7x + 1$

f $y = 2x^2 + x - 4$ and $y = 11x + 2$

g $y = x^2 - x$ and $y = 5x - 2$

h $y = 2x^3 + 3x^2$ and $y = 12x$

5 a Consider the derivatives of 1, x, x^2, x^3, x^4, and hence suggest a general rule about the derivative of x^n

b Consider the derivatives of 2, $4x$, $3x^2$, $5x^3$, $-2x^4$ and hence suggest a general rule about the derivative of ax^n

c Test your rule by considering the derivative of $2x^3$ and $3x^2$

d Can you say you have proved your rule?

Challenge

6 Can you show from first principles that

a The curve $y = \frac{1}{x}$ has a gradient of -4 when $x = \frac{1}{2}$

b The derivative of $\frac{1}{x}$ is $-\frac{1}{x^2}$?

4.2 Differentiating ax^n and Leibniz notation

Fluency and skills

You can use a simple rule to differentiate functions in the form ax^n

Key point

If $f(x) = ax^n$ then $f'(x) = nax^{n-1}$

If a function is a sum of two other functions, you can differentiate each function one at a time and then add the results.

Key point

If $h(x) = f(x) + g(x)$ then $h'(x) = f'(x) + g'(x)$

Isaac Newton is credited with having come up with the idea of Calculus first, but **Gottfried Leibniz**, a German mathematician and contemporary of Newton, also developed the concept and devised an alternative notation which is commonly used.

See p.382 For a list of mathematical notation.

For $y_Q - y_P$ he used the symbol δy and for $x_Q - x_P$ he used the symbol δx

So $\frac{y_Q - y_P}{x_Q - x_P} = \frac{\delta y}{\delta x}$ and he wrote that $\lim_{\delta x \to 0} \frac{\delta y}{\delta x} = \frac{dy}{dx}$

Key point

If $y = f(x)$ then $\frac{dy}{dx} = f'(x)$

Differentiating a function, or finding $\frac{dy}{dx}$, gives a formula for the gradient of the graph of the function at a point. This is also the gradient of the tangent to the curve at this point.

Try it on your calculator

You can use a calculator to evaluate the gradient of the tangent to a curve at a given point.

$\frac{d}{dx}(5X^2 - 2X)\Big|_{x=3}$

28

Activity

Find out how to calculate the gradient of the tangent to the curve $y = 5x^2 - 2x$ where $x = 3$ on *your* calculator.

Example 1

Differentiate $y = 3x^5 + 4x^2 + 2x + 3$

$y = 3x^5 + 4x^2 + 2x + 3 = 3x^5 + 4x^2 + 2x^1 + 3x^0$ — Write each term in the form ax^n

$\frac{dy}{dx} = 5 \times 3x^4 + 2 \times 4x^1 + 1 \times 2x^0 + 0 \times 3x^{-1}$ — Use $f'(x) = nax^{n-1}$ on each term.

$\frac{dy}{dx} = 15x^4 + 8x + 2$

Example 2

Work out the derived function when $f(x)=4+\frac{3}{x}+\sqrt{x}+2x^7$

$f(x)=4x^0+3x^{-1}+x^{\frac{1}{2}}+2x^7$ — Write each term in the form ax^n

$f'(x)=0\times4x^{-1}+(-1)\times3x^{-2}+\frac{1}{2}\times x^{-\frac{1}{2}}+7\times2x^6$ — Use $f'(x)=nax^{n-1}$ on each term.

$f'(x)=-3x^{-2}+\frac{1}{2}x^{-\frac{1}{2}}+14x^6$

$f'(x)=-\frac{3}{x^2}+\frac{1}{2\sqrt{x}}+14x^6$ — Simplify.

PURE

Exercise 4.2A Fluency and skills

1 Differentiate

a $3x^7$ b $6x^4$ c $5x$

d 8 e -4 f $\frac{1}{3}$

g $-2x^5$ h $-x^7$ i $2x^{-1}$

j $-x^{-3}$ k $x^{\frac{1}{2}}$ l $x^{\frac{2}{3}}$

m $5x^{\frac{5}{3}}$ n $-x^{-\frac{1}{4}}$ o 0

2 Work out the derivative of

a $\sqrt{x}$ b $\sqrt[3]{x}$ c $\sqrt[5]{x}$

d $\sqrt{x^3}$ e $\sqrt[3]{x^2}$ f $\sqrt[3]{3x}$

g $\frac{1}{x}$ h $\frac{3}{x^2}$ i $-\frac{6}{x^4}$

j $\frac{1}{\sqrt{x}}$ k $\frac{3}{\sqrt{x^3}}$ l $\frac{1}{\sqrt{5}}$

m π

3 Calculate $\frac{dy}{dx}$ when $y=$

a x^2+2x-3 b $1-2x-5x^2$

c x^3+2x^2-3x+1 d $x^2-x^4+\pi$

e $x-\frac{1}{x}$ f $3+x+\frac{2}{x^2}$

g $x^3-\frac{1}{x^3}+\frac{3}{x}+5$ h $\frac{10}{x}-1-\frac{x}{10}$

i $\sqrt{x}+\frac{1}{\sqrt{x}}$ j $x^5-\frac{5}{\sqrt[3]{x}}$

k $3+\frac{2}{\sqrt{x}}-\frac{5}{\sqrt[3]{x^2}}$ l $\frac{3}{\pi}-\frac{2}{x^2}-x^{\frac{3}{2}}$

4 For each question part **a** to **d**

i Find an expression in x for the gradient function,

ii Find the value of the gradient at the given point.

a Given that $f(x)=3x^2+4x-6$ work out the value of $f'(-2)$

b Given that $y=2x^3-5x^2-1$ work out the value of $\frac{dy}{dx}$ when $x=1$

c A curve has equation $y=10x+\frac{8}{x}$

Calculate the gradient of the curve at the point where $x=2$

d Calculate the gradient of the curve $y=\frac{8}{x}+\frac{4}{x^2}$ at the point $(2, 5)$

e i Expand $y=x(x-1)$

ii Hence evaluate $\frac{dy}{dx}$ when $x=4$ and $y=x(x-1)$

iii Hence state the gradient of the tangent to the curve at $(4, 12)$

5 Write an expression in x for $\frac{dy}{dx}$ and thus calculate the gradient of the tangent to each curve at the point given.

a $y=2x^2-5x+1$ at $(1, -2)$

b $y=1-5x+\frac{10}{x}$ at $(2, -4)$

c $y=x(2x+1)$ at $(-3, 15)$

d $y=\sqrt{x}+\frac{2}{\sqrt{x}}$ at $(4, 3)$

6 For each pair of functions, find which has the greater gradient at the given point.

a $y = x^2$ and $y = 20 - x$ at the point (4, 16)

b $y = x^2 + 3x - 8$ and $y = 6 - 2x$ at the point (2, 2)

c $y = 2x^2 + 13x - 18$ and $y = 2x + 3$ at the point (−7, −11)

d $y = 3x^2 - 5x - 2$ and $y = x^2 - 2x + 3$ at the point (−1, 6)

e $y = \sqrt{x}$ and $y = 2x - 15$ at the point (9, 3)

7 A curve is given by $y = x^3 + 5x^2 - 8x + 1$

Which of the following statements are true?

a $\frac{dy}{dx} > 4$ when $x = 1$

b $\frac{dy}{dx} < 0$ when $x = 0$

c $\frac{dy}{dx} = 0$ when $x = -4$

d $\frac{dy}{dx} > 0$ when $x = -2$

e $\frac{dy}{dx}$ when $x = -4$ is equal to $\frac{dy}{dx}$ when $x = \frac{2}{3}$

f $\frac{dy}{dx}$ when $x = -1$ is equal to $\frac{dy}{dx}$ when $x = 1$

Reasoning and problem-solving

Strategy

To solve differentiation problems involving polynomials with rational powers

1. Use the laws of algebra to make your expression the sum of terms of the form ax^n
2. Apply $f(x) = ax^n \Rightarrow f'(x) = nax^{n-1}$ to each term to find the derivative.
3. Substitute any numbers required and answer the question.

Example 3

Given that $f(x) = \frac{x+1}{\sqrt{x}}$, find the expression $f'(x)$ and hence find $f'(4)$

$f(x) = \frac{x+1}{\sqrt{x}}$

$= \frac{x}{x^{\frac{1}{2}}} + \frac{1}{x^{\frac{1}{2}}}$

$f(x) = x^{\frac{1}{2}} + x^{-\frac{1}{2}}$

$f'(x) = \frac{1}{2}x^{-\frac{1}{2}} - \frac{1}{2}x^{-\frac{3}{2}}$

$f'(x) = \frac{1}{2x^{\frac{1}{2}}} - \frac{1}{2x^{\frac{3}{2}}}$

$f'(x) = \frac{1}{2\sqrt{x}} - \frac{1}{2(\sqrt{x})^3}$

$f'(4) = \frac{1}{2\sqrt{4}} - \frac{1}{2(\sqrt{4})^3}$

$f'(4) = \frac{1}{4} - \frac{1}{16}$

$= \frac{3}{16}$

Write in index form.

1 Divide by $x^{\frac{1}{2}}$ to make your expression a sum of terms of the form ax^n

2 Apply $f'(x) = nax^{n-1}$ to each term.

3 Substitute $x = 4$

Exercise 4.2B Reasoning and problem-solving

1 Work out the derivative with respect to x of

a $f(x) = x(x+1)$

b $g(x) = (x-1)(x+1)$

c $h(x) = x^2(1-3x)$

d $k(x) = \sqrt{x}(x+3)$

e $m(x) = 3x(2x^2 + x - 3)$

f $n(x) = x(\sqrt{x} + 1)$

g $p(x) = x^{-1}(4 + 2x)$

h $q(x) = (x^{-1} - 2)(x + 1)$

i $r(x) = \frac{1}{x}(x^2 + 3x + 1)$

2 Given y, find $\frac{dy}{dx}$

a $y = x$

b $y = x + \sqrt{x}$

c $y = \frac{x + \sqrt{x}}{x}$

d $y = \frac{1 - x - 2\sqrt{x}}{x}$

e $y = \frac{x - x^2}{\sqrt{x}}$

f $y = \frac{1 + 2\sqrt{x}}{x}$

g $y = \frac{x^2 + 3x - 1}{x}$

h $y = \frac{1}{x} + \frac{1}{x^2}$

i $y = \frac{1}{\sqrt[3]{x}}$

3 Work out the derivative with respect to x of each of these functions.

a $f(x) = (x-3)(x+1)$

b $g(x) = (x-4)(x-5)$

c $h(x) = (1-x)(3+x)$

d $j(x) = (x+1)(2x+1)$

e $k(x) = x(x+1)(x-1)$

f $m(x) = (x+1)(\sqrt{x}+1)$

4 The sketch shows part of the curve $y = (x-1)(x-7)$ near the origin.

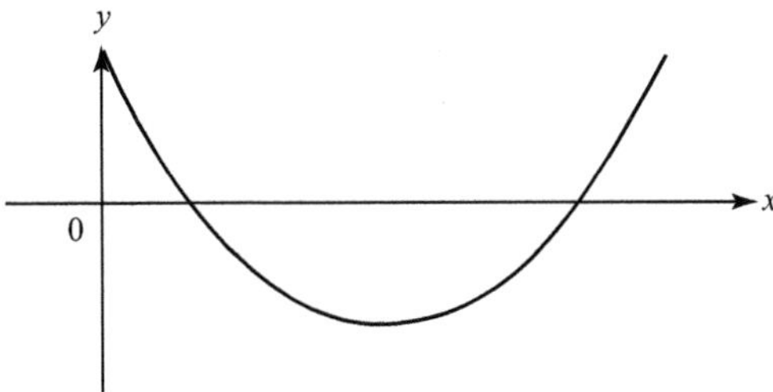

a Work out $\frac{dy}{dx}$

b Identify where the curve crosses the x-axis.

c Work out the point where the gradient of the curve is zero.

d i Where is $\frac{dy}{dx} < 0$?

ii Where is $\frac{dy}{dx} > 0$?

e Choose the correct word to make each statement true.

i When $\frac{dy}{dx} < 0$, the curve is (rising/falling) from left to right.

ii When $\frac{dy}{dx} > 0$, the curve is (rising/falling) from left to right.

5 The sketch shows the curve $y = \frac{10x^2 + 40}{x}$, $0.5 \le x \le 5.5$

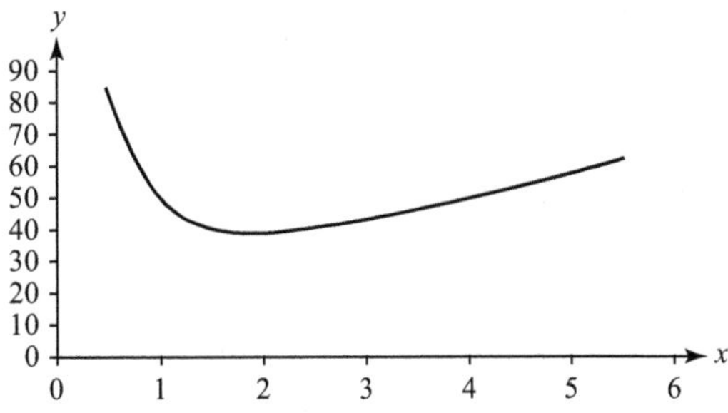

a Work out $\frac{dy}{dx}$

b i For which value of x is the gradient of the curve zero? Show your working.

ii What is the value of y at this point?

c For which values of x is the curve

i Decreasing, ii Increasing?

Challenge

6 Prove that the gradient of the function $f(x) = (x+1)^4$ at the point (1, 16) is 32

See Ch2.2

For a reminder on expanding brackets and the binomial theorem.

7 a Differentiate the function $f(x) = (x-1)(x+2)$

b Work out the gradient of the curve $y = f(x)$ at $x = 5$

c At which point is the gradient equal to zero?

d A line with equation $y = 2x - k$, where k is a constant, is a tangent to the curve.

i At what point does the tangent touch the curve?

ii What is the value of k?

4.3 Rates of change

Fluency and skills

When y is a function of x, the gradient of a graph of y against x tells you how the y-measurement is changing per unit x-measurement.

Key point

The rate of change of y with respect to x can be written $\frac{dy}{dx}$

See Ch7.2 For more on distance–time graphs.

The gradient of a distance-time graph is a measure of the rate of change of distance (r) with respect to time (t), this is called **velocity** (v).

Key point

$$v=\frac{dr}{dt}$$

If v metres per second represents velocity and t seconds represents time, then the gradient, $\frac{dv}{dt}$, is a measure of the rate at which velocity is changing with time, in metres per second per second.

Key point

The rate of change of velocity is called **acceleration**, $a=\frac{dv}{dt}$

See Ch7.4 For more on acceleration as a derivative.

Acceleration is the derivative of a derivative, which is called the **second derivative**.
A similar notation is used for the second derivative, $f''(x)$, as for the first derivative, $f'(x)$

Key point

$$y=f(x)\Rightarrow\frac{dy}{dx}=f'(x)\Rightarrow\frac{d^2y}{dx^2}=f''(x)$$

Example 1

A particle is moving on the y-axis such that its distance, r cm, from the origin is given by $r=t^3+2t^2+t$, where t is the time measured in seconds.

a Use the fact that the velocity $v=\frac{dr}{dt}$ to find an expression for the particle's velocity.

b Use the fact that the acceleration $a=\frac{dv}{dt}=\frac{d^2r}{dt^2}$ to find an expression for the particle's acceleration.

a $r=t^3+2t^2+t$ $\qquad \frac{dr}{dt}=3t^2+4t+1$ — Use $f'(t)=nat^{n-1}$

b $\frac{dr}{dt}=3t^2+4t+1$ $\qquad \frac{d^2r}{dt^2}=6t+4$

By differentiating a function and sketching a graph of the differential, you can find out whether

the function is increasing $\left(\frac{dy}{dx}>0\right)$,

the function is decreasing $\left(\frac{dy}{dx}<0\right)$,

or the function is neither increasing nor decreasing $\left(\frac{dy}{dx}=0\right)$, in which case you call it **stationary**.

A positive gradient means an increasing function. A negative gradient means a decreasing function.

PURE

Example 2

Consider the curve $y = 2x^3 - 3x^2 - 36x + 2$

a Work out $\frac{dy}{dx}$

b Use your equation from part **a** to sketch a graph of $\frac{dy}{dx}$. You must show your working.

c Using your sketch, determine where the value of y is increasing and decreasing.

d Work out where the value of y is stationary.

See Ch2.4 For a reminder on curve sketching.

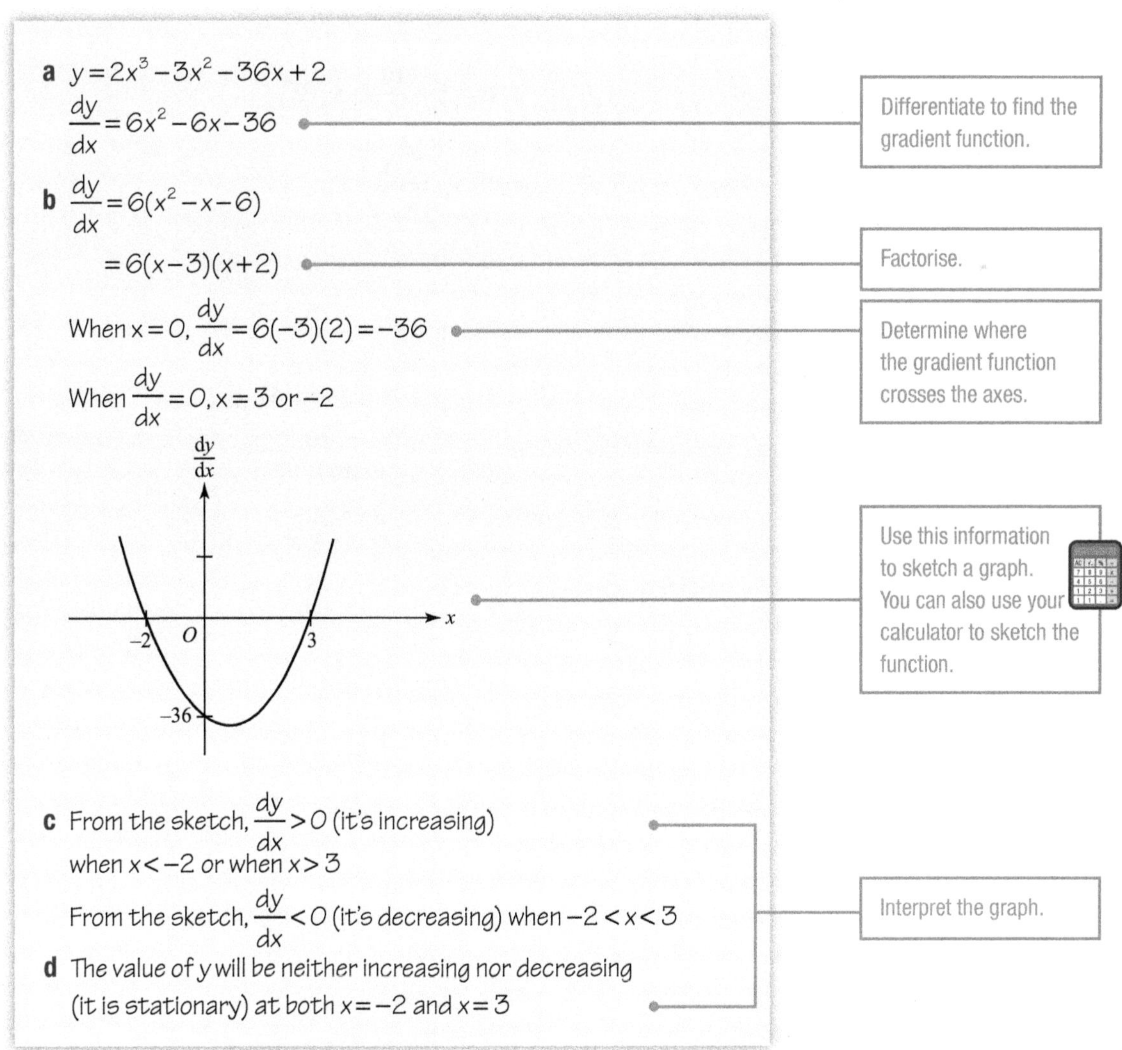

a $y = 2x^3 - 3x^2 - 36x + 2$
$\frac{dy}{dx} = 6x^2 - 6x - 36$
Differentiate to find the gradient function.
b $\frac{dy}{dx} = 6(x^2 - x - 6)$
$= 6(x-3)(x+2)$
Factorise.
When $x = 0$, $\frac{dy}{dx} = 6(-3)(2) = -36$
When $\frac{dy}{dx} = 0$, $x = 3$ or -2
Determine where the gradient function crosses the axes.
$\frac{dy}{dx}$
x
O
–2
3
–36
Use this information to sketch a graph. You can also use your calculator to sketch the function.
c From the sketch, $\frac{dy}{dx} > 0$ (it's increasing) when $x < -2$ or when $x > 3$
From the sketch, $\frac{dy}{dx} < 0$ (it's decreasing) when $-2 < x < 3$
d The value of y will be neither increasing nor decreasing (it is stationary) at both $x = -2$ and $x = 3$
Interpret the graph.

Calculator

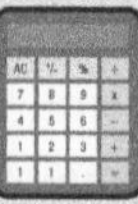

Try it on your calculator

You can use a calculator to sketch a function and its derivative.

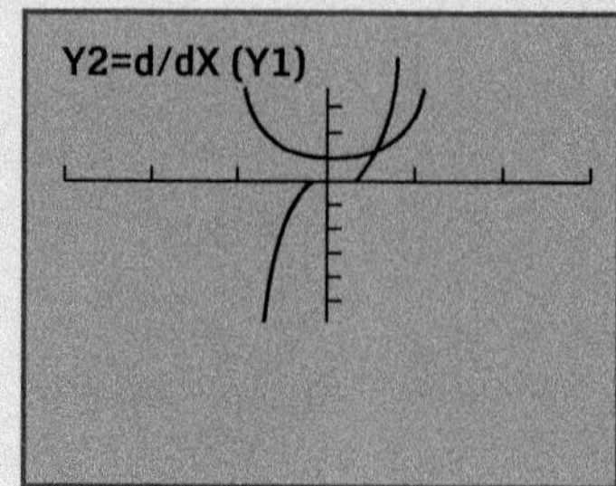

Activity

Find out how to sketch the curve $f(x) = 3x^4 - x^3 + 1$ and its derivative $f'(x)$ on *your* calculator.

Exercise 4.3A Fluency and skills

1 Calculate the rate of change of the following functions at the given points. You must show all your working.

- **a** $f(x)=x^2+10x$ at $x=4$
- **b** $g(x)=x^3+2x^2+x+1$ at $x=-1$
- **c** $h(x)=5x+6$ at $x=1$
- **d** $k(x)=x+\frac{1}{x}$ at $x=3$
- **e** $m(x)=9x^2+6x+1$ at $x=1$
- **f** $n(x)=x^{-1}+x^{-2}$ at $x=-2$
- **g** $p(x)=\frac{1}{\sqrt{x}}$ at $x=\frac{1}{9}$
- **h** $q(x)=x^{\frac{3}{2}}+1$ at $x=36$
- **i** $r(x)=x^4-8x^2$ at $x=-2$
- **j** $s(x)=\sqrt{x}-x$ at $x=4$
- **k** $t(x)=2\pi$ at $x=7$
- **l** $u(x)=4-3x$ at $x=-10$
- **m** $v(x)=\frac{1}{2x^4}$ at $x=-1$
- **n** $w(x)=1-3x-2x^2$ at $x=0$
- **o** $y(x)=\frac{162}{x^2}+2x^2$ at $x=3$
- **p** $z(x)=20\sqrt{x}+\frac{1000}{\sqrt{x}}$ at $x=25$

2 Work out the rate of change of the rate of change, $\left(\frac{d^2y}{dx^2}\right)$, of the following functions at the given points. You must show all your working.

- **a** $y=x^3+x$ at $x=3$
- **b** $y=\frac{10}{x}$ at $x=2$
- **c** $y=\frac{1}{\sqrt{x}}$ at $x=1$
- **d** $y=x^4-x^2$ at $x=-2$
- **e** $y=x^2-\frac{4}{x}$ at $x=2$
- **f** $y=x^3+4x^2+3x$ at $x=-1$
- **g** $y=\sqrt{x}$ at $x=0.25$
- **h** $y=3x-4\sqrt{x}$ at $x=0.16$
- **i** $y=4-ax$, where a is a constant, at $x=-3$
- **j** $y=x+2+\frac{1}{x}$ at $x=\frac{1}{2}$
- **k** $y=\frac{1}{x^4}$ at $x=0.5$
- **l** $y=1+x+\frac{x^2}{2}+\frac{x^3}{6}+\frac{x^4}{24}+\frac{x^5}{120}$ at $x=-2$

3 Using the gradient function of each curve, determine where the curve is

- **i** Stationary,
- **ii** Increasing,
- **iii** Decreasing.

a $y = x^2 + 4x - 12$

b $y = 3 - 5x - 2x^2$

c $y = x^2 - 1$

d $y = \frac{x^3}{3} - 4x + 1$

e $y = 1 + 21x - 2x^2 - \frac{1}{3}x^3$

f $y = \frac{x^3}{3} - \frac{3x^2}{2} + 2x - 1$

g $y = \frac{2}{3}x^{\frac{3}{2}}, x > 0$

h $y = 1 - \frac{1}{x}, x > 0$

i $y = \frac{x^3}{3}$

j $y = \frac{x^4}{4}$

4 By finding expressions for $\frac{dy}{dx}$, determine which of each pair of functions has the greater rate of change with respect to x at the given x-value.

a $y = x^2$ and $y = -x^2$ at $x = -1$

b $y = 5 - 3x$ and $y = \frac{1}{x}$ at $x = 0.5$

c $y = x^2 - 2$ and $y = x^2 - x - 2$ at $x = 3$

d $y = 1 + 10x - x^3$ and $y = x - x^2$ at $x = 2$

Reasoning and problem-solving

Strategy

To answer a problem involving rate of change

1. Read and understand the context, identifying any function or relationship.
2. Differentiate to get a formula for the relevant rate of change.
3. Evaluate under the given conditions.
4. Apply this to answer the initial question, being mindful of the context and units.

Example 3

A conical vat is filled with water. The volume in the vat at time t seconds is V ml.

V is a function of t such that the volume at time t is found by multiplying the cube of t by 50π

a Work out $V'(t)$, the rate at which the vat is filled, in millilitres per second.

b Calculate the rate at which the vat is filling with water after 3 seconds.

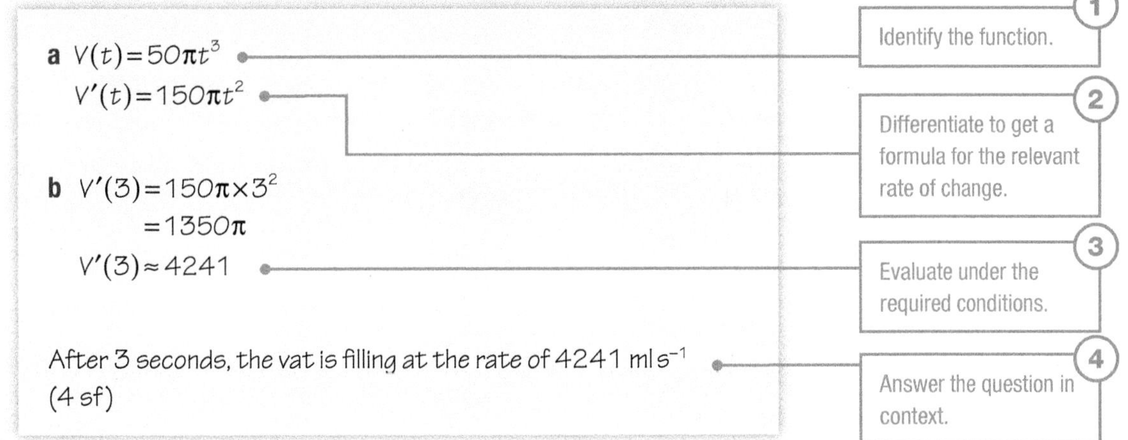

a $V(t) = 50\pi t^3$
$V'(t) = 150\pi t^2$
b $V'(3) = 150\pi \times 3^2$
$= 1350\pi$
$V'(3) \approx 4241$
After 3 seconds, the vat is filling at the rate of 4241 ml s^{-1} (4 sf)
1 Identify the function.
2 Differentiate to get a formula for the relevant rate of change.
3 Evaluate under the required conditions.
4 Answer the question in context.

Example 4

An object is thrown into the air. Its height after t seconds is given by $h = 1 + 30t - 5t^2$ where h is its height in metres.

a Write down an expression for the rate at which the object is climbing, in metres per second.

b Work out

i When the object is rising,
ii When the object is falling.

c After how many seconds does the object reach its maximum height?

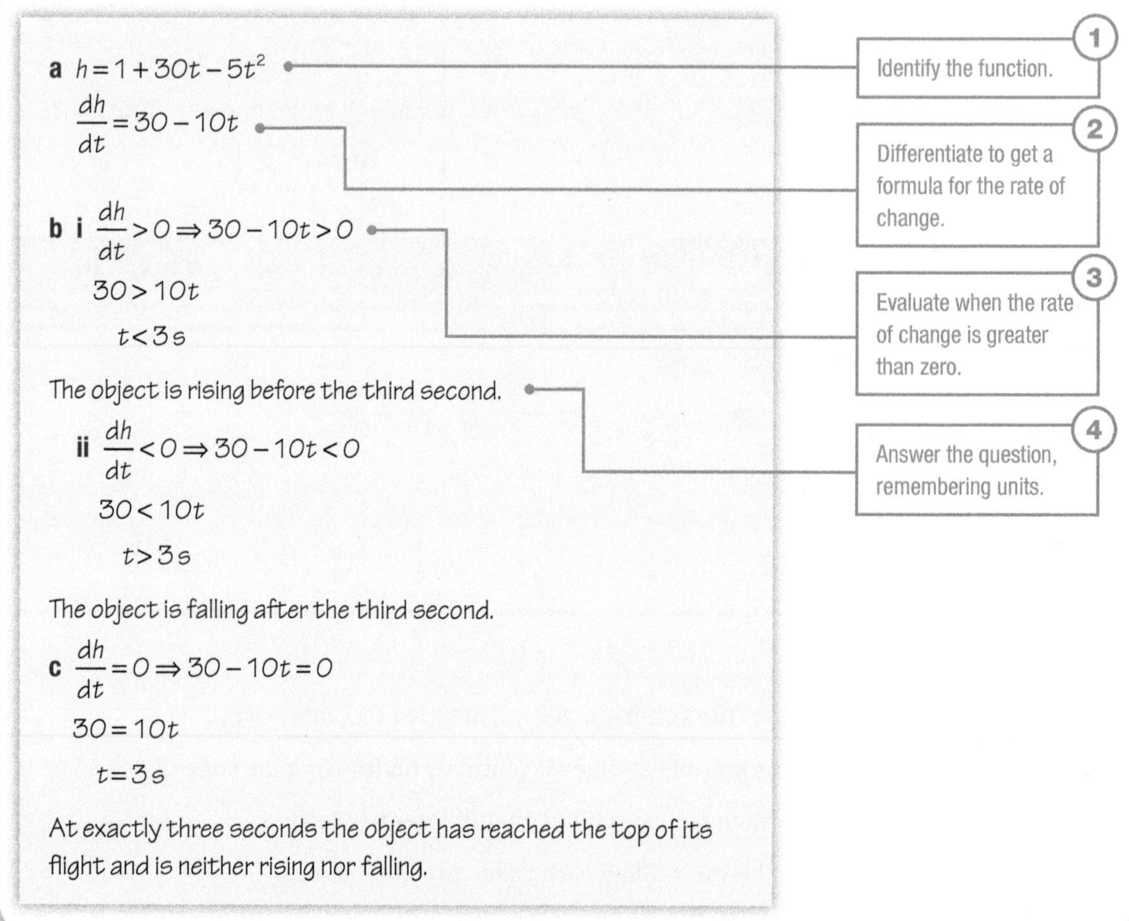

a $h = 1 + 30t - 5t^2$
$\frac{dh}{dt} = 30 - 10t$
b i $\frac{dh}{dt} > 0 \Rightarrow 30 - 10t > 0$
$30 > 10t$
$t < 3$ s
The object is rising before the third second.
ii $\frac{dh}{dt} < 0 \Rightarrow 30 - 10t < 0$
$30 < 10t$
$t > 3$ s
The object is falling after the third second.
c $\frac{dh}{dt} = 0 \Rightarrow 30 - 10t = 0$
$30 = 10t$
$t = 3$ s
At exactly three seconds the object has reached the top of its flight and is neither rising nor falling.
1 Identify the function.
2 Differentiate to get a formula for the rate of change.
3 Evaluate when the rate of change is greater than zero.
4 Answer the question, remembering units.

Exercise 4.3B Reasoning and problem-solving

1 A taxi driver charges passengers according to the formula $C(x) = 2.5x + 6$, where C is the cost in £ and x is the distance travelled in km.

Work out the rate in £ per km that the taxi driver charges.

2 A block of ice melts so that its dry mass M, at a time t minutes after it has been removed from the freezer, is given by $M(t) = 500 - 7.5t$

a Work out $\frac{dM}{dt}$, the rate at which the mass is changing.

b Your answer should have a negative sign in it. Explain why this is.

3 The volume of a sphere is $V=\frac{4}{3}\pi r^3$

The surface area of a sphere is $S=4\pi r^2$

A spherical bubble is expanding.

a Find an expression for the rate of change of the volume as r increases.

b Calculate the rate of change of the volume of the bubble when its radius is 3 cm.

c Calculate the rate of change of the surface area with respect to the radius when the radius is 3 cm.

4 A parabola has equation $y=35-2x-x^2$

a Work out the rate of change of y with respect to x when x is equal to

i -2 ii 2 iii 5

b When will this rate of change be zero?

5 A skydiver jumps from an ascending plane. His height, h m above the ground, is given by $h=4000+3t-4.9t^2$, where t seconds is the time since leaving the plane.

a Work out $\frac{dh}{dt}$, the rate at which the skydiver is falling.

b How fast is he falling after 5 seconds?

c How fast is he falling after 10 seconds?

d Calculate his acceleration at this time.

6 A golf ball was struck on the moon in 1974. Its height, h m, is modelled by $h=10t-1.62t^2$, where t seconds is the time since striking the ball.

a Calculate $\frac{dh}{dt}$, the rate at which the height of the ball is changing.

b After how much time will the ball be falling?

c Calculate the acceleration of the ball in the gravitational field of the moon.

7 A cistern fills from empty. A valve opens and the volume of water, V ml, in the cistern t seconds after the valve opens is given by $V=360t-6t^2$

a Write down an expression for the rate at which the cistern is filling after t seconds.

b Calculate the rate at which the cistern is filling after

i 10 seconds, ii 20 seconds.

c When the rate is zero, a ballcock shuts off the valve. At what time does this occur?

d What volume of water is in the tank when the valve closes?

8 For each function

i Work out an expression for the rate of change of y with respect to x

ii Evaluate this rate of change when $x=0$

iii By expressing the rate of change in the form $\frac{dy}{dx}=(x+a)^2+b$, establish that each function is increasing for all values of x

a $y=\frac{x^3}{3}+5x^2+30x+1$

b $y=\frac{x^3}{3}-3x^2+12x-4$

c $y=\frac{x^3}{3}-\frac{5}{2}x^2+8x+\frac{1}{2}$

9 The derivative of a function is

$\frac{dy}{dx}=8x-x^2-17$

Show that the function is always decreasing.

Challenge

10 A curve has equation

$y=x^4+\frac{1}{x^2}, x\geq 1$

a Work out i $\frac{dy}{dx}$ ii $\frac{d^2y}{dx^2}$

b Show that the gradient of the curve is an increasing function.

4.4 Tangents and normals

Fluency and skills

See Ch1.6
For a reminder on the equation of a straight line.

When lines with gradient m_1 and m_2 are **perpendicular** to each other, $m_1 \times m_2 = -1$

Key point

$m_1 = -\frac{1}{m_2}$ for perpendicular lines.

The **tangent** to the curve $y = f(x)$, which touches the curve at the point $(x, f(x))$, has the same gradient as the curve at that point, giving $m_T = f'(x)$

The **normal** to the curve $y = f(x)$, which passes through the point $(x, f(x))$, is perpendicular to the tangent at that point.

giving $m_N = -\frac{1}{m_T} = -\frac{1}{f'(x)}$

A line with gradient m passing through the point (a, b) has equation $(y - b) = m(x - a)$

Example 1

A curve has equation $y = 2x^2 - 3x - 10$

a Work out the equation of the tangent to the curve at the point (4, 10)

b Work out the equation of the normal to the curve when $x = -2$

a $y = 2x^2 - 3x - 10$ so $\frac{dy}{dx} = 4x - 3$ — Differentiate.

At the point (4, 10) the tangent has gradient

$\frac{dy}{dx} = 4 \times 4 - 3 = 13$ — Substitute.

The equation of the tangent is

$(y - 10) = 13(x - 4)$ — Use $(y - b) = m(x - a)$

$y = 13x - 42$

b When $x = -2$

$y = 2 \times (-2)^2 - 3 \times (-2) - 10$ — Substitute $x = -2$ into the original equation.

$y = 4$

So $(-2, 4)$ is a point on the normal.

At $(-2, 4)$ the tangent has gradient

$\frac{dy}{dx} = 4 \times (-2) - 3 = -11$

So the normal has a gradient of $\frac{1}{11}$ — Use $m_1 = -\frac{1}{m_2}$

The equation of the normal is

$(y - 4) = \frac{1}{11}(x + 2)$ — Use $(y - b) = m(x - a)$

$11y - x = 46$

Exercise 4.4A Fluency and skills

1 Work out the equation of the tangent to each of these curves at the given points. Show your working.

a $y = 2x^2 + 3x - 1$ at $(1, 4)$

b $y = 3 - x - x^2$ at $(-2, 1)$

c $y = 2x^3 + 3x^2$ at $(1, 5)$

d $y = 5x^2 - x^4$ at $(2, 4)$

e $y = \frac{3}{x}$ at $(3, 1)$

f $y = \frac{16}{x^2}$ at $(4, 1)$

g $y = \sqrt{x}$ at $(9, 3)$

h $y = \frac{25}{\sqrt{x}}$ at $(25, 5)$

i $y = \sqrt{x} + \frac{2}{\sqrt{x}}$ at $(4, 3)$

j $y = \frac{1}{\sqrt{x}} + \frac{1}{x}$ at $(1, 2)$

2 Work out the equation of the normal to each curve at the given points. Show your working.

a $y = x^2 + 2x - 7$ at $(2, 1)$

b $y = 4 - 5x - x^2$ at $(-3, 10)$

c $y = 3 - x^3$ at $(2, -5)$

d $y = x^4 + 2x^3 + x^2$ at $(1, 4)$

e $y = \frac{6}{x}$ at $(2, 3)$

f $y = \sqrt{x} + x$ at $(4, 6)$

g $y = \frac{1}{x} + \frac{1}{\sqrt{x}}$ at $(4, \frac{3}{4})$

h $y = \frac{1}{x^2} + \frac{\sqrt{x}}{x^2}$ at $(1, 2)$

3 A curve has equation $y = 3x^2$

a Work out the point on the curve with x-coordinate 3

b Work out the gradient of the curve at this point. Show your working.

c Work out the equation of the tangent to the curve at this point.

4 A line is a tangent to the curve $y = x^2 + 3x - 1$ at the point $(1, k)$

a What is the value of k?

b What is the gradient of the tangent at this point? Show your working.

c Work out the equation of the line.

5 A curve has equation $y = x^2 + 7x - 9$

a Calculate the point on this curve with x coordinate 1

b Calculate the gradient of the tangent to the curve at this point. Show your working.

c Hence state the gradient of the normal to the curve at this point.

d Work out the equation of the normal to the curve at this point.

6 A curve has equation $y = 2x + \frac{2}{x}$

a Calculate the point on the curve with x-coordinate 2

b Calculate the gradient of the tangent to the curve at this point. Show your working.

c Hence state the gradient of the normal to the curve at this point.

d Work out the equation of the normal to the curve at this point.

7 A parabola has equation $y = x^2 + 6x + 5$

a i Work out the gradient of the tangent at the point where $x = -3$. Show your working.

ii Give the equation of the tangent.

b Give the equation of the normal to the curve at this point.

c In a similar way, work out the equation of the tangent and normal to each curve at the given point. Show your working.

i $y = x^2 + 2x - 24$ at $x = -1$

ii $y = x^2 + 10x$ at $x = -5$

iii $y = 21 + 4x - x^2$ at $x = 2$

Reasoning and problem-solving

Strategy 1

To work out where a tangent or normal meets a curve

1. Differentiate the function for the curve.
2. Equate this to the gradient of the tangent or normal (remember $m_T = -\frac{1}{m_N}$).
3. Rearrange and solve for x
4. Substitute x in the function and solve for y

Example 2

The line $y = 3x + b$ is a tangent to the curve $y = 2x + 4\sqrt{x}$, $x > 0$

a Work out the point where the tangent meets the curve, thus find the value of the constant b

b Work out the equation of the normal to the curve at this point.

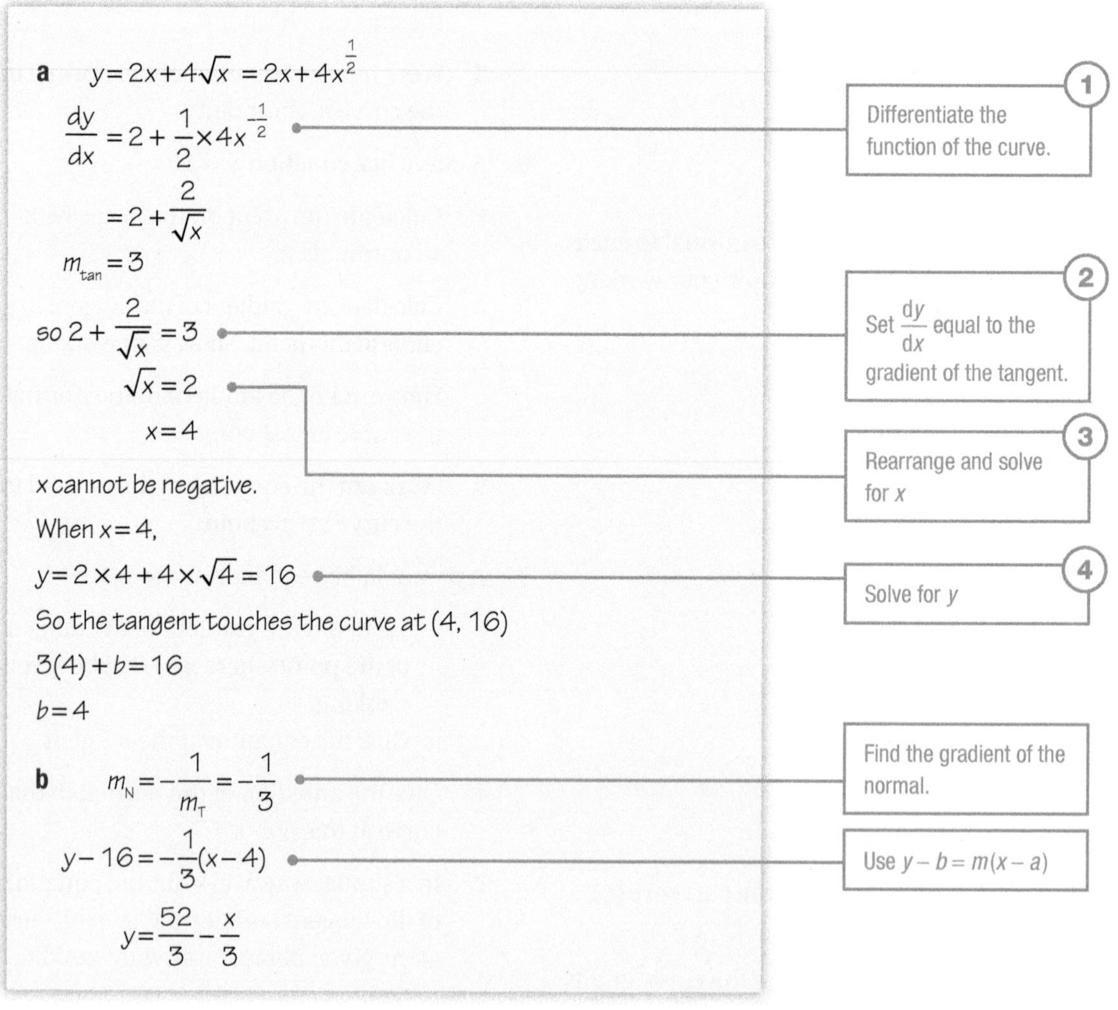

a $y = 2x + 4\sqrt{x} = 2x + 4x^{\frac{1}{2}}$

$\frac{dy}{dx} = 2 + \frac{1}{2} \times 4x^{-\frac{1}{2}}$ — Differentiate the function of the curve. (1)

$= 2 + \frac{2}{\sqrt{x}}$

$m_{tan} = 3$

so $2 + \frac{2}{\sqrt{x}} = 3$ — Set $\frac{dy}{dx}$ equal to the gradient of the tangent. (2)

$\sqrt{x} = 2$ — Rearrange and solve for x (3)

$x = 4$

x cannot be negative.

When $x = 4$,

$y = 2 \times 4 + 4 \times \sqrt{4} = 16$ — Solve for y (4)

So the tangent touches the curve at (4, 16)

$3(4) + b = 16$

$b = 4$

b $m_N = -\frac{1}{m_T} = -\frac{1}{3}$ — Find the gradient of the normal.

$y - 16 = -\frac{1}{3}(x - 4)$ — Use $y - b = m(x - a)$

$y = \frac{52}{3} - \frac{x}{3}$

PURE

Strategy 2

To work out the area bound between a tangent, a normal and the x-axis/y-axis

1. Work out the equation of the tangent and, from it, the equation of the normal.
2. Work out where each line crosses the required axis. Lines cut the x-axis when $y = 0$, and the y-axis when $x = 0$
3. Sketch the situation if required.
4. Use $A = \frac{1}{2} \times \text{base} \times \text{height}$, where the base is the length between the intercepts on the x-axis or y-axis and height is the y-coordinate or x-coordinate respectively.

Example 3

The point $T(1, 2)$ lies on the curve $y = x^3 + x$

Work out the triangular area trapped between the tangent and the normal to the curve at this point and the x-axis.

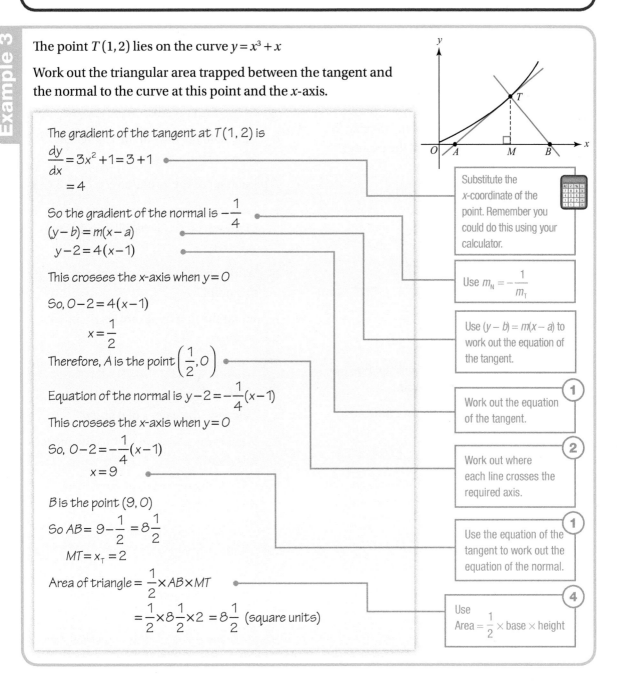

The gradient of the tangent at $T(1, 2)$ is

$\frac{dy}{dx} = 3x^2 + 1 = 3 + 1$

$= 4$ — Substitute the x-coordinate of the point. Remember you could do this using your calculator.

So the gradient of the normal is $-\frac{1}{4}$ — Use $m_N = -\frac{1}{m_T}$

$(y - b) = m(x - a)$ — Use $(y - b) = m(x - a)$ to work out the equation of the tangent.

$y - 2 = 4(x - 1)$ — (1) Work out the equation of the tangent.

This crosses the x-axis when $y = 0$

So, $0 - 2 = 4(x - 1)$

$x = \frac{1}{2}$

Therefore, A is the point $\left(\frac{1}{2}, 0\right)$ — (2) Work out where each line crosses the required axis.

Equation of the normal is $y - 2 = -\frac{1}{4}(x - 1)$

This crosses the x-axis when $y = 0$

So, $0 - 2 = -\frac{1}{4}(x - 1)$

$x = 9$ — (1) Use the equation of the tangent to work out the equation of the normal.

B is the point $(9, 0)$

So $AB = 9 - \frac{1}{2} = 8\frac{1}{2}$

$MT = x_T = 2$

Area of triangle $= \frac{1}{2} \times AB \times MT$ — (4) Use Area $= \frac{1}{2} \times \text{base} \times \text{height}$

$= \frac{1}{2} \times 8\frac{1}{2} \times 2 = 8\frac{1}{2}$ (square units)

Exercise 4.4B Reasoning and problem-solving

1 The line with equation $y=1-3x$ is a tangent to the curve $y=x^2-7x+k$ where k is a constant.

a Calculate the value of x at the point where the tangent meets the curve.

b Hence calculate the value of k

2 The curve $y=x+\frac{3}{x}, x>0$ has a normal that runs parallel to the line

$y=3x+4$

Work out the point where the normal crosses the curve at right angles.

3 a Work out the equation of the normal to the curve $y=x^2-3x+1$ at the point A, where $x=1$

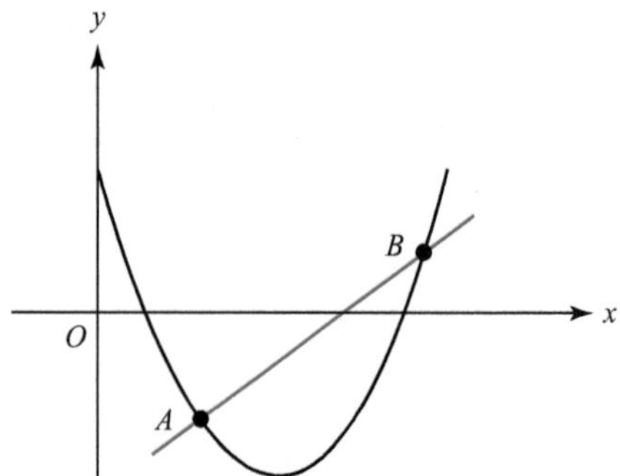

b Work out the point B where the normal at A crosses the curve again.

c Prove that the line AB is *not* normal to the curve at B

4 The curves $y=\frac{2x^2+1}{2}$ and $y=\frac{1+4x}{x}, x\neq 0$ intersect at the point (a, b). At this point, the line that is the tangent to one curve is the normal to the other line.

a Use two methods to work out an expression for the gradient of this line.

b Work out the point (a, b)

c Work out the equation of the line.

5 a Work out the equation of the tangent to the curve $y=2-\frac{9}{x}$ when $x=3$

b Work out the equation of the normal to the curve at the same point.

c Calculate the area of the triangle bound by the tangent, the normal and the y-axis.

6 A cubic curve $y=x^3+x^2+2x+1$ has a tangent at $x=0$

a Work out the equation of this tangent.

b Work out the coordinates of the point B where the tangent crosses the curve again.

c Work out the coordinates of the point C where the tangent at B crosses the x-axis.

d Calculate the area of the triangle BCO where O is the origin.

7 A parabola has the equation $y=2x^2-3x+1$

a Work out the equation of the tangent to the curve that is parallel to the line $y=5x$

b Work out the equation of the normal at this point.

8 A point in the first quadrant (p, q) lies on the curve $y=x^3+1$

The tangent at this point is perpendicular to the line $y=-\frac{x}{12}$

a Calculate the values of p and q

b What is the equation of the tangent at this point?

c What is the equation of the normal at this point?

9 A normal to the curve $y=x+\frac{18}{x}, x>0$, is parallel to $y=x$

a Work out the coordinates of the point where the normal crosses the curve at right angles.

b Work out the equation of the tangent at this point.

10 For each parabola

i Express the equation in the form $b \pm (x+a)^2$

ii Hence deduce the coordinates of the turning point on the parabola.

iii Work out the equation of the tangent and the normal at this point.

iv Comment on your answers.

a $y = x^2 + 6x - 1$

b $y = x^2 - 10x + 5$

c $y = x^2 - 3x - 7$

d $y = 4 - 8x - x^2$

e $y = x^2 + 4$

f $y = 1 - 2x - x^2$

11 A parabolic mirror has a cross-section as shown.

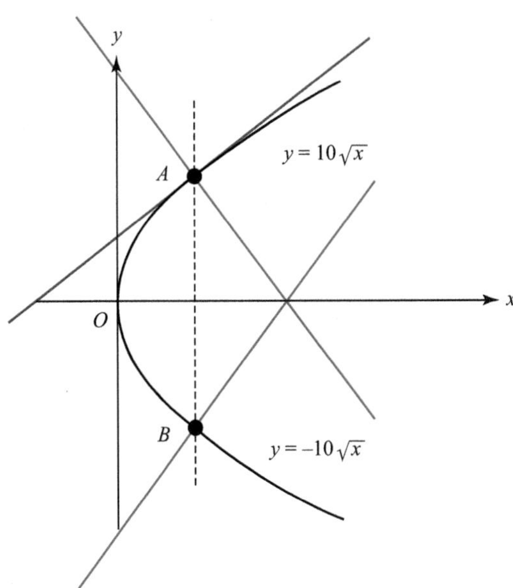

The branch above the x-axis has equation $y = 10\sqrt{x}$

The branch below it has equation $y = -10\sqrt{x}$

The line AB is parallel to the y-axis.
Let $x_A = x_B = p$

a Work out the equation of the normal to the curve

i At A ii At B

b Show that these two normals intersect on the x-axis.

12 The parabola $y = 4 + 2x - 2x^2$ crosses the y-axis at the point (p, q)

a State the values of p and q

b Work out the gradient of the tangent of the curve at this point.

c Work out the equation of the normal to this point.

d The curve crosses the x-axis at two points.

i Work out the coordinates of these points.

ii Work out the equation of the normal at both points.

e A related curve, $y = 4 - 2x^2$ crosses the y-axis at (0, 4)

Work out the equation of the normal to this curve at this point.

Challenge

13 a Work out the equation of the tangent to the curve $y = \frac{360}{x}$ $(x \geq 1)$ at the point where $x = 30$

b Work out the equation of the normal at the point where $x = 60$

c i At what point is the gradient of the normal $\frac{1}{10}$?

ii Give the equation of this normal.

d The line $y = -40x + k$ is a tangent to the curve.

i At what point does this line touch the curve?

ii What is the value of k?

MyMaths 2030 SEARCH

4.5 Turning points

Fluency and skills

When a curve changes from an increasing function to a decreasing function or vice versa, it passes through a point where it is stationary. This is called a **turning point** or **stationary point**.

Key point

At a turning point, the gradient of the tangent is zero. Therefore, you can work out the coordinates of the turning point by equating the derivative to zero.

A turning point is a stationary point, but a stationary point is not necessarily a turning point. You will learn about other types of stationary point in **Section 15.1**

Here are examples of a **maximum** turning point and a **minimum** turning point.

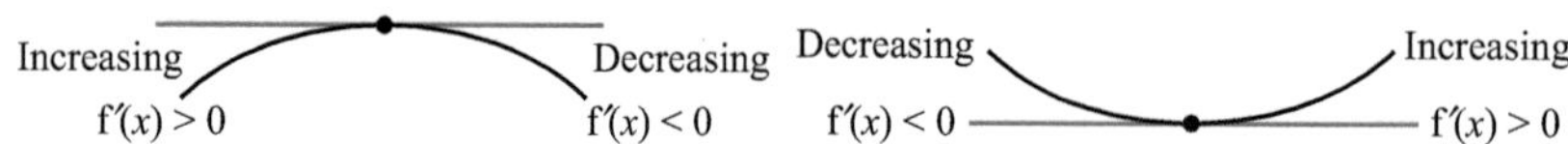

At a maximum turning point, as x increases, the gradient changes from positive through zero to negative.

At a minimum turning point, as x increases, the gradient changes from negative through zero to positive.

Example 1

Work out the coordinates of the turning point on the curve $y = x^2 + 4x - 12$ and determine its nature by inspection of the derivative either side of the point. Show your working.

$y = x^2 + 4x - 12$ so $\frac{dy}{dx} = 2x + 4$ — Differentiate.

At a turning point,

$\frac{dy}{dx} = 0 \Rightarrow 2x + 4 = 0 \Rightarrow x = -2$ — Find the value of x when the derivative is equal to zero.

$y = (-2)^2 + 4 \times (-2) - 12 \Rightarrow y = -16$ — Substitute into the original equation.

The turning point is at $(-2, -16)$

At $x = -2.1$, $\frac{dy}{dx} = 2(-2.1) + 4 = -0.2$

At $x = -1.9$, $\frac{dy}{dx} = 2(-1.9) + 4 = 0.2$ — Consider the value of the derivative either side of the turning point.

The gradient is increasing from negative to positive, so the point $(-2, -16)$ is a minimum turning point.

Calculator

PURE

Try it on your calculator

You can find turning points on a graphics calculator.

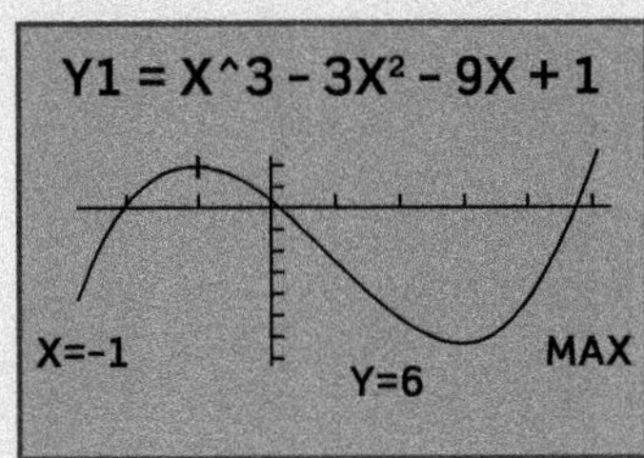

Activity

Find out how to find the turning points of $y=x^3-3x^2-9x+1$ on *your* graphics calculator.

As well as by inspection, the nature of a turning point can be determined by finding the second derivative with respect to x, $\frac{d^2y}{dx^2}$

If the gradient, $\frac{dy}{dx}$, is *decreasing* and the second derivative is negative, the turning point is a *maximum*.

Key point

At a *maximum* turning point, $\frac{d^2y}{dx^2}<0$

If the gradient, $\frac{dy}{dx}$, is *increasing* and the second derivative is positive, the turning point is a *minimum*.

Key point

At a *minimum* turning point, $\frac{d^2y}{dx^2}>0$

Example 2

Use calculus to work out the coordinates of the turning point on the curve $y=x+\frac{1}{x}, x>0$ and determine its nature.

$y=x+\frac{1}{x}$ so $\frac{dy}{dx}=1-\frac{1}{x^2}$

At a turning point, $\frac{dy}{dx}=0$ so $1-\frac{1}{x^2}=0$

Solving for x gives $x=\pm 1$

but $x>0$, so $x=1$

When $x=1$,

$y=1+\frac{1}{1}=2$ — Substitute into the original equation.

So the turning point is at $(1, 2)$

$\frac{dy}{dx}=1-\frac{1}{x^2}=1-x^{-2}$

$\frac{d^2y}{dx^2}=\frac{2}{x^3}$ — Find the second derivative to determine the nature of the turning point.

When $x=1$,

$\frac{d^2y}{dx^2}=\frac{2}{1^3}=2$

The second derivative is positive, so the turning point is a minimum.

MyMaths 2270 SEARCH

Exercise 4.5A Fluency and skills

1 For each curve, work out the coordinates of the stationary point(s) and determine their nature by inspection. Show your working.

a $y = x^2 + 4x - 5$ b $y = x^2 + 4x - 32$

c $y = x^2 - 6x - 7$ d $y = 1 - x^2$

e $y = 2x^2 + 7x + 6$ f $y = 20 - x - x^2$

g $y = 6x^2 - x - 1$ h $y = 2 - 13x - 7x^2$

i $y = 6 - x - 2x^2$ j $y = x + \frac{4}{x}, x > 0$

k $y = 2x + \frac{18}{x}, x \neq 0$

l $y = 10 - x - \frac{1}{x}, x \neq 0$

m $y = x - 10\sqrt{x}, x \geq 0$

n $y = x^2 - 32\sqrt{x}, x \geq 0$

2 The curve $y = x^3 - 6x^2$ has two turning points.

a Work out the coordinates of both turning points. Show your working.

b Use the second derivative to determine the nature of each.

3 Work out the coordinates of the turning points of $y = 2x^3 + 30x^2 + 1$ and determine their nature. Show your working.

4 $f(x) = 2x^3 - 9x^2 + 12x + 7$

a Differentiate $f(x)$

b The curve $y = f(x)$ has two turning points. Work out the coordinates of them both and determine their natures. Show your working.

c Repeat this process with the following functions.

i $f(x) = x^3 - 3x^2 - 24x + 1$

ii $f(x) = x^3 + 3x^2 - 45x - 45$

iii $f(x) = 1 - 36x - 21x^2 - 2x^3$

iv $f(x) = 2x^3 - 11x^2 - 8x + 2$

v $f(x) = 3 - 4x + 5x^2 - 2x^3$

vi $f(x) = 5 + x - 2x^2 - 4x^3$

5 The function $f(x) = 3x^4 + 8x^3 - 6x^2 - 24x - 1$ has three turning points.

a Show that stationary points can be found at $x = 1$, $x = -1$ and $x = -2$

b Work out the coordinates of the three stationary points.

c Use the second derivative to help you establish the nature of each.

6 The function $f(x) = 3x^4 - 4x^3 - 36x^2$ has three turning points. A sketch of the graph of the function is shown.

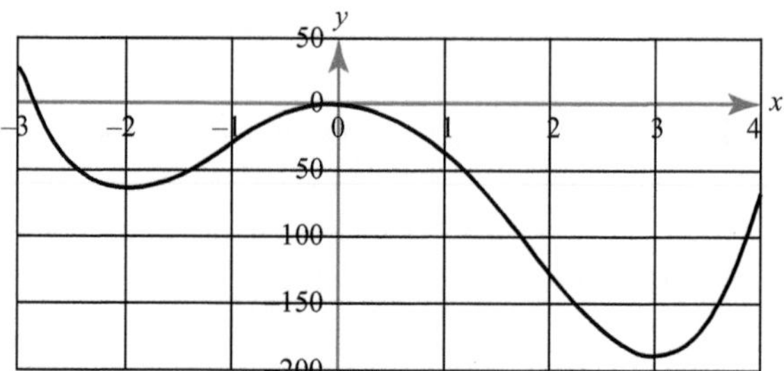

a The sketch suggests that these turning points can be found at $x = -2$, $x = 0$ and $x = 3$
Show that this is the case.

b Use the second derivative to verify the nature of each turning point.

7 A function is defined by $f(x) = 8x + \frac{72}{x}, x > 0$

a Show that the *only* stationary point on the curve $y = f(x)$ is at $x = 3$

b State the coordinates of the stationary point and determine its nature. Show your working.

c A related function is defined by $f(x) = 8x + \frac{72}{x}, x \neq 0$
It has a stationary point at $x = -3$
Determine the nature of this stationary point. Show your working.

Reasoning and problem-solving

Strategy 1

To identify the main features of a curve

1. Work out where it crosses the axes ($x = 0$ and $y = 0$)
2. Consider the behaviour of the curve as x tends to infinity and identify any asymptotes.
3. Work out the coordinates of the turning points $\left(\frac{dy}{dx} = 0\right)$ and determine their nature.
4. Use the information you have found to sketch the function.

See Ch2.4 For a reminder about sketching curves.

Example 3

$y = (x - 10)(x + 5)(x + 14)$

a Show that the curve crosses the y-axis at $(0, -700)$

b Show that there is a maximum turning point at $(-10, 400)$ and a minimum turning point at $(4, -972)$

c Sketch the curve.

a The curve crosses the y-axis when $x = 0$

So $y = -10 \times 5 \times 14$

$= -700$, giving $(0, -700)$ as a point on the curve. — *Find where the graph crosses the axes.* (1)

b $y = (x - 10)(x + 5)(x + 14)$

$y = x^3 + 9x^2 - 120x - 700$ — *Expand.*

$\frac{dy}{dx} = 3x^2 + 18x - 120 = 3(x - 4)(x + 10)$ — *Differentiate and factorise.*

Stationary points occur when $\frac{dy}{dx} = 0 \Rightarrow x = 4$ and $x = -10$

When $x = 4$

$y = (4 - 10)(4 + 5)(4 + 14) = -972$, giving $(4, -972)$

When $x = -10$

$y = (-10 - 10)(-10 + 5)(-10 + 14) = 400$, giving $(-10, 400)$ — *Work out the coordinates of the turning points.* (3)

$\frac{dy}{dx} = 3x^2 + 18x - 120$

So $\frac{d^2y}{dx^2} = 6x + 18$ — *Find the second derivative.* (3)

When $x = 4$, the second differential is positive so $(4, -972)$ is a minimum turning point.

When $x = -10$, it is negative so $(-10, 400)$ is a maximum turning point. — *Use the second derivative to determine the nature of the turning points.* (3)

c

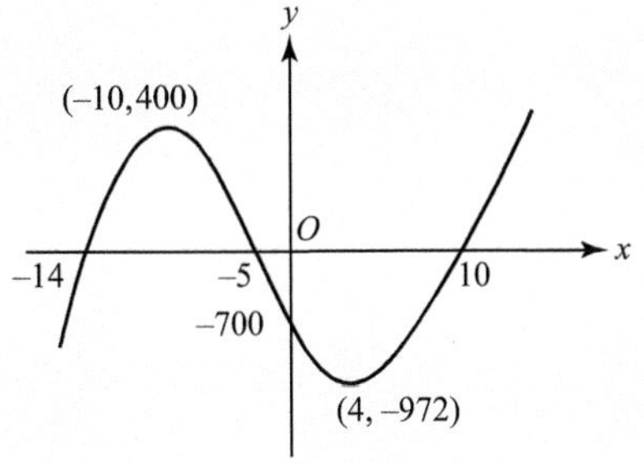

Use this information to sketch the function. You could use your graphics calculator to check your sketch. (4)

Strategy 2

To optimise a given situation

1. Express the dependent variable (say y) as a function of the independent variable (say x).
2. Differentiate y with respect to x
3. Let the derivative be zero and find the value of x that optimises the value of y
4. Examine the nature of the turning points to decide if it is a maximum or minimum.
5. Put your turning point in the context of the question.

Example 4

A piece of rope 120 m long is to be used to draw out a rectangle on the ground.

What is the biggest area that can be enclosed in the rectangle?

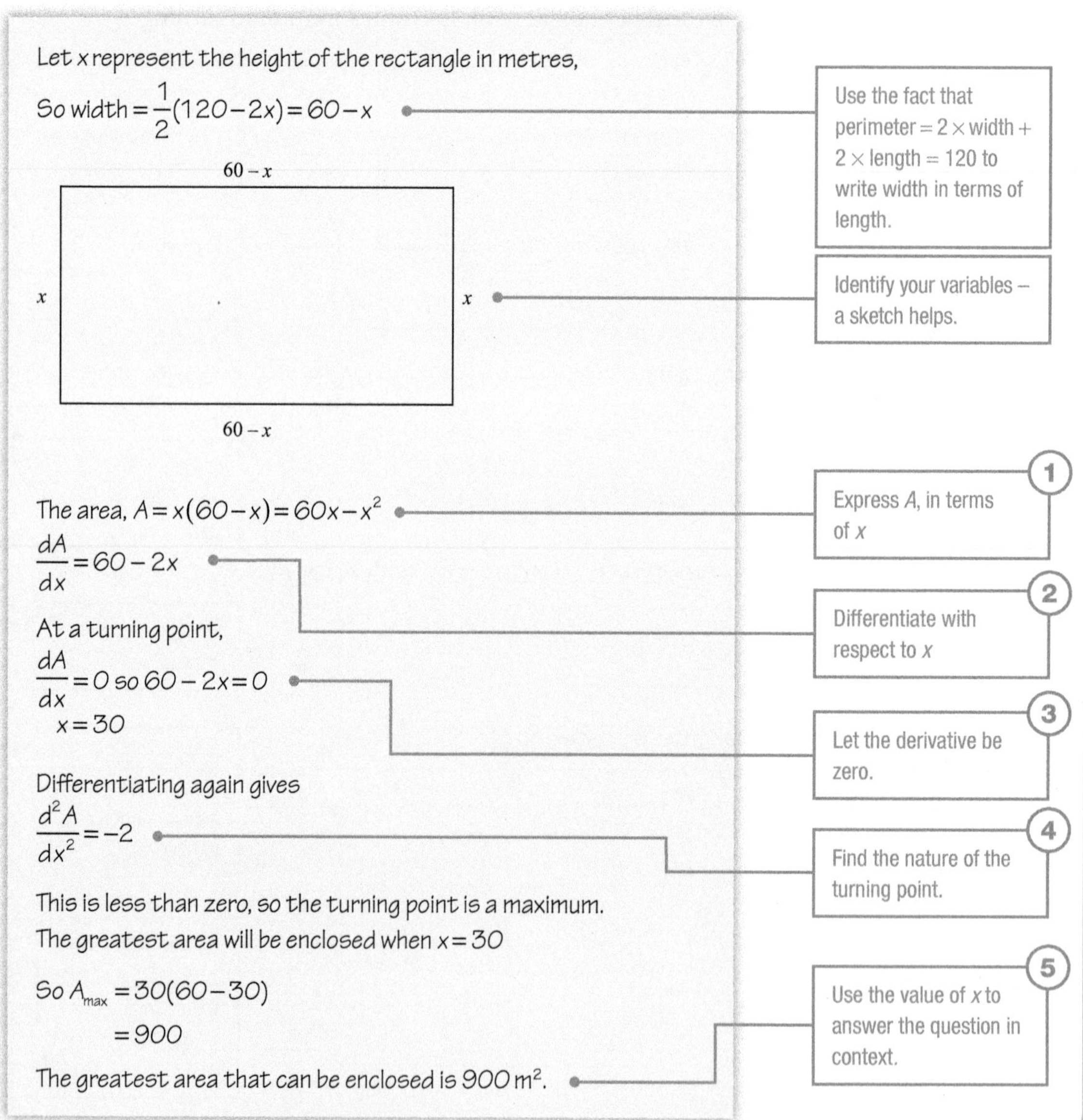

Let x represent the height of the rectangle in metres,
So width $= \frac{1}{2}(120 - 2x) = 60 - x$
Use the fact that perimeter $= 2 \times$ width $+ 2 \times$ length $= 120$ to write width in terms of length.
$60 - x$
x
x
$60 - x$
Identify your variables – a sketch helps.
The area, $A = x(60 - x) = 60x - x^2$
1 Express A, in terms of x
$\frac{dA}{dx} = 60 - 2x$
2 Differentiate with respect to x
At a turning point,
$\frac{dA}{dx} = 0$ so $60 - 2x = 0$
$x = 30$
3 Let the derivative be zero.
Differentiating again gives
$\frac{d^2A}{dx^2} = -2$
4 Find the nature of the turning point.
This is less than zero, so the turning point is a maximum.
The greatest area will be enclosed when $x = 30$
So $A_{max} = 30(60 - 30)$
$= 900$
The greatest area that can be enclosed is $900\ m^2$.
5 Use the value of x to answer the question in context.

Exercise 4.5B Reasoning and problem-solving

1 Two numbers, x and $1000 - x$, add to make 1000. What would they need to be to maximise their product?

2 The product of two positive whole numbers, x and $\frac{3600}{x}$, is 3600. Their sum is the smallest it can be. What are the two numbers?

3 Two numbers, x and y, have a sum of 12

What values of x and y will make x^2y a maximum?

4 A wire model of a cuboid is made. The total length of wire used is 600 cm.

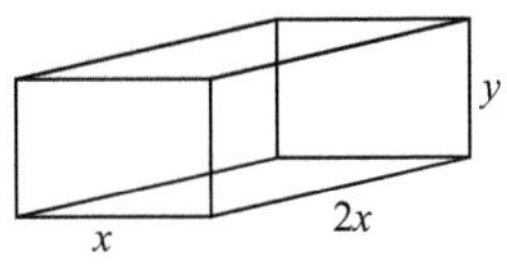

a Express y in terms of x

b Express the volume of the cuboid in terms of x

c Find the values of x and y which maximise the volume.

5 A rope of length 16 m is used to form three sides of a rectangular pen against a wall.

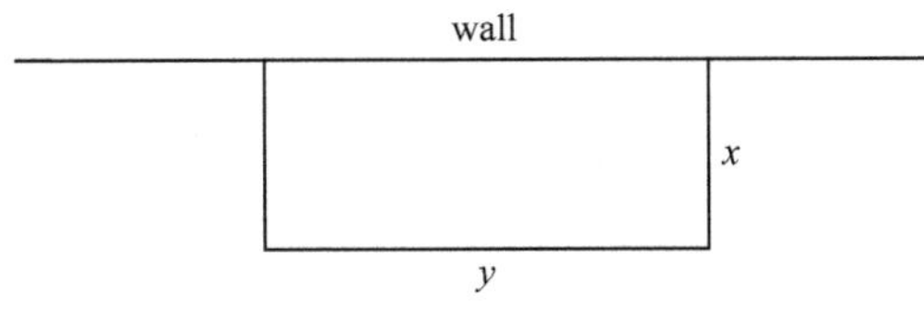

a How should the rope be arranged to maximise the area enclosed?

b Another length of rope is used to enclose an area of $50\,m^2$ in a similar way. What is the shortest length of rope that is needed?

6 The point $A(x, y)$ lies in the first quadrant on the line with equation $y = 6 - 5x$

A rectangle with two sides on the coordinate axes has A as one vertex.

a Work out an expression in x for the area of the rectangle.

b Work out the point for which this area is a maximum.

7 A small tray is made from a 12 cm square of metal.

Small squares are cut from each corner and the edges left are turned up.

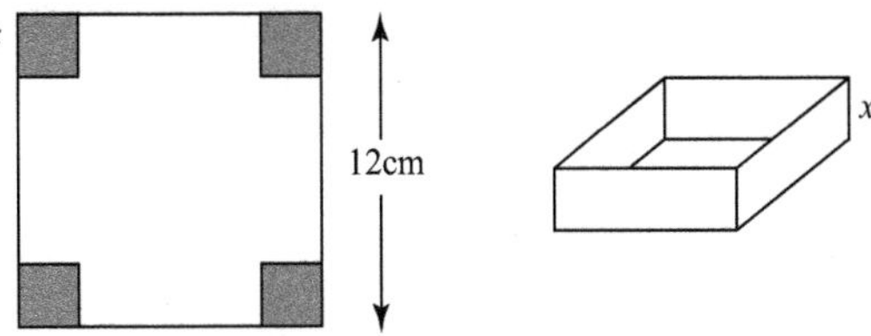

What should the side of the small square be so as to maximise the volume of the tray?

8 A fruit drink container is a cuboid with a square base. It has to hold 1000 ml of juice. Let one side of the square base be x cm and the height of the container be h cm.

a Express h in terms of x

b Find an expression for the surface area of the cuboid in terms of x

c Find the value of x that will minimise the surface area (and hence the cost of the container).

Challenge

9 A cylindrical container has to hold 440 ml of juice.

a Express its height h cm in terms of the radius x cm.

b The surface area of a cylinder with a lid and a base is given by

$$2\pi x^2 + 2\pi xh$$

Calculate the value of x that minimises this surface area.

Show your working and write your answer correct to 2 decimal places.

c If the container doesn't need a lid, how does this affect the answer?

MyMaths 2270 SEARCH

4.6 Integration

Fluency and skills

The reverse process of differentiation is known is **integration.**

To differentiate sums of terms of the form ax^n you multiply by the power then reduce the power by 1 to get nax^{n-1}

To **integrate** you do the exact opposite: you add 1 to the power then divide by the new power.

When you differentiate a constant, the result is zero. So when you perform an integration, you should add a constant, *c*, to allow for this. This is referred to as the **constant of integration**. The value of this constant can only be determined if further information is given.

When you perform an integration you can check your result by differentiating it – you should get back to what you started with.

Key point

Integrating x^n with respect to x is written as

$$\int x^n \mathrm{d}x = \frac{x^{n+1}}{n+1} + c,\ n \neq -1$$

When a function is multiplied by a constant, the constant can be moved outside the integral.

Key point

$\int a\mathrm{f}(x)\mathrm{d}x = a\int \mathrm{f}(x)\mathrm{d}x$ where a is a constant.

When integrating the sum of two functions, each function can be integrated separately.

You may see this rule referred to as the *sum rule.*

$$\int(\mathrm{f}(x)+\mathrm{g}(x))\mathrm{d}x = \int \mathrm{f}(x)\mathrm{d}x + \int \mathrm{g}(x)\mathrm{d}x$$

Given that displacement is a function of time $\mathrm{r}(t)$, then velocity $\mathrm{v}(t) = \mathrm{r}'(t)$. Reversing this, we get

Key point

$$\int \mathrm{v}(t)\mathrm{d}t = \mathrm{r}(t) + c$$

Given that velocity is a function of time $\mathrm{v}(t)$, then acceleration $\mathrm{a}(t) = \mathrm{v}'(t)$. Reversing this we get

Key point

$$\int \mathrm{a}(t)\mathrm{d}t = \mathrm{v}(t) + c$$

The Fundamental Theorem of Calculus shows how integrals and derivatives are linked to one another. The theorem states that, if $\mathrm{f}(x)$ is a continuous function on the interval $a \le x \le b$, then

Key point

$\int_a^b \mathrm{f}(x)\mathrm{d}x = \mathrm{F}(b) - \mathrm{F}(a)$ where $\frac{\mathrm{d}}{\mathrm{d}x}(\mathrm{F}(x)) = \mathrm{f}(x)$

Example 1

$\mathrm{f}'(x) = 6x^3 + 3x^2 + \frac{1}{\sqrt{x}} + 4$

a Integrate to find $\mathrm{f}(x)$

b Given that the point $(1, 4)$ lies on the curve $y = \mathrm{f}(x)$, find the constant of integration.

a $\mathrm{f}(x) = \int\left(6x^3 + 3x^2 + \frac{1}{\sqrt{x}} + 4\right)dx$

$= \int 6x^3 dx + \int 3x^2 dx + \int x^{-\frac{1}{2}} dx + \int 4x^0 dx$ — Express all the terms in index form and use the sum rule to isolate functions.

$= \frac{6x^4}{4} + \frac{3x^3}{3} + \frac{x^{\frac{1}{2}}}{\frac{1}{2}} + \frac{4x^1}{1} + c$ — Integrate using $\int ax^n\, dx = \frac{ax^{n+1}}{n+1} + c$

$= \frac{3}{2}x^4 + x^3 + 2x^{\frac{1}{2}} + 4x + c$ — Simplify.

b $y = \frac{3}{2}x^4 + x^3 + 2x^{\frac{1}{2}} + 4x + c$

$4 = \frac{3}{2} + 1 + 2 + 4 + c$ — Substitute in coordinate values for x and y

$c = -\frac{9}{2}$ — Rearrange and evaluate.

Exercise 4.6A Fluency and skills

1 Work out the integral of each function with respect to x, remembering the constant of integration.

- a 10
- b $3x$
- c $6x^2$
- d $12x^3$
- e $25x^4$
- f x^6
- g $3x+1$
- h $5-4x$
- i $3x^2+6x+2$
- j $12x^2+6$
- k $3-4x-6x^2$
- l x^3+x^2+x+1
- m $3-x-24x^3$
- n $2x^3+4x+1$
- o x^3+3x^{-2}
- p $2x^{-3}-x^{-2}$
- q $x^{\frac{1}{2}}-x^3+1$
- r $x^{\frac{1}{2}}-x^{-\frac{1}{2}}+x+6$
- s $3x^{\frac{1}{3}}-x^{\frac{2}{3}}$
- t $x^{\frac{1}{4}}-x^{\frac{3}{4}}+x^{-\frac{1}{2}}$
- u $1-x^{-2}-x^{-\frac{2}{3}}$
- v $\frac{1}{3}x^{-\frac{2}{3}}-x^4+4$
- w $\frac{4}{3}\pi x^3-\frac{1}{3}\pi x^{\frac{1}{3}}+2\pi x$
- x $x^2-x-1-x^{-2}$

2 Find

- a $\int 1\,dx$
- b $\int x\,dx$
- c $\int(6x+7)dx$
- d $\int(3-2x)dx$
- e $\int(x^2+x+1)dx$
- f $\int(1-4x-3x^2)dx$
- g $\int(4x^3+2x-7)dx$
- h $\int(2+9x^2-12x^3)dx$
- i $\int\sqrt{x}\,dx$
- j $\int\sqrt[3]{x}\,dx$
- k $\int\sqrt{\frac{1}{x}}\,dx$
- l $\int x^{\frac{2}{3}}\,dx$

3 Work out

- a $\int\pi\,dx$
- b $\int(3\pi+x)dx$
- c $\int(x^2\sin 30\ )dx$
- d $\int(x^2+6x)dx$
- e $\int 4x^2+4x-28dx$
- f $\int\left(x^2+\frac{1}{x^2}\right)dx$
- g $\int\left(\sqrt{x}-\frac{2}{\sqrt{x}}\right)dx$
- h $\int\left(8x-\frac{3}{x^2}\right)dx$
- i $\int\left(\frac{1}{x^3}-\frac{1}{x^4}\right)dx$
- j $\int\left(\frac{1}{x^2}-x-\frac{1}{x^3}\right)dx$
- k $\int x+\frac{1}{\sqrt{x}}dx$
- l $\int\left(x^2-\frac{3}{\sqrt{x}}+1\right)dx$
- m $\int\left(\frac{1}{x^2}+\frac{3}{x^3}\right)dx$
- n $\int x\sin\frac{\pi}{3}+x^2\sin\frac{\pi}{6}dx$
- o $\int 3x^2-\frac{1}{\sqrt{x^3}}dx$
- p $\int\left(\frac{x^2-x}{x}\right)dx$
- q $\int\frac{x}{\sqrt{x}}dx$
- r $\int\frac{x+1}{\sqrt{x}}dx$

4 The derivative and a point on the curve $y = f(x)$ are given.

Work out the function $f(x)$

a $f'(x) = 6x^2$ at $(1, 5)$

b $f'(x) = 12x^3$ at $(2, 18)$

c $f'(x) = 5\sqrt{x}$ at $(9, 100)$

5 Work out the function, $f(x)$ for the given $f'(x)$ and the point $(x, f(x))$

a $f'(x) = 4x + 3$; $(2, 4)$

b $f'(x) = 10$; $(1, 12)$

c $f'(x) = 3x^2 + 2x + 1$; $(-2, 1)$

d $f'(x) = 4x + 3\sqrt{x}$; $(1, 5)$

Reasoning and problem-solving

Strategy

To solve problems that require integration

1. Identify the variables and express the problem as a mathematical equation.
2. Integrate.
3. Use initial conditions to work out the constant of integration.
4. Substitute c into the equation and answer the question.

Example 2

A moving particle has acceleration -10 cm s^{-2}

The particle starts from rest 50 cm to the right of the origin.

a Express the velocity as a function of time.

b Express the displacement from the origin as a function of time.

c State the acceleration, velocity and displacement after 3 seconds.

'The particle starts from rest' means when $t = 0$, $v = 0$

a $v(t) = \int a(t)dt$

$= \int -10dt$

$= -10\int 1dt$ — (1) Identify the variables and express the rate of change as a mathematical equation.

$= -10t + c$ — (2) Integrate to work out the general form of the function.

$v(0) = -10 \times 0 + c = 0$

$c = 0$ — (3) Use the initial conditions to work out c

So $v(t) = -10t$ — (4) Use the value of c to answer the question.

b $s(t) = \int v(t)dt$

$= \int(-10t)dt$

$= -5t^2 + c$

$s(0) = -5 \times 0^2 + c = 50$

$c = 50$

So $s(t) = -5t^2 + 50$

c $a(3) = -10$ cm s^{-2}

$v(3) = -10 \times 3$

$= -30$ cm s^{-1}

$s(3) = -5 \times 3^2 + 50$

$= 5$ cm

Exercise 4.6B Reasoning and problem-solving

1 Given that the rate of change of P with respect to t is 7, and that when $t = 4$, $P = 2$

a Express P in terms of t

b Work out P when $t = 5$

c Work out t when $P = 16$

2 It is known that $f'(x) = 1 - 6x$ and that $f(3) = 6$

a Calculate $f(-2)$

b For what values of x does $f(x) = 0$?

3 Kepler was an astronomer who studied the relationship between the orbital period, in y years, of a celestial body and the mean distance from the Sun, R (measured in astronomical units AU). The rate at which the period increases as the distance from the Sun increases is given by $\frac{dy}{dr} = \frac{3}{2}r^{\frac{1}{2}}$

a Express y in terms of r and c, the constant of integration.

b The Earth is 1 astronomical unit from the Sun and takes one year to orbit it. Find the relation between y and r

c Mars is 1.5 AU from the Sun. How long does it take to orbit the Sun?

d Saturn takes 29.4 years to orbit the Sun. What is its mean distance form the Sun?

4 The rate at which the depth, h metres, in a reservoir drops as time passes is given by $\frac{dh}{dt} = \frac{8}{3}t^3 - 4t, t > 0$ where t is the time in days.
When $t = 0$, $h = 4$

a Express h as a function of t

b What is the depth after half a day?

c When is the depth 16 m?

5 The second derivative of a function is given by $\frac{d^2y}{dx^2} = 12x + 1$
When $x = 1$, $\frac{dy}{dx} = 4$

a Work out an expression in x for $\frac{dy}{dx}$

b When $x = 1$, $y = \frac{1}{2}$. What is the value of y when $x = 2$?

6 The second derivative of a function is given by $\frac{d^2y}{dx^2} = 6$
When $x = 2$, $y = 1$ and $\frac{dy}{dx} = 3$
What is the value of y when $x = 4$?

7 a A particle moves on the x-axis so that its acceleration is a function of time, $a(t) = 2t$. Initially (at $t = 0$) the particle was 2 cm to the right of the origin travelling with a velocity, v, of 2 cm s^{-1}.

i Express the velocity and the displacement, s, as a function of t

ii Calculate both velocity and displacement when $t = 1$

b Repeat part **a** for particles with the following initial conditions.

i $a(t) = 6$; $s(0) = -2$; $v(0) = 5$

ii $a(t) = t + 1$; $s(0) = 1$; $v(0) = 2$

iii $a(t) = t^2$; $s(0) = 0$; $v(0) = -1$

8 A piece of rock breaks away from the White Cliffs of Dover.

Its acceleration towards the shore, 90 m below, is 5 m s^{-2}.

Let t seconds be the time since the rock broke free. Let $v(t)$ m s^{-1} be its velocity at time t. Note that when $t = 0$, the velocity of the rock is 0 m s^{-1}. How long will it take for the rock to hit the shore?

Challenge

9 The velocity of a particle is given by $v(t) = 4 + 3t$ where distance is measured in metres and time in seconds. After one second the particle is 6 m to the right of the origin.

a Where was the particle initially?

b What is its acceleration at $t = 5$?

c How far has it travelled in the fifth second?

d Work out an expression in n for the distance travelled in the nth second.

4.7 Area under a curve

Fluency and skills

You can use integration to find the area between a curve and the x-axis. To do this, you perform a calculation using a **definite integral**.

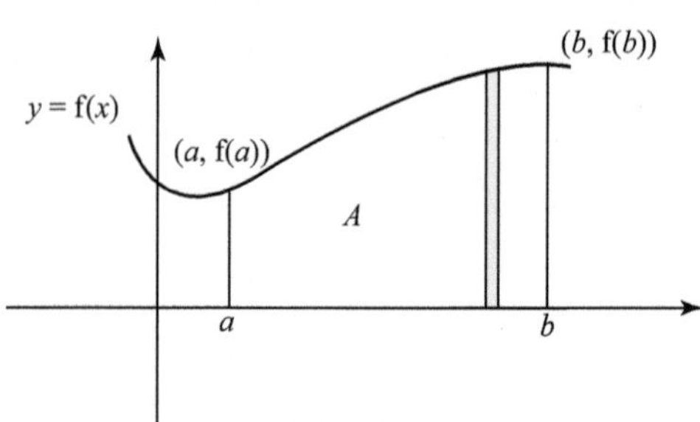

Key point

A definite integral is denoted by $\int_a^b f(x)\,dx$

b is called the **upper limit**, and a the **lower limit**.

Consider a continuous function $y = f(x)$ over some interval and where all points on the curve in that interval lie on the same side of the x-axis. The area, A, is bound by $y = f(x)$, the x-axis and the lines $x = a$ and $x = b$ where $a < b$

ICT Resource online

To experiment with numerical integration using rectangles, click this link in the digital book.

Consider a small change in x and the change in area, δA, that results from this change.

Use Leibniz notation where δx represents a small change in x and δy represents the corresponding change in y

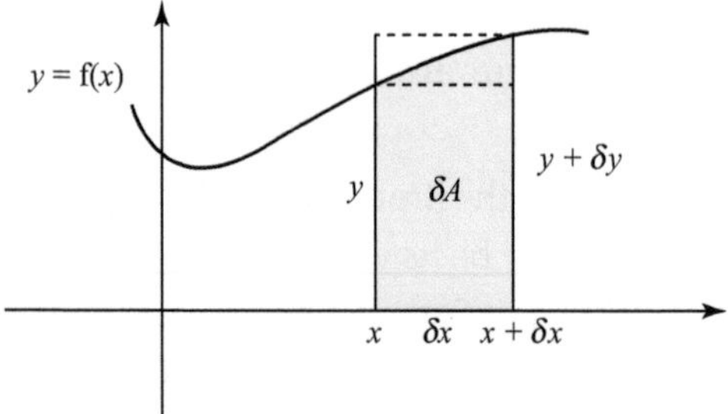

Zooming in you can see a small section of the area, trapped between vertical lines at x and at $x + \delta x$

The small area between the vertical lines at x and $x + \delta x$ (shaded) is denoted by δA

$$y\delta x \le \delta A \le (y + \delta y)\delta x$$

Dividing by δx: $$y \le \frac{\delta A}{\delta x} \le (y + \delta y)$$

As you let δx tend to zero:

$$y \le \lim_{\delta x \to 0} \frac{\delta A}{\delta x} \le \lim_{\delta x \to 0}(y + \delta y) \Rightarrow y \le \frac{dA}{dx} \le y$$

Remember

$\lim_{\delta x \to 0} \delta y = 0$ and $\lim_{\delta x \to 0} \frac{\delta y}{\delta x} = \frac{dy}{dx}$

So $$y = \frac{dA}{dx}$$

Integrating this will give you a formula for the area from the origin up to the upper bound, namely $A = \int y\,dx$. Note that, when calculating this small area, the lines $x = a$ and $x = b$ were not used so this has given us a *general* formula for calculating the area. A further calculation is needed to obtain the area between $x = a$ and $x = b$

If you wish to calculate the area between two vertical lines, $x = a$ and $x = b$, then you need only integrate to get the formula for area and substitute a and b for x. The difference between your results is the required area.

If $\int f(x)\,dx = F(x) + c$ then $(F(b) + c) - (F(a) + c) = F(b) - F(a)$

The constants of integration sum to zero. This means you don't need to worry about the constant of integration when calculating a definite integral.

Key point

The area under a curve between the x-axis, $x = a$, $x = b$ and $y = f(x)$, is given by $A = \int_a^b f(x)\,dx = F(b) - F(a)$

Working with areas below the x-axis will produce negative results. As area is a positive quantity you should use the magnitude of the answer only (ignore the negative sign).

Example 1

Evaluate the definite integral $\int_1^4 (3x^2 + 4x + 1)\,dx$. You must show your working.

$$\int_1^4 (3x^2+4x+1)\,dx$$

$$= \left[x^3+2x^2+x\right]_1^4$$ Integrate.

$$= (4^3+2\times4^2+4)-(1^3+2\times1^2+1)$$ Substitute the two values of x

$$= 100-4$$

$$= 96$$

Example 2

A parabola cuts the axes as shown.

Show that the area A is $10\frac{2}{3}$ units.

y, $y = x^2 - 8x + 12$, 12, O, 2, A, 6, x

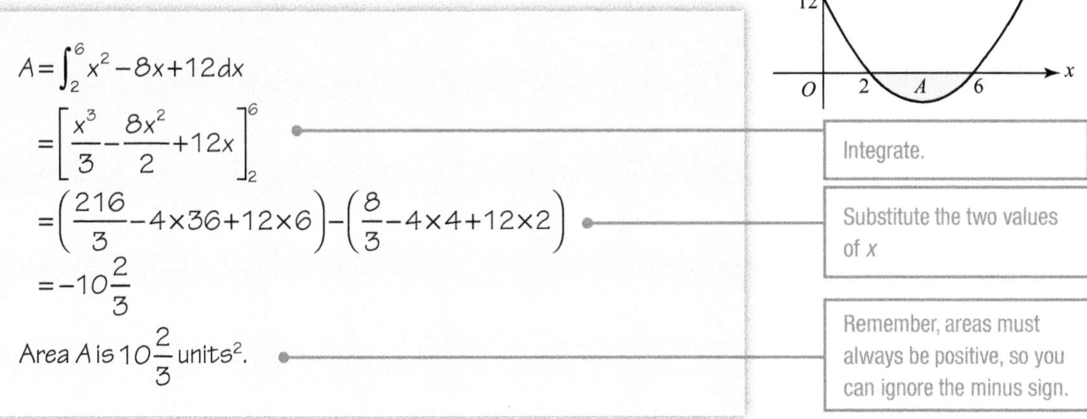

$$A = \int_2^6 x^2 - 8x + 12\,dx$$

$$= \left[\frac{x^3}{3} - \frac{8x^2}{2} + 12x\right]_2^6$$ Integrate.

$$= \left(\frac{216}{3} - 4\times36 + 12\times6\right) - \left(\frac{8}{3} - 4\times4 + 12\times2\right)$$ Substitute the two values of x

$$= -10\frac{2}{3}$$

Area A is $10\frac{2}{3}$ units^2. Remember, areas must always be positive, so you can ignore the minus sign.

If you had calculated the definite integral between 0 and 6, the answer would be zero. The positive and negative sections would cancel each other out.

Try it on your calculator

You can use a calculator to evaluate a definite integral.

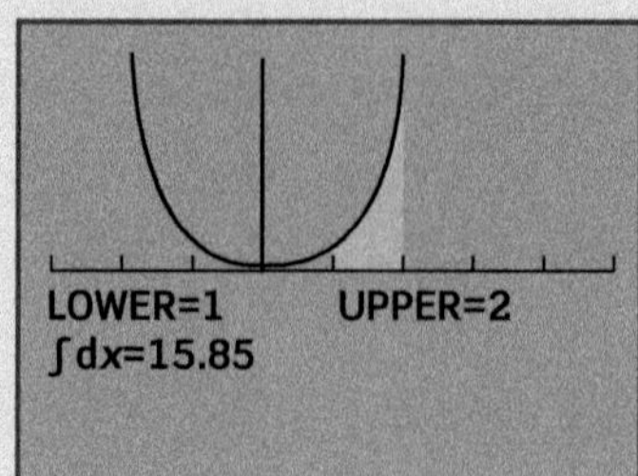

Activity

Find out how to evaluate $\int_1^2 3x^4 - x^3 + 1\,dx$ on *your* calculator.

Exercise 4.7A Fluency and skills

1 Evaluate these definite integrals. Show your working in each case.

a $\int_2^4 x+1\,dx$ b $\int_1^5 2x-1\,dx$

c $\int_0^3 4-x\,dx$ d $\int_{-2}^6 18-3x\,dx$

e $\int_{-2}^3 7\,dx$ f $\int_1^7 \pi\,dx$

g $\int_1^3 \pi+1\,dx$ h $\int_0^3 3x^2+4x+1\,dx$

i $\int_{-3}^3 x^2-6x+9\,dx$ j $\int_{-1}^1 3-x-x^2\,dx$

k $\int_{\sqrt{2}}^{\sqrt{3}} x^3\,dx$ l $\int_1^{1.5} x+\frac{1}{x^2}\,dx$

m $\int_{-1}^0 x^3+2x+3\,dx$ n $\int_4^{25} \frac{1}{\sqrt{x}}\,dx$

o $\int_2^5 4\pi x^2\,dx$ p $\int_{\sqrt{3}}^{\sqrt{5}} x^3+x\,dx$

q $\int_{-3}^3 x^3-2x-21\,dx$ r $\int_{\frac{1}{4}}^{\frac{1}{2}} \frac{\pi}{x^2}\,dx$

2 Work out the total shaded area for each of these graphs. Show your working in each case.

a
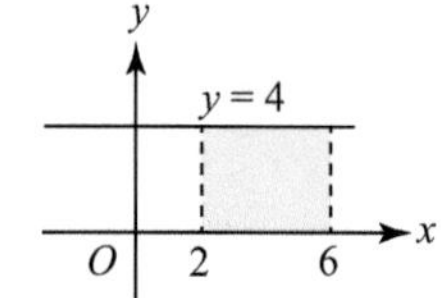

b
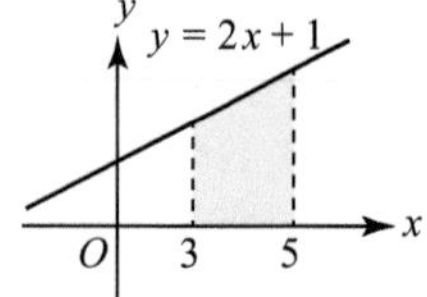

c
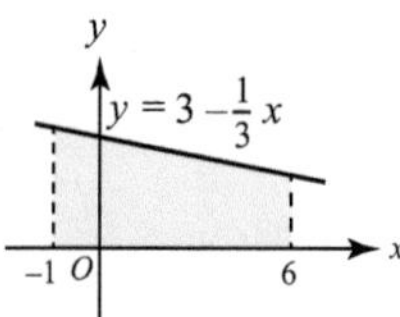

d
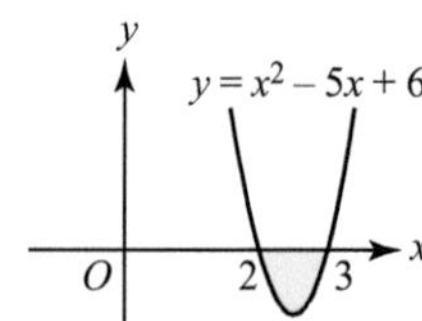

e
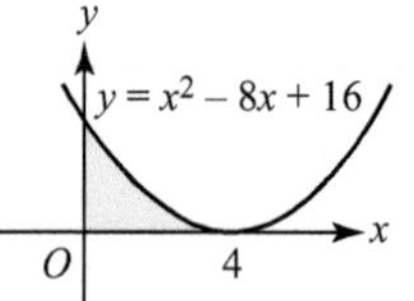

f
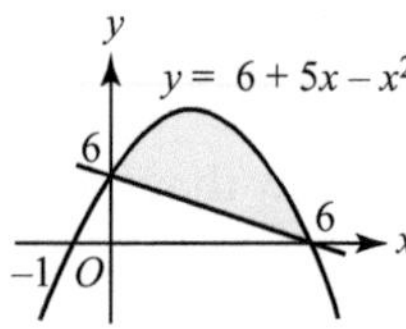

g
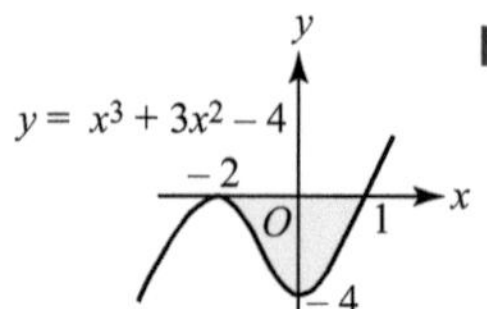

h
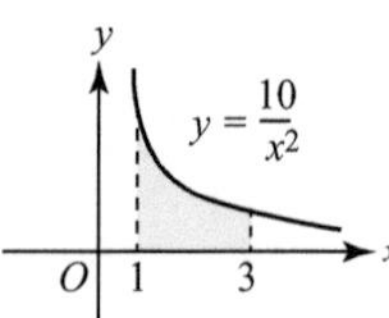

i
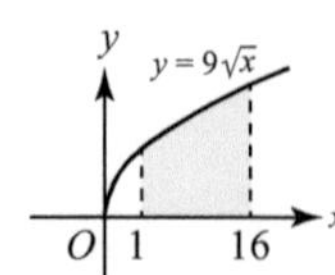

Reasoning and problem-solving

Strategy

To calculate the area under a curve

1. Make a sketch of the function, if there isn't one provided.
2. Identify the area that has to be calculated.
3. Write down the definite integral associated with the area.
4. Evaluate the definite integral and remember that the area is always positive.

Exercise 4.7B Reasoning and problem-solving

1 An area of 29 units^2 is bound by the line $y = x$, the x-axis and the lines $x = -3$ and $x = a$, $a < 0$. Use integration to help you calculate the value of a

2 Integrate the function $f(x) = x^2 - 6x + 8$ with respect to x and thus calculate the area bound by the curve $y = f(x)$, the x-axis, and the lines $x = 2$ and $x = 4$

3 A parabola has equation $y = 10 + 3x - x^2$

Calculate the area of the dome trapped between the curve and the x-axis.

4 The equation $y = \sqrt{x}$ is one part of a parabola with $y = 0$ as the axis of symmetry.

a Calculate $\int y \, dx$

b Calculate the area trapped between the curve, the x-axis and the lines $x = 0$ and $x = 1$

c P is the area below the curve between $x = 1$ and $x = 4$. Q is the area between $x = 4$ and $x = 9$. Calculate the ratio P : Q

d Calculate the area trapped between the curve $y = 1 + \sqrt{x}$, the x-axis and the lines $x = 0$ and $x = 1$ and explain the difference between this and the answer to part **b**.

5 **a** Area A is defined by $A(a) = \int_1^a 3x^2 \, dx$

For what value of a is the area 999 units^2?

b Area B is defined by $B(a) = \int_a^{2a} 3x^2 \, dx$

For what value of a is the area 875 units^2?

c Area C is defined by $C(a) = \int_a^{a+1} 3x^2 \, dx$

For what values of a is the area 19 units^2?

6 The area trapped between the x-axis and the parabola $y = (x+3)(x-9)$ is split in two parts by the y-axis. What is the ratio of the larger part to the smaller?

7 An area is bounded by the curve $y = 4x^3$, the x-axis, the line $x = 1$ and the line $x = a$, $a > 1$. What is the value of a if the area is 2400 units^2?

8 Sketch a graph of $y = \sin x$ for $0 \le x \le 360°$ and, using this, explain why the definite integral $\int_0^{360} \sin x \, dx$ equals zero.

9 On a velocity–time curve, the distance travelled can be obtained by calculating the area under the curve.

An object is thrown straight up. Its velocity, in m s^{-1}, after t seconds can be calculated using $v(t) = 24t - 5t^2$

What distance did the object travel between $t = 1$ and $t = 4$?

Challenge

10 Space debris is detected falling into the Earth's atmosphere. Its velocity in kilometres per second is modelled by $v(t) = 5 + 0.01t$ where t is the time in seconds measured from where the debris was detected. It completely burned up after 4 seconds. How far did the debris travel in the atmosphere?

MyMaths 2056, 2273 SEARCH

4 Summary and review

Chapter summary

- The gradient of the tangent to a curve at the point P can be approximated by the gradient of the chord PQ, where Q is a point close to P on the curve.
- The derivative at the point P on the curve can be calculated from first principles by considering the limiting value of the gradient of the chord PQ as Q tends to P
- The derivative is denoted by $f'(x)$

  $f'(x)=\lim_{h\to 0}\frac{f(x+h)-f(x)}{h}$

- If $y=f(x)$ then $\frac{dy}{dx}=f'(x)$
- The derivative gives the instantaneous rate of change of y with respect to x.
- The derivative of $y=ax^n$, where a is a constant, is $\frac{dy}{dx}=nax^{n-1}$
- $f(x)=ag(x)+bh(x)\Rightarrow f'(x)=ag'(x)+bh'(x)$, where a, b are constants.
- $f'(x)$ gives the rate of change of function f with respect to x. So,

  $f'(x)>0$ means the function is increasing,
  $f'(x)<0$ means the function is decreasing,
  $f'(x)=0$ means the function is stationary.

- By sketching the gradient function, you can determine whether a function is increasing, decreasing, or stationary.
- At a turning point, the tangent is horizontal, so $f'(x)=0$
- Where the function changes from being an increasing to a decreasing function, the turning point is referred to as a maximum.
- Where the function changes from being a decreasing to an increasing function, the turning point is referred to as a minimum.
- The derivative is itself a function that can be differentiated. The result is called the second derivative and is denoted by $f''(x)$ or $\frac{d^2y}{dx^2}$
- $\frac{d^2y}{dx^2}$ gives the rate of change of the gradient with respect to x. Assuming $\frac{dy}{dx}=0$, then $f''(x)>0$ means the gradient is increasing—it is a minimum turning point.

  $f''(x)<0$ means the gradient is decreasing—it is a maximum turning point.
- Since the derivative at a point $(a, f(a))$ gives the gradient of the curve at that point and the gradient of the tangent at that point then $m_{\text{tangent}}=f'(a)$ and the equation of the tangent is $y-f(a)=f'(a)(x-a)$
- The normal at $(a,f(a))$ has a gradient $m_{\text{normal}}=-\frac{1}{f'(a)}$ so the equation of the normal is

  $y-f(a)=-\frac{1}{f'(a)}(x-a)$

- The process of differentiation can be reversed by a process called integration.

  $F'(x) = f(x) \Rightarrow F(x) = \int f(x)\,dx + c$

  where c is the constant of integration.
- $\int ax^n\,dx = \frac{ax^{n+1}}{n+1}$ ... read as "the integral of ax^n with respect to x"
- $\int (af(x) + bg(x))\,dx = a\int f(x)\,dx + b\int g(x)\,dx$
- $\int_a^b f(x)\,dx$ is called a definite integral with upper limit b and lower limit a
- $\int f(x)\,dx = F(x) + c \Rightarrow \int_a^b f(x)\,dx = F(b) - F(a)$
- If $f(x)$ is continuous in the interval $a \le x \le b$ and all points on the curve in this interval are on the same side of the x-axis, the area bounded by the curve, the x-axis and the lines $x = a$ and $x = b$ is given by the positive value of $\int_a^b f(x)\,dx$

Check and review

You should now be able to...	Try Questions
✔ Differentiate from first principles.	1
✔ Differentiate functions composed of terms of the form ax^n	2, 3
✔ Use differentiation to calculate rates of change.	4
✔ Work out equations, tangents and normals.	5
✔ Work out turning points and determine their nature.	6
✔ Work out and interpret the second derivative.	7
✔ Work out the integral of a function.	8
✔ Understand and calculate definite integrals.	9
✔ Use definite integrals to calculate the area under a curve.	10

1 Differentiate each function from first principles.

a $y=3x^2$ at $(1, 3)$

b $y=x^3+1$ when $x=-1$

c $y=x^4-x-5$ at $(2,9)$

d $y=1-x^2$

e $y=x-x^2$

f $y=\pi x^2$

2 Work out the derivative of

a $f(x)=x^3+2x^2+3x+1$

b $y=4\sqrt{x}+x$

c $y=x+\frac{1}{x}+\frac{1}{x^2}$

d $f(x)=\sqrt[4]{x}-\sqrt[3]{x}$

e $y=\frac{x+3}{x^2}$

f $y=\frac{4}{x}+\frac{2}{\sqrt{x}}$

g $y=1-\frac{1}{x^2}+\frac{1}{\sqrt[3]{x}}$

h $y=\frac{x^2+2x+3}{x}$

i $y=\frac{x+2\sqrt{x}}{\sqrt{x}}$

j $y=\frac{\sqrt{x}+1}{\sqrt[3]{x}}$

3 Work out the value of the following functions. Show your working.

a $f'(2)$ when $f(x)=2x^2-5$

b $\frac{dy}{dx}$ when $x=3$ given $y=\frac{x+3}{x}$

c $f'(4)$ when $f(x)=1+\frac{1}{x}-\frac{33}{\sqrt{x}}$

d $\frac{dy}{dx}$ when $x=9$ given $y=\frac{x-9}{2\sqrt{x}}$

4 a Work out the rate of change of y with respect to x when $x=4$ given that $y=x(2x^2-5x)$ Show your working.

b A roll of paper is being unravelled. The volume of the roll is a function of the changing radius, r cm. $V=25\pi r^2$ [V cm³ is the volume].

Calculate the rate of change of the volume when the radius is 3 cm. Show your working.

c A particle moves along the x-axis so that its distance, D cm, from the origin at time t seconds is given by $D(t)=t^2-5t+1$

i Work out the particle's velocity at $t=3$ (the rate of change of distance with respect to time). Show your working.

ii Work out the particle's acceleration at this time (the rate of change of velocity with time). Show your working.

d A tourist ascends the outside of a tall building in a scenic elevator.

As she ascends, she can see further. The distance to her horizon, K km, can be calculated by the formula $K=\sqrt{\frac{hD}{1000}}$ where h is her height in metres and D is the diameter of the planet in kilometres. On Earth, $D=12\,742$ km

i Calculate the rate of change of the distance to her horizon with respect to her height

1 When she is 50 m up,

2 When she is 100 m up.

ii How high up will she be when the distance to the horizon is changing at a rate of 0.4 km per m of height?

iii Suppose the tourist were on a building on the Moon. The Moon has a diameter of 3474 km. Calculate the rate of change of the distance to her horizon with respect to her height when she is 50 m up.

5 Work out the equations of the tangent and normal to the given curves at the given points. Show your working.

a $y=x^2+4x+3$ at $(1,8)$

b $y=2x-\frac{3}{x}$ at $(3,5)$

c $y=5+4x-x^2$ at $(2,9)$

d $y=200\sqrt{x}$ at $(25,1000)$

6 Work out the turning points on each curve and determine their nature. Show your working.

a $y=x^2+4x-5$

b $y=3x^3-4x$

c $y=10x^4$

d $y=ax+\frac{a}{x}$ where a is a positive constant.

e $y=ax^2+bx+c$ where a, b and c are positive constants.

7 Work out the second derivative of

a $y=x^2$

b $y=x^3+x^2$

c $y=\frac{1}{x}$

d $y=1+x+\frac{x^2}{2}+\frac{x^3}{6}+\frac{x^4}{24}+\frac{x^5}{120}+\frac{x^6}{720}$

8 Calculate

a $\int 1-x-x^3\,dx$

b $\int \frac{1}{x^2}-4x^2\,dx$

c $\int \sqrt{x}-\frac{1}{\sqrt{x}}-\sqrt[3]{x}\,dx$

d $\int \frac{x^2-1}{\sqrt{x}}\,dx$

9 Calculate the following definite integrals. Show your working.

a $\int_1^4 x+6\,dx$

b $\int_1^{10} 2\pi\,dx$

c $\int_{-1}^4 x^2-4x-5\,dx$

d $\int_{\sqrt{2}}^{\sqrt{5}} 2x^3\,dx$

10 a Calculate the area bounded by $y=x^2-7x+10$, the x-axis, $x=2$ and $x=5$. Show your working.

b Calculate the area bounded by $y=x^2-5x+6$, the x-axis, $x=2$ and $x=3$. Show your working.

c Calculate the area bounded by $y=27-x^3$, the x-axis, and the y-axis. Show your working.

What next?

Score		
0 – 5	Your knowledge of this topic is still developing. To improve, search in MyMaths for the codes: 2028–2030, 2054–2056, 2269–2270, 2273	
6 – 8	You're gaining a secure knowledge of this topic. To improve, look at the InvisiPen videos for Fluency and skills (04A)	
9 – 10	You've mastered these skills. Well done, you're ready to progress! To develop your techniques, look at the InvisiPen videos for Reasoning and problem-solving (04B)	

Click these links in the digital book

History

Differentiation and integration belong to a branch of maths called calculus, which is the study of change. It took hundreds of years and the work of many mathematicians to develop calculus to the state in which we know it today.

The first person to write a book about both differentiation and integration, and the first woman to write any book about mathematics, was **Gaetana Maria Agnesi**. One of 21 children, Maria displayed incredible ability in a number of disciplines at an early age, and was an early campaigner for women's right to be educated – giving a speech on the topic when she was 9 years old.

Agnesi's book, **Istituzioni analitiche ad uso della gioventu italiana** (analytical institutions for the use of the Italian youth), was published in 1748 when she was just 30 years old.

Research

Before Maria, in the 17th century, the fundamental theorem of calculus was independently discovered by **Isaac Newton** and **Gottfried Leibniz**.

These two were both well-regarded by the mathematical community, and both published very similar works on the subject within only a few years of each other. This led to suspicions of copying and a bitter feud between the two lasted until Leibniz's death in 1716.

Find out the contributions that Newton and Leibniz made to the theorem of calculus and use your research to write a summary. In your summary, discuss any similarities and differences between the works of each mathematician, and how their results added to or built on what was previously established in the field of calculus.

Isaak Newton.

Gottfried Wilhelm Leibniz.

4 Assessment

1 $y = 5x^3 - 2x^{-3}$ Calculate and select the correct answer for

a $\frac{dy}{dx}$

A $15x^2 + 6x^{-2}$ B $15x^2 - 6x^{-4}$ C $15x^2 - 6x^{-2}$ D $15x^2 + 6x^{-4}$ **[1 mark]**

b $\int y\,dx$

A $\frac{5x^4}{4} - x^{-2} + c$ B $\frac{5x^4}{4} + x^{-2} + c$ C $\frac{5x^4}{4} - \frac{x^{-4}}{2} + c$ D $\frac{5x^4}{4} + \frac{x^{-4}}{2} + c$ **[1]**

2 Calculate the exact value of $\int_1^2 5x - 2\,dx$. Select the correct answer.

A 5.5 B 1.5 C 13 D 8 **[1]**

3 The radius (in cm) of a circle at time t seconds is given by $r = 20 - 2\sqrt{t}$

a Work out an expression for the rate of change of the radius. **[3]**

b Calculate the rate of change of the radius at time 25 s. State the units of your answer. **[2]**

4 $y = 2x^2 - x^{-3}$

a Calculate the gradient of the curve at the point (1, 1) **[3]**

b Work out the equation of the normal to the curve at the point (1, 1) **[3]**

5 $f(x) = \frac{1}{5}x^5 - x^2$

a Work out an expression for $f'(x)$ **[2]**

b Calculate $f'(2)$ **[2]**

c Work out the equation of the tangent to $y = f(x)$ at the point where $x = 2$

Give your equation in the form $ax + by + c = 0$ where a, b and c are integers. **[4]**

6 The curve C has equation $y = 6x^3 - 3x^2 - 12x + 5$

a Use calculus to show that C has a turning point when $x = -\frac{2}{3}$ **[4]**

b Work out the coordinates of the other turning point on C **[2]**

c Is this point is a maximum or a minimum? Explain your reasoning. **[2]**

7 Work out the range of values of x for which $y = x^3 + 5x^2 - 8x + 4$ is decreasing. **[5]**

8 Work out these integrals.

a $\int 2x + 3x^5\,dx$ **[2]** b $\int x^{-4} - 4x^3\,dx$ **[3]** c $\int 3\sqrt{x} + \frac{1}{\sqrt{x}}\,dx$ **[3]**

9 Calculate the exact values of these definite integrals. You must show your working.

a $\int_0^3 2\sqrt{x}\,dx$ **[4]** b $\int_1^2 \frac{2}{x^3} - 3x\,dx$ **[4]**

10 Find an expression for $f(x)$ when

a $f'(x) = 4x^2 + 5x - 1$ **[3]** **b** $f'(x) = 7x^{-3} - x + \sqrt{x}$ **[5]**

11 The region shown is bounded by the x-axis and the curve $y = -x^2 + 4x - 3$

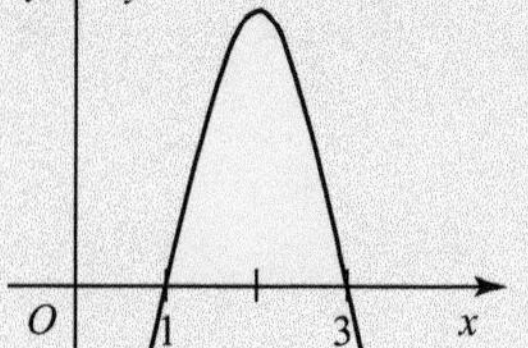

Show that the area of the shaded region is $1\frac{1}{3}$ square units. **[4]**

12 $y = 3x^2$

a Work out $\frac{dy}{dx}$ from first principles. **[5]**

b Calculate the gradient of the tangent where $x = 5$ **[2]**

13 $y = x^3 - 2x$

a Work out $\frac{dy}{dx}$ from first principles. **[5]**

b Calculate the gradient of the tangent where $x = 2$ **[2]**

14 Work out $\frac{dy}{dx}$ when

a $y = x(x+3)^2$ **[4]** **b** $y = \frac{x+2}{\sqrt{x}}$ **[4]**

15 The volume (in litres) of water in a container at time t minutes is given by

$$V = \frac{t^3 - 8t}{3t}$$

Calculate the rate of change of the volume after 4 minutes. **[6]**

16 Work out the equation of the normal to $y = x^2(2x+1)(x-3)$ at the point where $x = 2$ **[10]**

17 The curve C has equation $y = \frac{3x^3 + \sqrt{x}}{2x}$

a Work out the equation of the tangent to C at the point where $x = 1$ **[9]**

The tangent to C at $x = 1$ crosses the x-axis at the point A and the y-axis at the point B

b Calculate the exact area of the triangle AOB **[4]**

18 Show that the function $f(x) = (1+2x)^3$ is increasing for all values of x **[6]**

19 Work out the range of values of x for which $f(x) = 5\sqrt{x} + \frac{3}{\sqrt{x}}$ is a decreasing function. **[4]**

20 Work out the range of values of x for which $y = \frac{1}{3}x(x-1)(5-x)$ is an increasing function. **[7]**

21 Given that $f(x) = x^2(2x - \sqrt{x})$, work out expressions for

a $f'(x)$ **[4]** **ii** $f''(x)$ **[2]**

22 Given that $y = 3x^2 - 4\sqrt{x}$, work out $\frac{d^2y}{dx^2}$ **[4]**

23 Calculate the coordinates of the stationary point on the curve with equation

$$y = 32x + \frac{2}{x^2} - 15,\ x > 0$$

Show that this point is a minimum. **[8]**

24 A cylindrical tin is closed at both ends and has a volume of 200 cm³.

a Express the height, h in terms of the radius, x **[3]**

b Show that the surface area, A of the tin is given by

$$A = 2\pi x^2 + \frac{400}{x}$$ **[3]**

c Calculate the value of x for which A is a minimum. **[4]**

d Hence, work out the minimum value of A **[2]**

e Justify that the value found in part **d** is a minimum. **[2]**

25 A box has a square base of side length x

The volume of the box is 3000 cm^3.

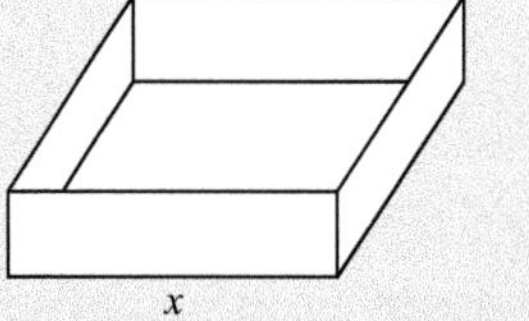

a Show that the surface area, A, of the box (not including the lid) is given by $A = x^2 + \frac{12000}{x}$ **[6]**

b Calculate the value of x for which A is a minimum. **[3]**

c Hence, work out the minimum value of A **[2]**

d Justify that the value found in part **c** is a minimum. **[2]**

26 Work out these integrals.

a $\int (2x+3)^2 \, dx$ **[3]** **b** $\int \sqrt{x}(5x-1) \, dx$ **[4]** **c** $\int \frac{2+x}{2\sqrt{x}} \, dx$ **[4]**

27 Calculate the exact values of these definite integrals. You must show your working.

a $\int_0^1 (3x-1)^3 \, dx$ **[5]** **b** $\int_{-1}^1 x^2(x-4)(x-5) \, dx$ **[5]**

28 Calculate the area of the region bounded by the x-axis and the curve

$y = x^2 - 3x - 10$ **[8]**

29 The shaded region shown is bounded by the x-axis, the line $x = \frac{1}{2}$ and the curve with equation $y = \frac{5}{x^2} - 3x^2 - 2\sqrt{x}$, $x > 0$

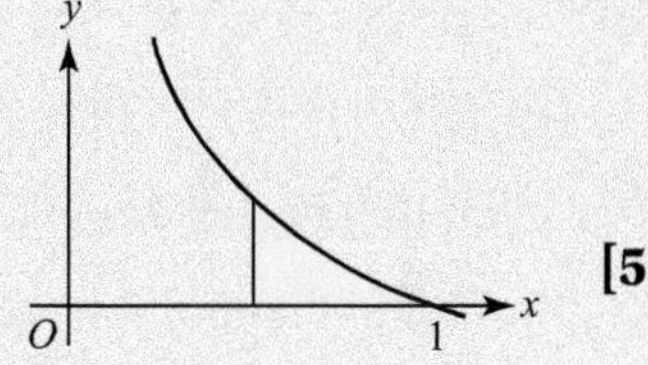

Calculate the area of the shaded region. **[5]**

30 The curve with equation $y = f(x)$ passes through the point $(1, 1)$

Given that $f'(x) = 5x^4 - \frac{2}{x^3}$

a Calculate $f(x)$ **[4]**

b Work out the equation of the normal to the curve at the point $(1, 1)$ **[4]**

31 **a** Differentiate with respect to x, where k is a constant.

i $kx + x^k$, $k \neq -1$ **ii** $\frac{1}{x^k} - k$, $k \neq 1$ **[4]**

b Integrate the functions in part **a** with respect to x **[5]**

32 $f(x) = x^3 - 2x$

P is the point of intersection of the tangent to the curve $y = f(x)$ at the point $x = 2$ and the normal to the curve $y = f(x)$ at the point $x = -1$. Calculate the coordinates of P **[13]**

33 The function $f(x)$ is given by $f(x) = x^2 + kx$ where k is a positive constant.

The tangent to $y = f(x)$ at the point where $x = k$ meets the x-axis at the point A and the y-axis at the point B

Given that the area of the triangle ABO is 36 square units, work out the value of k **[10]**

34 The normal to the curve $y = 2x^2 - x + 2$ at the point where $x = 1$ intersects the curve again at the point Q. Calculate the coordinates of Q **[9]**

35 Work out and classify all the stationary points of the curve with equation $y = x^4 - 2x^2 + 1$ **[8]**

36 Work out and classify all the stationary points of the curve with equation

$y = 3x^4 + 4x^3 - 12x^2 + 20$ **[8]**

37 Calculate the range of values of x for which $f(x) = 4x^4 - 2x^3$ is an increasing function. **[5]**

38 Calculate the range of values of x for which $f(x) = -3x^4 + 8x^3 + 90x^2 + 12$ is a decreasing function. **[6]**

39 A triangular prism has a cross-section with base twice its height.

The volume of the prism is 250 cm³.

Calculate the minimum possible surface area of the prism given that it is closed at both ends. **[12]**

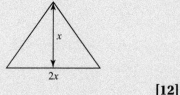

40 A cylinder of radius, r cm is open at one end. The surface area of the cylinder is 700 cm².

Calculate the maximum possible volume of the cylinder. **[11]**

41 The curve with equation $y = f(x)$ passes through the point $A(1, 4)$

Given that $f'(x) = 3x^2 - \dfrac{2}{x^3}$

a Work out the equation of the tangent to the curve at the point when $x = -1$ **[7]**

The tangent crosses the y-axis at the point B

b Calculate the area of the triangle ABO **[3]**

42 The curve with equation $y = f(x)$ passes through the point $(0, 0)$

Given that $f'(x) = 4x - 3x^2$, work out the area enclosed by the curve $y = f(x)$ and the x-axis. **[8]**

43 The shaded region is bounded by the curve with equation

$y = 12 - 7x - x^2$ and the line with equation $y = 4$

Calculate the area of the shaded region. **[9]**

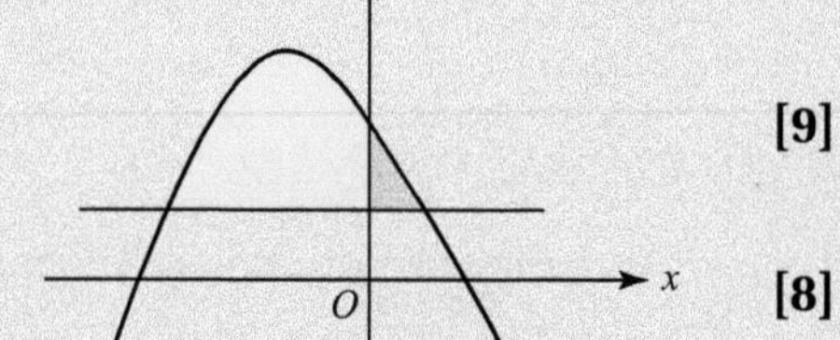

44 Calculate the area of the region bounded by the x-axis and the curve with equation $y = x(x+1)(x-2)$ **[8]**

45 The shaded region is bounded by the curve with equation $y = 3x^2 + 1$ and the lines $y = 4$ and $y = 2$

Calculate the area of the shaded region. **[11]**

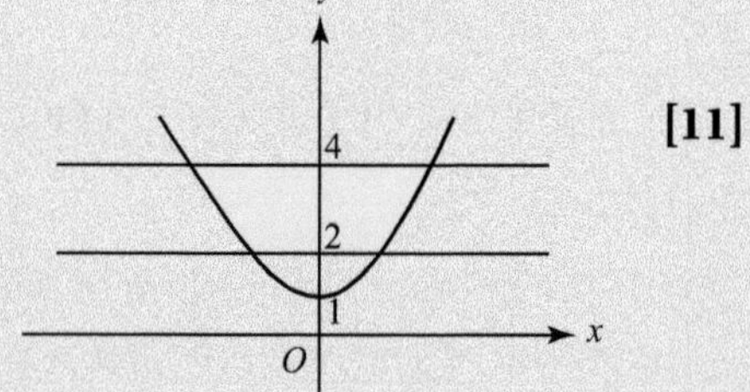

46 The region R is bounded by the x-axis and the curve with equation

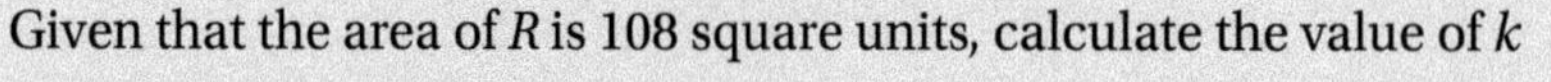

$y = x^2(k - x)$, where k is a positive constant.

Given that the area of R is 108 square units, calculate the value of k **[7]**

47 The region R is bounded by the curve with equation

$y = 13 - 2x - x^2$ and the line $y = 11 - x$

Calculate the area of R **[9]**

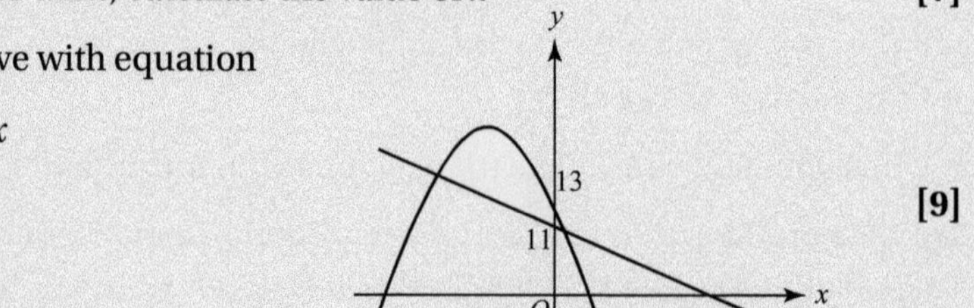

5 Exponentials and logarithms

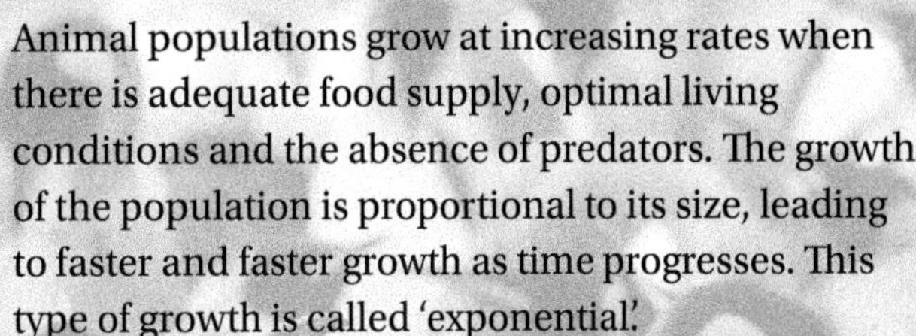

Animal populations grow at increasing rates when there is adequate food supply, optimal living conditions and the absence of predators. The growth of the population is proportional to its size, leading to faster and faster growth as time progresses. This type of growth is called 'exponential.'

Exponential functions and their inverses, logarithmic functions, appear in many areas of life as well as in many areas of mathematics. They are used to model growth and decline and can be used to calculate populations, radioactive decay and increasing bank balances as a result of interest. This means that these functions are an invaluable part of mathematical modelling.

Orientation

What you need to know

KS4

- Recognise, sketch and interpret graphs of exponential functions $y = k^x$ for positive values of k

p.58

Ch2.4 Curve sketching

- Graph transformations.
- Proportionality.

p.86

Ch4.1 Differentiation

- Differentiation from first principles.

What you will learn

- To convert between powers and logarithms.
- To manipulate and solve equations involving powers and logarithms.
- To use exponential functions and their graphs.
- To verify and use mathematical models and consider limitations of these models.

What this leads to

Ch15 Differentiation
Derivation for a^n and $\ln x$ and applications to problems.

Ch16 Integration
Integration of exponential functions.

Careers
Archeology.
Investment banking.
Wildlife biology.

Practise before you start

1070, 2024, 2028, 2258

5 Bridging Unit

Topic A: Exponentials and logarithms

Bridging to Ch5.1

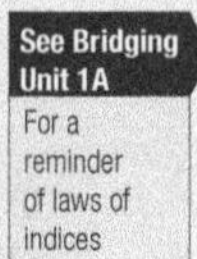
See Bridging Unit 1A For a reminder of laws of indices

You're likely to have come across **exponentials** before, but you may not have heard of **logarithms**. Section 5.1 will go into detail on what exactly a logarithm is, but for now you just need to know that it's the **inverse** of an exponential.

An exponential takes the form a^b whilst a logarithm takes the form $\log_a b$

You read "$\log_a b$" as "log base a of b" or "log of b to base a".

Key point

$a^x = b$ is equivalent to $x = \log_a b$

For example, $2^x = 19 \Rightarrow x = \log_2 19$. This expression, $\log_2 19$, is a logarithm with base 2. You can use your calculator to evaluate logarithms: typing in the numbers gives $x = \log_2 19 = 4.25$ to 3 significant figures.

Example 1

Solve the equation $3^x = 11$

$3^x = 11 \Rightarrow x = \log_3 11$ — The base of the logarithm will be the same as the base of the exponential. In this case, it's 3

$= 2.18$ (to 3 significant figures) — Calculate $\log_3 11$ on your calculator. The button will look like something like $\log_\square \square$

Try It 1

Use your calculator to solve these equations. **a** $5^x = 11$ **b** $6^x = \frac{1}{3}$

You can also use this technique of converting exponentials to logarithms when solving more complicated equations involving exponentials.

You may need to use laws of indices.

Example 2

Solve the equation $3^{2x} - 7(3^x) + 10 = 0$

$3^{2x} = (3^x)^2$ — Since $(3^a)^b = 3^{ab}$

Let $y = 3^x$ then the equation becomes

$y^2 - 7y + 10 = 0$ — Re-write the equation as a quadratic in y

The solutions to this are $y = 2$ and $y = 5$ — Use your calculator to solve the quadratic equation.

So $3^x = 2$ or $3^x = 5$ — Since $y = 3^x$

Therefore, $x = \log_3 2 = 0.631$

Or $x = \log_3 5 = 1.46$ — Give your answers to 3 significant figures.

See Ch1.4 For a reminder on how to solve quadratic equations with a calculator.

Bridging: PURE

Try It 2

Solve the equation $(3^x)^2 - 5(3^x) = 0$

Example 3

Solve the equation $2^{2x} - 2^{x+3} - 20 = 0$

$2^{2x} = (2^x)^2$ — Since $(2^a)^b = 2^{ab}$

$2^{x+3} = 2^3 \times 2^x$
$= 8(2^x)$ — Since $2^{a+b} = 2^a \times 2^b$

Let $y = 2^x$ so the equation becomes $y^2 - 8y - 20 = 0$ — Re-write the equation as a quadratic in y

The solutions to this are $y = 10$ and $y = -2$ — Use your calculator to solve the equation.

If $y = 10$ then $2^x = 10$ — Since $y = 2^x$

Therefore, $x = \log_2 10$
$= 3.32$ — Give your answer to 3 significant figures.

If $y = -2$ then there are no solutions for x because 2^x cannot be a negative value.

Try It 3

Solve the equation $2(5^{2x}) - 5^{x+2} + 50 = 0$

The graphs of exponential and logarithmic functions have characteristic curved shapes with an asymptote at one of the axes.

The graph of $y = a^x$ will always pass through the point (0, 1) for any base a since $a^0 = 1$

Example 4

Complete a table of values and plot the graph of $y = 6^x$ for values of x between $x = -2$ and $x = 3$

x	−2	−1	0	1	2	3
y	0.0278	0.167	1	6	36	216

Substitute each of the values for x, e.g. $6^{-2} = 0.0278$ (to 3 significant figures).

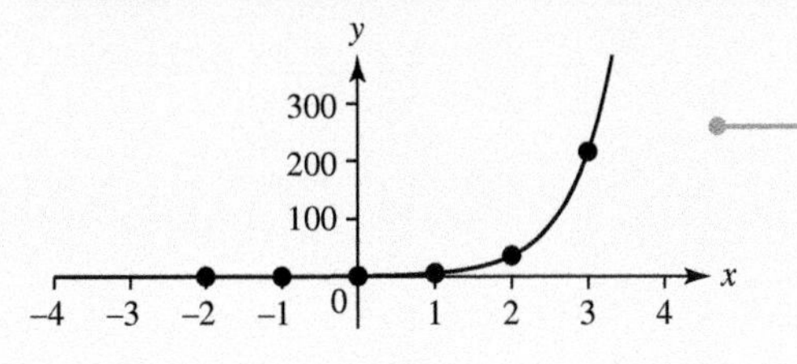

Note, the x-axis is an asymptote as the value of a^x can never be zero.

Try It 4

Complete a table of values and plot the graph of $y = 4^x$ between $x = -4$ and $x = 4$

As shown in Example 4, when a is positive, the graph of $y=a^x$ is only defined for positive values of y. So its inverse, the graph of $y=\log_a x$, is only defined for positive values of x

Example 5

Complete a table of values and plot the graph of $y=\log_2 x$ for $x=0$ to $x=10$

x	0.25	0.5	1	2	4	6	8
y	−2	−1	0	1	2	2.58	3

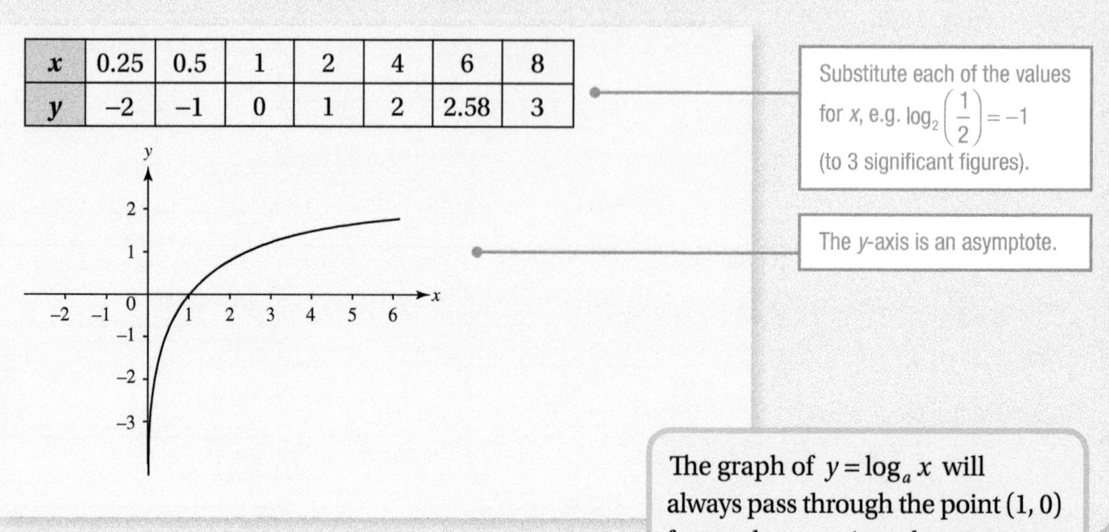

Substitute each of the values for x, e.g. $\log_2\left(\frac{1}{2}\right)=-1$ (to 3 significant figures).

The y-axis is an asymptote.

The graph of $y=\log_a x$ will always pass through the point (1, 0) for any base a, since $\log_a 1=0$

Complete a table of values and plot the graph of $y=\log_3 x$ for $x=0$ to $x=10$ **Try It 5**

See Ch2.4 For a reminder on graph transformations.

You can also transform the graphs of exponential and logarithmic functions using the usual rules for graph transformations. In the case of a translation, it may be helpful to think first about what happens to the asymptote. For example, the graph of $y=2^x+3$ is the graph of $y=2^x$ translated 3 units in the positive y-direction, therefore the asymptote of $y=0$ is translated to $y=3$

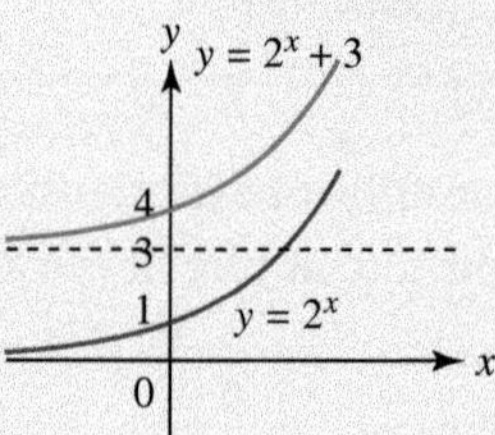

Example 6

The graph of $y=3^x$ is shown. Sketch the graphs of

a $y=3^x-2$ **b** $y=2(3^x)$

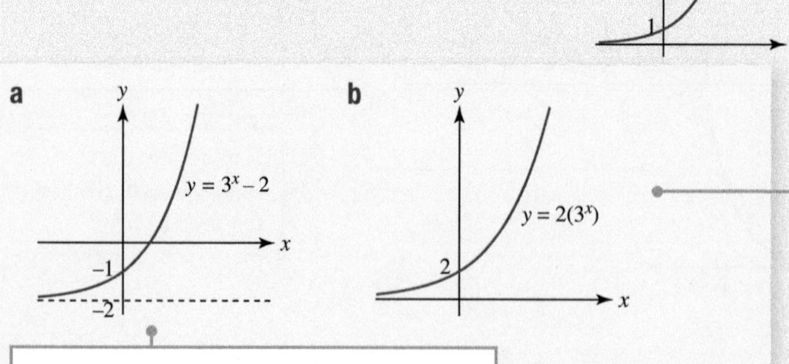

The asymptote is translated from $y=0$ to $y=-2$

The curve is stretched by scale factor 2 in the y-direction, so the y-intercept becomes (0, 2)

Sketch the graphs of **a** $y=4+3^x$ **b** $y=3^{-x}$ **Try It 6**

Example 7

log(x) is only defined when $x > 0$

The graph of $y = \log_4 x$ is shown.

Sketch the graphs of

a $y = \log_4(x+3)$, $x > -3$ **b** $y = \log_4(-x)$, $x < 0$

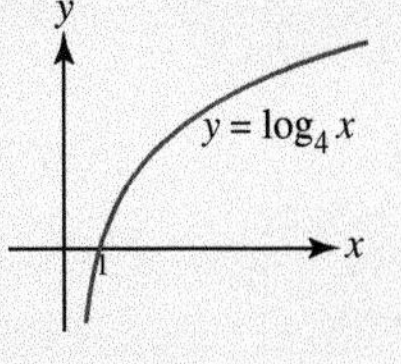

a

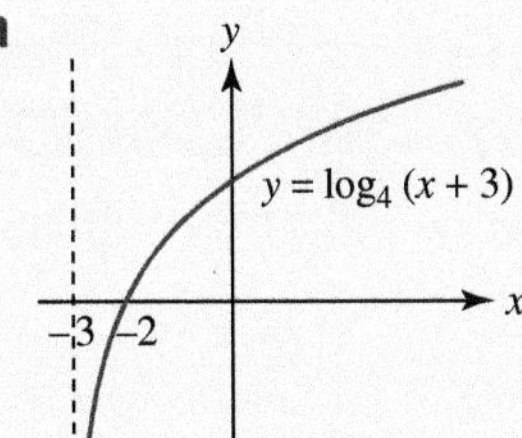

b

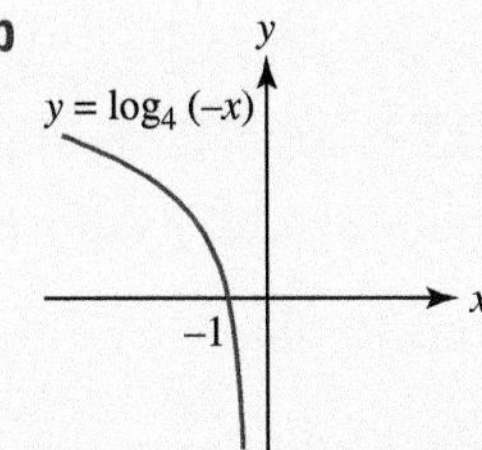

The curve is reflected in the y-axis so the x-intercept becomes (−1, 0)

Try It 7

Sketch the graphs of **a** $y = \log_4(x-2)$ **b** $y = \log_4\left(\frac{x}{3}\right)$

Bridging Exercise Topic A

Bridging to Ch5.1

1 Solve each of these equations, giving your answers to 3 significant figures where appropriate.

a $3^x = 27$ **b** $2^x = 16$ **c** $3^x = \frac{1}{9}$ **d** $2^x = \frac{1}{32}$

e $3^x = 13$ **f** $2^x = 29$ **g** $4^x = \frac{1}{256}$ **h** $5^x = 0.008$

i $6^x = 38$ **j** $7^x = 0.3$ **k** $5^x = 682$ **l** $6^x = 0.2$

m $2(4^x) = 85$ **n** $3(5^x) = 0.7$ **o** $6^{x+1} = 17$ **p** $7^{2x+3} = \frac{1}{49}$

2 Solve each of these equations, giving your answers to 3 significant figures where appropriate.

a $(3^x)^2 - 2(3^x) = 0$ **b** $(2^x)^2 - 5(2^x) + 6 = 0$ **c** $3^{2x} - 3(3^x) - 4 = 0$

d $2^{2x} + 2^x - 30 = 0$ **e** $2(6^{2x}) - 31(6^x) + 15 = 0$ **f** $3^{2x} - 11(3^x) + 24 = 0$

g $2^{2x} - 2^{x+1} - 120 = 0$ **h** $7^{2x} - 7^{x+1} - 8 = 0$ **i** $3^{2x} - 3^{x+2} + 20 = 0$

3 Copy and complete the table of values and plot the graph of

a $y = 5^x$ **b** $y = \left(\frac{1}{2}\right)^x$

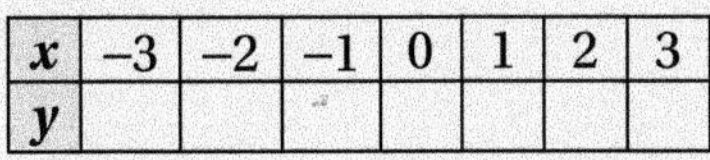

x	−3	−2	−1	0	1	2	3
y							

4 Copy and complete the table of values and plot, for $x > 0$, the graph of

a $y = \log_5(x)$ **b** $y = \log_4(x)$

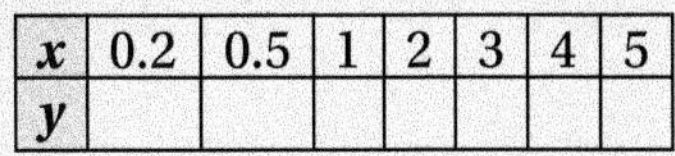

x	0.2	0.5	1	2	3	4	5
y							

5 The graph of $y = 6^x$ is shown. Sketch the graph of

a $y = 5 + 6^x$ **b** $y = 4(6^x)$

c $y = 6^x - 2$ **d** $y = -6^x$

6 The graph of $y = \log_7 x$, $x > 0$ is shown. Sketch the graph of

a $y = \log_7(x+3)$, $x > -3$ **b** $y = \log_7\left(\frac{x}{5}\right)$, $x > 0$

c $y = \log_7(x-5)$, $x > 5$ **d** $y = \log_7(-x)$, $x < 0$

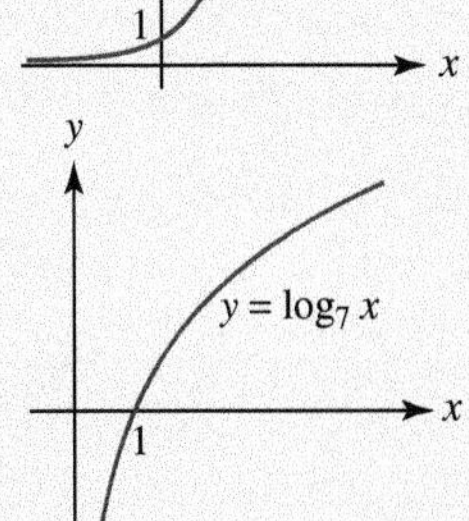

MyMaths 2062, 2063 SEARCH

5.1 The laws of logarithms

Fluency and skills

Logarithms allow you to perform certain calculations with a large number of digits more efficiently.

You read $10^4 = 10\,000$ as '10 to the **power** 4 equals 10 000'.

Similarly, you read $x = a^n$ as 'x is equal to a to the power n'.

Logarithms are a different way of working with powers. Another way of reading $x = a^n$ is to say 'n is the logarithm of x to **base** a'. The following three statements are equivalent.

Key point

$x = a^n$

$n = \log_a x$

n is the log of x to base a

For example, $10^2 = 100$ and $\log_{10} 100 = 2$ are equivalent statements.

The most common base to use is 10, but you can use any positive number.

You need to know the following three cases for $x = a^n$. They are true when $a > 0$ and $a \neq 1$

Key point

When $n = 1 \Rightarrow a^1 = a \Rightarrow \log_a a = 1$

When $n = 0 \Rightarrow a^0 = 1 \Rightarrow \log_a 1 = 0$

When $n = -1 \Rightarrow a^{-1} = \frac{1}{a} \Rightarrow \log_a \left(\frac{1}{a}\right) = -1$

There are three **laws of logarithms**:

Key point

Law 1 $\quad \log_a(xy) = \log_a x + \log_a y$

Law 2 $\quad \log_a\left(\frac{x}{y}\right) = \log_a x - \log_a y$

Law 3 $\quad \log_a(x^k) = k\log_a x$

See Ch1.2

For a reminder on index laws.

These laws are derived from the rules of indices.
For example, for Law 1, if $p = \log_a x$ and $q = \log_a y$, then $x = a^p$ and $y = a^q$
You know $xy = a^p \times a^q = a^{p+q}$ so $\log_a(xy) = p + q = \log_a x + \log_a y$

If the base of a logarithm is not given, then always use base 10

Key point

You can write $\log_{10} x$ as simply $\log x$

PURE

Example 1

Show that $\log_5 125 = 3$

$\log_5 125 = \log_5 5^3$ — Write 125 using powers.

$= 3 \times \log_5 5$ — Use Law 3 to simplify the log.

$= 3 \times 1 = 3$

Example 2

Write as a single logarithm $5\log_{10} 2 - \log_{10} 4 + \frac{1}{2}\log_{10} x$. Show your working.

$5\log 2 - \log 4 + \frac{1}{2}\log x = \log 2^5 - \log 4 + \log x^{\frac{1}{2}}$ — Use Law 3 to write 5 log 2 as $\log 2^5$ and $\frac{1}{2}\log x$ as $\log x^{\frac{1}{2}}$

$= \log\left(\frac{2^5}{4} \times x^{\frac{1}{2}}\right)$ — Use Laws 1 and 2 to combine logs.

$= \log\left(8\sqrt{x}\right)$ — Simplify the expression.

Exercise 5.1A Fluency and skills

1 Write each of the following in logarithmic form.

a $2^3 = 8$ **b** $3^2 = 9$

c $10^3 = 1000$ **d** $2 = 16^{\frac{1}{4}}$

e $0.001 = 10^{-3}$ **f** $\frac{1}{4} = 4^{-1}$

g $\frac{1}{8} = 2^{-3}$ **h** $\frac{1}{4} = 8^{-\frac{2}{3}}$

2 Write each of the following in index notation.

a $\log_2 32 = 5$ **b** $\log_2 16 = 4$

c $\log_3 81 = 4$ **d** $\log_5 1 = 0$

e $\log_3\left(\frac{1}{9}\right) = -2$ **f** $1 = \log_6 6$

g $\log_{16} 2 = \frac{1}{4}$ **h** $\log_2\left(\frac{1}{64}\right) = -6$

3 Find the value of each of these expressions. Show your working.

a $\log_3 9$ **b** $\log_2 16$

c $\log_4 16$ **d** $\log_{16} 16$

e $\log_5 125 + \log_5 5$ **f** $\log_3 81 - \log_3 27$

g $\log_2 8 + \log_2 2$ **h** $\log_3 9 - \log_3 27$

4 Write each of these as a single logarithm.

a $\log 4 + \log 3$

b $\log 12 - \log 2$

c $\log 2 + \log 6 - \log 3$

d $3\log 2 + \log 4$

e $3\log 3 + 2\log 2$

f $3\log 6 - 2\log 3 + \frac{1}{2}\log 9$

g $4\log 2 - 5\log 1$

h $2\log 4 + \log\frac{1}{2}$

i $\frac{1}{2}\log 9 + \frac{1}{3}\log 8$

j $\frac{1}{2}\log 4 + \frac{2}{3}\log 27$

k $4\log x + 2\log y$

l $2\log x + \log y$

5 Write each of these in terms of $\log a$, $\log b$ and $\log c$, where a, b and c are greater than zero.

a $\log(a^2b)$

b $\log\left(\frac{a}{b}\right)$

c $\log\left(\frac{a^2}{b^3}\right)$

d $\log(a\sqrt{b})$

e $\log\left(\frac{\sqrt{ab}}{c}\right)$

f $\log(\sqrt{abc})$

g $\log\left(a\sqrt{\frac{b}{c}}\right)$

6 Write each of these logarithms in terms of $\log 2$ and $\log 3$. By taking $\log 2 \approx 0.301$ and $\log 3 \approx 0.477$, find their approximate values.

a $\log 12$ b $\log 18$

c $\log 4.5$ d $\log 13.5$

e $\log 5$ f $\log 0.125$

7 Prove that

a $\frac{\log 125}{\log 25} = 1.5$ b $\frac{\log 27}{\log 243} = 0.6$

8 Write $\log 40$ in terms of $\log 5$. If $\log 5 = 0.698\,970\,00\ldots$ find the value of $\log 40$ correct to 6 decimal places.

Reasoning and problem-solving

Strategy

To solve problems with logarithms

1. Convert between index notation and logarithmic notation.
2. Apply the laws of logarithms if necessary and any results for special cases.
3. Manipulate and solve the equation. Check your solution by substituting back into the original equation.

Example 3

Solve the equation $2^{3x+1} = 36$. Show your working.

$3x + 1 = \log_2 36$

$3x = \log_2 36 - 1$

$x = \frac{1}{3}(\log_2 36 - 1) = 1.39$ (to 3 sf)

Check: $2^{3x+1} = 2^{4.17+1} = 2^{5.17} = 36.0$

(1) Convert to logarithmic notation using the equivalent statements $x = a^n$ and $n = \log_a x$

(3) Manipulate and solve for x. Use the original equation to check your solution.

Calculator

Try it on your calculator

You can use a calculator to solve equations with exponents.

Eq: $3^{2X-1} = 10$
X = 1.547951637
Lft = 10
Rgt = 10

REPT

Activity

Find out how to solve $3^{2x-1} = 10$ on *your* calculator.

Exercise 5.1B Reasoning and problem-solving

1 Solve each of these equations. Give your answers to 3 sf and show your working.

a $2^x = 5$ b $3^x = 10$

c $5^x = 4$ d $2^{x+1} = 7$

e $3^{x-1} = 80$ f $2^x \times 2^{x-1} = 20$

g $\frac{1}{2^x} = 9$ h $\left(\frac{3}{4}\right)^x = 2$

i $2^{x+1} = 3^{2x}$ j $5^{3x-1} = 2^{x+3}$

2 Solve each of these equations. Show your working.

a $\log_{10}(2x-40) = 3$

b $\log_5(3x+4) = 2$

c $\log_3(x+2) - \log_3 x = \log_3 8$

d $\log_3(x+2) + \log_3 x = 1$

3 Show that $x = 316$ and $y = 3.16$ is the solution to these simultaneous equations to 3 sf.

$\log x + \log y = 3$
$\log x - \log y = 2$

4 Let $y = a^x$ and use the fact that $a^{2x} = (a^x)^2$ to solve each of these quadratic equations. Give your answers to 3 sf where appropriate.

a $2^{2x} - 3 \times 2^x + 2 = 0$

b $3^{2x} - 12 \times 3^x + 27 = 0$

c $2^{2x} + 6 = 5 \times 2^x$

d $2^{2x} - 2^{x+1} = 8$

e $5^{2x} + 5^{x+1} - 50 = 0$

f $3^{2x+1} - 26 \times 3^x = 9$

g $2^{2x+1} - 13 \times 2^x + 20 = 0$

h $9^x + 8 = 2(3^{x+1})$

i $25^x + 4 = 5^{x+1}$

j $2^x \times 2^{x+1} = 10$

5 a Solve $3^{2x-1} - 5 \times 3^{x-1} + 2 = 0$

b Hence, find the point of intersection of the graphs of $y = 3^{2x-1} + 2$ and $y = 5 \times 3^{x-1}$

6 Find the point of intersection of the graphs of $y = 2^{2x} - 5$ and $y = 4 \times 2^x$. Show your working.

7 a Solve

i $3^x > 10\,000$ ii $0.2^x < 0.001$

b Calculate the smallest integer for which $2^n > 1\,000\,000$

c Calculate the largest integer for which $0.2^n > 0.000\,005$

8 Calculate the smallest positive integer value of x such that

a $\left(1 + \frac{x}{100}\right)^6 > 1.25$

b $1 - 0.8^x > 0.95$

9 Given that $x = a^p$ and $y = a^q$, prove that

a $\log_a\left(\frac{x}{y}\right) = \log_a x - \log_a y$

b $\log_a(x^k) = k\log_a x$

10 a Prove that $\log_a b = \frac{1}{\log_b a}$

b Solve

i $\log_5 x = 4\log_x 5$

ii $\log_3 x + 8\log_x 3 = 6$

Challenge

11 a Starting with $x = a^y$, prove that $\log_a x = \frac{\log_b x}{\log_b a}$

b Express $\log_2 12$ as a logarithm in base 10, so that $\log_2 12 = k \times \log_{10} 12$ and find the value of k

c Write these as logarithms in base 10 and use a calculator to find their values.

i $\log_6 37$ ii $\log_4 6$ iii $\log_3 25$

12 If $x = \log_y z$, $y = \log_z x$ and $z = \log_x y$, prove that $xyz = 1$

MyMaths 2062, 2063, 2257 SEARCH

5.2 Exponential functions

Fluency and skills

Key point

The general equation of an **exponential function** is $y = a^x$ where a is a positive constant.

The table shows values for the graphs of $y = 2^x$ and $y = 2^{-x}$

x	-2	-1	0	1	2
$y = 2^x$	$2^{-2} = \frac{1}{4}$	$2^{-1} = \frac{1}{2}$	$2^0 = 1$	$2^1 = 2$	$2^2 = 4$
$y = 2^{-x} = \left(\frac{1}{2}\right)^x$	$2^2 = 4$	$2^1 = 2$	$2^0 = 1$	$2^{-1} = \frac{1}{2}$	$2^{-2} = \frac{1}{4}$

Note that $a^{-x} = \left(\frac{1}{a}\right)^x$

The diagram shows how the shape of the graph varies for different values of a. It also shows the relationship between $y = a^x$ and $y = a^{-x}$

There is one special value of a where the gradient of the curve $y = a^x$ is equal to the value of a^x for all values of x

This is called ***the* exponential function** and is written $y = e^x$, where $\mathbf{e = 2.71828}$ (to 5 dp).

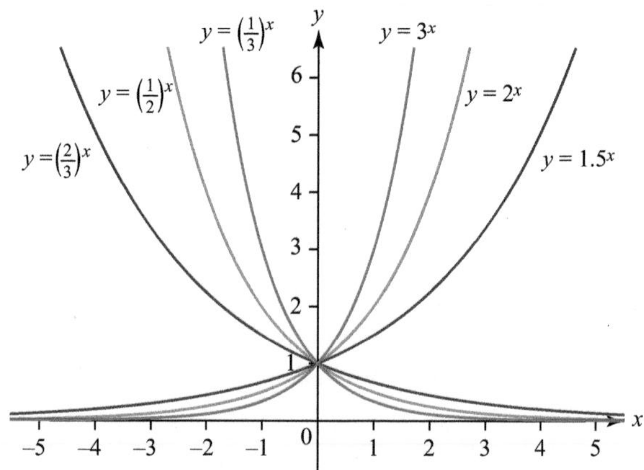

Key point

The graph of $y = e^x$ has a gradient of e^x at any point (x, y)

e is irrational (like π) and e^x is sometimes written $\exp(x)$

See Ch2.4 For a reminder on transformations of graphs and proportionality.

You can transform the graph of $y = e^x$ in various ways.

In particular, a stretch parallel to the x-axis by a scale factor $\frac{1}{k}$ results in a curve with the equation $y = e^{kx}$

The tangent at P_2 is steeper by a factor k than the tangent at P_1

The graph of $y = e^{kx}$ has a gradient of $k \times e^{kx}$ at the point (x, y)

This means that the gradient of $y = e^{kx}$ is proportional to y at the point (x, y), where the constant of proportionality is k

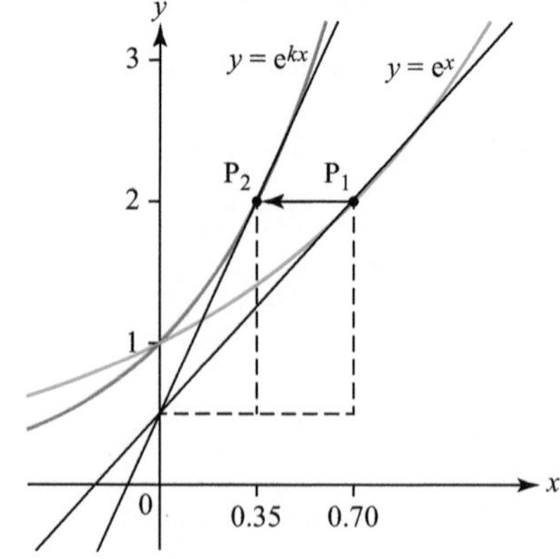

▲ This diagram shows $k = 2$

The inverses of exponential functions are logarithmic functions. You can find the inverse of an exponential function, $y = a^x$, using the following method.

$y = a^x$

$\log_a y = \log_a a^x$

$\log_a y = x \log_a a$

$\log_a y = x$

$y = \log_a x$

To find the inverse:

Take logs to base a on both sides.

Use log laws to find x in terms of y

Remember $\log_a a = 1$

Then interchange x and y

Key point

The inverse of $y = a^x$ is the logarithmic function, $y = \log_a x$

The diagram shows the shape of some exponential functions and their inverses, and the relationships between them.
In particular

- a^x is positive for all values of x
- $\log_a x$ does not exist for negative values of x
- all the exponential graphs pass through $(0, 1)$ because $a^0 = 1$ for all a $(a \neq 0)$
- all the logarithmic graphs pass through $(1, 0)$ because $\log_a 1 = 0$ for all $a > 0$

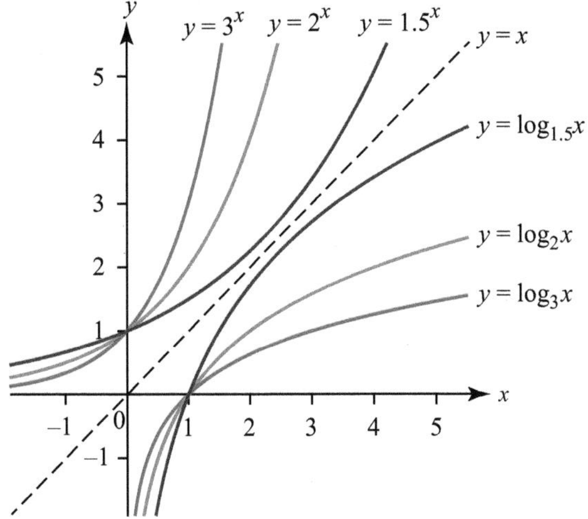

ICT Resource online

To experiment with the graph of $y = a^x$ and its inverse, click this link in the digital book.

Key point

The inverse of $y = e^x$ is $y = \log_e x$ which can be written $y = \ln x$

$\ln x$ is called the **natural (or Naperian) logarithm**.

See p.382
For a list of mathematical notation.

Note that $y = e^x$ passes through the point $(0, 1)$ and its inverse $y = \ln x$ passes through the point $(1, 0)$
The x-axis is an asymptote for the graph $y = e^x$
The y-axis is an asymptote for the graph $y = \ln x$

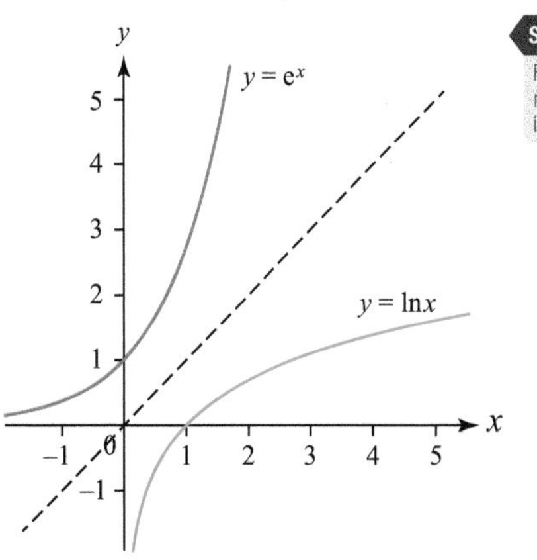

MyMaths 2061, 2133, 2134 SEARCH

Example 1

a This is a table of values for the function $y = a^x$
Find the value of a. Hence, find the values of p and q

x	0	1	2	3
y	1	p	9	q

b The graph of the function $y = e^{kx}$ contains the point $(3, e^6)$
Find the value of k and the gradient of the graph at this point.

c Using your value of a from part **a**, find the inverse of the function $y = a^x$

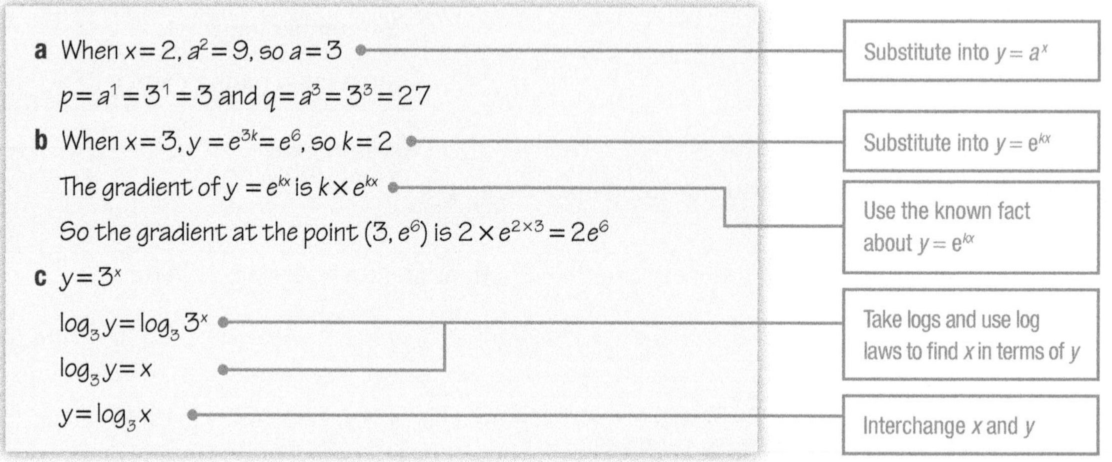

a When $x = 2$, $a^2 = 9$, so $a = 3$
$p = a^1 = 3^1 = 3$ and $q = a^3 = 3^3 = 27$
b When $x = 3$, $y = e^{3k} = e^6$, so $k = 2$
The gradient of $y = e^{kx}$ is $k \times e^{kx}$
So the gradient at the point $(3, e^6)$ is $2 \times e^{2\times3} = 2e^6$
c $y = 3^x$
$\log_3 y = \log_3 3^x$
$\log_3 y = x$
$y = \log_3 x$
Substitute into $y = a^x$
Substitute into $y = e^{kx}$
Use the known fact about $y = e^{kx}$
Take logs and use log laws to find x in terms of y
Interchange x and y

Exercise 5.2A Fluency and skills

1 Draw the graphs of $y = 3^x$ and $y = 3^{-x}$ for $-2 \le x \le 2$ on the same axes.
Explain why the graph of $y = \left(\frac{1}{3}\right)^x$ is identical to one of the graphs you have drawn.

2 These two tables give the values (x, y) for three relationships. Values are given to 4 sf where appropriate. For each table

i Write the equation for each of the three relationships,

ii State which one of the three relationships is exponential,

iii Write the three y-values when $x = 5$

a

x	1	2	3	4
y_1	2	4	6	8
y_2	1	4	9	16
y_3	2	4	8	16

b

x	1	2	3	4
y_1	12	6	4	3
y_2	1	1.414	1.732	2
y_3	$\frac{1}{2}$	$\frac{1}{4}$	$\frac{1}{8}$	$\frac{1}{16}$

3 These two tables give the values of x and y for two exponential relationships.

Copy and complete the tables and write the equations for the relationships.

a

x	−2	−1	1	2	3	4
y			3	9	27	

b

x	−2	−1	1	2	3	4
y			$\frac{1}{5}$	$\frac{1}{25}$	$\frac{1}{125}$	

4 a Copy and complete this table for the curve $y = e^x$. Give your answers to 3 sf.

x	y	Gradient at (x, y)
0		
1		
2		
3		
4		

b Repeat for the curve $y = e^{3x}$

5 Each of the points W, X, Y, Z lies on one of the curves A, B, C or D. Match each point to a curve. The coordinates are given correct to 2 dp.

$W(2.10, 0.23)$ $X(0.80, 0.33)$ $Y(1.20, 2.30)$ $Z(0.50, 1.73)$

$A\ y = 2^x$ $B\ y = 3^x$ $C\ y = 2^{-x}$ $D\ y = 4^{-x}$

6 a Write down the gradient of the graph of

i $y = e^x$ at the point $(2, e^2)$

ii $y = e^{-x}$ at the point $(2, e^{-2})$

b Find the equation of the tangent to the curves at the given points.

7 Determine the equation of the tangent to the curve $y = e^x$ at the point where

a $x = 3$ **b** $x = \frac{1}{2}$

8 Describe the transformation that maps the curve $y = e^x$ onto each of these curves. Sketch the graph of $y = e^x$ and its image in each case. You can check your sketches on a graphics calculator.

a $y = e^{-x}$ **b** $y = -e^x$ **c** $y = 2e^x$ **d** $y = e^{2x}$

e $y = e^x + 1$ **f** $y = e^x - 1$ **g** $y = e^{x+1}$ **h** $y = e^{x-1}$

9 a Draw the graph of $y = 3^x$ on graph paper with both axes labelled from –2 to 9. On the same axes, draw the graph of its inverse by reflecting in the line $y = x$

Answer parts **b** to **d** using your graphs. You can use your calculator to check your answers.

b Write the coordinates of the image of the point (1, 3) under the reflection.

c Write the coordinates of the point where $x = 1.2$ and also of its image point. Hence write the value of $\log_3 3.74$

d Write the values of

i $3^{1.5}$ **ii** $\log_3 5.20$

iii $3^{0.5}$ **iv** $\log_3 1.73$

v $\log_3 6$ **vi** $\log_3 7$

10 a The graph of $y = e^x$ passes through the points $(3, p)$, (q, e^2) and $(r, 9)$. Write the values of p, q and r

b The graph of $y = e^{-x}$ passes through the points $\left(\frac{1}{2}, a\right)$, (b, e^3) and $(c, 9)$. Write the values of a, b and c

MyMaths 2061, 2133, 2134 SEARCH

Reasoning and problem-solving

Strategy

To solve problems involving exponential functions

1. Draw or sketch a graph if it is helpful.
2. Use what you know about the gradients of $y = e^x$ and $y = e^{kx}$
3. Use the relationship between an exponential function and its inverse.

Example 2

The tangent to the curve $y = e^x$ at point P, where $x = 2$, intersects the y-axis at point Q
Find point Q

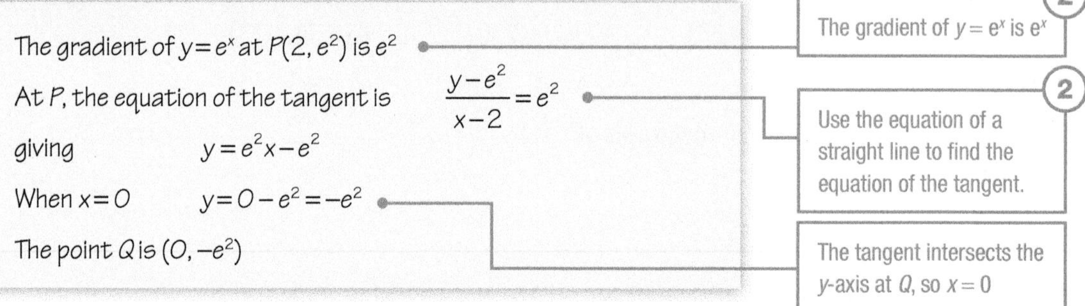

The gradient of $y = e^x$ at $P(2, e^2)$ is e^2
At P, the equation of the tangent is $\frac{y - e^2}{x - 2} = e^2$
giving $y = e^2x - e^2$
When $x = 0$ $y = 0 - e^2 = -e^2$
The point Q is $(0, -e^2)$
2 The gradient of $y = e^x$ is e^x
2 Use the equation of a straight line to find the equation of the tangent.
The tangent intersects the y-axis at Q, so $x = 0$

Example 3

State the inverse of the function $y = \ln x$. Find the equation of the normal to this inverse at $x = 3$, giving values to 2 dp.

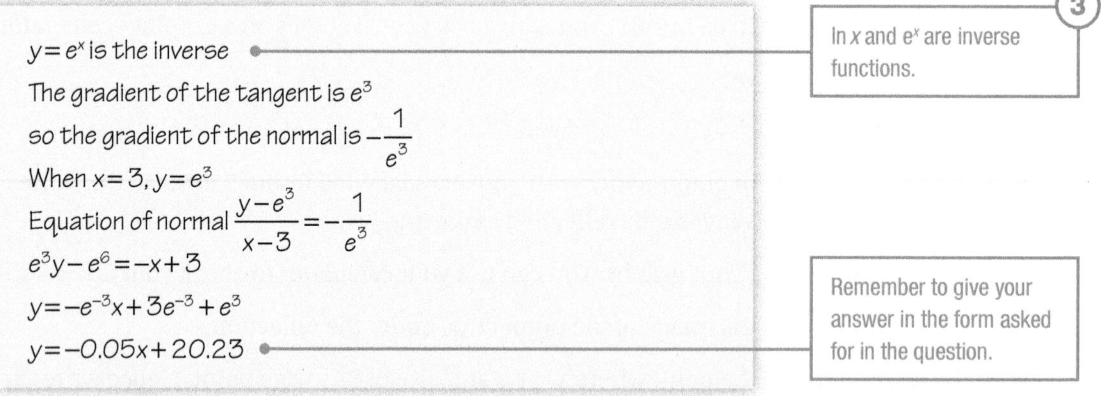

$y = e^x$ is the inverse
The gradient of the tangent is e^3
so the gradient of the normal is $-\frac{1}{e^3}$
When $x = 3$, $y = e^3$
Equation of normal $\frac{y - e^3}{x - 3} = -\frac{1}{e^3}$
$e^3y - e^6 = -x + 3$
$y = -e^{-3}x + 3e^{-3} + e^3$
$y = -0.05x + 20.23$
3 $\ln x$ and e^x are inverse functions.
Remember to give your answer in the form asked for in the question.

Exercise 5.2B Reasoning and problem-solving

1 a Draw the graphs of the curve $y = 2^x$ and the line $y = 6$ on the same axes for $0 \le x \le 3$

Write an estimate of their point of intersection and hence deduce an approximate solution to the equation $2^x = 6$

b Draw the graphs of $y = 3^x$ and $y = x + 3$ on the same axes for $0 \le x \le 2$. Find an approximate solution to the equation $3^x - x = 3$ from your diagram.

2 a Sketch the graphs of $y = 2^x$ and $y = x^2$ on the same axes and explain why there is more than one possible solution to the equation $2^x = x^2$

b Draw the two graphs for $-2 \le x \le 4$ and find solutions to the equation in part **a**, giving answers to 2 dp where appropriate. You can check your sketches on a graphics calculator.

3 Use an algebraic method to find the point of intersection for each of these pairs of curves.

a $y = 4^{3x}$ and $y = 4^{x+6}$

b $y = 4^x$ and $y = 2^{x+1}$

c $y = 3^x$ and $y = 9^{x-2}$

d $y = 4 \times 2^x$ and $y = 2^{-x}$

4 The curves of $y = e^x$ and $y = e^{-x}$ intersect the straight line $x = 2$ at points P and Q respectively. Calculate the distance PQ to 3 sf.

5 Prove that the tangent to the curve $y = e^{3x}$ at the point where $x = 1$ passes through the point $\left(\frac{2}{3}, 0\right)$

6 Find the point where the tangent to the curve $y = e^{\frac{1}{2}x}$ at the point $(4, e^2)$ intersects the straight line $x = 6$. Show your working.

7 Find the equation of the normal to the curve $y = e^{-2x}$ when $x = 1$, giving values to 3 dp.

8 Find the x-intercept of the normal to the curve $y = e^{2x}$ at the point where $x = 1$

9 a Write the equation of the graph of the inverse of the function $y = a^x$

b Show that the inverse of $y = 2^x$ is approximately $y = 1.44 \times \ln x$. Hence, find the value of x if $2^x = 17$

PURE

Challenge

10 The diagram shows how to investigate the gradient of a curve $y = a^x$ at P(0, 1) using differentiation from first principles.

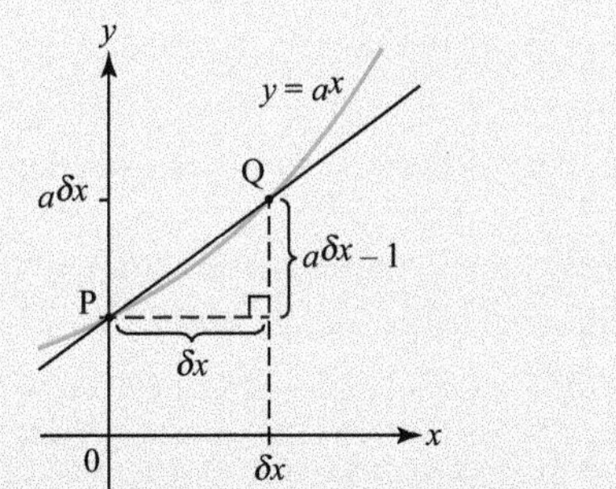

See Ch4.1

For a reminder on differentiation from first principles.

a Taking $\delta x = 0.0001$, use a spreadsheet to copy and complete the table of values for the gradient of the chord $PQ = \frac{a^{\delta x} - 1}{\delta x}$ for curves with different values of a

	A	B	C	D	E	F	G
1	a	2.0	2.2	2.4	2.6	2.8	3.0
2	$\frac{a^{\delta x} - 1}{\delta x}$						

b Between which pair of values does e lie? Give a reason for your answer.

c Create further tables for narrower ranges of a to find the value of e correct to 2 decimal places.

d Why will this method only give an approximate value of e?

(You can take increasingly small values for δx to convince yourself that e does lie between the values you have found.)

11 The graph of $y = e^x$ has a gradient of e^x at the point (x, y)

a Justify this statement by copying and completing this table of values. Take $\delta x = 0.0001$ and choose your own values of x for $x > 3$. You could use a computer spreadsheet.

	A	B	C	D	E
1	x	1	2	3	
2	e^x	2.71828			
3	$\frac{e^{x+\delta x} - e^x}{\delta x}$				

b Why is this method not a proof of the statement?

MyMaths 2061, 2133, 2134 SEARCH

5.3 Exponential processes

Fluency and skills

Mathematical models are used to describe and make predictions about real-life events using mathematical language and symbols.

The gradient of $y = e^{kx}$ is proportional to y at the point (x, y), where the constant of proportionality is k. This property of proportionality occurs frequently in the natural world and allows you to create mathematical models of events such as radioactive decay and population growth.

For example, if the rate of increase of a population of bacteria is directly proportional to the number of bacteria, y, then $\frac{dy}{dt} = ky$. The rate of change of $y = e^{kt}$ is proportional to y, so an exponential function is a good model for this situation.

Key point

An equation of the form $y = Ae^{kt}$ gives an exponential model where A and k are constants.

Example 1

The population, P hundreds of cells, of an organism grows exponentially over time, t hours, according to $P = Ae^{\frac{1}{20}t}$

t	0	5	10	15
P	4			

a Find the value of A, copy and complete the table and draw a graph to represent the data.

b What is the rate of increase in the population when $t = 5$? Show your working.

a When $t = 0$, $P = A \times e^0 = A \times 1 = A$, so $A = 4$

t	0	5	10	15
P	4	$4e^{0.25} = 5.14$	$4e^{0.5} = 6.59$	$4e^{0.75} = 8.47$

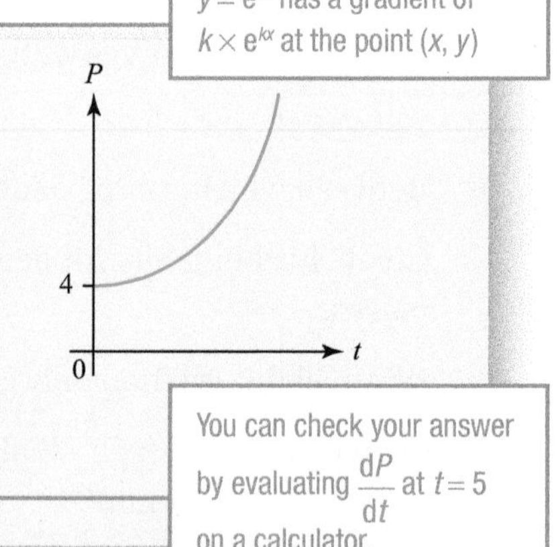

b When $t = 5$, the rate of change of P is

$A \times \frac{1}{20}e^{\frac{5}{20}} = 4 \times \frac{1}{20}e^{\frac{1}{4}} = 0.257$ hundreds

$= 25.7$ cells per hour

$y = e^{kx}$ has a gradient of $k \times e^{kx}$ at the point (x, y)

You can check your answer by evaluating $\frac{dP}{dt}$ at $t = 5$ on a calculator.

Exercise 5.3A Fluency and skills

1 The value of n, after t hours, is given by $n = 2^t$

a Copy and complete this table.

t	0	1	2	3	4
n					

b Calculate the value of n after 6 hours.

2 A patient is injected with 10 units of insulin. The drug content in the body decreases exponentially and the number of units, n, left in the patient after t minutes is modelled by $n = A \times 0.95^t$

The actual number of units in the patient is measured at the times given in this table.

t	0	2	4	6	8
Actual units	10.0	9.1	8.2	7.4	6.7
n					

a State the value of A

b Copy and complete the table. Does the model predict reasonably accurate values of n for $t \le 8$?

c Does the model indicate that less than half the initial insulin is present in the patient after 16 minutes?

3 The variables x, y and z change with time t. Calculate the rate of change of each variable at the instant when **i** $t=0$ **ii** $t=2$

a $x=e^{2t}$ **b** $y=5e^{3t}$ **c** $z=100e^{-\frac{t}{20}}$

Show your working.

Do x, y and z grow or decay over time?

4 **a** The height h cm of a bean shoot t hours after germination is given by $h=0.3e^{0.1t}$

What is the rate of growth of the bean shoot when $t=5$? Show your working.

b A radioactive chemical decays so that, after t hours, its mass m is given by $m=2\times e^{-0.01t}$ kg. Calculate its rate of decay after 100 hours.

5 The temperature θ°C of water in a boiler rises so that $\theta=\frac{1}{5}e^{t}$ after t minutes.

a What is the temperature

i Initially, **ii** After 2 minutes?

b What is the average rate of change of temperature over the first 2 minutes?

c What is the instantaneous rate of change of temperature when $t=2$? Show your working.

d The boiler switches off after 6 minutes. What is the temperature at this time?

6 The population of a country at the start of a given year, P millions, is growing exponentially so that $P=15e^{0.06t}$ where t is the time in years after 2000. Calculate

a The size of the population at the start of 2006

b The rate of increase in the population at the start of 2006

c The average rate of increase in the population from the start of 2000 to the start of 2006

7 An injected drug decays exponentially. Its concentration, C mg per ml, after t hours is given by $C=2\times e^{-0.45t}$

a What is the initial concentration?

b Calculate the concentration after 4 hours.

c What is the rate of decay of the drug when **i** $t=0$ **ii** $t=4$?

Show your working.

8 The cost of living in Exruria increased by 5% each year from 2000 to 2010. The weekly food bill per family was £P at the start of 2000

a Show that by the start of 2002 this bill was £$P\times 1.05^2$

b If $P=102$, find the weekly food bill at the start of 2010

9 £P invested for n years at r% p.a. increases to £A where $A=P\left(1+\frac{r}{100}\right)^n$

Two friends have £1000 each. They place their money in different accounts and make no withdrawals for 4 years. Both accounts have annual fixed rates; one at 6% p.a., the other at 4% p.a. What is the difference in their savings at the end of the 4 years?

10 The mass m of a radioactive material at a time t is given by $m=m_0e^{-kt}$ where k and m_0 are constants. If $m=\frac{9}{10}m_0$ when $t=10$, find the value of k. Also find the time taken for the initial mass of material to decay to half that mass.

MyMaths 2136 SEARCH

Reasoning and problem-solving

Strategy

To solve modelling problems involving rates of change

1. Calculate data using the model.
2. Consider sketching or using a graphical calculator to graph the model.
3. Use your knowledge of exponential functions and logarithms to find rates of change and solve equations.
4. Compare actual data with your model and, where necessary, comment on any limitations.

Example 2

An area of fungus, A cm^2, grows over t days such that $A = 2 + 6e^{0.1t}$

a Sketch a graph of A against t by calculating A for $t = 0, 10, 20$ and 30

b How long does it take for the area of the fungus to double?

c What is the initial rate of change of A?

d Why might this model not be realistic for large values of t?

a

t	A
0	$2+6e^0=8$
10	$2+6e^1=18.3$
20	$2+6e^2=46.3$
30	$2+6e^3=122.5$

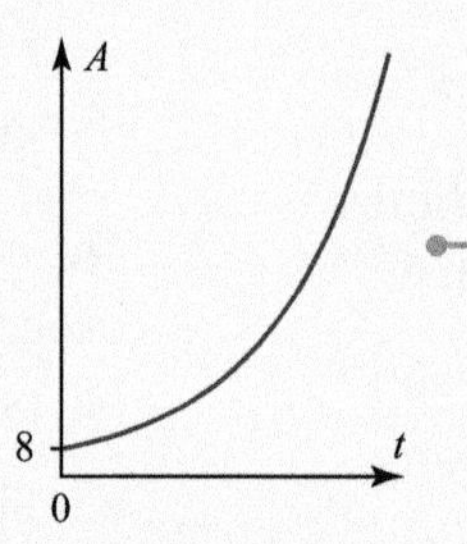

b Initially $A = 8$, so area doubles when $A = 16$

$16 = 2 + 6e^{0.1t}$

so $e^{0.1t} = \frac{7}{3}$

$\ln(e^{0.1t}) = \ln\frac{7}{3}$

$0.1t \times 1 = \ln\frac{7}{3}$

$t = 8.47$ to 3sf

It takes approximately 8 days for the fungus to double its area.

c Rate of change is $0 + 6 \times 0.1e^{0.1t}$

When $t = 0$, initial rate of change $= 0.6$ cm^2 per day.

d There is no limit placed on the area of fungus—it would increase to infinity according to the model. The model ignores factors such as limited space and changes in conditions.

(1) Substitute the given values of t to calculate A over time.

(2) Sketch a graph, starting with $t = 0$

The table shows that t is less than 10

Form an equation using the model.

(3) Take natural logs of both sides and simplify using log laws. Remember $\ln e = 1$

(3) The rate of change of $y = e^{kx}$ is ke^{kx}. The constant 2 has zero rate of change.

(4) Comment on limitations of the model.

Exercise 5.3B Reasoning and problem-solving

1 A population of insects, n, increases over t days, and can be modelled by $n = 100 - 80e^{-\frac{1}{5}t}$

a What was the initial number of insects?

b If there are 72 insects after 5 days, does this data fit the model?

c How many insects are there after 9 days?

d Does the model predict a limiting number of insects? If so, what is it?

2 A population of bacteria grows so that the actual number of bacteria is n after t hours. The number, p, predicted by a model of the growth, is given by $p = 2 + Ae^{\frac{1}{20}t}$

t	0	20	40	60	80	100
n	5	11	25	60	140	265
p	5					

a Calculate A and copy and complete the table.

b According to the model, how long does it take for the initial number of bacteria to triple?

c Draw graphs of n and p against t on the same axes. Is the model accurate at predicting the actual numbers? What limitations would you place on the model?

d What rate of growth is predicted at $t = 40$?

3 A doctor injects a patient with a drug that decays exponentially. Its concentration, C mg per ml, after a time, t minutes, is modelled by $C = Ae^{-kt}$ where A and k are constants.

a At the time of injecting, $C = 1.5$. After 5 minutes, $C = 0.25$. Calculate A and k

b The drug becomes ineffective if $C < 0.2$ For how long is the drug effective?

c Calculate the rate of decay of the drug 5 minutes after injection.

4 Trees in a local wood are infected by disease. The number of unhealthy trees, N, was observed over t years and modelled by $N = 200 - Ae^{-\frac{1}{20}t}$

a If there are 91 unhealthy trees after 10 years, calculate the value of A

b What is the initial number of unhealthy trees and the initial rate of change in the number of trees?

c How long does it takes for the initial number of unhealthy trees to triple?

d Explain why the model predicts a limit to the number of unhealthy trees and state its value.

5 A block of steel leaves a furnace and cools. Its temperature, θ°C, is given by $\theta = c + 500e^{-0.005t}$ after it has been cooling for t minutes.

a If the steel leaves the furnace at 530 °C, how long does it take for the temperature to drop to 100 °C?

b Calculate the rate of decrease of temperature (to 3 sf) when $t = 20$

c Explain why the model predicts a minimum temperature below which the steel does not cool. State its value.

d Why might this model not be appropriate?

Challenge

6 A rare plant has been monitored over time. Its population, n, after t years of monitoring has been modelled as $n = \frac{Ae^{\frac{1}{4}t}}{e^{\frac{1}{4}t} + k}$, where A and k are positive constants.

a There were 250 plants initially and 528 plants after 4 years. Calculate the values of A and k to 2 sf.

b Show that, according to this model, the population cannot exceed a certain number and calculate this number.

MyMaths 2136 SEARCH

5.4 Curve fitting

Fluency and skills

Two common, non-linear relationships take the form $\boldsymbol{y = ax^n}$ and $\boldsymbol{y = kb^x}$, where a, n, b and k are constants. These are called **polynomial** and **exponential relationships** respectively. Given values for x and y, we can use logs to determine the form of the relationship and find values for the constants.

When you plot these relationships on axes where the scales are linear, the range between plotted y-values gets larger and larger, and the graph becomes harder to read accurately. By using logs, you can scale down the y-values and convert the relationships into linear ones. This means that their graphs are transformed into straight line graphs.

Key point

$y = ax^n$ becomes $Y = nX + c$, where $Y = \log y$, $X = \log x$ and $c = \log a$

$y = kb^x$ becomes $Y = mx + c$, where $Y = \log y$, $m = \log b$ and $c = \log k$

These straight lines are **lines of best fit** through the data points.

Example 1

The relationship between x and y is given by $y = ax^n$ where a and n are constants.

x	10	20	30	50	100
y	88	232	409	837	2208

Use the data in this table to find the values of a and n, and write the relationship.

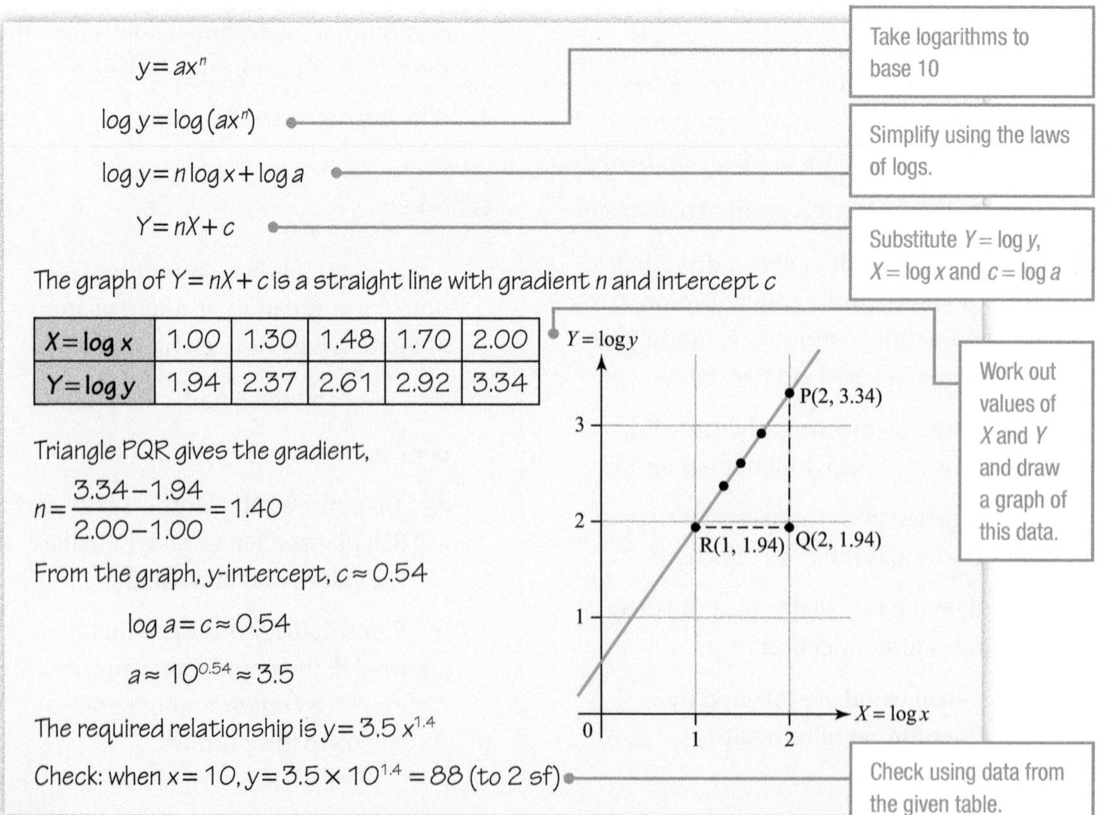

$y = ax^n$

$\log y = \log(ax^n)$ — Take logarithms to base 10

$\log y = n\log x + \log a$ — Simplify using the laws of logs.

$Y = nX + c$ — Substitute $Y = \log y$, $X = \log x$ and $c = \log a$

The graph of $Y = nX + c$ is a straight line with gradient n and intercept c

$X = \log x$	1.00	1.30	1.48	1.70	2.00
$Y = \log y$	1.94	2.37	2.61	2.92	3.34

Work out values of X and Y and draw a graph of this data.

Triangle PQR gives the gradient,

$$n = \frac{3.34 - 1.94}{2.00 - 1.00} = 1.40$$

From the graph, y-intercept, $c \approx 0.54$

$\log a = c \approx 0.54$

$a \approx 10^{0.54} \approx 3.5$

The required relationship is $y = 3.5\,x^{1.4}$

Check: when $x = 10$, $y = 3.5 \times 10^{1.4} = 88$ (to 2 sf) — Check using data from the given table.

Exercise 5.4A Fluency and skills

1 The relationship $y = ax^n$ is graphed as $Y = nX + c$ where $Y = \log y$, $X = \log x$ and $c = \log a$ as shown. Take readings from the graph to find the values of a and n. Write the relationship between x and y

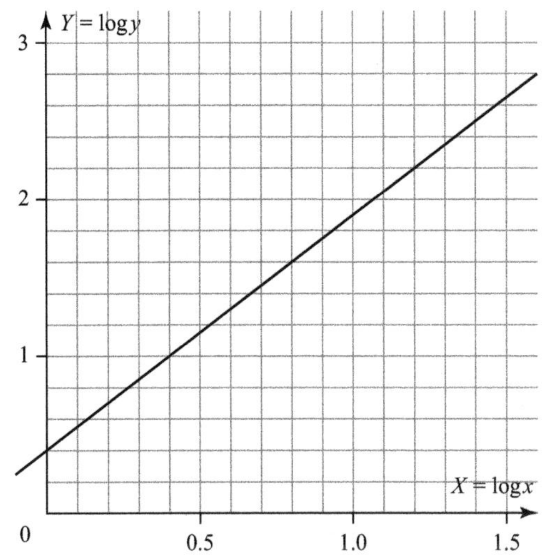

2 The relationship $y = kb^x$ is graphed as $Y = mx + c$ where $Y = \log y$, $m = \log b$ and $c = \log k$ as shown. Use the graph to find the values of b and k. Write the relationship between x and y

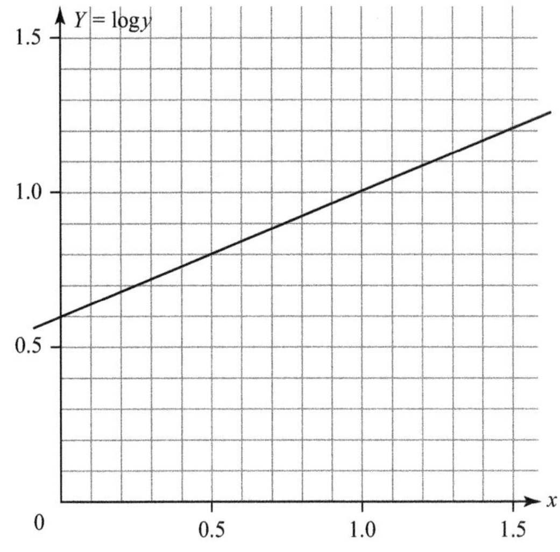

3 For each set of data, $y = ax^n$ where a and n are constants. For each table

i Draw an appropriate straight-line graph,

ii Find a and n and write the relationship.

Check your answers using the data.

a

x	2	3	4	5	6
y	17	40	74	117	172

b

x	1	2	3	4	5
y	1.6	2.3	2.9	3.4	3.8

c

x	10	15	20	25	30
y	376	333	305	286	270

d

x	2	4	6	8
y	1.72	0.57	0.30	0.19

4 For each set of data, $y = kb^x$ where k and b are constants. For each table

i Draw an appropriate straight-line graph,

ii Find k and b and write the relationship.

Check your answers using the data.

a

x	3	5	7	9
y	40	176	774	3415

b

x	0.2	0.3	0.4	0.5	0.6
y	8.2	9.2	10.4	11.6	13.1

c

x	2.0	2.5	3.0	3.5	4.0
y	176	139	110	87	69

d

x	0.3	0.4	0.5	0.6
y	0.33	0.39	0.46	0.54

5 This set of data gives four points on the graph of the function $y = f(x)$

x	0.5	1.0	2.0	3.0
y	0.66	1.08	2.92	7.87

Draw appropriate graphs to investigate whether the function has the form $f(x) = ax^n$ or $f(x) = kb^x$ and work out the function $f(x)$

Reasoning and problem-solving

Strategy

To solve problems involving polynomial and exponential relationships

1. Transform the non-linear functions $y = ax^n$ and $y = kb^x$ to linear functions using logarithms.
2. Use the transformed data to draw a straight-line graph, using a line of best fit when necessary.
3. Use your graph to calculate the constants and work out the relationship between x and y

When experimental data is gathered, it is often not exact. Data points plotted on a scatter diagram may not lie exactly on a straight line. You have to judge by eye where to draw the **line of best fit.**

ICT Resource online

To investigate logarithmic graphs and lines of best fit, click this link in the digital book.

Example 2

A patch of algae grows so that its area is y cm² after x days.

x	2	4	6	8
y	4.9	7.4	12.6	18.7

a Use the data in this table to find the relationship between x and y in the form $y = kb^x$

b What was the initial area of algae and what would you expect the area to be after 10 days?

a $\log y = \log(kb^x)$ — Take logarithms to base 10 (1)

$\log y = x\log b + \log k$ — Simplify using the laws of logs.

$Y = (\log b) \times x + \log k$ — Substitute $Y = \log y$ (1)

This is a straight line with gradient $\log b$ and intercept $\log k$ — Compare with $y = mx + c$

x	$Y = \log y$
2	0.69
4	0.87
6	1.10
8	1.27

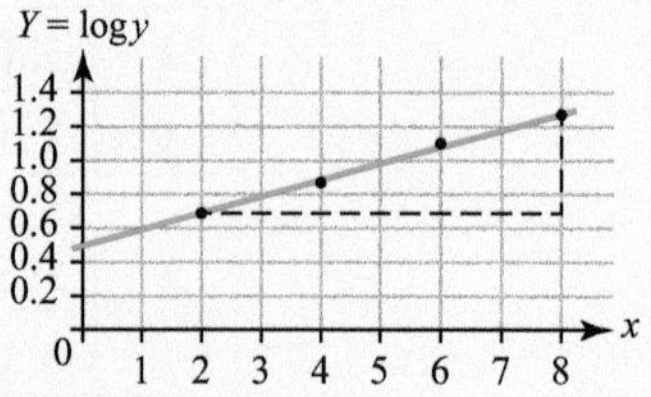

Work out values of Y. Plot the data and draw a line of best fit by eye. (2)

$$\text{Gradient} = \log b \approx \frac{1.27 - 0.69}{8 - 2} = 0.0966$$

$$b \approx 10^{0.0966\ldots} \approx 1.25$$

y-intercept $= \log k \approx 0.49$

$$k = 10^{0.49} \approx 3.1$$ — Use the graph to find the gradient and y-intercept. (3)

The required relationship is $y = 3.1 \times 1.25^x$

b When $x = 0$, the initial area $= 3.1 \times 1.25^0 = 3.1$ cm²

After 10 days, the expected area $= 3.1 \times 1.25^{10} = 28.9$ cm²

Exercise 5.4B Reasoning and problem-solving

1 A pipe empties water into a river. The table shows the volume of water, $y\,\text{m}^3$, leaving the pipe each second, for different depths of water, $x\,\text{m}$, in the pipe. Engineers expect x and y to be connected by the relationship $y = ax^n$

x	1.2	1.3	1.4	1.5
y	5.8	7.6	8.7	10.0

a Show, by drawing a line of best fit on appropriate axes, that the engineers are right and find approximate values of a and n

b What volume of water (in m^3) would you expect to leave the pipe each second when the depth of water is

i 1.35 metres   ii 3.2 metres?

c Which of your two answers in part **b** is the more reliable? Explain why.

2 When an oil droplet is placed on the surface of water it forms a circle. As time, t seconds, increases, the area, A, of the oil increases.

t	2	3	4	5
A	17	40	73	117

a Show that there is a relationship between A and t given by $A = k \times t^n$ and find the values of the constants k and n

b Explain why this algebraic model is better at estimating the area of the oil slick after 1 second than after 1 minute.

3 A small town has had a growing population, p, for many years, x, since 1980

Year	1980	1985	1995	2000
x	0	5	15	20
p	4140	5000	8400	10200

a Show that the population can be approximately related to the number of years since 1980 using $p = k \times b^x$. State the value of k and find the value of b using an appropriate straight-line graph.

b Estimate the size of the population in 1990 and in 2010. Which of your two answers is the better estimate? Explain why.

4 When 200 cm^3 of gas being held under high pressure suddenly expands, its pressure p (measured in thousands of pascals) decreases rapidly as its volume, $v\,\text{cm}^3$, increases. The values are given in the table.

v	200	250	300	350	400
p	78.0	56.5	43.7	36.0	29.7

a Show that $p = k \times v^n$ is a good approximation for this data and calculate the values of the constants k and n to 3 sf.

b What is the volume of the gas when its pressure has fallen to

i Half its initial pressure,

ii A tenth of its initial pressure?

c Which of your two answers to part **b** is more reliable? Explain why.

Challenge

5 The mass, m grams, of a radioactive material decreases with time, t hours. The mass predicted to remain radioactive after various times is given in the table.

t	5	10	15	30
m	47.6	45.3	43.1	38.0

a Draw an appropriate straight-line graph and find the relationship between m and t

b According to this model, what mass of radioactive material is present when $t = 0$?

c How long does the model predict it will take for the radioactive substance to decay to half its original mass? This is its half-life.

MyMaths 2268 SEARCH

5 Summary and review

Chapter summary

- $x = a^n$ and $n = \log_a x$ are equivalent statements.
- The inverse of $y = a^x$ is $y = \log_a x$
- Logarithmic expressions can be manipulated using the following three laws of logarithms.
  Law 1: $\log_a(xy) = \log_a x + \log_a y$
  Law 2: $\log_a\left(\frac{x}{y}\right) = \log_a x - \log_a y$
  Law 3: $\log_a(x^k) = k\log_a x$
- Logs can have different bases. Bases 10 and e are the most common.
- The value of e is 2.71828 (to 5 dp).
- The general exponential function is a^x. *The* exponential function is e^x
- The inverse of $y = e^x$ is $y = \log_e x$ which can be written as $y = \ln x$
- The gradient of $y = e^x$ is e^x and the gradient of $y = e^{kx}$ is ke^{kx}
- The graph of $y = e^x$ can be transformed in a variety of ways.
- Mathematical models using exponential functions can be used to describe real-life events. A common model is $y = Ae^{kt}$. Contextual limitations of these models should be considered.
- Two common, non-linear relationships take the form $y = ax^n$ and $y = kb^x$
- The graphs of these relationships can be transformed into straight lines using logs.
  $y = ax^n$ becomes $Y = nX + c$, where $Y = \log y$, $X = \log x$ and $c = \log a$
  $y = kb^x$ becomes $Y = mx + c$, where $Y = \log y$, $m = \log b$ and $c = \log k$
- These straight lines are lines of best fit through the data points.

Check and review

You should now be able to...	Try Questions
✔ Convert between powers and logarithms.	1–2
✔ Manipulate expressions and solve equations involving powers and logarithms.	3–7
✔ Use the exponential functions $y = a^x$, $y = e^x$, $y = e^{kx}$ and their graphs.	8–14
✔ Verify and use mathematical models, including those of the form $y = ax^n$ and $y = kb^x$	15–17
✔ Consider limitations of exponential models.	18

1 Express each of these in logarithmic form.

a $2^5 = 32$ **b** $0.0001 = 10^{-4}$

2 Express each of these in index notation.

a $\log_2 16 = 4$ **b** $\log_4\left(\frac{1}{16}\right) = -2$

3 Evaluate

a $\log_2 16 + \log_2 2$ **b** $\log_2 16 - \log_2 2$

4 Write $2\log 9 + \log\left(\frac{1}{3}\right)$ as a single logarithm.

5 Write $\log\left(\frac{a^2}{\sqrt{b}}\right)$ in terms of $\log a$ and $\log b$

6 Solve these equations, giving your answers to 3 sf.

a $2^x \times 2^{x+1} = 24$

b $2^{2x} - 9 \times 2^x + 8 = 0$

c $\log_{10}(2x-15) = 2$

7 Calculate the smallest positive integer x such that

a $\left(1+\frac{x}{100}\right)^4 > 2$ b $1-0.9^x > 0.8$

8 a Find the equation of the tangent to the curve $y = e^x$ at the point $(3, e^3)$

b Find the equation of the tangent to $y = e^{\frac{1}{2}x}$ at the point where $x = 2$. What are the coordinates of the point where this tangent intersects the line $x = 1$?

9 Describe the transformation that maps $y = e^x$ onto $y = e^{2x}$

10 The graph of $y = e^x$ passes through the points $(3, a)$ and $(b, 4)$
Find the exact values of a and b

11 Calculate the value of k (to 3 sf) where the inverse of $y = 3^x$ is $y = k\ln x$

12 Find the equation of the normal to the graph of $y = \frac{1}{2}e^x + 2x^2$ at the point $\left(0, \frac{1}{2}\right)$

13 Prove that the normal to the curve $y = 1 - x + e^x$ at the point $(1, e)$ passes through the point $(e^2, -1)$

14 Sketch the graph of $y = x^2 + e^x$

Find the minimum point.

15 A bank account has £250 invested in it at a compound rate of r% p.a.

a Prove that, after n years, the account holds £A where $A = 250\left(1+\frac{r}{100}\right)^n$

b Find A when $r = 5$ and $n = 4$

c How long does it take for the amount invested to double in value when $r = 6$?

16 Two variables are measured. The relationship between them is likely to be $y = k \times b^x$

x	1.0	2.0	3.0	4.0
y	15.0	23.7	43.4	70.2

a Plot a suitable straight-line graph and estimate the values of the constants b and k

b Further measurements give the values $x = 7.5$, $y = 372$. Show whether the model predicts these results sufficiently accurately or not.

17 The variables x and y satisfy the relationship $y = a \times x^b$, where a and b are constants. By drawing a line of best fit, show that this is approximately correct and estimate the values of a and b

x	4.0	4.8	5.6	6.4
y	90	140	198	270

18 A scientist proposes that the population of deer, D, on an island can be modelled by the formula $D = 25e^{0.6t}$. Why might this model not be appropriate?

What next?

Score		
	0–9	Your knowledge of this topic is still developing. To improve, search in MyMaths for the codes: 2061–2063, 2133, 2134, 2136, 2257, 2268
	10–14	You're gaining a secure knowledge of this topic. To improve, look at the InvisiPen videos for Fluency and skills (05A)
	15–18	You've mastered these skills. Well done, you're ready to progress! To develop your techniques, look at the InvisiPen videos for Reasoning and problem-solving (05B)

Click these links in the digital book

Exploration

Going beyond the exams

History

Towards the end of 16th century, rapid advances in science demanded calculations to be carried out on a scale never seen before. This put pressure on mathematicians to devise methods for simplifying the processes involved in multiplication, division and finding powers and roots.

In 1614 the Scottish mathematician **John Napier** published his work on **logarithms**, which scientists, navigators and engineers quickly adopted in their calculations.

Swiss mathematician **Leonhard Euler** did further work on logarithms in the 18th century, creating the system that we still use today.

John Napier

Have a go

Try the following calculations by hand.

$$5438 \times 2149$$

$$4769604 \div 567$$

Now try these calculations using logarithms.

"For the sake of brevity, we will always represent this number 2.718281828459... by the letter e."
- Leonhard Euler

Research

Find the dates for each of the following events and plot them all on a single timeline.

1. Today
2. Your date of birth
3. The moon landing
4. The first powered flight
5. The first use of a printing press
6. The first use of coins
7. The end of the Neolithic age
8. The extinction of dinosaurs

Explain the need for a logarithmic timescale. Have a go at plotting a timeline of the same events with a logarithmic scale.

Research

Find a method for 'extracting square roots' by hand.

Use the method to find the square root of 126 736

5 Assessment

1 Express each of these as a single logarithm. Select the correct answer.

a $\log_5 7 + \log_5 8$

A $\log_5 15$ B $\log_{10} 15$ C $\log_5 56$ D $\log_{25} 56$ **[1 mark]**

b $3\log_5 2 - \log_5 10$

A $3\log_5(-8)$ B $\log_{-5}(-2)$ C $3\log_5 0.2$ D $\log_5 0.8$ **[1]**

2 Write down the value of each logarithm. Select the correct answer.

a $\log_3 81$

A 27 B 4 C 4.32 D 3 **[1]**

b $\log_3\left(\frac{1}{3}\right)$

A $\frac{1}{9}$ B 1 C -1 D 3 **[1]**

3 Sketch, on the same axes, the graphs of $y = 2^x$, $y = 5^x$ and $y = 0.5^x$ **[4]**

4 Given $f(x) = \log_3 x$

a Sketch the graph of $y = f(x)$ and write down the equation of any asymptote. **[3]**

b Solve the equations

i $f(x) = 3$ ii $f(x) = -2$ iii $f(x) = 0.5$ **[3]**

5 Express as a single logarithm

a $2\log_n 5 + \log_n 3$ **[2]** b $1 - \log_n 5$ **[2]**

6 Solve these equations, giving your answers to 3 significant figures.

a $6^x = 13$ **[2]** b $e^x = 5$ **[2]**

7 Solve these equations.

a $\log_2 x - \log_2 3 = \log_2 4$ **[2]** b $\log_2 9 + 2\log_2 x = \log_2 x^4$ **[2]**

8 The number of bacteria on a dish is given by the equation $B = 200e^{3t}$ where t is the time in hours.

a How many bacteria are there originally? **[1]**

b How long will it be until there are 10 000 bacteria? **[3]**

c Give one reason why this model might not be appropriate. **[1]**

9 The value, £V, of an investment after t years is given by the formula

$V = V_0 e^{0.07t}$ where V_0 is the amount originally invested.

How long will it take for the investment to double in value? **[3]**

10 The graph of $y = k^x$ passes through (2, 15)

Find an exact value of k **[2]**

11 The graph of $y = ax^n$ passes through the points (1, 2) and (−2, 32)

Find exact values of a and n **[3]**

12 **a** Sketch the graph of $y=2+e^x$, and write down the equation of the asymptote. **[3]**

b Solve the equation $e^{2x}=5$, giving your answer in terms of logarithms. **[2]**

13 Given that $\log_n 3 = p$

a Write in terms of p

i $\log_n 9$ **ii** $\log_n \frac{1}{3}$ **iii** $\log_n 3n$ **[6]**

b Find the value of n given that $\log_n 3n - \log_n 9 = 2$ **[3]**

14 Solve these equations.

a $\log_3 x + \log_3(x-1) = \log_3 12$ **[5]**

b $\log_3 x - \log_3(x-1) = 2$ **[3]**

c $3\ln x - \ln 2x = 5$ **[4]**

15 The population of a particular species on an island t years after a study began is modelled as $P=\frac{1500a^t}{2+a^t}$, where a is a positive constant.

a What was the population at the beginning of the study? **[2]**

b Given that the population after two years was 600

i Find the value of a

ii Calculate, to the nearest whole year, how long it takes for the population to double its initial size. **[8]**

c Explain why, according to this model, the population cannot exceed 1500 **[2]**

d Give one reason why this model might not be appropriate. **[1]**

16 A radioactive isotope has mass, M grams, at time t days given by the equation

$M = 50e^{-0.3t}$

a What is the initial mass of the isotope? **[1]**

b What is the half-life of the isotope? **[3]**

17 The graph of $y=a(b)^x$ passes through the points $(0, 5)$ and $(2, 1.25)$

a Find exact values for a and b **[3]**

b Sketch the curve. **[2]**

18 Find the rate of change of the function $f(t)=e^{2t}$ at time $t=1.5$ **[3]**

19 Solve the equation

$2^{2x}+7(2^x)-18=0$ **[3]**

20 The concentration of a drug, C mg per litre, in the blood of a patient at time t hours is modelled by the equation

$C=C_0e^{-rt}$

where C_0 is the initial concentration and r is some measure of the removal rate.

The concentration after 1 hour is 9.2 mg/litre and after 2 hours is 8.5 mg/litre.

a Calculate the initial concentration and the value of r. Give your answers to 2 significant figures. **[7]**

The drug becomes ineffective when the concentration falls below 3.6 mg/litre.

b What is the maximum time that can elapse before a second dose should be given? Give your answer to the nearest hour. **[3]**

1-5

Assessment

Chapters 1-5: Pure

1 **a** Simplify these expressions. Select the correct answer for each part.

i $(5^n)^m$

A 5^{n+m} **B** $m5^n$ **C** 5^{mn} **D** $n5^m$

ii $\dfrac{3^m \times 3^n}{3}$

A $\dfrac{3^{mn}}{3}$ **B** $\dfrac{3^{m+n}}{3}$ **C** 1^{m+n} **D** 3^{m+n-1} **[2 marks]**

b Solve the equation $a^{x+1} = (a^2)^3$. Select the correct answer.

A $x = 31$ **B** $x = 5$ **C** $x = 7$ **D** $x = 4$ **[1]**

2 Solve these simultaneous equations. You must show your working.

$$x + 12y = 5$$

$$x^2 + 2xy - 5 = 0$$ **[5]**

3 $f(x) = x^2 + 3x + k - 4$

a Show that when $k = 10$ the equation $f(x) = 0$ has no real solutions. **[2]**

b Solve the equation $f(x) = 0$ given that

i $k = 6$ **ii** $k = 5$ **[4]**

c Find the value of k for which the equation $f(x) = 0$ has precisely one solution. **[2]**

4 **a** Write $x^2 - 4x + 12$ in the form $(x + p)^2 + q$ where p and q are integers to be found. **[2]**

b Show algebraically that the equation $x^2 - 4x + 12 = 0$ has no real solutions. **[2]**

c Sketch the graph of $y = x^2 - 4x + 12$ and write down the coordinates of the minimum point. **[3]**

5 **a** Express $\dfrac{2+\sqrt{3}}{1-\sqrt{3}}$ in the form $p + q\sqrt{3}$ and write down the values of p and q. You must show your working. **[4]**

b Solve the equation $x\sqrt{8} = 5\sqrt{2} - \sqrt{32}$, giving your answer in its simplest form. You must show your working. **[3]**

6 A circle has equation $x^2 + y^2 - 6x + 14y + 33 = 0$

a Work out the centre and the radius of the circle. Show your working. **[4]**

b Sketch the circle, labelling its points of intersection with the coordinate axes. **[3]**

7 The line p_1 has equation $2x + 8y + 14 = 0$

a Find the gradient of p_1 **[2]**

The line p_2 is perpendicular to p_1 and passes through the point (3, 6)

b Find the equation of the line p_2 in the form $ax + by + c = 0$ **[3]**

The lines p_1 and p_2 intersect at the point A

c Calculate the length OA, showing all your working. **[5]**

8 Prove by counter-example that the following statement is false.

"The product of two consecutive integers is odd." **[2]**

9 a Write down the value of ${}^9C_5$ **[1]**

b Find the binomial expansion in ascending powers of x of

i $(1-x)^6$ ii $(2x+1)^4$ **[6]**

10 Given $p(x)=x^3-2x^2-5x+6$

a Show that $x-3$ is a factor of $p(x)$ **[2]**

b Find all the solutions to the equation $p(x)=0$. Show your working. **[3]**

c Sketch the curve of $y=p(x)$, labelling where the curve crosses the axes. **[3]**

11 a Sketch these graphs on the same axes, giving the coordinates of their intersections with the x- and y-axis.

i $y=\frac{1}{x+3}$ ii $y=3^x$ **[5]**

b State how many solutions there are to the equation $\frac{1}{x+3}=3^x$. Explain your answer. **[1]**

12 A triangle has sides 4 cm and 9 cm and angle $y°$ as shown. Find the range of values of y for which the area of the triangle is less than or equal to $9\sqrt{2}$ cm². **[5]**

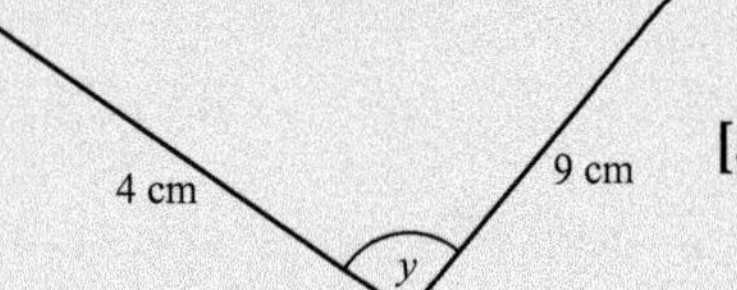

13 a Solve these equations for $0 \le x \le 180°$
Give your answers to 1 dp where appropriate and show all your working.

i $\sin x = 0.5$ ii $\cos 2x = 0.75$ **[6]**

b Sketch the graph of $y=\sin(x+40)$ for $0 \le x \le 360°$
Label the x-intercepts. **[2]**

14 A triangle has sides of 12.3 cm and 13.7 cm and angles of 34° and θ as shown.

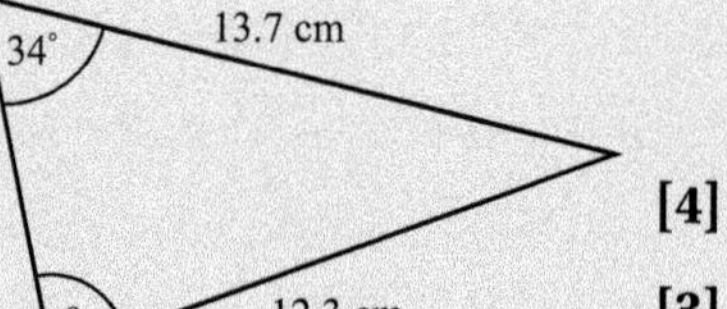

a Calculate the possible values of θ **[4]**

b Calculate the area of the triangle, given θ is acute. **[3]**

15 a Differentiate the following expressions with respect to x

i x^3-2x^5+3 ii $\sqrt{x}+\frac{1}{x}$ **[5]**

b For what values of x do the curves $y=\frac{4}{x}$ and $y=-\sqrt{x}$ have the same gradient? **[4]**

16 Find the equation of the normal to the curve $y=x^3-2x$ at the point (2, 4). Give your answer in the form $ax+by=c$ where a, b and c are integers. **[6]**

17 Work out these integrals.

a $\int (2x+3)^2\,dx$ **b** $\int \sqrt{x}(5x-1)\,dx$ **c** $\int \frac{2+x}{2\sqrt{x}}\,dx$ **[11]**

18 The shaded area is enclosed by the x-axis, the line $x=5$ and the curve with equation $y=-2+\sqrt{x}$

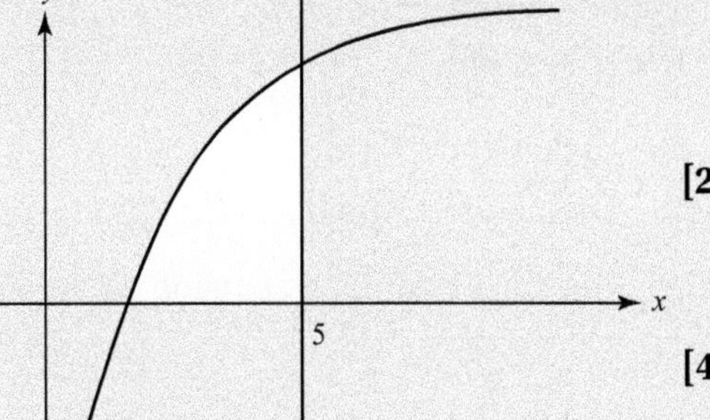

a Find the coordinates of the point where the curve crosses the x-axis. **[2]**

b Calculate the shaded area, giving your answer in surd form. You must show your working. **[4]**

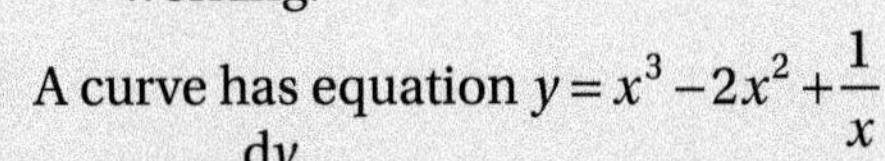

19 A curve has equation $y=x^3-2x^2+\dfrac{1}{x}$

a Find $\dfrac{dy}{dx}$  **[2]**

b The tangent to the curve at the point where $x=-1$ crosses the x-axis at the point A and the y-axis at the point B. Calculate the area of triangle OAB. **[6]**

20 The graph of $y=a(x^b)$ passes through the points (3, 16) and (2, 4)

Find an approximate equation for the curve. **[4]**

21 a Write in the form $\log_2 a$

i $\log_2 15-\log_2 3$ **ii** $\log_2 6+2\log_2 3$ **iii** $1+\log_2 5$ **[5]**

b Solve the equation $3^x=17$. Show your working and give your answer to 3 sf. **[2]**

22 a Sketch the graph of $y=\log_4 x$. Label any points of intersection with the coordinate axes and give the equations of any asymptotes. **[3]**

b Solve these equations, giving your answers as exact fractions. You must show your working.

i $\log_4 x=-3$ **ii** $\ln x+\ln 5=\ln(x+1)$ **[4]**

23 a Solve these inequalities.

i $10-2x\geq 4-5x$ **ii** $x^2+x<6$ **[5]**

b Give the range of values of x that satisfies both $10-2x\geq 4-5x$ and $x^2+x<6$ **[2]**

24 The points $A(3, 7)$ and $B(5, -3)$ are points on the circumference of a circle.

AB is a diameter of the circle.

a Find the equation of the circle. **[4]**

b Find an equation of the tangent to the circle at the point A **[4]**

25 a Work out the points of intersection, A and B, between the line with equation $x+2y=5$ and the curve with equation $x^2-xy-y^2=5$
You must show your working. **[5]**

b Find an equation of the chord AB in the form $ax+by+c=0$ **[3]**

26 Prove that n^2-m^2 is odd for any consecutive integers m and n **[3]**

27 You are given $f(x)=2x^2+2kx-4k$

Find the range of values of k for which $f(x)=0$ has real solutions. **[4]**

28 $f(x) = 2x^3 + 3x^2 - 23x - 12$
$x = -4$ is a solution of $f(x) = 0$

a Fully factorise $f(x)$ **[2]**

b Sketch the graph of $y = f(x)$, giving the coordinates of the x and y-intercepts. **[3]**

29 a Use the binomial expansion to expand then fully simplify these expressions.

i $(1+\sqrt{5})^5$ **ii** $(1-\sqrt{3})(2-\sqrt{3})^4$ **[7]**

b Use binomial expansions to prove that $(x+\sqrt{5})^5 \geq (x-\sqrt{5})^5$ for all values of x. **[5]**

30 a Find the expansion of $(2+x)^6$ in ascending powers of x **[3]**

b Use your expansion to find the value of 2.001^6 correct to 3 decimal places. **[2]**

31 a Show that the equation $\sin 3x + 2\cos 3x = 0$ can be written $\tan 3x = -2$ **[2]**

b Solve the equation $\sin 3x + 2\cos 3x = 0$ for $0 \leq x \leq 180°$. Show your working. **[4]**

32 a Sketch, for $-180° \leq x \leq 180°$, the following graphs. In each case, give the coordinates of intersection with the x- and y-axes, the equations of any asymptotes, and the coordinates of any maximum or minimum points.

i $y = a\sin x,\ a > 0$ **ii** $y = \tan(x+b),\ 0 < b < 90°$ **[8]**

b Solve the equation $2\sin(x+30) = \sqrt{3}$ for values of x in the range $-180° < x < 180°$
You must show your working. **[4]**

33 Given that $\cos x = \frac{1}{3}$ for an acute angle x, find the exact value of the following expressions. Give your answers in the form $a\sqrt{2}$

a $\sin x$ **b** $\tan x$ **[5]**

34 A triangle has side lengths 19 mm, 24 mm and 31 mm.

a Calculate the size of the smallest angle in the triangle. **[3]**

b Calculate the area of the triangle. **[3]**

35 A triangle has side lengths 12 cm and 17 cm with an angle of 30° as shown.

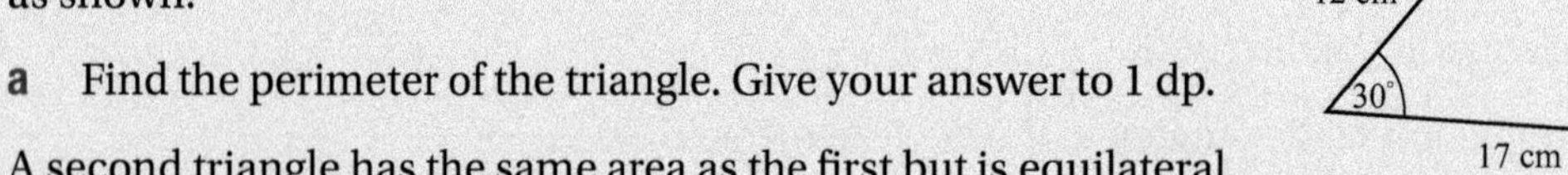

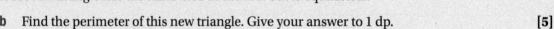

a Find the perimeter of the triangle. Give your answer to 1 dp. **[3]**

A second triangle has the same area as the first but is equilateral.

b Find the perimeter of this new triangle. Give your answer to 1 dp. **[5]**

36 $y = 4x^3$

a Work out $\frac{dy}{dx}$ from first principles. **[5]**

b Calculate the gradient of the tangent to the curve when $x = -2$ **[2]**

37 Find the range of values of x for which $f(x) = -\frac{x^2}{2} - \frac{16}{\sqrt{x}}$ is an increasing function. **[4]**

38 Find and classify all the stationary points of the curve with equation

$$3x^4 + 4x^3 - 12x^2 + 20$$

Show your working. **[8]**

39 A box has a square base with side x cm and height h cm.

The volume of the box is 3000 cm^3

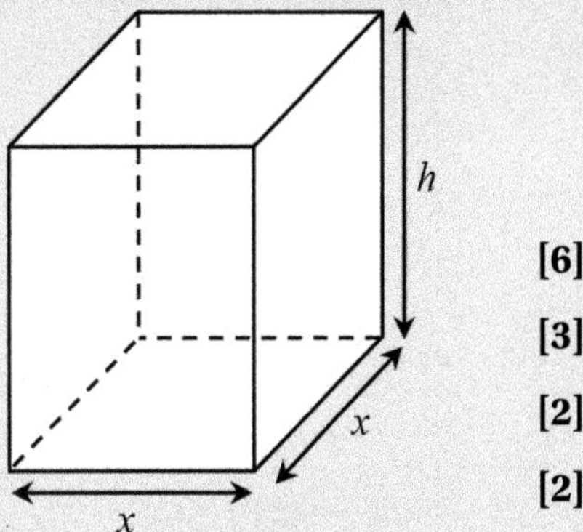

a Show that the surface area, A, of the box (not including the lid) is given by $A = x^2 + \frac{12\,000}{x}$ **[6]**

b Calculate the value of x for which A is a minimum. **[3]**

c Hence, find the minimum value of A. **[2]**

d Justify that the value found in part **b** is a minimum. **[2]**

40 A curve has equation $y = f(x)$

Given that $\frac{dy}{dx} = 10x^4 - 12x^2 + 1$, and that $y = f(x)$ passes through the point (2, 9),

a Find the equation of the curve, **[5]**

b Show that the curve passes through the point (–1, –24) **[2]**

41 Use the data in the table to approximate the equation of the curve, given that it is in the form $y = ax^b$

x	5	10	15	20
y	25	70	130	200

Give your values of a and b to one decimal place. **[4]**

42 The price in £ of a car that is t years old is modelled by the formula

$$P = 12000e^{-\frac{t}{5}} - 1000$$

a When will the car first be worth less than £3000? Give your answer in years and months to the nearest month. **[3]**

b Comment on the appropriateness of this model. **[1]**

43 a Sketch on the same axes the graphs of

i $y = \log_5 x$ ii $y = \ln x$ **[3]**

b Solve the equation $\ln x - 2\ln 3 = 5$, giving your answer in terms of e. You must show your working. **[3]**

44 a Use a suitable method of proof to prove or disprove the following statements.

i "$2^n + 1$ is a prime number for all positive integers n"

ii "$3^n + 1$ is even for all integers n in the range $1 \le n \le 4$"

iii "$5^n + 10$ is a multiple of 5 for all positive integers n" **[6]**

b For each proof in part a, state the name of the method of proof you have used. **[3]**

45 The curve shown has equation $y = A + \frac{B}{x+C}$

Find the values of A, B and C **[4]**

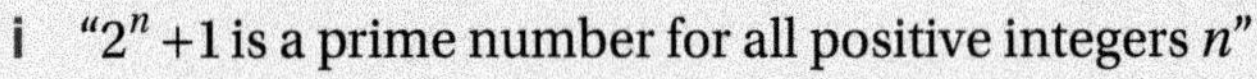
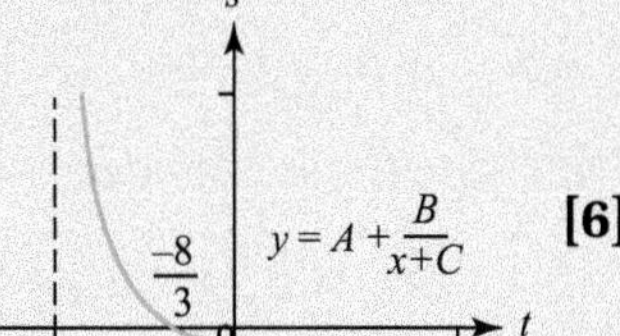

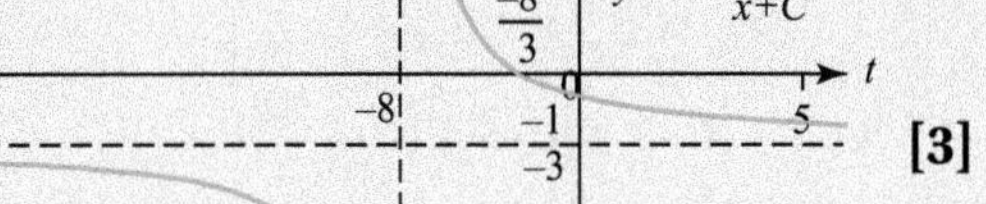

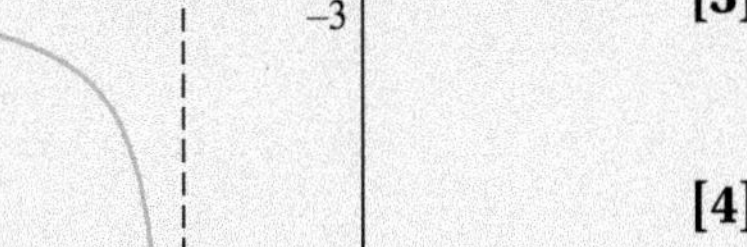

46 The first three terms in the expansion of $(1+ax)^n$ are $1-8x+30x^2$

Given that $n \in \mathbb{Z}^+$, find the value of

a n **[4]** **b** a **[1]**

47 Solve these equations for $-360 \le x \le 360°$. Show your working and give your answers to 3 sf.

a $\sin^2 x = 0.6$ **[4]** **b** $\cos^2 x - \cos x - 2 = 0$ **[3]**

48 **a** Prove that $3\sin^2 x - 4\cos^2 x \equiv 7\sin^2 x - 4$ **[3]**

b Solve the equation $3\sin^2 x - 4\cos^2 x = 1$ for $-180° \le x \le 180°$. Show your working. **[5]**

49 Solve the quadratic equation $\cos x + 3\sin^2 x = 2$ for x in the range $0 \le x \le 360°$
Show your working. **[6]**

50 Given that $g(x) = x^3 - 9x^2 + 11x + 21$

a Solve the equation $g(x) = 0$. Show your working. **[3]**

b Calculate the area bounded by the curve with equation $y = g(x)$ and the x-axis.
Show your working. **[6]**

51 A closed cylinder is such that its surface area is 50π cm^2

a Calculate the radius of the cylinder that gives the maximum volume. **[7]**

b Find the maximum volume and prove it is a maximum. **[4]**

52 A population of an organism grows such that after t hours the number of organisms is N thousand, where N is given by the equation $N = A - 8e^{-kt}$

Initially there are 3000 organisms and this number doubles after 5 hours.

a Find the value of

i A **ii** k **[5]**

b Sketch the graph of N against t for $t > 0$ **[2]**

c How many organisms are there after 3 hours? **[2]**

d What is the rate of change of N after 3 hours? **[3]**

e What is the limiting value of N as $t \to \infty$? **[2]**

53 Solve the equation $2e^{2x} - 13e^x + 15 = 0$. Show your working. **[4]**

54 Solve these equations. You must show your working.

a $\log_4(x+6) - \frac{1}{2} = 2\log_4 x$ **b** $\log_3 y + \log_9 3y = 2$ **c** $(\ln x)^2 - 3\ln x = 1$ **[13]**

55 Solve these simultaneous equations. You must show your working.

$$xy = 24$$

$$\ln y + 2\ln x = \frac{1}{2}\ln\left(\frac{64}{9}\right)$$ **[6]**

6 Vectors

In a basketball match, a player tries to shoot a ball through the hoop. The success of this shot depends on the distance from the hoop and the direction in which the ball is thrown. The player might be 10 m from the end of the court and want to aim the ball at a space 3.5 m above the ground in order to place it just above the basket. This information about the displacement and direction of the ball could be described as components of a vector.

Vectors can be used to describe anything that has both magnitude and direction, and are used in many fields including physics and engineering. They crop up over and over again in descriptions of motion. Velocity, acceleration, force and momentum are vector quantities. The concept is simple but fundamental to mathematics.

Orientation

What you need to know

Ch1 Algebra 1 p.22

- Simultaneous equations.

Ch3 Trigonometry p.76

- Sine and cosine rules and the area formula.

What you will learn

- To identify vector and scalar quantities.
- To solve geometric problems in 2D using vector methods.
- To solve problems involving displacements, velocities and forces.
- To find the magnitude and direction of a vector and use the components.
- To use the position vectors to find displacements and distances.

What this leads to

Ch7 Units and kinematics
Motion in a straight line.
Distance and speed as scalar quanitites.
Position, displacement and velocity as vector quantites.

Ch8 Forces and Newton's laws
Resolving in two directions.
Magnitude and direction of a resultant force.

MyMaths Practise before you start 2005, 2018, 2045, 2046

6.1 Definitions and properties

Fluency and skills

Key point

A **scalar** quantity has **magnitude** (size) *only*.
Examples include distance (100 m) and speed (10 m s^{-1}).
A **vector** quantity has *both* magnitude and **direction**. Examples include displacement (100 m north) and velocity (10 m s^{-1} on a bearing of 217°).

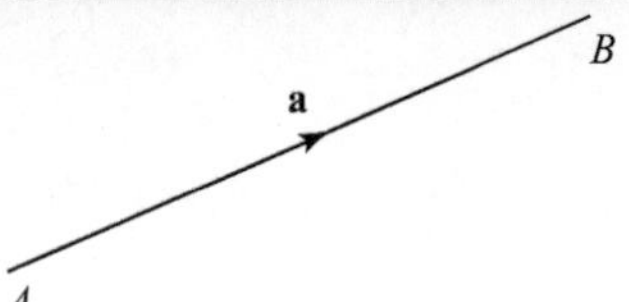

Vector **a** represents the directed line segment from A to B

You can represent any vector between points A and B by the magnitude and direction of a directed line segment; written $\overrightarrow{AB}$, **a**, $\underset{\sim}{a}$ or $\underline{a}$. When you write a vector by hand, using only a single letter, make sure to underline it.

You write the magnitude as $|\overrightarrow{AB}|$ or AB or $|\mathbf{a}|$ or a

See p.382
For a list of mathematical notation.

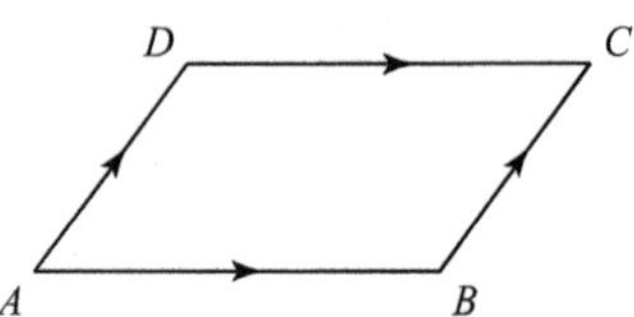

Two pairs of equal vectors are shown.
$\overrightarrow{AB} = \overrightarrow{DC}$ and $\overrightarrow{AD} = \overrightarrow{BC}$

Key point

Vectors are **equal** if they have the same magnitude and direction.

Multiplying a vector by a number (a **scalar**) changes its magnitude but not its direction.

Key point

$k\mathbf{a}$ is a vector parallel to **a** and with magnitude $k|\mathbf{a}|$
a is parallel to **b** only if $\mathbf{a} = k\mathbf{b}$

Combining vectors $\overrightarrow{AB}$ and $\overrightarrow{BC}$ gives the vector $\overrightarrow{AC}$
You add vectors by placing them 'nose to tail.'

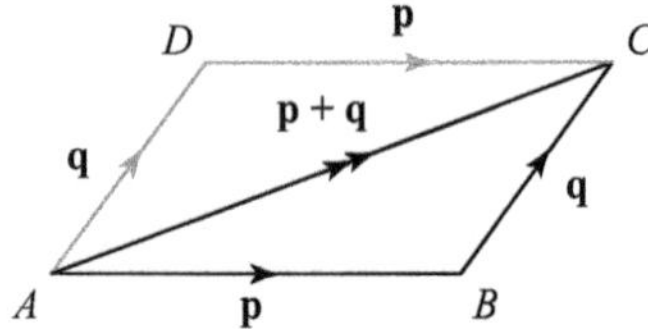

$\overrightarrow{AC}$ is the resultant of $\overrightarrow{AB}$ and $\overrightarrow{BC}$
$\overrightarrow{AB} + \overrightarrow{BC} = \overrightarrow{AC}$

Key point

The vector sum of two or more vectors is called their **resultant**.

Two or more points are **collinear** if a single vector, or multiple parallel vectors, can pass through the points to form a single, straight line segment.

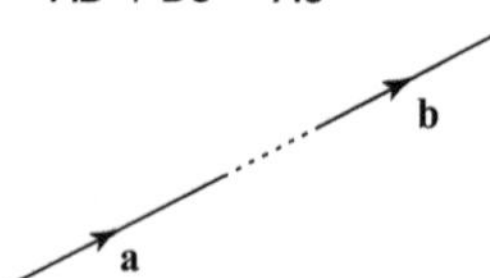

Vectors **a** and **b** are collinear

Key point

A vector with a magnitude of 1 is called a **unit vector.**

The unit vector in the direction of **a** is $\hat{\mathbf{a}} = \frac{\mathbf{a}}{|\mathbf{a}|}$

$\overrightarrow{AB} + \overrightarrow{BA} = \mathbf{0}$ (the **zero vector**. It has zero magnitude and undefined direction). This means that $\overrightarrow{BA} = -\overrightarrow{AB}$

Key point

a and −**a** have the same magnitude but opposite directions.

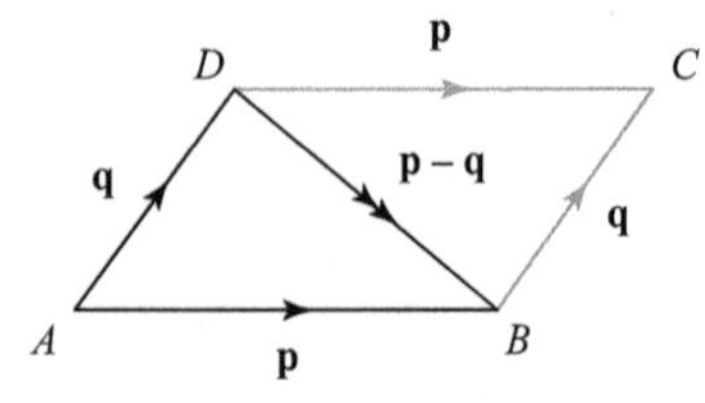

In the diagram $\overrightarrow{DB} = \overrightarrow{DA} + \overrightarrow{AB} = -\mathbf{q} + \mathbf{p} = \mathbf{p} - \mathbf{q}$

Key point

Subtracting a vector is the same as adding its negative.

Example 1

Vector **p** has magnitude 10 and direction due east. Vector **q** has magnitude 8 and direction 055°. Find the resultant of **p** and **q**

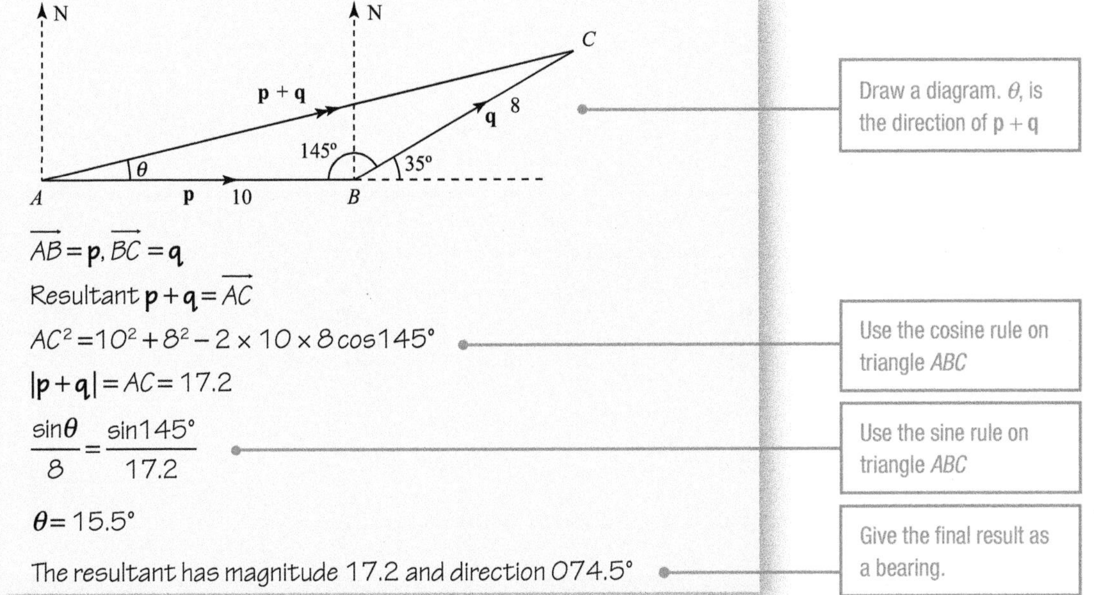

$\overrightarrow{AB} = \mathbf{p}$, $\overrightarrow{BC} = \mathbf{q}$

Resultant $\mathbf{p} + \mathbf{q} = \overrightarrow{AC}$

$AC^2 = 10^2 + 8^2 - 2 \times 10 \times 8 \cos 145°$ — Use the cosine rule on triangle *ABC*

$|\mathbf{p} + \mathbf{q}| = AC = 17.2$

$\frac{\sin\theta}{8} = \frac{\sin 145°}{17.2}$ — Use the sine rule on triangle *ABC*

$\theta = 15.5°$

The resultant has magnitude 17.2 and direction 074.5° — Give the final result as a bearing.

MECH

See Ch3.2

For a reminder of the sine and cosine rules.

Exercise 6.1A Fluency and skills

1 The diagram shows parallelogram *ABCD*, where $\overrightarrow{AB} = \mathbf{p}$ and $\overrightarrow{AD} = \mathbf{q}$. Express these vectors in terms of **p** and **q**

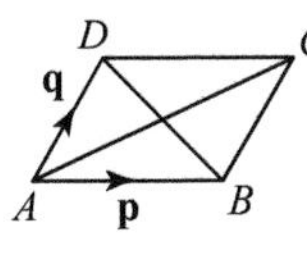

a $\overrightarrow{BC}$ b $\overrightarrow{DC}$ c $\overrightarrow{BA}$ d $\overrightarrow{CB}$

e $\overrightarrow{AC}$ f $\overrightarrow{BD}$ g $\overrightarrow{DB}$

2 The diagram shows two squares *ABEF* and *BCDE*, where $\overrightarrow{AB} = \mathbf{p}$ and $\overrightarrow{AF} = \mathbf{q}$. Express these vectors in terms of **p** and **q**

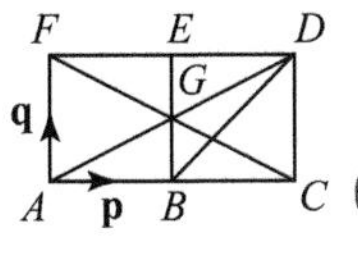

a $\overrightarrow{BD}$ b $\overrightarrow{AD}$ c $\overrightarrow{CF}$ d $\overrightarrow{AG}$

3 The diagram shows a trapezium *ABCD* in which *BC* is parallel to *AD* and twice as long. Express these vectors in terms of **p** and **q**

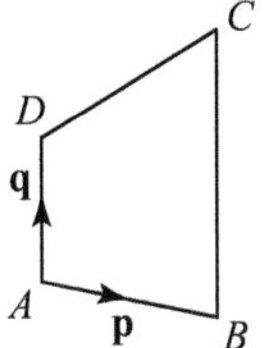

a $\overrightarrow{BC}$ b $\overrightarrow{DC}$

c $\overrightarrow{AC}$ d $\overrightarrow{BD}$

4 In each part of this question you are given the magnitude and direction (bearing) of two vectors **p** and **q**. In each case, work out the magnitude and direction of their resultant.

a **p**: 16, east **q**: 20, north

b **p**: 5, 100° **q**: 6, 060°

c **p**: 12, north **q**: 18, 200°

5 Vectors **p** and **q** have magnitudes 8 and 5 respectively and the angle between their directions is 45°. Find the magnitude and direction from **p** of these vectors.

a $\mathbf{p} + \mathbf{q}$ b $\mathbf{p} - \mathbf{q}$

6 Find the magnitude and direction of the resultant of each pair of vectors. Show your working.

a A displacement of 3.5 km on a bearing of 050° and a displacement of 5.4 km on a bearing of 128°

b A displacement of 26 km on a bearing of 175° and a displacement of 18 km on a bearing of 294°

c Velocities of 15 km h^{-1} due north and 23 km h^{-1} on a bearing of 253°

d Forces of 355 N on a bearing of 320° and 270 N on a bearing of 025°

MyMaths 2206 SEARCH

Reasoning and problem-solving

Strategy

To solve problems involving vectors

1. Sketch a diagram using directed line segments, to show all the information given in the question.
2. Look for parallel, collinear and equal vectors.
3. Break down vectors into a route using vectors you already know.

Example 2

ABCD is a parallelogram. *E* is the midpoint of *AC*

Vector $AB = \mathbf{p}$ and vector $AD = \mathbf{q}$

Prove that *E* is the midpoint of *BD*

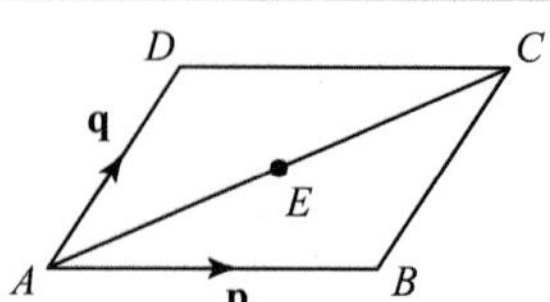

$\overrightarrow{AC} = \mathbf{p} + \mathbf{q}$ so $\overrightarrow{AE} = \frac{1}{2}(\mathbf{p} + \mathbf{q})$

$\overrightarrow{BE} = \overrightarrow{BA} + \overrightarrow{AE} = -\mathbf{p} + \frac{1}{2}(\mathbf{p} + \mathbf{q}) = \frac{1}{2}(\mathbf{q} - \mathbf{p})$ — (3) Find the position of *E* in relation to *B*

But $\overrightarrow{BD} = \mathbf{q} - \mathbf{p}$

So $\overrightarrow{BE} = \frac{1}{2}\overrightarrow{BD}$ — (2) $\overrightarrow{BE}$ and $\overrightarrow{BD}$ are collinear and *BE* is half as long as *BD*

Hence *BED* is a straight line and *E* is the midpoint of *BD*

Example 3

The diagram shows parallelogram *ABCD*. *E* lies on *DC*, and $DE : EC = 1 : 3$

AE and *BC*, when extended (produced), meet at *F*

$AB = \mathbf{p}$ and $AD = \mathbf{q}$

$AF = \lambda AE$ and $BF = \mu BC$

a Express *AF* in terms of λ, **p** and **q**

b Express *AF* in terms of μ, **p** and **q**

c Hence find the values of λ and μ

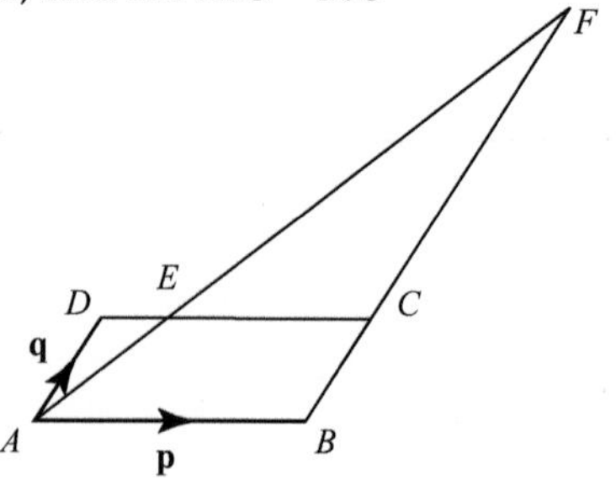

a $\overrightarrow{AE} = \overrightarrow{AD} + \overrightarrow{DE} = \mathbf{q} + \frac{1}{4}\mathbf{p}$ — (3) Break down vectors and apply the rules you know.

$\overrightarrow{AF} = \lambda\overrightarrow{AE} = \lambda\left(\mathbf{q} + \frac{1}{4}\mathbf{p}\right)$

b $\overrightarrow{AF} = \overrightarrow{AB} + \overrightarrow{BF} = \overrightarrow{AB} + \mu\overrightarrow{BC} = \mathbf{p} + \mu\mathbf{q}$

c $\lambda\left(\mathbf{q} + \frac{1}{4}\mathbf{p}\right) = \mathbf{p} + \mu\mathbf{q}$ gives $\left(\frac{1}{4}\lambda - 1\right)\mathbf{p} = (\mu - \lambda)\mathbf{q}$ — Use the answers from parts **a** and **b**.

$\frac{1}{4}\lambda - 1 = 0$ and $\mu - \lambda = 0$ — **p** and **q** are not parallel, so the equation is only possible if both sides are the zero vector.

$\lambda = 4$ and $\mu = 4$

Exercise 6.1B Reasoning and problem-solving

1 In triangle ABC, D and E are the midpoints of AB and AC. $\overrightarrow{BC} = \mathbf{p}$ and $\overrightarrow{BD} = \mathbf{q}$
Express these vectors in terms of $\mathbf{p}$ and $\mathbf{q}$

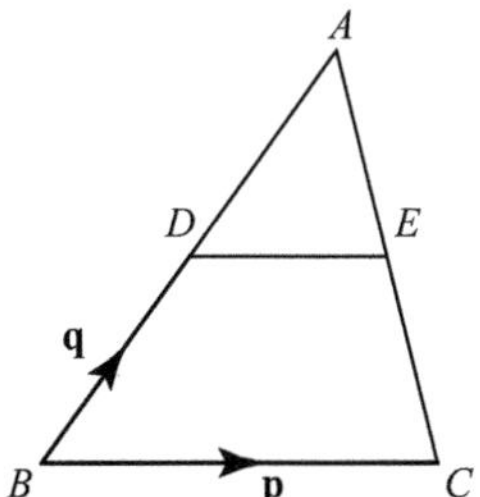

a $\overrightarrow{AC}$

b $\overrightarrow{DE}$

c What does this tell you about DE and BC? Explain why.

2 In the triangle shown, D and E are the midpoints of BC and AC respectively. The point G lies on AD and AG is twice GD

Let $\overrightarrow{AB} = \mathbf{p}$ and $\overrightarrow{AC} = \mathbf{q}$

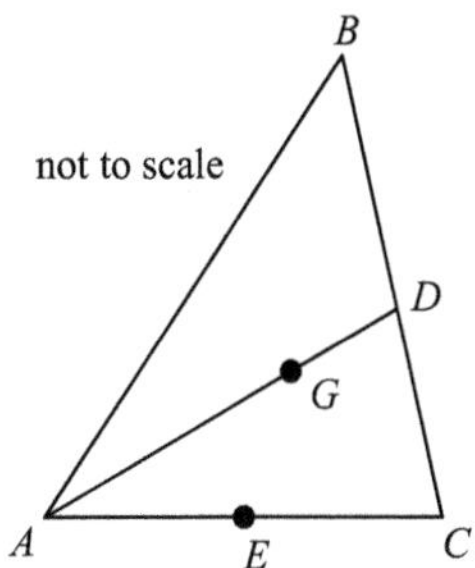

Show that BGE is a straight line and find the ratio $BG : GE$

3 In a triangle ABC, the points D, E and F are the midpoints of AB, BC and CA respectively. Point P lies in the plane of the triangle. By setting $\overrightarrow{PA} = \mathbf{a}$, $\overrightarrow{PB} = \mathbf{b}$ and $\overrightarrow{PC} = \mathbf{c}$, show that $\overrightarrow{PD} + \overrightarrow{PE} + \overrightarrow{PF} = \overrightarrow{PA} + \overrightarrow{PB} + \overrightarrow{PC}$

4 $ABCD$ is a quadrilateral. E, F, G and H are the midpoints of AB, BC, CD and DA respectively. By setting $\overrightarrow{AB} = \mathbf{p}$, $\overrightarrow{BC} = \mathbf{q}$ and $\overrightarrow{CD} = \mathbf{r}$, show that $EFGH$ is a parallelogram.

5 In the diagram, D is the midpoint of AE and $AB : BC = 1 : 2$
$\overrightarrow{AB} = \mathbf{p}$ and $\overrightarrow{AD} = \mathbf{q}$
$\overrightarrow{CF} = \lambda\overrightarrow{CD}$ and $\overrightarrow{EF} = \mu\overrightarrow{EB}$

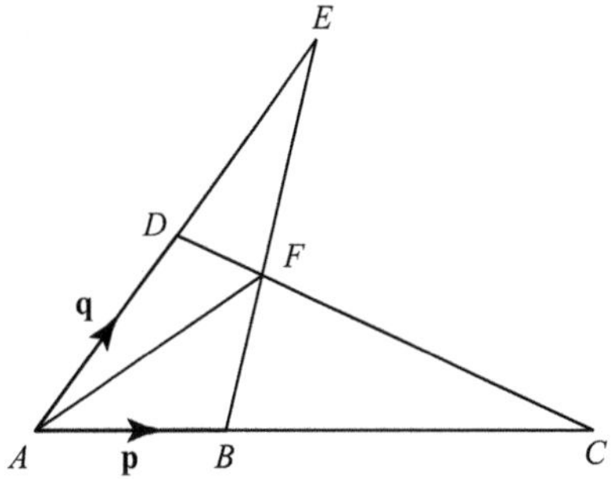

a Express $\overrightarrow{AF}$ in terms of λ, $\mathbf{p}$ and $\mathbf{q}$

b Express $\overrightarrow{AF}$ in terms of μ, $\mathbf{p}$ and $\mathbf{q}$

c Hence determine the values of λ and μ

Challenge

6 Points A and B are directly opposite each other across a river that is 100 m wide and flowing at 2 m s^{-1}. A boat, which can travel at 4 m s^{-1} in still water, leaves A to cross the river.

a If the boat is steered directly across the river, how far downstream of B will it reach the other bank?

b In what direction should it be steered so that it travels directly to B?

7 An aircraft has a speed in still air of 300 km h^{-1}. A wind is blowing from the south at 80 km h^{-1}. The pilot must fly to a point south-east of his present position.

a On what bearing should the pilot steer the aircraft?

b At what speed will the aircraft travel?

8 A ship, which can travel at 12 km h^{-1} in still water, is steered due north. A current of 9 km h^{-1} from west to east pushes the ship off course.

a Find the ship's resultant velocity.

b The ship is turned around with the intention of returning to its starting point. On what bearing should it be steered?

6.2 Components of a vector

Fluency and skills

Vectors **i** and **j** are unit vectors in the x and y directions.

The vector $\overrightarrow{OP}$ shown has an ***x*-component** of 3 and a ***y*-component** of 2. You can write it in **component form** in terms of **i** and **j**, or as a **column vector**: $\overrightarrow{OP} = 3\mathbf{i} + 2\mathbf{j} = \begin{pmatrix} 3 \\ 2 \end{pmatrix}$

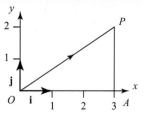

▲ **i** and **j** are unit vectors, $|\mathbf{i}| = 1 = |\mathbf{j}|$

Suppose that a vector $\overrightarrow{OP}$ has magnitude r, direction θ and components x and y, as shown.

If you know r and θ, you can **resolve** the vector into components.

If you are given x and y, you can find the magnitude and direction.

Key point

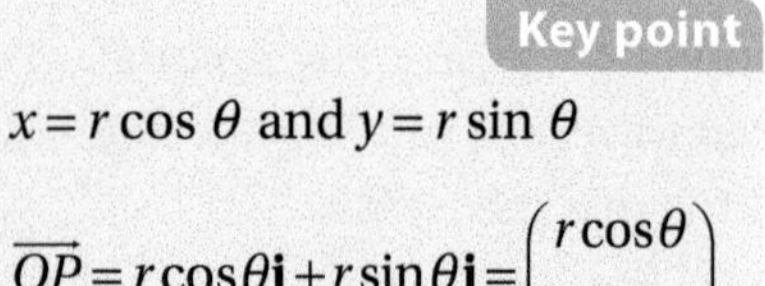

$x = r\cos\theta$ and $y = r\sin\theta$

$\overrightarrow{OP} = r\cos\theta\mathbf{i} + r\sin\theta\mathbf{j} = \begin{pmatrix} r\cos\theta \\ r\sin\theta \end{pmatrix}$

Key point

$r = \sqrt{x^2 + y^2}$ and $\tan\theta = \dfrac{y}{x}$

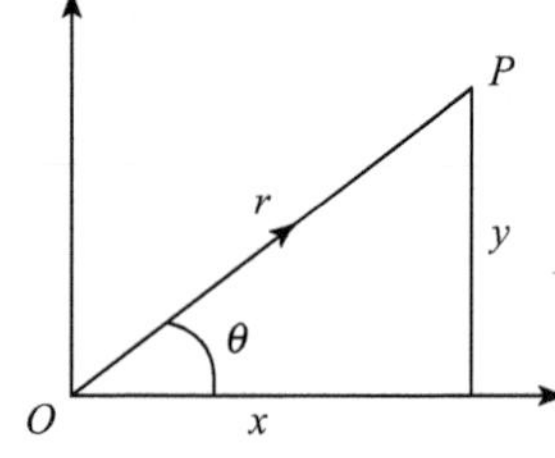

It is a good idea to sketch a diagram when finding θ because, for example, $(3\mathbf{i} - 2\mathbf{j})$ and $(-3\mathbf{i} + 2\mathbf{j})$ would give the same value of $\tan\theta$

Example 1

Express these vectors in component form.

a

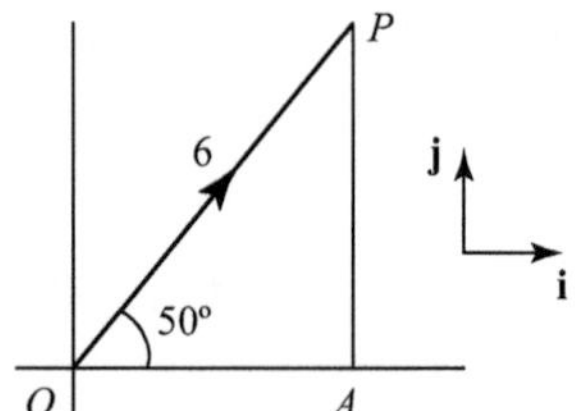

b

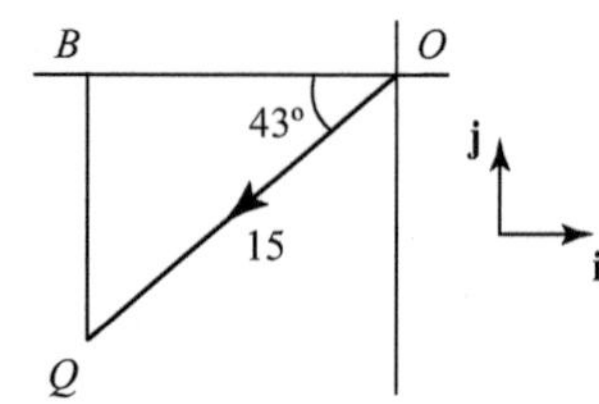

a $OA = 6\cos 50° = 3.86$ and $AP = 6\sin 50° = 4.60$

$\overrightarrow{OP} = 3.86\mathbf{i} + 4.60\mathbf{j}$ or $\begin{pmatrix} 3.86 \\ 4.60 \end{pmatrix}$

b $OB = 15\cos 43° = 11.0$ and $BQ = 15\sin 43° = 10.2$

$\overrightarrow{OQ} = -11.0\mathbf{i} - 10.2\mathbf{j}$ or $\begin{pmatrix} -11.0 \\ -10.2 \end{pmatrix}$

ICT Resource online

To experiment with converting vectors between different forms, click this link in the digital book.

Example 2

Work out the magnitude and direction of these vectors.

a $\mathbf{p} = 5\mathbf{i} + 2\mathbf{j}$ **b** $\mathbf{q} = -\mathbf{i} - 2\mathbf{j}$

a

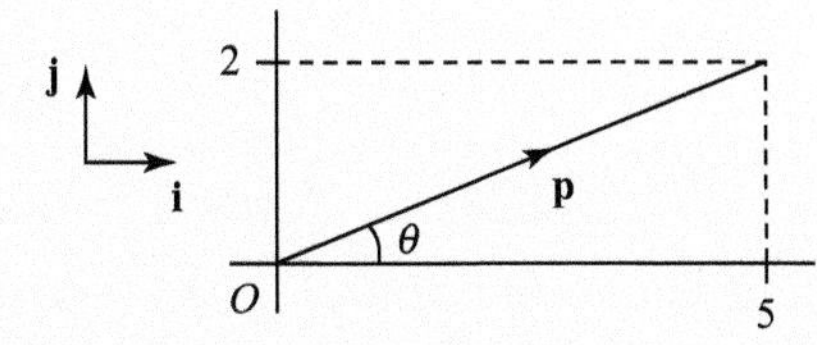

b

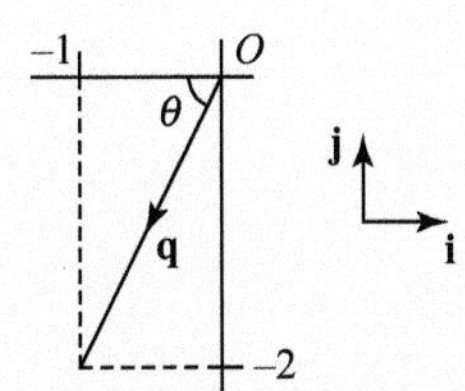

$|\mathbf{p}| = \sqrt{5^2 + 2^2} = 5.39$

$\tan\theta = \frac{2}{5}$ given $\theta = 21.80°$

$|\mathbf{q}| = \sqrt{(-1)^2 + (-2)^2} = 2.24$

$\tan\theta = 2$ given $\theta = 63.4°$

You should always make the direction clear - in this case by marking the angle in a diagram. More formally, you can state the direction as a rotation θ from the positive x-direction, where $-180° < \theta \leq 180°$

The answer to part **b** would then be given as $-116.6°$

When vectors are expressed in component form

- to add (or subtract) two vectors, you add (or subtract) the two **i**-components and add (or subtract) the two **j**-components,
- to multiply a vector by a scalar, you multiply both components by that scalar.

Key point

$(a\mathbf{i} + b\mathbf{j}) + (c\mathbf{i} + d\mathbf{j}) = (a + c)\mathbf{i} + (b + d)\mathbf{j}$

$(a\mathbf{i} + b\mathbf{j}) - (c\mathbf{i} + d\mathbf{j}) = (a - c)\mathbf{i} + (b - d)\mathbf{j}$

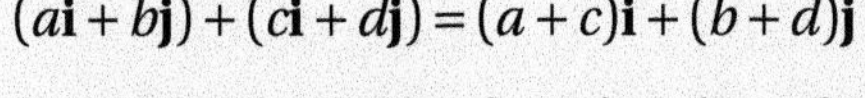

$k(a\mathbf{i} + b\mathbf{j}) = ka\mathbf{i} + kb\mathbf{j}$ where k is a scalar.

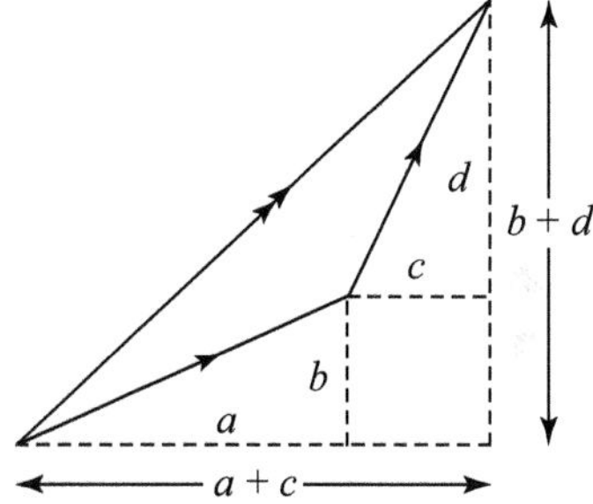

Example 3

Given $\mathbf{p} = 12\mathbf{i} + 5\mathbf{j}$ and $\mathbf{q} = 3\mathbf{i} - 4\mathbf{j}$, work out

a **i** $\mathbf{p} - \mathbf{q}$ **ii** $2\mathbf{p} + 3\mathbf{q}$ **b** A vector parallel to **p** and with magnitude 39

c The unit vector $\hat{\mathbf{q}}$

a **i** $\mathbf{p} - \mathbf{q} = 12\mathbf{i} + 5\mathbf{j} - (3\mathbf{i} - 4\mathbf{j}) = (12 - 3)\mathbf{i} + (5 - (-4))\mathbf{j} = 9\mathbf{i} + 9\mathbf{j}$

ii $2\mathbf{p} + 3\mathbf{q} = 2(12\mathbf{i} + 5\mathbf{j}) + 3(3\mathbf{i} - 4\mathbf{j}) = (24\mathbf{i} + 10\mathbf{j}) + (9\mathbf{i} - 12\mathbf{j}) = 33\mathbf{i} - 2\mathbf{j}$

b $|\mathbf{p}| = \sqrt{12^2 + 5^2} = 13$

A parallel vector with magnitude 39 is $3\mathbf{p} = 36\mathbf{i} + 15\mathbf{j}$

Vector $k\mathbf{p}$ is parallel to **p** and has k times the magnitude.

c $|\mathbf{q}| = \sqrt{3^2 + (-4)^2} = 5$

$\hat{\mathbf{q}} = \frac{1}{5}\mathbf{q} = 0.6\mathbf{i} - 0.8\mathbf{j}$

MECH

You can separate a two-dimensional vector equation into two equations, one for each of the x- and y-components.

Key point

Two vectors are **equal** if and only if *both* their **i**- and **j**-components are equal.

If $a\mathbf{i} + b\mathbf{j} = c\mathbf{i} + d\mathbf{j}$ then $a = c$ and $b = d$

Example 4

Given that $\mathbf{p} = x\mathbf{i} + y\mathbf{j}$ and $\mathbf{q} = (2y + 5)\mathbf{i} + (1 - x)\mathbf{j}$, work out the values of x and y for which $\mathbf{p} = \mathbf{q}$

$x\mathbf{i} + y\mathbf{j} = (2y + 5)\mathbf{i} + (1 - x)\mathbf{j}$

So $x = 2y + 5$ [1]

and $y = 1 - x$ [2] — Equate components.

$x = 2\frac{1}{3}$ and $y = -1\frac{1}{3}$ — Solve [1] and [2].

To describe the position of a point A relative to an origin O, you use the vector $\overrightarrow{OA}$. This is called the **position vector** of A. It is often labelled **a** or $\mathbf{r}_A$

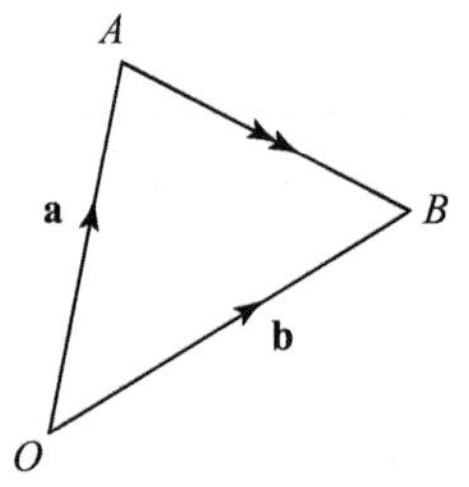

Key point

If points A and B have position vectors **a** and **b** then

vector $\overrightarrow{AB} = \mathbf{b} - \mathbf{a}$ distance $AB = |\mathbf{b} - \mathbf{a}|$

As A has position vector $(2\mathbf{i} + \mathbf{j})$, you can say that its coordinates are (2, 1)

Example 5

Points A and B have position vectors $\mathbf{a} = 2\mathbf{i} + \mathbf{j}$ and $\mathbf{b} = 5\mathbf{i} - 6\mathbf{j}$
Evaluate the distance AB

$\overrightarrow{AB} = (5\mathbf{i} - 6\mathbf{j}) - (2\mathbf{i} + \mathbf{j}) = 3\mathbf{i} - 7\mathbf{j}$ — Vector $\overrightarrow{AB} = \mathbf{b} - \mathbf{a}$

$|\overrightarrow{AB}| = \sqrt{3^2 + (-7)^2} = 7.62$ — Distance $AB = |\mathbf{b} - \mathbf{a}|$

Example 6

Points A, B and C lie on a straight line, with C between A and B, and $AC : CB = 2 : 3$

A and B have position vectors $\mathbf{a} = 2\mathbf{i} + 3\mathbf{j}$ and $\mathbf{b} = 12\mathbf{i} + 5\mathbf{j}$ respectively. Find the position vector **c** of C

$\overrightarrow{AB} = \mathbf{b} - \mathbf{a} = (12\mathbf{i} + 5\mathbf{j}) - (2\mathbf{i} + 3\mathbf{j}) = 10\mathbf{i} + 2\mathbf{j}$

$\overrightarrow{AC} = \frac{2}{5}\overrightarrow{AB} = 4\mathbf{i} + 0.8\mathbf{j}$ — This comes from $AC : CB = 2 : 3$

$\mathbf{c} = \mathbf{a} + \overrightarrow{AC} = (2\mathbf{i} + 3\mathbf{j}) + (4\mathbf{i} + 0.8\mathbf{j})$
$= 6\mathbf{i} + 3.8\mathbf{j}$

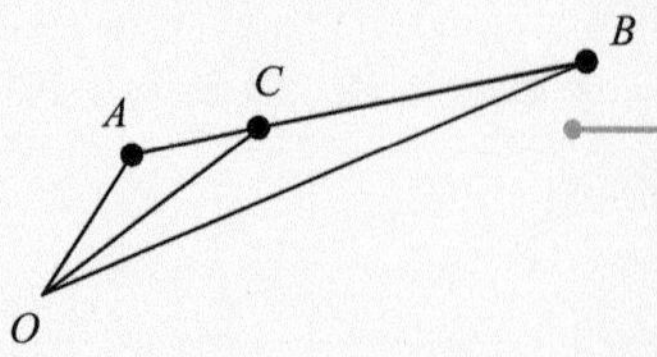

Sketch a diagram to help you answer the question.

Exercise 6.2A Fluency and skills

1 Write these vectors in the form $x\mathbf{i} + y\mathbf{j}$. Show your working.

a 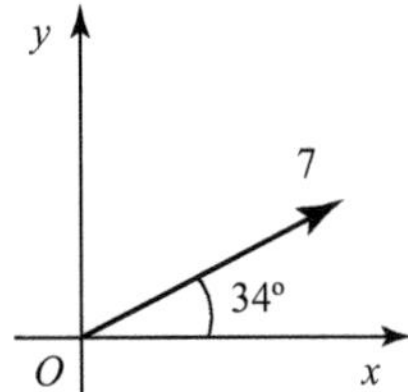

b 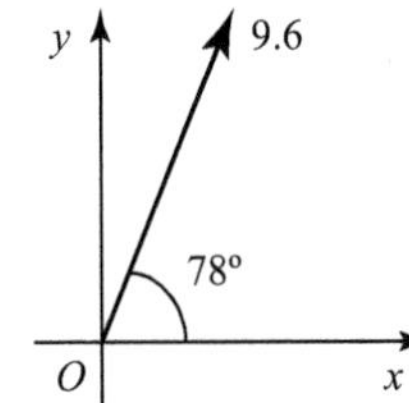

c 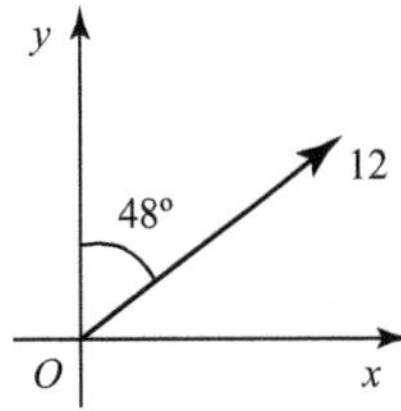

d 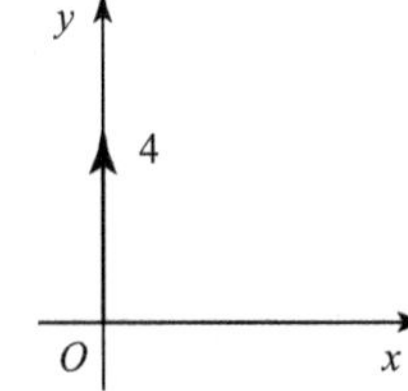

e 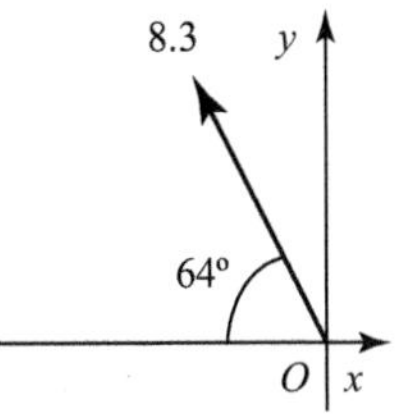

f 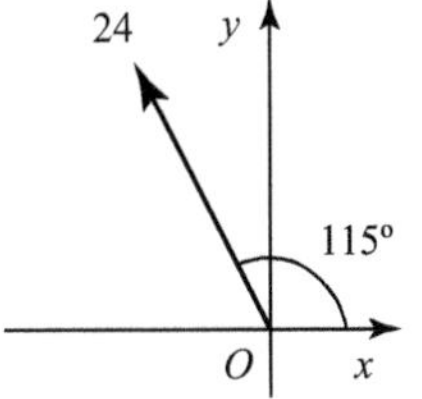

g 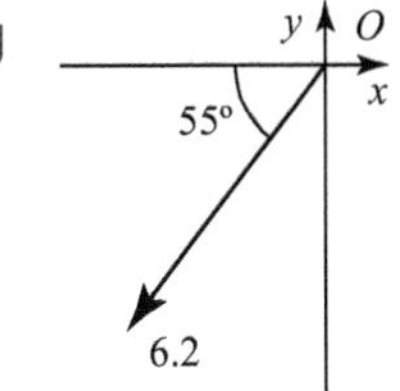

h 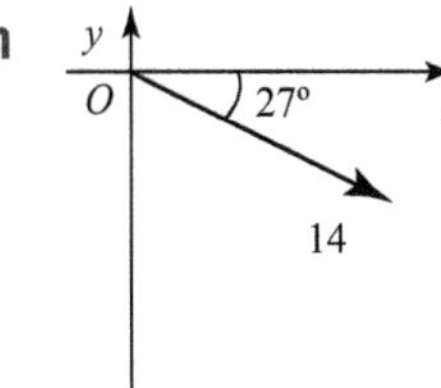

2 Evaluate the magnitude, r, and the direction, θ, of these vectors, where θ is the anticlockwise rotation from the positive x-direction and $-180° < \theta \le 180°$
Show your working.

a $5\mathbf{i} + 2\mathbf{j}$ b $7\mathbf{i} + 9\mathbf{j}$ c $-5\mathbf{j}$

d $-2\mathbf{i} + 3\mathbf{j}$ e $3\mathbf{i} - 5\mathbf{j}$ f $-6\mathbf{i} - 5\mathbf{j}$ g $-2\mathbf{i}$

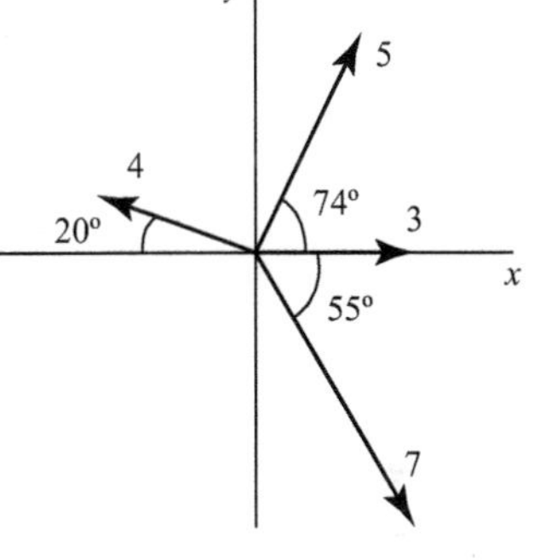

3 The diagram shows the magnitude and direction of four vectors. By writing each of these vectors in component form, find the magnitude and direction of their resultant.

4 Given vectors $\mathbf{p} = 2\mathbf{i} - \mathbf{j}$, $\mathbf{q} = -2\mathbf{i} + 3\mathbf{j}$ and $\mathbf{r} = 4\mathbf{i} + \mathbf{j}$, calculate each of these vectors.

a $\mathbf{p} + \mathbf{q}$ b $\mathbf{p} - \mathbf{r}$ c $2\mathbf{q} - \mathbf{p}$ d $2\mathbf{p} + 3\mathbf{r}$ e $|\mathbf{p}|$ f $|\mathbf{q} + \mathbf{r}|$

5 Given vectors $\mathbf{p} = 3\mathbf{i} + u\mathbf{j}$, $\mathbf{q} = v\mathbf{i} - 4\mathbf{j}$ and $\mathbf{r} = 4\mathbf{i} - 6\mathbf{j}$, work out

a The values of u and v if $\mathbf{p} - \mathbf{q} = \mathbf{r}$

b The value of u if $\mathbf{p}$ and $\mathbf{r}$ are parallel.

6 Given $\mathbf{p} = -3\mathbf{i} + 4\mathbf{j}$, write down

a A vector parallel to $\mathbf{p}$ with magnitude 20

b The unit vector $\hat{\mathbf{p}}$ in the direction of $\mathbf{p}$

7 Evaluate the values of a and b which satisfy these equations.

a $(2a + 1)\mathbf{i} + (a - 2)\mathbf{j} = (b - 1)\mathbf{i} + (2 - b)\mathbf{j}$

b $(a^2 - b^2)\mathbf{i} + 2ab\mathbf{j} = 3\mathbf{i} + 4\mathbf{j}$

8 Point A has position vector $(8\mathbf{i} - 15\mathbf{j})$ m.

Work out the distance of point A from the origin. Show your working.

9 Points A and B have position vectors $(-2\mathbf{i}+3\mathbf{j})$ m and $(4\mathbf{i}+7\mathbf{j})$ m respectively. Evaluate

a The length AB

b The angle made by AB with the x-direction.

10 A particle is at point A, position vector $(3\mathbf{i}+7\mathbf{j})$ m. It undergoes a displacement of $(6\mathbf{i}-3\mathbf{j})$ m, ending up at B

a Calculate the particle's final distance from the origin O

b Find the angle AOB

11 $ABCD$ is a parallelogram. The vertices A, B and C have position vectors $\mathbf{a}=\mathbf{i}+2\mathbf{j}$, $\mathbf{b}=3\mathbf{i}+6\mathbf{j}$ and $\mathbf{c}=-4\mathbf{i}+7\mathbf{j}$ respectively.

a Find the vector $\overrightarrow{BC}$

b Hence find the position vector $\mathbf{d}$ of D

12 $ABCD$ is a parallelogram. A, B and D have position vectors $\mathbf{a}=\begin{pmatrix}-10\\-4\end{pmatrix}$, $\mathbf{b}=\begin{pmatrix}15\\6\end{pmatrix}$ and $\mathbf{d}=\begin{pmatrix}-4\\12\end{pmatrix}$. Find the position vector $\mathbf{c}$ of C

13 Points A, B and C lie on a straight line in that order, with $AB:BC=7:3$

A and C have position vectors $\mathbf{a}=-5\mathbf{i}+4\mathbf{j}$ and $\mathbf{c}=7\mathbf{i}+12\mathbf{j}$ respectively. Find the position vector $\mathbf{b}$ of B

Reasoning and problem-solving

Strategy

To solve a problem involving vector components

1. Draw a diagram if appropriate.
2. Convert vectors to components if they're not already in component form.
3. When solving vector equations remember that you can equate x- and y-components separately.

Example 7

Show that the points $A\,(0, 2)$, $B\,(2, 5)$ and $C\,(6, 11)$ are collinear.

$A=\begin{pmatrix}0\\2\end{pmatrix}\quad B=\begin{pmatrix}2\\5\end{pmatrix}\quad C=\begin{pmatrix}6\\11\end{pmatrix}$ — Express each point as a vector from the origin. (2)

$\overrightarrow{AB}=\begin{pmatrix}2\\5\end{pmatrix}-\begin{pmatrix}0\\2\end{pmatrix}=\begin{pmatrix}2\\3\end{pmatrix}$

$\overrightarrow{BC}=\begin{pmatrix}6\\11\end{pmatrix}-\begin{pmatrix}2\\5\end{pmatrix}=\begin{pmatrix}4\\6\end{pmatrix}$ — Determine two vectors that join a common point (B).

$\overrightarrow{BC}=2\overrightarrow{AB}$ — Show that the vectors are parallel (vectors **a** and **b** are parallel if $\mathbf{a}=k\mathbf{b}$).

So the points A, B and C must be collinear.

Exercise 6.2B Reasoning and problem-solving

1 A canoeist takes part in a race across a lake. They must pass through checkpoints, whose positions on a grid map are given by the x- and y-coordinates (1, 11), (7, 6) and (13, 1) respectively. Show that the canoeist will pass through all three checkpoints if they paddle in a straight line.

2 Particles A and B have position vectors $\mathbf{a} = (2\mathbf{i} + 5\mathbf{j})$ m and $\mathbf{b} = (6\mathbf{i} + 3\mathbf{j})$ m respectively. Particle A undergoes a displacement of $(2\mathbf{i} - 3\mathbf{j})$ m and particle B moves in the opposite direction and three times as far. Calculate the distance between the particles after these displacements.

3 Points A and B have position vectors $\mathbf{a} = 3\mathbf{i} + \mathbf{j}$ and $\mathbf{b} = 11\mathbf{i} + 6\mathbf{j}$ respectively. Point C lies on the same straight line as A and B
The lengths AC and BC are in the ratio 3 : 2
Show that there are two possible positions for point C, and find the position vector of each.

4 A town contains four shops A, B, C and D
Shop B is 200 m west of A. Shop C is 100 m north of A. Shop D is 283 m north-east of A
Show that the positions of shops B, C and D are collinear, given that the distances are rounded.

5 Particle A is stationary at the point $(2\mathbf{i} + 3\mathbf{j})$ m, particle B is stationary at the point $(3\mathbf{i} - \mathbf{j})$ m and particle C is stationary at the point $(4\mathbf{i} + 13\mathbf{j})$. Particle B undergoes a displacement of $k(\mathbf{i} + 2\mathbf{j})$ and particle C undergoes a displacement of $k(4\mathbf{i} - \mathbf{j})$ so that all three particles are aligned in a straight line.
Determine the possible values of k

6 Particle A starts at the point $(3\mathbf{i} + \mathbf{j})$ m and travels for 2 s along a track, finishing at the point $(7\mathbf{i} + 4\mathbf{j})$ m. A second particle, B, starts at the same time from the point $2\mathbf{i}$ m. It travels along a parallel track for a distance of d m.

  a Work out the final position vector of B in component form, in terms of d

  b If $d = 15$, evaluate the final distance from A to B

7 The road from P to Q makes a detour round a mountain. It first goes 6 km from P on a bearing of 080°, then 7 km on a bearing of 020° and finally 5 km on a bearing of 295° to reach Q. There is a plan to bore a tunnel through the mountain from P to Q. It will be considered cost effective if it reduces the journey by more than 10 km. Determine whether the tunnel should be built based on this information.

Challenge

8 If a particle moves with constant velocity $\mathbf{v}$, its displacement after time t is $\mathbf{v}t$
Particle A is initially at the point with position vector $(\mathbf{i} + 4\mathbf{j})$ m and moving with constant velocity $(3\mathbf{i} + 3\mathbf{j})$ m s^{-1}. At the same time, particle B is at the point $(5\mathbf{i} + 2\mathbf{j})$ m and moving with constant velocity $(2\mathbf{i} + 3.5\mathbf{j})$ m s^{-1}. After t seconds the particles are at A' and B' respectively.

  a Find the position vector of A' in terms of t

  b Find the vector $\overrightarrow{A'B'}$ in terms of t

  c Decide if the particles will collide, and, if so, find the position vector of the point of collision.

9 If a particle moves with constant velocity $\mathbf{v}$, its displacement after time t is $\mathbf{v}t$. A particle starts from the point with position vector $4\mathbf{j}$ m and moves with constant velocity $(2\mathbf{i} - \mathbf{j})$ m s^{-1}. At the same time a second particle starts from the point $(6\mathbf{i} + 8\mathbf{j})$ m and moves with constant velocity $(-\mathbf{i} - 3\mathbf{j})$ m s^{-1}
Show that the particles collide, and find the time at which they do so.

10 Particle A starts from the point with position vector $5\mathbf{j}$ m and moves with constant velocity $(2\mathbf{i} + \mathbf{j})$ m s^{-1}. Five seconds later a second particle, B, leaves the origin, moving with constant velocity, and collides with A after a further 2 s. Find the velocity of B

2207

6 Summary and review

Chapter summary

- A vector quantity has both magnitude and direction.
  A scalar quantity has magnitude only.
- Equal vectors have the same magnitude *and* direction.
- $k\mathbf{a}$ is parallel to $\mathbf{a}$ and has magnitude $k|\mathbf{a}|$
- Two or more points are collinear if a single vector, or multiple parallel vectors, pass through those points.
- The unit vector has a magnitude of 1 in the direction of $\mathbf{a}$ is $\hat{\mathbf{a}} = \frac{\mathbf{a}}{|\mathbf{a}|}$
- $\overrightarrow{AC} = \overrightarrow{AB} + \overrightarrow{BC}$, $\overrightarrow{AC}$ is the resultant of $\overrightarrow{AB}$ and $\overrightarrow{BC}$
- Making a vector negative reverses its direction, so $\overrightarrow{BA} = -\overrightarrow{AB}$
- A vector is written in component form as $x\mathbf{i} + y\mathbf{j}$ or $\begin{pmatrix} x \\ y \end{pmatrix}$
- For magnitude r and direction θ (the positive rotation from the x-direction), $x = r\cos\theta$ and $y = r\sin\theta$
- For components x and y, $r = \sqrt{x^2 + y^2}$ and $\tan\theta = \frac{y}{x}$
- Treat components separately in calculations and equations
  $(a\mathbf{i} + b\mathbf{j}) \pm (c\mathbf{i} + d\mathbf{j}) = (a \pm c)\mathbf{i} + (b \pm d)\mathbf{j}$
  $k(a\mathbf{i} + b\mathbf{j}) = ka\mathbf{i} + kb\mathbf{j}$
  if $a\mathbf{i} + b\mathbf{j} = c\mathbf{i} + d\mathbf{j}$ then $a = c$ and $b = d$
- The position vector of a point A relative to an origin O is $\overrightarrow{OA}$, often labelled $\mathbf{a}$ or $\mathbf{r}_A$
- If points A and B have position vectors $\mathbf{a}$ and $\mathbf{b}$ then vector $\overrightarrow{AB} = \mathbf{b} - \mathbf{a}$ and distance $AB = |\mathbf{b} - \mathbf{a}|$

Check and review

You should now be able to...	Try Questions
✔ Identify vector quantities and scalar quantities.	1
✔ Solve geometric problems in two dimensions using vector methods.	2–4
✔ Solve problems involving displacements, velocities and forces.	5, 6
✔ Find and use the components of a vector.	6
✔ Find the magnitude and direction of a vector expressed in component form.	6
✔ Use position vectors to find displacements and distances.	7

1 State whether each quantity is vector or scalar.

a Speed **b** Displacement **c** Force **d** Velocity **e** Distance

2 In the diagram, C is the midpoint of BD, F is the midpoint of AC and $EB = 2AE$. The vectors $\overrightarrow{AE}$ and $\overrightarrow{EF}$ are $\mathbf{p}$ and $\mathbf{q}$ respectively.

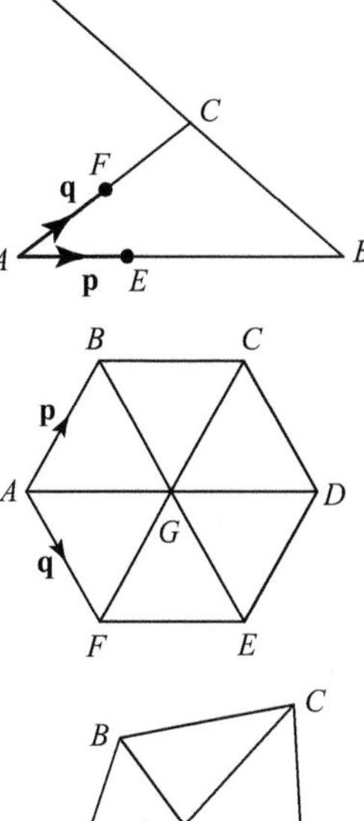

a Express these vectors in terms of $\mathbf{p}$ and $\mathbf{q}$

i $\overrightarrow{AB}$ ii $\overrightarrow{BC}$ iii $\overrightarrow{DF}$

b Show that E, F and D are collinear.

3 The diagram shows a regular hexagon $ABCDEF$. $\overrightarrow{AB} = \mathbf{p}$ and $\overrightarrow{AF} = \mathbf{q}$

Express these vectors in terms of $\mathbf{p}$ and $\mathbf{q}$

a $\overrightarrow{BC}$ **b** $\overrightarrow{BE}$ **c** $\overrightarrow{FD}$ **d** $\overrightarrow{CE}$

4 The diagram shows a quadrilateral $ABCD$

$\overrightarrow{AB} = \mathbf{p}$, $\overrightarrow{AC} = \mathbf{q}$ and $\overrightarrow{AD} = \mathbf{r}$. E is the midpoint of CD

The point F lies on BD and $BF : FD = 3 : 1$

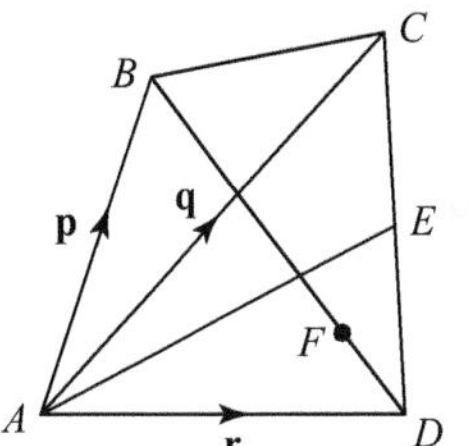

Express these vectors in terms of $\mathbf{p}$, $\mathbf{q}$ and $\mathbf{r}$

a $\overrightarrow{DC}$ **b** $\overrightarrow{DE}$ **c** $\overrightarrow{AE}$ **d** $\overrightarrow{BD}$ **e** $\overrightarrow{BF}$ **f** $\overrightarrow{AF}$

5 $\mathbf{V}$ is the resultant of two velocities, $\mathbf{v}_1$ and $\mathbf{v}_2$

$\mathbf{v}_1$ has magnitude $3\,\text{m s}^{-1}$ and direction $150°$ (anticlockwise rotation from the x-direction). $\mathbf{v}_2$ has magnitude $10\,\text{m s}^{-1}$ and direction θ. Given that the direction of $\mathbf{V}$ is $90°$, calculate

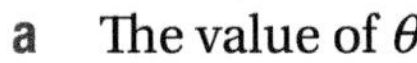

a The value of θ **b** The magnitude of $\mathbf{V}$

6 Forces $\mathbf{P}$, $\mathbf{Q}$ and $\mathbf{R}$ have directions $50°$, $100°$ and $-20°$ respectively (measured from the positive x-direction) $|\mathbf{P}| = 8$ N, $|\mathbf{Q}| = 10$ N and $|\mathbf{R}| = 6$ N

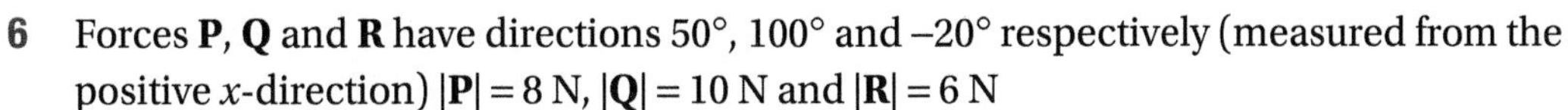

a Express $\mathbf{P}$, $\mathbf{Q}$ and $\mathbf{R}$ in component form.

b Calculate the magnitude and direction of the resultant of the three forces.

7 In relation to origin O, points A and B have position vectors $\mathbf{a} = 2\mathbf{i} + 5\mathbf{j}$ and $\mathbf{b} = 6\mathbf{i} - 2\mathbf{j}$ respectively. Find

a The distance OA **b** The displacement $\overrightarrow{AB}$ **c** The distance AB

What next?

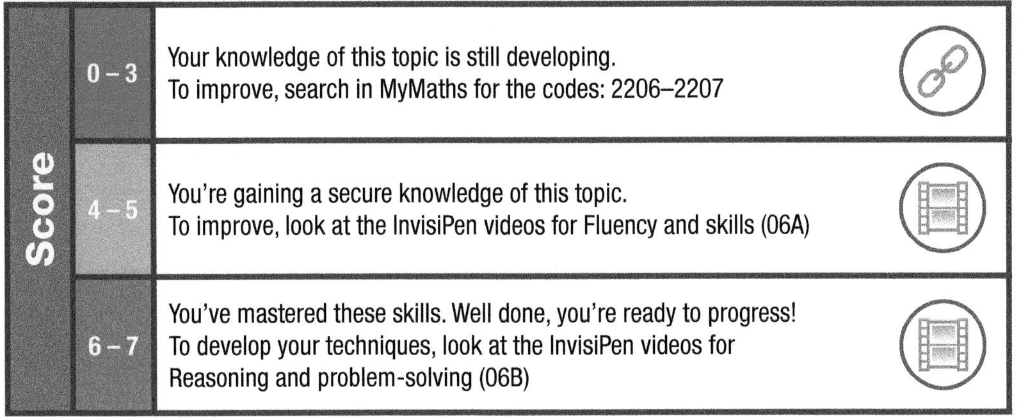

Score			
Score	0 – 3	Your knowledge of this topic is still developing. To improve, search in MyMaths for the codes: 2206–2207	
	4 – 5	You're gaining a secure knowledge of this topic. To improve, look at the InvisiPen videos for Fluency and skills (06A)	
	6 – 7	You've mastered these skills. Well done, you're ready to progress! To develop your techniques, look at the InvisiPen videos for Reasoning and problem-solving (06B)	

Click these links in the digital book

MECH

History

J Willard Gibbs was an American mathematical physicist. Between 1881 and 1884, while teaching at Yale University, he produced lecture notes to help his students understand the electromagnetic theory of light. In his notes he replaced something called **quaternions** with a simpler representation that we now know as **vectors**.

These vector analysis notes proved to be a great success and were later adapted and published as a textbook by one of Gibbs' students, **Edwin Bidwell Wilson**, in 1901. This means that vectors are a relatively recent part of mathematics.

"Mathematics is a language."
- J Gibbs

Research

The theory of quaternions was developed by **Sir William Rowan Hamilton**. What is the connection with Broom Bridge in Dublin?

A British mathematician also developed a theory of vector calculus independently around the same time as Gibbs. Who was it?

Did you know?

Vectors are used extensively in video game development. They are required in order to control or determine the position of objects on the screen.

Programmers combine the language of mathematics with the computer programming language to create the game.

Most people would think that computer programming is a modern invention that didn't truly come about until the 2nd half of the 20th century.

However, the title of 'first computer programmer' is often given to **Ada Lovelace**, a mathematician born in 1815.

Lovelace met **Charles Babbage**, a mathematician and engineer who was working on an 'analytical engine', which was essentially what we would now describe as a programmable computer. She took interest in his work and developed what we now consider to be the first ever computer programme.

The analytical engine never did get built, but Lovelace laid the foundations for the type of computer programming that we encounter in so many aspects of our lives today.

Ada Lovelace

6 Assessment

1 Vector $\mathbf{p} = \begin{pmatrix} 6 \\ -1 \end{pmatrix}$ and vector $\mathbf{q} = \begin{pmatrix} -3 \\ 4 \end{pmatrix}$

a Evaluate $3\mathbf{p} + 5\mathbf{q}$. Select the correct answer.

A $\begin{pmatrix} 33 \\ 7 \end{pmatrix}$ **B** $\begin{pmatrix} 21 \\ 7 \end{pmatrix}$ **C** $\begin{pmatrix} 3 \\ 17 \end{pmatrix}$ **D** $\begin{pmatrix} 3 \\ 23 \end{pmatrix}$ **[1 mark]**

b Calculate the unit vector $\hat{\mathbf{q}}$, in the direction of $\mathbf{q}$. Select the correct answer.

A $\begin{pmatrix} -0.6 \\ 0.8 \end{pmatrix}$ **B** $\begin{pmatrix} -15 \\ 20 \end{pmatrix}$ **C** $\begin{pmatrix} -1.5 \\ 1.5 \end{pmatrix}$ **D** $\begin{pmatrix} 0.6 \\ 0.8 \end{pmatrix}$ **[1]**

2 Four vectors, **a**, **b**, **c** and **d** have a resultant of **0**. $\mathbf{a} = 2\mathbf{i} + 7\mathbf{j}$, $\mathbf{b} = 3\mathbf{i} - 10\mathbf{j}$ and $\mathbf{c} = 5\mathbf{i} - 21\mathbf{j}$

a Evaluate **d**. Select the correct answer.

A $\begin{pmatrix} 10 \\ 24 \end{pmatrix}$ **B** $-10\mathbf{i} + 24\mathbf{j}$ **C** $10\mathbf{i} - 24\mathbf{j}$ **D** $\begin{pmatrix} 24 \\ 10 \end{pmatrix}$ **[1]**

b Evaluate $|\mathbf{d}|$. Select the correct answer.

A 676 **B** $\pm\sqrt{676}$ **C** 26 **D** $\sqrt{476}$ **[1]**

3 A sailing boat starts from buoy A and sails 800 metres on a bearing of 032° to buoy B. It then sails 1200 metres on a bearing of 294° to buoy C. Work out

a The distance AC **[6]**

b The bearing of C from A **[4]**

4 Given vectors $\mathbf{p} = x\mathbf{i} + 2\mathbf{j}$, $\mathbf{q} = 3\mathbf{i} + y\mathbf{j}$ and $\mathbf{r} = 4\mathbf{i} + 6\mathbf{j}$, evaluate

a x and y if $\mathbf{r} = \mathbf{p} - \mathbf{q}$ **[3]**

b x if **p** and **r** are parallel. **[2]**

5 The diagram shows two vectors, **a** and **b**, where $\overrightarrow{OA} = \mathbf{a}$ and $\overrightarrow{OB} = \mathbf{b}$. The point X lies on OA where $OX:XA = 2:1$. The point Y lies on OB where $OY:YB = 3:1$. M is the midpoint of XB

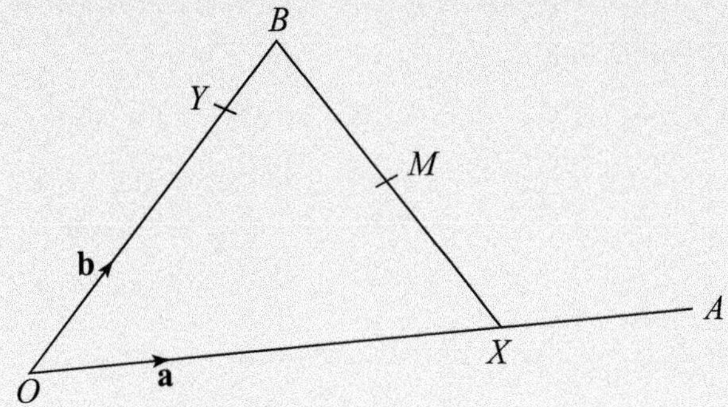

a Express these vectors in terms of **a** and **b** **[11]**

i $\overrightarrow{OX}$ **ii** $\overrightarrow{OY}$ **iii** $\overrightarrow{OM}$ **iv** $\overrightarrow{AM}$ **v** $\overrightarrow{MY}$

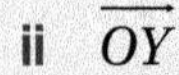

b Use your answer to **a** to prove that A, M and Y are collinear, and work out the ratio $AM:MY$ **[2]**

6 Vector $\mathbf{a} = \begin{pmatrix} 2 \\ -5 \end{pmatrix}$ and vector $\mathbf{b} = \begin{pmatrix} 4 \\ 2 \end{pmatrix}$

Work out the magnitude and direction of the resultant of **a** and **b** **[6]**

7 A is the point $(-2, 5)$, B is the point $(1, 3)$ and C is the point $(10, -3)$

a Write down **i** $\overrightarrow{AB}$ **ii** $\overrightarrow{BC}$ **[2]**

b Prove that A, B and C are collinear. **[2]**

8 Two vectors, **a** and **b**, are given by $\mathbf{a} = \begin{pmatrix} -3 \\ 8 \end{pmatrix}$ and $\mathbf{b} = \begin{pmatrix} 2 \\ 1 \end{pmatrix}$

Given $|\mathbf{a} + \lambda\mathbf{b}| = 13$, work out the possible values of the scalar λ **[6]**

9 The diagram shows a trapezium $OABC$ where $\overrightarrow{OA} = \mathbf{a}$, $\overrightarrow{OC} = \mathbf{c}$, and $\overrightarrow{CB} = 2\mathbf{a}$
M is the midpoint of AB, and N is the midpoint of BC

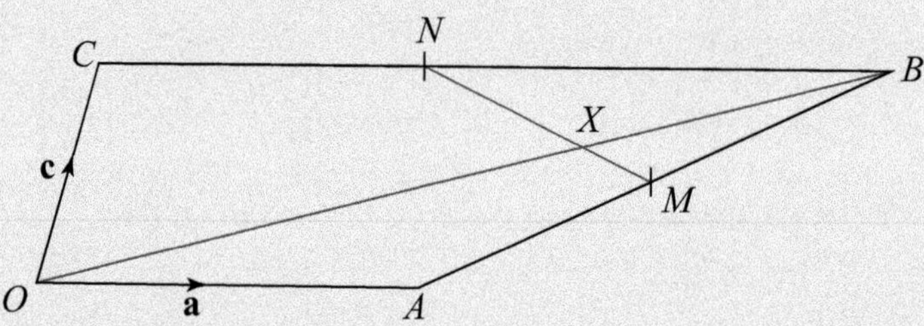

a Express these vectors in terms of **a** and **c** **[7]**

i $\overrightarrow{OB}$ **ii** $\overrightarrow{ON}$ **iii** $\overrightarrow{AB}$ **iv** $\overrightarrow{OM}$ **v** $\overrightarrow{MN}$

The line OB meets the line MN at the point X

b Find $\overrightarrow{OX}$ in terms of **a** and **c** **[8]**

10 Points P, Q, R and S have position vectors $\mathbf{p} = \begin{pmatrix} 6 \\ 3 \end{pmatrix}$, $\mathbf{q} = \begin{pmatrix} -3 \\ -5 \end{pmatrix}$, $\mathbf{r} = \begin{pmatrix} 1 \\ -3 \end{pmatrix}$ and $\mathbf{s} = \begin{pmatrix} 10 \\ 5 \end{pmatrix}$ **[7]**

Prove that the quadrilateral $PQRS$ is a parallelogram.

11 Vectors **x**, **y** and **z** are given by $\mathbf{x} = \begin{pmatrix} 1 \\ -1 \end{pmatrix}$, $\mathbf{y} = \begin{pmatrix} 5 \\ 3 \end{pmatrix}$ and $\mathbf{z} = \begin{pmatrix} 2 \\ 1 \end{pmatrix}$

a Prove that $\mathbf{x} + 3\mathbf{y}$ is parallel to **z** **[3]**

b Work out the value of the integer c, for which $\mathbf{x} + c\mathbf{z}$ is parallel to **y** **[4]**

12 Two forces act on an object: a 2 N force on a bearing of 030° and a 3.5 N force.
The resultant of the two forces acts in a northerly direction. Work out

a the direction of the 3.5 N force, **[6]**

b the magnitude of the resultant force. **[2]**

7 Units and kinematics

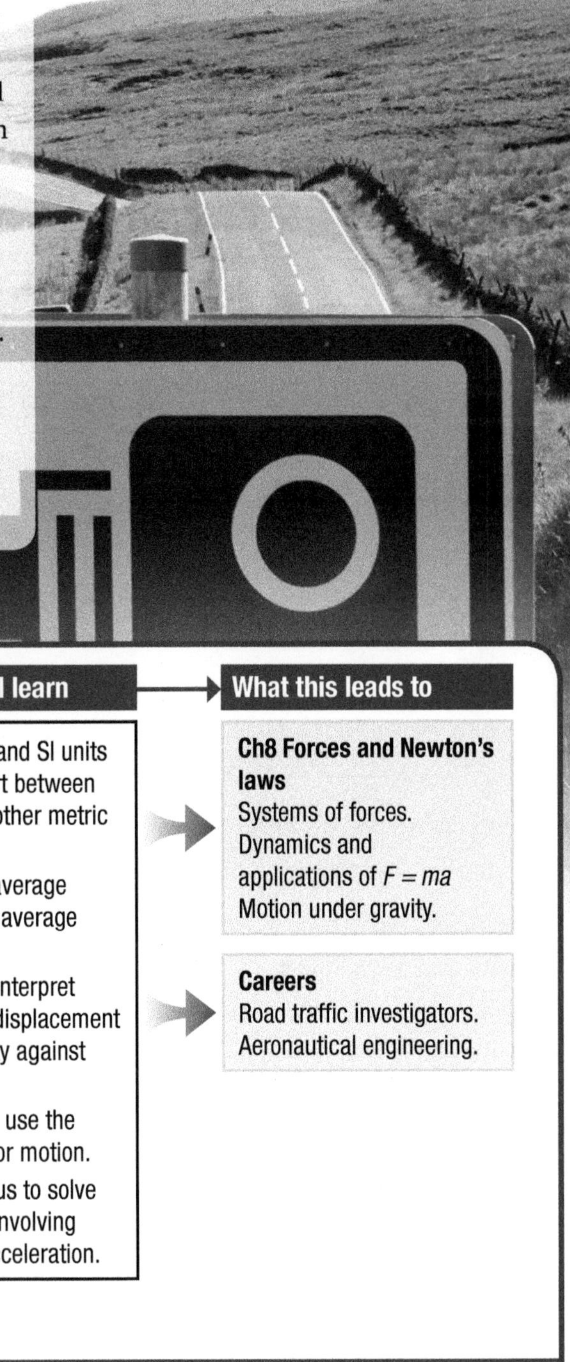

Speed cameras use a variety of techniques for detecting speeding cars. One method involves taking two photos of a car at the start and end of a known time period, and then measuring the distance that the car has travelled in that time. These numbers can then be used to calculate the average speed of the car within that timeframe. This is an application of the equations of motion.

In the speed camera example, only the average speed is considered and so acceleration isn't taken into account. However, in many scenarios, acceleration is important, and it can be both constant (e.g. the effect of gravity) and variable (e.g. racing cars during a Formula One race). This chapter will explore the equations of motion for both constant and variable acceleration.

Orientation

What you need to know

KS4

- Change freely between related standard units and compound units.
- Plot and interpret graphs to solve simple kinematic problems.

Ch4 Integration

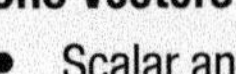

p.116

- Area under a curve.
- Velocity-time graphs.

Ch6 Vectors

p.160

- Scalar and vector definitions.

What you will learn

- To understand SI units and convert between them and other metric units.
- Calculate average speed and average velocity.
- Draw and interpret graphs of displacement and velocity against time.
- Derive and use the formulae for motion.
- Use calculus to solve problems involving variable acceleration.

What this leads to

Ch8 Forces and Newton's laws
Systems of forces.
Dynamics and applications of $F = ma$
Motion under gravity.

Careers
Road traffic investigators.
Aeronautical engineering.

MyMaths Practise before you start 1322, 1970, 2056, 2206, 2269

7 Bridging Unit

Topic A: Kinematics

Bridging to Ch7.2

Kinematics is the study of motion, and this chapter covers the motion of objects in all kinds of different situations. The most basic equation for motion is the one that links speed, distance and time:

$$\text{speed} = \frac{\text{distance}}{\text{time}}$$

Whilst this equation relates only to an object moving with constant speed, you can also use a more general equation for an object whose speed is changing:

$$\text{average speed} = \frac{\text{total distance}}{\text{total time}}$$

You can rearrange this to distance = speed × time or $\text{time} = \frac{\text{distance}}{\text{speed}}$

Example 1

a A car travels at a constant speed of 47 km h^{-1} for 12 minutes. How far does it travel in this time?

b A cyclist rides at a constant speed of 8 m s^{-1}. How long does it take her to cycle 7.2 km?

km h^{-1} means 'kilometres per hour' and m s^{-1} means 'metres per second'.

a 12 minutes = 0.2 hours — The speed is given in km h^{-1}, so convert the time to hours.

Distance = 47 × 0.2 — Use distance = speed × time

= 9.4 km

b 7.2 km = 7200 m — The speed is given in m s^{-1}, so convert the distance to metres.

$\text{Time} = \frac{7200}{8}$ — Use $\text{time} = \frac{\text{distance}}{\text{speed}}$

= 900 seconds (= 15 minutes)

Try It 1

A car travels at a constant speed. It takes 9 minutes to travel 12 km. Calculate its speed in km h^{-1}

The **displacement** of an object is its distance *from a fixed point in a given direction*. This might be the same as the distance, but not necessarily. For example, if a dog runs across a 100 m field, its distance and displacement will both be 100 m. However, if it runs across the field, turns round, and comes back to its starting point, it will have travelled a *distance* of 200 m, but its *displacement* from the starting point will be 0 m.

Displacement and velocity are vector quantities because they have magnitude and direction.

Velocity is the rate of change of displacement, much like speed is the rate of change of distance. So velocity also takes into account the direction from a fixed point, whilst speed does not. Again, they may be the same, but not if there's a change of direction or if the object moves backwards.

Displacement-time graphs are useful tools for studying the motion of an object. Positive gradients show the object is moving in the positive (or forwards) direction, and negative gradients show motion in the negative (or backwards) direction.

Key point

On a **displacement–time graph**, the gradient represents the velocity.

When the displacement is negative, the object is 'behind' its starting point, but be careful not to confuse this with the gradient. If the graph shows a negative displacement but a positive gradient, it's behind the starting point but moving 'forwards', that is, back towards the starting point.

Displacement-time graphs	Positive value	Negative value
Positive gradient	In front of fixed point, moving forwards away from it.	Behind fixed point, moving forwards towards it.
Negative gradient	In front of fixed point, moving backwards towards it.	Behind fixed point, moving backwards away from it.

Example 2

The motion of a particle is described by this displacement–time graph.

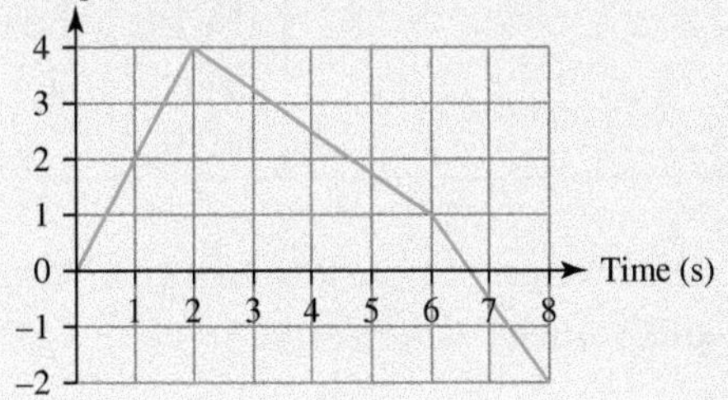

a Find the velocity of the particle during the first 2 seconds.

b Find the velocity of the particle during the next 4 seconds.

c Find the displacement of the particle after 8 seconds.

d Calculate the total distance travelled during these 8 seconds.

e Calculate the average speed of the particle over the whole 8 seconds.

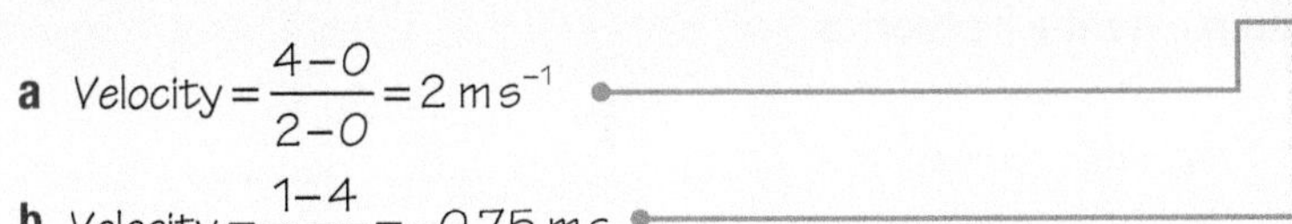

a Velocity $= \frac{4-0}{2-0} = 2\ \text{m s}^{-1}$ — Use the gradient of the line.

b Velocity $= \frac{1-4}{6-2} = -0.75\ \text{m s}$ — The velocity is negative as the particle is travelling 'backwards'. The speed would be $0.75\ \text{m s}^{-1}$

c Displacement $= -2\text{m}$ — You can just read this off the graph.

d Total distance $= 4 + 3 + 3 = 10\ \text{m}$ — This is different to the final displacement, because the direction doesn't matter. Split the motion into distinct sections each time the gradient changes, and add them together.

e Average speed $= \frac{10}{8} = 1.25\ \text{m s}^{-1}$ — Use average speed $= \frac{\text{total distance}}{\text{total time}}$

Try It 2

For the graph in Example 2,

a Calculate the velocity in the final 2 s,

b State the speed in the final 2 s.

MyMaths

1322 SEARCH

Velocity–time graphs are another useful tool for showing the motion of an object. Just like displacement-time graphs, their gradient is important when performing kinematics calculations.

Key point

On a **velocity–time graph**, the gradient represents the **acceleration** and the displacement is given by the area under the graph.

Be careful not to confuse displacement–time and velocity–time graphs. Despite looking very similar, they don't display the same information. In particular, the gradient and value of velocity–time graphs don't tell you anything about where the object is in relation to the fixed point, which is why you need to find the area.

Velocity-time graphs	Positive value	Zero value	Negative value
Positive gradient	Moving in forwards direction, speeding up.	Changing direction, from backwards to forwards.	Moving in backwards direction, slowing down.
Zero gradient	Moving in forwards direction at steady speed.	Not moving.	Moving in backwards direction at steady speed.
Negative gradient	Moving in forwards direction, slowing down.	Changing direction, from forwards to backwards.	Moving in backwards direction, speeding up.

Example 3

The motion of a particle in a given direction is given by this velocity-time graph.

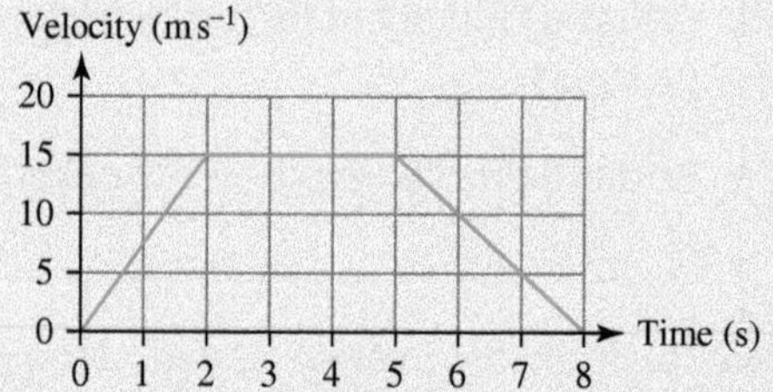

a Describe the motion of the particle during these 8 seconds.

b State the acceleration of the particle in the final 3 seconds.

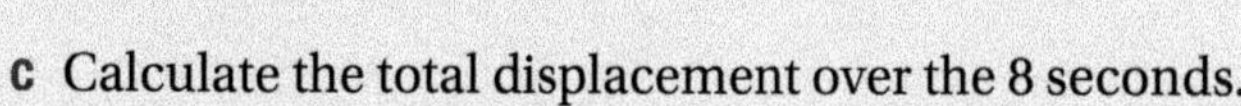

c Calculate the total displacement over the 8 seconds.

a The particle accelerates over 2 seconds to a velocity of 15 m s^{-1} then moves at a constant velocity of 15 m s^{-1} for 3 seconds, before decelerating to rest over a further 3 seconds.

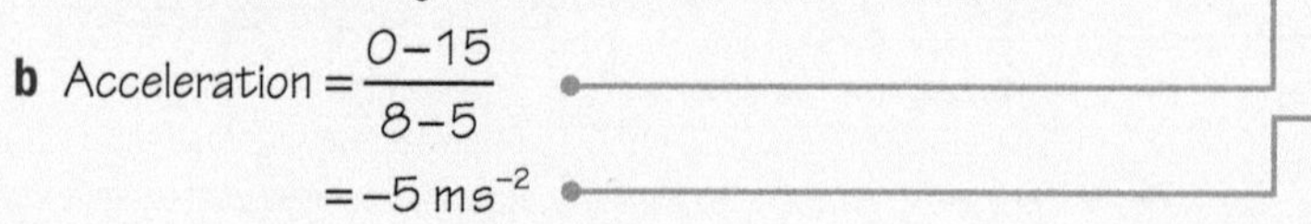

b Acceleration $= \dfrac{0-15}{8-5}$ — Use acceleration = gradient.

$= -5\text{ m s}^{-2}$ — This is the same as a deceleration of 5 m s^{-2}

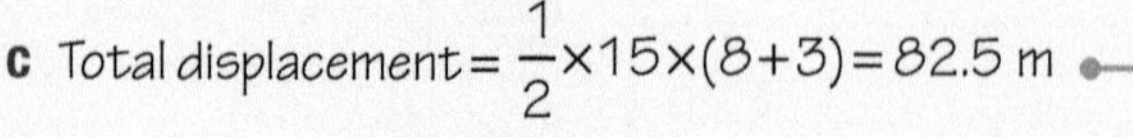

c Total displacement $= \dfrac{1}{2} \times 15 \times (8+3) = 82.5\text{ m}$ — Using the formula for the area of a trapezium to calculate the area under the graph.

Try It 3

For the graph in Example 3,

a Calculate the acceleration over the first 2 s,

b Calculate the total displacement of the particle whilst it is moving at a constant velocity.

1 Calculate the speed of an object that travels

a 18 km in 4 hours (in km h^{-1}),

b 30 m in 48 s (in m s^{-1}),

c 12 km in 32 minutes (in km h^{-1}).

2 Calculate the distance travelled by a particle moving at

a 13 km h^{-1} for 2.5 hours,

b 9.2 m s^{-1} for 75 seconds,

c 68 km h^{-1} for 3 hours 24 minutes.

3 Calculate the time taken from a particle to travel

a 27 km at 30 km h^{-1},

b 13 m at 0.8 m s^{-1},

c 36 cm at 0.3 m s^{-1}.

4 The displacement-time graph describes the motion of a car travelling along a straight road.

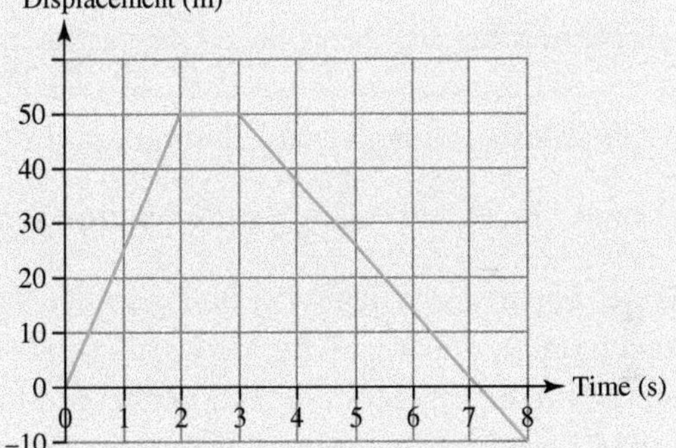

a Calculate the velocity during the first 2 s.

b For how long was the car stationary?

c Calculate the velocity and the speed during the final 5 s.

d Calculate the total distance travelled by the car.

e Calculate the average speed over the whole 8 s.

5 The speed-time graph describes the motion of a particle moving in a straight line.

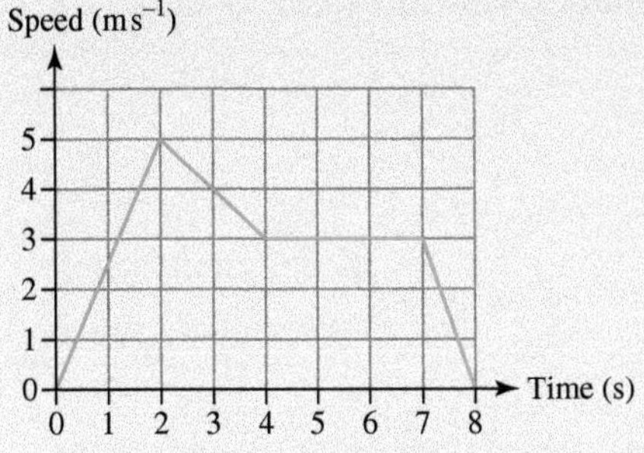

As long as the motion is in a straight line and the object is always either stationary or moving forwards, then a speed-time graph is the same as a velocity-time graph.

a Calculate the acceleration over the first 2 s.

b Calculate the acceleration over the next 2 s.

c Describe the motion of the particle between 4 s and 7 s.

d Calculate the deceleration over the final 1 s.

e Calculate the distance travelled by the particle over the final 4 s.

  1322  SEARCH

7.1 Standard units and basic dimensions

Fluency and skills

All quantities in mechanics are defined in terms of three **fundamental quantities** or **dimensions:** mass, length and time. Quantities, or dimensions, are measured in units.

Some SI (Système International d'Unités) base units you'll have come across before are kilogram / kg (mass), metre / m (length), and second / s (time).

Kinematics is the study of motion. In kinematics, you will meet distance, displacement, speed, velocity and time. These are derived quantities that you can describe in terms of the fundamental quantities (mass, length and time).

See Ch6.1 For a reminder on vectors and scalars.

Vector	Scalar	Fundamental quantities	SI Units
Displacement	Distance	length	metres (m)
Velocity	Speed	$\frac{\text{length}}{\text{time}}$	metres per second (m s^{-1} or m/s)
Acceleration		$\left(\frac{\text{velocity}}{\text{time}}\right) = \frac{\text{length}}{(\text{time})^2}$	metres per second squared (m s^{-2} or m/s^2)

Mechanics also involves the derived quantities force and weight.

Force = mass × acceleration. The SI unit is the newton (N). **Key point**

Weight is the force of gravity on an object. An object with mass m kg has weight mg N, where g is the acceleration due to gravity. On Earth, this is 9.81 m s^{-2} to 3 sf. If you were on the Moon, your mass would be the same but your weight would be less. In common speech, you might use mass and weight to mean the same thing, but make sure you don't do this in Maths.

Correct formulae are dimensionally consistent. If, for example, $a = b + c$ and a is a velocity, then b and c also have the dimensions of a velocity. You must also use the same units throughout, and so you may need to convert some units before carrying out calculations for a formula to work.

Example 1

Express a speed of 15 km h^{-1} in m s^{-1}

$15\text{ km h}^{-1} = 15000\text{ m h}^{-1}$ — 1 km = 1000 m

$15000\text{ m h}^{-1} = \frac{15000}{3600}\text{ m s}^{-1} = 4\frac{1}{6}\text{ m s}^{-1}$ — 1 h = 60 × 60 = 3600 s

Exercise 7.1A Fluency and skills

1 State the quantity described by these units.

a Newtons, N

b Kilograms, kg

c Metres per second, m s^{-1}

d Metres per second squared, m s^{-2}

2 Convert

a 8.5 km to m

b 2.3 m to mm

c 482 cm to m

d 1650 m to km

e 72 km h^{-1} to m s^{-1}

f 14 m s^{-1} to km h^{-1}

g 25 cm s^{-1} to km h^{-1}

h 2.4 m^2 to cm^2

i 1.4 kg to g

j 1.6 tonnes to kg

3 A car travels 70 km in 35 minutes. Evaluate its speed in

a km h^{-1} b m s^{-1}

4 Work out the distance, in km, travelled in a quarter of an hour by a car that has a constant speed of 20 m s^{-1}

5 A particle has an acceleration of 200 km h^{-2} Express this in m s^{-2}

6 The force F (in N) on an object is related to its mass m (in kg) and its acceleration a (in m s^{-2}) by $F = ma$. Work out, in kg, the mass of an object if a force of 0.25 kN (kilonewtons) accelerates it at 20 km min^{-2}

MECH

Reasoning and problem-solving

Strategy

When answering a question involving units

1. Convert units if they're inconsistent and perform any necessary calculations.
2. Check that dimensions have been conserved and that your final answer is in the correct units.

Example 2

u and v are velocities, a is acceleration and s is displacement. Use the formula $v^2 = u^2 + 2as$ to work out s if $u = 24 \text{ km h}^{-1}$, $v = 32 \text{ km h}^{-1}$ and $a = 0.005 \text{ m s}^{-2}$. Give your answer in kilometres.

$u = \frac{24 \times 1000}{3600} = 6.67 \text{ m s}^{-1}$ and $v = \frac{32 \times 1000}{3600} = 8.89 \text{ m s}^{-1}$ — Convert velocities to m s^{-1} (1)

$8.89^2 = 6.67^2 + 2 \times 0.005s$

$s = \frac{8.89^2 - 6.67^2}{0.01} = 3456.8 \text{ m}$ — Substitute values and solve for s (2)

$s = 3.46 \text{ km}$ (to 3 sf) — Convert your answer to km. (2)

Exercise 7.1B Reasoning and problem-solving

1 A runner travels 3900 m at 8 km h^{-1} Find, in minutes, the time she takes.

2 In the formula $s = ut + \frac{1}{2}at^2$, u is velocity, a is acceleration, t is time and s is displacement. Find the value of s if $u = 4 \text{ km h}^{-1}$, $a = 0.01 \text{ m s}^{-2}$ and $t = 40$ minutes. Give your answer in km.

3 A station platform is 180 m long. A train of length 120 m passes it at 30 km h^{-1} How long will it take for the train to pass completely?

Challenge

4 A liquid of density 1.2 g cm^{-3} is flowing at 3 km h^{-1} through a cylindrical pipe of radius 5 cm. Given that density $= \frac{\text{mass}}{\text{volume}}$, and that for a cylinder with radius r and height h its volume is given by $\pi r^2 h$, calculate the mass, in kg, of the liquid emerging from the pipe in 30 seconds.

2183

7.2 Motion in a straight line – definitions and graphs

Fluency and skills

You use these terms to describe location and movement.

> **Key point**
>
> **Position** is a vector: the distance and direction from the origin O
>
> **Displacement** is a vector: the change of position.
>
> **Distance** is a scalar: the magnitude of displacement.
>
> **Velocity** is a vector: the rate of change of displacement.
>
> **Speed** is a scalar: the magnitude of velocity.

The diagram shows displacement $\overrightarrow{PQ} = -6$ from position 4 to position –2, then displacement $\overrightarrow{QO} = 2$ from position –2 to position 0 (the origin).

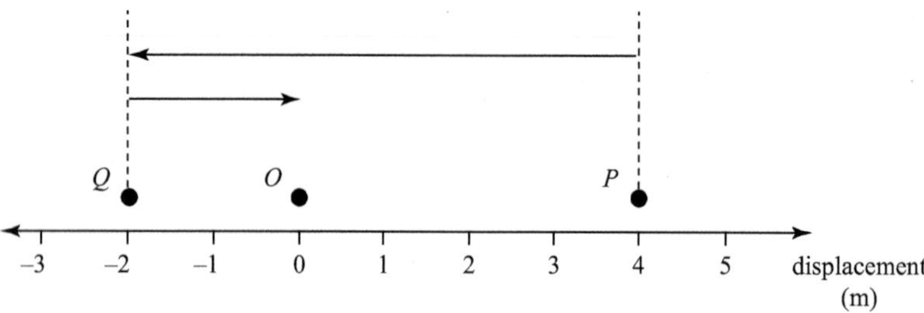

The resultant displacement $\overrightarrow{PO}$ is –4 m but the total distance moved is 8 m.

You can see from this that it's important to distinguish between displacement and distance. Similarly, it's important to distinguish between the average velocity during motion and the average speed during motion. Average speed will not take into account the direction of the motion.

> **Key point**
>
> $$\text{Average velocity} = \frac{\text{resultant displacement}}{\text{total time}}$$

> **Key point**
>
> $$\text{Average speed} = \frac{\text{total distance}}{\text{total time}}$$

You can illustrate motion with a **displacement–time (or s–t) graph.**

For an s–t graph, gradient $= \frac{\text{change of displacement}}{\text{change of time}}$, which is velocity.

Displacement is usually represented by the letter s

> **Key point**
>
> The **gradient** of a displacement–time graph is the velocity.

For straight-line s–t graphs, you should assume that any changes of gradient are instantaneous. This makes calculations easier, but in reality the velocity would change over a given period of time.

Example 1

The graph shows the motion of a particle along a straight line between 0 and 11 seconds.

a Find the displacement and velocity for the first 6 seconds and for the final 5 seconds.

b Find **i** The resultant displacement,

ii The average velocity,

iii The average speed.

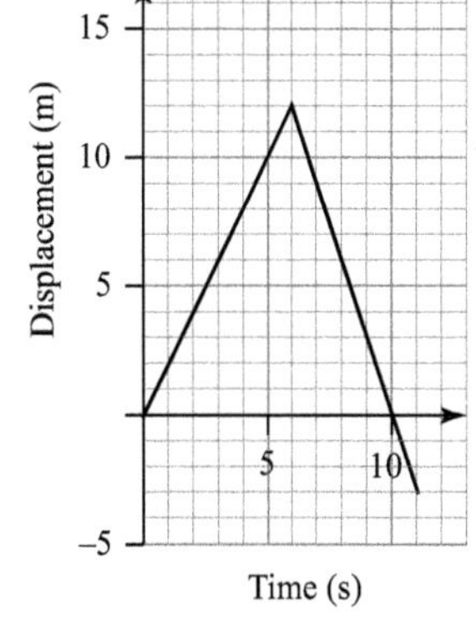

a First 6 seconds:

displacement = 12 m

$\text{Velocity} = \dfrac{12 - 0}{6 - 0} = 2\text{ m s}^{-1}$

Velocity = gradient

Final 5 seconds:

displacement = (–3) – 12 = –15 m

$\text{Velocity} = \dfrac{(-3) - 12}{11 - 6} = -3\text{ m s}^{-1}$

b i Resultant displacement = 12 + (–15) = –3 m

ii $\text{Average velocity} = \dfrac{-3}{11}\text{ m s}^{-1}$

Average velocity = $\dfrac{\text{Resultant displacement}}{\text{Total time}}$

iii $\text{Average speed} = \dfrac{12 + 15}{11} = 2\dfrac{5}{11}\text{ m s}^{-1}$

Average speed = $\dfrac{\text{Total distance}}{\text{Total time}}$

Speed is a scalar, so all motion is positive.

You can also draw a **velocity–time graph** (a ***v–t* graph**).

If velocity changes from u m s^{-1} to v m s^{-1} in t s, as shown, then

$$\text{gradient} = \frac{v-u}{t}\text{ m s}^{-2} = \text{rate of change of velocity} = \text{acceleration.}$$

Key point

Acceleration is the rate of change of velocity.

The gradient of a velocity–time graph is the acceleration.

Key point

$$\text{Acceleration} = \frac{\text{change in velocity}}{\text{time}} = \frac{v-u}{t}$$

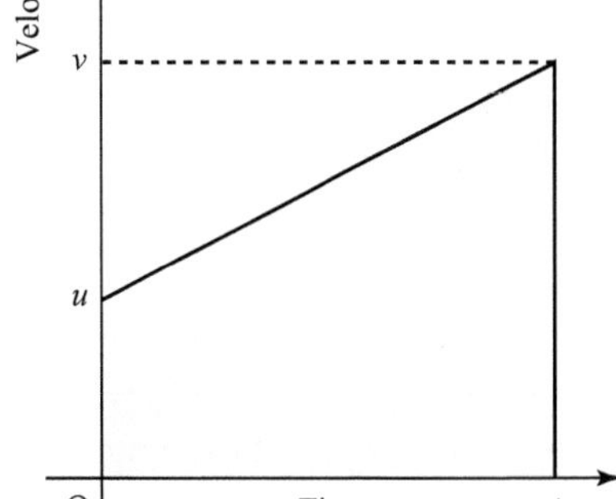

The shaded area $= \frac{1}{2}(u+v)t$ = average velocity × time = displacement

Key point

The area between the v–t graph and the t-axis is the displacement.

Example 2

Use this velocity–time graph to answer the questions.

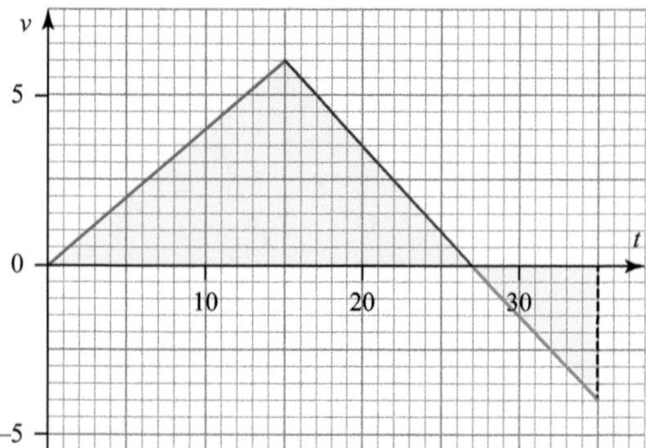

a Calculate the acceleration during

i The first 15 s **ii** The remaining 20 s

b Calculate **i** The resultant displacement, **ii** The total distance travelled.

c Sketch the corresponding *s–t* graph.

a **i** Acceleration $= \frac{6-0}{15-0} = 0.4\ \text{m s}^{-2}$

Acceleration = gradient

ii Acceleration $= \frac{(-4)-6}{35-15} = -0.5\ \text{m s}^{-2}$

b **i** Object moves forward for the first 27 s

The velocity is positive.

Displacement $= \frac{1}{2} \times 27 \times 6 = 81\ \text{m}$

Displacement = shaded area

In final 8 s, the object moves backwards:

displacement $= \frac{1}{2} \times 8 \times (-4) = -16\ \text{m}$

Resultant displacement $= 81 - 16 = 65\ \text{m}$

ii Total distance $= 81 + 16 = 97\ \text{m}$

c

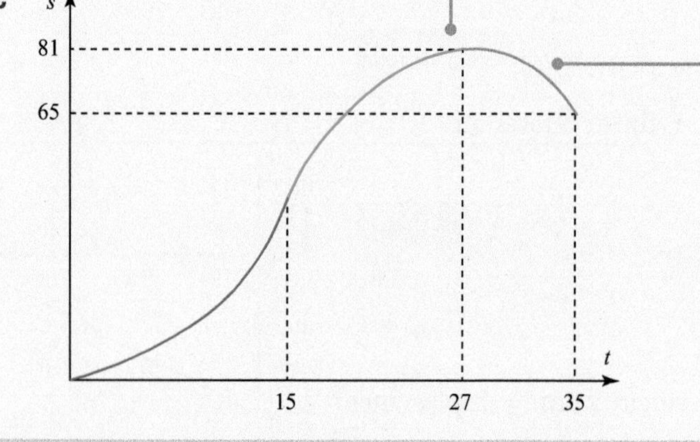

Gradient (velocity) increases (in red) to be the steepest at $t = 15$ s, then decreases (in blue) to zero at $t = 27$ s, when $s = 81$ m

Gradient then becomes negative (in green) and increasingly steep, ending at $s = 65$ m

Negative acceleration is sometimes called **deceleration**. Though it often means slowing down, in the final stage of Example 2 the acceleration is negative but the *speed* is increasing.

Exercise 7.2A Fluency and skills

1 The graph shows the motion of an object in a straight line over 20 seconds.

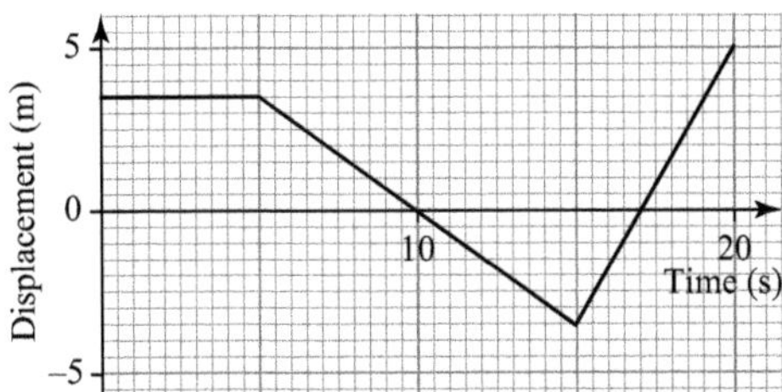

a During what time period(s) is the particle moving backwards?

b During what time period(s) is the object not moving?

c Without performing any calculations, state with reason the time period during which the object's speed is greatest.

2 The graph shows a particle moving in a straight line.

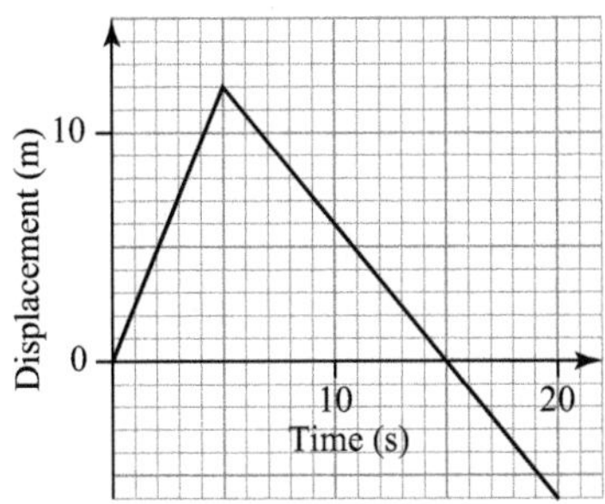

a Describe the motion of the particle.

b Evaluate its velocity in each phase.

c Work out i Its average speed,
ii Its average velocity.

3 The graph shows a particle moving along a straight line.

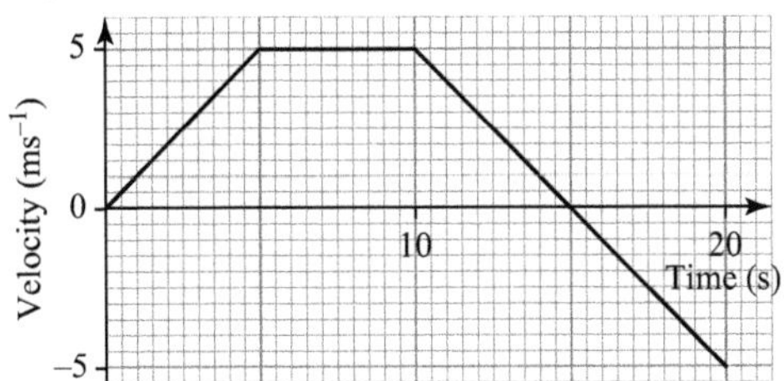

a Calculate the distance travelled by the particle during the first 10 seconds.

b Calculate the acceleration of the particle between 10 and 20 seconds.

4 The graph shows a particle moving along a straight line.

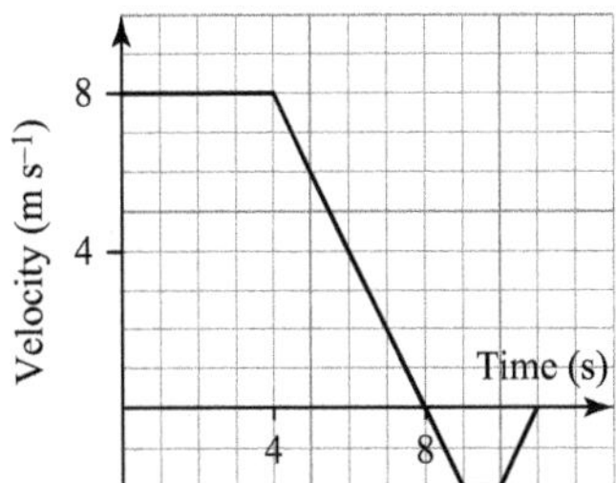

a During what time period is the particle moving backwards?

b Evaluate the average acceleration of the particle between 0 and 11 seconds.

c Work out the acceleration of the particle between 4 and 9 seconds.

5 A particle, travelling at 36 m s^{-1}, is brought uniformly to rest with acceleration -1.2 m s^{-2}

a Sketch the velocity-time graph for this motion.

b Calculate the distance the particle travels before it stops.

6 A man stands 2.5 m from the wall of an ice rink and hits a hockey puck towards the wall. The puck hits the wall after 1.4 s before travelling straight back in the opposite direction. After another 3.2 s the puck stops suddenly in a goal, 1.8 m behind the man.

a Draw a displacement-time graph for the motion of the puck.

b Calculate the puck's

i Velocity during the first 1.4 s,

ii Velocity during the final 3.2 s,

iii Average speed between being hit and coming to a stop.

7 A particle, travelling at 15 m s^{-1}, accelerates uniformly to a velocity of 45 m s^{-1} in 12 s. Sketch a velocity-time graph for this motion and use it to

a Work out the particle's acceleration,

b Calculate how far the particle travels while accelerating.

MyMaths 2183 SEARCH

Reasoning and problem-solving

Strategy

To solve problems using motion graphs

1. Be clear whether you are being asked for displacement or distance, and velocity or speed.
2. Use gradient to calculate velocity from an s–t graph, and acceleration from a v–t graph.
3. Use area under a v–t graph to calculate displacement. Keep in mind that area below the t-axis is negative displacement.

Example 3

The graph shows the acceleration of an object during a period of 7 seconds. At the start of that period, the velocity of the object is 2 m s^{-1}

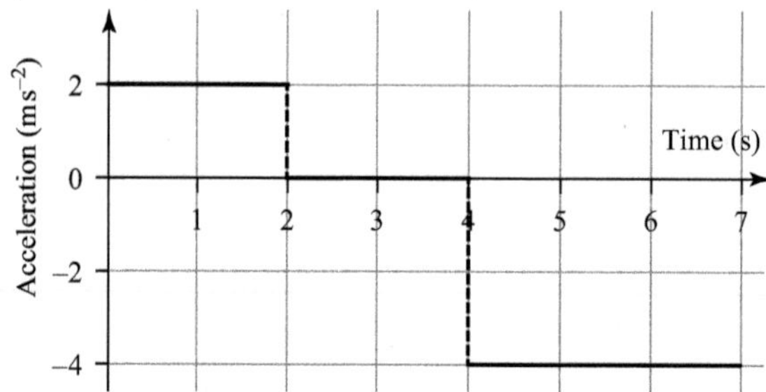

By sketching the velocity-time graph, calculate the resultant displacement of the object and the total distance it travels.

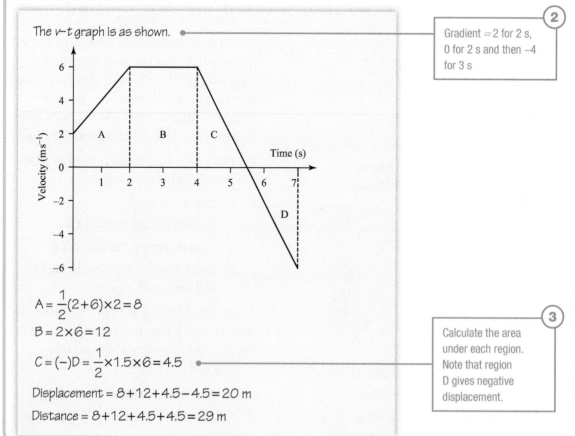

The v–t graph is as shown.

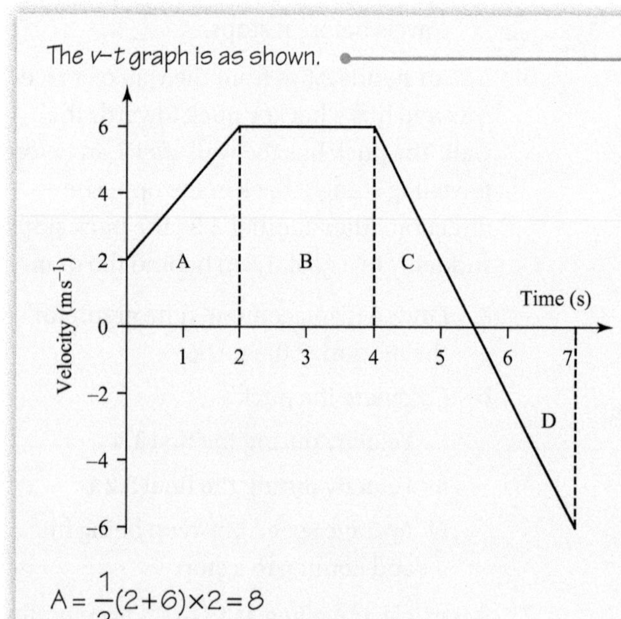

$A = \frac{1}{2}(2+6)\times 2 = 8$

$B = 2\times 6 = 12$

$C = (-)D = \frac{1}{2}\times 1.5\times 6 = 4.5$

Displacement $= 8+12+4.5-4.5 = 20$ m

Distance $= 8+12+4.5+4.5 = 29$ m

(2) Gradient = 2 for 2 s, 0 for 2 s and then −4 for 3 s

(3) Calculate the area under each region. Note that region D gives negative displacement.

Exercise 7.2B Reasoning and problem-solving

1 Ben rolls a ball at a constant speed of 4 m s^{-1}. After 6 m, it hits a wall and rebounds along the same line at a constant speed of 2.5 m s^{-1}. A further 3 s later he stops the ball.

a Draw a displacement–time graph.

b Work out

i The average speed of the ball,

ii The average velocity of the ball.

2 The graph shows the motion of a cat along the top of a fence.

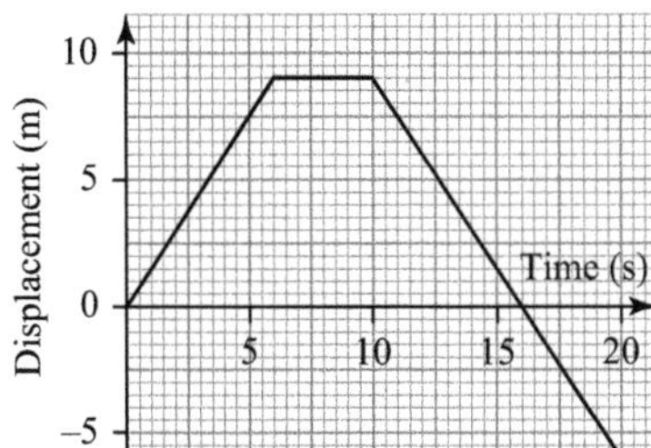

a Describe the cat's motion during this period of 20 s

b Calculate the cat's velocity during each phase.

c Work out

i The average speed of the cat,

ii The average velocity of the cat during this period of 20 s

d Explain why, in reality, the motion of a cat could not be represented by a straight-line graph like the one shown.

3 A car accelerates from rest at point O for 6 s at 2.5 m s^{-2}, then brakes uniformly to rest in 4 s. It immediately reverses, accelerating uniformly and passing point O at a speed of 6 m s^{-1}

a Sketch a velocity–time graph.

b Calculate the car's greatest positive displacement from O

c At what time was the car back at O?

d Calculate the car's acceleration whilst reversing.

4 A runner and a cyclist start together from rest at the same point. The runner accelerates at 0.6 m s^{-2} for 10 s, then continues at uniform velocity. The cyclist accelerates uniformly for 8 s then immediately slows uniformly to rest after a further 22 s. At the moment the cyclist stops they have both gone the same distance.

a Draw a velocity–time graph for the runner and for the cyclist.

b Work out

i How far each travels,

ii The cyclist's greatest speed.

5 Car P is at rest when car Q passes it at a constant speed of 20 m s^{-1}. Immediately, P sets off in pursuit, accelerating at 2 m s^{-2} until it reaches 30 m s^{-1} and continues at that speed.

a Sketch a velocity–time graph for this situation.

b How far is P behind Q when it reaches full speed?

c How much longer after this point does it take for P to draw level with Q?

d In reality, the graph showing this motion would not be made up of perfect straight lines. Suggest why this is.

Challenge

6 A car can accelerate and decelerate at 2.5 m s^{-2} and has a cruising speed of 90 km h^{-1}. It travels 8 km along a road, but 2 km of road is affected by road works with a speed limit of 36 km h^{-1}. The total journey time will depend on where the road-works occur. By sketching suitable graphs, work out the maximum and minimum journey times. Assume the car starts and finishes at rest.

7.3 Equations of motion for constant acceleration

Fluency and skills

All motion problems involve some or all of these quantities.

See p.382
For a list of mathematical notation.

Key point

s = displacement u = initial velocity v = final velocity
a = acceleration t = time

For constant acceleration, these quantities satisfy five equations, sometimes called the ***suvat* equations**. You need to memorise these equations and know how to derive them.

Each one involves four of the five variables, and when solving a problem you should list the variables you know and the variables you want to find. This will let you choose the correct equation to use—find the one with 3 variables that you know and 1 that you don't.

Acceleration = gradient of v-t graph, so $a = \frac{v-u}{t}$

1 **Key point**

$$\Rightarrow v = u + at$$

Displacement = area under graph

2 **Key point**

$$\Rightarrow s = \frac{1}{2}(u+v)t$$

Substituting for v from 1 into 2, $s = \frac{1}{2}(u+u+at)t$

3 **Key point**

$$\Rightarrow s = ut + \frac{1}{2}at^2$$

Rearranging 1 you get $u = v - at$. Substituting this into 2, $s = \frac{1}{2}(v - at + v)t$

4 **Key point**

$$\Rightarrow s = vt - \frac{1}{2}at^2$$

Rearranging 1, you get $t = \frac{v-u}{a}$. Substituting this into 2, $s = \frac{(u+v)(v-u)}{2a}$

5 **Key point**

$$\Rightarrow v^2 = u^2 + 2as$$

ICT Resource online

To practise choosing which equation of motion to use, click this link in the digital book.

In problems involving these equations, the moving objects are usually treated as 'particles', with a small or irrelevant size.

In most cases this won't affect the results. However, when you're dealing with a large object moving across a relatively short distance, such as a bus moving between street junctions, the true distance may be shorter than that used in calculations—you might need to consider this when discussing the limitations of mathematical models.

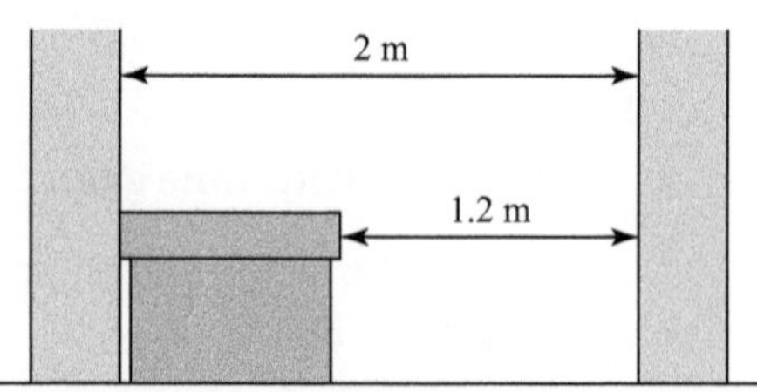

If this box were treated as a particle with irrelevant size, it would move 2 m to travel between the walls of the corridor. In reality, it will only move 1.2 m

Example 1

An object travelling at $15\,\text{m s}^{-1}$ accelerates at $3\,\text{m s}^{-2}$ for 5 s. Work out

a Its final velocity, **b** Its displacement during this period.

a $u = 15, a = 3, t = 5$, find v — List what you know and your 'target' variable.

$v = 15 + 3 \times 5 = 30$ — Use $v = u + at$

$\therefore$ final velocity is $30\,\text{m s}^{-1}$

b $u = 15, a = 3, t = 5$, find s

$s = 15 \times 5 + \frac{1}{2} \times 3 \times 5^2 = 112.5$ — Use $s = ut + \frac{1}{2}at^2$

$\therefore$ displacement $= 112.5\,\text{m}$

In part **b** you could use $s = \frac{1}{2}(u + v)t$ with the value of v found in part **a**.

Exercise 7.3A Fluency and skills

1 You are given the initial velocity u and the uniform acceleration a of a particle during time t. Write down the equation you could use to find its displacement during this time.

2 **a** Evaluate v if $u = 4\,\text{m s}^{-1}$, $a = 2\,\text{m s}^{-2}$, $t = 7\,\text{s}$

b Evaluate u if $v = 12\,\text{m s}^{-1}$, $a = -3\,\text{m s}^{-2}$, $t = 4\,\text{s}$

c Evaluate t if $u = 8\,\text{m s}^{-1}$, $v = 35\,\text{m s}^{-1}$, $a = 3\,\text{m s}^{-2}$

3 **a** Evaluate s if $u = 5\,\text{m s}^{-1}$, $a = 2\,\text{m s}^{-2}$, $t = 4\,\text{s}$

b Evaluate a if $s = 30\,\text{m}$, $u = -4\,\text{m s}^{-1}$, $t = 6\,\text{s}$

c Evaluate t if $u = 12\,\text{m s}^{-1}$, $a = -6\,\text{m s}^{-2}$, $s = 9\,\text{m}$

4 **a** Evaluate s if $u = 4\,\text{m s}^{-1}$, $v = 6\,\text{m s}^{-1}$, $a = 5\,\text{m s}^{-2}$

b Evaluate t if $u = 3\,\text{m s}^{-1}$, $v = 5\,\text{m s}^{-1}$, $s = 20\,\text{m}$

c Evaluate s if $v = 8\,\text{m s}^{-1}$, $a = 2\,\text{m s}^{-2}$, $t = 3\,\text{s}$

5 $v = 12.1\,\text{m s}^{-1}$, $a = -1.20\,\text{m s}^{-1}$, $s = 73.6\,\text{m}$

a Calculate u

b Calculate t

6 A particle starts from rest and accelerates uniformly for 10 s. It moves 150 m in that time. Work out

a Its acceleration,

b Its speed at the end of the period.

7 A particle accelerates uniformly for 8.1 s. After this time, it has a displacement of 80 cm and its speed is $14\,\text{cm s}^{-1}$

a Calculate its acceleration during this time in m s^{-2}

b Calculate its speed before it underwent the acceleration in m s^{-1}

8 A particle travelling at $144\,\text{km h}^{-1}$ is brought uniformly to rest in a distance of 200 m. Work out

a Its acceleration,

b The time it takes to stop.

9 A particle, accelerating at $4\,\text{m s}^{-2}$, takes 5 s to travel from A to B. Its speed at B is $24\,\text{m s}^{-1}$ Work out

a The distance AB **b** Its speed at A

10 A driver notices the speed limit change and applies the brakes suddenly, causing a uniform deceleration. The car's speed after braking is $60\,\text{km h}^{-1}$, the car's deceleration while braking is $(-)10\,\text{m s}^{-2}$ and the distance it covers while braking is 26 m.

a How fast, in km h^{-1}, was the car moving before applying the brakes?

b For how long did the driver apply the brakes?

SEARCH

Reasoning and problem-solving

Strategy

When solving a problem with the equations of motion for constant acceleration

1. Use the information in the question to list the known values and the variable you need to find. Be careful to distinguish between displacement and distance, and between velocity and speed.
2. Choose the correct equation to use.
3. Apply the equation to find the numerical value and use it to answer the original question.

Example 2

A hot-air balloon is drifting in a straight line at a constant velocity of 3 m s^{-1}. A sudden head-wind gives it an acceleration of –0.5 m s^{-2} for a period of 16 s. For this period, calculate

a The resultant displacement of the balloon, **b** The distance travelled.

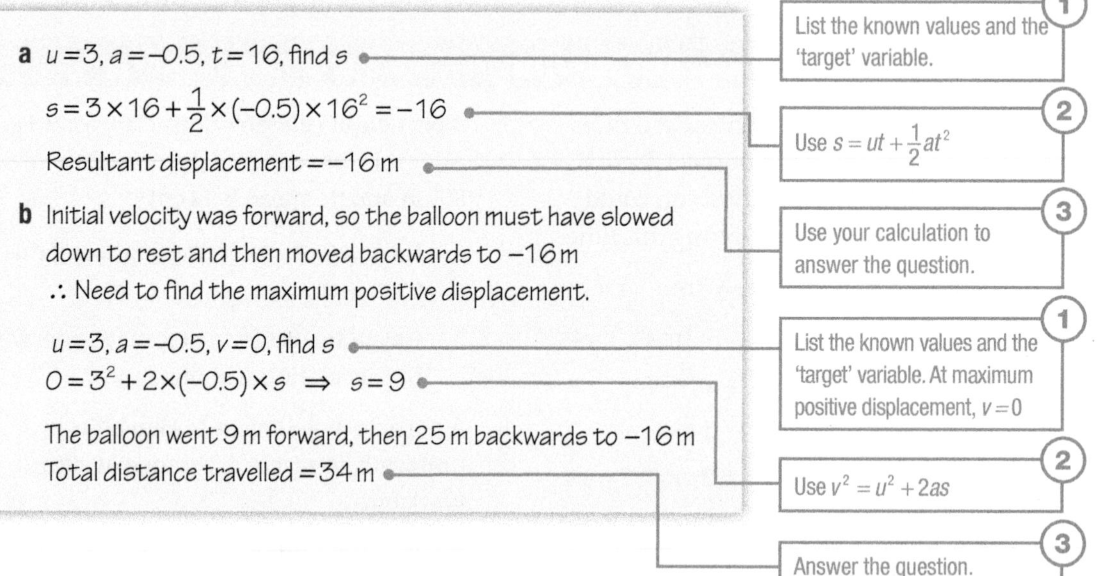

a $u=3, a=-0.5, t=16$, find s
$s=3\times16+\frac{1}{2}\times(-0.5)\times16^2=-16$
Resultant displacement $=-16$ m
b Initial velocity was forward, so the balloon must have slowed down to rest and then moved backwards to -16 m
∴ Need to find the maximum positive displacement.
$u=3, a=-0.5, v=0$, find s
$0=3^2+2\times(-0.5)\times s \Rightarrow s=9$
The balloon went 9 m forward, then 25 m backwards to -16 m
Total distance travelled $=34$ m
1 List the known values and the 'target' variable.
2 Use $s=ut+\frac{1}{2}at^2$
3 Use your calculation to answer the question.
1 List the known values and the 'target' variable. At maximum positive displacement, $v=0$
2 Use $v^2=u^2+2as$
3 Answer the question.

Example 3

Car P is accelerating at 2 m s^{-2}. When its velocity is 10 m s^{-1}, it is overtaken by car Q, which is travelling at 16 m s^{-1} and accelerating at 1 m s^{-2}. How long will P take to catch up with Q?

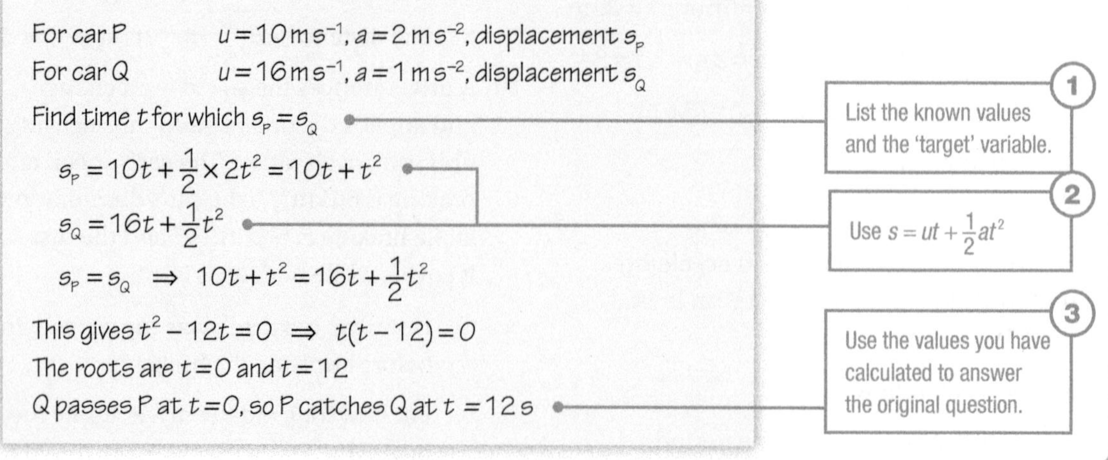

For car P $u=10$ m s^{-1}, $a=2$ m s^{-2}, displacement s_P
For car Q $u=16$ m s^{-1}, $a=1$ m s^{-2}, displacement s_Q
Find time t for which $s_P=s_Q$
$s_P=10t+\frac{1}{2}\times2t^2=10t+t^2$
$s_Q=16t+\frac{1}{2}t^2$
$s_P=s_Q \Rightarrow 10t+t^2=16t+\frac{1}{2}t^2$
This gives $t^2-12t=0 \Rightarrow t(t-12)=0$
The roots are $t=0$ and $t=12$
Q passes P at $t=0$, so P catches Q at $t=12$ s
1 List the known values and the 'target' variable.
2 Use $s=ut+\frac{1}{2}at^2$
3 Use the values you have calculated to answer the original question.

Exercise 7.3B Reasoning and problem-solving

1 A lorry starts from rest and accelerates uniformly at $3\,m\,s^{-2}$. It passes point A after 20 s and point B after a further 10 s. Calculate the distance AB

2 A train leaves station A from rest with constant acceleration $0.2\,m\,s^{-2}$. It reaches maximum speed after 2 minutes, maintains this speed for 4 minutes, then slows down to stop at station B with acceleration $-1.5\,m\,s^{-2}$. Calculate the distance AB

3 A ferry carries passengers between banks of a river, which are 20 m apart. After setting off, the ferry accelerates at $0.2\,m\,s^{-2}$ for 12 seconds before turning off the engine and decelerating at a constant rate and coming to a stop at the opposite bank.

  a Calculate the speed of the ferry after the first 12 seconds.

  b Calculate the distance the ferry travels during these 12 seconds.

  c Calculate the value of the ferry's deceleration after the engines are turned off.

  d State any assumptions you made in part **c** and explain why, in reality, this value would be higher.

4 A train accelerates uniformly from rest for 1 minute, at which time its velocity is $30\,km\,h^{-1}$. It maintains this speed until it is 500 m from the next station. It then decelerates uniformly and stops at the station. Calculate the train's acceleration during the first and last phases of this journey.

5 A boat is travelling at $4\,m\,s^{-1}$. Its propeller is then put into reverse, giving it an acceleration of $-0.4\,m\,s^{-2}$ for 25 seconds.

  a Work out the displacement of the boat during this period.

  b Work out the distance travelled by the boat during this period.

6 An object travelling forwards along a straight line travels 10 m during one second and 15 m during the next second. Work out the acceleration of the object, assuming it to be constant.

7 Lorry A is travelling along a straight road, with lorry B following 40 m behind. They are both travelling at a constant $25\,m\,s^{-1}$. Lorry A then brakes to a halt, with acceleration $-5\,m\,s^{-2}$ The driver of lorry B takes 0.2 seconds to react, then brakes with acceleration $-4\,m\,s^{-2}$

  a Do the lorries collide?

  b State any assumptions you made, and explain how this may affect the answer.

Challenge

8 There are five equations of constant motion often referred to as the *suvat* equations.

Use the fact that the gradient of a v–t graph gives the acceleration, and the area under a v–t graph gives the displacement, to show how each of these equations can be derived.

9 A car crosses a speed hump with a velocity of $4\,m\,s^{-1}$. It then accelerates at $2.5\,m\,s^{-2}$ to $9\,m\,s^{-1}$. The driver then brakes, causing an acceleration of $-3\,m\,s^{-2}$, reducing the speed to $4\,m\,s^{-1}$ to cross the next hump.

  a How far apart are the humps?

  b How long does the car take to travel from one hump to the next?

  c The question implies that the car is being modelled as a particle. In what way, if any, does this assumption affect your results?

MECH

2184

7.4 Motion with variable acceleration

Fluency and skills

You know that velocity is equal to the gradient of a displacement–time (s–t) graph. If the graph is not linear, the gradient changes, but at a particular point on the graph you can say

Key point

gradient of s–t graph = velocity *at that instant*

Similarly, for a non-linear velocity–time (v–t) graph

Key point

gradient of v–t graph = acceleration *at that instant*

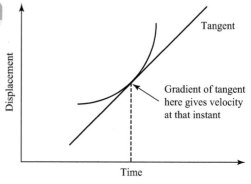

Displacement
Tangent
Gradient of tangent here gives velocity at that instant
Time

See Ch4.3 For a reminder of finding gradients with differentiation.

If you know s as a function of t, you can work out the gradient (velocity) by differentiation. The velocity is the rate of change of displacement.

Key point

$$v = \frac{\mathrm{d}s}{\mathrm{d}t} \quad \text{or} \quad v = s$$

Similarly, acceleration is the rate of change of velocity. You can obtain acceleration by differentiating velocity with respect to time, which is the same as differentiating displacement twice.

Key point

$$a = \frac{\mathrm{d}v}{\mathrm{d}t} = \frac{\mathrm{d}^2 s}{\mathrm{d}t^2} \quad \text{or} \quad a = v = s$$

A dot above a variable denotes the derivative with respect to time. Two dots show the second derivative.

Example 1

A particle moves in a straight line so that at time t seconds its displacement s, in metres, from an origin O is given by $s = t^4 - 32t$. Evaluate its velocity and acceleration when $t = 2$. Show your working.

$$v = \frac{ds}{dt} = 4t^3 - 32$$

Differentiate displacement to find velocity.

When $t = 2$, $v = 4 \times 2^3 - 32 = 0$ so the particle is at rest when $t = 2$

$$a = \frac{dv}{dt} = 12t^2$$

Differentiate velocity to find acceleration.

When $t = 2$, $a = 12 \times 2^2 = 48$ so $a = 48\,\mathrm{m\,s^{-2}}$ when $t = 2$

You can check your answer by finding $\frac{d^2s}{dt^2}$ at the point where $t = 2$ on a calculator.

See Ch4.7 For a reminder of integration and areas under graphs.

You can also write the relationship between s, v and a using integrals.

Key point

$$v = \frac{\mathrm{d}s}{\mathrm{d}t} \Rightarrow s = \int v\,\mathrm{d}t \quad \text{and} \quad a = \frac{\mathrm{d}v}{\mathrm{d}t} \Rightarrow v = \int a\,\mathrm{d}t$$

Remember to include the constant of integration.

You know that integration corresponds to area under a graph, and so the first of these relationships confirms that displacement is the area under the v–t graph.

Example 2

A particle starts with velocity $2\,\text{m s}^{-1}$ and moves with acceleration $a = (6t + 4)\,\text{m s}^{-2}$

a Calculate its velocity after 5 s **b** Calculate the distance it travels in that time.

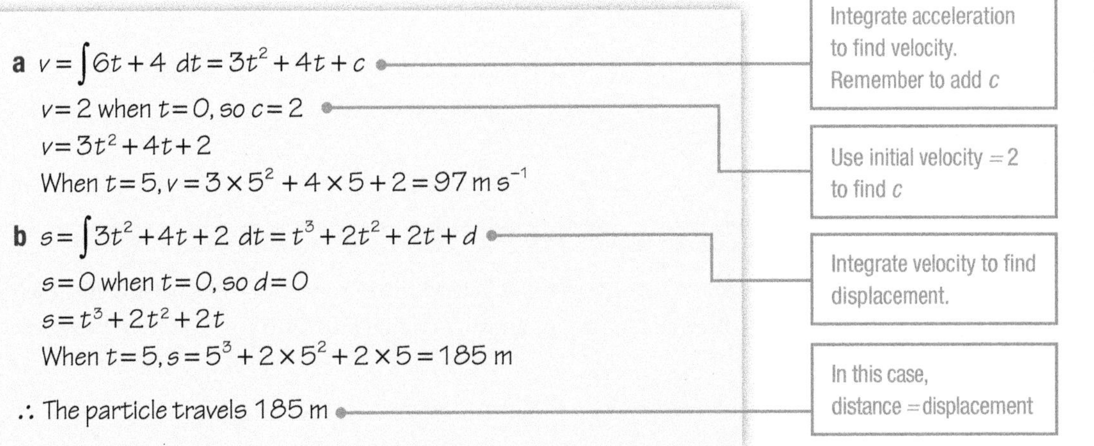

a $v = \int 6t + 4\,dt = 3t^2 + 4t + c$
$v = 2$ when $t = 0$, so $c = 2$
$v = 3t^2 + 4t + 2$
When $t = 5$, $v = 3 \times 5^2 + 4 \times 5 + 2 = 97\,\text{m s}^{-1}$
b $s = \int 3t^2 + 4t + 2\,dt = t^3 + 2t^2 + 2t + d$
$s = 0$ when $t = 0$, so $d = 0$
$s = t^3 + 2t^2 + 2t$
When $t = 5$, $s = 5^3 + 2 \times 5^2 + 2 \times 5 = 185$ m
$\therefore$ The particle travels 185 m
Integrate acceleration to find velocity. Remember to add c
Use initial velocity = 2 to find c
Integrate velocity to find displacement.
In this case, distance = displacement

MECH

Exercise 7.4A Fluency and skills

Show your working for these questions.

1 Given $s = 4t^2 - t^3$

a Write an expression for v

b Evaluate v when $t = 2$ s

2 Given $s = (t + 2)(t - 6)$

a Write an expression for v

b Evaluate v when $t = 10$ s

c Evaluate v when $t = 100$ s

3 Given $v = 6t^2 - 8t$

a Write an expression for a

b Evaluate a when $t = 3$ s

4 Given $v = t(t^2 + 7)$

a Write an expression for a

b Evaluate a when $t = 0$ s

5 Given $s = t^4 + 5t^2$, evaluate

a v when $t = 1$ s

b a when $t = 0.5$ s

6 Given $s = 3t^2 - 12t + 5$

a Write an expression for v,

b Evaluate t when $v = 0\,\text{m s}^{-1}$

7 Given $v = 14t - t^2$

a Write an expression for a,

b Evaluate t when $a = 6\,\text{m s}^{-2}$

8 If $v = 24 - 6t$ and $s = 0$ m when $t = 0$ s

a Write an expression for s,

b Evaluate s when $t = 2$ s

9 If $a = 6t^2 + 3$ and $v = 4\,\text{m s}^{-1}$ when $t = 0$ s

a Write an expression for v

b Evaluate v when $t = 2$ s

10 If $v = 9 - t^2$ and $s = 4$ m when $t = 0$ s, evaluate

a s when $t = 1$ s

b s when $v = 0\,\text{m s}^{-1}$

11 If $a = 6t - 12$ and $v = 2\,\text{m s}^{-1}$ when $t = 0$ s, evaluate

a v when $t = 3$ m,

b v when $a = 0\,\text{m s}^{-2}$

12 $a = 24t^2 + 6t - 10$

a Given that $v = 2\,\text{m s}^{-1}$ when $t = 0$ s, work out v when $t = 1$ s

b Given that $s = 5$ m when $t = 0$ s, work out s when $t = 1$ s

Reasoning and problem-solving

Strategy

When solving a motion problem with differentiation or integration

1. Identify what dimension(s) you're dealing with (speed, velocity, distance, displacement) and differentiate or integrate as appropriate.
2. (When integrating) include the constant of integration and calculate its value.
3. Use the result of your differentiation or integration to answer the original question.

Example 3

A particle moves with acceleration $(2t-3)\,\text{m s}^{-2}$. It is initially at a point O and is travelling with velocity $2\,\text{m s}^{-1}$. Show that its direction of travel changes twice and find the distance between the points where this occurs.

$v=\int 2t-3\,dt=t^2-3t+c$

When $t=0$, $v=2$, so $c=2$

$\therefore v=t^2-3t+2=(t-1)(t-2)$

The particle changes direction when $v=0$

$\Rightarrow (t-1)(t-2)=0 \Rightarrow t=1$ and $t=2$

$v>0$ for $t<1$ or $t>2$

$v<0$ for $1<t<2$

$s=\int t^2-3t+2\,dt=\frac{1}{3}t^3-\frac{3}{2}t^2+2t+c_1$

When $t=0$, $s=0$, so $c_1=0$

$\therefore s=\frac{1}{3}t^3-\frac{3}{2}t^2+2t$

When $t=1$, $s=\frac{1}{3}-\frac{3}{2}+2=\frac{5}{6}$ m

When $t=2$, $s=\frac{8}{3}-6+4=\frac{2}{3}$ m

Distance between the points $=\frac{5}{6}-\frac{2}{3}=\frac{1}{6}$ m

(1)(2) Integrate a to obtain v and include the constant of integration.

As the particle starts with positive velocity, it must be moving backwards between 1 and 2 seconds, and forwards the rest of the time.

(1)(2) Integrate v to obtain s and include the constant of integration.

(3) Find the positions at which $v=0$ and use this information to answer the question.

Exercise 7.4B Reasoning and problem solving

1 A particle, moving in a straight line, starts from rest at O. Its acceleration (in m s^{-2}) at time t is given by $a=30-6t$

- a Calculate its velocity and position at time t
- b How long does the particle take to return to O?
- c What is its greatest positive displacement from O?

2 A particle is initially traveling at $18\,\text{m s}^{-1}$ It is acted upon by a braking force that brings it to rest. At time t s during braking, its acceleration has magnitude $t\,\text{m s}^{-2}$ Calculate

- a The length of time it takes to stop,
- b The distance it travels before coming to rest.

3 A particle, initially at rest, is acted on for 3 s by a variable force that gives it acceleration of $(8t+2)\,\text{m s}^{-2}$. Work out the distance it travels while the force is acting.

4 A ball is thrown straight up from an open window. Its height, at time t s, is h m above the ground, where $h = 4 + 8t - 5t^2$

a How high is the window above the ground?

b For how long is the ball in the air?

c At what speed does it hit the ground?

d Evaluate the maximum height it reaches above the ground.

5 A particle starts from point O and moves so that its displacement s, in metres, at time t seconds is given by $s = t^4 - 32t$. Work out

a i Its position at the moment it is instantaneously at rest,

ii Its acceleration at the moment it is instantaneously at rest,

b Its speed at the moment that it returns to O

c The distance it travels in the first 3 s

6 A particle travels in a straight line through a point O with constant acceleration. Its velocity is given by $v = 2t - 3\,\text{m s}^{-1}$

If its displacement from O is –4 m at $t = 0$ s, work out how long before the particle passes point O

7 A particle moving along a straight line through a point O has acceleration given by $a = (2t - 5)\,\text{m s}^{-2}$. When $t = 4$ s, the particle has velocity $2\,\text{m s}^{-1}$ and displacement from O of +8 m. Work out the two positions where the particle is at rest.

8 A particle is moving in a straight line. Its velocity at time t s is given by $v = 6t - 3t^2\,\text{m s}^{-1}$

a Evaluate its velocity at times $t = 1$ and $t = 3$

b What distance does it travel from time $t = 1$ to time $t = 3$?

Challenge

9 A car travels from rest at a set of traffic lights until it stops at the next set of lights. The car's displacement x m from its starting position at time t s is given by $x = \dfrac{1}{13500}t^3(t^2 - 75t + 1500)$

a Write an expression for the car's velocity.

b Work out the time it takes for the car to travel between the two sets of lights.

c What is the distance between the two sets of lights?

d Work out the car's maximum speed.

10 The *suvat* equations can be derived using differentiation and integration.

Acceleration is the derivative of velocity, so $a = \dfrac{dv}{dt}$

For a constant acceleration between $t = 0$ and $t = t$, this gives $\int_0^t a\,dt = \int_u^v dv$

a Use this information to derive the *suvat* equation that uses a, t, u and v

Similarly, $v = \dfrac{ds}{dt}$ so, between displacement $s = 0$ and $s = s$,

$$\int_0^s ds = \int_0^t v(t)\,dt$$

b Use this information to derive the suvat equation that uses s, u, a and t

MECH

7 Summary and review

Chapter summary

- The SI base units used in this chapter are kilograms / kg (mass), metres / m (length) and seconds / s (time).
- In kinematics, the quantities in the table are used.

Vector Quantity	Scalar Quantity	SI Units
Displacement	Distance	m
Velocity	Speed	m s^{-1} or m/s
Acceleration	—	m s^{-2} or m/s^2

- Formulae should be dimensionally consistent and you must use the same units throughout, converting if necessary.
- Position is a vector – the distance and direction from the origin.
  Displacement is a vector – the change of position.
  Distance is a scalar – the magnitude of displacement.
  Velocity is a vector – the rate of change of displacement.
  Speed is a scalar – the magnitude of velocity.
  Acceleration is a vector – the rate of change of velocity.
- Average velocity $= \dfrac{\text{resultant displacement}}{\text{total time}}$ and average speed $= \dfrac{\text{total distance}}{\text{total time}}$
- The gradient of a displacement–time (s–t) graph is the velocity.
- The gradient of a velocity–time (v–t) graph is the acceleration.
- The area between a v–t graph and the t-axis is the displacement.
- In straight-line graphs, changes in motion are assumed to be instantaneous. In reality, this is usually not possible.
- s = displacement, u = initial velocity, v = final velocity, a = acceleration, t = time
  For constant acceleration: $v = u + at$; $s = ut + \frac{1}{2}at^2$; $v^2 = u^2 + 2as$; $s = \frac{1}{2}(u+v)t$; $s = vt - \frac{1}{2}at^2$
- The equations of motion for constant acceleration assume objects to be particles with tiny or irrelevant size. In reality, the size of an object may affect calculations.
- For variable acceleration, use calculus: $v = \dfrac{\mathrm{d}s}{\mathrm{d}t}$ and $a = \dfrac{\mathrm{d}v}{\mathrm{d}t} = \dfrac{\mathrm{d}^2 s}{\mathrm{d}t^2}$; $s = \int v \, \mathrm{d}t$ and $v = \int a \, \mathrm{d}t$

Check and review

You should now be able to...	Try Questions
✔ Understand and use standard SI units and convert between them and other metric units.	1
✔ Calculate average speed and average velocity.	2, 4
✔ Draw and interpret graphs of displacement and velocity against time.	3, 4
✔ Derive and use the formulae for motion in a straight line with constant acceleration.	5
✔ Use calculus to solve problems involving variable acceleration.	6, 7

1 Using the formula $s = vt - \frac{1}{2}at^2$, calculate s if $v = 5\,\text{km h}^{-1}$, $a = 0.002\,\text{m s}^{-2}$ and $t = 10$ minutes. Give your answer in kilometres to 3 sf.

2 An object travels for 10 s at $8\,\text{m s}^{-1}$ and then for a further distance of 120 m at $18\,\text{m s}^{-1}$ Work out its average speed for the whole journey.

3 A log flume climbs a slope for 20 s at a constant $1\,\text{m s}^{-1}$. It then pauses for 5 s before moving rapidly downhill at $10\,\text{m s}^{-1}$ for 5 s

a Sketch a displacement-time graph for this motion.

b Explain why it is unrealistic that a log flume would be able to move with the motion described by your graph.

4 The velocity-time graph shows the motion of a particle for a period of 10 s

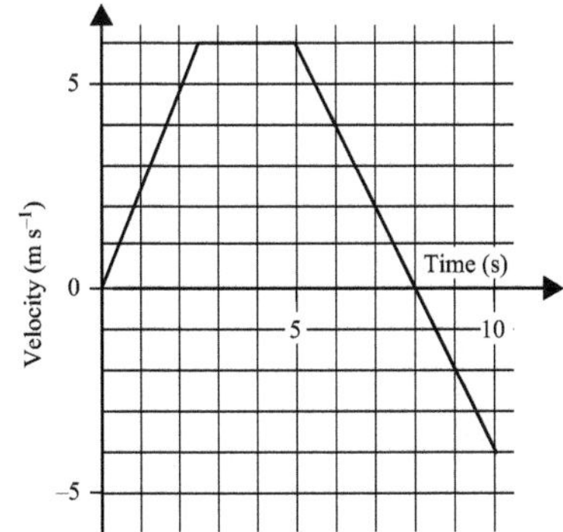

a Work out the acceleration of the object for each of the three stages of the motion.

b Calculate i The resultant displacement,
ii The distance travelled.

c Calculate the average velocity of the particle.

5 A particle moves with constant acceleration along a straight line. It passes the origin, O, at $2\,\text{m s}^{-1}$ and travels 15 m in the next 5 s

a Calculate its acceleration.

b Work out its position when its velocity is $8\,\text{m s}^{-1}$

6 A particle, P, is moving along a straight line. At time t seconds, its displacement, s metres from the origin, O, is given by $s = t(t^2 - 16)$

a Write down an expression for the velocity of P at time t

b Work out the acceleration of P when $t = 5$

7 A particle moves along a straight line with acceleration $a = (4t - 3)\,\text{m s}^{-2}$ at time t s. Initially it has a velocity of $5\,\text{m s}^{-1}$

a Write down an expression for its velocity at time t

b If the particle started at the origin, O, work out its displacement from O after 4 s

MECH

What next?

Score			
	0 – 4	Your knowledge of this topic is still developing. To improve, search in MyMaths for the codes: 2183–2184, 2289	
	5 – 6	You're gaining a secure knowledge of this topic. To improve, look at the InvisiPen videos for Fluency and skills (07A)	
	7	You've mastered these skills. Well done, you're ready to progress! To develop your techniques, look at the InvisiPen videos for Reasoning and problem-solving (07B)	

Click these links in the digital book

History

If you have ever struggled with mechanics, you may take some comfort from the fact that the laws and equations that are used today took many centuries to establish.

Galileo (1564 - 1642) has often been credited with establishing the principles governing uniformly accelerated motion. Much of the work, however, had already been done almost three centuries earlier, by a group known as The Oxford Calculators. **The Oxford Calculators** were a group of 14th century thinkers associated with Oxford University's Merton College.

The group distinguished between **kinematics** and **dynamics** and they were the first to formulate the **mean speed theorem**.

This theorem states that an object which uniformly accelerates from rest will cover the same distance as an object travelling at a constant velocity if this velocity is half the final speed of the accelerated object.

Investigation

Explain how this diagram illustrates the mean speed theorem.

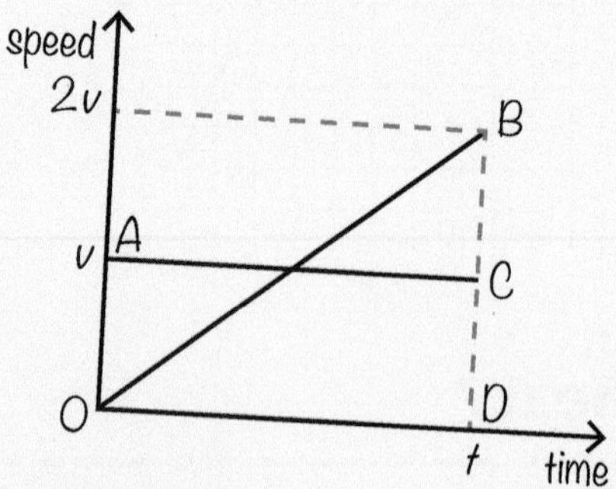

Applications

The theory of kinematics continues to be developed all the time. **Ferdinand Freudenstein** was a 20th century American physicist and engineer, regarded by many as the father of modern kinematics.

The **Freudenstein equation**, which he presented in his 1954 doctoral thesis, can be applied to a number of machines used in daily life such as the braking system of a vehicle. Designers of modern mechanical systems test them using apps that build on Freudenstein's ideas.

The derivation of Freudenstein's equation relies on the theory of vectors and trigonometry that you have already studied.

7 Assessment

1 The velocity of a particle at time t seconds is given by the formula $v = 3t^2 - 3t + 16$. Choose the correct expression for the acceleration of the particle.

A $t^3 - 1.5t^2 + 16t + c$ **B** $6t + 13$ **C** $6t - 3$ **D** $9t^3 - 6t^2 + 16t + c$ **[1 mark]**

2 At time $t = 0$ s, a body passes the origin with a velocity of $60\,\text{m s}^{-1}$, and decelerates uniformly at a rate of $4\,\text{m s}^{-2}$

a Determine the time at which the body is at rest. Select the correct answer.

A $t = 15$ s **B** $t = -15$ s **C** $t = 240$ s **D** $t = -240$ s **[1]**

b Determine the times at which the body is 400 metres from the origin.

A $t = 20$ s and $t = 40$ s **B** $t = -10$ s and $t = -20$ s

C $t = -20$ s and $t = -40$ s **D** $t = 10$ s and $t = 20$ s **[1]**

3 A train leaves a station, P, and accelerates from rest with a constant acceleration of $0.4\,\text{m s}^{-2}$ until it reaches a speed of $24\,\text{m s}^{-1}$. It maintains this speed for 6 minutes. It then decelerates uniformly with a deceleration of $0.2\,\text{m s}^{-2}$, until it comes to rest at station Q

a Draw a velocity-time graph of the journey from P to Q, labelling all the relevant times. **[5]**

b Calculate the distance PQ **[2]**

4 Points A, B and C lie on a straight line. The distances AB and BC are 80 m and 96 m respectively. A particle moves in a straight line from A to C with an acceleration of $4\,\text{m s}^{-2}$. It takes 5 seconds to travel from A to B. Work out

a The speed of the particle at A **[3]**

b The time taken for the particle to travel from B to C **[6]**

5 The displacement, s metres, of a particle, at a time t seconds, is given by the formula $s = t^3 - 9t^2 + 24t$

a Write an expression for the velocity of the particle. **[2]**

b Calculate the times at which the particle is at rest. **[3]**

c Work out the distance travelled by the particle between $t = 0$ s and $t = 5$ s **[5]**

6 A speeding van passes a police car. The van is travelling at $27\,\text{m s}^{-1}$, and the police car is travelling at $15\,\text{m s}^{-1}$. From the instant when the van is level with the police car, the police car accelerates uniformly at $3\,\text{m s}^{-2}$ in order to catch the van. Work out

a The time taken until the police car is level with the van, **[5]**

b The speed of the police car at this time. **[2]**

7 A man on a bicycle accelerates uniformly from rest to a velocity of $10\,\text{m s}^{-1}$ in 5 seconds. He maintains this speed for 20 seconds, and then decelerates uniformly to rest. His journey takes a total of T seconds.

a Draw a velocity-time graph of his journey. **[3]**

Given that he cycles a total of 265 metres, calculate

b The value of T **[3]**

c The acceleration for the final stage of his journey. **[2]**

8 A car moving along a straight road with constant acceleration passes points A and B with velocities $10\,\mathrm{m\,s^{-1}}$ and $40\,\mathrm{m\,s^{-1}}$ respectively. Work out the velocity of the car at the instant when it passes M, the midpoint of AB **[6]**

9 A bus travels on a straight road with a constant acceleration of $0.8\,\mathrm{m\,s^{-2}}$. A and B are two points on the road, a distance of 390 metres apart. The bus increases its velocity by $12\,\mathrm{m\,s^{-1}}$ in travelling from A to B

a What is the speed of the bus at A? **[5]**

b Work out the time taken for the bus to travel from A to B **[2]**

10

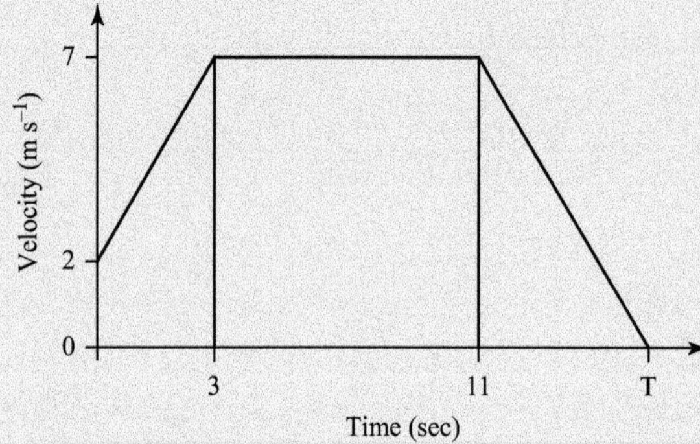

The diagram shows the velocity–time graph of the motion of a runner over a time period of T seconds. During that time, the runner travels a distance of 76 metres.

a Write down the initial speed of the runner. **[1]**

b Work out the value of her initial acceleration. **[2]**

c Describe her motion from $t = 3$ to $t = 11$ **[1]**

d Calculate the value of T **[6]**

11 The acceleration, $a\,\mathrm{m\,s^{-2}}$, of a particle moving in a straight line is given by the formula $a = 2t - 6$

At time $t = 0$, the particle is moving through the origin with a velocity of $10\,\mathrm{m\,s^{-1}}$

a Write an expression for the velocity of the particle at time t **[4]**

b At which times is the particle moving with a velocity of $2\,\mathrm{m\,s^{-1}}$? **[4]**

c Write an expression for the displacement of the particle at time t **[4]**

d Work out the displacement of the particle when $t = 6$ **[2]**

12 A jogger is running along a straight road with a velocity of $4\,\mathrm{m\,s^{-1}}$ when she passes her friend who is stationary with a bicycle. Three seconds after the jogger is level with her friend, her friend sets off in pursuit. Her friend accelerates from rest with a constant acceleration of $2\,\mathrm{m\,s^{-2}}$ When the cyclist has been riding for T seconds, the cyclist and her friend are level.

a Draw a velocity–time graph for $t = 0$ to $t = T + 3$ **[3]**

b Write down an equation for T **[4]**

c Solve this equation to find the value of T **[3]**

d How fast is the cyclist travelling when they draw level? **[2]**

8 Forces and Newton's laws

The dog in this picture is in equilibrium. The leash is taut but the dog is not moving, which means that all forces acting on the dog are balanced. The force of the person pulling on the leash balances with the force of the dog pulling forward. The dog's weight balances with the normal reaction upwards from the floor. This simple example illustrates that systems of forces surround us every day.

Drawing diagrams and labelling forces is a useful way to visualise these systems. It is important to remember that any object that is not accelerating is in a state of equilibrium. This means that the forces can be balanced even if the object is moving. There are many different types of forces, and an understanding of how they work is vital for modelling.

Orientation

What you need to know

p.159
Ch6 Vectors
- Scalar and vector definitions.
- Components of a vector.

p.175
Ch7 Units and kinematics
- Formulae for motion.
- Acceleration under gravity.

What you will learn
- To resolve forces in two perpendicular directions.
- To calculate the magnitude and direction of a resultant force.
- To resolve for particles with constant acceleration, including those which are connected by string over pulleys and 'connected' particles.
- Understand mass and weight.

What this leads to

Ch18 Motion in two dimensions
SUVAT equations in 2D.
The significance of g

Ch19 Forces
Modelling friction.
Inclined planes.
Moments.

Practise before you start — 2206, 2183, 2184

8 Bridging Unit

Topic A: Forces and Newton's laws

Bridging to Ch8.3, 8.4

Most situations covered in this chapter will involve one or more forces acting on an object. Often, it can be useful to draw a diagram showing these forces.

For example, consider a ball held up by a string. There are two forces acting on the ball: its weight (W) and the tension (T) in the string. If the tension in the string is equal to the weight, then the ball is in **equilibrium**. If the ball is simply hanging, it will continue to do so, and if it's being lifted by the string, it will continue moving upwards at constant speed.

Key point

The **resultant force** in a direction is the sum of all forces acting in that direction. If the resultant force in all directions is zero then an object is in **equilibrium**.

Example 1

The diagram shows the forces acting on a particle.

Given that the particle is in equilibrium, state the value of the forces T and P.

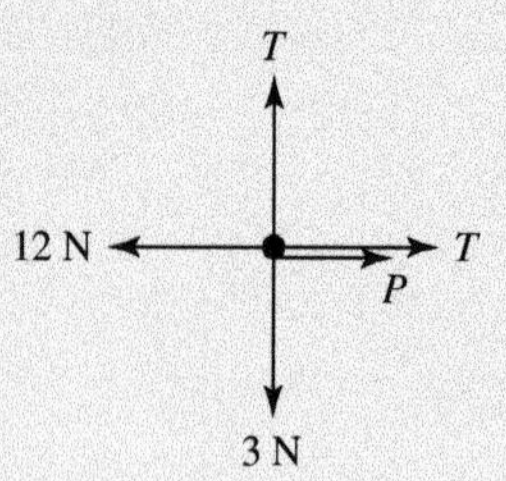

Resolving vertically gives $T - 3 = 0$ so $T = 3$ N

Take forces upwards as positive and forces downwards as negative. Since the system is in equilibrium, the forces must equal zero.

Resolving horizontally gives $P + T - 12 = 0$

$P + 3 - 12 = 0$ so $P = 9$ N

Substitute $T = 3$ and take upwards forces as positive.

Forces are measured in newtons (N).

Try It 1

Given that this system is in equilibrium, state the values of the forces T and P

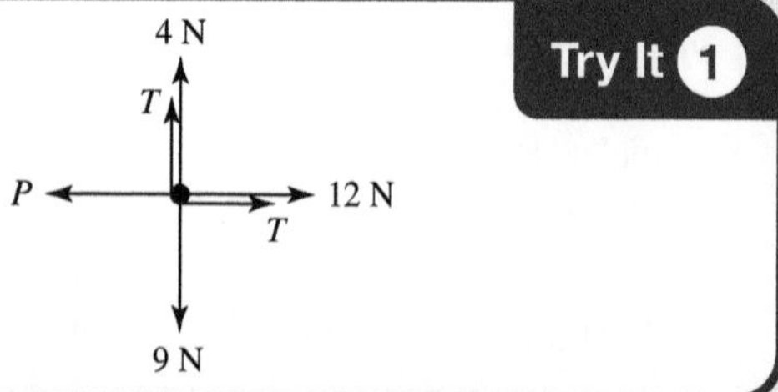

If the resultant force acting on an object is > 0, the object is not in equilibrium and will accelerate in the direction of the resultant force. Accelerate could mean speed up, slow down, or start moving from rest.

Key point

The resultant force, F, in any direction is given by $F = ma$ where m is the mass (in kg) and a is the acceleration (in m s^{-2}) in that direction. This is Newton's 2nd law

The weight of an object is its mass multiplied by the acceleration due to gravity, $W = mg$. In Topic A, use the estimate $g = 9.8$ m s^{-2} for the acceleration due to gravity on Earth.

Example 2

The forces acting on a particle are shown in the force diagram.

The mass of the particle is 5 kg.

Calculate the acceleration of the particle.

Resolving vertically gives $5g - 5g = 0$

So the particle will not accelerate vertically.

Resolving horizontally gives $12 - 8 = 4$ N

So the particle will accelerate horizontally to the right.

Since the resultant force acts in this direction.

$F = ma$ gives $4 = 5a$

The resultant force is 4 and the mass is 5

So $a = 0.8\ \text{m s}^{-2}$ to the right.

Remember to give the units and direction of your answer.

As this is a force diagram, the weight is shown not the mass.

Try It 2

A particle of mass 0.8 kg is acted on by the forces shown in the diagram. Find the size and direction of the acceleration of the particle.

You can indicate acceleration on a force diagram using double arrows.

Example 3

A particle of mass 7 kg is acted upon by the forces shown in the diagram and moves horizontally with acceleration $1.2\ \text{m s}^{-2}$

Calculate the values of R and T

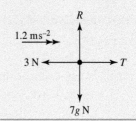

Resolving vertically gives $R - 7g = 0$ since the particle does not accelerate vertically.

So $R = 7g = 69$ N (to 2 significant figures)

Since we have used the estimate $g = 9.8\ \text{m s}^{-2}$, answers should only be given to 2 significant figures.

Using $F = ma$ horizontally gives

$T - 3 = 7 \times 1.2$

Remember you need the mass, which is 7 (*not* $7g$, which is the weight).

$T - 3 = 8.4$

$T = 11.4$ N (to 3 significant figures)

This calculation did not involve an approximation of g, so give your answer to 3 significant figures unless directed otherwise.

Try It 3

A particle of mass 13 kg is acted upon by the forces shown in the diagram and moves vertically with acceleration 0.6 m s^{-2}

Calculate the values of X and Y

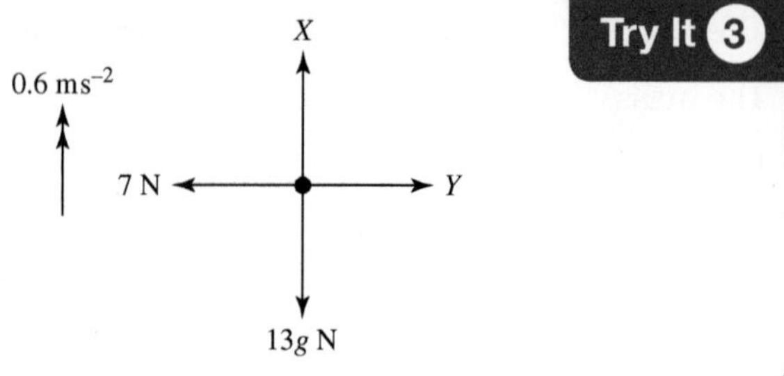

If you have two particles connected by a string, you can calculate their acceleration and the tension in the string using simultaneous equations.

Example 4

Two particles, one of mass 3 kg and the other of mass 5 kg, are connected by a light, inextensible string, which passes over a smooth pulley as shown.

Calculate the tension in the string and the acceleration of the particles.

T T a 3 kg 5 kg a 29.4 N 49.0 N

First consider the 3 kg particle:

$F = ma$ gives $T - 29.4 = 3a$ (1)

> The resultant force is $T - 29.4$ acting vertically *upwards*.

Now consider the 5 kg particle:

$F = ma$ gives $49 - T = 5a$ (2)

> The resultant force is $49 - T$ acting vertically *downwards*.

(1) + (2): $(T - 29.4) + (49 - T) = 3a + 5a$

$$19.6 = 8a$$

$$\Rightarrow a = \frac{19.6}{8} = 2.45 \text{ ms}^{-2}$$

> Solve simultaneously to find the values of a and T

So both particles move with acceleration 2.45 m s^{-2}

> The 3 kg particle accelerates vertically upwards and the 5 kg particle accelerates vertically downwards.

Substitute the value of a into (1):

$T - 29.4 = 3(2.45)$

$T = 29.4 + 3(2.45) = 36.8$ N

The tension in the string is 36.8 N

See Ch1.6

For a reminder on simultaneous equations.

If the string in a question is described as 'light and inextensible' then you can ignore its mass and assume that the tension is the same throughout, and the acceleration is the same for all particles attached to it. If an object is described as 'smooth' then you can assume there is no friction.

Try It 4

Two particles, one of mass 2 kg and the other of mass 12 kg, are connected by a light, inextensible string, which passes over a smooth pulley as shown.

Using $F = ma$ for each of the particles gives the simultaneous equations

$T - 19.6 = 2a$ and $117.6 - T = 12a$

Calculate the tension in the string and the acceleration of the particles.

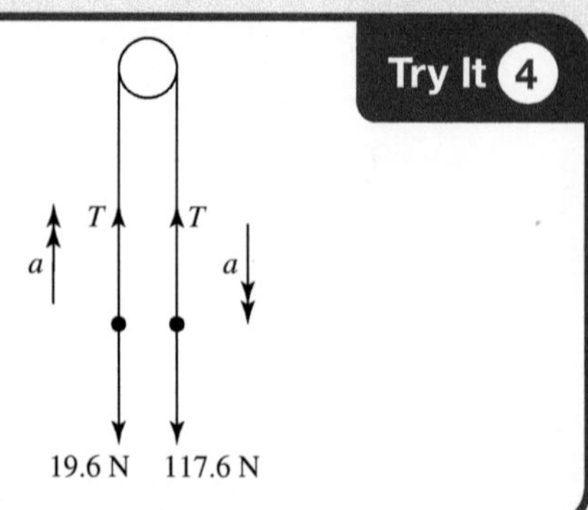

1 Given that the particle in each of these force diagrams is in equilibrium, calculate the values of the forces X and Y in each case.

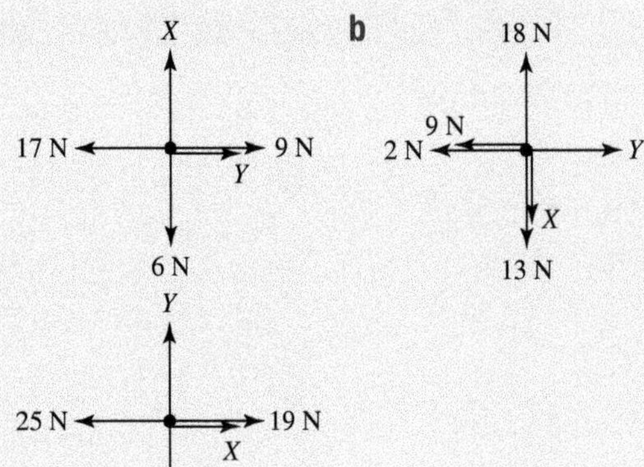

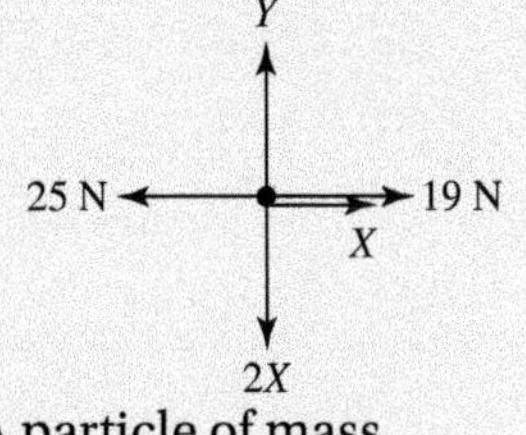

2 A particle of mass 3.5 kg is acted on by the forces shown in the diagram. Find the size and direction of the acceleration of the particle.

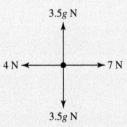

3.5g N
4 N
7 N
3.5g N

3 A particle of mass 2.8 kg is acted on by the forces shown in the diagram. Find the size and direction of the acceleration of the particle.

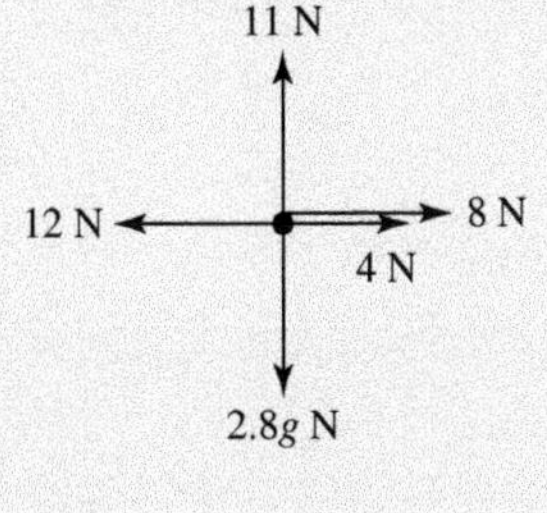

4 Use the force diagram and the acceleration given to find the mass of each particle.

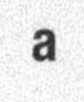

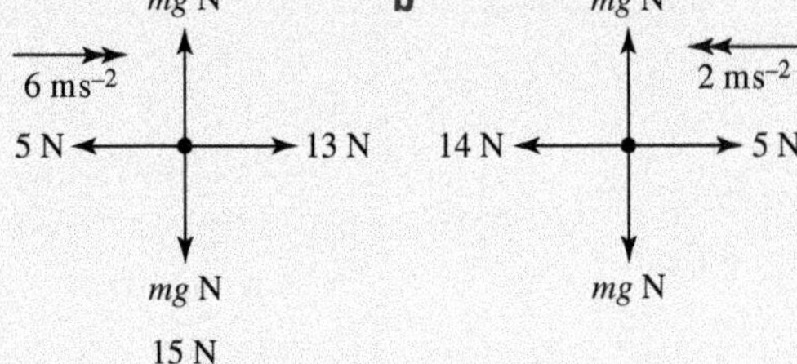

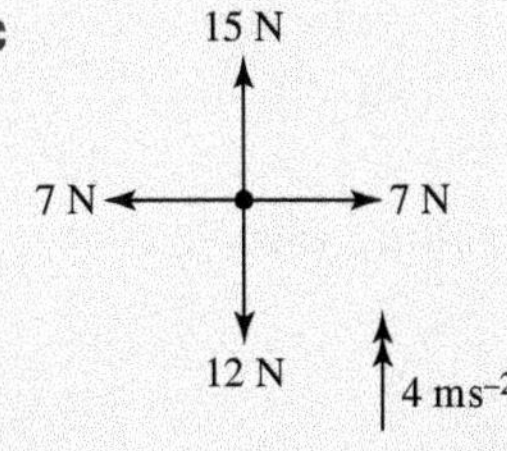

5 A particle of mass 1.7 kg moves under the forces shown with a horizontal acceleration of 5.2 m s^{-2}

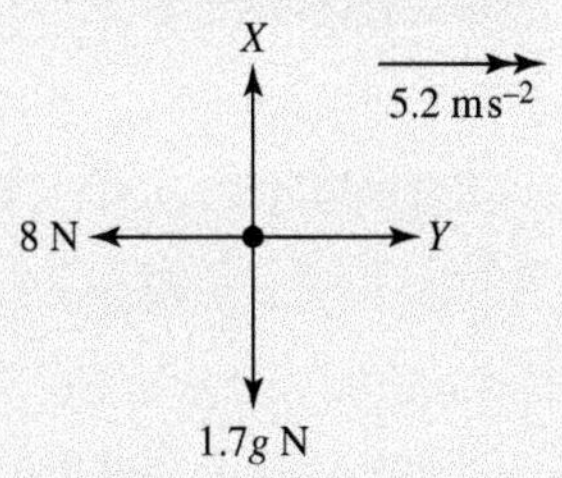

Calculate the values of X and Y

6 A particle of mass 2.5 kg moves under the forces shown with a vertical acceleration of 3.7 m s^{-2} downwards.

X
Y
X
3.7 ms⁻²
2.5g N

Calculate the values of X and Y

7 Two particles, one of mass 4 kg and the other of mass 16 kg are connected by a light, inextensible string, which passes over a smooth pulley as shown. Using $F = ma$ for each of the particles gives the simultaneous equations $T - 39.2 = 4a$ and $156.8 - T = 16a$

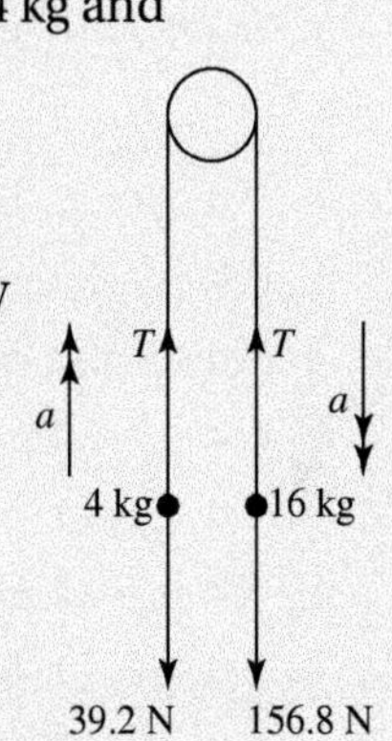

Calculate the tension in the string and the acceleration of the particles.

8 Two particles, one of mass 0.7 kg and the other of mass 0.2 kg, are connected by a light, inextensible string which passes over a smooth pulley as shown. Using $F = ma$ for each of the particles gives the simultaneous equations $6.86 - T = 0.7a$ and $T - 1.96 = 0.2a$

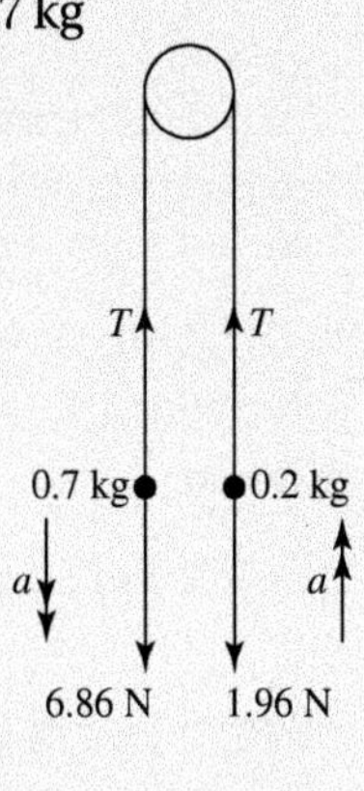

Calculate the tension in the string and the acceleration of the particles.

8.1 Forces

Fluency and skills

When you hit a tennis ball, you are applying a **force** to it. When you pick up a book, you are applying a force to it.

To describe a force you give both its **magnitude** and the **direction** in which it acts, so force is a vector quantity. The magnitude of a force is measured in newtons (N). The direction can be given as an angle or bearing.

See Ch6 For a reminder on vectors.

Examples of forces include:

Kicking a ball with a force of magnitude 200 N in the easterly direction.

A force $\mathbf{F} = (3\mathbf{i} + 4\mathbf{j})$ N acting on a particle.

All objects accelerate downwards towards the ground due to the gravitational pull of the Earth, also known as an object's **weight**.

Objects stop accelerating downwards when a contact force pushes upwards to counteract their weight. The contact force is called the **normal reaction**. It is always perpendicular (or normal) to the surface and if the object is at rest then the forces must balance each other.

Example 1

A box, with weight W N, is at rest on a horizontal table. Explain why it does not move even though its weight is pulling it directly down.

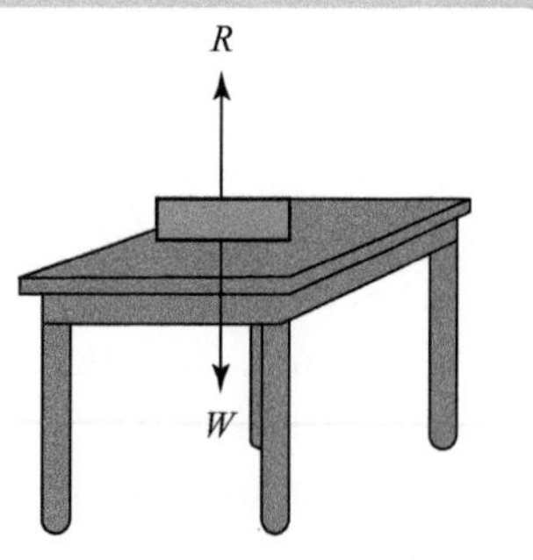

The weight of the box is balanced by the reaction from the table, which pushes directly upwards with the same strength as the weight.

The box in Example 1 is said to be in **equilibrium**.

> **Key point**
> An object that is at rest or moving with constant velocity is in equilibrium.

In these situations you can **resolve** the forces in a particular direction. This means you find the overall force acting in that direction.

Resolve vertically $\qquad R - W = 0$

> **Key point**
> When you resolve in any direction for an object that is in equilibrium, the overall force in that direction will be zero.

Bracketed arrows are sometimes used to denote the direction in which you are resolving. $(\uparrow)$ indicates that you are resolving vertically and $(\rightarrow)$ indicates that you are resolving horizontally.

If you apply a horizontal force P to a box laying on a rough surface, a **frictional force** acts in the opposite direction. If the box doesn't move, the frictional force must equal P

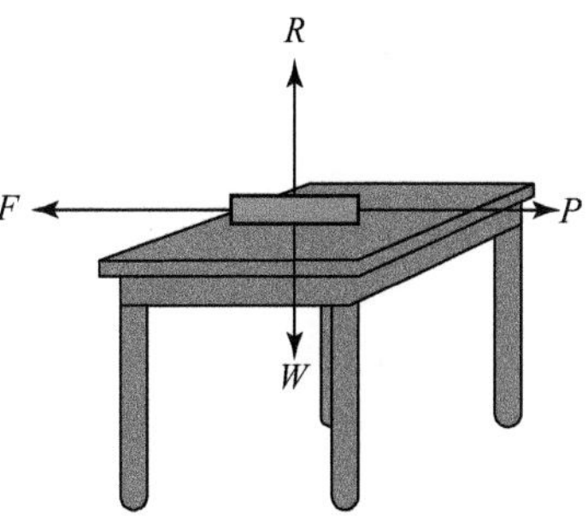

Resolve horizontally $\quad P - F = 0$

Resolve vertically $\quad R - W = 0$

> **Key point**
>
> The frictional force always acts in a direction which opposes motion. There is no friction on smooth surfaces. Friction appears when rough surfaces try to move relative to each other.

An object does not have to be at rest to be in equilibrium. It could be moving at a constant velocity.

Newton's first law of motion states that an object will remain at rest or continue to move with constant velocity unless an external force is applied to it, i.e. a moving object remains in equilibrium as long as the resultant force on it is zero.

A string is attached to a box and pulled vertically so that the box moves upwards with a constant speed of $1\,\text{m s}^{-1}$

Resolve vertically $\quad T - W = 0$

The force T in the string is a **tension.**

The box is pushed upwards at a constant speed of $5\,\text{m s}^{-1}$ by two rods.

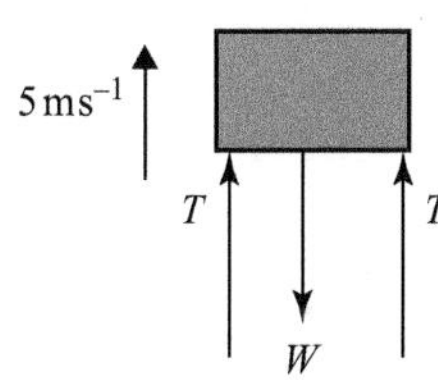

Resolve vertically $\quad 2T - W = 0$

The force T in each rod is called **thrust.**

Example 2

A bag of weight W rests on a table. It is pulled by two horizontal strings with tensions T and S, both pulling in the same direction, so that the box moves at a constant speed of $1\,\text{m s}^{-1}$. The frictional force is F and the normal reaction of the table on the box is R

a Draw a diagram showing all the forces acting on the box.

b Resolve forces horizontally and vertically to form two equations.

c The resistance to motion is 5 N and the tension in string T is 1.5 N. Find the tension S

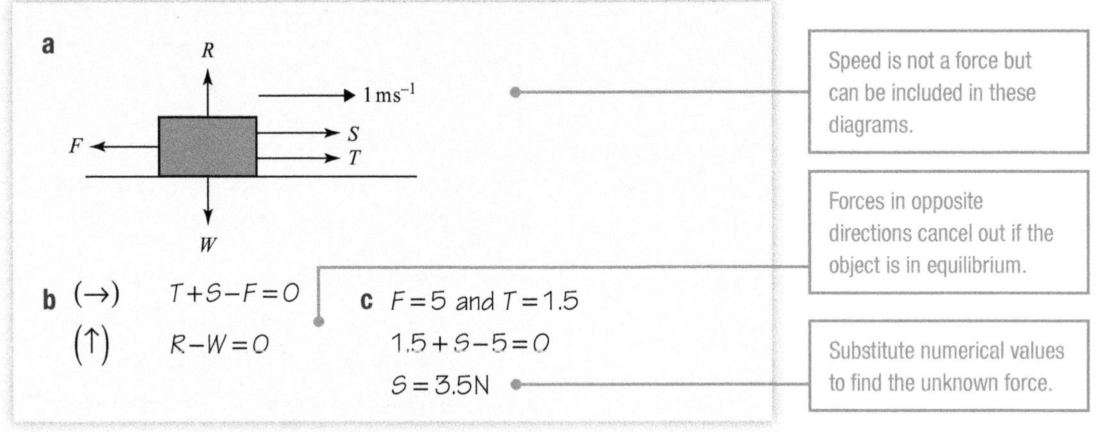

b $(\rightarrow) \quad T + S - F = 0$

$(\uparrow) \quad R - W = 0$

c $F = 5$ and $T = 1.5$

$1.5 + S - 5 = 0$

$S = 3.5\text{N}$

Speed is not a force but can be included in these diagrams.

Forces in opposite directions cancel out if the object is in equilibrium.

Substitute numerical values to find the unknown force.

Exercise 8.1A Fluency and skills

1 Draw diagrams showing all the forces acting on the blocks in the following situations. Use F for the frictional force and R for the normal reaction.

a A block of weight W is placed on a smooth horizontal table.

b A block of weight $4W$ is placed on a rough horizontal table and is pushed with a horizontal force of P which causes it to move at $2\,\text{m s}^{-1}$

c A block of weight W is placed on a rough horizontal table and a string is attached to the block which pulls horizontally, at constant velocity, with tension T

d A block of weight W is placed on a rough horizontal table and a string, attached to the block, pulls the block along the table with tension T at an angle of $40°$ to the horizontal.

2 a Resolve forces vertically in question **1 a**.

b Resolve forces horizontally and vertically in question **1 b**.

c Resolve forces horizontally and vertically in question **1 c**.

3 A girl holds a string, which is attached to a box of weight 40 N. The box hangs vertically below the girl's hand.

a Find the magnitude of the tension in the string.

She then pulls the box upwards with constant velocity.

b Find the magnitude of the tension in the string.

4 Resolve forces horizontally and vertically for the following objects, which are in equilibrium.

a

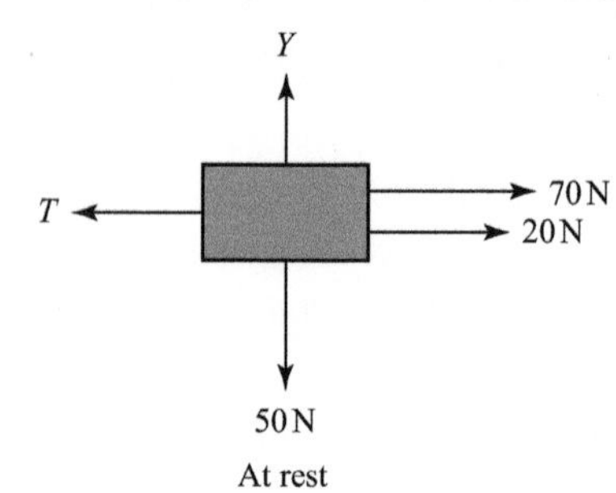

b

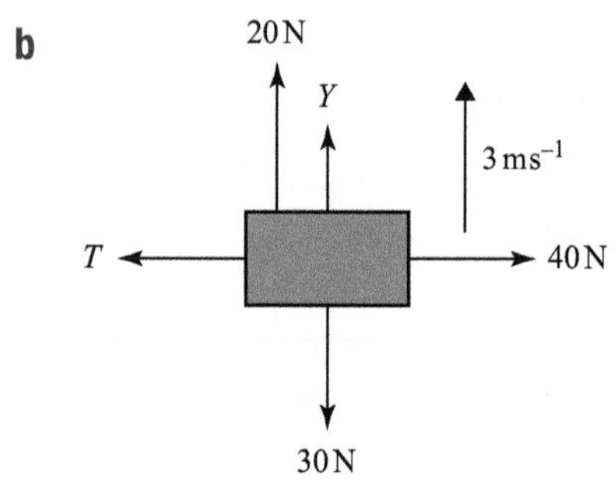

c

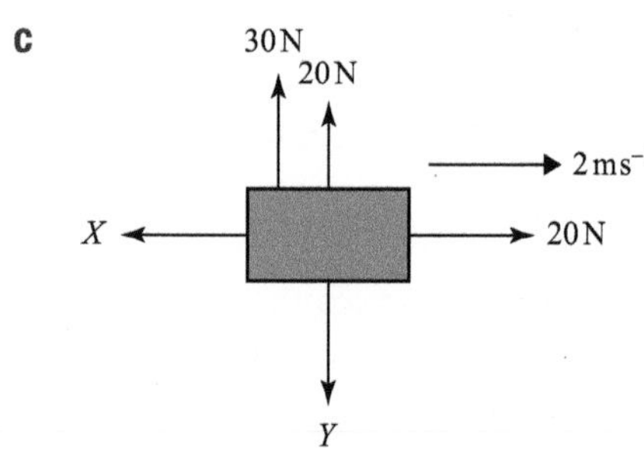

d

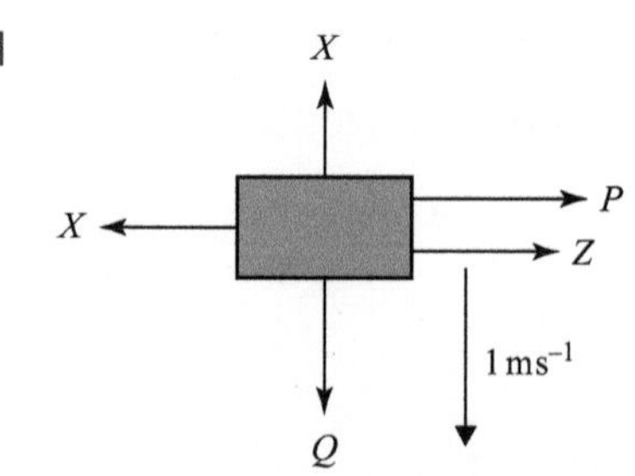

5 A car travels with constant velocity of $8\,\text{m s}^{-1}$ in a straight line along a horizontal road. The resistance to its motion is 900 N.

a What is the driving force of the engine?

b As the car travels, keeping the same driving force, the resistance to its motion increases. What happens to the speed of the car?

Reasoning and problem-solving

When more than one force acts on an object, the **resultant force** is the single force that is equivalent to all the forces acting on the object.

Key point

If forces $\mathbf{F}_1, \mathbf{F}_2, \ldots, \mathbf{F}_n$ act on an object then the resultant force is $\mathbf{R} = \mathbf{F}_1 + \mathbf{F}_2 + \quad \mathbf{F}_n$

MECH

Strategy

To solve questions involving the resultant force

1. Resolve in two perpendicular directions (always in the direction of one of the forces) to find the sum of the components of all the forces in these two directions. Label the components P and Q
2. Draw a right-angled triangle with P and Q as the two shorter sides.
3. To calculate the resultant $R = \sqrt{P^2 + Q^2}$ is the magnitude and α gives the direction.

   If the resultant R is known, use $P = R\cos\alpha$ and $Q = R\sin\alpha$ to find the components.

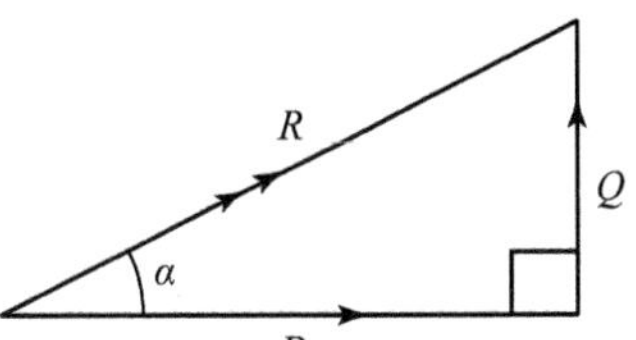

Example 3

The forces $(3\mathbf{i}+14\mathbf{j})$N, $(5\mathbf{i}-2\mathbf{j})$N and $(7\mathbf{i}-\mathbf{j})$N act on an object.

a Calculate the resultant force in the form $(a\mathbf{i}+b\mathbf{j})$N.

b Find the magnitude and direction of the resultant force. Give the direction as the angle between the resultant force and the unit vector **i**.

a $(3\mathbf{i}+14\mathbf{j})\text{ N}+(5\mathbf{i}-2\mathbf{j})\text{ N}+(7\mathbf{i}-\mathbf{j})\text{ N}=(15\mathbf{i}+11\mathbf{j})\text{ N}$ — (1) Work out the sum of forces in both directions. See Ch6.2 for a reminder on components of vectors.

b Magnitude:

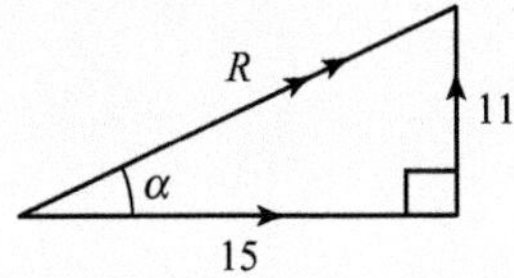

(2) Draw a sketch.

$\sqrt{15^2+11^2} = 18.6\text{ N (to 3 sf)}$ — (3) Use Pythagoras' theorem to find the magnitude.

Direction:

$\tan\alpha = \dfrac{11}{15}$

$\alpha = \tan^{-1}\left(\dfrac{11}{15}\right) = 36.3$ (to 1 dp) — (3) Use trigonometry to find the angle.

The angle the resultant makes with the horizontal is 36.3°

2186, 2293

Example 4

Work out X and Y if the resultant force on this object has magnitude 24 N and makes an angle of 30° above the rightward horizontal.

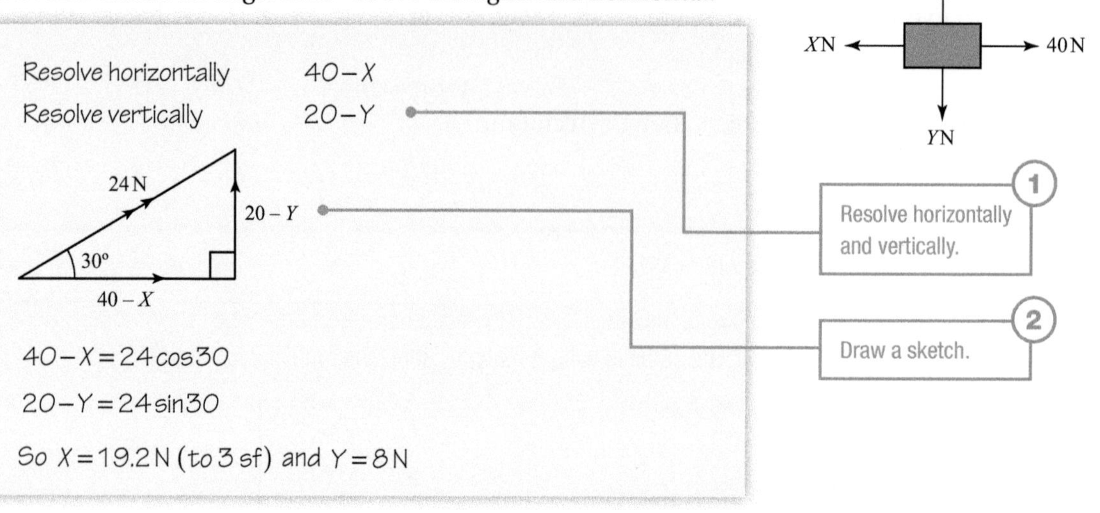

Exercise 8.1B Reasoning and problem-solving

1 The forces $(10\mathbf{i}+15\mathbf{j})$ N, $(25\mathbf{i}+7\mathbf{j})$ N and $(13\mathbf{i}-4\mathbf{j})$ N act on an object.

  a Work out the resultant force in the form $(a\mathbf{i}+b\mathbf{j})$ N.

  b Find the magnitude and direction of the resultant force. Show your working.

2 The forces $(200\mathbf{i}+350\mathbf{j})$ N, $(125\mathbf{i}+75\mathbf{j})$ N and $(-200\mathbf{i}-300\mathbf{j})$ N act on an object.

  a Work out the resultant force in the form $(a\mathbf{i}+b\mathbf{j})$ N.

  b Find the magnitude and direction of the resultant force. Show your working.

3 The forces $\begin{pmatrix}90\\-25\end{pmatrix}$ N, $\begin{pmatrix}-75\\60\end{pmatrix}$ N and $\begin{pmatrix}a\\b\end{pmatrix}$ N act on an object.

  a Calculate the values of a and b if the resultant force is $\begin{pmatrix}-50\\120\end{pmatrix}$ N.

  b Find the magnitude and direction of the resultant force. Show your working.

4 These objects are all in equilibrium. Find the values of the lettered forces.

a

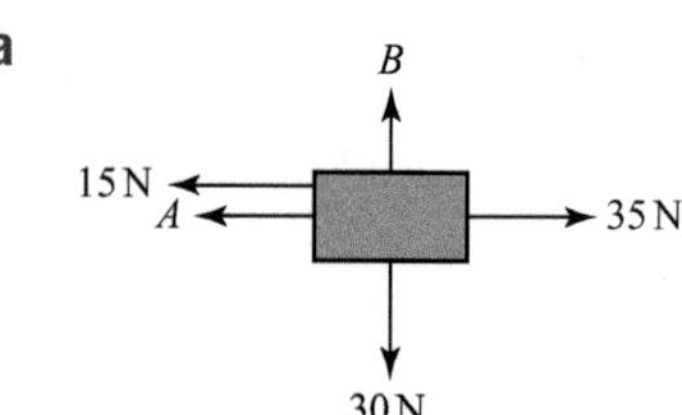

b

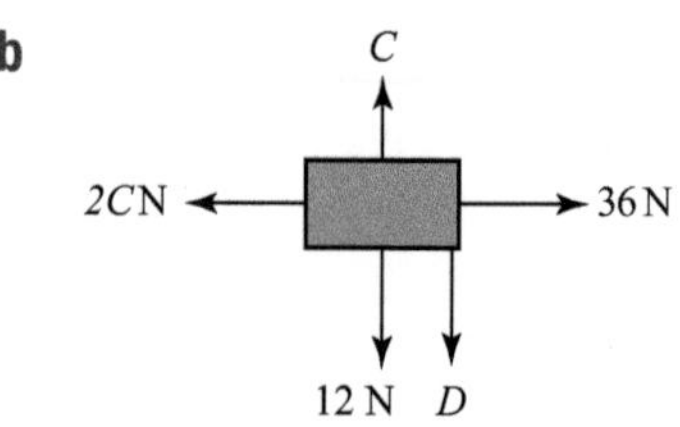

c

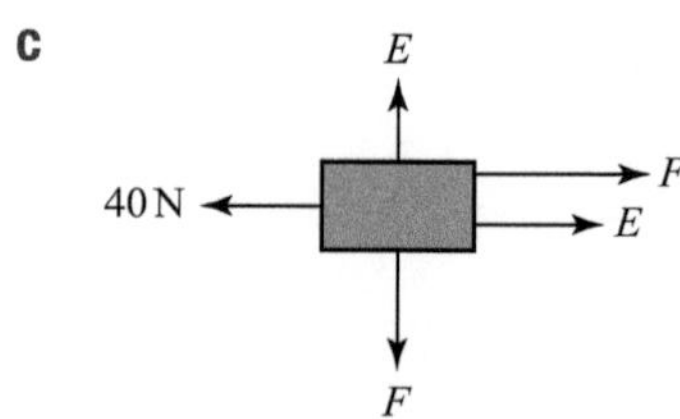

d

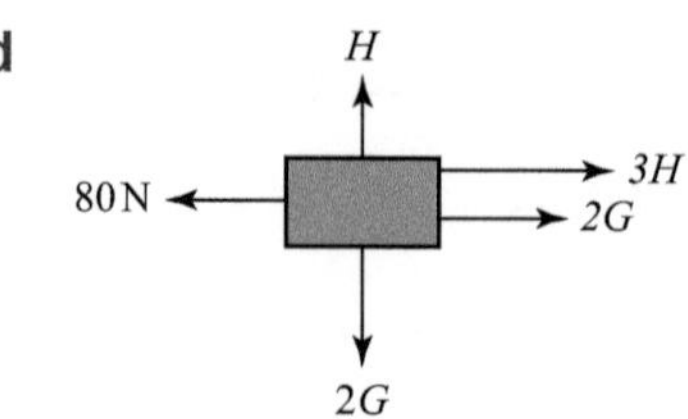

5 Find the magnitude and direction of the resultant force on this object. Show your working.

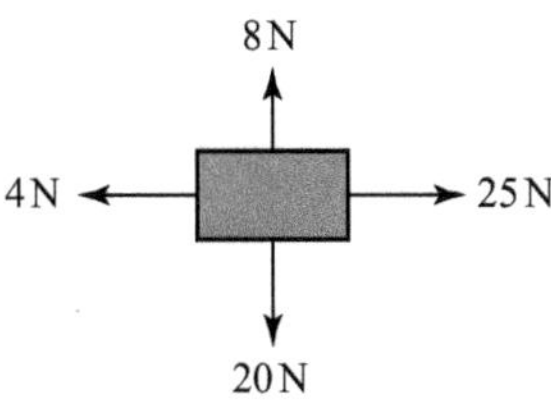

6 Find the magnitude and direction of the resultant force on the following objects. Show your working.

a

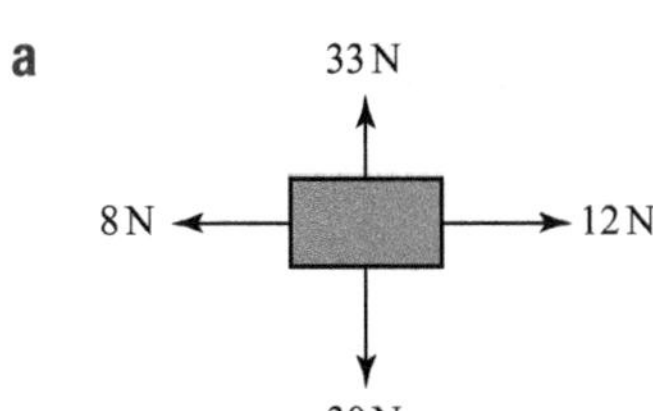

b

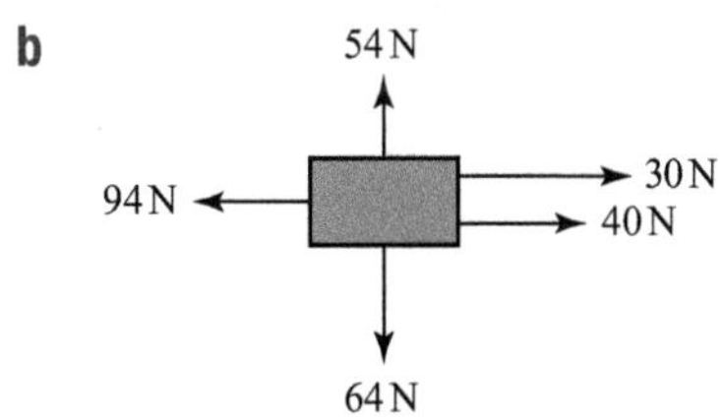

7 State why the following objects cannot be in equilibrium.

a

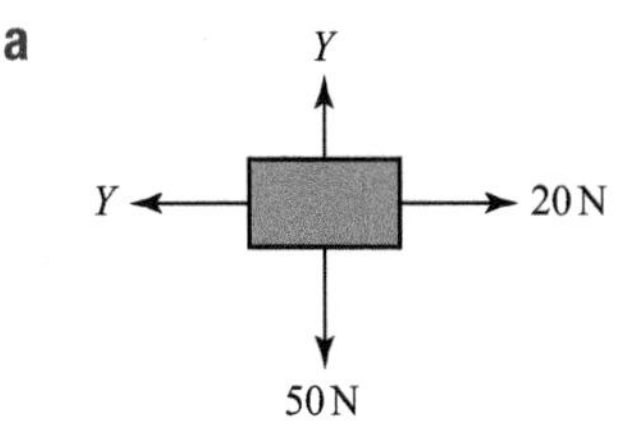

b

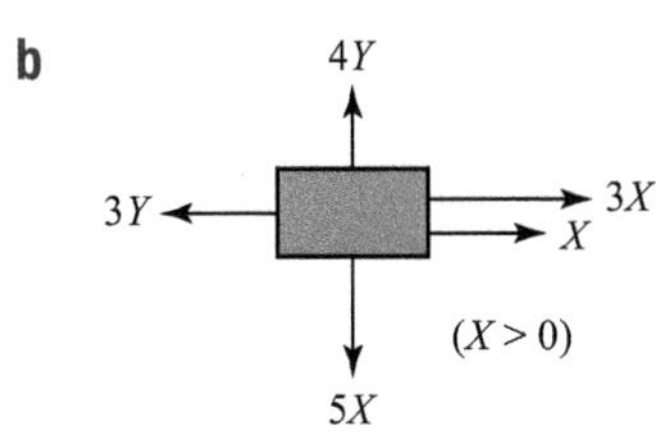

8 A bicycle travels with a constant velocity of $6\,\text{m s}^{-1}$ in a straight line along a horizontal road. The resistance to motion is 300 N.

a What is the magnitude of the force applied by the cyclist?

b At a certain point in the journey the cyclist accelerates without increasing the force she applies to the bicycle. Give a possible reason for this.

9 A block of weight 50 N is placed on a rough, horizontal table. A horizontal force of magnitude 5 N is applied to the block but the block does not move. Find

a The magnitude of the vertical force exerted by the table on the block,

b The magnitude of the horizontal force exerted by the table on the block.

10 A train moves over rough tracks at a constant velocity. One carriage follows the engine, attached to the engine by a light, inextensible tow bar.

a Is the tow bar in thrust or tension? Give a reason for your answer.

b The train then travels in the opposite direction. Does your answer to part **a** change? Give a reason for your answer.

11 The underside of a block of weight 50 N is attached to a vertical rod which applies an upward vertical force of 60 N to the block. State whether the rod is in tension or thrust.

Challenge

12 Work out the positive values of X and Y given that the resultant force has magnitude 30 N on a 315° bearing and Y acts in a northerly direction.

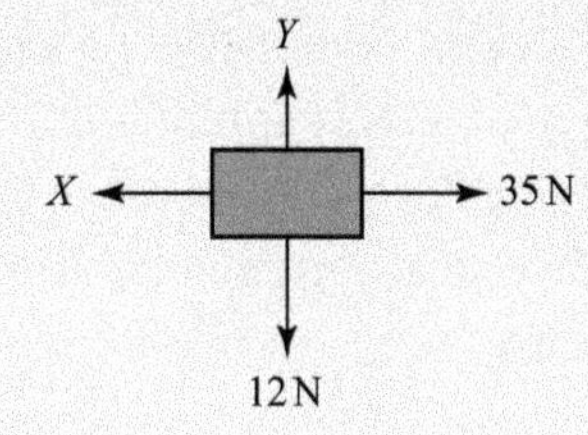

MECH

8.2 Dynamics

Fluency and skills

When you give a large push to a sledge on ice, the sledge begins to accelerate. The bigger the force that you apply, the bigger the acceleration. Acceleration is directly proportional to the force that you apply.

If you had a sledge of twice the mass, you would need a force twice the size to create the same acceleration. The force you apply is directly proportional to the mass.

See Ch2.4

For a reminder on proportionality.

Newton's second law of motion states that the resultant force acting on a particle is proportional to the product of the mass of the particle and its acceleration, $\mathbf{F} \propto m\mathbf{a}$

You measure force in newtons; one newton is the force required to give a 1 kg mass an acceleration of 1 m s^{-1}. Using these units, Newton's second law is $\mathbf{F} = m\mathbf{a}$

Key point

If a resultant force **F** N acts on an object of mass m kg giving it an acceleration $\mathbf{a}\text{ m s}^{-2}$ then $\mathbf{F} = m\mathbf{a}$

Example 1

Calculate the acceleration if the forces acting on an object of mass 25 kg are $(40\mathbf{i}+15\mathbf{j})$ N, $(20\mathbf{i}-7\mathbf{j})$ N and $(31\mathbf{i}+23\mathbf{j})$ N.

Remember that acceleration is a vector not a scalar.

The resultant force

$\mathbf{F} = (40\mathbf{i}+15\mathbf{j})+(20\mathbf{i}-7\mathbf{j})+(31\mathbf{i}+23\mathbf{j}) = 91\mathbf{i}+31\mathbf{j}$ N

$91\mathbf{i}+31\mathbf{j} = 25\mathbf{a}$ — Use $\mathbf{F} = m\mathbf{a}$

$\therefore \mathbf{a} = (3.64\mathbf{i}+1.24\mathbf{j})\text{ m s}^{-2}$

Key point

The equation $F = ma$ can be used in any direction where F is the resultant force in that direction and a is the acceleration in that direction.

Example 2

A box of mass 5 kg and weight 49 N rests on a horizontal floor. A horizontal force of 30 N is applied to the box. The box is subject to a frictional force of 10 N when it is moving. Resolve horizontally and vertically to calculate the normal reaction R and the acceleration a

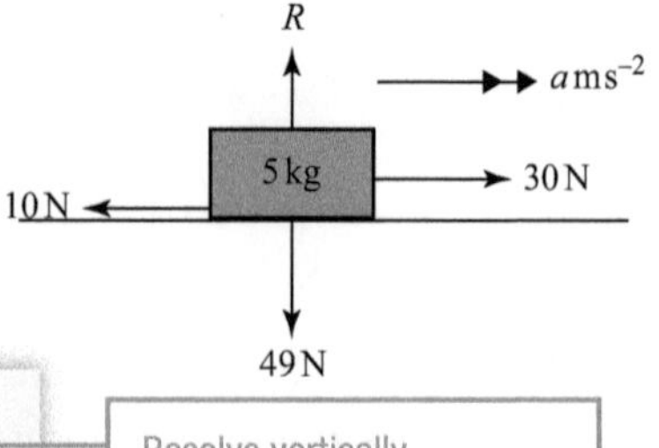

$R - 49 = 0$, so $R = 49$ N — Resolve vertically.

$30 - 10 = 5a$ — Apply Newton's second law horizontally and solve.

$\therefore a = 4\text{ m s}^{-2}$

Exercise 8.2A Fluency and skills

1 A box of mass 5 kg lies on a horizontal table. A horizontal force of magnitude 20 N is applied to the box. Find the magnitude of the resistive force if the acceleration is $1\,\text{m s}^{-2}$.

2 A crate of mass 60 kg lies on a horizontal floor. A horizontal force of magnitude 300 N is applied to the box. Find the acceleration of the crate if the resistive force has magnitude 210 N.

3 A car of mass 1000 kg is travelling along a horizontal road. The total resistance to motion is 400 N and the driving force is 1600 N. Calculate the acceleration of the car.

4 A rope is attached to a block of mass 250 kg, which lies on horizontal ground. The rope is pulled horizontally with tension T. The magnitude of the resistive force is 650 N. Find T if the block accelerates at $0.25\,\text{m s}^{-2}$.

5 Calculate the acceleration acting on an object if

a The resultant force is $(9\mathbf{i}+18\mathbf{j})$ N and the mass is 5 kg,

b The resultant force is $\begin{pmatrix}3\\5\end{pmatrix}$ N and the mass is 2 kg,

c The resultant force is 7 N and the mass is 25 kg,

d The forces acting on the object are $(5\mathbf{i}-\mathbf{j})$ N and $(3\mathbf{i}-4\mathbf{j})$ N and the mass is 4 kg,

e The forces acting on the object are $(3\mathbf{i}+8\mathbf{j})$ N, $(9\mathbf{i}+11\mathbf{j})$ N, $(-\mathbf{i}-7\mathbf{j})$ N and $(4\mathbf{i}+9\mathbf{j})$ N and the mass is 0.25 kg.

6 A car of mass 1200 kg is at rest on a horizontal road. Work out the force needed to give the car an acceleration of $3\,\text{m s}^{-2}$ if the total resistance to motion is 300 N.

7 Work out the missing values in the following diagrams.

a

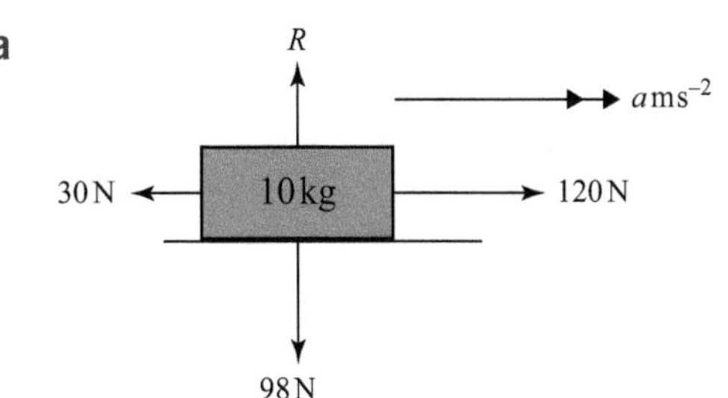

b

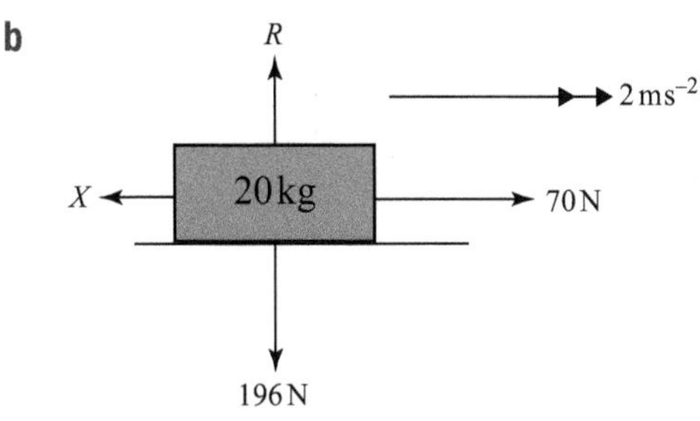

8 A truck of mass 2000 kg is travelling on a horizontal road. The total resistance to motion is 500 N. A horizontal braking force of magnitude 900 N is applied to the truck. Work out the deceleration of the truck.

9 A box of mass m kg rests on a horizontal floor. A horizontal force 40 N is applied to the box which gives it an acceleration of $2\,\text{m s}^{-2}$. Calculate the value of m if the total resistance to motion is 12 N.

10 A car of mass 800 kg is moving along a straight level road with a velocity of $30\,\text{m s}^{-1}$, when the driver spots an obstacle ahead. The driver immediately applies the brakes, providing a net braking force of 3000 N. Calculate

a The deceleration,

b The time taken for the car to come to rest,

c The distance travelled by the car in coming to rest.

MyMaths 2187, 2293 SEARCH

Reasoning and problem-solving

You saw earlier that the equation $F = ma$ can be used in any direction where F is the overall force in that direction and a is the acceleration in that direction.

Strategy

To solve questions involving acceleration

1. Draw a clear diagram, marking on all the forces which act on the object and the acceleration.
2. Use $F = ma$ to write an equation of motion where F is the sum of the components of all the forces in the direction of a
3. Solve the equation to calculate the unknown force.

Example 3

A box of mass 10 kg and weight 98 N is pulled upwards by a vertical string. The block is decelerating at a rate of $2\,\mathrm{m\,s^{-2}}$.

a Find the tension in the string.

b State any assumptions you made in part **a**.

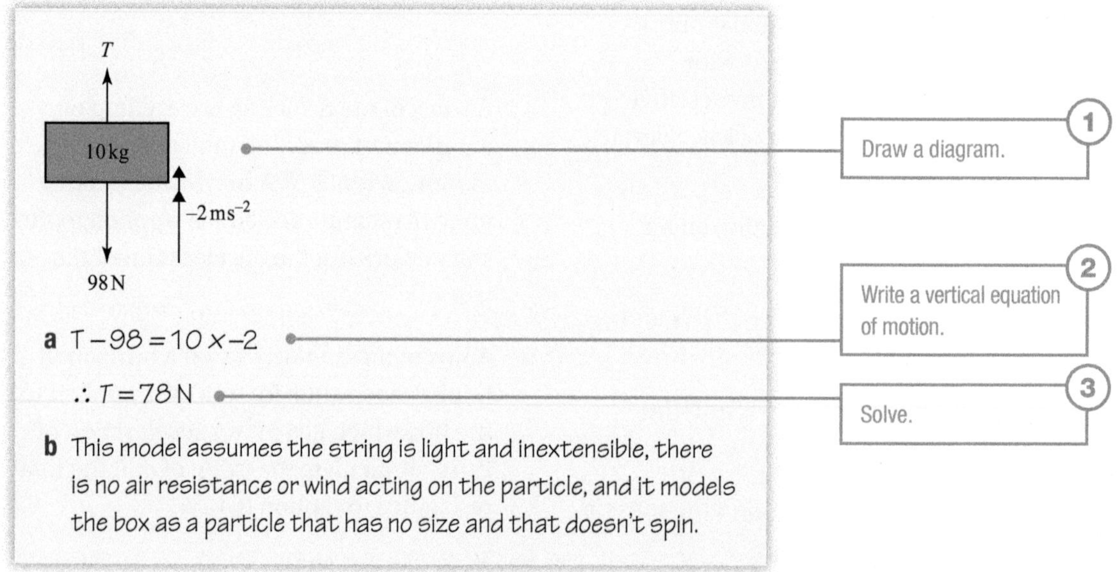

a $T - 98 = 10 \times -2$

$\therefore T = 78\,\mathrm{N}$

b This model assumes the string is light and inextensible, there is no air resistance or wind acting on the particle, and it models the box as a particle that has no size and that doesn't spin.

In mechanics, you often model strings as inextensible which means they do not stretch under tension.

Exercise 8.2B Reasoning and problem-solving

1 Work out the mass of a car if a resultant force of magnitude 700 N causes a constant acceleration that brings the car from rest to $10\,\mathrm{m\,s^{-1}}$ in 200 m.

2 Work out the magnitude of the resultant force on a bike of mass 100 kg that goes from $15\,\mathrm{m\,s^{-1}}$ to $12\,\mathrm{m\,s^{-1}}$ in 20 m.

3 Work out how far an initially stationary object of mass 0.5 kg will travel in 1 second if a resultant force of 30 N is applied to it for that one second.

4 A train of mass 50 tonnes accelerates from rest to a speed of $10\,\mathrm{m\,s^{-1}}$ over a distance of 50 m. If the total resistance to motion is 3000 N then work out the driving force acting on the train.

5 A cyclist and her bike have a combined mass of 80 kg. She travels on a horizontal road at $12\,\text{m s}^{-1}$ and the total resistance to motion is 25 N.

a What is the magnitude of the force that she is applying?

The cyclist sees a problem ahead so immediately stops pedalling and applies the brake. Her braking distance is 10 m.

b Assuming that the resistance to motion stays at 25 N, find the braking force that she applies.

c State, with a reason, what will actually happen to the resistance to motion and what effect this will have on your answer to part **b**.

6 A box of mass m kg has the following forces acting on it: $(2\mathbf{i} + 7\mathbf{j})$ N, $(3\mathbf{i} - 2\mathbf{j})$ N, $(11\mathbf{i} - 2\mathbf{j})$ N, $(11\mathbf{i} + 3\mathbf{j})$ N and $(5\mathbf{i} + p\mathbf{j})$ N. The resultant acceleration is $(7\mathbf{i} + p\mathbf{j})$. Find the values of m and p

7 A parachutist of mass 80 kg and weight 800 N is falling to the ground. Her speed changes from $50\,\text{m s}^{-1}$ to $10\,\text{m s}^{-1}$ in 2 seconds.

a Assuming constant acceleration over these two seconds, find the resistive force of the parachute.

b Comment on the assumption that the acceleration will be constant over these two seconds.

8 A lorry of mass m kg accelerates from rest to a speed of $20\,\text{m s}^{-2}$ over 16 seconds. The total resistance to motion of the lorry is 1200 N and the driving force of the lorry is 3700 N. Find m

9 A string is attached to the top of a box of mass 5 kg and weight 49 N. The box is at rest on the ground. The string is held vertically above the box and the tension in the string is slowly increased until the box is just about to lift off the ground.

a Find the tension in the string at this point.

The tension in the string is then increased to 60 N.

b Find how long the box takes to reach a height of 1 m.

When the box is at a height of 1 m above the ground the tension is reduced to 40 N.

c Find the speed at which the box hits the ground.

10 A force $\begin{pmatrix} 3 \\ 4 \end{pmatrix}$ N is applied to a box of mass 4 kg. Another force $\begin{pmatrix} a \\ b \end{pmatrix}$ N of magnitude 10 N is to be applied in order to give the box the greatest possible acceleration. Find a and b and find the magnitude of the acceleration of the box.

Challenge

11 Calculate the two possible values of x if the forces $\begin{pmatrix} x \\ 1 \end{pmatrix}$ N and $\begin{pmatrix} 8 \\ x \end{pmatrix}$ N applied to a box of mass 5 kg cause an acceleration of magnitude $2.6\,\text{m s}^{-2}$. Assume there is no resistance to motion. Assume both forces act only in directions parallel to the ground.

12 Work out R and m in this diagram.

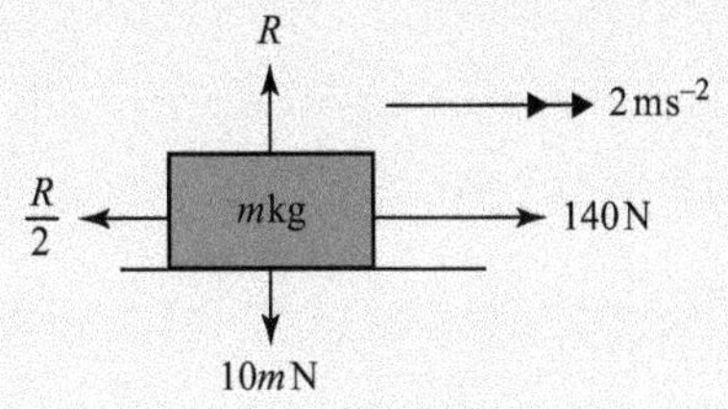

MECH

MyMaths 2187, 2293 SEARCH

8.3 Motion under gravity

Fluency and skills

If a ball of mass m kg is dropped and air resistance is ignored then the only force acting on the ball is the gravitational pull of the Earth. The acceleration of the ball is entirely due to the weight of the ball.

The weight of the ball depends on where the ball is. If the ball is in space its weight is much less than if it is near the surface of the Earth. Note however that even if the ball's weight changes based on where it is, its mass never changes.

Using Newton's second law,
$F = ma$ gives $W = mg$

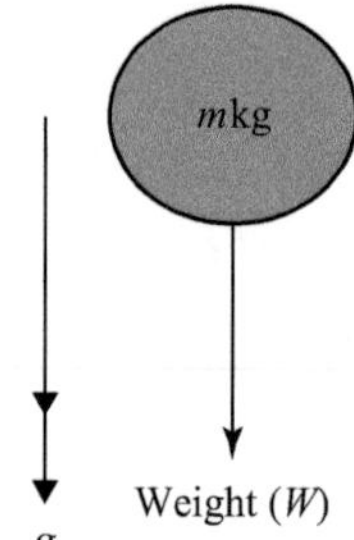

Key point

An object of mass m kg has weight mg newtons.

In questions involving motion on Earth, you will be told which approximation to use for g. You may be told to use g as 9.81 m s^{-2}, 9.8 m s^{-2} or even 10 m s^{-2}. In your answers use the same number of significant figures as the value of g you are given.

Example 1

Work out the following quantities.

a The weight of a car of mass 1230 kg. You may assume that g is 9.8 m s^{-2}.

b The weight of a piece of paper of mass 4.54 g. You may assume that g is 9.81 m s^{-2}.

c The mass of a cup of water of weight 3.2 N. You may assume that g is 10 m s^{-2}.

a $W = 1230 \times 9.8 = 12054$ N

So $W = 12000$ N (to 2 sf) — *g is given to 2 sf so give your answer to 2 sf.*

b $W = 0.00454 \times 9.81 = 0.0445374$ N — *Convert the mass to kilograms.*

$\therefore W = 0.0445$ N (to 3 sf)

c $3.2 = m \times 10$

$m = 0.32$ kg $= 320$ g

$m = 300$ g (to 1 sf)

Example 2

A block is pulled along a horizontal surface by a horizontal string. The tension in the string is 40 N. The frictional force is 30 N and the normal reaction is 20 N. Work out the mass and the acceleration of the block. You may assume that g is $10\,\text{m s}^{-2}$.

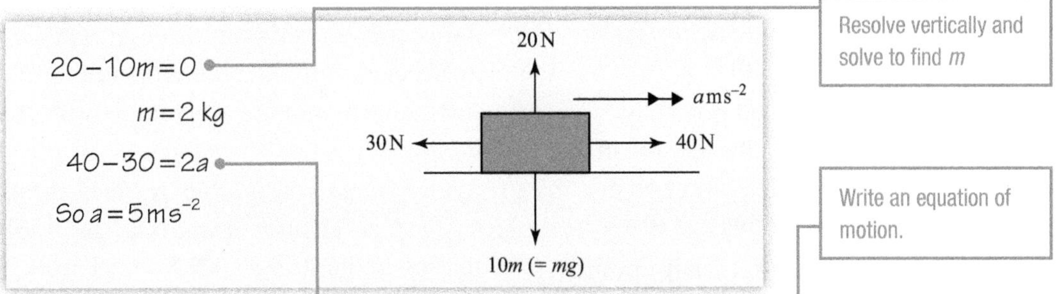

Example 3

A stone of mass 50 g is dropped from the top of a cliff, which is 20 m above the sea. You may assume that g is $9.8\,\text{m s}^{-2}$.

a Find the speed at which the stone hits the sea below.

b State any assumptions you have made.

a The acceleration of the stone is $9.8\,\text{m s}^{-2}$.

$u = 0 \qquad s = 20 \qquad a = 9.8$

$v^2 = u^2 + 2as$

$v^2 = 2 \times 20 \times 9.8$

$v = 20\,\text{m s}^{-1}$ (to 2 sf)

b The model assumes that there is no air resistance or wind and it models the stone as a particle that has no size and does not spin.

Write down the quantities that you know. The acceleration is equal to g. See Ch7.3 for a reminder on equations of motion for constant acceleration.

Select an appropriate equation and solve for v

Exercise 8.3A Fluency and skills

1 Work out the following. You may assume that g is $9.8\,\text{m s}^{-2}$.

  a The weight of a 10 g piece of paper.

  b The weight of a 10 tonne lorry.

  c The mass of a 100 N weight.

2 A stone is dropped from a height of 2m above the ground. You may assume that g is $9.8\,\text{m s}^{-2}$.

  a How long does it take for the stone to reach the ground?

  b State any assumptions you made in your calculation for part **a**.

  c With what velocity does the stone hit the ground?

3 A block of mass 75 kg is pulled up by a vertical rope. The tension in the rope is 1200 N. Calculate the acceleration of the block. You may assume that g is $9.8\,\text{m s}^{-2}$.

4 A ball is thrown vertically upwards. It returns to its starting point after 3 seconds. Find the speed with which it was thrown. You may assume that g is $9.81\,\text{m s}^{-2}$.

5 A particle is thrown vertically downwards with a speed of $18\,\text{m s}^{-1}$ from the top of a building, which is 5 m high. Find the speed with which the particle hits the ground. You may assume that g is $10\,\text{m s}^{-2}$.

6 A block is pulled along a horizontal surface by a horizontal string. The tension in the string is 450 N. The frictional force is 300 N and the normal reaction force is 250 N. Work out the mass and the acceleration of the block. You may assume that g is $10\,\text{m s}^{-2}$.

MyMaths 2185, 2187 SEARCH

7 A block is pulled along a horizontal surface by a horizontal string. The tension in the string is 70 N. The frictional force is 50 N and the normal reaction force is 40 N. Work out the mass and the acceleration of the block. You may assume that g is $10\,\text{m s}^{-2}$.

8 A crate of mass 10 kg is pulled vertically upwards by a vertical rope. The tension in the rope is 130 N. Find the acceleration of the crate. You may assume that g is $10\,\text{m s}^{-2}$.

9 A box of mass m kg is pulled vertically upwards by a vertical string. The tension in the string is 80 N and the acceleration of the box is $0.2\,\text{m s}^{-2}$. Find m. You may assume that g is $9.8\,\text{m s}^{-2}$.

10 Work out the tension in the cable attached to the top of a lift of mass 400 kg when the lift is

a Stationary,

b Accelerating at $1\,\text{m s}^{-2}$ vertically upwards,

c Accelerating at $2\,\text{m s}^{-2}$ vertically downwards.

You may assume that g is $9.8\,\text{m s}^{-2}$.

11 A box of mass 10 kg is lifted by a light string so that it accelerates upwards at $2.0\,\text{m s}^{-2}$. Work out the tension in the string. You may assume that g is $9.8\,\text{m s}^{-2}$.

12 A box of mass 2 kg sinks through water with an acceleration of $2\,\text{m s}^{-2}$. Work out the resistance to motion. You may assume that g is $9.8\,\text{m s}^{-2}$.

13 A box of mass m kg is lifted by a light string so that it accelerates upwards at $3\,\text{m s}^{-2}$. If the tension in the string is 256 N calculate m. You may assume that g is $9.8\,\text{m s}^{-2}$.

14 A block of mass 200 kg is pulled vertically upwards by a vertical cable. The block is accelerating upwards at $0.5\,\text{m s}^{-2}$. Find the magnitude of the tension in the cable. You may assume that g is $9.81\,\text{m s}^{-2}$.

15 A container of mass 60 kg is being lowered vertically downwards by a vertical cable. The container is accelerating downwards at $0.1\,\text{m s}^{-2}$. Find the magnitude of the tension in the cable. You may assume that g is $9.81\,\text{m s}^{-2}$.

Reasoning and problem-solving

Strategy

To solve questions involving weight

1 Draw a clear diagram, marking on all the forces which act on the object and the acceleration.

2 Resolve vertically and use $F = ma$ to create an equation.

3 Solve the equation for the acceleration of the object or any unknown force or mass.

Example 4

A bag of mass $10b$ kg is lowered by a light, inextensible string so that it accelerates downwards at $3b\,\text{m s}^{-2}$. Calculate the possible values of b if the tension is 41.5 N. You may assume that g is $9.8\,\text{m s}^{-2}$.

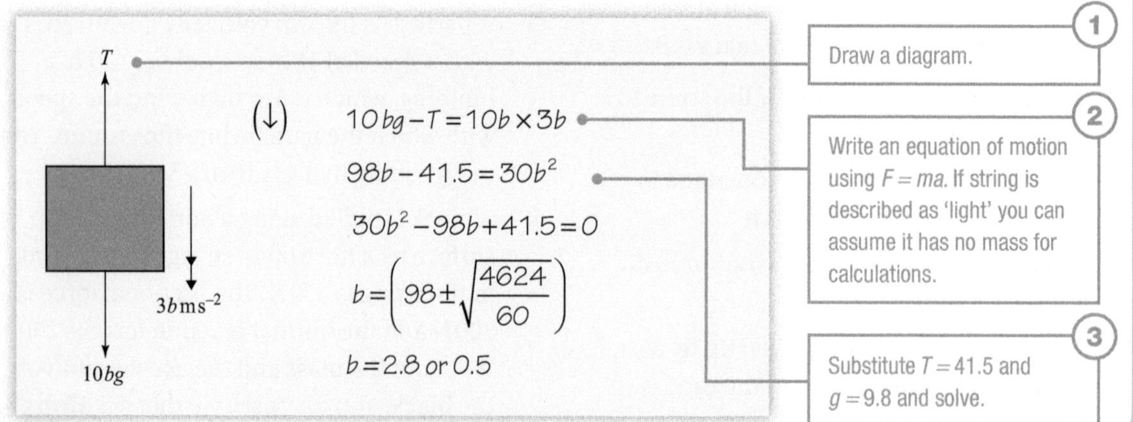

$(\downarrow)\quad 10bg - T = 10b \times 3b$

$98b - 41.5 = 30b^2$

$30b^2 - 98b + 41.5 = 0$

$b = \left(98 \pm \sqrt{4624}\right) / 60$

$b = 2.8$ or 0.5

1 Draw a diagram.

2 Write an equation of motion using $F = ma$. If string is described as 'light' you can assume it has no mass for calculations.

3 Substitute $T = 41.5$ and $g = 9.8$ and solve.

Exercise 8.3B Reasoning and problem-solving

1 A crane lifts up a crate of bricks with mass 1200 kg from the ground. The acceleration is constant, and after 1 second the crate is 0.5 m off the ground. Work out the acceleration of the crate and the tension in the crane's cable. You may assume that g is $9.8\,\text{m s}^{-2}$.

2 A block of mass 5 kg is pulled along a horizontal surface by a horizontal string. The tension in the string is 50 N. The normal reaction force is R and the frictional force is $\frac{R}{2}$. Work out R and the acceleration of the block. You may assume that g is $10\,\text{m s}^{-2}$.

3 A man who weighs 780 N on the surface of the Earth weighs only 130 N on the surface of the moon. Work out the ratio of the value of g on the moon to the value of g on the surface of the Earth.

4 A crate of mass 20 kg is pulled upwards from rest by a light rope to a speed of $4\,\text{m s}^{-1}$. If the tension in the rope is 260 N then work out how far the crate travelled before getting to $4\,\text{m s}^{-1}$. You may assume that g is $9.8\,\text{m s}^{-2}$.

5 A ball is thrown vertically upwards from the ground. Its highest point is 20 m above the ground. Find how long the ball is in the air.

6 A box of mass 8.5 kg is pulled vertically upwards by a rope.

a Calculate the tension in the rope, in terms of g, when

i The basket is stationary,

ii The basket is accelerating at $2\,\text{m s}^{-2}$ upwards.

b What assumptions have you made about the rope?

7 A box of mass 25 kg is being lifted. The box starts from rest and then a rope attached to the top of the box pulls it vertically upwards. The tension in the rope is 400 N. Another rope is attached to the bottom of the box and a man pulls lightly down to stop the box from swaying to the side. The tension in this rope is 50 N. Find the acceleration of the box and the height it will be at after 5 seconds. You may assume that g is $10\,\text{m s}^{-2}$.

8 A man on the surface of the moon lifts a box of mass 10 kg by pulling on a string. The tension in the string is 30 N. Find how long it takes to lift the box to a height of 1 m. You may assume that g is $1.6\,\text{m s}^{-2}$.

9 A box of mass 3 kg rests on a tray of mass 2 kg. They are lifted up by two vertical strings each attached to the tray so that the tension in each string is the same. The box and the tray accelerate upwards at $1\,\text{m s}^{-2}$. You may assume that g is $9.8\,\text{m s}^{-2}$.

a Find the tension in each of the strings.

b The 3 kg box falls off the tray but the tension in the string stays the same. Find the new acceleration of the tray.

Challenge

10 A block of mass m kg is pulled along a horizontal surface by a horizontal string. The normal reaction force is R. The tension in the string is half the normal reaction force and the frictional force is one fifth of the normal reaction force. You may assume $g = 10\,\text{m s}^{-2}$

a Calculate the acceleration of the block.

b Calculate the acceleration of the block assuming there is no friction.

11 A bag of mass 4 kg hangs stationary from a taut string at a height of 2 m. It is lowered to the ground at constant acceleration, hitting the ground after twice the time it would have taken if it had been in freefall. Work out the tension in the string in terms of g

8.4 Systems of forces

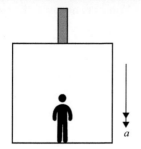

Fluency and skills

A person standing in a lift exerts a force, acting vertically downwards, on the floor of the lift. The lift exerts a force, acting vertically upwards, on the person. As the lift moves up and down, the size of this force changes but the force that the person exerts on the lift is always equal and opposite to the force the lift exerts on the person.

Newton's third law states that for every action there is an equal and opposite reaction.

> **Key point**
>
> Newton's third law states that when an object A exerts a force on an object B, object B exerts an equal and opposite force on object A.

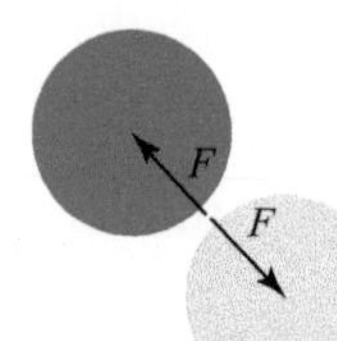

If a red ball hits a yellow ball, the force that the red ball exerts on the yellow ball has equal magnitude to the force that the yellow ball exerts on the red ball. These forces act in opposite directions.

Example 1

A man of mass 80 kg is in a lift of mass 300 kg, which is accelerating upwards at 4.1 m s^{-2}. You may assume that $g = 9.8\text{ m s}^{-2}$

a Work out the tension in the cable pulling the lift by

  i Considering the set of forces acting on the man and the set of forces acting on the lift,

  ii Considering the man and the lift as one object.

b Use Newton's third law to explain why you can consider the man and the lift as one object.

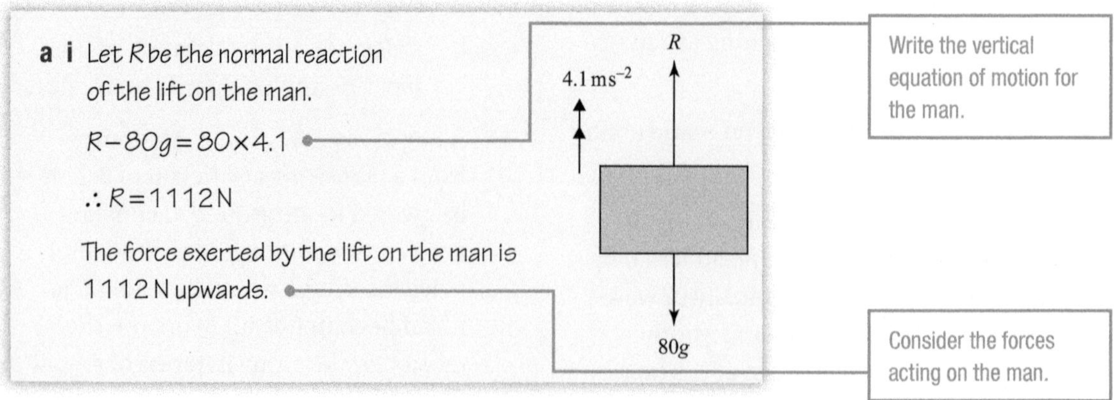

a i Let R be the normal reaction of the lift on the man.

$R - 80g = 80 \times 4.1$ — Write the vertical equation of motion for the man.

$\therefore R = 1112\text{ N}$

The force exerted by the lift on the man is 1112 N upwards. — Consider the forces acting on the man.

(Continued on the next page)

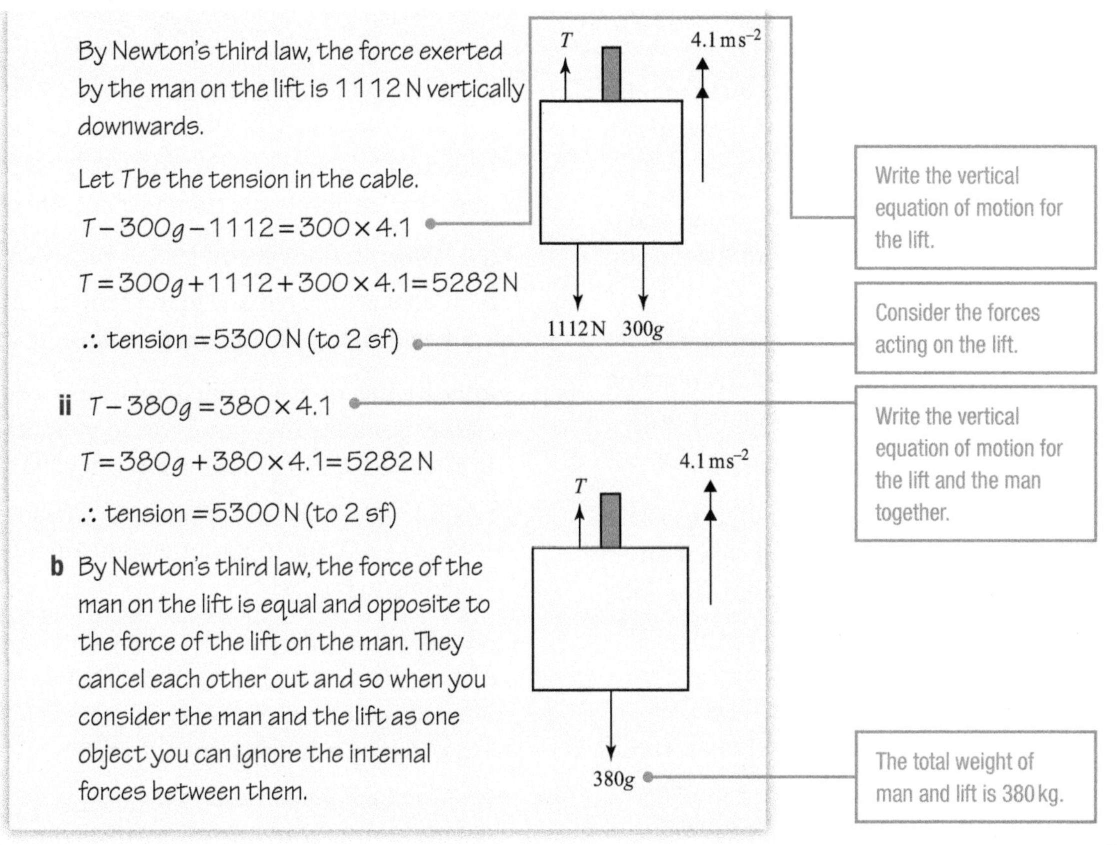

By Newton's third law, the force exerted by the man on the lift is 1112 N vertically downwards.

Let T be the tension in the cable.

$T - 300g - 1112 = 300 \times 4.1$

$T = 300g + 1112 + 300 \times 4.1 = 5282$ N

$\therefore$ tension $= 5300$ N (to 2 sf)

ii $T - 380g = 380 \times 4.1$

$T = 380g + 380 \times 4.1 = 5282$ N

$\therefore$ tension $= 5300$ N (to 2 sf)

b By Newton's third law, the force of the man on the lift is equal and opposite to the force of the lift on the man. They cancel each other out and so when you consider the man and the lift as one object you can ignore the internal forces between them.

Write the vertical equation of motion for the lift.

Consider the forces acting on the lift.

Write the vertical equation of motion for the lift and the man together.

The total weight of man and lift is 380 kg.

MECH

Example 2

A van of mass 1250 kg tows a trailer of mass 250 kg along a horizontal road. The driving force from the engine is D N, the van experiences air resistance of 300 N and the trailer experiences air resistance of 100 N.

The main source of resistance to motion for a moving vehicle is usually air resistance.

If the acceleration of the van and trailer is 2.4 m s^{-2}, calculate the tension in the connection between the van and the trailer and calculate the driving force D

Let T be the tension in the connection between the van and the trailer

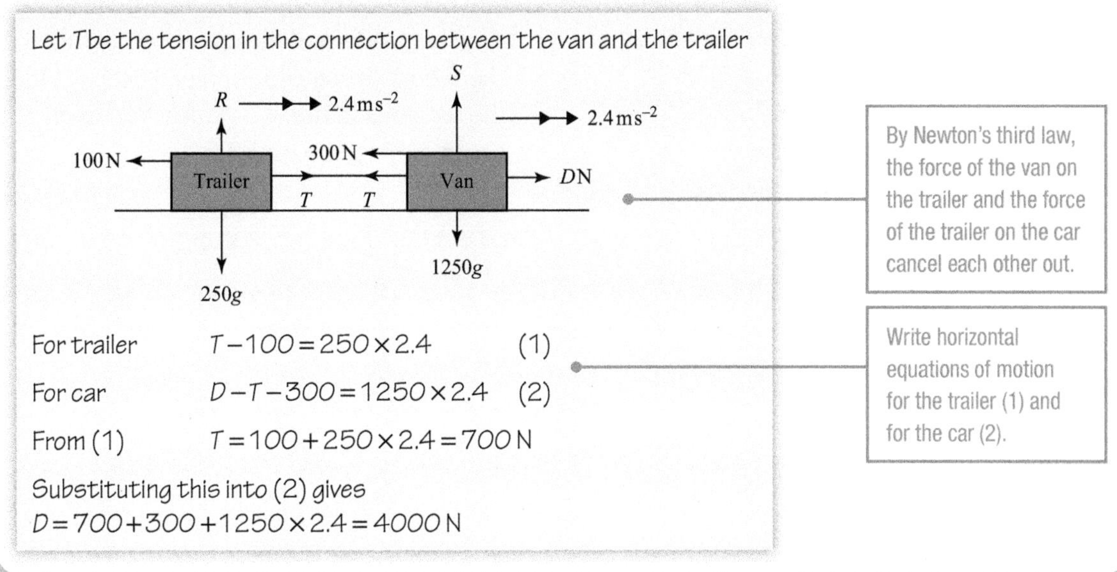

For trailer	$T - 100 = 250 \times 2.4$	(1)
For car	$D - T - 300 = 1250 \times 2.4$	(2)
From (1)	$T = 100 + 250 \times 2.4 = 700$ N	

Substituting this into (2) gives

$D = 700 + 300 + 1250 \times 2.4 = 4000$ N

By Newton's third law, the force of the van on the trailer and the force of the trailer on the car cancel each other out.

Write horizontal equations of motion for the trailer (1) and for the car (2).

Exercise 8.4A Fluency and skills

1 A woman of mass 60 kg is in a lift of mass 250 kg which is accelerating downwards at $3.2\,\text{m s}^{-2}$.

  a Resolve vertically for the woman and the lift together to work out the tension in the cable.

  b Resolve vertically for the woman to work out the magnitude of the reaction force between the woman and the lift.

  You may assume that $g = 9.8\,\text{m s}^{-2}$

2 A box of mass 20 kg is in a lift of mass 200 kg. If the tension in the cable is 2800 N then work out the acceleration of the lift and the reaction force between the box and the lift. You may assume that $g = 9.8\,\text{m s}^{-2}$

3 A car of mass 1200 kg tows a caravan of mass 800 kg along a horizontal road. The driving force is 4000 N, the car experiences air resistance of 150 N and the caravan experiences air resistance of 100 N. Calculate

  a The acceleration of the car,

  b The force transmitted through the tow bar.

4 A tray of mass 500 g has a box of mass 750 g placed on it. A string is attached to the tray and the string is pulled upwards to cause the tray and box to accelerate. The tension in the string is 15 N. Find the acceleration of the tray and the box. Find also the force exerted by the tray on the box. You may assume that g is $10\,\text{m s}^{-2}$.

5 A car of mass 1500 kg tows a caravan of mass 400 kg along a horizontal road. The driving force is D N, the car experiences air resistance of 250 N and the trailer experiences air resistance of 120 N. The car and caravan accelerate from rest at $1.8\,\text{m s}^{-2}$.

  a Work out the value of D

  b Calculate the tension in the connection between the car and the caravan.

  This driving force is applied for 10 seconds. At that point the driver puts his foot on the brake and applies a braking force of 2000 N.

  c Calculate how far the car and caravan travel with the brake applied before they come to rest.

6 A man of mass 80 kg carries a bag of mass 2 kg in a lift of mass 500 kg. The lift is moving upwards and decelerates from a speed of $5\,\text{m s}^{-1}$ to rest in a distance of 20 m.

  a Calculate the tension in the cable and the normal reaction force between the man and the lift.

  b Calculate the force that the man feels from carrying the bag.

  You may assume that $g = 9.81\,\text{m s}^{-2}$

7 The cable from a crane is attached to a crate of mass 220 kg. Another crate of mass 150 kg is connected to the 220 kg crate by a cable and hangs vertically below it. The crane then begins to pull the crates up, giving them both an acceleration of $0.6\,\text{m s}^{-2}$. Find the tensions in the two cables. You may assume that g is $9.8\,\text{m s}^{-2}$.

Reasoning and problem-solving

Strategy

To solve questions involving connected objects

1. Draw a clear diagram marking on all the forces which act on the objects and the acceleration.
2. Consider the whole system or isolate one of the objects and create an equation of motion, using $F = ma$
3. Use equations of motion for constant acceleration to solve the equations for the unknown quantity.

Example 3

The two ends of a light, inextensible string are attached to two objects of mass 4 kg and 9 kg. The string passes over a smooth, fixed pulley. The 9 kg mass is initially 0.2 m above the ground and the 4 kg mass is initially 0.6 m above the ground. They are released from rest.

a **i** Write an equation of motion for each mass.

ii Use your equations of motion to work out the acceleration of the two masses and the tension in the string.

b Work out the speed at which the 9 kg mass hits the ground.

c Work out the greatest height of the 4 kg mass above the ground in subsequent motion.

You may assume that $g = 9.8\,\text{m}\,\text{s}^{-2}$

a **i** 4 kg mass: $(\uparrow)\ T - 4g = 4a$ (1)

9 kg mass: $(\downarrow)\ 9g - T = 9a$ (2)

ii Adding up (1) and (2) gives $5g = 13a$

$\therefore a = 3.8\,\text{m}\,\text{s}^{-2}$ (to 2 sf)

Using (1) $T = 54\,\text{N}$ (to 2 sf)

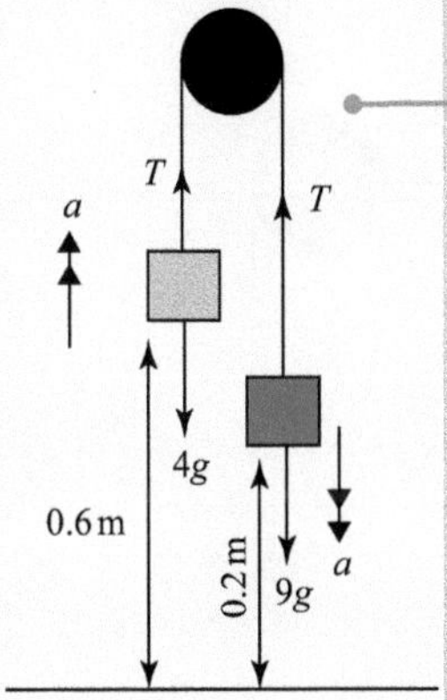

b $u = 0 \quad a = 3.8 \quad s = 0.2$

$v^2 = u^2 + 2as = 1.52$

So $v = 1.2\,\text{m}\,\text{s}^{-1}$ (to 2 sf)

c For upward motion,

$a = -9.8\,\text{m}\,\text{s}^{-2}$ (to 2 sf)

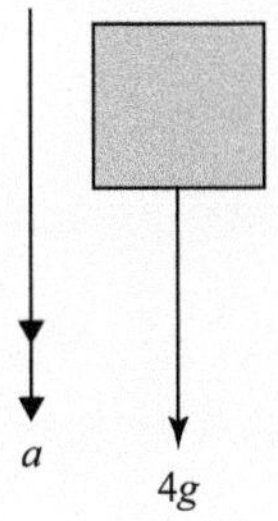

At highest point $v = 0$

$u = 1.23 \quad v = 0 \quad a = -9.8$

$v^2 = u^2 + 2as$ gives

$$s = \frac{0^2 - 1.23^2}{2 \times (-9.8)} = 0.077$$

So greatest height of 9 kg mass above the ground is $0.6 + 0.2 + 0.077 = 0.88\,\text{m}$ (to 2 sf)

1. Draw a diagram.

When the 9 kg mass hits the ground, the string becomes slack. There is no upwards force on the 4 kg mass and it begins to decelerate due to gravity.

2. Consider the 4 kg mass.

3. Use values from your answer from part **b** and identify the equation of motion needed.

The particle is initially 0.6 m above the ground, travels 0.2 m as the 9 kg mass falls and then travels a further 0.077 m.

MECH

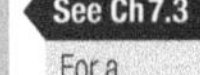
See Ch7.3

For a reminder of the equations of motion.

ICT Resource online

To experiment with the motion of connected masses, click this link in the digital book.

2188 SEARCH

Example 4

A 2 kg mass rests on a rough horizontal table. It is attached to a long string that passes over a smooth pulley at the end of the table, and is tied to a 3 kg mass held still in the air, 1 m above the ground. The 3 kg mass is released. The 2 kg mass experiences a constant frictional force of magnitude 10 N. You may assume that $g = 9.8\text{ m s}^{-2}$

a Work out the acceleration of the two masses and the tension in the string.

b Work out the speed at which the 3 kg mass hits the ground.

c Work out how far the 2 kg mass travels before it comes to rest.

d Work out the magnitude of the force exerted by the string on the pulley before the 3 kg mass hits the floor.

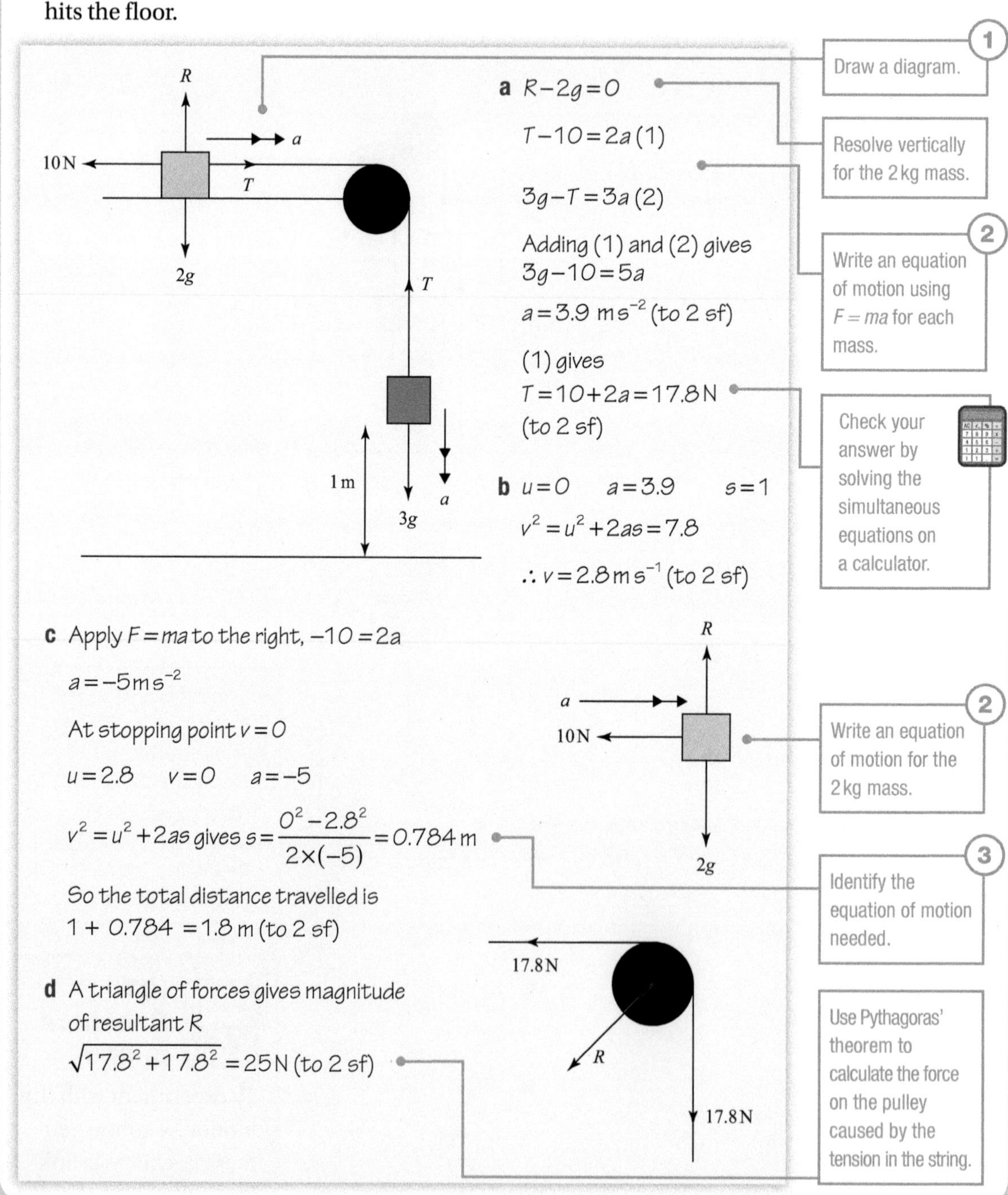

R
a
10 N
T
2g
T
1 m
a
3g
a R−2g=0
T−10=2a (1)
3g−T=3a (2)
Adding (1) and (2) gives
3g−10=5a
a=3.9 m s^-2 (to 2 sf)
(1) gives
T=10+2a=17.8 N
(to 2 sf)
b u=0 a=3.9 s=1
v^2=u^2+2as=7.8
∴ v=2.8 m s^-1 (to 2 sf)
c Apply F=ma to the right, −10=2a
a=−5 m s^-2
At stopping point v=0
u=2.8 v=0 a=−5
v^2=u^2+2as gives s=(0^2−2.8^2)/(2×(−5))=0.784 m
So the total distance travelled is
1 + 0.784 = 1.8 m (to 2 sf)
R
a
10 N
2g
d A triangle of forces gives magnitude
of resultant R
√(17.8^2+17.8^2) = 25 N (to 2 sf)
17.8 N
R
17.8 N
1
Draw a diagram.
Resolve vertically
for the 2 kg mass.
2
Write an equation
of motion using
F = ma for each
mass.
Check your
answer by
solving the
simultaneous
equations on
a calculator.
2
Write an equation
of motion for the
2 kg mass.
3
Identify the
equation of motion
needed.
Use Pythagoras'
theorem to
calculate the force
on the pulley
caused by the
tension in the string.

Exercise 8.4B Reasoning and problem-solving

1 A 2 kg mass and a 3 kg mass are connected by a light inextensible string and hang either side of a smooth, fixed pulley. Calculate the tension in the string and the acceleration of the particles. You may assume that $g = 9.81\,\text{m s}^{-2}$

2 A 3 kg mass rests on a rough horizontal table. It is attached to a long string that passes over a smooth pulley at the end of the table and is tied to a mass of 5 kg, which is held at rest in the air 0.2 m above the ground. The 5 kg mass is released from rest. The 3 kg mass experiences a constant frictional force of magnitude 12 N. You may assume that $g = 9.8\,\text{m s}^{-2}$

- **a** Work out the tension in the string and the magnitude of the acceleration of the two masses.
- **b** Work out the speed at which the 5 kg mass hits the ground.
- **c** Work out how far the 3 kg mass travels before it comes to rest.
- **d** Work out the magnitude of the force exerted by the string on the pulley while the 5 kg mass is falling.

3 Two blocks, A and B, of masses 5 kg and 10 kg respectively are connected by a light, inextensible string that passes over a smooth pulley, P. Initially A is at rest on a smooth horizontal platform, and B hangs freely, as shown in the diagram. The system is released from rest. You may assume that $g = 10\,\text{m s}^{-2}$

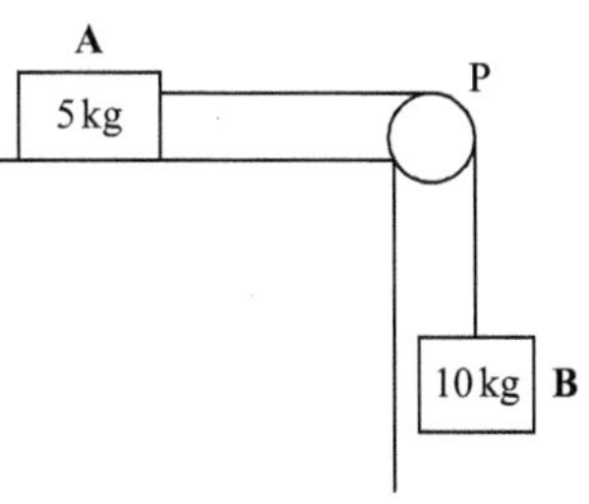

- **a** Calculate the acceleration of B.
- **b** Calculate the tension in the string.

After three seconds, A is still moving freely when B hits the floor.

- **c** Calculate the velocity of A at this time.
- **d** Calculate the initial height of B above the floor.

4 Two blocks, A and B, of mass 3 kg and 4 kg respectively, are connected by a light, inextensible string, passing over a fixed smooth, light pulley. The blocks are released from rest with the string taut, and the hanging parts vertical. You may assume that $g = 9.81\,\text{m s}^{-2}$. Find

- **a** The acceleration of B,
- **b** The tension in the string.

After 5 seconds, B strikes the floor. Block A continues upwards, and does not hit the pulley. Find

- **c** The velocity of A at the instant when B strikes the floor,
- **d** The greatest height above its initial starting height reached by A.

Challenge

5 Two trays, each of mass 1 kg, are connected by a long light, inextensible string that passes over a smooth, fixed pulley. Both trays are 1 m above the horizontal ground. A mass of 2 kg and 3 kg respectively is placed in each tray and the system is released from rest. When the tray with the 3 kg mass is 0.5 m above the ground, the 3 kg mass slips out. Find the time from the system initially being released from rest and the tray with the 2 kg mass hitting the ground. You may assume that $g = 9.8\,\text{m s}^{-2}$

6 An x kg and a y kg mass are connected by a light, inextensible string and hang either side of a smooth fixed pulley. The masses are initially both s metres above the horizontal ground. They are released from rest. If $y > x$, work out the tension in the string and the acceleration in terms of x, y and g

2188

8 Summary and review

Chapter summary

- Newton's first law of motion states that an object will remain at rest or continue to move with constant velocity unless an external force is applied to it.
- Force is a vector. It has both magnitude and direction.
- An object is in equilibrium if it is at rest or moving at constant velocity.
- The resultant force is the single force equivalent to all the forces acting on the object.
- If an object is in equilibrium, the resultant force is zero.
- You can summarise Newton's second law of motion as:
  If a resultant force $\mathbf{F}$ N acts on an object of mass m kg, giving it an acceleration $\mathbf{a}$, then $\mathbf{F} = m\mathbf{a}$
- You can use the equation $F = ma$ in any direction, where F is the overall force in that direction and a is the acceleration in that direction.
- Friction always opposes motion. If a particle is moving to the right, friction will act to the left.
- Deceleration of $a\,\text{m s}^{-2}$ in one direction is acceleration of $-a\,\text{m s}^{-2}$ in the opposite direction.
- An object of mass m kg has weight mg N. g is approximately $9.81\,\text{m s}^{-2}$ on the Earth's surface. It decreases as the object moves further from the Earth's surface.
- Newton's third law states that for every action there is an equal and opposite reaction. So when an object A exerts a force on an object B, object B exerts an equal and opposite force on object A.
- If two objects are connected, the internal forces between them can be ignored when the two objects are considered as a whole. E.g. a man standing in a lift or a van towing a trailer.
- A number of assumptions are often made in questions involving forces and Newton's laws:
  - Objects are particles. There is no turning effect and mass acts at one point.
  - Strings are light and inextensible.
  - The acceleration is constant through the string.
  - The tension is constant through the string.
  - The tension in the string is the same on both sides of a pulley.
  - Pulleys and smooth surfaces are perfectly smooth. There is no resistance force acting.

Check and review

You should now be able to...	Try Questions
✔ Resolve in two perpendicular directions for a particle in equilibrium.	3
✔ Calculate the magnitude and direction of the resultant force acting on a particle.	1, 2
✔ Resolve for a particle moving with constant acceleration. Work out acceleration or forces.	4
✔ Understand the connection between the mass and the weight of an object. Know that weight changes depending on where the object is.	5
✔ Resolve for "connected" objects, such as an object in a lift.	6
✔ Resolve for particles moving with constant acceleration connected by string over pulleys.	7

1 Express, as a vector, the resultant force acting on this box.

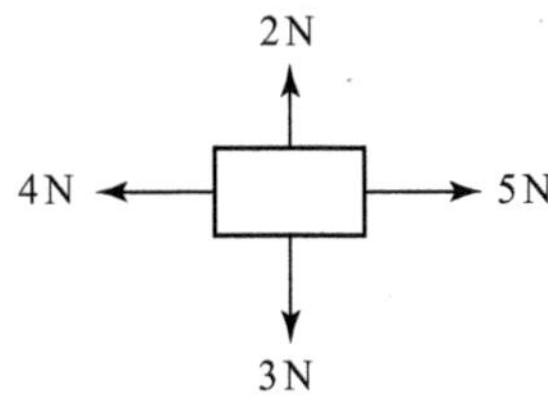

2 The forces $(13\mathbf{i}-5\mathbf{j})$N, $(-5\mathbf{i}+7\mathbf{j})$N and $(12\mathbf{i}+\mathbf{j})$N act on an object.
Work out the magnitude and direction of the resultant force.

3 A bird is held in equilibrium so that it is not moving in any direction. Its mass is 3 kg and it is flying with a forwards driving force parallel to the ground of 20 N. If the only external force exerted on the bird is caused by the wind, find the vector that describes the force exerted on the bird by the wind.

4 A car of mass 1250 kg is travelling at 40 m s^{-1} on a horizontal road. There is no resistance force. Work out the braking force needed to bring the car to rest in 100 m.

5 A man lifts up a bucket of mass 200 g. Inside the bucket is a brick of mass 500 g. The acceleration of the bucket and the brick is 0.3 m s^{-2}. Find the force exerted by the man on the bucket and the force exerted by the bucket on the brick if

a The man is on the surface of the Earth. You may assume that g is 10 m s^{-2}

b The man is on the surface of the Moon. You may assume that g is 1.6 m s^{-2}.

6 A woman of mass 40 kg is in a lift of mass 450 kg, which is moving upwards but decelerating at 0.9 m s^{-2}. Work out the tension in the cable that holds up the lift and the magnitude of reaction force between the woman and the lift.

You may assume that $g = 9.8$ m s^{-2}

7 Two masses, A and B, of 5 kg and 10 kg respectively are attached to the ends of a light, inextensible string. Mass A lies on a rough, horizontal platform. The string passes over a small smooth pulley fixed on the edge of the platform. The pulley is initially 3 m from A. The particle B hangs freely below the pulley. The particles are released from rest with the string taut. There is a constant resistive force of 20 N on particle A as it is moving.

You may assume $g = 9.8$ m s^{-2}

a Work out the tension in the string and the magnitude of the acceleration of the masses.

After 0.5 seconds, a 9 kg section of the 10 kg mass falls off.

b Work out how long it takes for the 10 kg mass to come to rest after the 9 kg section falls off.

c Work out the closest distance between A and the pulley.

d Calculate the magnitude of the force on the pulley from the strings after the 9 kg section falls off.

What next?

Score		
0 – 3	Your knowledge of this topic is still developing. To improve, search in MyMaths for the codes: 2185–2188, 2293	
4 – 5	You're gaining a secure knowledge of this topic. To improve, look at the InvisiPen videos for Fluency and skills (08A)	
6 – 7	You've mastered these skills. Well done, you're ready to progress! To develop your techniques, look at the InvisiPen videos for Reasoning and problem-solving (08B)	

Click these links in the digital book

History

Galileo Galilei was born in Italy in 1564 and throughout his lifetime he made contributions to astronomy, physics, engineering, philosophy and mathematics.

Before Galileo, the description of **forces** and their influence, given by **Aristotle**, had remained unchallenged for almost two thousand years. Galileo exposed the errors in Aristotle's work through experiment and logic.

Galileo's work in mechanics paved the way for **Newton** to define his **three laws of motion** in 1687

Galileo Galilei

"All truths are easy to understand, once discovered; the point is to discover them."
- Galileo

Investigation

"A ship is sailing on a calm sea when a cannon ball is dropped from the crow's nest.

Where does the cannon ball land?"

* Describe the path of the cannon ball as seen by someone on the ship.
* Describe the path of the cannon ball as seen by someone on the shore, as the ship passes by.
* Why is the path of the cannonball different in each case?

Find out how Galileo used the dropping of a cannon ball to demonstrate a flaw in Aristotle's understanding of the way forces behave.

Research

The work of Galileo and Newton stood the test of time until the 20th century and is now referred to as **classical mechanics**.

In the late 1800s and early 1900s, **Einstein** and other scientists showed that the laws of classical mechanics don't always hold. They needed a more general theory, and this gave rise to a new field of study called **quantum mechanics**.

In 1905, Einstein introduced the concept of the speed of light, and stated that nothing can travel faster this speed. What is the value of the speed of light, and do we still believe his theory to be true?

8 Assessment

1 The forces $(4\mathbf{i}+6\mathbf{j})$ N, $(12\mathbf{i}-9\mathbf{j})$ N and $(-9\mathbf{i}+\mathbf{j})$ N act on an object. Work out the magnitude of the resultant force. Choose the correct answer.

A 7.28 N B 9 N C 29.2 N D 6.71 N **[1 mark]**

2 Calculate the acceleration of an object of mass 15 kg if the forces acting on it are $(55\mathbf{i}+13\mathbf{j})$ N and $(-17\mathbf{i}+22\mathbf{j})$ N.

A $(0.395\mathbf{i}+0.429\mathbf{j})\,\text{m s}^{-2}$ B $4.87\,\text{m s}^{-2}$ C $(2.53\mathbf{i}+2.33\mathbf{j})\,\text{m s}^{-2}$ D $3.44\,\text{m s}^{-2}$ **[1]**

3 A car of mass 750 kg moves along a level straight road at a constant velocity of $20\,\text{m s}^{-1}$. The engine produces a driving force of 3000 N.

a Write the magnitude of the resisting force. **[1]**

The car increases the driving force to 6000 N. Assuming that the resisting force remains constant,

b Find the acceleration of the car, **[2]**

c Calculate the distance travelled by the car as it increases its speed from $20\,\text{m s}^{-1}$ to $30\,\text{m s}^{-1}$. **[2]**

4 Two particles, P and Q, of mass 20 kg and 30 kg respectively, are connected by a light inextensible string, passing over a fixed smooth light pulley. The particles are released from rest with the string taut, and the hanging parts vertical. You may assume that $g = 9.8\,\text{m s}^{-2}$ Find

a The acceleration of P, **[6]**

b The tension in the string. **[1]**

5 A small block of mass 5 kg is released from rest at the surface of a lake of still water. The water offers a constant resisting force of 29 N. You may assume that $g = 9.8\,\text{m s}^{-2}$

a Calculate the acceleration of the block. **[3]**

After 8 seconds the block hits the bottom of the lake.

b How fast is the block moving when it hits the bottom of the lake? **[2]**

c How deep is the lake at that point? **[2]**

6 A car of mass 1200 kg tows a caravan of mass 800 kg along a horizontal road. The car and the caravan experience resistances of 500 N and 300 N respectively. The constant horizontal force driving the car forwards is 1500 N.

Set up equations of motion for the car and the caravan and solve to find

a The acceleration of the car and the caravan, **[6]**

b The tension in the tow bar connecting the car and the caravan. **[1]**

7 The upwards motion of a lift between two floors is in three stages. Firstly, the lift accelerates from rest at $2\,\text{m s}^{-2}$ until it reaches a velocity of $6\,\text{m s}^{-1}$. It maintains this velocity for 5 seconds, after which it slows to rest with a deceleration of $3\,\text{m s}^{-2}$. You may assume that $g = 9.8\,\text{m s}^{-2}$

a Draw a velocity-time graph for the motion of the lift between the two floors. **[3]**

b Calculate the reaction force between a man of mass 100 kg and the floor of the lift during each of the three stages of the motion. **[5]**

8

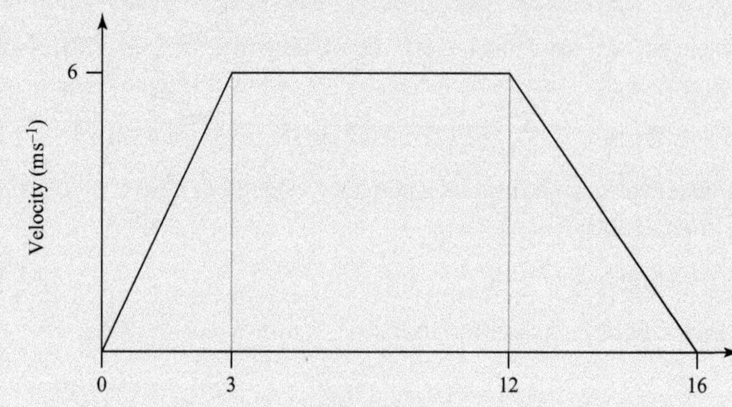

The diagram shows the velocity-time graph for the motion of a lift moving up between two floors in a tall building. A parcel of mass 40 kg rests of the floor on the lift. You may assume that $g = 9.8\,\text{m s}^{-2}$

Calculate the vertical force exerted by the floor of the lift on the parcel between

a $t = 0$ and $t = 3$ **[3]**

b $t = 3$ and $t = 12$ **[2]**

c $t = 12$ and $t = 15$ **[3]**

9 A lorry of mass 1900 kg tows a trailer of mass 800 kg along a straight horizontal road. The lorry and the trailer are connected by a light horizontal tow bar. The lorry and the trailer experience resistances to motion of 700 N and 400 N respectively. The constant horizontal driving force on the lorry is 2900 N.

a Set up equations of motion for the lorry and the trailer. **[4]**

b Use your equations to work out

i The acceleration of the lorry and the trailer,

ii The tension in the tow bar. **[3]**

When the speed of the vehicles is $12\,\text{m s}^{-1}$ the tow bar breaks. The resistance to the motion of the trailer remains 400 N.

c Find the distance moved by the trailer from the moment the tow bar breaks to the moment the trailer comes to rest. **[4]**

10 Two boxes, A and B, of masses 0.2 kg and 0.3 kg respectively are connected by a light, inextensible string that passes over a smooth pulley, P. Initially A is at rest on a rough horizontal platform, a distance 4 m from the pulley, and B hangs freely. The system is released from rest. A experiences a constant frictional resisting force of $0.15g$. In this question give your answers in terms of g

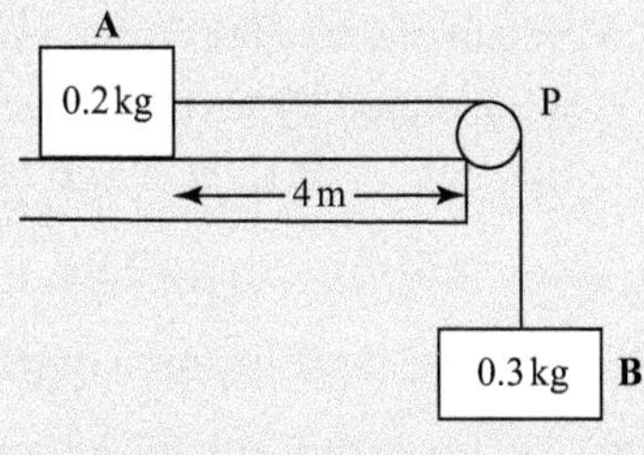

a Calculate the acceleration of A. **[5]**

When A is 1 metre from the pulley, the string breaks.

b Calculate the velocity of A at this instant. **[2]**

c Calculate the deceleration of A after the string has broken. **[2]**

d Show that A is moving at a speed of $\sqrt{\frac{3g}{10}}$ m s^{-1} when it hits the pulley. **[2]**

11 Two particles, A and B, of masses 2 kg and 3 kg respectively are attached to the ends of a light inextensible string. The string passes over a smooth fixed pulley. The system is released from rest with both masses a height of 72 cm above a horizontal table. You may assume that $g = 9.8$ m s^{-2}. Calculate

a The speed with which B hits the table, **[8]**

b How long it takes for B to hit the table. **[2]**

When B has hit the table, particle A continues upwards without hitting the pulley.

c Calculate the greatest height above the table reached by A. **[3]**

12

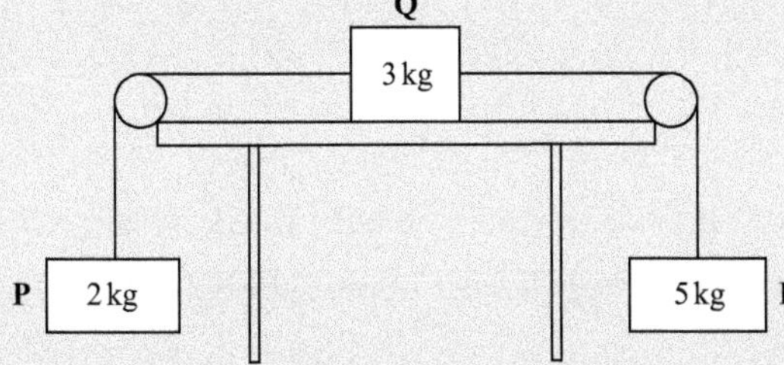

The diagram shows three bodies, P, Q and R, connected by two light inextensible strings, passing over smooth pulleys. Q lies on a smooth horizontal table, and P and R hang freely. The system is released from rest. You may assume that $g = 9.8$ m s^{-2}

a Calculate the acceleration of Q. **[8]**

b Calculate the tension in the string joining P to Q. **[2]**

c Calculate the tension in the string joining Q to R. **[2]**

13 A lift of mass 820 kg transports a woman of mass 80 kg. The lift is accelerating upwards at 4 m s^{-2} You may assume that $g = 9.8$ m s^{-2}

a Calculate the tension in the lift cable. **[2]**

b Calculate the vertical force exerted on the woman by the floor of the lift. **[2]**

Some time later the tension in the lift cable is 8640 N.

c Calculate the acceleration of the lift. **[2]**

d Calculate the vertical force exerted on the woman by the floor of the lift. **[2]**

14 A tug of mass 8000 kg is pulling a barge of mass 6000 kg along a canal. The tug and the barge are connected by an inextensible horizontal tow rope. The tug and the barge experience resistances to motion of 1200 N and 600 N respectively. The tug is accelerating at $0.2\,\text{m}\,\text{s}^{-2}$. Find

a The force in the tow rope, **[2]**

b The tractive force of the tug. **[2]**

The tow rope can operate safely up to a maximum force of 2100 N.

c Calculate the maximum safe tractive force for the tug. **[4]**

15 A car of mass 1000 kg tows a trailer of mass 400 kg along a horizontal road. The engine of the car exerts a forward force of 4.9 kN. The car and the trailer experience resisting forces that are each proportional to their masses. Given that the car accelerates at $3\,\text{m}\,\text{s}^{-2}$ find

a The tension in the tow bar, **[7]**

b The resisting force on the trailer. **[1]**

16

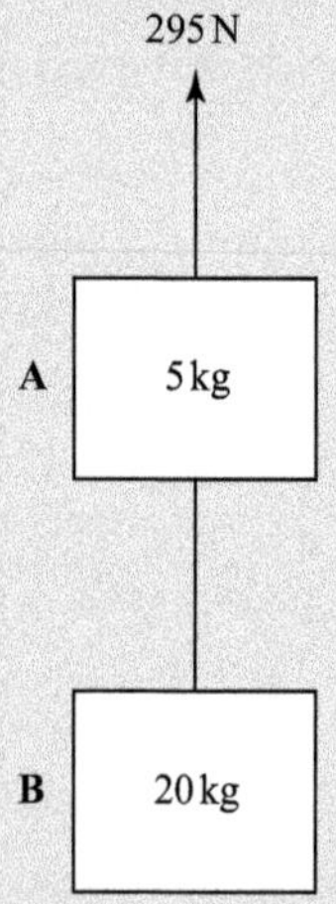

295 N

A 5 kg

B 20 kg

A body, B, of mass 20 kg, hangs below a mass, A, of 5 kg, connected by a light inextensible string. The system is lifted by a vertical force of 295 N, applied to A. You may assume that $g = 9.8\,\text{m}\,\text{s}^{-2}$

a Calculate the acceleration of A. **[6]**

b Calculate the tension in the string between A and B. **[2]**

17 A locomotive of mass 10 tonnes pushes a carriage of mass 5 tonnes along straight, horizontal rails.

The locomotive and the carriage are joined by a horizontal coupling. The locomotive and the carriage experience resisting forces of 3 kN and 2 kN respectively. They accelerate at $0.3\,\text{m}\,\text{s}^{-2}$. Find

a The force in the coupling, **[2]**

b The force of the engine on the locomotive. **[3]**

When the locomotive and the carriage are travelling at $20\,\text{m}\,\text{s}^{-1}$, the locomotive turns off its engine.

c Calculate the new force in the coupling. **[8]**

d Calculate the time until the locomotive and the carriage come to rest. **[2]**

Assessment

Chapters 6-8: Mechanics

1 The diagram shows the triangle ABC where $\overrightarrow{AB} = \mathbf{p}$ and $\overrightarrow{AC} = \mathbf{q}$
The point D is the midpoint of BC

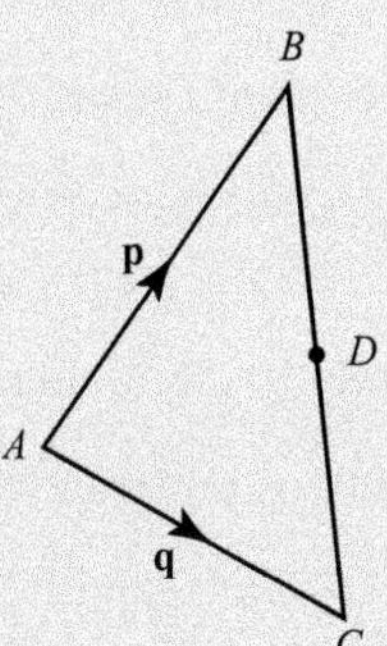

a Express the vector $\overrightarrow{BD}$ in terms of $\mathbf{p}$ and $\mathbf{q}$

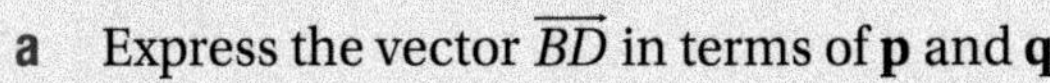

Select the correct answer.

A $\frac{1}{2}(\mathbf{p}+\mathbf{q})$ **B** $\frac{1}{2}(\mathbf{p}-\mathbf{q})$

C $\frac{1}{2}(-\mathbf{p}-\mathbf{q})$ **D** $\frac{1}{2}(\mathbf{q}-\mathbf{p})$ **[1 mark]**

b Calculate the magnitude of $\overrightarrow{BD}$ given that $\mathbf{p} = \begin{pmatrix} 2 \\ 3 \end{pmatrix}$ and $\mathbf{q} = \begin{pmatrix} 3 \\ -2 \end{pmatrix}$

A 2.45 **B** 2.55 **C** 1.73 **D** 1.41 **[1]**

2 $\mathbf{a} = 2\mathbf{i} - 3\mathbf{j}$ and $\mathbf{b} = 8\mathbf{i} + \mathbf{j}$ where $\mathbf{i}$ and $\mathbf{j}$ are unit vectors in a due east and due north direction respectively.

a Find the vector $\mathbf{c} = \mathbf{a} - 3\mathbf{b}$ **[2]**

b Calculate the magnitude and direction of vector $\mathbf{c}$ **[4]**

c Describe the geometric relationship between the vector $\mathbf{c}$ and the vector $11\mathbf{i} + 3\mathbf{j}$

Explain how you determined this relationship. **[2]**

3 A boat is sailing with a velocity of $\begin{pmatrix} -1 \\ 4 \end{pmatrix}$ m s^{-1}

a Calculate the speed of the boat to 1 decimal place. **[2]**

b Find the direction the boat is sailing as a bearing. **[2]**

4 A train travels at a constant speed of 50 km h^{-1} for 5 minutes before decelerating at a constant rate for one minute until it comes to a stop. After waiting for 2 minutes the train goes in the opposite direction, accelerating at a constant rate for 1 minute to reach a speed of 30 km h^{-1}

a Sketch a velocity-time graph for the journey described, using km h^{-1}. **[3]**

b Find the total distance travelled in km. **[2]**

c Calculate the resultant displacement of the train, in km, after 9 minutes. **[2]**

5 An object starts from rest and accelerates at a constant rate for 3 seconds until it is moving at 12 m s^{-1}

a Calculate

i The acceleration of the particle, **ii** The distance taken to reach 12 m s^{-1} **[3]**

The acceleration of the object is caused by a force of 5 N

b Calculate the mass of the object. State the unit of your answer. **[3]**

6 The displacement, s, in metres of a particle from its start point at time t seconds is given by $s=2t^3-t^2+\frac{t}{3}$

Calculate

a The velocity after 3 seconds, **[3]** **b** The acceleration after 3 seconds. **[3]**

7 Calculate the acceleration of an object of mass 4 kg when acted on by forces $(5\mathbf{i}-\mathbf{j})$ N and $(3\mathbf{i}+4\mathbf{j})$ N, where **i** and **j** are perpendicular unit vectors.

Give your answer in the form $a\mathbf{i}+b\mathbf{j}$ **[3]**

8 A book falls from a shelf and takes 0.6 seconds to hit the ground.

a Using $g=9.8\text{ m s}^{-2}$, calculate

i The height of the shelf, **ii** The speed at which the book hits the ground. **[4]**

b What assumptions did you make in your answers to part **a**? **[2]**

9 A basket of mass 5 kg is pulled vertically upwards by a rope.

a Calculate the tension in the rope, in terms of g, when

i The basket is stationary, **ii** The basket is accelerating at 2 m s^{-2} upwards. **[3]**

b What assumptions have you made about the rope? **[2]**

10 $\overrightarrow{OA}=\begin{pmatrix}3\\-1\end{pmatrix}$ and $\overrightarrow{OB}=\begin{pmatrix}-2\\5\end{pmatrix}$. Calculate the area of triangle OAB to 1 decimal place. **[6]**

11 A ball is kicked from the ground at an angle of 30° to the horizontal at a speed of 15 m s^{-1}

Write down the initial velocity of the ball as a vector. **[3]**

12 What vector describes the translation of the curve $y=\sqrt{x}$ onto the curve $y=\sqrt{x+3}$? **[1]**

13 A jet ski travels 200 m in a straight line on a bearing of 200°, then 600 m in a straight line on a bearing of 060°

Calculate the distance the jet ski must travel, to the nearest m, and the bearing on which it needs to travel to return directly to its start point. **[6]**

14 A toy train moves back and forwards along a straight section of track.

The displacement-time graph shows the motion of the toy train.

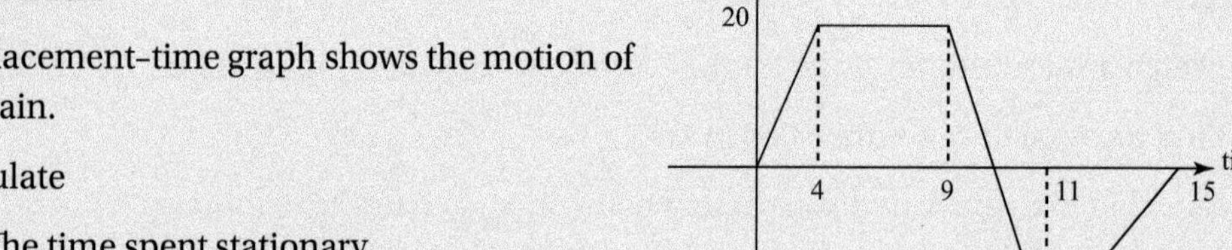

a Calculate

i The time spent stationary,

ii The velocity during the first 3 seconds,

iii The total distance travelled. **[4]**

b Find the average speed of the train during the final 6 seconds. State the units of your answer. **[3]**

15 The displacement, s m of a particle at time t s is given by the formula $s = t^3 - 3t + 2$

a Calculate the displacement of the particle after 5 seconds. **[2]**

b Find an expression for the velocity of the particle after t seconds. **[2]**

c Calculate the acceleration of the particle after 5 seconds. **[3]**

d Sketch a velocity–time graph for this situation. **[3]**

16 A motorbike starts from rest and moves along a straight horizontal road.

It accelerates at 3 m s^{-2} until it reaches a speed of 24 m s^{-1}, then it decelerates at 2 m s^{-2} until it comes to a stop.

a Find the length of the journey. **[4]**

b Calculate the average speed over the whole journey. **[4]**

17 A family of total mass 180 kg are in a lift of mass 500 kg. The lift is accelerating upwards at 2.4 m s^{-2}. Use 9.81 m s^{-2} as an approximation of g.

a Find the tension in the cable. **[3]**

A child in the family has a mass of 22 kg

b Calculate the reaction force between the child and the lift. **[3]**

c Without further calculation, state how the reaction force will vary for the heavier members of the family. **[1]**

18 A car of mass 1000 kg is travelling along a horizontal road. The total resistance to motion is 400 N and the driving force is 1600 N

a Calculate the acceleration of the car and state the units of your answer. **[3]**

b Work out the time taken for the car to accelerate from 10 km h^{-1} to 30 km h^{-1}

Give your answer to the nearest tenth of a second. **[3]**

19 A box of mass 0.5 kg starts from rest and is pulled along a horizontal table by a string. The frictional force is 2 N and the tension in the string is 3.5 N

a Calculate the distance travelled by the box in the first 2 seconds. **[4]**

After 4 seconds the string breaks.

b Calculate the length of time until the box comes to rest. **[3]**

20 A train consists of a carriage of mass 3000 kg pulled by an engine of mass 4000 kg with a driving force of 12 000 N. The train moves along a straight, horizontal track. The carriage and the engine are joined by a horizontal light bar and their total resistances to motion are 600 N and 900 N respectively.

a Calculate

i The acceleration of the train, **ii** The tension in the bar. **[4]**

When the speed of the train is 13 m s^{-1} the carriage becomes disconnected from the engine.

b Calculate the distance travelled by the carriage until it comes to rest. **[3]**

21 This graph shows the acceleration of a particle from rest.

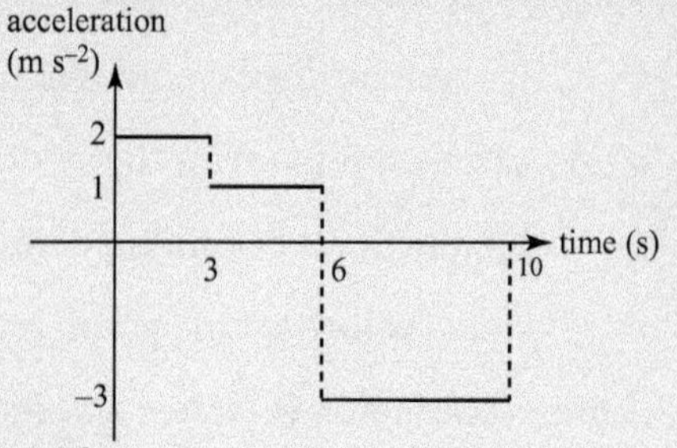

a Sketch a velocity–time graph. **[4]**

b Work out the displacement of the particle after 10 seconds. **[3]**

c Calculate the average speed over the 10 seconds. **[4]**

acceleration ($m\,s^{-2}$)
2
1
3 6 10 time (s)
−3

22 The acceleration (in $m\,s^{-2}$) of a particle at time t s is given by the formula $a = 2t - 4$
Find the distance travelled by the particle in the first 7 seconds, given that it is initially travelling at $5\,m\,s^{-1}$ **[5]**

23 A diver of mass 65 kg jumps vertically upwards with a speed of $3.5\,m\,s^{-1}$ from a board 3 m above a swimming pool. By modelling the diver as a particle moving vertically only (but not hitting the board on the way down), and using $g = 9.8\,m\,s^{-1}$, calculate

a The time taken for the diver to reach the water, **[4]**

b The speed of the diver when they reach the water. **[2]**

Give your answers to an appropriate degree of accuracy.

24 A particle has acceleration $(4t - 5)\,m\,s^{-2}$ and is initially travelling at $3\,m\,s^{-1}$

a Find the times at which the particle changes direction. **[5]**

b Calculate the displacement of the particle after 2 seconds. **[4]**

c Calculate the distance travelled in the first 2 seconds. **[4]**

25 Two remote-control cars are on a smooth, horizontal surface.

One of the cars passes the other at point A whilst travelling at a constant speed of $3\,m\,s^{-1}$

Two seconds after the first car passes, the second car accelerates from rest at a rate of $1.8\,m\,s^{-2}$ until it catches up with the first car at point B

a Calculate the time the second car has been moving when they meet at point B **[5]**

b Calculate the distance from A to B **[2]**

26 A ball is dropped from a height of 2 m and at the same time a second ball is projected vertically from the ground at a speed of $6\,m\,s^{-1}$

When and where will the two balls collide? Give your answers to 2 decimal places. **[6]**

27 Two boxes of mass 3 kg and 2.5 kg are connected by a light inextensible string that passes over a smooth pulley. The boxes both hang 2 m above the ground before the system is released from rest. Assuming $g = 9.81\,m\,s^{-2}$,

a Calculate

i The initial acceleration of the boxes, **ii** The tension in the string. **[5]**

b Assuming it does not reach the pulley, work out the greatest height reached by the 2.5 kg box. **[4]**

Consider instead that the 2.5 kg box lies on a smooth horizontal surface and that the 3 kg box hangs vertically below the pulley. The system is once again released from rest and the 2.5 kg box moves horizontally towards the pulley.

c Without further calculation, explain how the initial acceleration of the system will differ from that in the original scenario. **[3]**

9 Collecting, representing and interpreting data

If an ice-cream seller wants to investigate whether temperatures affect his ice-cream sales, he can record the mean temperature and his income each day. The data can then be analysed to determine whether a correlation exists. This is an example of bivariate data — temperature and ice-cream sales are two variables which may or may not affect each other. It is important to recognise, however, that correlation does not always imply causation. This means that two sets of data can show correlation without one affecting the other.

Data collection and analysis is the foundation of many different kinds of research. Being able to collect relevant data accurately and without bias, effectively represent it, and then interpret the results in a meaningful way is very important when undertaking investigations and testing hypotheses.

Orientation

What you need to know

KS4

- Apply statistics to describe a population.
- Construct and interpret tables, charts and diagrams for numerical data.
- Recognise appropriate measures of central tendency and spread.
- Use and interpret scatter graphs of bivariate data.
- Recognise correlation.

What you will learn

- To distinguish a population and its parameters from a sample and its statistics.
- To identify and name sampling methods and highlight sources of bias.
- Read discrete and continuous data from a variety of diagrams.
- Plot and use scatter diagrams.
- Summarise raw data.

What this leads to

Ch10 Probability and discrete random variables
Binomial distribution.

Ch11 Hypothesis testing 1
Formulating a test.
The critical region.

MyMaths Practise before you start

1192, 1194, 1195, 1213, 1248

9 Bridging Unit

Topic A: Averages

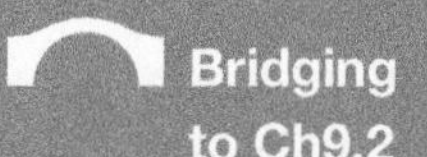
Bridging to Ch9.2

You can summarise data using **summary statistics**, which are key values relating to the data. The three averages (that is, the **mean, median** and **mode**) are three examples of summary statistics, and you should make sure you know how to define and calculate them.

Key point

The **mean** is the sum of the values divided by the number of values.

The **median** is the middle value when the data is listed in order of size. For n values, listed in order of size,

- If $\frac{n+1}{2}$ is an integer, then this is the position of the median.
- If $\frac{n+1}{2}$ is a not an integer, then find the mean of the two values either side of the $\frac{n+1}{2}$ position.

The **mode** is the most commonly occurring value or class.

The modal class, (or group or category) refers to when data is grouped – see Example 3. If values or classes are tied for most common, you get multiple modes.

Example 1

Calculate the mean, median and mode of this data.

5 12 17 4 14 19 12 12 24 14

$$\text{Mean} = \frac{133}{10} = 13.3$$

Since the sum of the values is 133 and there are 10 values.

To find the median, write the numbers in size order first:

4 5 12 12 12 14 14 17 19 24

There are 10 numbers, and $\frac{n+1}{2} = \frac{10+1}{2} = 5.5$ is not a whole number, so take the mean of the 5th and 6th values.

The 5th value is 12 and the 6th value is 14

Therefore the median is $\frac{12+14}{2} = 13$

Mode = 12

Since 12 occurs three times which is more than any other value.

Try It 1

Calculate the mean, median and mode of this data.

5, 6, 2, 8, 4, 6, 9, 5, 6

Data can also be presented in a frequency table. In these cases, you need to multiply each value by its frequency when working out the mean and median. The mode is just the value(s) with the highest frequency.

Example 2

Calculate the mean, median and mode of this data.

Number	4	5	6	7	8
Frequency	24	25	17	4	1

Your calculator may be able to find the mean, mode and median of data – make sure you know how to use it.

$$\text{Mean} = \frac{(4\times24)+(5\times25)+(6\times17)+(7\times4)+(8\times1)}{24+25+17+4+1}$$

The sum of the values is the sum of each number multiplied by its frequency.

$$= \frac{359}{71} = 5.06$$

Divide the sum by the total frequency.

The total frequency is 71

$\frac{71+1}{2} = 36$ so the median is the 36th value.

Median = 5

The data is already listed in order. The 36th value would lie in the 2nd column.

Mode = 5

The frequency of 5 is highest.

Try It 2

Calculate the mean, median and mode of this data.

x	6	7	8	9
f	2	6	12	11

If data is grouped, use the midpoint of each group as an approximation of all the values in the group.

Example 3

Estimate the mean of this data and state the modal class.

Number, x	$0 \le x < 10$	$10 \le x < 15$	$15 \le x < 20$	$20 \le x < 25$	$25 \le x < 30$
Frequency	2	8	16	20	15

Since you don't know the exact values, the mean is only an estimate.

The modal class is $20 \le x < 25$

Since it has the highest frequency (20).

To estimate the mean, use the midpoints of each group and multiply by the frequency:

$(5\times2)+(12.5\times8)+(17.5\times16)+(22.5\times20)+(27.5\times15)$
$= 1252.5$

To find the midpoint, add the bounds of the interval and divide by 2, e.g. $\frac{25+30}{2} = 27.5$

$$\text{Then mean} \approx \frac{1252.5}{2+8+16+20+15} = \frac{1252.5}{61} = 20.5$$

Divide by the total frequency.

Try It 3

Estimate the mean of this data and state the modal class.

x	$0 \le x < 2$	$2 \le x < 5$	$5 \le x < 10$	$10 \le x < 20$
Frequency	14	16	11	8

To estimate the median of grouped data you can use a method called **linear interpolation**. You assume that the values in each group are evenly spread out and work out how far through the group the median will lie.

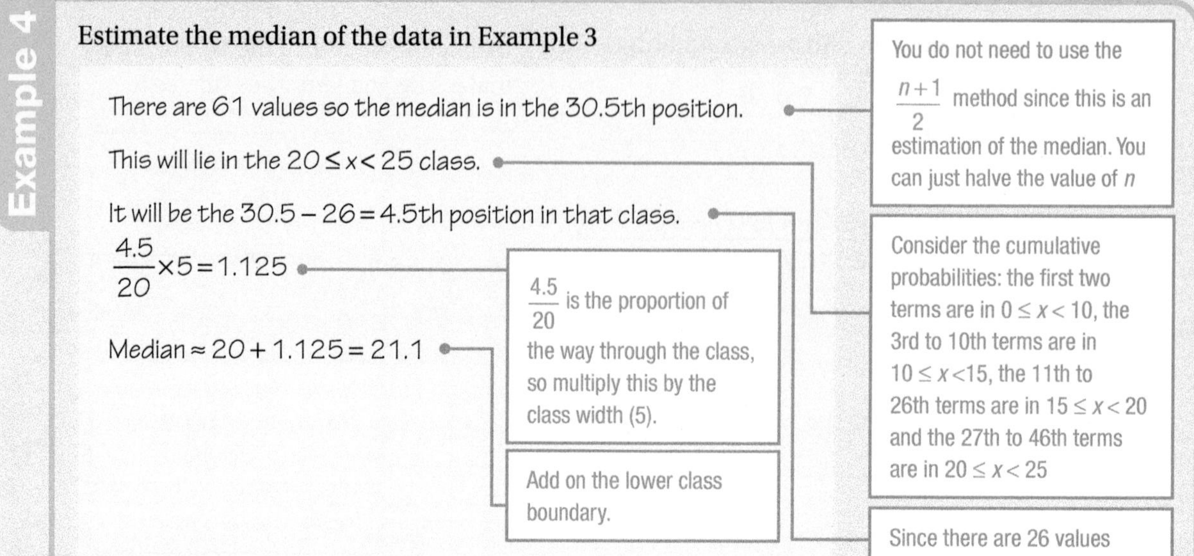

Example 4
Estimate the median of the data in Example 3
There are 61 values so the median is in the 30.5th position.
This will lie in the $20 \le x < 25$ class.
It will be the $30.5 - 26 = 4.5$th position in that class.
$\frac{4.5}{20} \times 5 = 1.125$
Median $\approx 20 + 1.125 = 21.1$
You do not need to use the $\frac{n+1}{2}$ method since this is an estimation of the median. You can just halve the value of n
Consider the cumulative probabilities: the first two terms are in $0 \le x < 10$, the 3rd to 10th terms are in $10 \le x < 15$, the 11th to 26th terms are in $15 \le x < 20$ and the 27th to 46th terms are in $20 \le x < 25$
$\frac{4.5}{20}$ is the proportion of the way through the class, so multiply this by the class width (5).
Add on the lower class boundary.
Since there are 26 values smaller than this class.

Try It 4

Estimate the median of the data in **Try It 3**

Data is sometimes **coded** to make the numbers easier to work with.

For example, the distance from the first 4 planets in our solar system to the Sun is given in the table in millions of kilometres. So in this case, the true values are actually a million times larger.

Planet	Distance to Sun (millions of km)
Mercury	58
Venus	108
Earth	150
Mars	228

Mean $= \frac{544}{8} = 68$

So the mean distance is $68 \times 1\,000\,000 = 68\,000\,000$ km

Example 5

Raw data has been coded by subtracting 20 from each value, giving the values

0.5, 0.8, −0.2, 0.9, −0.5, −0.4, −0.8, 0.4, −0.1, 0.5

Calculate the mean of the raw data.

The mean of the coded data is $\frac{1.1}{10} = 0.11$ — Use the sum of the values divided by the number of values.

So the mean of the raw data is $20 + 0.11 = 20.11$ — Add the 20 back on to find the mean of the uncoded data.

Try It 5

Raw data has been coded by dividing it by 10, giving the values:

7, 5, 2, 8, 5, 1, 7

Calculate the mean of the raw data.

Bridging Exercise Topic A

Bridging to Ch9.2

1 Calculate the mean, median and mode of each set of data.

a 3, 7, 1, 8, 1, 4

b 3, 9, 13, 7, 6, 5, 6

c 1.3, 0.3, 1.6, 1.3, 2.4, 0.5, 1.3, 0.1

d 12, 19, 23, 19, 24, 18, 23, 19, 23

e

Number	2	3	4	5	6
Frequency	12	13	10	7	2

f

Number	2	4	6	8
Frequency	5	8	12	7

g

Number	5	10	15	20	25
Frequency	1	5	7	0	2

h

Number	0.5	1	1.5	2
Frequency	8	6	6	5

2 Estimate the mean and median and state the modal class of each set of data.

a

x	$0 \le x < 10$	$10 \le x < 20$	$20 \le x < 30$	$30 \le x < 40$	$40 \le x < 50$
f	4	17	12	6	1

b

x	$0 \le x < 2$	$2 \le x < 4$	$4 \le x < 6$	$6 \le x < 8$	$8 \le x < 10$
f	7	5	6	12	8

c

x	$10 \le x < 15$	$15 \le x < 20$	$20 \le x < 25$	$25 \le x < 30$	$30 \le x < 35$
f	3	8	11	9	4

d

x	$1 \le x < 5$	$5 \le x < 10$	$10 \le x < 15$	$15 \le x < 25$	$25 \le x < 40$
f	16	22	13	8	3

3 In a class of 30 students, the mean score of the 12 boys in a test is 75% and the mean test score of the girls is 80%. Calculate the mean score for the whole class. **Hint:** calculate the total score for the boys and the total for the girls and add to give the total for the whole class.

4 In another class of 25 students, the mean score for the whole class is 70% and the mean score for the 10 boys is 72%. What is the mean score for the girls?

5 Raw data is coded by multiplying each of the values by 10. If the coded data has a mean of 14, find the mean of the original data.

6 Raw data is coded by subtracting 5 from each of the values. If the coded data has a mean of 0.4, find the mean of the original data.

7 Raw data is coded by dividing each of the values by 2. If the coded data has a mean of 32, find the mean of the original data.

8 Raw data is coded by dividing each of the values by 10^9. If the coded data has a mean of 1.8, find the mean of the original data in standard form.

Topic B: Measures of spread

Bridging to Ch9.2

As well as the three averages covered in Topic A, summary statistics also include measures of **spread** of the data. These include the **range, interquartile range** and **standard deviation.**

The interquartile range gives you an idea of how spread out the data points are, and whether they concentrate around the middle value. The standard deviation gives you an idea of how much the data points differ from the mean.

Key point

The **range** is the smallest value subtracted from the largest value.

The **interquartile range (IQR)** is the lower quartile subtracted from the upper quartile.

The upper quartile, Q_3, is the value that $\frac{3}{4}$ of the data is less than.

The lower quartile, Q_1, is the value that $\frac{1}{4}$ of the data is less than.

You can calculate the quartiles using similar methods as for the median. Find the value of $\frac{n+1}{4}$ for Q_1 or $\frac{3(n+1)}{4}$ for Q_3. If it is an integer then this is the position of the quartile, if it is not an integer then use the mean of the values either side of that position.

If the data were grouped, you would need to use linear interpolation to find the quartiles - like in Example 4 of Topic A.

Example 1

Calculate the range and interquartile range of this data: 5, 7, 10, 3, 5, 9, 3, 9, 6, 1, 4, 3

Range $= 10 - 1 = 9$

Subtract the smallest value from the largest.

Data in order: 1, 3, 3, 3, 4, 5, 5, 6, 7, 9, 9, 10

To find the quartiles, first put the data in order.

$\frac{n+1}{4} = \frac{12+1}{4} = 3.25$ so Q_1 is the mean of the 3rd and 4th values:

$Q_1 = \frac{3+3}{2} = 3$

$\frac{3(n+1)}{4} = \frac{3(12+1)}{4} = 9.75$ so Q_3 is the mean of the 9th and 10th values:

$Q_3 = \frac{7+9}{2} = 8$

IQR $= 8 - 3 = 5$

Subtract the lower quartile from the upper quartile.

Try It 1

Calculate the range and interquartile range of this data:
2, 7, 3, 9, 1, 1, 4, 8, 9

Key point

The **standard deviation** is a measure of the average distance between each data point and the mean.

You can calculate the standard deviation by subtracting the mean from each value then squaring the result to give a positive number.

Take the numbers 2, 3, 6 and 9. Their mean is $\frac{2+3+6+9}{4}=5$

Subtract the mean from each value and square:

$(2-5)^2=(-3)^2=9$
$(3-5)^2=(-2)^2=4$
$(6-5)^2=1^2=1$
$(9-5)^2=4^2=16$

Then find the mean of these values: $\frac{9+4+1+16}{4}=7.5$

Finally, square-root this to give the standard deviation.
Standard deviation $=\sqrt{7.5}=2.74$ (to 3 significant figures)

Example 2

Calculate the standard deviation of this data. 3, 7, 2, 4, 9, 6, 11

Mean $=\frac{42}{7}=6$ — First find the mean.

$(3-6)^2=9$ $(7-6)^2=1$
$(2-6)^2=16$ $(4-6)^2=4$
$(9-6)^2=9$ $(6-6)^2=0$ $(11-6)^2=25$

Subtract the mean from each data point then square the result.

standard deviation $=\sqrt{\frac{64}{7}}=3.02$ (3 significant figures) — Calculate the mean of these values then square-root to find the standard deviation.

Try It 2

Calculate the standard deviation of this data. 3, 6, 12, 2, 1, 10, 9, 13

If you are given the sum of the values Σx, or the sum of the squares Σx^2, you can save time by using formulae to calculate the mean and standard deviation.

Key point

Mean $=\frac{\sum x}{n}$

Standard deviation $=\sqrt{\frac{\sum x^2}{n}-\left(\frac{\sum x}{n}\right)^2}$

where $\sum x$ means the sum of all the data points and $\sum x^2$ means the sum of the squares of all the data points.

Σx and Σx^2 can also be referred to as summary statistics.

2087

Example 3

A sample of 8 people complete a task and their times, x, are recorded. Given that $\sum x = 46$ and $\sum x^2 = 354$, calculate the mean and standard deviation of the times taken.

$$\text{Mean} = \frac{46}{8} = 5.75$$

Use mean $= \frac{\sum x}{n}$

$$\text{Standard deviation} = \sqrt{\frac{354}{8} - \left(\frac{46}{8}\right)^2}$$
$$= 3.34 \text{ (to 3 significant figures)}$$

Use standard deviation $= \sqrt{\frac{\sum x^2}{n} - \left(\frac{\sum x}{n}\right)^2}$

Try It 3

A sample of 12 people complete a task and their times, x, are recorded. Given that $\sum x = 78$ and $\sum x^2 = 650$, calculate the mean and standard deviation of the times taken.

If you are given the raw data you can calculate the sums of x and x^2 yourself.

Example 4

Calculate the standard deviation of this data.

x	4	5	6	7	8
Frequency	24	25	17	4	1

$$\sum x = (4\times24)+(5\times25)+(6\times17)+(7\times4)+(8\times1)$$
$$= 359$$

Multiply each value by its frequency.

$$\sum x^2 = (4^2\times24)+(5^2\times25)+(6^2\times17)+(7^2\times4)+(8^2\times1)$$
$$= 1881$$

Square each value first then multiply by its frequency.

$$n = 24+25+17+4+1$$
$$= 71$$

Add together the frequencies to find n

$$\text{Standard deviation} = \sqrt{\frac{1881}{71} - \left(\frac{359}{71}\right)^2}$$
$$= 0.962 \text{ (to 3 significant figures)}$$

Use standard deviation $= \sqrt{\frac{\sum x^2}{n} - \left(\frac{\sum x}{n}\right)^2}$

Your calculator might be able to calculate the standard deviation using either raw data or the summary statistics – make sure you know how to do that.

Try It 4

Calculate the standard deviation of this data.

x	6	7	8	9
f	2	6	12	11

Bridging Exercise Topic B

Bridging to Ch9.1

1 Calculate the range and interquartile range of each set of data.

a 3, 7, 1, 8, 1, 4

b 3, 9, 13, 7, 6, 5, 6

c 1.3, 0.3, 1.6, 1.3, 2.4, 0.5, 1.3, 0.1

d 12, 19, 23, 19, 24, 18, 23, 19, 23

e

Number	2	3	4	5	6
Frequency	12	13	10	7	2

f

Number	2	4	6	8
Frequency	5	8	12	7

g

Number	5	10	15	20	25
Frequency	1	5	7	0	2

h

Number	0.5	1	1.5	2
Frequency	8	6	6	3

2 Calculate the standard deviation of each set of data.

a 4, 8, 2, 5, 1

b 16, 22, 35, 18, 4, 15, 3, 7

c 12, 17, 3, 27, 33, 64, 19

d 2, 5, 3, 7, 3, 8, 2, 1, 3, 8

3 A sample of 10 people complete a task and their times, x, are recorded. Given that $\sum x = 273$ and $\sum x^2 = 7561$, calculate the mean and standard deviation of the times taken.

4 A sample of 9 people complete a task and their times, x, are recorded. Given that $\sum x = 25$ and $\sum x^2 = 163$, calculate the mean and standard deviation of the times taken.

5 A sample of 16 people complete a task and their times, x, are recorded. Given that $\sum x = 155$ and $\sum x^2 = 2301$, calculate the mean and standard deviation of the times taken.

6 A sample of 20 people complete a task and their times, x, are recorded. Given that $\sum x = 66.2$ and $\sum x^2 = 388.1$, calculate the mean and standard deviation of the times taken.

7 For each set of data, calculate $\sum x$, $\sum x^2$ and n, then use these values to find the mean and standard deviation.

a

Number	1	2	3	4	5
Frequency	14	11	9	2	1

b

Number	5	10	15	20
Frequency	6	4	8	2

2087 SEARCH

Topic C: Histograms

You can use a **histogram** to display continuous data. In a histogram, the **area** of the bar represents the frequency. The height of the bar is called the **frequency density**.

Key point

The **frequency density** for a class is given by the frequency divided by the range of the values in that class (called the class width). That is, Frequency density = $\frac{\text{Frequency}}{\text{Class width}}$

Example 1

Draw a histogram to display this data for the heights of a group of children.

Height, h (cm)	Frequency
$60 \le h < 80$	8
$80 \le h < 90$	8
$90 \le h < 100$	7
$100 \le h < 150$	10

Height, h (cm)	Frequency	Class width	Frequency density
$60 \le h < 80$	8	20	$8 \div 20 = 0.4$
$80 \le h < 90$	8	10	$8 \div 10 = 0.8$
$90 \le h < 100$	7	10	$7 \div 10 = 0.7$
$100 \le h < 150$	10	50	$10 \div 50 = 0.2$

Calculate the frequency density using the formula

Frequency density = $\frac{\text{Frequency}}{\text{Class width}}$

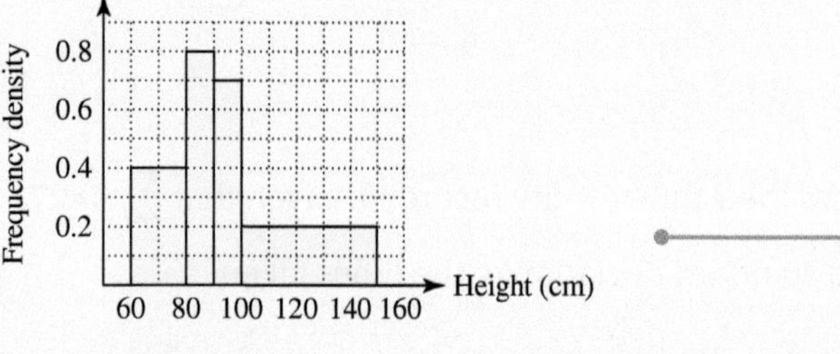

Note that the vertical axis is labelled 'frequency density' **not** 'frequency' and the horizontal axis has a continuous scale. There should be no gaps between the bars.

Try It 1

Draw a histogram to display this data about the heights of a group of children.

Height, h (cm)	$60 \le h < 100$	$100 \le h < 110$	$110 \le h < 130$
Frequency	16	9	4

In Example 1, the area of each bar is *equal* to the frequency of its class. However, often the area is only *proportional* to the frequency. You need to find the number k such that Area $= k \times$ frequency.

Example 2

The table gives the masses of some books. The bar representing the $200 \le m < 250$ class has width 5 and height 9

a Calculate the dimensions of the $100 \le m < 200$ bar.

b Estimate the number of books with a mass > 268 g.

Mass, m (g)	Frequency
$100 \le m < 200$	8
$200 \le m < 250$	15
$250 \le m < 300$	10
$300 \le m < 400$	6

(*Continued on the next page*)

a The $200 \le m < 250$ class has class width 50 and bar width 5

The $100 \le m < 200$ class has class width 100, so the width of the bar will be 10

First find the width of the bar.

The $200 \le m < 250$ class has frequency 15 and area $= 5 \times 9 = 45$

So area $= 3 \times$ frequency for this histogram

Find the relationship between area and frequency for this histogram using the information given.

Therefore the $100 \le m < 200$ bar has area $= 3 \times 8 = 24$

So the height of the bar is $24 \div 10 = 2.4$

Use height = area ÷ width.

b All 6 of the books in the $300 \le m < 400$ class have a mass greater than 268 g

Estimate of number of books in $250 \le m < 300$ class with mass greater than 268 g is $\frac{32}{50} \times 10 = 6.4$

Since the class width is 50 and you want values of m between 268 and 300. Multiply by 10 as that's the frequency of this class

Total estimate $= 6 + 6.4 = 12.4$ which is 12 books to the nearest book.

Try It 2

The table gives the masses of some books. A histogram is created from this data. The bar for the $250 \le m < 300$ class has width 5 and height 2

Mass, m (g)	Frequency
$250 \le m < 300$	20
$300 \le m < 340$	25
$340 \le m < 360$	16
$360 \le m < 500$	12

a Calculate the dimensions of the $340 \le m < 360$ bar.

b Estimate the number of books with a mass less than 308 g.

Bridging Exercise Topic C

Bridging to Ch9.3

1 **a** Draw a histogram to display this data.

x	$0 \le x < 20$	$20 \le x < 30$	$30 \le x < 40$	$40 \le x < 80$
f	6	12	10	8

b Estimate the probability that x is less than 22

2 The table gives the time it took 30 people to wash their cars. The bar representing people taking between 30 and 40 minutes has width 2 and height 3

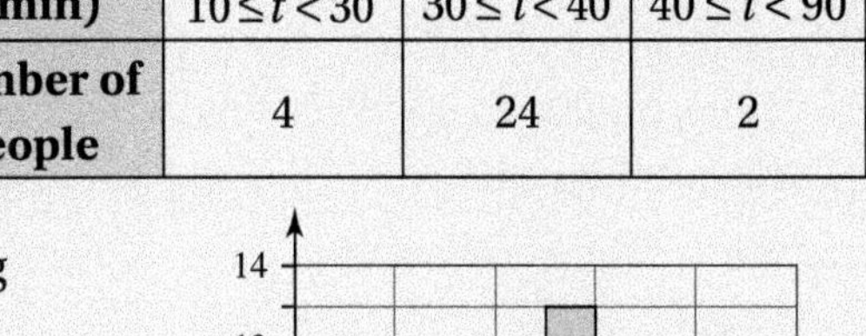

t (min)	$10 \le t < 30$	$30 \le t < 40$	$40 \le t < 90$
Number of people	4	24	2

a Calculate the dimensions of the bar representing people taking between 10 and 30 minutes.

b Estimate how many people took over an hour.

3 The histogram shows the distances thrown in a shot-put competition.

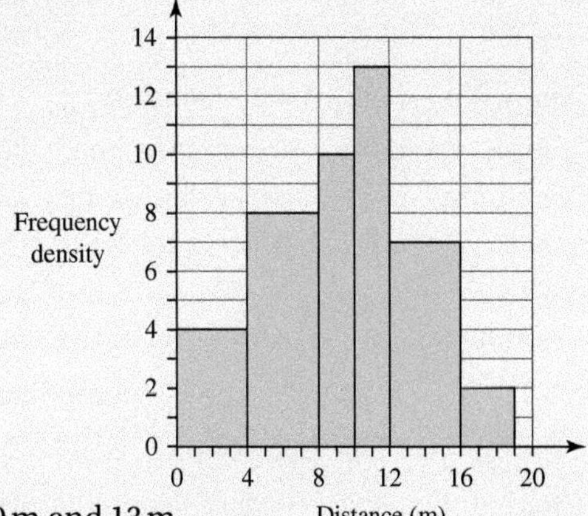

There were 64 throws recorded in the competition.

a Calculate the number of shots that were

i Less than 4 m, **ii** More than 12 m.

b Estimate the number of shots that were between 10 m and 13 m.

9.1 Sampling

Fluency and skills

Sampling from a population can provide extremely useful information, if done effectively.

Key point

The **population** is the set of things you are interested in.
A **sample** is a subset of the population.

The population may be finite, like the current top-selling pop artists, or infinite, such as the range of locations at which an archer's arrow might land. Parameters for a population can be almost impossible to accurately measure, so it is best to take a statistic from a sample and use the result to say something about the population.

Key point

A **parameter** is a number that describes the entire population. A **statistic** is a number taken from a single sample—you can use one or more of these to estimate the parameter.

You can use a statistic to estimate a parameter. For example, the mean of a sample is an estimate of the population mean.

Example 1

A student wants to know the mean number of sweets in a packet of their favourite snack.

They open 10 packets and count the number of sweets in each. They find the mean of these totals.

Identify the population, the parameter, the sample(s) and the statistic(s) in this example, and say which statistic can be used to estimate the parameter.

The population is all packets of the snack. The mean number of sweets per packet is the parameter. The 10 packets of sweets opened by the student is the sample. The statistics are the number of sweets in each of these 10 packets and the sample mean. The sample mean is an estimator for the parameter.

When deciding on the best way to produce a sample, it is useful to know if you would, at least in principle, be able to list every single member of the population.

The table shows some typical methods of sampling that you can use if you *are* able to list every member of the population.

Sampling method	Description
Simple Random Sampling	Every member of the population is equally likely to be chosen. For example, allocate each member of the population a number. Then use random numbers to choose a sample of the desired size.
Systematic Sampling	Find a sample of size n from a population of size N by taking one member from the first k members of the population at random, and then selecting every k^{th} member after that, where $k=\frac{N}{n}$
Stratified Sampling	When you know you want distinct groups to be represented in your sample, split the population into these distinct groups and then sample within each group in proportion to its size.

Often, you're not able to list every member of a population. In this case, you have to generate a sample to represent the population in the best way you can.

Sampling method	Description
Opportunity sampling	Take samples from members of the population you have access to until you have a sample of the desired size.
Quota sampling	When you know you want distinct groups to be represented in your sample, decide how many members of each group you wish to sample in advance and use opportunity sampling until you have a large enough sample for each group.
Cluster sampling	Split the population into clusters that you expect to be similar to each other, then take a sample from each of these clusters.

STATS

Exercise 9.1A Fluency and skills

1 A child wants to know the average height of an adult in their family. They measure the heights of all their adult relatives who live in their city. Identify the population, the parameter, the sample(s) and the statistic(s) in this situation.

2 A conservationist is interested to know the maximum height that various species of trees reach in their country's forests. They choose a random forest and measure the heights of all the trees there. Identify the population, the parameter, the sample(s) and the statistic(s) in this situation.

3 You wish to find out the favourite bands of students in your school. Describe how to take a sample of size 40

  a Using simple random sampling,

  b Using cluster sampling.

4 Systematic sampling is used to find a sample from a population of 3783 people. A random number between 1 and 13 is generated, then that numbered member of the population and every 13th member thereafter is chosen for the sample. How large is the sample?

5 A school wishes to know how popular its new after-school club is with the 1000 students in the school. The school has three year-groups of 150 students each and two year-groups of 275 students each. It comes up with three methods of sampling a group of 40 students to get an idea.

State the name of each sampling method.

  a 40 students are chosen at random from the list of 1000 students using random numbers between 1 and 1000

  b The school lists the students by year and by class and gives each student a number. The school randomly chooses one student from the first 25 listed and then every 25th student from the list, until they've picked 40 students.

  c The school lists the students by year and by class and gives each student a number. Six students are chosen at random from each of the three smaller year groups and 11 students are chosen at random from each of the two larger year groups.

6 A manufacturer of bean bags produces 2772 bags, amongst which are 1001 red bags, 1309 blue bags and 462 green bags. 36 bags are chosen to check for defects. Calculate how many of each colour bag are chosen if a stratified sample is taken.

Reasoning and problem-solving

Strategy

When deciding on a sampling method

1. Consider whether or not you can list every member of a population.
2. Identify any sources of bias and any difficulties you might face in taking certain samples.
3. Compare the different sampling methods you have available and choose the one that best suits your needs and limitations.

Large data set

Know your dataset

The various makes of car in the LDS have their own characteristics. For example, the average engine size of a BMW car is 2160 cm^3 whilst for a Vauxhall car it is 1450 cm^3.

Click this link in the digital book for more information about the Large data set.

Taking a sample that accurately reflects the population is not a simple job. It is all too easy to bias your sample or get results that may not accurately reflect the population.

Key point

A sampling method is **biased** if it creates a sample that does not represent the population.

When deciding on a sampling method, you should aim to produce as unbiased a sample as possible, but you may need to factor in the difficulty and cost of any sampling method chosen.

Example 2

A BMW car dealer believes that given a free choice customers would choose a car with a larger engine size over the same model with a smaller engine. To test out his idea he asks the next 40 people that come into his showroom whether they would choose a larger engine model. 24 people say yes.

a Give a reason why any conclusions drawn by the car dealer may not be valid.

b Suggest an alternative sampling method that he could use to get a better representation of the whole population's views. Explain why it is a better method.

a *On average the engines of BMW cars are larger than the engines in other cars. You can therefore expect that people interested in BMW cars prefer cars with above average engine sizes. This would bias the sample towards people who prefer larger engines.*

(2) Use your knowledge of the LDS to help identify possible sources of bias.

b *He could use quota sampling, in which he also asks dealers for other makes of car to carry out surveys. The dealers should sample pre-determined numbers of customers and then combine the results. In this way he can match the proportions of people who are interested in particular makes of cars with the known numbers of registered cars of that make.*

(3) You need to choose a sampling method that suits the situation.

Exercise 9.1B Reasoning and problem-solving

1 A meteorologist wants to know the average temperature in their town. The values of temperature, taken hourly over a period of three months, give an average result of 14 °C, but the values taken hourly over a period of three years give a result of 21 °C.

a How reliable are the two methods and what might you conclude about the actual average temperature?

The meteorologist takes monthly average values and orders them by time, from first to last. They then use systematic sampling to generate a random number between 1 and 12 for the first monthly value used for a sample, and every 12th monthly value is used after that.

b Discuss why this method would not give a fair representation of the average temperature.

2 A newspaper is writing a report on the average cost of housing in the UK. It sends a reporter to estate agencies near to the newspaper's offices in central London to collect details. It finds that the average cost to buy a house from the agencies asked is £1.7 million. What might the newspaper conclude from this sample, and how reliable is that conclusion?

3 A student owns 50 books. 20 are in a pile in the living room and 30 in their bedroom. They want to take a sample to see how many pages the average book in their collection has, so they take two books at random from the living room and two from their bedroom. Why is this not a simple random sample? Why might it not matter?

4 A teacher is organising a conference in their city but is not sure where best to hold it so that everyone can attend. There are 293 students spread across 12 schools. The teacher wants to obtain a sample of 36 students to take into account their views. The teacher comes up with two ideas:

Idea 1. The teacher could list all 293 students and number them from 000 to 292. The teacher randomly generates 36 different 3-digit numbers in this range and the corresponding students are included in the sample.

Idea 2. Two schools are selected at random and 18 students from each school are selected at random. If there aren't enough students then the teacher randomly chooses another school and draws random students until enough are chosen.

a For each idea, decide if every possible sample is equally likely to be chosen.

b Name the sampling method used in each idea.

c State, with reason, which idea is best from a statistical point of view.

d Suggest another suitable method the teacher could use based on stratified sampling.

Challenge

Large data set

5 a In the large data set 38.7% of cars are registered to men, 22.9% are registered to women and 35.3% are registered to companies. Explain why the total is not 100%.

b If you wanted a stratified sample of 1000 car owners, how many owners should be men, women and companies?

Large data set

6 A researcher wishes to investigate how tolerant British car owners are of high levels of car pollution.

a He considers carrying out a census. Explain what a census means in this situation and give one advantage and one disadvantage of using a census.

b He also considers using stratified sampling.

i Explain what stratified sampling means.

ii Suggest two ways in which the researcher could stratify his sample.

c A colleague warns the researcher that he needs to avoid bias in his sample. Explain what is meant by bias.

d The researcher decides to quota sample car owners living in London, grouping people according to the engine sizes of their cars. Suggest how this might give a biased sample and how the bias may affect the researcher's results.

9.2 Central tendency and spread

Fluency and skills

Discrete data can take any one of a finite set of categories (non-numeric) or values (numeric), but nothing in between those values. Often, the values are different categories. **Continuous data** is always numeric, and it can take any value between two points on a number line.

Statistical investigations generate large quantities of **raw data**. It is useful to reduce this data to some key values, called **summary statistics**. These can be categorised as **measures of central tendency** (also known as 'averages') and **measures of dispersion** (also known as 'spread'). There are three measures of central tendency: **mode**, **median** and **mean**.

Key point

The mode of a set of data is the value or category that occurs most often or has the largest frequency. For grouped data, the **modal interval** or **modal group** is normally given.

See p.382

For a list of mathematical notation.

Key point

To work out the **mean** $\bar{x}$ of a set of n observations, calculate their sum and divide the result by n $\bar{x}=\frac{\sum x}{n}$

The mean of a set of data given in the form of a frequency distribution is given by $\bar{x}=\frac{\sum fx}{\sum f}$

The symbol 'Σ' means 'the sum of' whatever follows it. The bar on top of the x indicates the mean of the x-values.

For a grouped frequency distribution, you can only calculate an estimate for the mean rather than the exact value. In this case, x is the middle value of each group.

f denotes the frequency, so Σf is the number of observations and Σfx is the sum of the x-values.

Key point

The **median** of a set of data is the middle value of data listed in order of size.

To calculate the position of the median of a set of n observations, work out the value of $\frac{n+1}{2}$

If the value of $\frac{n+1}{2}$ is a whole number, then the median is the value in that position.

If the answer is not a whole number, then the median is the mean of the two values in the positions on either side of $\frac{n+1}{2}$

Calculator

Try it on your calculator

You can calculate the mode, mean ($\bar{x}$) and median for a set of data using your calculator.

SUB	List 1	List 2	List 3	List 4
1	0	3		
2	1	5		
3	2	6		
4	3	4		

4

1VAR 2VAR REG SET

1-Variable
$\bar{x}$ =1.61111111
Σx =29
Σx^2 =65
σx =1.0076865
sx =1.03690086
n =18

1-Variable
Q1 =1
Med =2
Q3 =2
maxX =3
Mod =2
Mod:n =1

Activity

Find out how to calculate the mean, median and mode of the data set

x	0	1	2	3
f	3	5	6	4

on *your* calculator.

Example 1

Work out the mode, median and mean of this data set.

x	0	1	2	3	4
f	3	5	6	4	1

Mode = 2

Median = 2

Mean = 1.74

The mode is the value of x with the highest frequency.

$\Sigma f = 19$ so the median is the $\frac{19+1}{2} = 10$th value.

Check the value given by your calculator is sensible.

Example 2

Write the modal interval, the interval containing the median and an estimate of the mean for this data set.

x	$0 \le x < 4$	$4 \le x < 8$	$8 \le x < 12$	$12 \le x < 16$	$16 \le x < 20$
f	2	3	7	10	1

Modal interval: $12 \le x < 16$

Median lies in $8 \le x < 12$

Midpoints: 2, 6, 10, 14, 18

Estimate for mean = 10.9

$\Sigma f = 23$ so the median is the $\frac{23+1}{2} = 12$th value.

Use the midpoint of each interval and your calculator to find an estimate for the mean.

STATS

In order to summarise data, you can also use measures of **spread** or **dispersion.**

Key point

You should be familiar with four measures of spread: range, interquartile range, variance and standard deviation.

Key point

The **range** of a set of data is the largest value minus the smallest value.

The median, i.e. the middle value of ordered data, is sometimes called the **second quartile**. The **first quartile** (or **lower quartile**) is the middle value between the lowest value and the median, and the **third quartile** (or **upper quartile**) is the middle value between the median and the largest value.

Quartiles divide the ordered data into four groups, with equal numbers of observations in each.

To evaluate the first and third quartiles for a set of n ungrouped observations, work out the values of $\frac{n+1}{4}$ and $\frac{3(n+1)}{4}$ respectively.

If the value of $\frac{n+1}{4}$ or $\frac{3(n+1)}{4}$ is a whole number, then the quartile is the value in that position.

If the answer is not a whole number, then the quartile is the mean of the two values in the positions on either side of $\frac{n+1}{4}$ or $\frac{3(n+1)}{4}$

Key point

If the first and third quartiles of a set of observations are denoted by Q_1 and Q_3, the interquartile range is $\text{IQR} = Q_3 - Q_1$

ICT Resource online

To investigate standard deviation and mean, click this link in the digital book.

Example 3

Find the interquartile range for this set of data: 1, 6, 9, 23, 4, 19, 2, 7, 13, 24, 2, 7, 14, 19, 14, 20

1, 2, 2, 4, 6, 7, 7, 9, 13, 14, 14, 19, 19, 20, 23, 24 — Write the data in ascending order.

$n = 16$, $\frac{n+1}{4} = 4.25$ so $Q_1 = \frac{4+6}{2} = 5$ — Calculate Q_1

$\frac{3(n+1)}{4} = 12.75$ so $Q_3 = \frac{19+19}{2} = 19$ — You can check the values of Q_1 and Q_3 on your calculator.

$IQR = Q_3 - Q_1 = 19 - 5 = 14$

The **variance** of a set of data measures how spread-out the values are from the mean. To find the variance of a population you calculate the difference between each data value (x) and the population mean (μ). You then find the mean of the squares of these values. This is the mean of $(x - \mu)^2$, and is called the variance, denoted σ^2. The square root of the variance, σ, is the **standard deviation.**

The values are squared to remove minuses.

Key point

The population variance of n observations with mean μ is defined as

$$\sigma^2 = \frac{\sum(x-\mu)^2}{n} = \frac{\sum x^2}{n} - \mu^2$$

The population standard deviation of n values with mean μ is defined as

$$\sigma = \sqrt{\frac{\sum(x-\mu)^2}{n}} = \sqrt{\frac{\sum x^2}{n} - \mu^2}$$

The prefix 'population' is often dropped in practice. When a population is large it may be more practical to use a sample from the population to estimate the variance of the whole population.

Key point

An **unbiased estimate of the population variance** using a sample of n observations with sample mean $\bar{x}$ is given by the **sample variance**, s^2. The divisor in s^2 is $n - 1$

$$s^2 = \frac{\sum(x-\bar{x})^2}{n-1} = \frac{\sum x^2}{n-1} - \frac{\left(\sum x\right)^2}{n(n-1)}$$

The sample standard deviation of a set of n values is defined as

$$s = \sqrt{\frac{\sum(x-\bar{x})^2}{n-1}} = \sqrt{\frac{\sum x^2}{n-1} - \frac{\left(\sum x\right)^2}{n(n-1)}}$$

Calculator

Try it on your calculator

You can calculate the standard deviation of a population, or estimate the standard deviation using a sample, on your calculator.

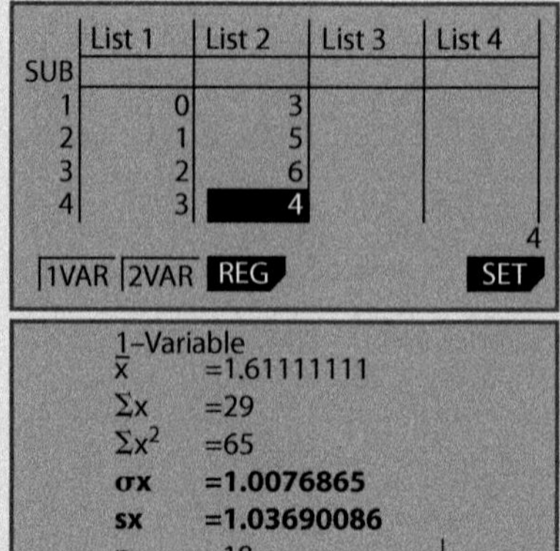

Activity

Find out how to calculate the population standard deviation of the data set

x	0	1	2	3
f	3	5	6	4

on *your* calculator.

Example 4

In each of the following situations

i Decide which standard deviation formula is appropriate,

ii Calculate the standard deviation.

a A researcher is investigating the variation in alcohol consumption of pensioners. He asks a sample of 50 pensioners how many alcoholic drinks they consumed on a given evening. The results are summarised $\sum(x-\bar{x})^2 = 68.4$

b A census is taken in a village to gather information about the spread of ages of its 1000 residents. The ages, x, are summarised $\sum x = 47000$, $\sum x^2 = 2\,900\,000$

c A researcher is looking at the variation in litter sizes for a particular pack of wolves. She records the size of every litter in the pack. The results are
6, 7, 3, 4, 4, 5, 6, 6, 8, 9, 4, 5, 6, 7, 8

d A researcher is interested in the masses of eggs for a particular species of bird. He takes a sample of 34 eggs and records their masses.

x	$7 \le x < 8$	$8 \le x < 9$	$9 \le x < 10$	$10 \le x < 11$	$11 \le x < 12$
f	3	10	11	7	3

a i Unbiased estimate of the population standard deviation. — This is a sample representing a bigger population.

ii $s = \sqrt{\dfrac{68.4}{49}} = 1.18$ — Use the formula.

b i Population standard deviation. — A census includes all the population data.

ii $\sigma = \sqrt{\dfrac{2900000}{1000} - \left(\dfrac{47000}{1000}\right)^2} = 26.3$

c i Population standard deviation.

ii $\sigma = 1.67$ — Find σ using your calculator.

d i Unbiased estimate of the population standard deviation.

ii Midpoints: 7.5, 8.5, 9.5, 10.5, 11.5 — The data is grouped so use midpoints and find s using your calculator.

$s = 1.11$

Exercise 9.2A Fluency and skills

1 For these sets of data, give

i The mode(s), **ii** The mean, **iii** The median, **iv** The range,
v The standard deviation.

a 6, 8, 9, 2, 5, 6, 10, 8, 5, 7, 4, 8, 11

b 68, 71, 72, 75, 68, 65, 69, 70, 71, 68, 62, 64, 71

2 For this data, calculate

a The modal interval,

b An estimate for the mean.

x	4–9	10–15	16–21	22–27	28–33
f	1	3	7	4	2

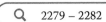

3 The graph shows the number of times a sample of households went out for a meal in September 2010

a Write the modal class of the sample.

b Estimate the mean of the sample.

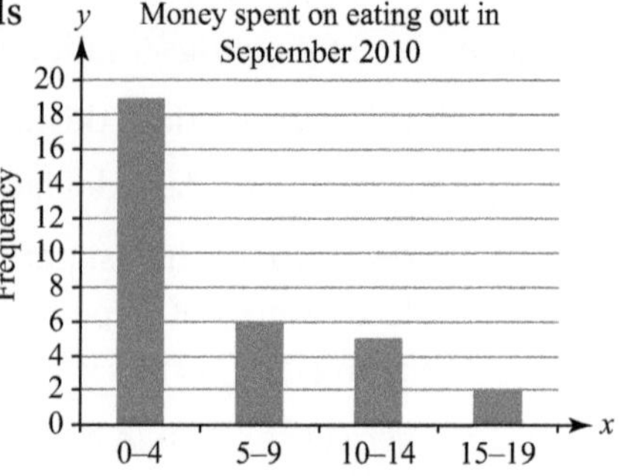

4 50 milk samples, each of mass 100 g, were analysed as part of a food quality inspection and the potassium content, in mg, was recorded. The results are shown in the table.

Potassium (x mg)	$140 \le x < 144$	$144 \le x < 148$	$148 \le x < 152$	$152 \le x < 156$	$156 \le x < 160$
Frequency	5	12	16	11	6

Estimate the mean of this data.

5 Work out the 1st and 3rd quartiles and the interquartile range for these observations.

a 12, 14, 15, 19, 21, 25, 26, 29, 32, 36

b 1.2, 1.7, 1.9, 2.3, 2.5, 2.9, 3.1, 3.6, 3.7, 3.9, 4.0, 4.1

6 Work out the mean and variance of this set of x-values: 7.1, 6.4, 8.5, 7.5, 7.8, 7.2, 6.9, 8.1, 8.0, 6.6

7 For these sets of data, work out the standard deviation.

a $\sum(x-\bar{x})^2 = 141.4$, $n = 10$

b $n = 9$, $\sum x = 295$, $\sum x^2 = 9779$

8 This sample was taken from a large population. Use it to work out

a An estimate for the population mean,

b An unbiased estimate for the population variance.

x	4	5	6	7	8
f	1	5	7	3	2

9 A researcher is interested in the masses of a species of bird, in grams. A sample of 25 birds are weighed, giving the summary statistics $\sum x = 20.4$, $\sum(x-\bar{x})^2 = 52.1$

a Estimate the population mean.

b i Is it appropriate to use s^2 or σ^2 in this situation?

ii Calculate the value you have chosen in part **b i**.

10 A researcher is interested in the heights of seedlings of a certain species of plant. The heights, in centimetres, of a sample of 40 seedlings are given in the table.

a Explain why it is appropriate to use s rather than σ in this situation.

b Use the data from this sample to find an unbiased estimation for the population standard deviation. Give your answer to 2 dp.

x	$16 \le x < 18$	$18 \le x < 20$	$20 \le x < 22$	$22 \le x < 24$	$24 \le x < 26$
f	8	13	11	6	2

Reasoning and problem-solving

Strategy

To solve a problem about summary statistics

1. Identify the summary statistics appropriate to the problem.
2. Calculate values of the required statistics, using a calculator where appropriate.
3. Use the statistics to describe key features of the data set and make comparisons.
4. If not already done, identify any outliers and remove them, then see how this affects the calculations.

Key point

Outliers are values that lie significantly outside the typical set of values of the variable.

An outlier can be defined in several ways. For example, any value that lies outside the interval $(Q_1 - 1.5 \times IQR, Q_3 + 1.5 \times IQR)$

This is not the only rule to identify outliers. You will be told which rule to apply if another is needed.

Outliers may indicate natural variation in the data set or may be the result of measurement and recording errors. If an outlier is due to an error it should be removed. This is one way in which you can clean data.

STATS

Example 5

Define an outlier as a value more than two standard deviations from the mean.
A group of 15 students complete a timed test for their homework. Their times (in minutes) are recorded: 32, 34, 33, 37, 39, 39, 42, 45, 41, 40, 40, 44, 13, 36, 36

a Calculate the mode, median and mean of the data.

b Show that there is exactly one outlier in the data.

c Give one possible reason for **i** Removing the outlier, **ii** *Not* removing the outlier.

d A teacher investigates the outlier and decides to remove it. Without further calculation explain how removing this value would affect your answers to part **a**.

a Modes = 36, 39 and 40, median = 39, mean = $\mu = 36.7$

b $\sigma = 7.33$ — (1) You need the standard deviation and the mean to identify outliers.

$\mu - 2\sigma = 36.7 - 2 \times 7.33 = 22.04$

$\mu + 2\sigma = 36.7 + 2 \times 7.33 = 51.36$ — (2) Calculate upper and lower bounds for outliers.

13 is the only value in the list that is less than 22.04; no value in the list is more than 51.36 — (3) Compare the listed values with the bounds.

c **i** The outlier could be an error, for instance it could have been recorded incorrectly; or a parent could have helped the student, meaning this is not a valid test result. In both cases including it distorts the results.

ii If this is a true value, removing it gives a false picture, underestimating the variation of results.

d The mode and median are not affected, the mean increases. — (4) $13 < \mu$ so the remaining values sum to a value greater than 14μ

2279 – 2282 SEARCH

You should choose appropriate statistics for central tendency and spread. You can determine which statistics to use by considering the properties of the data.

Statistic		Pros	Cons
Measure of central tendency	Mode	Useful for non-numerical data. Not usually affected by outliers. Not usually affected by errors or omissions. Is always an observed data point.	Doesn't use all the data. May not be representative if it has a low frequency. There may be other values with similar frequency.
	Median	Not affected by outliers. Not significantly affected by errors.	Doesn't make use of all the data.
	Mean	When the data set is very large a few extreme values have negligible impact.	When the data set is small a few extreme values or errors have a big impact.
Measure of spread	Range	Reflects the full data set.	Distorted by outliers.
	IQR	Not distorted by outliers.	Does not reflect all the data.
	Standard deviation	When the data set is very large a few outliers have negligible impact.	When the data set is small a few outliers have a big impact.

Example 6

An environmentalist in Bristol records the level of hydrocarbon emissions, in g km^{-1}, for 7 cars.
0.003, 0.008, 0.067, 0.014, 0.003, 0.017, 0.007

a Calculate an appropriate measure of spread. Explain your choice of measure.

b Explain why the mode is not an appropriate measure of central tendency and suggest a better measure.

a IQR = 0.014

Both the range and standard deviation are very dependent on the outlier value, 0.067, whilst the IQR is robust.

b The data set is too small with only two occurrences of 0.003 for the mode to be safely considered as representative of the data. The median, 0.008, is a better measure as it is not very sensitive to the presence of the outlier.

Consider the effects of the outlier. (4)

In a small data set, the mean is sensitive the presence of outliers. (1)

Exercise 9.2B Reasoning and problem-solving

1 The lifetimes of a batch of 100 batteries are measured and have the following distribution, where x is the lifetime measured to the nearest half-hour.

x	15.5	16.0	16.5	17.0	17.5	18.0	18.5	19.0	19.5
f	13	21	28	22	15	0	0	0	1

a Work out the median lifetime, the range and the interquartile range.

b Explain why the range is *not* an appropriate measure of spread.

c Outliers are defined for this data as being outside the interval $(Q_1 - 2 \times \text{IQR}, Q_3 + 2 \times \text{IQR})$. Calculate the limits of this interval.

d To test the quality of another batch of batteries, one is chosen at random and its lifetime is measured. If its lifetime is outside the interval found in part **c**, the whole batch is tested. Otherwise the batch is accepted. If the lifetime of the selected battery was 14.5 hours, what decision should be taken about a test of the whole batch?

2 Tom records the number of emails he receives per day over a period of 16 days:
13, 15, 19, 17, 14, 27, 19, 9, 10, 17, 18, 14, 20, 18, 15, 10

a Work out the mean, μ, and standard deviation, σ, for this data.

b An outlier is defined as any observation less than $\mu - 2\sigma$ or more than $\mu + 2\sigma$
Show that one of the observations is an outlier.

c Give one reason for
  i Removing the outlier,
  ii Not removing the outlier.

d The decision is made to remove the outlier. State the effect this has on μ and σ

3 The table shows the average mass of a car for each of the top five car makes in 2002.

Car make	BMW	Ford	Toyota	Vauxhall	Volkswagen
Mean car mass (kg)	1585	1215	1260	1650	1315

a Find the mean and variance for this data.

b The cars produced by the top five car manufacturers in 2002 are grouped together to make a set. Does the mean found in part **a** give the mean mass of the cars in this set? Give your reasons.

4 The table shows the carbon dioxide emissions for diesel-fuelled, Ford cars registered in north-west England in 2002 in the large data set.

CO_2 (g km^{-1})	173	159	156	145	143	121
Frequency	5	3	3	7	2	1

a Calculate the median and interquartile range for this data.

b The median and IQR for the Vauxhall cars registered in north-west England in 2002 in the large data set are 152.5 g km^{-1} and 45 g km^{-1} respectively.
Compare the CO_2 emissions for the two samples.

5 9 students were asked how much they spent, in £ per week, on travel. The results were as follows:
£21 £31 £19.84 £19.32 £22.54 £21.48 £20.73 £16.10 £21.40

a Find the interquartile range (IQR).

b Using the rule that outliers are values less than $Q_1 - 1.5 \times \text{IQR}$ or more than $Q_3 + 1.5 \times \text{IQR}$, where Q_1 is the lower quartile and Q_3 is the upper quartile, identify any outliers within the data.

c If the outlier(s) in part **b** were removed, what effect would that have on the median of the data?

Challenge

6 11 members of a golf team record their scores on an 18-hole course.
108, 110, 114, 101, 99, 98, 107, 103, 109, 145, 105

a Calculate the mode, median and mean of the scores.

b Explain which of the measures you calculated in part **a** is the most representative measure of central tendency in this case.

A competing team has the same mean score. The team has seven members and their scores are: 110, 112, 115, 108, 111, a, 105

c Find the value of a and compare the scores of the two teams. Justify your choice of statistics.

STATS

9.3 Single-variable data

Fluency and skills

Data is often summarised by five main statistics.

Key point

The **five-number summary** gives the **minimum value, lower quartile, median, upper quartile and maximum value.**

You can use a **box-and-whisker plot** to display these values. The values are marked along a linear scale and the points are joined to form a central box and two whiskers. One quarter of the data values in the sample lie between each consecutive pair of vertical lines on the diagram. Lines placed further apart, i.e. longer whiskers and box, show a greater spread of the data but do not show more data values.

You display outliers on a box plot as crosses (×), they are not included in whiskers.

If you have sufficient information you should use the most extreme value that is not an outlier as the end of the whisker. Otherwise, use the boundary for outliers as the end of the whisker.

Suppose you define an outlier as a value less than $Q_1 - 1.5 \times \text{IQR}$ or more than $Q_3 + 1.5 \times \text{IQR}$. This box plot represents the set of data 20, 22, 23, 25, 25, 26, 27, 27, 27, 28, 39

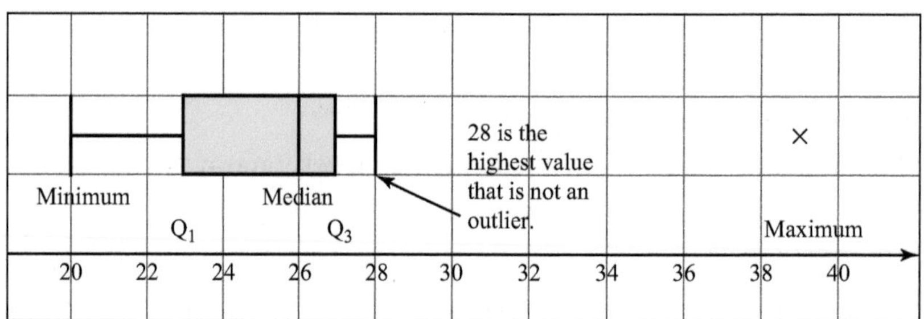

This box plot represents the set of data summarised by minimum = 43, $Q_1 = 46$, median = 49, $Q_3 = 50$, maximum = 65

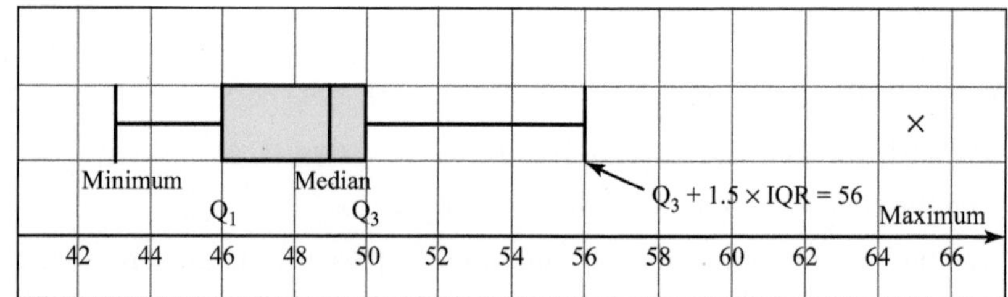

Note that, if you don't know the values of the quartiles or interquartile range, you should also be able to identify outliers by direct observation. In the diagrams above, it should be clear even without calculation that the maximum value is set apart from the rest.

Key point

Box-and-whisker plots are useful for comparing sets of data.

Example 1

A building company works with two plumbers. Over a period of time, they assess how long it takes each plumber to fix leaking pipes.
This data is displayed in box-and-whisker plots.

An outlier is defined as a value less than $Q_1 - 1.5 \times IQR$ or more than $Q_3 + 1.5 \times IQR$.

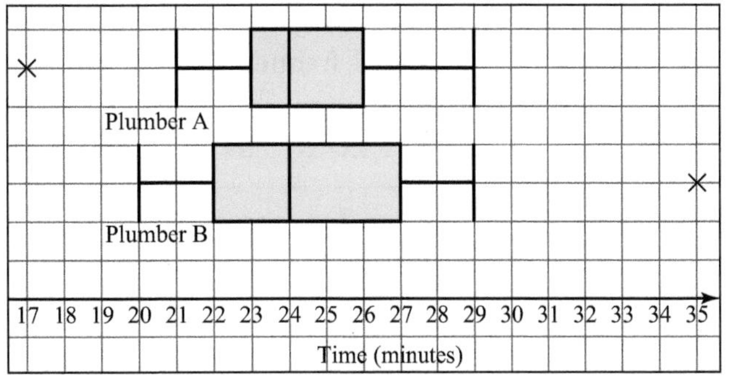

a Write down the minimum, lower quartile, median, upper quartile and maximum for each set of data.

b Recommend a choice of plumber given that no outliers are deleted.

a Plumber A's times: minimum = 17 minutes, Q_1 = 23 minutes, median = 24 minutes, Q_3 = 26 minutes, maximum = 29 minutes

Plumber B's times: minimum = 20 minutes, Q_1 = 22 minutes, median = 24 minutes, Q_3 = 27 minutes, maximum = 35 minutes

b It would be most sensible to choose Plumber A. Although both plumbers have a median time of 24 minutes. Plumber A's data shows less variation – it has a smaller IQR of 3 compared with 5 for Plumber B's data, plumber A's data has one outlier representing a quick time and Plumber B's data has one outlier representing a slow time.

Use a measure of central tendency and a measure of spread.

Another way to display continuous data is a **cumulative frequency diagram** (or graph, or curve). You can use a cumulative frequency diagram (or graph, or curve) to estimate values, including those in the five-number summary.

A cumulative frequency diagram consists of points whose x-coordinates are the upper boundary of each interval, and whose y-coordinates are the sums of the frequencies up to those points. Or, in other words, the y-coordinates are the cumulative frequencies.

You join the points by a smooth curve, which is always increasing from zero to the size of the sample. Drawing dotted line from the y-axis to the curve at the appropriate point allows you to estimate the value any required quartile.

For data given in intervals, you can use $\frac{n}{2}$, $\frac{n}{4}$ and $\frac{3n}{4}$ for the median, lower quartile and upper quartile. There is no need to use $n + 1$ in each case.

Know your dataset

The LDS contains many data values that are the same. This is because it contains many cars of the same make and model and these will all have, for example, the same engine size.

Click this link in the digital book for more information about the Large data set.

STATS

Example 2

The heights of a sample of a species of plant are recorded. Copy and complete the frequency table and use the cumulative frequency graph to fill in the missing values.

Height in cm	$0 \le x < 1$	$1 \le x < 3$	$3 \le x < 7$	$7 \le x < 10$	$10 \le x < 16$	$16 \le x < 24$	$24 \le x < 31$	$31 \le x$
Frequency f	1	2			9			0

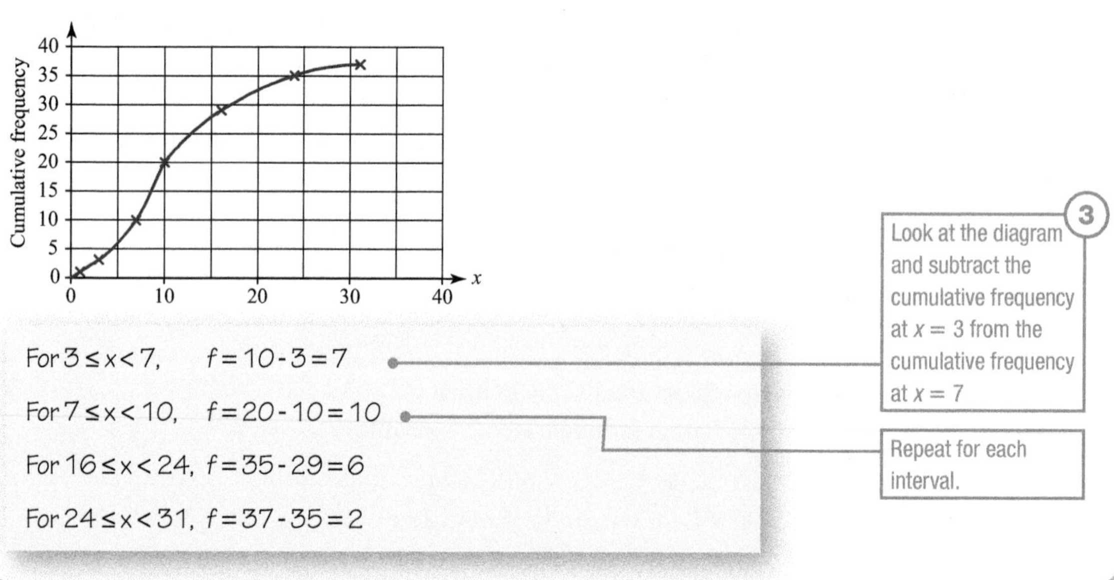

Cumulative frequency
40
35
30
25
20
15
10
5
0
0
10
20
30
40
x
For 3 ≤ x < 7, f = 10 - 3 = 7
For 7 ≤ x < 10, f = 20 - 10 = 10
For 16 ≤ x < 24, f = 35 - 29 = 6
For 24 ≤ x < 31, f = 37 - 35 = 2
3
Look at the diagram and subtract the cumulative frequency at x = 3 from the cumulative frequency at x = 7
Repeat for each interval.

Sometimes your interest is in the way in which the data is distributed rather than the individual values.

> **Key point**
>
> The **distribution** of data is how often each outcome occurs. Each outcome occurs with a given **frequency**.

When representing grouped data, the groups must be consecutive, non-overlapping ranges and do not have to be equal in width.

Distributions for continuous variables can be complicated because data values may be rounded measurements of true values. You may need to apply a **continuity correction** to ensure that the intervals meet but don't overlap.

> **Key point**
>
> A continuity correction involves altering the endpoints of an interval of rounded data to include values which would fall in the interval when rounded.

> **Key point**
>
> You can use a **histogram** to display continuous data.

A histogram consists of rectangles whose areas are proportional to the frequencies of the groups. The width of each rectangle is the size of the interval. A histogram often displays **frequency density** on its vertical axis.

$$\text{Frequency density} = \frac{\text{Frequency}}{\text{Class width}}$$

When working with grouped data you can make estimates by assuming that data points are distributed equally throughout a group.

ICT Resource online

To investigate single variable data using graphs, click this link in the digital book.

Example 3

24 students in a gym class balance on one leg for as long as they can, and their times are recorded to the nearest second.

a **i** Find the missing frequency.
  ii Give the width and height of the missing bar on the histogram.

b Estimate the number of students who could stand on one leg for

  i Less than 5.5 seconds,
  ii More than 10 seconds.

Time (seconds)	Continuity correction	Frequency
0-4	$0 \le t < 4.5$	
5-7	$4.5 \le t < 7.5$	3
8-11	$7.5 \le t < 11.5$	6
12-16	$11.5 \le t < 16.5$	6
≥ 17	$16.5 \le t$	0

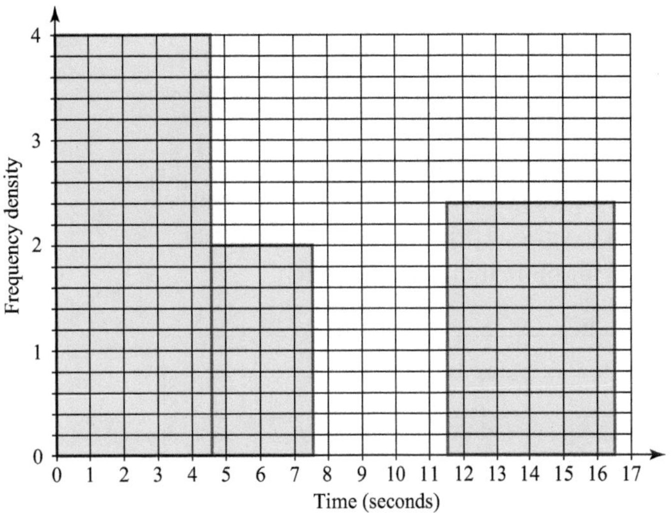

a **i** When area $= 2(7.5 - 4.5) = 6$, frequency $= 3$ — Find the constant of proportionality using a known value.

So area $= 4 \times 4.5 = 18$ means frequency $= 9$ — In this diagram area $= 2 \times$ frequency

ii Frequency $= 6$ so area $= 12$ — Use the constant of proportionality, 2

Width $= 11.5 - 7.5 = 4$

Height $= 12 \div 4 = 3$ — Use the known area and width of the missing bar to find the height.

b **i** $\frac{5.5-4.5}{7.5-4.5} = \frac{1}{3}$ — Assume the data in the 2nd bar is equally distributed throughout the interval. Find the fraction of the 2nd bar that lies to the left of 5.5

$\frac{1}{3} \times 3 = 1$ — Multiply by the frequency of the group.

$1 + 9 = 10$ people — Add the frequency of the 1st group to $\frac{1}{3}$ of the frequency of the 2nd group.

ii $\frac{11.5-10}{11.5-7.5} \times 6 = 2.25$ — Find the fraction of the 3rd bar that lies to the right of 10 and multiply by frequency.

$2.25 + 6 = 8.25$

8 people — Round your answer to a sensible degree of accuracy.

STATS

Exercise 9.3A Fluency and skills

1 A maths teacher compares the results of two classes in a test. The teacher creates box-and-whisker plots to compare the percentage scores gained. Compare briefly the scores achieved by the two sets.

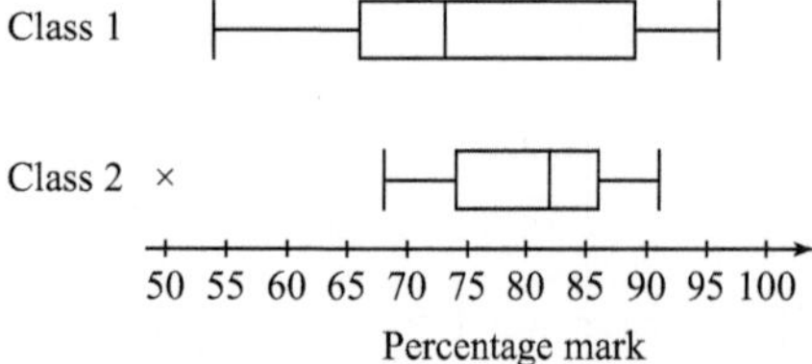

2 An estate agent collects data on the houses sold in various price brackets. 100 houses sold for more than £500 000 each.

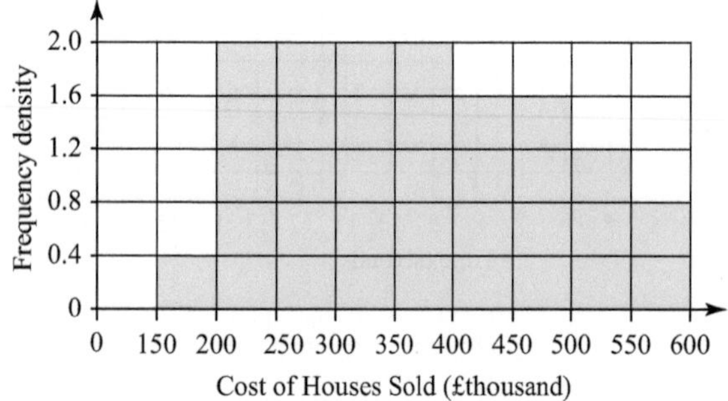

a How many houses sold for less than £300 000?

b Estimate the number of houses that sold for between £360 000 and £480 000

3 The IQ scores of 30 people are measured and a cumulative frequency graph is plotted.

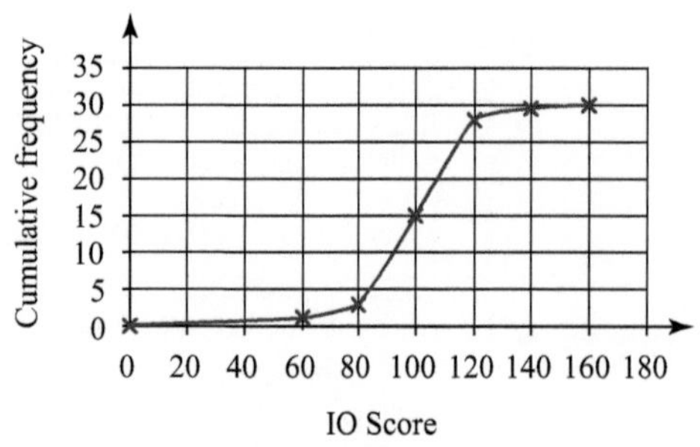

a Estimate the number of people with an IQ between 80 and 100

b Use the graph to estimate the upper quartile and lower quartile.

c An outlier is defined as a value less than $Q_1 - 1.5 \times \text{IQR}$ or greater than $Q_3 + 1.5 \times \text{IQR}$. Show that there are three outliers in this data.

4 A sample of 80 people was asked how much whole milk they consume each week. The same group of people were asked how much pure fruit juice they consume each week. The results are shown in cumulative frequency curves.

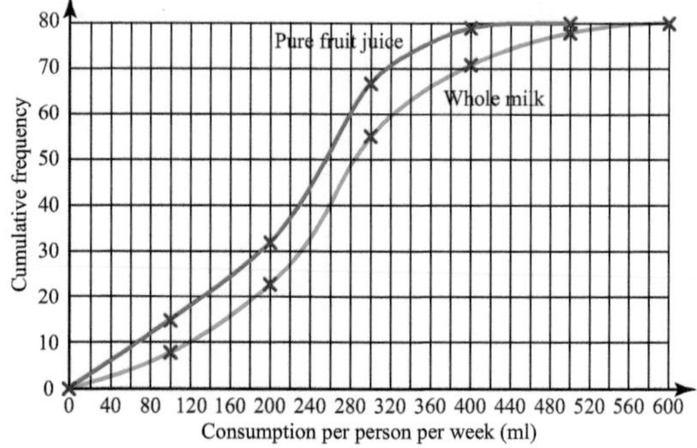

a Estimate the number of people who consume

i Less than 200 ml of whole milk per week,

ii Less than 200 ml of pure fruit juice per week,

iii More than 400 ml of milk per week.

b Make two comparisons between the consumption of whole milk and pure fruit juice. Use a measure of spread and a measure of central tendency in your comparison.

5 A set of continuous data is recorded to one decimal place. The results are summarised in a histogram.

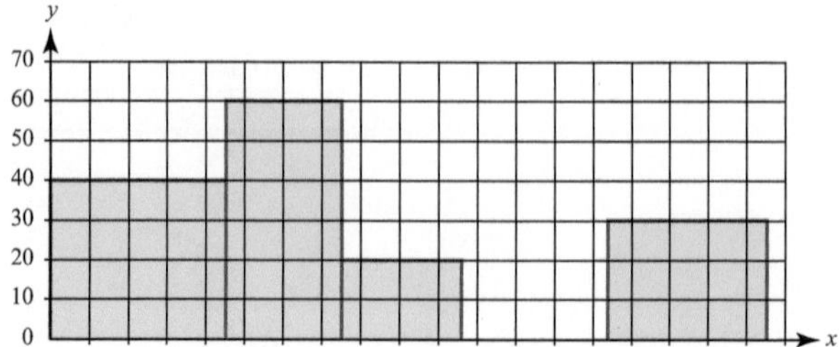

x	Frequency
0-0.4	18
0.5-0.7	
0.8-1.0	
1.1-1.4	15
1.5-1.8	12

a Find the missing frequencies.

b Calculate the width and height of the missing bar.

c Estimate the percentage of values which are

i Below 0.6

ii Above 1.2

Reasoning and problem-solving

Strategy 1

When deciding on the most appropriate diagram to represent data

1. Consider whether you need to be able to display all of the values, including outliers.
2. Consider whether you need to be able display relative or absolute frequencies.
3. Consider whether you are more interested in displaying the distribution or the summary statistics.

STATS

It is important to use an appropriate diagram to represent your data.

	Advantages	Disadvantages
Box plot	Highlights outliers. Makes it easy to compare data sets.	Data is grouped into only four categories so some detailed analysis is not possible.
Histogram	Clearly shows shape of distribution.	Doesn't always highlight outliers. It is possible but not easy to estimate Q_1, Q_2 and Q_3
Cumulative frequency curve	Makes it easy to find the five number summary.	Doesn't always highlight outliers. If interval boundaries are not shown the degree of detail is not clear.

Example 4

The manager of a company's fleet of BMW cars in north-west England recorded the carbon monoxide emissions, in g km^{-1}, of the cars during one week in 2016.

0.086, 0.098, 0.101, 0.128, 0.128, 0.128, a, 0.144, 0.144,
0.148, 0.151, 0.157, 0.172, 0.172, 0.187, b, 0.312
The data is given in ascending order but two values, a and b, have been lost.
An outlier is defined as any value outside the interval $[Q_1 - 1.5 \times IQR, Q_3 + 1.5 \times IQR]$

a Explain why it is not possible to draw a box-and-whisker plot to display this data.

b Suggest an appropriate diagram to represent the data giving your reasons for the choice.

(*Continued*)

a $Q_1 = 0.128$ and $Q_3 = 0.172 \Rightarrow \text{IQR} = 0.044$ and the interval is $[0.062, 0.238]$ 0.187 is not an outlier but 0.312 is an outlier. It is not possible to say if b is an outlier so it is not possible to accurately draw the upper whisker for the data.

b A histogram.
This allows the shape of the data's distribution to be shown and, provided that class boundaries avoid the values a and b, you do not need to know their exact values to draw an accurate histogram.

1 Outliers should be plotted as crosses (×).

If a class interval was 0.125 – 0.150 then it would contain 7 data points independent of the exact value of a.

When interpreting a diagram displaying data

1. Consider what is being represented and whether your data is discrete or continuous.
2. If necessary, identify any outliers or missing/incorrect data, and consider the effects of removing them.
3. Read what is being asked for in the question and use the diagram to answer it.

See Ch10.1 For information about probability.

You can use statistical diagrams to find probabilities of events.

If an experiment can have any one of N equally likely outcomes and n of those outcomes result in event A, then the theoretical probability of event A happening is $P(A) = \frac{n}{N}$

You can estimate probabilities of certain events happening by using diagrams that show grouped data. This often involves assuming data is equally distributed on each interval.

Example 5

This box -and-whisker plot shows the masses (in grams) of 52 eggs from a certain species of bird.

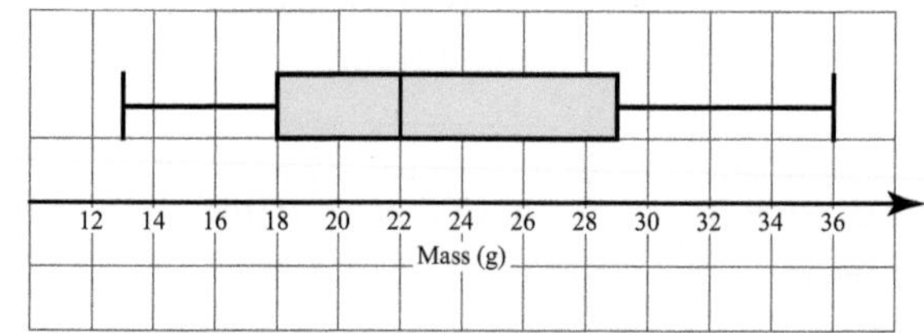

An egg is picked at random from this set. Estimate the probability that its mass is

a More than 22 g, **b** Less than 18 g,

c Less than 20 g, **d** Between 16 g and 21 g

a 50%

b 25%

c $25\% + 0.5 \times 25\% = 37.5\%$

d

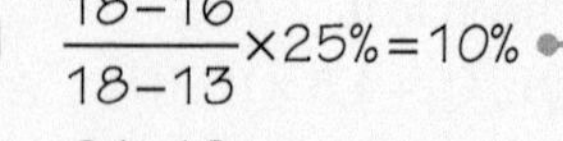

$\frac{18-16}{18-13} \times 25\% = 10\%$

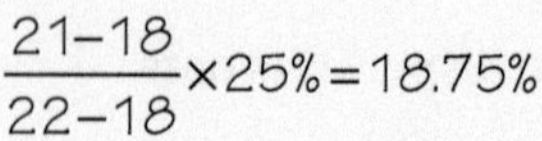

$\frac{21-18}{22-18} \times 25\% = 18.75\%$

$10\% + 18.75\% = 28.75\%$

22 g is the median, so 26 of the 52 eggs weigh more than 22 g.

One quarter of the eggs weigh less than the lower quartile value of 18 g.

25% of the eggs weigh between 18 g and 22 g. 20 g is halfway through this interval so you assume half of the eggs between 18 g and 22 g weigh less than 20 g.

Sum to find the total probability.

Find the proportion of the interval you want and multiply by the proportion of the whole data that the interval represents.

Exercise 9.3B Reasoning and problem-solving

1 A doctor is investigating obesity in his local area. He records the masses (in kg) of 1000 patients and creates a box-and-whisker plot. The lightest patient weighs 5 kg.

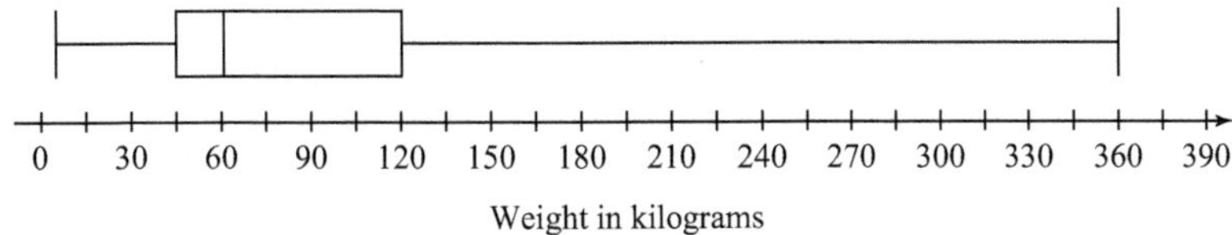

a A patient is selected at random. Estimate the probability that the patient weighs

  i More than 45 kg   **ii** Less than 90 kg

b The two largest masses measured are 360 kg and 165 kg, but it was later decided that the 360 kg mass should be considered an outlier. If it is removed, how will the box-and-whisker plot change?

2 An art student wants to know how much money she could make from selling her work. Over a one-year period, she collects data on how much money pieces of art sold for at auction. Would a box-and-whisker plot or a histogram be better to display this data?

3 The box-and-whisker plot shows the Min, Q_1, Q_2, Q_3 and Max engine sizes for the estate cars in the large data set for 2002.

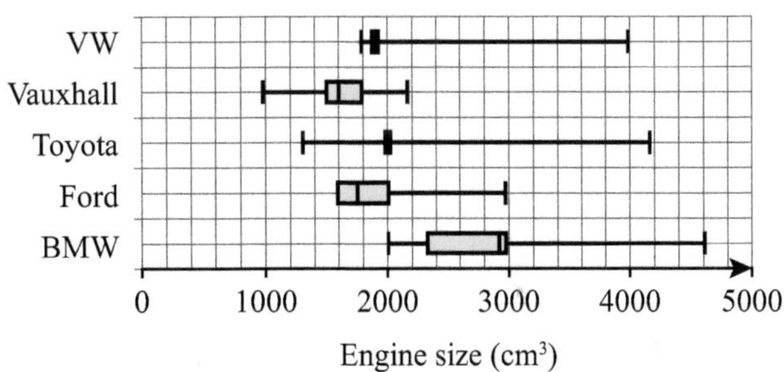

a Suggest a reason why Q_1, the median and Q_3 all have the same value for Volkswagen estates.

b A car is chosen at random from the sample of Ford estates, what is the probability that its engine size is less than 1750 cm^3?

c A car is chosen at random from the sample of Toyota estates, what is the probability that its engine size is greater than 2000 cm^3?

d A journalist claims that nearly all BMW estates have bigger engines than Vauxhall estates. Explain why this claim may not be true.

4 A garage in south-west England that specialises in hybrid electric-petrol cars records the carbon monoxide emissions of the 37 cars it serviced in one month.

CO (g km^{-1})	Frequency
$0 \le x < 0.050$	0
$0.050 \le x < 0.075$	4
$0.075 \le x < 0.100$	11
$0.100 \le x < 0.150$	4
$0.150 \le x < 0.200$	
$0.200 \le x < 0.300$	
$0.300 \le x$	0

Distribution of number of cars CO emissions

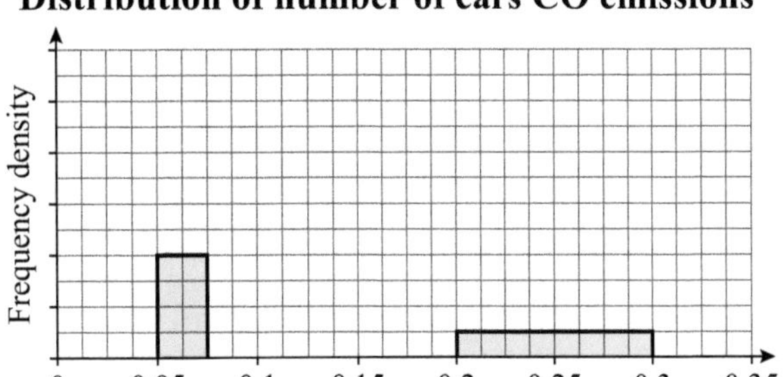

a Find the missing values in the table and complete a copy of the histogram.

b Using the garage's data, estimate the proportion of hybrid cars that had CO emissions

  i above 0.100 g km^{-1} **ii** below 0.250 g km^{-1}

2276 – 2278 SEARCH

9.4 Bivariate data

Fluency and skills

It is often useful to look for connections between two sets of data. Data relating to *pairs* of variables is called **bivariate data.**

> **Key point**
>
> Variables that are statistically related are described as **correlated**.
>
> There are three types of correlation: positive, negative and zero.

If the variables increase together, they have **positive correlation.**

If one variable increases as the other decreases, they have **negative correlation.**

Two variables can also be uncorrelated and the data is then said to have **zero correlation.**

You can **identify correlation** by plotting a **scatter diagram,** which shows each pair of data values as a point on a graph. A scatter diagram shows both the type and strength of the relationship between two variables.

The **independent variable** goes on the x-axis. This is the variable that doesn't depend on other things—the values for it are usually selected by the person gathering data. The **dependent variable** goes on the y-axis. This is the variable that is expected to change in response to a change in the other variable.

> Data that lies exactly on a straight line has perfect correlation. Otherwise, the correlation may be described as strong, moderate or weak.

Types and strength of correlation

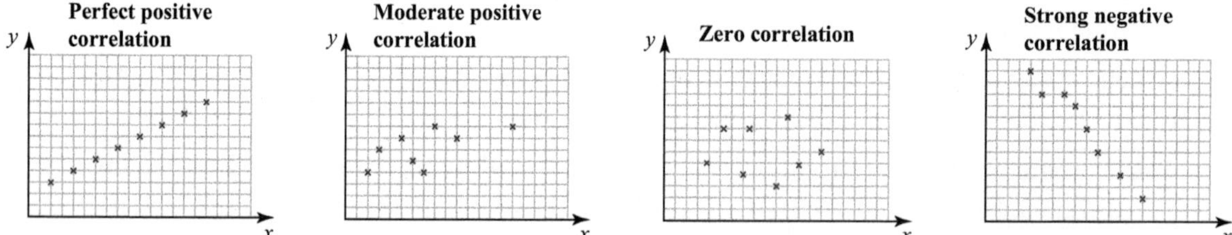

There is no exact definition of the terms strong, moderate or weak. It is therefore useful to quantify the correlation between two variables.

> **Key point**
>
> The correlation coefficient, known as r, can have a value between -1 and $+1$ inclusively.
>
> - For perfect positive correlation, $r = +1$
> - For perfect negative correlation, $r = -1$
> - For no correlation, $r = 0$

> You do not need to be able to calculate r at this stage.

Example 1

The data in the table relates to the width and length of the petals of 9 roses. Plot this data on a scatter diagram and describe the type and strength of the correlation.

Width, cm	4.9	4.7	4.6	5	5.4	4.6	5	4.4	4.9
Length, cm	3	3.2	3.1	3.6	3.9	3.4	3.4	2.9	3.1

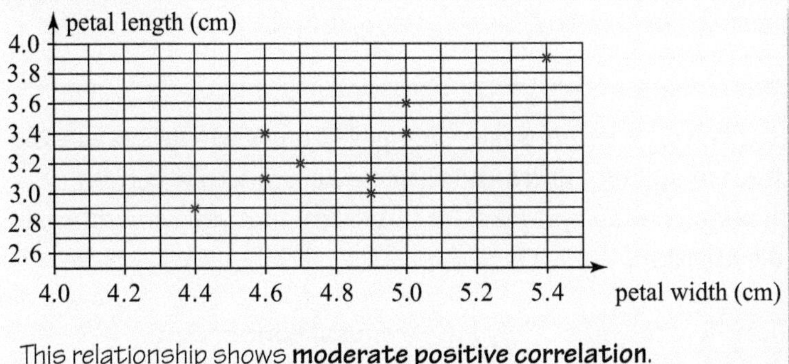

This relationship shows **moderate positive correlation**.

ICT Resource online

To investigate correlation, click this link in the digital book.

STATS

Exercise 9.4A Fluency and skills

1 Give the most likely type and strength of correlation (if any) for these pairs of variables.

  a A town's unemployment rates and a standard of living index.

  b The heights of fathers and their adult sons.

  c Cooking time and weight of a chicken.

  d Adults' shoe sizes and salaries.

2 This diagram relates two variables, x and y

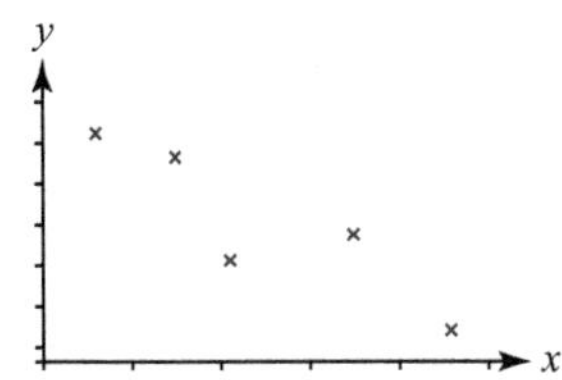

  a Describe the correlation.

  b What is the most likely value of r: +0.6, -0.2 or -0.7

3 a For each of these data sets, draw a scatter diagram and use it to describe the type and degree of correlation.

  i

x	2	4	4	7	8	10	12	14
y	3	4	7	7	9	10	12	12

  ii

x	-2	0	2	4	3	5	7	7
y	9	8	5	6	1	1	2	6

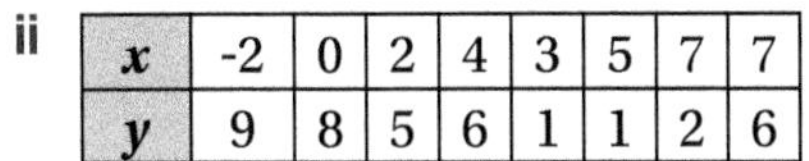

  b Somebody claims that r for the data in part **i** is either 1 or -1. Explain why they're incorrect.

4 The length of service L (years) against annual salary S (£1000) is shown for 10 workers in a company.

L	5	15	2	12	11	1	3	6	4	12
S	22	26	16	21	22	18	23	23	18	27

  a Plot a scatter diagram for this data.

  b The correlation coefficient is either -0.63, +0.73 or +1.1

    i State which is the most likely value.

    ii Explain why each other value cannot be correct.

5 This table gives data for a sample of 10 individuals from a bivariate population.

x	51	62	47	53	71	65	55	69	57	61
y	68	55	68	64	50	58	65	50	61	54

  a Draw a scatter diagram for this data.

  b Describe the correlation and estimate the value of the correlation coefficient.

MyMaths

2283 SEARCH

Reasoning and problem-solving

Strategy

To solve a problem about bivariate data and correlation

1. Draw a scatter diagram to identify any correlation between two variables.
2. Identify data points that don't fit the general pattern shown by the data.
3. Use correlation in the scatter diagram to determine the value of missing data points.

You should not assume that, because two variables correlate, changes in one variable are *causing* changes in the other. For example, rates of diabetes and annual income correlate for certain groups, but this is because they both relate to dietary intake. It's important that you remember the difference between **correlation** and **causation**.

Key point

When a change in one variable *does* affect the other, they have a **causal connection.** Correlation without a causal connection is known as **spurious correlation.**

When data is correlated, a scatter diagram shows a pattern. **Outliers** are points that do not fit the pattern . They are easy to identify on a scatter diagram.

Example 2

The following data gives an ice-cream parlour's daily sales figures, y, in hundreds of pounds, against the number of hours of sunshine, x, on six consecutive Saturdays.

x	2.2	3.5	4.7	5.2	6.6	7.8
y	7.2	9.3	13.8	8.1	4.1	13.1

a Plot a scatter diagram to represent this data and write down the type and strength of correlation.

b One of the points relates to a day when there was a power cut and the ice-cream parlour was closed for several hours. Write the most likely coordinates of this point.

c For the day identified in part **b**, give a sales figure that could be expected if the power cut had not occurred.

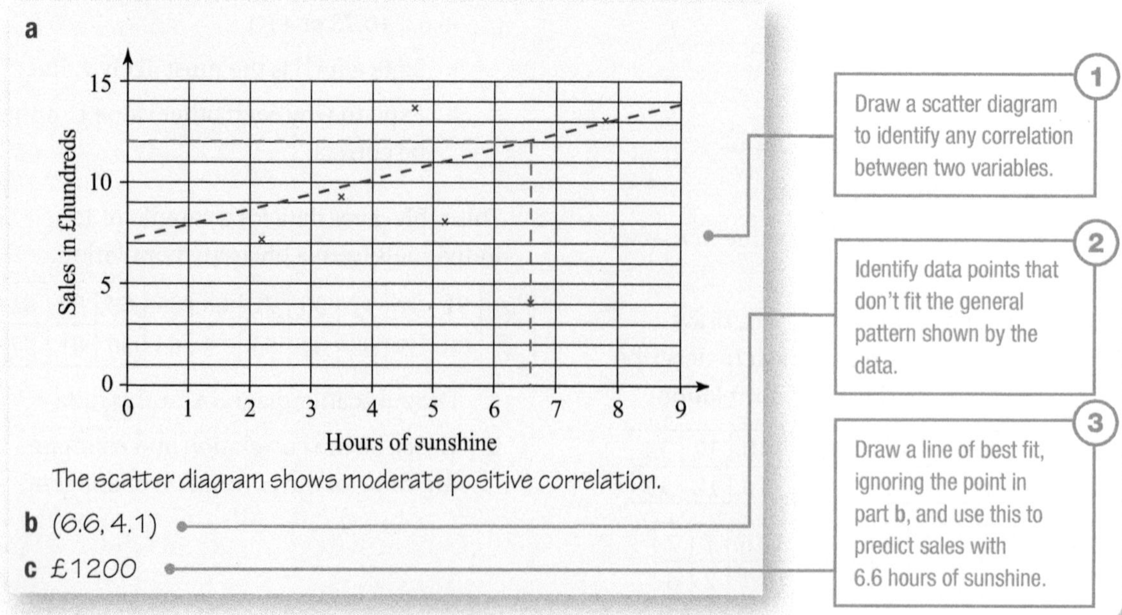

a
Sales in £hundreds
15
10
5
0
0 1 2 3 4 5 6 7 8 9
Hours of sunshine
1
Draw a scatter diagram to identify any correlation between two variables.
2
Identify data points that don't fit the general pattern shown by the data.
3
Draw a line of best fit, ignoring the point in part b, and use this to predict sales with 6.6 hours of sunshine.

The scatter diagram shows moderate positive correlation.

b (6.6, 4.1)

c £1200

Exercise 9.4B Reasoning and problem-solving

1 For each of these examples of bivariate data, state whether the correlation is likely to be positive, negative or zero. If non-zero, state whether you think there is a causal relationship between the variables.

- a Daytime temperature at a seaside resort and number of deckchairs hired out.
- b Plant growth and amount of fertilizer applied.
- c A person's annual income and weight.
- d Unemployment rates and measures of the standard of living.

2 This data shows the number of customer arrivals in a five-minute interval, and queue length at the end of the interval, in a small grocery store. One mistake was made in recording the data.

Number of arrivals, x	16	5	12	8	4	9	17
Queue length, y	4	0	5	3	1	2	1

- a Draw a scatter diagram and identify which reading was most likely recorded incorrectly.
- b Estimate the true queue length for that point.

3 The table shows a car's engine size and hydrocarbon emissions, in g km^{-1}, for a sample of electric-petrol hybrid cars in north-west England.

	A	B	C	D	E	F	G
Engine size	1798	1997	1998	1497	1798	2494	647
hc emissions	0.036	0.014	0.017	0.043	0.033	0.049	0003

- a Plot a scatter diagram for cars A – E.
- b Describe, in context, any correlation shown by the data.
- c Cars F and G were initially excluded as outliers. Add these points to your scatter diagram and explain how this changes your answer to part **b**.

4 According to Department for Transport statistics, in the period 2006 - 2016 the numbers of licensed cars and vans in Britain has steadily increased whilst the number of licensed buses has steadily decreased.

Large data set

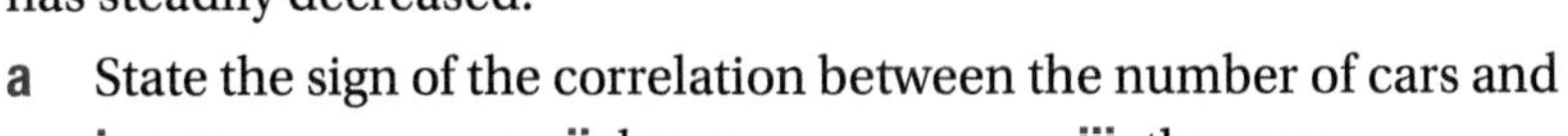

- a State the sign of the correlation between the number of cars and
  - i vans
  - ii buses
  - iii the year.
- b Explain what is meant by the statement 'the correlation between the numbers of cars and vans is an example of a spurious correlation.'

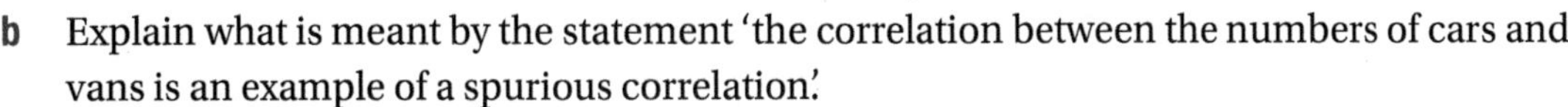

Challenge

5 A sample of seven data items is taken from a bivariate population and the following values of the variables, X and Y, are obtained:

X	3	9	10	12	13	15	18
Y	2	9	9	11	11	19	13

- a Plot a scatter diagram for this data and hence describe the type and strength of the correlation.

The researcher leading this investigation expected the correlation coefficient to be about +0.95 and decided to check her data. She found that one y-value had been recorded incorrectly.

- b Identify the most likely value to be incorrectly recorded.
- c This error is corrected and a line of best fit is given by $y = 0.8x + 0.9$. Draw this line on your graph and estimate the value of y when $x = 7$.

9 Summary and review

Chapter summary

- You can estimate a population and its parameters using statistics taken from a sample.
- There are many sampling methods, including simple random sampling, systematic sampling, stratified sampling, opportunity sampling, quota sampling and cluster sampling.
- Sampling methods can produce biased samples in some situations.
- The mode of a set of observations is the value that occurs with the largest frequency.
- To find the mean of a set of n observations, calculate their sum and divide the result by n
- The median of a set of data is the middle value once the data is in order of size.
- The range of a set of observations is the largest observation minus the smallest.
- The interquartile range is the difference between the first and third quartiles.
- The variance of a set of data measures the degree of spread.
- Outliers are points that don't fit the pattern of the data.
- Data is either discrete or continuous.
- Continuous data needs summarising for representation.
  - You use the five-number summary to form a box-and-whisker plot.
  - You group data to draw a histogram or a cumulative frequency diagram.
- Variables whose values are linearly related are said to be correlated. The type of correlation can be positive, negative or zero. If non-zero, it can be weak, moderate or strong.
- A scatter diagram shows both the type and strength of the relationship between two variables.
- The correlation coefficient, r, takes all values from -1 to $+1$ inclusive.
- Variables that have a non-zero correlation are not necessarily causally connected.

Check and review

You should now be able to...	Try Questions
✔ Distinguish a population and its parameters from a sample and its statistics.	1
✔ Identify and name sampling methods.	2
✔ Highlight sources of bias in a sampling method.	2
✔ Read continuous data given in box-and-whisker plots, histograms and cumulative frequency diagrams.	3
✔ Plot scatter diagrams and use them to identify types and strength of correlation.	4
✔ Use scatter diagrams and rules using quartiles to identify outliers.	4, 6
✔ Summarise raw data using appropriate measures of location and spread.	5

1 Identify the population, sample, parameter and statistics in these situations.

a 23 LED light bulbs made in a certain factory are tested to see how warm they get when lit.

b Nine matured casks of whiskey from a certain brewery have their alcohol levels measured.

2 A sample of 15 families is to be taken to determine the average number of children per household in a given city. Name the sampling methods in parts **a** to **c**.

   a A list of all house or flat numbers and names is found. The numbers and names are ordered and labelled 1, 2, 3, … , n. 15 random numbers are generated between 1 and n. Those labels give the numbers or names of the households that are sampled.

   b A surveyor spends three hours on a Saturday in the high street asking passers-by until he records enough responses.

   c 15 random streets in the city are picked and one random house from each street is chosen to be asked.

   d Which of the sampling methods in parts **a** to **c** could be biased?

3 Create a table of frequencies that could produce this histogram.

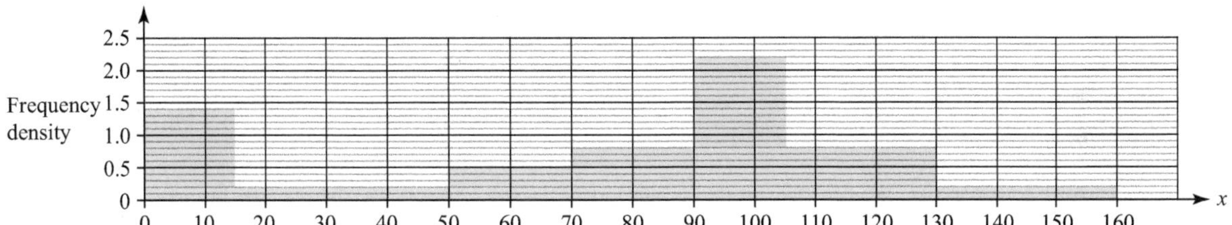

4 **a** Draw a scatter diagram for this data and comment on the correlation between the variables.

x	3	9	5	14	10	7
y	12	13	9	1	8	8

   b The y-value of one point was recorded incorrectly. Given that there is strong negative correlation between the variables, identify the most likely incorrect point and state a more plausible y-value for this point.

5 For this data set, find the modal interval and estimates for the median, mean and variance.

x	$0 \le x < 4$	$4 \le x < 8$	$8 \le x < 12$	$12 \le x < 16$	$16 \le x < 20$
f	1	4	8	4	3

6 **a** For this set of observations, find the interquartile range (IQR).

   14, 16, 15, 26, 10, 9, 12, 17, 15, 18, 11, 10

   b Using the rule that outliers are values less than $Q_1 - 1.5 \times \text{IQR}$ or more than $Q_3 + 1.5 \times \text{IQR}$, where Q_1 is the first quartile and Q_3 is the third quartile, identify any outliers within the data.

What next?

Score		
	0 – 3	Your knowledge of this topic is still developing. To improve look at MyMaths: 2275 – 2283
	4 – 5	You're gaining a secure knowledge of this topic. To improve, look at the InvisiPen videos for Fluency and skills (09A)
	6	You have mastered these skills. Well done, you're ready to progress! To develop your techniques, look at the InvisiPen videos for Reasoning and problem-solving (09B)

Click these links in the digital book

Exploration

Going beyond the exams

History

Florence Nightingale (1820 - 1910) is best known for her contributions to modern nursing, but she was also a very influential statistician.

Nightingale worked to make statistical data more accessible by developing innovative graphs and charts. She used them to present information about soldier mortality in order to convince governments that changes to hospital conditions were needed.

Nightingale's work transformed the way we can represent statistics and highlighted the importance of statistics in government.

Research

Sir Francis Galton (1822 – 1911) is credited with inventing the concepts of **standard deviation** and **correlation**, and was the first to recognise the phenomenon of **regression towards the mean**.

What is regression towards the mean?

Why does it need to be taken into account when designing experiments?

Have a go

In 1973, the English statistician **Francis Anscombe** constructed the four sets shown, known as Anscombe's quartet, in order to make a point about the value of statistical diagrams.

Calculate for each dataset

1. The mean value of x
2. The mean value of y
3. The variance of x
4. The variance of y

The correlation between x and y is 0.816 for each dataset, and in each case the line of best fit is $y = 0.5x + 3$

Comment on what all of the summary statistics appear to tell us about the datasets.

Plot the datasets on four diagrams, using the same scales.

What do you observe? What point do you think Anscombe wanted to make?

Did you know?

Sir Francis Galton used information from weather stations in England to produce the world's first weather map.

I		II		III		IV	
x	y	x	y	x	y	x	y
10.0	8.04	10.0	9.14	10.0	7.46	8.0	6.58
8.0	6.95	8.0	8.14	8.0	.77	8.0	5.76
13.0	7.58	13.0	8.74	13.0	12.74	8.0	7.11
9.0	8.81	9.0	8.77	9.0	7.11	8.0	8.84
11.0	8.33	11.0	9.26	11.0	7.81	8.0	8.47
14.0	9.96	14.0	8.10	14.0	8.84	8.0	7.04
6.0	7.24	6.0	6.13	6.0	6.08	8.0	5.25
4.0	4.26	4.0	3.10	4.0	5.399	19.0	12.50
12.0	10.84	12.0	9.13	12.0	8.15	8.0	5.56
7.0	4.82	7.0	7.26	7.0	6.42	8.0	7.91
5.0	5.68	5.0	4.74	5.0	5.73	8.0	6.89

9 Assessment

1 The estimated mean of the data in the table is 11

x	$0 \le x < 4$	$4 \le x < 8$	$8 \le x < 12$	$12 \le x < 16$	$16 \le x < 20$
Frequency	5	2	13	a	8

Calculate the value of the missing frequency, a. Select the correct answer.

A 0.857 **B** 4 **C** 12 **D** 0.182 **[1 mark]**

2 The maximum temperature, t, was recorded every day one year in July.

You are given $\sum t = 686$ and $\sum t^2 = 15598$

Calculate the standard deviation to 3 sf. Select the correct answer.

A 13.5° **B** 21.9° **C** 3.67° **D** 4.70° **[1]**

3 A battery manufacturer needs to test the average life of its batteries.

a What is the population in this case? **[1]**

b Aside from cost, give a reason why only a sample should be tested. **[1]**

c What would be a sensible statistic to calculate? **[1]**

4 The scatter diagram shows the carbon monoxide and oxides of nitrogen emissions, in g km^{-1}, for a sample of Vauxhall, 5-door hatchbacks registered in south-west England.

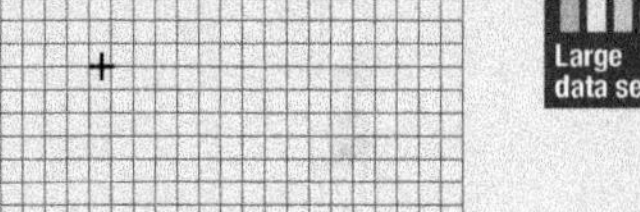

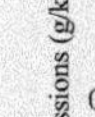

a Describe the relationship between the CO and and NOX emissions. **[2]**

b Which of these values is most likely to be the correlation coefficient between CO and NOX emissions? **[1]**

−0.6 −0.3 0.0 0.3 0.6

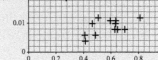

5 This histogram shows the total expenditure on petrol in a sample of 45 households during one week.

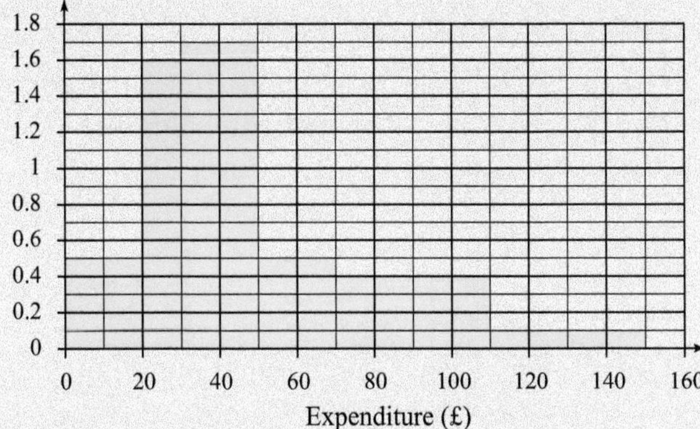

a How many households spent less than £20? **[4]**

b Estimate the percentage of households that spent over £100 **[3]**

c Use the histogram to estimate the mean expenditure on petrol. **[4]**

6 The large data set contains data on 20 BMW, petrol fuelled, convertible cars registered in London. The table shows a copy of the measured values of hydrocarbon emissions, in g km^{-1}, for the cars.

0.040	0.048	0.035	0.086	0.120	0.120	0.125	0.035	0.097	0.062
0.040	0.035	0.125	0.048	0.038	0.034	0.020	0.172	0.079	0.037

a Calculate **i** the mean, $\bar{x}$ **ii** the sample standard deviation, s **[2]**

b An outlier is defined as any value greater than $\bar{x} + 2s$ or less than $\bar{x} - 2s$
Identify any outliers in the data set. **[2]**

c Find the median of the data set. **[1]**

d Explain whether the mean, median or mode is a better representative of a typical value for this data set. **[2]**

e Another 10 cars were sampled but their emission measurements failed. Assuming that the emissions were 0.050 for each of these cars, how do your answers to parts **a** - **d** change? **[7]**

7 A teacher wants to find out the average number of homework assignments that students have been given each week. He takes a sample of 30 students. The results are shown in this graph.

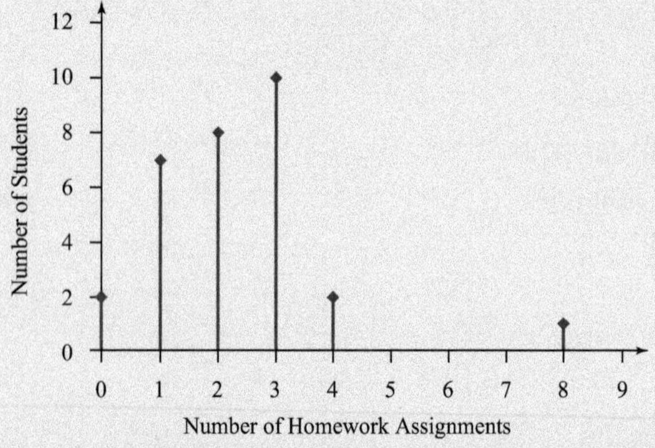

a What is the modal number of assignments? **[1]**

b Calculate the mean and standard deviation of the sample. **[4]**

c State the median and first and third quartiles. **[2]**

d Advise which measure of average would be best to use, stating your reasons clearly. **[1]**

8 This cumulative frequency diagram shows the weekly expenditure on public transport in a sample of 100 households in the year 2000

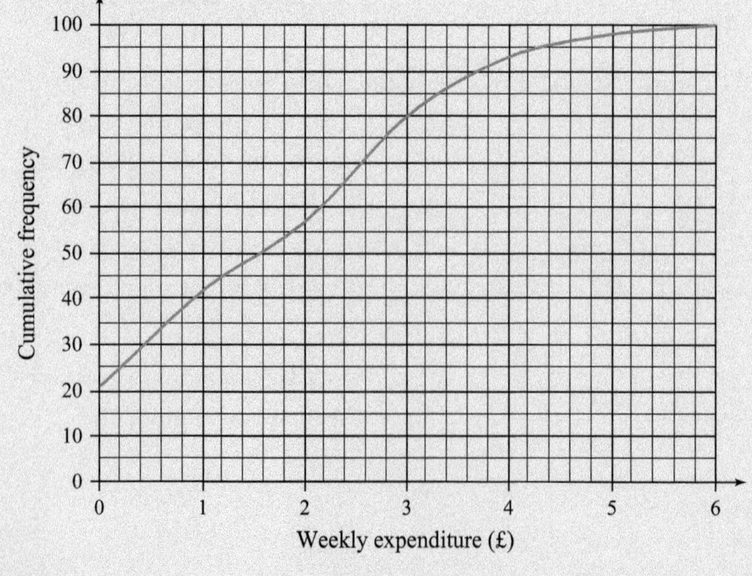

a What is the median expenditure? **[1]**

b Calculate the interquartile range. **[3]**

c Estimate the percentage of households that spent over £3 on public transport. **[2]**

9 A sample of students in a sixth form is to be taken and surveyed regarding their use of the library. The numbers of boys and girls in each year is given in the table.

	Year 12	Year 13
Girls	60	51
Boys	39	30

a It is suggested that a sample of size 60 should be taken by randomly selecting 15 boys and 15 girls from each year. State a disadvantage of taking a sample in this way. **[1]**

b It is instead decided to use a stratified sample of 60 students. Calculate the number of boys and the number of girls that should be sampled in each year. **[4]**

c In order to select individuals for the survey, an interviewer will randomly choose students as they leave their common room. Explain why the results of the survey could be biased. **[1]**

10 An outlier is an observation greater than $Q_3 + 1.5(Q_3 - Q_1)$ or less than $Q_1 - 1.5(Q_3 - Q_1)$

A box-and-whisker plot is drawn for a large volume of data. Four extra observations are then recorded. Which of the extra observations A, B, C and D are outliers? Show your working. **[4]**

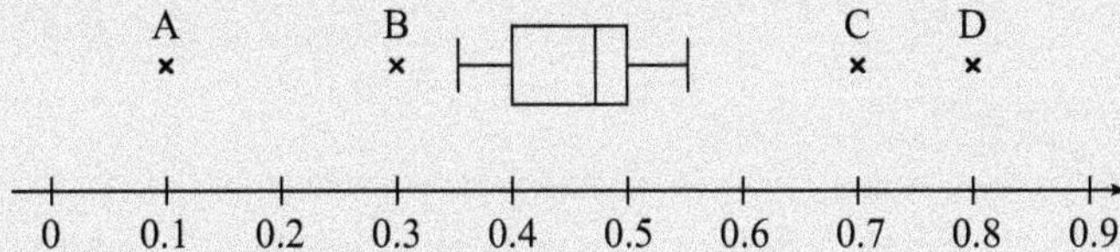

11 An experiment was carried out using tomato plant seeds. Trays of seeds were planted and each tray was placed in a controlled environment with a different temperature for each tray. All other variables, such as light and water, were the same for each of the trays. After 10 days, the number of seeds that had germinated in each was counted. The results are shown in the scatter diagram.

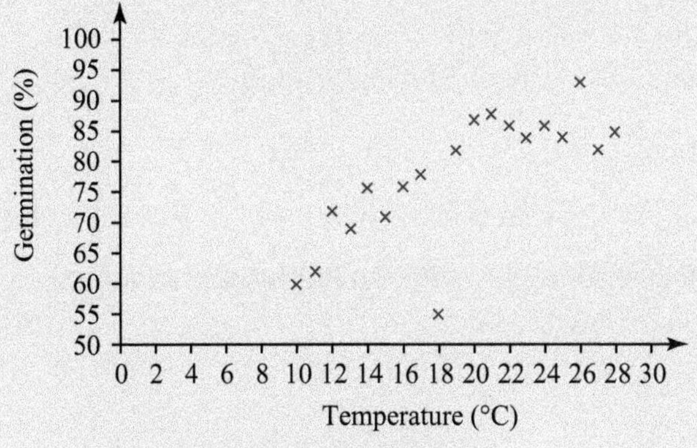

a Which is the independent variable? **[1]**

b Describe the relationship observed. **[2]**

c Without calculation, estimate the value of the correlation coefficient. **[2]**

d It is suggested that a temperature of 35 °C would result in almost all seeds germinating. Comment on whether this is a sensible suggestion. **[2]**

12

NOX emissions x g/km	Frequency
$0.00 \le x < 0.05$	667
$0.05 \le x < 0.10$	253
$0.10 \le x < 0.15$	31
$0.15 \le x < 0.20$	0
$0.20 \le x < 0.25$	0
$0.25 \le x < 0.30$	1
$0.30 \le x < 0.35$	49
$0.35 \le x < 0.40$	89
$0.40 \le x < 0.45$	98
$0.45 \le x < 0.50$	32
$0.50 \le x < 0.55$	0
$0.55 \le x < 0.60$	0
$0.60 \le x < 0.65$	4
$0.65 \le x < 0.70$	2

Large data set

A sample of 1226 cars was surveyed in 2002 and the oxides of nitrogen emissions recorded. The table shows the results.

a Estimate, to 3 significant figures,

i the mean

ii the standard deviation. **[4]**

b In 2016 a second sample of 2527 cars was taken and the following results obtained:

$\sum x = 83.947$, $\sum x^2 = 3.959833$

Calculate, to three significant figures,

i the mean

ii the standard deviation. **[3]**

c Compare the results for 2002 with those for 2016. **[2]**

13 The police in a town wish to survey members of a number of "Neighbourhood Watch" schemes. They wish to survey people from each scheme and the number selected is to be in proportion to the size of the scheme. Two possible methods of selecting the sample are suggested.

Method A: 10 *people are randomly selected from each scheme from a list of all members.*
Method B: The person who manages each scheme is asked to choose a proportionally sized sample from their population.

a State the name of each of these methods of sampling. **[2]**

b Which method is preferable? Clearly explain why this method is better. **[2]**

14 The box-and-whisker diagram summarises information from the large data set. It shows the mass of cars in the sample that are registered to men and women in London in 2002.

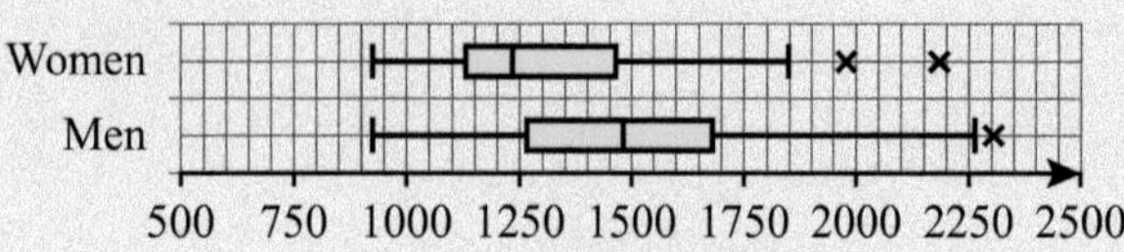

a Write down the median mass of cars registered to men and women. **[1]**

b Estimate the interquartile ranges for men and women. **[2]**

c Compare the masses of cars registered to men and women in the sample. **[3]**

d Eric claims that 'if you pick a car with a mass greater than 1500 kg then it is twice as likely to be registered to a man as a woman.' Comment on Eric's claim. **[2]**

15 The lengths of 40 fish caught in a competition are recorded to the nearest cm. Unfortunately, some of the numbers are now illegible. These values have been labelled a and b

Length (cm)	Frequency
18–22	3
23–25	10
26–28	14
29–31	a
32–40	b

Given that $\bar{x} = 27.3$

a Calculate the values of a and b **[5]**

b Find an unbiased estimation for the population standard deviation. **[3]**

16 The large data set contains records for 106 Coupe cars registered in 2016. The table shows the distribution of their carbon dioxide emissions to the nearest g/km.

CO_2 emissions, x g/km	Frequency
0 – 50	4
51 – 100	0
101 – 120	28
121 – 140	33
141 – 160	20
161 – 200	12
201 – 300	9

a A histogram is to be drawn to show this data. The 101 – 120 bar is 40 mm wide and 56 mm high.

Calculate the width and height of the 161 – 200 bar. **[2]**

b For this sample of Coupe cars, estimate

i the median **[3]**

ii the interquartile range. **[3]**

10 Probability and discrete random variables

Probability is a big part of genetics. Consider a child's chance of inheriting a specific eye colour. The basic model involves two possibilities: B (brown) and b (blue), where brown is dominant over blue. Each parent passes down either B or b and only the combination bb will result in blue eyes. The combinations Bb, bB and BB will all result in brown eyes. So if both parents have the combination Bb, they both have brown eyes and the probability of their child having brown eyes is 75%.

The example above treats eye colour as a discrete variable: either blue or brown in this case. Age and blood type are also typically treated as discrete variables. Probability is important in subjects such as statistics, research, medicine, weather forecasting and business, and this chapter provides an introduction to the terminology and techniques used in calculating and expressing probabilities.

Orientation

What you need to know

KS4

- Record, describe and analyse the frequency of experiment outcomes.
- Understand and draw probability tree diagrams.
- Construct and interpret diagrams for discrete data.

What you will learn

- To use the vocabulary of probability theory.
- To solve problems involving mutually exclusive and independent events.
- To use a probability function or given context to find a probability distribution and probabilities for particular events.
- To recognise and solve problems related to the binomial distribution.

What this leads to

Ch20 Probability and continuous random variables
Conditional probability.
Modelling with probability.
Connecting the binomial distribution with the normal distribution.

Careers
Weather forecasting.
Actuarial science.
Genetics.

Practise before you start

1193, 1211, 1935

10.1 Probability

Fluency and skills

A **random experiment** is any repeatable action with a collection of clearly defined outcomes that cannot be predicted with certainty. For example, the experiment 'an ordinary dice is thrown once'.

The **sample space** for an experiment is the collection of all possible outcomes of the experiment. **Events** are groups of outcomes within the sample space. In the diagram, S is the sample space for the experiment 'an ordinary dice is thrown once'. A is the event 'getting an odd number'.

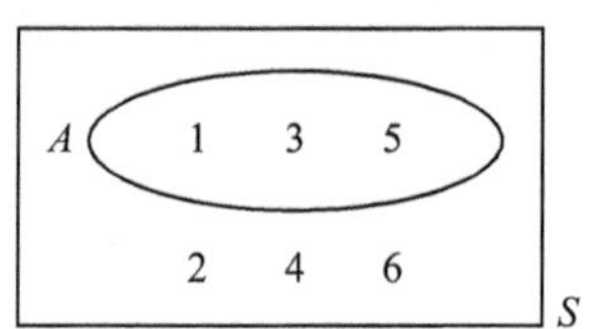

▲ Sample space S with event A

> **Key point**
> The total probability associated with a sample space is 1

If, in a trial where 60 dice are thrown, twelve 6s and seven 5s are thrown, then an estimate of the probability of a 5 or a 6 is $\frac{12+7}{60}=\frac{19}{60}$

See p.382
For a list of mathematical notation.

> **Key point**
> The probability of an even happening is written 'P(event)'.

Probabilities can be added like this when events are **mutually exclusive**. This means they cannot both happen in one trial.

> **Key point**
> If A and B are mutually exclusive, $\text{P}(A \text{ or } B)=\text{P}(A)+\text{P}(B)$

> Events are mutually exclusive if they cannot occur together. You cannot roll a dice once and get 5 *and* 6

Two events, A and not-A, are known as **complementary events**. They are mutually exclusive and **exhaustive**. Exhaustive means that no other outcome exists.

> The event not-A is written as A'

> **Key point**
> $\text{P}(A')=1-\text{P}(A)$

Example 1

Show that the probability of obtaining a score greater than 4 when a fair dice is thrown is $\frac{1}{3}$

$\text{P}(1)=\text{P}(2)=\ldots=\text{P}(6)=\frac{1}{6}$ — The dice is fair, so all outcomes are equally likely.

$\text{P}(\text{score}>4)=\text{P}(5 \text{ or } 6)=\text{P}(5)+\text{P}(6)=\frac{1}{6}+\frac{1}{6}=\frac{1}{3}$

For a sample space with N equally likely outcomes, the probability of any one occurring is $\frac{1}{N}$

> **Key point**
> If event A occurs in $\text{n}(A)$ of the equally probable outcomes, the probability of A is given by $\text{P}(A)=\frac{\text{n}(A)}{N}$

Two events, A and B, are **independent** if the fact that A has occurred does not affect the probability of B occurring.

Key point

If A and B are independent, $P(A \text{ and } B) = P(A) \times P(B)$

The $\cup$ and $\cap$ symbols are called the union and intersection symbols respectively.

$P(A \text{ or } B)$ is sometimes written $P(A \cup B)$. $P(A \text{ and } B)$ is sometimes written $P(A \cap B)$.

Example 2

Two boxes contain counters. Box 1 contains four red and five white counters; box 2 contains two red and three white counters. One counter is taken from each box. Find the probability that

a They are both red, **b** One is red and the other is white.

Let R_1, R_2, W_1, W_2 be the events 'red from box 1, …'

a $P(\text{two reds}) = P(R_1 \cap R_2) = P(R_1) \times P(R_2)$

$= \frac{4}{9} \times \frac{2}{5} = \frac{8}{45}$

R_1 and R_2 are independent events.

b $P(\text{red and white}) = P((R_1 \cap W_2) \text{ or } (W_1 \cap R_2))$

$= P(R_1) \times P(W_2) + P(W_1) \times P(R_2)$

$= \frac{4}{9} \times \frac{3}{5} + \frac{5}{9} \times \frac{2}{5} = \frac{22}{45}$

Red and white in either order.

A **probability distribution** for a random experiment shows how the total probability of 1 is distributed between all the possible outcomes.

Example 3

A fair coin is thrown 3 times and the number of heads noted. By calculating the probabilities of the possible outcomes, copy and complete the table.

Number of heads	0	1	2	3
Probability				

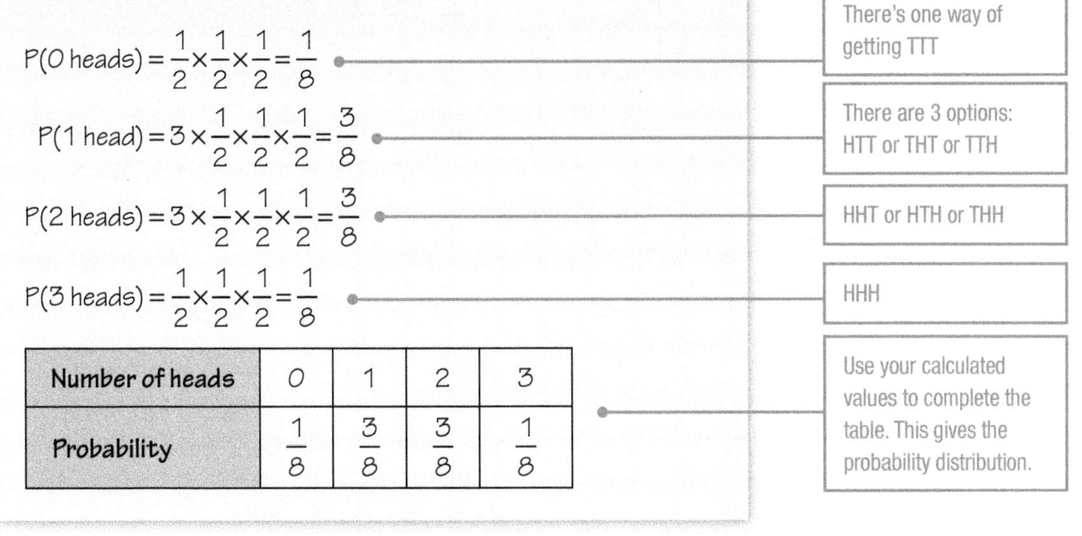

$P(0 \text{ heads}) = \frac{1}{2} \times \frac{1}{2} \times \frac{1}{2} = \frac{1}{8}$

There's one way of getting TTT

$P(1 \text{ head}) = 3 \times \frac{1}{2} \times \frac{1}{2} \times \frac{1}{2} = \frac{3}{8}$

There are 3 options: HTT or THT or TTH

$P(2 \text{ heads}) = 3 \times \frac{1}{2} \times \frac{1}{2} \times \frac{1}{2} = \frac{3}{8}$

HHT or HTH or THH

$P(3 \text{ heads}) = \frac{1}{2} \times \frac{1}{2} \times \frac{1}{2} = \frac{1}{8}$

HHH

Number of heads	0	1	2	3
Probability	$\frac{1}{8}$	$\frac{3}{8}$	$\frac{3}{8}$	$\frac{1}{8}$

Use your calculated values to complete the table. This gives the probability distribution.

STATS

MyMaths 2091, 2093, 2094, 2114 SEARCH

Sometimes probabilities are found using a **probability distribution function**. This is a function that gives the probabilities of all possible outcomes of an experiment.

A **discrete random variable** X takes values x_i with probabilities $P(X=x_i)$.

Example 4

A random variable X has probability distribution function given by $P(X=x)=kx^2; x=1, 2, 3, 4$

a Find the value of k **b** Show that $P(X>1)=\frac{29}{30}$

a $P(X=x)=kx^2; x=1,2,3,4$

$k+4k+9k+16k=1$ — Sum of all probabilities equals 1

$k=\frac{1}{30}$

b $P(X=x)=\left(\frac{1}{30}\right)x^2; x=1,2,3,4$

$P(X>1)=1-P(X=1)=1-\frac{1}{30}=\frac{29}{30}$

Exercise 10.1A Fluency and skills

1 A bag contains four counters numbered 1 to 4. A counter is chosen at random, not replaced, and then another counter is chosen. List all the pairs of numbers that make the sample space for this experiment.

2 A small health and fitness club offers gym facilities, personal training and Pilates classes. Its 45 members do only one of these activities each, as shown in the table.

Activity	Gym	Training	Pilates	Total
Male	7	a	1	20
Female	10	b	c	d
Total	17	20	8	e

a Find the values of a to e

b Find the probability that a randomly chosen member

i Does training,

ii Is female and does Pilates,

iii Is male.

3 Cara records the number of emails she receives per day throughout March. The results are shown in the bar chart.

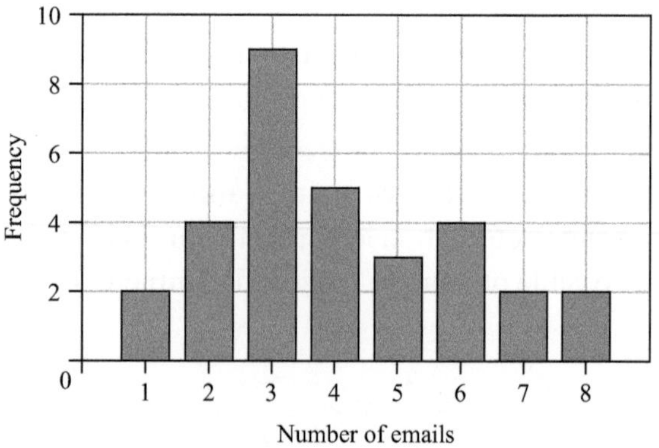

Find the probability that, on a randomly chosen day, the number of emails she receives is

a 5 **b** More than 5

c Between 1 and 4 inclusive.

4 One Saturday afternoon, Susie decides to either go to the cinema, visit her friends, go shopping or stay in to do her homework. The probabilities of each of these activities are shown in the table.

Activity	Probability
Cinema	$2h$
Friends	$4h$
Shopping	$3h$
Homework	h

a Find the value of h

b What is the probability that she doesn't do her homework?

5 The three mutually exclusive and exhaustive events, P, Q and R are associated with a random experiment.
$P(Q) = 0.2$ and $P(R) = 0.3$. Find

a $P(P)$ **b** $P(P')$

c $P(P \cup Q)$ **d** $P(P \cap R)$

6 A fair dodecahedral dice has integers 1 to 6, each occurring twice, on its 12 faces. The dice is rolled once. What is the probability that the uppermost face shows a square number or a prime number?

7 You throw two fair six-sided dice.

a What is the probability that one score is less than 4 and the other score is greater than 4?

b X is a variable for the difference (biggest – smallest) of the scores. By drawing a two-way table showing all possible outcomes, write down the probability distribution of X

8 A random variable X has probability distribution function given by

$$P(X = x) = \begin{cases} 0.1(x+1) & \text{for } x = 1, 2, 3 \\ b & \text{for } x = 4 \end{cases}$$

a Show that $b = 0.1$

b Find $P(X > 1)$

STATS

Reasoning and problem-solving

Strategy

To solve a probability problem

1. Identify mutually exclusive events and use the addition rule.
2. Identify independent events and use the multiplication rule.
3. For unknown probabilities, consider using the "probabilities total 1" result.

Example 5

a A and B are two events associated with a random experiment. Find the probability of B if $P(A) = \frac{1}{3}$, $P(A \text{ and } B) = 0$ and $P(A \text{ or } B) = \frac{3}{5}$

b R and S are two events associated with a random experiment. Given that $P(R) = 0.4$, $P(S) = 0.7$ and $P(R \text{ and } S) = 0.3$, show that R and S are *not* independent.

a $P(A \text{ or } B) = P(A) + P(B)$ — **1** $P(A \text{ and } B) = 0$, so A and B are mutually exclusive.

$P(B) = P(A \text{ or } B) - P(A) = \frac{3}{5} - \frac{1}{3} = \frac{4}{15}$

b $P(R \text{ and } S) = 0.3$ $\quad P(R) \times P(S) = 0.4 \times 0.7 = 0.28$

R and S are not independent. — **2** $P(R \text{ and } S) \neq P(R) \times P(S)$

Example 6

A sample of Ford cars, registered in London in June 2002 is taken. The mass of the car is recorded as 'light', mass ≤ 1100 kg or 'heavy', mass > 1100 kg. The CO_2 emission of the car is recorded as 'low', emissions ≤ 175 g/km or 'high' emissions > 175 g/ km. The table shows the results.

	Low emission	High emission
Light mass	30	4
Heavy mass	45	a

A car is selected at random from the sample. The probability that the car is either heavy or the car is light and has high CO_2 emissions is 0.75

a **i** Find the value of a

ii What is the probability that the chosen car is heavy and has high CO_2 emissions?

b A car is selected at random from the sample then replaced and a second car is selected at random from the sample. What is the probability that both cars are light?

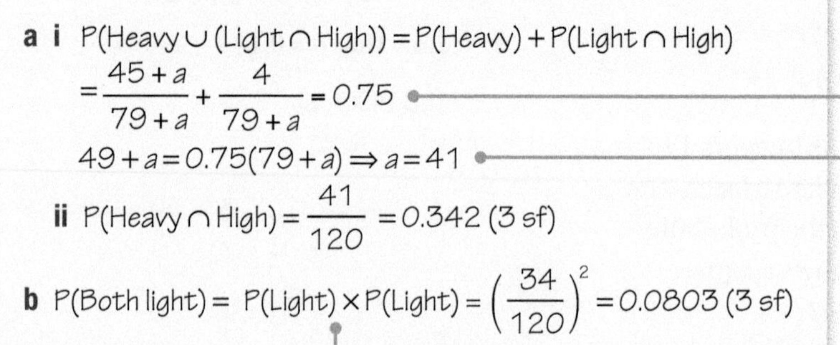

a i $P(\text{Heavy} \cup (\text{Light} \cap \text{High})) = P(\text{Heavy}) + P(\text{Light} \cap \text{High})$

$$= \frac{45+a}{79+a} + \frac{4}{79+a} = 0.75$$

$49 + a = 0.75(79 + a) \Rightarrow a = 41$

ii $P(\text{Heavy} \cap \text{High}) = \frac{41}{120} = 0.342$ (3 sf)

b $P(\text{Both light}) = P(\text{Light}) \times P(\text{Light}) = \left(\frac{34}{120}\right)^2 = 0.0803$ (3 sf)

1 'Heavy' and 'light, high emissions' are mutually exclusive events.

41 cars are classed as heavy and with high emissions.

2 Since the car is replaced in the sample before the second car is selected then the two events are independent.

Large data set

Know your data set

Emissions data for carbon dioxide, carbon monoxide, nitrous oxides, particulates and hydrocarbons are not available for all cars. These appear as blank entries in the large data set. Click this link in the digital book for more information about the Large data set.

Exercise 10.1B Reasoning and problem-solving

1 In a game between 2 players, a calculator is set to generate random numbers from 0 to 9. If the number is less than 3 you score 1 point, if it is between 3 and 8 inclusive you score 2 points, more than 8 scores 3 points.

a Copy and complete the following probability distribution table.

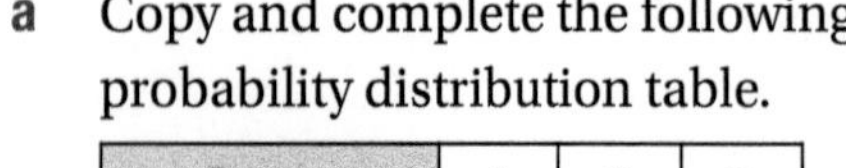

Score	1	2	3
Probability			

b Find the probability that the score is

i Less than 3 **ii** 1 or 3

2 The large data set records the car make and where it was registered. The table shows the results for the cars registered in 2002.

	BMW	Ford	Toyota	Vauxhall
North-west	21	132	48	82
South-west	46	155	40	104

a A car is selected at random from the sample. Find the probability that the car is
i a Ford, ii from the south-west.

b Are the events being a Ford and being from the south-west independent events? Give your reasons.

c Three cars are chosen at random from the sample. What is the probability that two of the choices are Ford cars from the south-west?

3 The large data set records the make of car for 2528 cars registered in 2016.

Make	BMW	Ford	Toyota	Vauxhall	VW
Proportion	0.16	0.27	0.05	a	b

The proportion of Toyota and Vauxhall cars was 0.37

a Find a and b

b A car is chosen at random from the sample. What is the probability that the car's make is

i BMW or Ford

ii Neither BMW nor Ford

iii Not Ford

4 Two events A and B are associated with a random experiment. $P(A \cap B) = 0$, $P(A \cup B) = 0.8$ and $P(A) = 0.6$

a Describe the relationship between the events A and B

b Find i $P(B)$ ii P(neither A nor B)

5 Two six-sided dice are each thrown once and their scores added. Draw a two-way table to show all possible outcomes. The probability that the sum of the scores is greater than 10 is $\frac{5}{36}$. Are both dice fair? You must give reasons for your answer.

6 P and Q are two events in an experiment, E The probability of P occurring when E is performed is 0.2 and the probability of both P and Q occurring is 0.06. If P and Q are independent of each other, what is the probability of

a Q occurring,

b P and not Q occurring,

c Neither P nor Q occurring?

7 A market researcher questions shoppers about their mobile phone provider. The probability of a shopper using Speak Mobile is p. The probability distribution function of X, the number of shoppers approached, is

$$P(X = x) = \begin{cases} (1-p)^{x-1}p; & x = 1, 2, 3 \\ (1-p)^3; & x = 4 \end{cases}$$

a Given that $P(X = 2) = 0.21$ and $p < 0.5$, show that $p = 0.3$

b Find the probability that at most 3 people are approached.

8 For any family of three children, A is the event 'there is at least one boy and one girl' and B is the event 'there are more girls than boys'. Assuming all combinations of children are equally likely, are the events A and B independent? Show your working.

9 A box contains blue, yellow and black beads. There are four blue beads and twice as many yellow as black beads. When a bead is chosen at random, the probability that it is black is $\frac{3}{11}$. How many yellow beads were initially in the box?

Challenge

10 Two events, E and F, are associated with a random experiment. $P(E \text{ and } F') = 0.4$, $P(E' \text{ and } F) = 0.3$,? $P(E \text{ and } F) = 0.1$

a Find i $P(E)$, $P(F)$, $P(E \text{ or } F)$

ii P(neither E nor F)

b Verify the equation
$P(E \text{ or } F) = P(E) + P(F) - P(E \text{ and } F)$

MyMaths 2091, 2093, 2094, 2114 SEARCH

10.2 Binomial distribution

Fluency and skills

Probability often involves sets of identical, independent trials. When trials have two outcomes: 'success' and 'failure', you can use a **binomial probability distribution** to model the situation.

Key point

Conditions for a binomial probability distribution:

- Two possible outcomes in each trial.
- Fixed number of trials.
- Independent trials.
- Identical trials (p is the same for each trial).

You can see from the tree diagram that if a biased coin is thrown three times and the probability of a head in any throw is 0.4, there are three ways to obtain exactly two heads.

P(two heads) $= 3 \times 0.4^2 \times 0.6$

The trials are independent of each other and there are three ways to obtain exactly two heads.

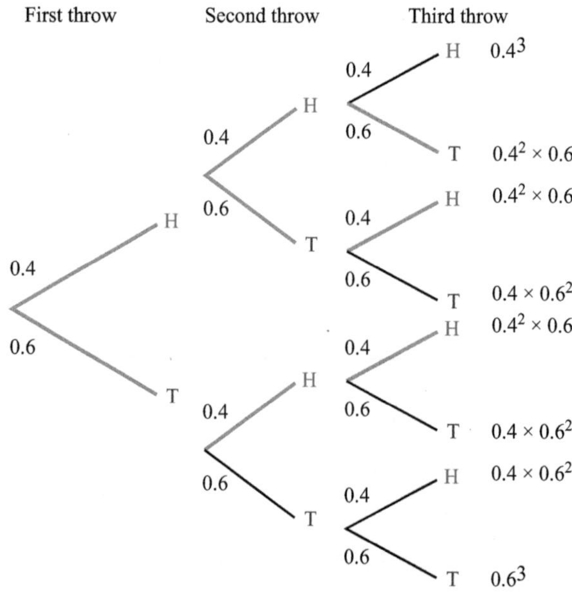

Although tree diagrams will not be assessed at AS level, they can be a helpful way to record the possible outcomes of a small number of trials. When the number of trials is greater than three this becomes unwieldy in practice and the ${}^nC_r$ formula is more appropriate.

In general, for a binomial probability distribution where X is a **random variable** for the number of successes, the probability of x successes is given by:

See Ch2.2

For a reminder on the ${}^nC_r$ function.

Key point

$P(X=x) = {}^nC_x\, p^x(1-p)^{n-x}$

where n is the number of trials and p is the probability of success in any given trial.

${}^nC_x$ is the number of ways of getting x successes in n trials.

A random variable, X, with this distribution function is said to follow a binomial distribution with parameters n and p. This is written $X \sim B(n, p)$.

Individual probabilities, $P(X=x)$, and cumulative probabilities, $P(X \le x)$, can be found directly on a calculator. Related probabilities can also be calculated.

ICT Resource online

To practise calculating binomial distribution values, click this link in the digital book.

Key point

If X can only take integer values, $P(X < x) = P(X \le x - 1)$ and $P(X > x) = 1 - P(X \le x)$

Calculator

Try it on your calculator

You can calculate binomial probabilities on a calculator.

```
Binomial    C.D
Data        : Variable
x           : 5
Numtrial    : 30
P           : 0.2
Save Res    : None
Execute
Calc
```

```
Binomial    C.D
      p=0.42751243
```

Activity

Find out how to calculate $P(X \le 5)$ for a random variable $X \sim \text{Bin}(30, 0.2)$ on *your* calculator.

Unless told otherwise, you should work out probabilities using a calculator.

Example 1

Two fair six-sided dice are thrown 24 times. X represents the number of double sixes.

a Write down the probability distribution of X and its distribution function.

b Using the distribution function, find $P(X = 1)$

c Find the value of $P(X < 5)$

d Find the probability of at least three double sixes.

a $X \sim B\left(24, \frac{1}{36}\right)$ — 24 independent and identical trials.

$P(X = x) = {}^{24}C_x \left(\frac{1}{36}\right)^x \left(\frac{35}{36}\right)^{24-x}$ — $n = 24$, $p = \frac{1}{6} \times \frac{1}{6} = \frac{1}{36}$

b $P(X = 1) = 0.3488$ (4 dp) — Check this using the binomial probability distribution function option on your calculator.

c $P(X < 5) = P(X \le 4) = 0.9995$ (4 dp)

d $P(X \ge 3) = 1 - P(X \le 2) = 0.0281$ — Change to $P(X \le 5 - 1)$ and use your calculator.

The shape of a binomial distribution has some interesting features.

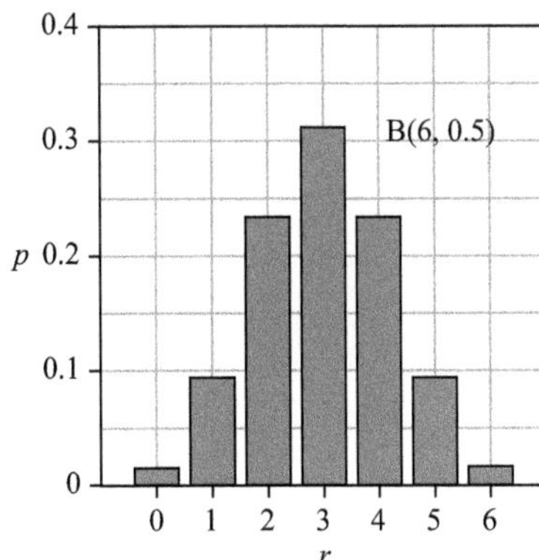

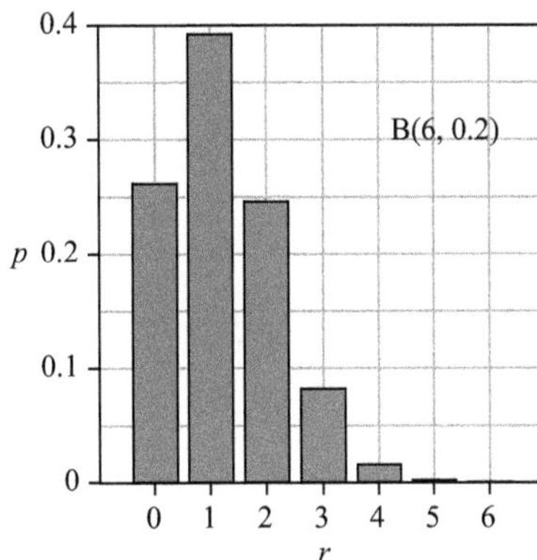

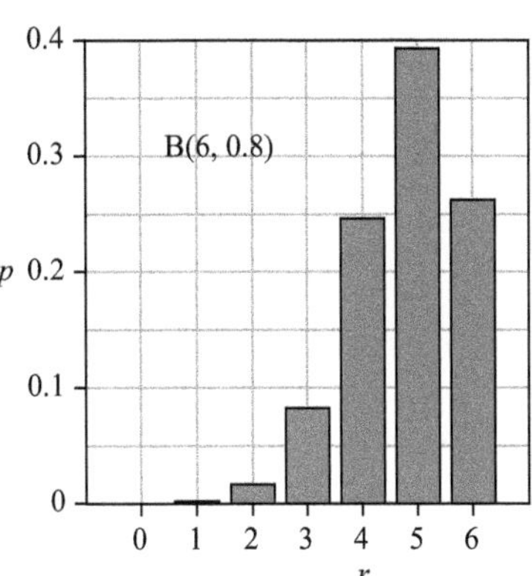

If $p = \frac{1}{2}$, the distribution shows symmetry. This is because the binomial coefficients have symmetry.

The distribution of $B(n, p)$ shows reflective symmetry with $B(n, 1 - p)$. For example, $P(X = 4)$ for distribution $X \sim B(6, 0.2)$ is equal to $P(X = 2)$ for $X \sim B(6, 0.8)$.

MyMaths 2110, 2111 SEARCH

Exercise 10.2A Fluency and skills

1 Given that $X \sim B(5,0.3)$, find

a $P(X=3)$ b $P(X\leq 2)$ c $P(X\neq 0)$

2 $X \sim B(8, 0.6)$. Find, to 2 sf

a $P(x\leq 0)$ b $P(x\leq 3)$

c $P(x<5)$ d $P(x>2)$

3 The random variable T has a binomial distribution, $n=8$, $p=\frac{1}{4}$. Find, to 2 sf

a $P(T=4)$ b $P(T\geq 7)$ c $P(3\leq T<5)$

4 Given that $X \sim B(5,0.4)$

a Write an expression for $P(X=x)$

b Copy and complete the probability distribution table.

x	0	1	2	3	4	5
$P(X=x)$	0.078			0.230	0.077	0.010

5 A fair six-sided dice is thrown 4 times and the random variable X denotes the number of 6s obtained.

a Give the distribution of X

b Find, giving your answers to 3 dp

i $P(X=4)$ ii $P(X>2)$ iii $P(1\leq X<3)$

6 A bag contains 12 counters. Three are red and the rest are black. A sample of five counters is taken, placing each back in the bag after it is chosen. Find the probability that the sample contains more than 3 red counters.

Reasoning and problem-solving

Strategy

To solve a probability problem involving the binomial distribution

1. Check the conditions for a binomial distribution are met. List any assumptions.
2. Identify the random variable and the corresponding values of n and p
3. Calculate probabilities using the addition and multiplication rules if necessary.

Example 2

A bag of sweets contains fudges, toffees and caramels in the ratio 4:3:3. 12 sweets are selected from the bag with replacement. Stating any assumptions you make, calculate the probability of

a 4 fudges being chosen, b At least 6 sweets in the sample being either toffees or caramels.

a Let X be the number of fudges. Assume the selections are independent of each other and that each sweet has an equal chance of being chosen. — (1) 12 identical, independent trials with two outcomes.

$X \sim B(12, 0.4)$ — (2) 12 trials. $P(\text{fudge}) = \frac{4}{4+3+3} = 0.4$

$P(X=4)=0.21$ (2 dp) — (3) Use your calculator to find $P(X=4)$

b Let Y be the number of toffees or caramels.

$p = P(\text{sweet is a toffee or a caramel}) = 0.3+0.3=0.6$ — The outcomes are mutually exclusive.

$Y \sim B(12, 0.6)$ — (2) 12 trials, $p=0.6$

$P(Y\geq 6) = 1-P(Y\leq 5) = 1-0.1582 = 0.8418$

Exercise 10.2B Reasoning and problem-solving

1 a In a repeated set of trials, X is a random variable for the total number of 'successes.' State three conditions required to be able to model X by a binomial distribution.

In 2016 the proportion of cars having oxides of nitrogen emissions above 0.015 g km^{-1} was 7%. An independent random sample of size 200 cars was taken.

b Find the probability that

i Exactly 16

ii fewer than 16

iii between 10 and 16 (inclusive)

cars had NOX emissions above 0.015 g km^{-1}.

2 For each of the following random variables, state whether the binomial distribution can be used as a good probability model. If it can, state the values of n and p; if it can't, or if its use is questionable, give reasons.

a The number of black counters obtained when 4 counters are chosen, with each being returned before the next is chosen, from a bag containing 6 black and 8 white counters.

b The number of patients in an independent random sample of size 8 at a GP practice who are prescribed antibiotics. You are given that 12% of patients are prescribed antibiotics.

c The number of heads in 5 throws of a biased coin where the probability of a head is 0.6

d The number of throws of a fair coin up to and including the first head.

3 A calculator claims it can randomly generate a digit from 0–9. For any 4 digits generated, the probability of 2 zeros is 0.03. Is the calculator's claim correct? Show your working.

Large data set

4 According to the Department for Transport, in 2016, 60% of all cars registered in Great Britain used petrol as sole fuel for propulsion.

a In a random sample of 40 cars, find the probability that

i 27

ii fewer than 22

cars were petrol fuelled.

b Write down the probability that of the 40 cars

i more than 22

ii more than 26

cars were petrol fuelled.

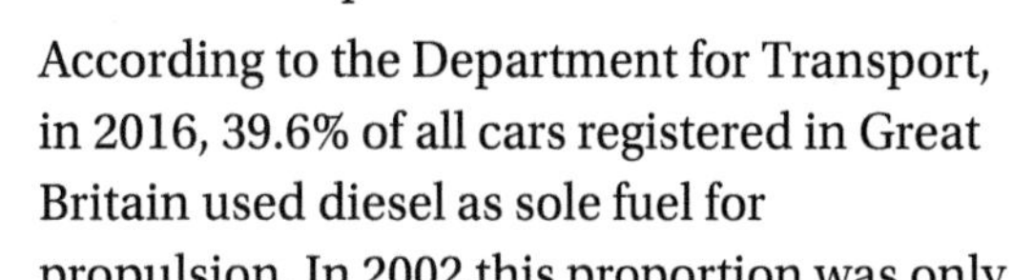

Large data set

5 According to the Department for Transport, in 2016, 39.6% of all cars registered in Great Britain used diesel as sole fuel for propulsion. In 2002 this proportion was only 15.2%.

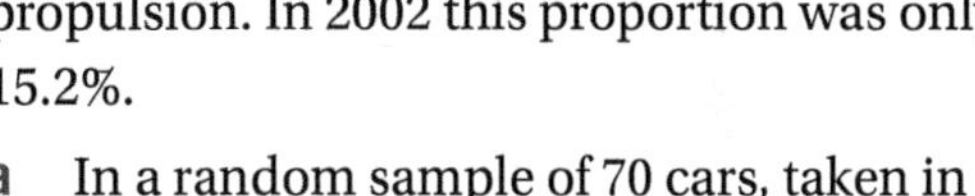

a In a random sample of 70 cars, taken in 2016, find the probability that more than 30 cars would be diesel fuelled.

b How would your answer to part **a** change if the same investigation had been carried out in 2002? Give a reason for your answer.

6 A pair of fair six-sided dice is thrown eight times. Find the probability that a score greater than 7 is scored no more than five times.

7 Somebody claims they can tell the difference between two different brands, A and B, of tea. They are given 5 pairs of cups, where in each pair 1 cup contains brand A and 1 contains brand B. Assuming that they are guessing, find the probability that they correctly identify at least 3 pairs.

Challenge

8 For any family of 5 children, A is the event 'there is at least 1 boy and 1 girl' and B is the event 'there are more girls than boys.' A symmetrical binomial probability distribution can model X, the number of girls in a family of 5 children.

Are the events A and B independent of each other? Show your working.

10 Summary and review

Chapter summary

- $P(A \cup B) = P(A) + P(B)$ if A and B are mutually exclusive.
- The total probability of all mutually exclusive events is 1
- If A is an event associated with a random experiment, then $P(A') = 1 - P(A)$
- For any sample space with N equally probable outcomes, if an event A can occur in $n(A)$ of these outcomes, the probability of A is given by $P(A) = \frac{n(A)}{N}$
- If A and B are independent, $P(A \cap B) = P(A) \times P(B)$
- If A and B are mutually exclusive, $P(A \cap B) = 0$
- A probability distribution for a random experiment shows how the total probability of 1 is distributed between all the possible outcomes.
- The conditions for a binomial probability distribution are:
  - Two possible outcomes in each trial.
  - Fixed number of trials.
  - Independent trials.
  - Identical trials (p is the same for each trial).
- If all the conditions are met, the binomial probability distribution can be used to calculate the probabilities of events expressed in terms of the number of 'successes' in a set of trials.
- If $X \sim B(n, p)$ then $P(X = x) = {}^nC_x p^x (1-p)^{n-x}$, where n is the number of trials and p is the probability of success in any given trial.
- Individual probabilities, $P(X = x)$, and cumulative probabilities, $P(X \le x)$, can be found directly on a calculator.

Check and review

You should now be able to...	Try Questions
✔ Use the vocabulary of probability theory, including the terms random experiment, sample space, independent events and mutually exclusive events.	1, 2
✔ Solve problems involving mutually exclusive and independent events using the addition and multiplication rules.	1, 6, 8
✔ Use a probability function or a given context to find the probability distribution and probabilities for particular events.	1, 2
✔ Recognise and solve problems relating to experiments which can be modelled by the binomial distribution.	3–10

1 You throw a coin until a head or three tails has occurred. If the probability of a head is p, find the probability distribution of the number of throws made. Check that the sum of the probabilities equals 1

2 A random variable X has probability distribution function given by $P(X=x)=k(1-2x)^2$; $x=1, 2, 3, 4$. Find

a The value of k

b $P(X \geq 2)$

3 Given that $X \sim B(16, 0.4)$, find

a $P(X=12)$

b $P(X>6)$

Give your answers to 3 dp.

4 A fair six-sided dice is thrown 5 times and the random variable X denotes the number of times an odd value is obtained.

a Give the distribution of X

b Find, giving your answers to 2 dp,

i $P(X=3)$

ii $P(X>0)$

iii $P(1 \leq X < 3)$

5 Telephone calls to an online bank are held in a queue until an advisor is available. Over a long period the bank has found that 8% of callers have to wait more than 4 minutes for a response. In a random sample of 20 callers, find the probability that fewer than 3 have to wait more than 4 minutes.

6 A pair of fair six-sided dice is thrown five times. Find the probability that a double six is scored no more than once.

7 A box contains 15 coloured beads. Three of the beads are red, 2 are black and the rest are white. A sample of 6 beads is taken, with each bead being returned to the box before the next is chosen. Find the probability that the sample contains more than 3 beads that are not white.

8 In a game of dice, you commit to getting at least one double six in 24 throws of two fair six-sided dice. Your friend says that you won't succeed. Who is more likely to be right? You must show your working.

9 A large packet of mixed flower seeds produce flowers which are either red, blue or white. These colours occur in the ratio 2 : 3 : 5. Find the probability that, when 12 plants are grown, at least 6 of them are white.

10 At a postal sorting office, 32% of letters are classified as large and the rest are standard.

A random sample of 12 letters is taken. The random variable X is the number of standard letters in the sample.

a Write down the probability distribution of X

b Find the probability that more than two-thirds of the letters are standard.

STATS

What next?

Score		
0 – 4	Your knowledge of this topic is still developing. To improve, search in MyMaths for the codes: 2091, 2093, 2094, 2110, 2111, 2114	
5 – 7	You're gaining a secure knowledge of this topic. To improve, look at the InvisiPen videos for Fluency and skills (10A)	
8 – 10	You've mastered these skills. Well done, you're ready to progress! To develop your techniques, look at the InvisiPen videos for Reasoning and problem-solving (10B)	

Click these links in the digital book

History

A popular gambling game in 17th century France was to wager that there would be at least one 6 in every four throws of a standard dice.

Antoine Gombaud, a gambler, reasoned that an equivalent wager was that there would be at least one double 6 in every 24 throws of a pair of standard dice.

Over a period of time, however, Gombaud lost money on the wager. He decided to ask his friend **Blaise Pascal** for help. Pascal, in turn, contacted **Pierre de Fermat** and between them they formulated the fundamental principles of **probability** for the first time.

Blaise Pascal

Pierre de Fermat

Have a go

Find the probability of rolling at least one 6 in four throws of a standard dice.

Find the probability of rolling at least one double 6 in 24 throws of a standard pair of dice.

Explain why Gombaud won money on the first wager but lost money on the second.

Did you know?

Assuming the probability of having a birthday on any given day is the same, you only need 23 people in a room to make it more likely than not that at least two of them share a birthday. If there are 70 people in a room, the chance of two or more of them sharing a birthday is 100% if rounded to three significant figures.

This feels counterintuitive because 23 is such a small number compared to the number of days in a year (365, or 366 on a leap year).

The maths, however, is actually fairly simple. There are ${}^{23}C_2$ (or 253) ways of choosing a pair of random people out of a group of 23. To find the chances of finding *at least one* match, you just need to subtract the chance of no matches from 1.

$$P(\text{2 or more shared birthdays}) = 1 - \left(\frac{364}{365}\right)^{253} = 0.5004\ldots\ (>50\%)$$

Birthdays actually come in clusters, so in reality this number may actually be lower than 23

10 Assessment

1 R and S are two independent events associated with a random experiment.
Given $P(R) = 0.6$ and $P(R \cap S) = 0.15$, work out $P(R \text{ or } S)$. Choose the correct answer.

A 0.85 **B** 0.7 **C** 0.09 **D** 0.75 **[1 mark]**

2 Given $X \sim B(18, 0.7)$, calculate $P(10 \le X < 14)$ to 3 sf. Choose the correct answer.

A 0.695 **B** 0.776 **C** 0.527 **D** 0.608 **[1]**

3 Data is collected on the number of students nationally who like two given brands of fast food.

		Brand B	
		Like	Dislike
Brand A	Like	11713	19981
	Dislike	9061	15457

a How many students are polled in total? **[1]**

b What is the probability a randomly chosen student likes

i Brand A, **ii** Brand B, **iii** Both brands? **[9]**

c Are opinions of the two brands independent? **[2]**

4 On any given day, the probability that a commuter misses their bus to work is $\frac{1}{10}$ and the probability that they miss the bus home is $\frac{1}{12}$. The probability that they accidentally overcook their dinner is $\frac{1}{7}$. These events are independent.

a What does it mean for events to be independent? **[2]**

b Calculate the probability that the commuter misses their bus home and accidentally overcooks their dinner. **[2]**

c Calculate the probability that the commuter misses both buses but doesn't overcook their dinner. **[3]**

5 Two bags contain balls of various colours. A ball is drawn at random from a bag. The probabilities of drawing a specific colour from each bag are given in the table.

Event	**Probability for first bag**	**Probability for second bag**
White	$3k$	?
Blue	$6k$	0.15
Black	$4k$	0.1
Red	$2k$	0.25
Green	$5k$	0.15

a Calculate the value of k **[3]**

b Calculate the probability of drawing a white ball from the second bag. **[2]**

c Tia wants to maximise the probability of drawing a white or a blue ball. Which bag should she choose? **[5]**

6 A satsuma must meet a minimum size requirement in order to be suitable for packaging. Each packet contains 8 satsumas. The grower finds that the probability of a randomly chosen satsuma not being large enough is 0.01

a Find the probability that a random set of 8 satsumas contains at least one that is not suitable for packaging. **[4]**

b Find the probability that a random set of 8 satsumas contains at most one that is not suitable for packaging. **[2]**

A batch is accidentally sent out without being checked for the minimum size. A supermarket receives 60 packets.

c Find the probability that the supermarket has received at least one packet which contains at least one undersized satsuma. **[5]**

7 At a factory, sweets are automatically discarded if they are misshapen. An inspector picks five discarded sweets at random to check that the right decisions are being made. If at least four of the discarded sweets are misshapen, then the inspector is satisfied.

a What conditions must be true for the binomial distribution to be a suitable model for this situation? **[2]**

On average, 84 out of 360 discarded sweets are not misshapen.

b Find the probability that the first four inspected sweets are misshapen but then the fifth is fine. **[2]**

c Find the probability that exactly one sweet is not misshapen. **[2]**

d Find the probability that the inspector is satisfied. **[4]**

8 In a football tournament, only two teams have a chance of winning. Team B will only win the tournament if they win all three of their remaining matches *and* Team A fails to win any of its four remaining matches to win the league. All match results are independent. The probability that Team A wins any of their matches is 0.56 per match and the probability that Team B wins any of their matches is 0.61 per match. Find the probability that Team A wins the tournament. **[4]**

9 Data is collected on the amount of money people spend per week on clothes and their relative levels of income in England.

		Household level of income	
Spending on clothes		Below average	Above average
	Above average	25	61
	Below average	40	58

a How many people are included in the survey? **[1]**

b What is the probability a randomly chosen person has

i Above-average spending on clothes,

ii Below-average household level of income,

iii Both above-average spending on clothes and below-average household level of income? **[3]**

c Are household level of income and amount of spending on clothes independent in this sample? Justify your answer. **[2]**

10 During each day in 2016, a Volkswagen car was randomly selected and its exhaust emissions measured. The probability that the car had 'high' carbon dioxide emissions, above 150 g km^{-1}, was 0.107

Assume that this probability was constant throughout the year.

a In one given week, calculate the probability that at least one car had high CO_2 emissions. **[3]**

b In one given week, calculate the probability that at most one car had high CO_2 emissions. **[2]**

c Find the probability that on at least 20 weeks in the year a car with high emissions is selected. **[2]**

d What is the expected number of weeks in the year that at least one car with high emissions is selected? **[2]**

11 Every week, three car makers test the exhaust emissions of a random sample of their cars. The probabilities that car makers A, B and C measure the highest average emission in a given week are 0.41, 0.36 and 0.23 respectively. Find the probability that over a five week period car maker A records the highest emissions on more weeks than the other two car makers. **[4]**

11 Hypothesis testing 1

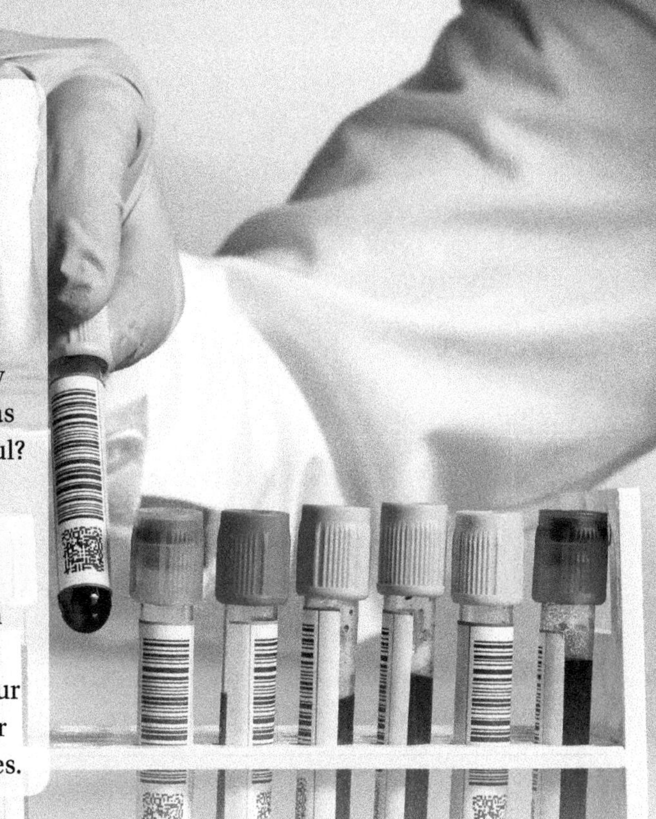

Before a new medicine can be released for public use, it has to go through a rigorous testing process. These tests include medical trials on people who are candidates for the treatment. This allows for the monitoring of side effects, the observation of how many trial patients actually respond to the treatment, and consideration for what level of improvement is deemed sufficient. For example, if a drug significantly improves the condition of 40% of trial patients, but has no effect on the rest, would it be considered successful?

In order to test a premise, you must have a clearly defined hypothesis as well as unbiased methods for collecting and analysing data. Does the data follow any kind of pattern, or fit a known model? How much error is there in the measurements and how much error is acceptable? What would lead you to reject your hypothesis? This is all part of the scientific method for investigating phenomena and making new discoveries.

Orientation

What you need to know

KS4
- Interpret, analyse and compare the distributions of data sets.
- Recognise appropriate measures of central tendency and spread.

p.261 **Ch10 Probability and discrete random variables**
- Binomial distributions.

What you will learn
- To understand the terms null hypothesis, alternative hypothesis, critical value, critical region and significance level.
- To calculate the critical region and the p-value.
- To decide whether to reject or accept the null hypothesis and make conclusions.

What this leads to

Ch21 Hypothesis testing 2
Testing correlation.
Testing a normal distribution.

Careers
Scientific research.
Quality control.

MyMaths Practise before you start 2111

11 Bridging Unit

Topic A: Hypothesis testing

Bridging to Ch11.1

A hypothesis test is used to determine whether or not a statement or assumption is likely to be true.

Imagine you have a coin and you wish to determine whether it is 'fair'; in other words, is it a reasonable assumption that the probability of 'heads' is 0.5? You flip the coin 20 times and record the number of 'heads.' The coin lands on heads only 8 times. You may want to say the coin is biased and that the probability of it landing on heads is less than 0.5. But does the evidence support this claim? The assumption that the coin is fair ($p = 0.5$) is called the null hypothesis.

Key point

The **null hypothesis** is your starting assumption. It states that a parameter, such as the probability, takes a certain value, and you assume this to be true in your calculations.

The claim that the coin is biased towards tails ($p < 0.5$) is called the alternative hypothesis.

Key point

The **alternative hypothesis** contradicts the null hypothesis. It might state that the true value is greater than, less than or simply different to the value stated in the null hypothesis.

If the coin is fair, then you expect 10 heads out of 20 flips. However, this is just an average, and you won't actually get 10 heads every time you repeat the experiment. 8 isn't far off 10 so it's quite possible that you will only get 8 heads even if the coin is fair. So this result does not provide evidence that the coin is biased.

What about if you only got 7 heads? Or 6? It all depends on how certain you want to be that the coin is biased and that the result wasn't just a fluke with a fair coin.

Eventually, you get to a point where you consider the chance of randomly getting the observed result *or lower* to be so small that you can reasonably assume the null hypothesis is wrong, and therefore you can reject it in favour of the alternative hypothesis.

The probability at this point is equal to the significance level.

Key point

The **significance level** is the probability of rejecting the null hypothesis incorrectly when it's actually true.

In other words, you reject the null hypothesis if the chance of randomly getting the observed value, or a more extreme value, is less than or equal to the significance level.

Common significance levels are 1%, 5% and 10%, but in this section you can assume it is always 5%.

A significance level of 5% means that if the null hypothesis is true you have a 5% chance of incorrectly rejecting it. If you use a significance level of 1% then your chance of incorrectly rejecting the null hypothesis goes down to 1%.

Key point

To determine the result of a hypothesis test, assume the null hypothesis is true in your calculations and work out the probability of getting an observed result or something more extreme by fluke. If the probability you calculate is less than or equal to the significance level, you should reject the null hypothesis in favour of the alternative hypothesis.

Example 1

A coin is flipped 20 times and lands on heads 7 times.

The probability that a fair coin when flipped 20 times lands on heads 7 or fewer times is 0.1316

a Do you think the coin is fair or biased towards tails?

The table gives the probability of different results when a fair coin is flipped 20 times.

Number of heads	Probability
7 or fewer	0.1316
6 or fewer	0.0577
5 or fewer	0.0207
4 or fewer	0.0059

b Use the table to suggest how few heads would need to be flipped to support the claim that the coin is biased towards tails.

a The probability that a fair coin lands on heads 7 or fewer times out of 20 is 13.16%

The probability is much bigger than the significance level, which is 5%

This is higher than the significance level, so we can't say with much confidence that the coin is biased. Therefore, we should continue to assume that it is a fair coin.

b The probability that a fair coin lands on heads 5 or fewer times out of 20 is 2.07%

The probability is smaller than 5%

So there is only a 2.07% chance of getting this result if the coin is fair.

Therefore, if you get 5 heads, this is low enough to support the claim that the coin is biased towards tails.

If you get fewer than 5 heads then this is even better evidence to support the claim of bias towards tails.

Try It 1

A coin is flipped 30 times and lands on heads 12 times. The probability that a fair coin when flipped 30 times lands on heads 12 or fewer times is 0.1808

a Do you think the coin is fair or biased towards tails? Explain your answer by referring to the probability.

The table gives the probability of different results when a fair coin is flipped 30 times.

b Use the table to suggest how few heads would need to be flipped to support the claim that the coin is biased towards tails.

Number of heads	Probability
8 or fewer	0.0081
9 or fewer	0.0214
10 or fewer	0.0494
11 or fewer	0.1002

If a coin is fair, then as the number of flips increases the number of heads will tend toward 50% of the total. So, with a smaller number of flips, a more 'extreme' result is needed to imply the coin is biased. For example, if you only flip the coin 4 times then you can never get a result that supports a claim that the coin is biased. As you can see in the table, even if you don't get any heads at all, the probability of this is 6.25%, which is more than 5%.

Number of heads	Probability
0	0.0625
1 or fewer	0.3125
2 or fewer	0.6875
3 or fewer	0.9375

However, if you flip the coin 1000 times, then the probability of 473 or fewer heads is 0.0468. This is less than 5%, so it provides strong evidence that the coin is biased towards tails - even if 473 seems relatively close to 500.

Example 2

A coin is flipped 8 times and lands on heads only once. The probability of a fair coin landing on heads 1 or fewer times out of 8 is 0.0352

a Do you think the coin is fair or biased towards tails?

A second coin is flipped 16 times and lands on heads twice.

b Use your result from part **a** to suggest whether the second coin is fair.

a The probability of 1 or fewer heads is 3.52%

This is less than 5%.

So there is evidence that the coin is biased since there is only a 3.52% chance that a fair coin would give a result this extreme.

b 16 is double 8 so 2 out of 16 is the same proportion as 1 out of 8

1 head of 8 implied that the coin is biased.

So 2 heads out of 16 will definitely also imply that the coin is biased, since a larger number of trials requires a less 'extreme' result to imply the coin is biased.

In fact, it can be calculated that only 4 or fewer heads are required to imply the coin is biased in this case.

A coin is flipped 12 times and a result of 2 or fewer heads is sufficient to imply that the coin is biased. A second coin is flipped 120 times and lands on heads 20 times. What can you conclude about this coin?

1 The table gives the probability of different results when a fair coin is flipped 80 times.

Number of heads	Probability
28 or fewer	0.0048
29 or fewer	0.0092
30 or fewer	0.0165
31 or fewer	0.0283
32 or fewer	0.0465
33 or fewer	0.0728

Use the table to suggest how few heads would need to be flipped to support the claim that the coin is biased towards tails.

You should compare the results to 5% as in the examples.

2 The table gives the probability of different results when a fair coin is flipped 50 times.

Number of heads	Probability
17 or fewer	0.0164
18 or fewer	0.0325
19 or fewer	0.0595
20 or fewer	0.1013

For each of these experiments, explain what conclusion you would draw about whether the coin is fair or biased towards tails.

You should compare the results to 5% as in the examples.

a A coin is flipped 50 times and lands on heads

  i 20 times,

  ii 18 times,

  iii 15 times,

  iv 22 times.

b A coin is flipped 100 times and lands on heads 36 times.

c A coin is flipped 25 times and lands on heads 10 times.

Bridging: STATS

MyMaths 2115 SEARCH

11.1 Formulating a test

Fluency and skills

A dietitian claims that 85% of his clients consume less than 42 g of butter per week. You think this proportion is too high and take a random sample of 10 clients. If only 8 of the sample consume less than 42 g butter per week, would you conclude that 85% is too high? What if only 6 or 7 meet this target?

You use a **hypothesis test** to determine whether to accept or reject the dietitian's claim.

Key point

The **null hypothesis**, H_0, is a statistical statement representing your basic assumption.

Don't reject the null hypothesis until there is sufficient evidence to do so.

Key point

The **alternative hypothesis**, H_1, is a statement that contradicts the null hypothesis.

Example 1

A car maker claims that '75% of our cars have carbon monoxide emission below 0.5 g km^{-1}'.
Let p equal the probability that a randomly selected car, of the given make, has emissions below 0.5 g km^{-1}.
Write down the null and alternative hypotheses for these three cases.

a The claim is believed to be an overestimate.

b The claim is believed to be an underestimate.

c The claim is believed to be wrong.

a $H_0: p = 0.75$ and $H_1: p < 0.75$

b $H_0: p = 0.75$ and $H_1: p > 0.75$

c $H_0: p = 0.75$ and $H_1: p \neq 0.75$

H_1 expresses the alternative hypothesis.

If H_0 is an underestimate then the true value of p must be higher.

The true value of p could be higher or lower

Key point

The null hypothesis always includes the equality sign.

In parts **a** and **b** of Example 1, you can test the null hypothesis against the alternative hypothesis with a **one-tailed test**. This only tests either *below* the value stated in H_0 or only *above* the value stated in H_0

In part **c**, you can test the hypothesis with a **two-tailed test**. This tests both below and above the value stated in H_0 ($p < 0.85$ and $p > 0.85$)

Let X be the random variable representing the number of clients, in the chosen group of 10, who meet the target. X is called the **test statistic.**

The value of X determines whether you accept or reject the null hypothesis.

Assuming H_0 is true, X is binomially distributed: $X \sim B(10, 0.85)$

See Ch 10.2

For a reminder of the binomial distribution.

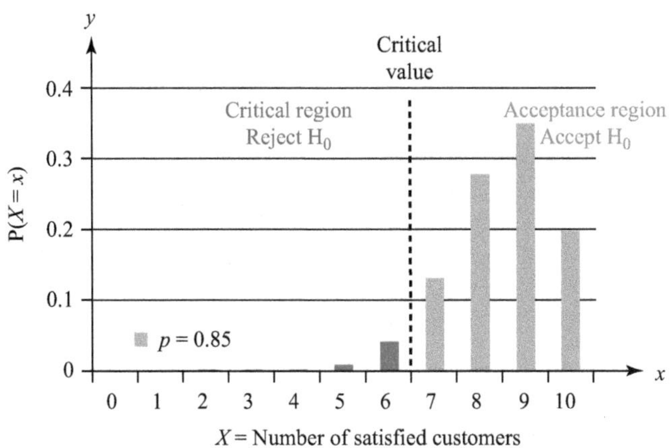

For every hypothesis test, there is a set of values of X for which you accept H_0 and a set for which you reject H_0

Key point

The **critical region** is the set of values that leads you to *reject* the null hypothesis. The **acceptance region** is the set of values that leads you to *accept* the null hypothesis.

The **critical value** lies on the border of the critical region. It depends on the **significance level** of the test. The critical region includes the **critical value** and all values that are more extreme than that.

In a discrete distribution such as the binomial distribution, the probability of incorrectly rejecting H_0 is the probability represented in the critical region and is therefore **less than or equal to** the significance level.

Key point

Every hypothesis test has a significance level. This is equal to the probability of *incorrectly* rejecting the null hypothesis.

If you use a *low* significance level, e.g. 1%, you can be fairly certain of your result if you reject the null hypothesis. If you reject the null hypothesis when using a *high* significance level, e.g. 20%, you may need to carry out further tests to be sure of your result. Significance levels of 10%, 5% and 1% are often used, but the significance level could be any value.

After carrying out a hypothesis test, you must state whether you accept H_0 or reject H_0

You must also state a **conclusion** that relates directly to the problem.

There is always a chance that your decision to accept H_0 or reject H_0 is wrong. Your conclusion must use language that reflects this.

A different significance level may produce a different critical region and so may lead to a different conclusion.

The lower the significance level, the smaller the critical region, and vice versa.

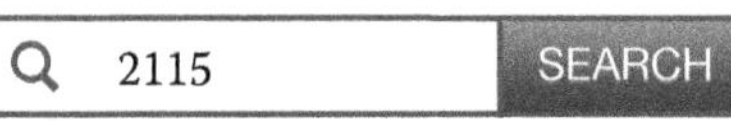

Example 2

Let X be the number of clients who meet a dietary target in a sample of 10 clients of a dietitian. Let p be the probability that a client, chosen at random, meets the target. A hypothesis test is carried out to assess the dietitian's claim $p = 0.85$ against the alternative hypothesis $p < 0.85$

At a significance level of 5%, the critical region is $X \le 6$

If the significance level is changed to a%, the critical region is $X \le 4$

a Write an inequality for a

b A sample is taken and 6 clients out of 10 meet the target. Write down the conclusion if the significance level is

i 5% **ii** a%

a $a < 5$

The size of the critical region is reduced so, assuming H_0 is correct, the probability of getting a value in the critical region is reduced.

b i Reject H_0, as there is enough evidence to suggest that the dietitian's claim is too high.

6 falls into the critical region $X \le 6$. State your answer in context.

ii Accept H_0, as there is not enough evidence to suggest that the dietitian's claim is too high.

6 falls into the acceptance region $X \ge 5$

Key point

As you lower the significance level, you need more evidence to reject the null hypothesis and you lower the chance of making an incorrect conclusion.

Example 3

A driving instructor claims that 70% of his candidates pass first time. An inspector thinks that this is inaccurate, so he does a survey of 25 former candidates and records the number who passed first time. The significance level of his test is 10% and the critical values are 14 and 21

The null hypothesis is that the driving instructor's claim is correct, so H_0: $p = 0.7$ where p is the probability that a candidate passes first time.

The alternative hypothesis is that the driving instructor's claim is wrong, so H_1: $p \ne 0.7$

a State the critical region and the acceptance region for the test.

b State whether the inspector would accept or reject the null hypothesis if he found that

i 14 of the former candidates passed first time,

ii 20 of the former candidates passed first time.

X is the number of the 25 former candidates who passed first time.

a Critical region is $X \le 14$ and $X \ge 21$

The critical region consists of the critical value, 14, and more extreme values. Here, that is values less than 14

Acceptance region is $15 \le X \le 20$

This is everything not in the critical region. X can only be an integer in this situation.

b i Reject the null hypothesis.

14 is the critical value and so lies in the critical region.

ii Accept the null hypothesis.

20 lies in the acceptance region.

Exercise 11.1A Fluency and skills

In questions **1** to **3**, p is the probability of an event occurring.

1 State the critical region and acceptance region for the following tests.

a H_0: $p = 0.6$ and H_1: $p < 0.6$
The critical value is 5

b H_0: $p = 0.6$ and H_1: $p > 0.6$
The critical value is 5

c H_0: $p = 0.6$ and H_1: $p \neq 0.6$
The critical values are 5 and 12

2 State whether to accept or reject the null hypothesis for the following tests.

a H_0: $p = 0.4$ and H_1: $p < 0.4$
The critical value is 3 and in the sample taken, X takes a value of 4

b H_0: $p = 0.3$ and H_1: $p > 0.3$
The critical value is 17 and in the sample taken, X takes a value of 18

c H_0: $p = 0.7$ and H_1: $p \neq 0.7$
The critical values are 2 and 11 and in the sample taken, X takes a value of 10

3 a H_0: $p = 0.4$ and H_1: $p < 0.4$
In the sample taken, X takes a value of 18. H_0 is rejected at the 10% significance level. State whether you accept or reject the null hypothesis at a 20% significance level.

b H_0: $p = 0.3$ and H_1: $p > 0.3$
In the sample taken, X takes a value of 18. H_0 is accepted at the 10% significance level. State whether you accept or reject the null hypothesis at a 5% significance level.

4 A football coach claimed that he lost only 15% of his games. One of his players thinks that this claim is inaccurate and decides to test it at the 5% significance level.

A random sample of 50 games is taken. The critical values for the number of losses are 2 and 14

a Write a null and alternative hypothesis to represent this situation.

b State the critical region and the acceptance region for the test.

c State whether you accept or reject the null hypothesis if in the sample

i 3 of the games were lost,

ii 15 of the games were lost,

iii 14 of the games were lost.

5 A courier company states on its website that 90% of its parcels arrive within 24 hours. An investigator thinks that this claim is too high, so he tests it by taking a random sample of 40 parcels.

The critical value at the 10% significance level is 33

a State the critical region and the acceptance region for the test.

b State whether you accept or reject the null hypothesis if

i 32 of the parcels arrive within 24 hours,

ii 35 of the parcels arrive within 24 hours.

6 A factory worker estimates that 1 in 20 products on the production line are faulty. A supervisor thinks that his estimate is too low, so she tests it at the 10% significance level.

A random sample of 100 products is taken. The critical value is 9

a State the critical region and the acceptance region for the test.

b The supervisor says that the critical value for the 20% significance level is more than 9. Is she correct?

7 A vehicle inspector thinks that 1 in 25 cars do not have the correct tyre pressure. A colleague believes that the proportion suggested by the inspector may be higher, so she tests it at the 10% significance level.

A random sample of 60 cars is taken. She measures tyre pressure and notes the number of cars which do not have the correct pressure. The critical value is 5

a State the critical region and the acceptance region for the test.

b The critical value for the p% significance level is more than 5. Find an inequality for p

Reasoning and problem-solving

Strategy

To interpret the result of a hypothesis test

1. Identify the critical region and draw a diagram if it helps.
2. If the value from the sample lies in the critical region then you reject the null hypothesis. If the value does not lie in the critical region then you accept the null hypothesis.
3. End with a conclusion that relates back to the situation described in the question.

Know your data set

The heaviest car in the LDS weighs 2450 kg, a 2002 Vauxhall, and the lightest car weighs 925 kg, a 2016 Toyota. There are 92 zero mass entries in the 2002 data. Excluding these, the average car in the sample weighs 1404 kg. Click this link in the digital book for more information about the Large data set.

Example 4

An environmentalist claims that only 10% of cars weighing more than 2300 kg are Toyotas. A researcher believes that this claim is not true. To test the claim she takes a random sample of 60 cars weighing more than 2300 kg and counts the number of Toyota cars in the sample. The critical values at the 5% significance level are 1 and 12.

a State the hypotheses needed to test this claim.

b If 15 cars were found to be Toyotas, say whether you would accept or reject the null hypothesis. State your conclusion.

c Assuming that the environmentalist's claim is rejected, write an inequality for n, the possible number of cars in the sample that are not Toyotas.

a Let p = the probability that a car in the sample is a Toyota. *(State what p represents.)*

$H_0: p = 0.1 \quad H_1: p \neq 0.1$ *(H_1 expresses the alternative hypothesis. Since the researcher does not believe the claim, p could be higher or lower. It is a two-tailed test.)*

b Let X = number of Toyota cars in the sample.

The critical region is $X \leq 1$ and $X \geq 12$ *(1 The critical values are included in the critical region.)*

Reject the null hypothesis. *(2 15 lies in the critical region.)*

There is evidence, at the 5% significance level, to say that it is not true that 10% of cars weighing more than 2300 kg are Toyotas. *(3 State your conclusion.)*

c Number of Toyotas in sample = $60 - n$

Reject $\Rightarrow 60 - n \leq 1$ or $60 - n \geq 12$

$n \geq 59$ or $n \leq 48$ *(3 State your conclusion.)*

Exercise 11.1B Reasoning and problem-solving

1 State H_0 and H_1 in each of the following cases.

  a The probability of an event occurring is stated as 0.6 but this is thought to be too low.

  b The probability of an event occurring is stated as 0.7 but this is thought to be too high.

  c The probability of an event occurring is stated as 0.4 but this is thought to be inaccurate.

2 A car dealer estimates that a quarter of the cars he sold in 2016 had engine sizes above 1975 cm^3. He looks at a sample of 20 sold cars. The critical values at the 5% confidence level are 1 and 9

  a State the hypotheses needed to investigate this claim.

  b If the null hypothesis is accepted then state the inequality for n, the number of cars with engine sizes above 1975 cm^3.

3 Using the large data set for 2016, 202 out of 809 Vauxhall cars in the survey had carbon monoxide emissions above 0.5 g km^{-1}. An environmentalist thinks that at a particular dealership the proportion of Vauxhall cars with 'high' CO emissions was higher than this. The claim is tested using the records for 150 Vauxhall cars sold at the dealership in 2016.

The critical value at the 15% confidence level is 44.

  a State the hypotheses needed to investigate this claim.

  b If 41 cars had CO emissions above 0. 5 g km^{-1}, decide whether you would accept or reject the null hypothesis and state your conclusion.

  c Suppose the conclusion is 'there is evidence that the proportion of 2016 Vauxhall cars with high CO emissions is higher at the dealership than the survey would suggest'. What is the lowest possible number of cars in the sample that had CO emissions above 0.5 g km^{-1}?

4 James claims that he can hit the bullseye with a dart on 25% of his throws. Mia thinks that James hits the bullseye less often than that. The critical value at the 10% significance level is 6

  a State the hypotheses needed to investigate the claim.

  b If James hit the bullseye with 5 of his darts state whether you would accept or reject the null hypothesis. State the conclusion.

  c James hit the bullseye with n darts and the conclusion was 'there is not sufficient evidence, at the 10% significance level, to say that his rate of hitting the bullseye is less than 25%'. State an inequality for n

  d The critical value at the x% significance level is 5. State an inequality for x

Challenge

5 $H_0: p = 0.2$ $\quad$ $H_1: p \neq 0.2$

A random sample of 80 is taken.

The critical values at the 5% significance level are 8 and 24

  a Find inequalities for a and b where a and b are the critical values at the 10% significance level and $a < b$

  b Find an inequality for x if 7 and 26 are the critical values at the x% significance level.

STATS

2115 SEARCH

11.2 The critical region

Fluency and skills

Suppose somebody claims that 40% of people prefer apples to pears. You disagree with this claim, and believe the true value to be lower than 40%. You plan to investigate this claim by asking a sample of n people whether they prefer apples to pears. The hypotheses are then:

$H_0: p = 0.4$ $\qquad$ $H_1: p < 0.4$

where p is the probability of a person, chosen at random, saying they prefer apples to pears.

In this kind of hypothesis test, you can find the critical values or critical region using the binomial distribution, provided certain conditions are true: samples are chosen at random, there are two possible outcomes (in this case, yes or no), the outcomes are independent of one another, and the same chance of each outcome can be assumed each time.

You can therefore assume that the number of people, X, who say they prefer apples to pears follows a binomial distribution with n trials. Thus $X \sim B(n, p)$

Say you ask a random sample of 50 people. Assuming the null hypothesis to be true, you would expect $x \approx 20$, where x is the number who say yes. The probabilities for any observed value of x are binomially distributed around $x = 20$, as shown in the graph for $X \sim B(50, 0.4)$

Always assume H_0 to be true in your calculations.

The alternative hypothesis is $p < 0.4$. That means this is a one-tailed test and the critical region involves values equal to or lower than the critical value. Say you are testing at a significance level of 5%: that gives a critical value such that $P(X \le x) = 0.05$. Using your calculator, you can see that $P(X \le 14) = 0.054$ and $P(X \le 13) = 0.028$

See Ch10.2
To check how to calculate probabilities using the binomial theorem.

Because the number of people can only take an integer value, the critical value must also be an integer. So you choose the first value where the probability is more extreme than the significance level, that is, $X = 13$. So if 13 or fewer people say they prefer apples to pears, you would reject H_0

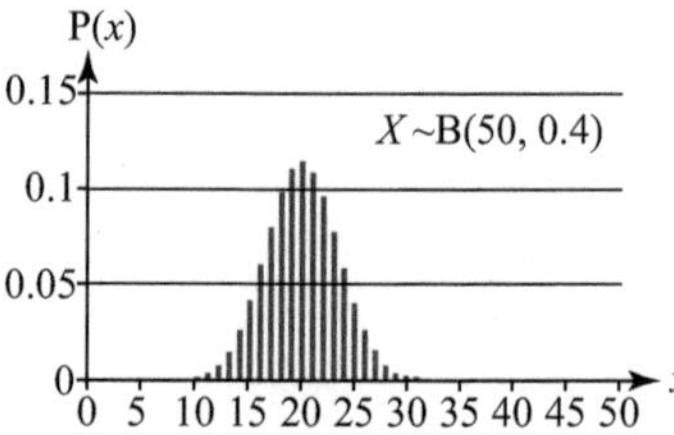

▲ If you assume p to be 0.4, then the most likely outcome is for 20 of the 50 people to prefer apples. But that doesn't mean other numbers are impossible: as you can see, the probability of exactly 20 people is only just over 0.1

Key point

If H_0 is assumed to be true, for discrete random variables, a value lies within the critical region if the probability of X being *equal to or more extreme* than that value is *equal to or less than* the significance level.

In many cases, instead of finding the critical region, it is easier to simply calculate the probability that X is *equal to or more extreme* than the observed value. If the probability of any observed value is *less than or equal to* the significance level, you have grounds to reject the null hypothesis.

Key point

The **p-value** is the probability that X is *equal to or more extreme* than an observed value. If the p-value is *greater* than the significance level, you accept H_0. If the p-value is *less than or equal to* the significance level, you reject H_0

So, going back to the case of people's preference for apples and pears: the p-value for 13 people preferring apples is 0.028 (=2.8%) and the p-value for 14 people preferring apples is 0.054 (=5.4%). At a significance level of 5%, you would reject the null hypothesis for $x = 13$ and accept the null hypothesis for $x = 14$

If you had been testing at a higher significance level, say 10%, you would reject the null hypothesis in both cases, because the p-value for both is lower than 10%

Example 1

You are told that $X \sim B(40, p)$ and $H_0: p = 0.25$, $H_1: p \neq 0.25$

For each of the following values, **i** State the p-value to 4 decimal places and **ii** State, with clear reason, your conclusion for the hypothesis test at a 10% significance level.

a $X = 7$ **b** $X = 16$

$0.25 \times 40 = 10$

a i $P(X \leq 7) = 0.1820$

ii The p-value is greater than 0.05, so accept H_0

b i $P(X \geq 16) = 1 - P(X \leq 15)$

$= 1 - 0.9738$

$= 0.0262$

ii The p-value is less than 0.05, so reject H_0

10 is the most likely outcome. This is a two-tailed test, so you must consider values much higher and much lower than 10

Use your calculator to find the p-value for $X = 7$

STATS

Exercise 11.2A Fluency and skills

1 $X \sim B(30, p)$ and

$H_0: p = 0.2$ $H_1: p < 0.2$

a Find the critical region for X if the significance level is 2%

b State the conclusion if $X = 1$

2 $X \sim B(20, p)$ and

$H_0: p = 0.4$ $H_1: p \neq 0.4$

Find the critical region for X if the significance level is 10%

3 $X \sim B(30, p)$ and

$H_0: p = 0.15$ $H_1: p < 0.15$

The significance level is 10%

Find the p-value (to 4 decimal places) and state the conclusion if

a $X = 1$ **b** $X = 2$

4 $X \sim B(25, p)$ and

$H_0: p = 0.2$, $H_1: p \neq 0.2$

The significance level is 5%

Find the p-value (to 4 decimal places) and state the conclusion if

a $X = 1$ **b** $X = 9$

5 $X \sim B(60, p)$ and

$H_0: p = 0.45$ $H_1: p \neq 0.45$

The significance level is 10%.

Find the p-value (to 4 decimal places) and state the conclusion if

a $X = 34$ **b** $X = 21$

6 $X \sim B(30, p)$ and

$H_0: p = 0.15$, $H_1: p > 0.15$

Find the critical region for X if the significance level is 15%

7 $X \sim B(40, p)$ and

$H_0: p = 0.35$, $H_1: p \neq 0.35$

A test has significance level 1%

Without finding the critical region, state the conclusion if

a $X = 6$ **b** $X = 22$

8 $X \sim B(50, p)$ and $H_0: p = 0.3$, $H_1: p > 0.3$

a Find the critical region for X if the significance level is

i 0.2% ii 2% iii 20%

b State the conclusion in each case if $X = 22$

c The critical value for the p% significance level is 21. Use your answer to part **a** to find an inequality for p

9 $X \sim B(80, p)$ and

$H_0: p = 0.7$ $H_1: p > 0.7$

a Find the critical region for X if the significance level is

i 1% ii 5% iii 10%

b State the conclusion in each case if $X = 64$

c The null hypothesis is rejected at the p% significance level when $X = 64$. Find the smallest possible whole number value of p

10 $X \sim B(40, p)$ and $H_0: p = 0.4$, $H_1: p < 0.4$

The critical region for X is $X \le 8$ at the y% significance level. Find the greatest possible whole number value of y

11 $X \sim B(50, p)$ and $H_0: p = 0.05$, $H_1: p < 0.05$

a Find the critical region for X if the significance level is 10%

b Find the smallest integer value for x such that the critical region at the x% significance level is different to your answer to **a**.

12 $X \sim B(60, p)$ and $H_0: p = 0.2$, $H_1: p \ne 0.2$

The critical region for X is $X \le 6$, $X \ge 18$ at the $2a$% significance level.
Find the two possible integer values of a

Reasoning and problem-solving

Strategy

To find a critical region and interpret the result when X is binomially distributed

1. Define X, state its distribution and write down H_0 and H_1
2. Assume that H_0 is true and either find the critical region or calculate the p-value.
3. Decide whether to accept or reject H_0 and interpret the result in the context of the question.

To experiment with hypothesis testing, click this link in the digital book.

Example 2

The Department for Transport says that 19% of new cars sold in 2017 was made by Ford or Vauxhall. A journalist thinks that this is no longer true and randomly selects the records of 28 new cars sold this year. She uses a hypothesis test at the 10% significance level to test the claim. Formulate the hypothesis test and state the journalist's conclusion if she found that 26 of the cars in her sample were made by neither Ford nor Vauxhall.

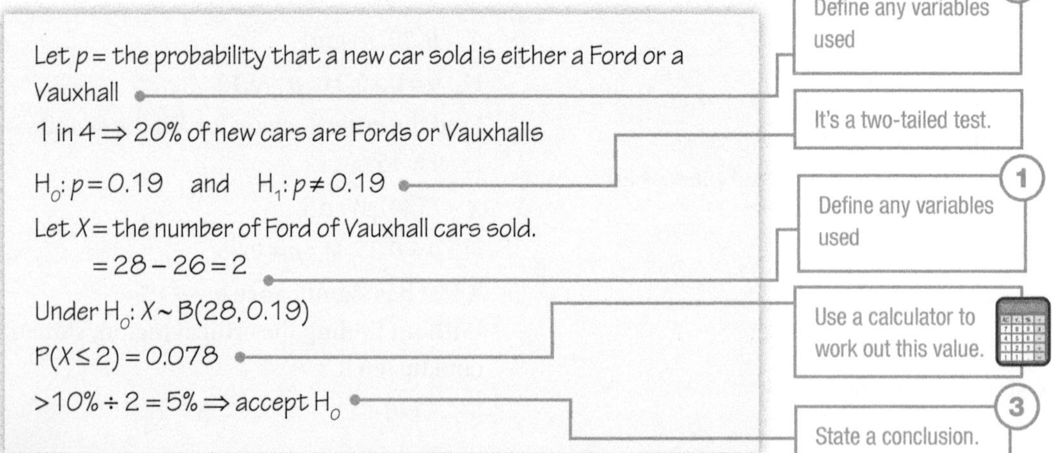

Let p = the probability that a new car sold is either a Ford or a Vauxhall
1 in 4 ⇒ 20% of new cars are Fords or Vauxhalls
H0: p = 0.19 and H1: p ≠ 0.19
Let X = the number of Ford of Vauxhall cars sold.
= 28 − 26 = 2
Under H0: X ~ B(28, 0.19)
P(X ≤ 2) = 0.078
>10% ÷ 2 = 5% ⇒ accept H0
1 Define any variables used
It's a two-tailed test.
1 Define any variables used
Use a calculator to work out this value.
3 State a conclusion.

Exercise 11.2B Reasoning and problem-solving

1 A BMW dealer estimates that 17 in 20 BMWs are owned by men. To test this claim is not an over estimate, 40 records for BMW cars are randomly selected and the gender of the registered owner noted. The test uses a 5% significance level.

a State the hypotheses for this test and the critical region.

b What is the conclusion if 32 of the owners are male?

2 A car maker estimates that 65% of new cars are bought by companies. An employee believes this claim is inaccurate and looks at a randomly selected set of records for 70 cars bought in 2016 in order to carry out a hypothesis test at the 1% significance level. She found that 54 of the cars were company owned.

a State the hypotheses for this test and calculate the p-value, to five decimal places.

b What conclusion should the employee make?

3 In a large container of sweets, 15% are blackcurrant-flavoured. After a group of children have eaten a lot of the sweets, one of the children wants to see whether the proportion of blackcurrant sweets in the container has changed. She selects a random sample of 60 sweets and finds that 4 of them are blackcurrant-flavoured.

a Stating your hypotheses clearly, test at the 10% level of significance whether or not there is evidence that the proportion of blackcurrant-flavoured sweets has changed.

b If the child discovers that she miscounted and that actually there are 5 blackcurrant-flavoured sweets, would your conclusion change?

4 When a coin is tossed a number of times, more than twice as many heads as tails are recorded.

a Use the p-value to test, at the 5% level, whether a coin is biased towards heads if

i The coin is tossed 4 times,

ii The coin is tossed 24 times.

b If the conclusion is 'there is sufficient evidence to say that the coin is biased towards heads', find the smallest number of times that the coin could have been tossed.

5 The discrete random variable X has the following probability distribution.

x	0	1	2	3	4	5	6
$P(X=x)$	$5k^2$	$2k$	0.1	$3k$	$10k^2$	0.2	0.05

a State the value of k

b If the critical values are 0 and 6 then find a lower bound for the significance level.

6 A grocer found that 1 in 10 crates of strawberries from a supplier contained rotten fruit. She suspects that over time this proportion has reduced. She carries out a hypothesis test at the 10% significance level on the next 40 crates that are brought in.

a State clearly H_0 and H_1

She initially thought that n of these crates contained rotten fruit and concluded that she should reject H_0. She then found that one more crate contained rotten fruit and concluded that she should accept H_0

b Find the value of n

Challenge

7 A coin is tossed n times. X represents the number of heads recorded. It is tested at the 2% level to see whether the coin is fair. The critical region is $X \leq 12$ and $X \geq 28$

a Find the value of n

b Find the critical region for a 1% significance level.

STATS

11 Summary and review

Chapter summary

- The null hypothesis, H_0, is the basic assumption in a statistical test. Do not reject the null hypothesis until there is sufficient evidence to do so.
- If H_1 has the form $p <$.... only look at the left tail of the distribution (green in the diagram).
- If H_1 has the form $p >$.... only look at the right tail of the distribution (red in the diagram).
- If H_1 has the form $p \neq$.... look at both tails of the distribution (green and red in the diagram).
- In a two-tailed test you always consider half of the significance level at the left tail and half of the significance level at the right tail.

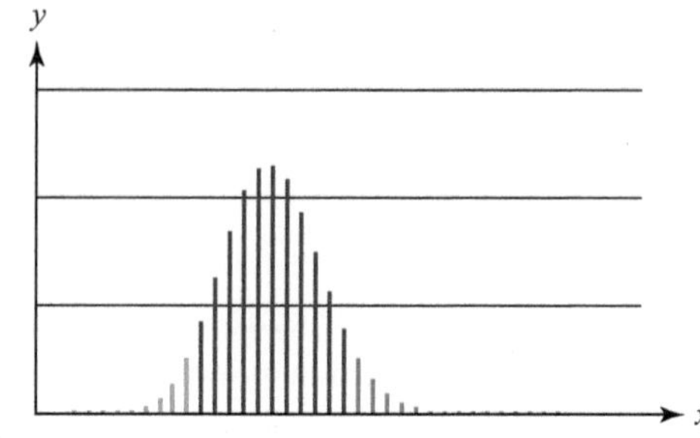

- The critical value lies in the critical region and is the cut-off point for whether values are in the critical region or not. The lower the significance level, the smaller the critical region.
- If you are given a value and asked to test it then it is often quicker to use the p-value method.
- In every hypothesis test you must state whether you accept or reject the null hypothesis.
- If the context is given for a hypothesis test then you must state a conclusion that relates directly to the context of the question.

Check and review

You should now be able to...	Try Questions
✔ Understand the terms null hypothesis and alternative hypothesis.	1–6
✔ Understand the terms critical value, critical region and significance level.	1–6
✔ Calculate the critical region.	2, 3, 5, 6
✔ Calculate the p-value.	4, 6
✔ Decide whether to reject or accept the null hypothesis.	4, 6
✔ Make a conclusion based on whether you reject or accept the null hypothesis.	4, 6

1 A proportion, p, of individuals in a large population have a characteristic, C. An independent random sample of size n is taken from the population and the number, X, of individuals with C is noted. Write down the distribution of X

2 A grocer sells apples in bags of 25. 20% of the apples are thought to be wind-damaged. A manager thinks that the figure of 20% may be wrong so she selects a bag at random and carries out a hypothesis test at the 5% significance level.

a State the hypotheses clearly.

b Find the critical region for the test.

c Using this critical region, state the probability of concluding the figure of 20% is incorrect when in fact it is correct.

d Explain why the answer to part **c** is not equal to 5%.

3 Lydia flips a coin to predict whether or not her school meal will include at least two vegetables. Olivia says that her analysis is better, so she records the number of vegetables in the next 18 meals and carries out a hypothesis test at the 10% significance level.

  a State the hypotheses clearly.

  b What is the lowest number of predictions that Olivia would have to get right in order to justify her claim?

4 Students took a multiple choice maths exam. Each of the 50 questions had 5 possible answers. One student did the test very quickly, didn't seem to read the questions at all and got 15 answers right. The teacher thought that this student had simply guessed all the answers. She conducts a hypothesis test at the 5% significance level to see if there is any evidence to suggest that the student performed better than if he had simply been guessing.

  a State the hypotheses clearly.

  b Use the p-value method to carry out the test and state the conclusion.

5 **a** State the conditions under which the binomial distribution provides a good model for a statistical experiment and state the probability distribution function.

  b Using the binomial distribution function, copy and complete the following table for binomial probabilities with $n = 6$, $p = 0.34$

x	0	1	2	3	4	5	6
$P(X = x)$	0.0827	0.2555	0.3290				0.0015

  c A random variable X is known to have a binomial distribution with $n = 6$. A test with a significance level of 5% is performed to investigate whether the parameter p equals 0.34 against the hypothesis that it is greater than this value. Use your answer to part **b** to find the set of x-values which would suggest that $p > 0.34$ and give the probability of incorrectly rejecting H_0

6 A scratch-card company claims that 10% of cards win prizes. Peter conducts a hypothesis test at the 10% significance level to see if the claim is inaccurate. He collects 45 cards and only wins with one card.

  a State the hypotheses clearly.

  b Peter says that the probability of winning with only one card is less than 5%, so he concludes that the company's claim is too high. Is he correct?

  c Peter carried out another hypothesis test on 90 cards. He won with two cards. He said that as both the total number of cards and the number of winning cards had doubled he should make the same conclusion as before. Is he right?

STATS

What next?

Score		
0 – 3	Your knowledge of this topic is still developing. To improve, search in MyMaths for the code: 2115	
4 – 5	You're gaining a secure knowledge of this topic. To improve, look at the InvisiPen videos for Fluency and skills (11A)	
6	You've mastered these skills. Well done, you're ready to progress! To develop your techniques, looks at the InvisiPen videos for Reasoning and problem-solving (11B)	

Click these links in the digital book

Exploration

Going beyond the exams

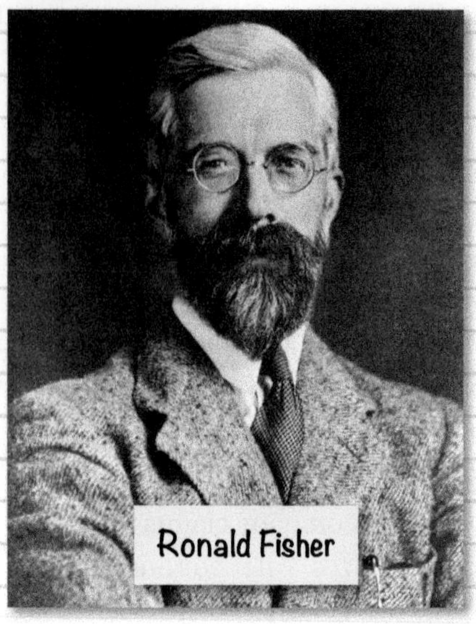
Ronald Fisher

History

Ronald Fisher (1890 - 1962) was born in England and studied at Cambridge University.

In 1925 he published a book on statistical methods, in which he defined the **statistical significance test**. The book was well received by the mathematical and scientific community and significance testing became an established method of analysis in experimental science.

Two other statisticians, **Jerzy Neyman** and **Egon Pearson**, also published a number of papers on **hypothesis testing**. The papers introduced much of the terminology used today, though their approach was different to Fisher's. Neyman and Fisher argued over the relative merits of their work for many years.

"Statistics is the servant to all sciences."
- Jerzy Neyman

Did you know?

Fisher applied his methods to test the claim, made by **Muriel Bristol**, that she could tell whether the tea or the milk was added to a cup first. This became known as the **lady tasting tea** experiment.

Have a go

In the lady tasting tea experiment, the lady was offered 8 cups of tea at the same time. 4 were prepared with milk first and 4 were prepared with tea first.

The lady was then asked to choose the 4 cups which had been prepared with tea first.

How many possible combinations of 4 cups could she give?

(You could use the nCr button on your calculator.)

In the test, Fisher would only consider the lady's claim as valid if she identified all 4 cups correctly. What was the significance level of this test?

How many cups would Fisher have had to use if he wanted to test the lady's claim at a significance level of less than 0.1%?

Muriel Bristol successfully identified all 4 cups, passing Fisher's test.

11 Assessment

1 A call centre manager claims that 45% of calls they receive are complaints. An employee thinks this claim is inaccurate. Select the correct null and alternative hypotheses.

A H_0: $p > 0.45$ and H_1: $p < 0.45$ **B** H_0: $p = 0.45$ and H_1: $p < 0.45$

C H_0: $p = 0.45$ and H_1: $p > 0.45$ **D** H_0: $p = 0.45$ and H_1: $p \neq 0.45$ **[1 mark]**

2 $X \sim B(25, p)$ and H_0: $p = 0.3$ H_1: $p \neq 0.3$
Find the critical region for X if the significance level is 10%
Select the correct answer.

A $X \le 3$ and $X \ge 11$ **B** $X \le 3$ and $X \ge 12$

C $X \ge 3$ and $X \le 12$ **D** $X \ge 4$ and $X \le 10$ **[1]**

3 A certain variety of pepper produces vegetables of various colours. Plants with yellow peppers are particularly prized. A random sample of n plants is chosen to test, at the 5% significance level, whether or not the proportion of plants with yellow peppers is $\frac{1}{2}$

a If n is the sample size and p is the proportion of plants producing yellow peppers, write down the null and alternative hypotheses for this test. **[1]**

b If $n = 6$, find the critical region, giving the probability of incorrectly rejecting H_0 **[5]**

4 A random variable, X, has a binomial distribution with parameters $n = 10$ and p a constant. The value of p was 0.5 but is believed to have increased.

a Write down the null and alternative hypotheses that should be used to test this belief. **[1]**

b Use a significance level of 5% to determine the values of X that would lead you to conclude that the belief is incorrect. **[4]**

In 2002, 50% of registered cars were five-door hatchbacks. By 2016 it is thought that this proportion had increased. Ten cars were selected at random and 9 of them were found to be five-door hatchbacks.

c Using your answers to parts **a** and **b**, state whether you think that the proportion of five-door hatchbacks has increased or not. **[1]**

5 You wish to investigate whether a coin is biased towards heads. You toss the coin 5 times and note the number of heads showing.

a Given that the number of heads is 4, perform a 5% significance test, stating clearly your null and alternative hypotheses. **[5]**

b Would your conclusion change if the number of heads showing was 5? **[1]**

6 It is estimated that 40% of ready-meals in a supermarket contain more than the recommended daily allowance of salt. A sample of 20 meals were examined and 6 were found to contain more than the recommended daily allowance of salt.

a State a condition on the method of choosing the sample so that a binomial probability model can be used to test the estimate. **[1]**

b Assuming that the condition in part **a** is met, test at a 5% significance level whether the data suggests that 40% is an overestimate. You should state clearly your null and alternative hypotheses. **[4]**

7 A researcher defines high particulate emissions from cars as a value above 0.002 g km^{-1}. In 2016, 45% of petrol cars had high particulate emissions. In a random sample of 12 diesel cars, 2 had high particulate emissions.

Is there evidence that a smaller proportion of diesel cars have high particulate emissions? You should use a 10% significance level and describe the critical region(s) for the test. **[5]**

8 A journalist thinks that people drove heavier cars in London in the past than they do now. According to the large data set, in 2002, 81% of cars had a mass of less than or equal to 1600 kg. The journalist took a random sample of 14 cars registered in London in 2016 and found that 8 of them had a mass less than or equal to 1600 kg.

a Explain why the binomial probability distribution provides a good model for this data. **[2]**

b Copy and complete this table for X, a random variable for the number of cars with mass less than or equal to 1600 kg, assuming that 81% of cars have a mass in this range. **[2]**

x	6	7	8	9
$P(X \le x)$	0.00169			

c Carry out a hypothesis test of the journalist's belief, using a 10% significance level. **[5]**

9 A survey found that 60% of documents printed in an office were printed single-sided. Employees were asked not to print single-sided in order to save paper. A fortnight later the manager of the office wanted to see if there had been any reduction in the rate of single-sided printing. He tested 40 documents and found that 18 of them had been printed single-sided.

a State the hypotheses clearly. **[1]**

b If he concluded that there had been an improvement, using a significance level of a%, what is the lowest possible whole number value of a? **[3]**

10 In a survey, 35% of adults in urban areas of England are found to spend more than £13.20 per week on petrol. A sample of 30 adults in rural areas found that 15 spent more than £13.20 per week on petrol. A statistical test is to be carried out to determine whether the spending on petrol is higher in rural areas than in urban areas.

a State a condition for the sample to be suitable for use in the test and state why the condition is necessary. **[2]**

Let X be a random variable for the number of people who spend more than £13.20 per week on petrol in the sample of size 30

b State the null and alternative hypotheses to be used in the test and, assuming that the null hypothesis is true, give the distribution of X **[2]**

c Perform this test at a significance level of 5%. **[3]**

11 $X \sim B(n,p)$ and $H_0: p = k$, $H_1: p > k$. The critical value is $n-1$

a Find the probability that X is in the critical region in terms of k and n **[4]**

b If $n = 2$ and the significance level is at most 19% then find an inequality for k **[5]**

9-11 Assessment

Chapters 9-11: Statistics

1 This cumulative frequency graph shows the results of a general knowledge test taken by 80 people.

For each question part, choose the correct answer.

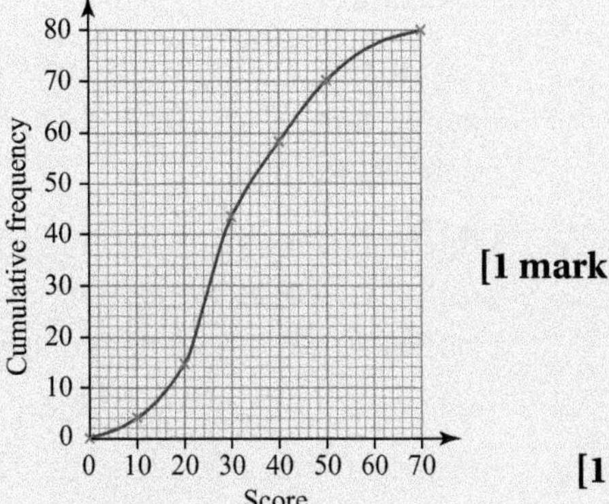

a What is the interquartile range?

A 20 **B** 40 **C** 51 **D** 62 **[1 mark]**

The test was out of 80 marks.

b How many people scored over 60%?

A 48 **B** 12 **C** 32 **D** 68 **[1]**

2 A market research company wants to survey British car owners using a sample of size 200.

Large data set

a In each of the cases below, state the name of the sampling method.

i Numbering the entries in a register of car owners and using a random number generator to select 200 owners. **[1]**

ii Standing outside a car park and asking people to complete a survey until you have 200 records **[1]**

iii Selecting a particular town and splitting its population into different income groups, then sampling each group in proportion to its size to obtain a total sample of size 200. **[1]**

b For each of the sampling methods in part **a**, give one advantage and one disadvantage of the method. **[6]**

3 The blood glucose levels of a group of adults is recorded immediately before and 2 hours after eating a meal. The results are summarised in these box and whisker plots.

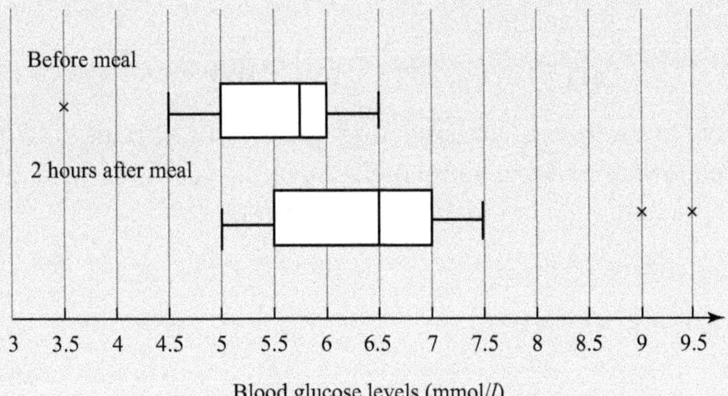

a What is the interquartile range 2 hours after eating the meal? **[2]**

b Compare the blood glucose levels immediately before eating with those 2 hours after eating. **[2]**

A blood glucose level of below 6 mmol/l before eating is considered normal.

c What percentage of the group have a normal blood glucose level before the meal? **[1]**

Two adults from the group are selected at random 2 hours after the meal.

d Calculate the probability that both adults have a blood glucose level of above 5.5 mmol/l. **[3]**

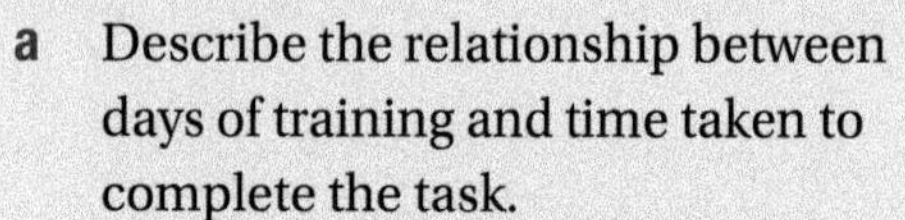

4 A group of trainees are timed to see how quickly they can complete a specific task. The results are shown in this scatter diagram.

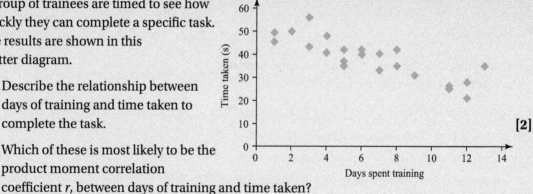

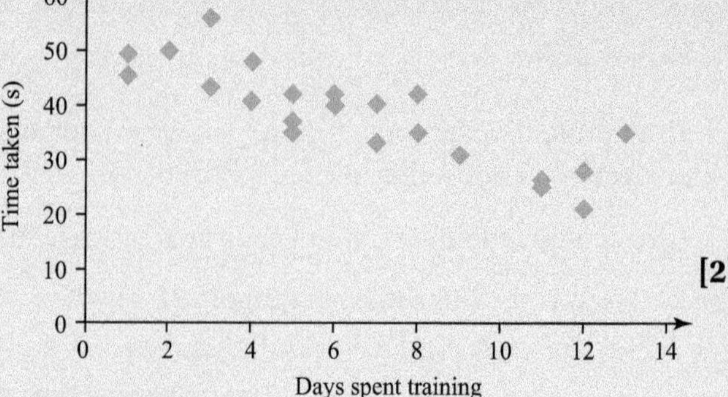

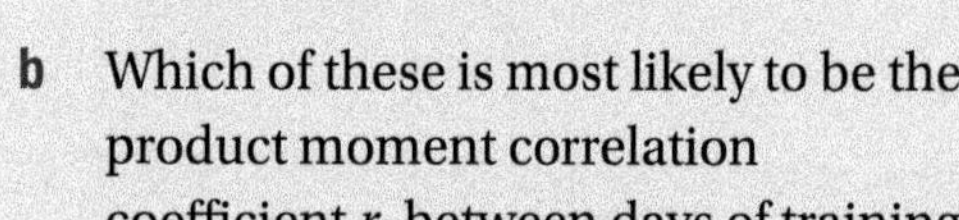

a Describe the relationship between days of training and time taken to complete the task. **[2]**

b Which of these is most likely to be the product moment correlation coefficient r, between days of training and time taken?

-0.5 $\quad 1.3 \quad$ -0.8 $\quad 0.7$ **[1]**

c Another trainee completes the task in 20 seconds after only 1 day of training.
What effect would including this trainee have on the value of r? **[1]**

Large data set

5 The large data set contains 12 coupe cars registered to either men or women in south-west England in 2016. These cars are further classified as made by BMW or not made by BMW. The table shows the results.

	Male	Female	Total
BMW	4		7
Not BMW			
Total		4	

a A car is selected at random from the sample of coupe cars. What is the probability that

i it is not made by BMW?

ii it is made by BMW and registered to a man? **[3]**

b Are the events 'made by BMW' and 'registered to a man' independent?
You must give your reasons. **[3]**

6 Use the fact that $X \sim B(20, 0.4)$ to find these probabilities.

a $P(X = 13)$ **[1]** b $P(X \leq 5)$ **[1]** c $P(X > 10)$ **[2]**

7 In a 2016 sample of Toyota car owners in south-west England, one-quarter of cars registered to men and two-thirds of cars registered to women were petrol fuelled. Four-sevenths of the cars in the sample were registered to men. The remainder were registered to women.

a For a randomly selected car in the sample, calculate the probability that the car is

i registered to a woman and not petrol fuelled,

ii petrol fuelled. **[4]**

b In a sub-sample of 16 cars, registered to men, calculate the probability that

i no cars are petrol fuelled,

ii exactly 6 cars are petrol fuelled,

iii at least 3 cars are petrol fuelled. **[4]**

8 The resting pulse rates (in beats per minute, bpm) of a group of people were measured. The mean for the group is 71 bpm.

71	67	22	99	68	67
63	56	76	65	68	75

The data is given in the box but one of the numbers has been entered incorrectly.

a Identify the error and calculate the correct value. **[3]**

b Calculate the variance of the correct data. **[1]**

An outlier is defined as a value which is further than 2 standard deviations from the mean.

c Are there any outliers in this data? Explain how you know. **[3]**

d Calculate the median of the data and explain why it might be a better average to use than the mean. **[2]**

9 A smallholder records the mass of all eggs laid by her hens over a week. The table shows her results.

Size	Mass e, of egg	Frequency
Small	$43 \le e < 53$	42
Medium	$53 \le e < 63$	59
Large	$63 \le e < 73$	24
Very large	$73 \le e < 83$	18

a What is the modal size of egg? **[1]**

b Estimate the mean and standard deviation of the mass of these eggs. **[3]**

c Estimate the median and interquartile range of the mass of the eggs. **[6]**

The smallholder decides to keep all the small eggs to use at home or give to friends.

d Without further calculation, state the effect this will have on the mean and standard deviation of the remaining eggs. **[2]**

10 $P(A) = 0.3$, $P(B) = 0.75$ and $P(A \cap B) = 0.15$

a Calculate

i $P(A \cup B)$ **ii** $P(A' \cap B)$ **[3]**

Event C is mutually exclusive to event A and to event B

b Given that $P(C) \le x$ calculate the value of x **[2]**

11 Based on a 2016 sample of Ford cars in London, the probability that a car has 'low' oxides of nitrogen emissions, below 0.03 g km^{-1}, is 20%. Large data set

a Assuming a binomial distribution, what is the probability that out of ten randomly selected Ford cars

i no cars have low NOX emissions.

ii fewer than three cars have low NOX emissions.

iii over half the cars have low NOX emissions. **[4]**

b Comment on the assumptions being made in the use of the binomial distribution. **[2]**

12 Using the large data set, in 2002 the probability that a Ford car in London had 'low' oxides of nitrogen emissions, below 0.03 g km^{-1}, was 10%. A Ford spokesman claims that other cars have significantly higher NOX emissions. A test of 30 non-Ford cars showed that 5 had low NOX emissions. Large data set

a Use a hypothesis test, with a 10% significance level, to test whether the Ford spokesman's claim is correct. **[6]**

b What is the probability of incorrectly rejecting the null hypothesis? **[1]**

13 A group of four car makers commission a statistician to research customer satisfaction with their makes of car. The statistician needs to create a sample of car owners.

a For each method described below, state the name of the sampling method and explain one disadvantage of the method.

i Visiting car parks around the country and asking people to participate. **[2]**

ii Using a local phone directory and a random number generator to select a sample. **[2]**

b The statistician decides to take a stratified random sample. The table shows the 2016 percentages of registered cars for the four car makers. If the sample size is 10 000, calculate the number of car owners that need to be selected for each car maker. **[3]**

Car maker	Percentage of all registered cars
BMW	5%
Ford	14%
Vauxhall	11%
Volkswagen	9%

14 In 2016, the population of countries in the European Union, to the nearest million, is summarised in the table.

A histogram is drawn to illustrate the data.

The bar for 6–15 million is 1 cm wide and 13.5 cm tall.

Population (millions)	Number of Countries
0–5	11
6–15	9
16–60	4
61–85	4

a Calculate the height and width of the 61–85 bar. **[5]**

b Estimate how many counties have a population, to the nearest million, of under 10 million. **[4]**

c Estimate the probability that a person chosen at random from the European Union lives in a country with a population of 0–5 million. **[4]**

15 You are given that $X \sim B(25, 0.45)$

Find the largest value of x such that $P(X \geq x) > 0.8$ **[4]**

16 Using the large data set, in 2002, 19% of all cars had 'low' carbon dioxide emissions, less than or equal to 145 g km^{-1}.

Joe wants to see whether the proportion of cars with low emissions is higher in 2016. He measures the CO_2 emissions of 21, petrol fuelled Multi Purpose Vehicles that visit his London garage in 2016 and finds 12 have emissions below 145 g km^{-1}.

a Comment on whether you think this sample provides a fair comparison between the level of CO_2 emissions from cars in 2002 and 2016. **[1]**

b Carry out a hypothesis test of Joe's suggestion using his sample of MPVs and a 1% significance level. **[6]**

c What range of significance levels could Joe use without changing his conclusion? Give your answer to three significant figures. **[2]**

Mathematical formulae

For AS Level Maths

The following mathematical formulae will be provided for you.

Pure Mathematics

Binomial series

$$(a+b)^n = a^n + \binom{n}{1}a^{n-1}b + \binom{n}{2}a^{n-2}b^2 + \ldots + \binom{n}{r}a^{n-r}b^r + \ldots + b^n \qquad (n \in \mathbb{N})$$

where $\binom{n}{r} = {}^nC_r = \dfrac{n!}{r!(n-r)!}$

Mechanics

Constant acceleration

$$s = ut + \frac{1}{2}at^2$$

$$s = vt - \frac{1}{2}at^2$$

$$v = u + at$$

$$s = \frac{1}{2}(u+v)t$$

$$v^2 = u^2 + 2as$$

Probability and Statistics

Discrete distributions

Standard discrete distributions

Distribution of X	$\mathrm{P}(X=x)$
Binomial $B(n, p)$	$\binom{n}{x}p^x(1-p)^{n-x}$

Sampling distributions

For a random sample $X_1, X_2, \ldots, X_n$ of n independent observations from a distribution having mean μ and variance σ^2:

$\overline{X}$ is an unbiased estimator of μ, with $Var(\overline{X}) = \dfrac{\sigma^2}{n}$

s^2 is an unbiased estimator of σ^2, where $s^2 = \dfrac{\sum(X_i - \overline{X})^2}{n-1}$

Mathematical formulae - to learn

For AS Level Maths

You are expected to know the following formulae for AS Level Mathematics.

Pure Mathematics

Quadratic equations

$ax^2 + bx + c = 0$ has roots $\dfrac{-b \pm \sqrt{b^2 - 4ac}}{2a}$

Laws of indices

$a^x a^y \equiv a^{x+y}$

$a^x \div a^y \equiv a^{x-y}$

$(a^x)^y \equiv a^{xy}$

Laws of logarithms

$x = a^n \Leftrightarrow n = \log_a x$ for $a > 0$ and $x > 0$

$\log_a x + \log_a y \equiv \log_a xy$

$\log_a x - \log_a y \equiv \log_a \left(\dfrac{x}{y}\right)$

$k \log_a x \equiv \log_a (x)^k$

Coordinate geometry

A straight line graph, gradient m passing through (x_1, y_1) has equation $y - y_1 = m(x - x_1)$

Straight lines with gradients m_1 and m_2 are perpendicular when $m_1 m_2 = -1$

Trigonometry

In the triangle ABC

Sine rule $\dfrac{a}{\sin A} = \dfrac{b}{\sin B} = \dfrac{c}{\sin C}$

Cosine rule $a^2 = b^2 + c^2 - 2bc \cos A$

Area $\dfrac{1}{2} ab \sin C$

$\cos^2 A + \sin^2 A \equiv 1$

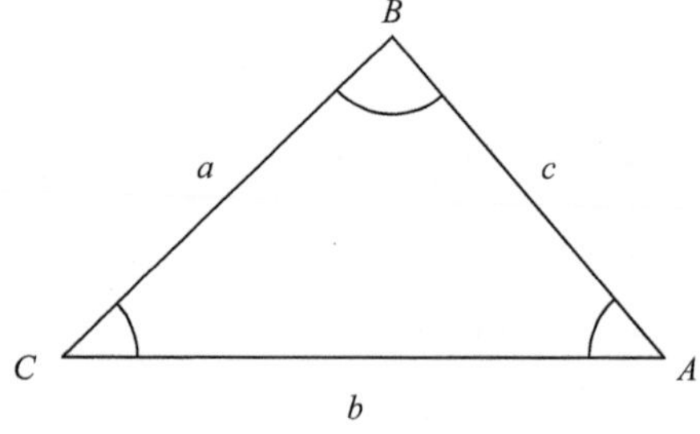

Mensuration

Circumference, C, and area, A, of circle with radius r and diameter d:

$C = 2\pi r = \pi d$ $\qquad A = \pi r^2$

Pythagoras' Theorem:

In any right-angled triangle where a, b and c are the lengths of the sides and c is the hypotenuse, $c^2 = a^2 + b^2$

Area of a trapezium $= \frac{1}{2}(a+b)h$, where a and b are the lengths of the parallel sides and h is their perpendicular separation.

Volume of a prism = area of cross section × length

Calculus and differential equations

Differentiation		**Integration**	
Function	Derivative	Function	Integral
x^n	nx^{n-1}	x^n	$\frac{1}{n+1}x^{n+1}+c, n \neq -1$
e^{kx}	ke^{kx}		
$f'(x)+g(x)$	$f'(x)+g'(x)$	$f'(x)+g(x)$	$f'(x)+g(x)+c$

Area under a curve $= \int_a^b y\,dx \quad (y \geq 0)$

Vectors

$|x\mathbf{i}+y\mathbf{j}| = \sqrt{x^2+y^2}$

Statistics

The mean of a set of data: $\bar{x} = \frac{\sum x}{n} = \frac{\sum fx}{\sum f}$

Mechanics

Forces and equilibrium

Weight = mass × g

Newton's second law in the form: $F = ma$

Kinematics

For motion in a straight line with variable acceleration:

$v = \frac{dr}{dt} \qquad a = \frac{dv}{dt} = \frac{d^2r}{dt^2}$

$r = \int v\,dt \qquad v = \int a\,dt$

Mathematical notation

For AS and A Level Maths

You should understand the following notation without need for further explanation.
Anything highlighted is only used in the full A Level, and so will not be needed at AS Level.

Set Notation

$\in$	is an element of
$\notin$	is not an element of
$\subseteq$	is a subset of
$\subset$	is a proper subset of
$\{x_1, x_2, \dots\}$	the set with elements $x_1, x_2, \dots$
$\{x: \dots\}$	the set of all x such that ...
$\text{n}(A)$	the number of elements in set A
$\varnothing$	the empty set
ε	the universal set
A'	the complement of the set A
$\mathbb{N}$	the set of natural numbers, $\{1, 2, 3, \dots\}$
$\mathbb{Z}$	the set of integers, $\{0, \pm1, \pm2, \pm3, \dots\}$
$\mathbb{Z}^+$	the set of positive integers, $\{1, 2, 3, \dots\}$
$\mathbb{Z}_0^+$	the set of non-negative integers, $\{0, 1, 2, 3, \dots\}$
$\mathbb{R}$	the set of real numbers
$\mathbb{Q}$	the set of rational numbers, $\left\{\frac{p}{q} : p \in \mathbb{Z},\ q \in \mathbb{Z}^+\right\}$
$\cup$	union
$\cap$	intersection
(x, y)	the ordered pair x, y
$[a, b]$	the closed interval $\{x \in \mathbb{R} : a \le x \le b\}$
$[a, b)$	the interval $\{x \in \mathbb{R} : a \le x < b\}$
$(a, b]$	the interval $\{x \in \mathbb{R} : a < x \le b\}$
(a, b)	the open interval $\{x \in \mathbb{R} : a < x < b\}$

Miscellaneous Symbols

$=$	is equal to
$\neq$	is not equal to
$\equiv$	is identical to or is congruent to
$\approx$	is approximately equal to
∞	infinity
$\propto$	is proportional to
$<$	is less than
$\leqslant, \le$	is less than or equal to, is not greater than
$>$	is greater than
$\geqslant, \ge$	is greater than or equal to, is not less than
$\therefore$	therefore
$\because$	because
$\angle A$	angle A
$p \Rightarrow q$	p implies q (if p then q)
$p \Leftarrow q$	p is implied by q (if q then p)
$p \Leftrightarrow q$	p implies and is implied by q (p is equivalent to q)

a	first term for an arithmetic or geometric sequence
l	last term for an arithmetic sequence
d	common difference for an arithmetic sequence
r	common ratio for a geometric sequence
S_n	sum to n terms of a sequence
S_∞	sum to infinity of a sequence

Operations

$a+b$	a plus b
$a-b$	a minus b
$a\times b$, ab, $a\cdot b$	a multiplied by b
$a\div b$, $\frac{a}{b}$	a divided by b
$\sum_{i=1}^{n} a_i$	$a_1 + a_2 + \ldots + a_n$
$\prod_{i=1}^{n} a_i$	$a_1 \times a_2 \times \ldots \times a_n$
$\sqrt{a}$	the positive square root of a
$\lvert a \rvert$	the modulus of a
$n!$	n factorial: $n! = n\times(n-1)\times\ldots\times 2\times 1$, $n\in\mathbb{N}$; $0! = 1$
$\binom{n}{r}$, ${}^nC_r$, ${}_nC_r$	the binomial coefficient $\frac{n!}{r!(n-r)!}$ for $n, r \in \mathbb{Z}_0^+$, $r\le n$ or $\frac{n(n-1)\ldots(n-r+1)}{r!}$ for $n\in\mathbb{Q}$, $r\in\mathbb{Z}_0^+$

Functions

$\mathrm{f}(x)$	the value of the function f at x
$\mathrm{f}: x \mapsto y$	the function f maps the element x to the element y
f^{-1}	the inverse function of the function f
gf	the composite function of f and g which is defined by $\mathrm{gf}(x) = \mathrm{g}(\mathrm{f}(x))$
$\lim_{x\to a} \mathrm{f}(x)$	the limit of $\mathrm{f}(x)$ as x tends to a
Δx, δx	an increment of x
$\frac{\mathrm{d}y}{\mathrm{d}x}$	the derivative of y with respect to x
$\frac{\mathrm{d}^n y}{\mathrm{d}x^n}$	the nth derivative of y with respect to x
$\mathrm{f}'(x) \ldots, \mathrm{f}^{(n)}(x)$	the first, ..., nth derivatives of $\mathrm{f}(x)$ with respect to x
$\dot{x}, \ddot{x}, \ldots$	the first, second, ... derivatives of x with respect to t
$\int y\,\mathrm{d}x$	the indefinite integral of y with respect to x
$\int_a^b y\,\mathrm{d}x$	the definite integral of y with respect to x between the limits $x=a$ and $x=b$

Exponential and Logarithmic Functions

e	base of natural logarithms
e^x, $\exp x$	exponential function of x
$\log_a x$	logarithm to the base a of x
$\ln x$, $\log_e x$	natural logarithm of x

Trigonometric Functions

sin, cos, tan, cosec, sec, cot	the trigonometric functions
$\sin^{-1}$, $\cos^{-1}$, $\tan^{-1}$, arcsin, arccos, arctan	the inverse trigonometric functions
$^\circ$	degrees
rad	radians

Vectors

$\mathbf{a}$, $\underline{a}$, $\underset{\sim}{a}$	the vector $\mathbf{a}$, $\underline{a}$, $\underset{\sim}{a}$
$\overrightarrow{AB}$	the vector represented in magnitude and direction by the directed line segment AB
$\hat{\mathbf{a}}$	a unit vector in the direction of $\mathbf{a}$
$\mathbf{i}$, $\mathbf{j}$, $\mathbf{k}$	unit vectors in the directions of the Cartesian coordinate axes
$\lvert\mathbf{a}\rvert$, a	the magnitude of $\mathbf{a}$
$\lvert\overrightarrow{AB}\rvert$, AB	the magnitude of $\overrightarrow{AB}$
$\begin{pmatrix} a \\ b \end{pmatrix}$, $a\mathbf{i} + b\mathbf{j}$	column vector and corresponding unit vector notation
$\mathbf{r}$	position vector
$\mathbf{s}$	displacement vector
$\mathbf{v}$	velocity vector
$\mathbf{a}$	acceleration vector

Probability and Statistics

A, B, C, etc.	events
$A \cup B$	union of the events A and B
$A \cap B$	intersection of the events A and B
$\mathrm{P}(A)$	probability of the event A
A'	complement of the event A
$\mathrm{P}(A \mid B)$	probability of the event A conditional on the event B
X, Y, R, etc.	random variables
x, y, r, etc.	values of the random variables X, Y, R etc.
$x_1, x_2, \ldots$	observations
$f_1, f_2, \ldots$	frequencies with which the observations $x_1, x_2, \ldots$ occur
$\mathrm{p}(x)$, $\mathrm{P}(X = x)$	probability function of the discrete random variable X
$p_1, p_2, \ldots$	probabilities of the values $x_1, x_2, \ldots$ of the discrete random variable X

$\mathrm{E}(X)$	expectation of the random variable X
$\mathrm{Var}(X)$	variance of the random variable X
$\sim$	has the distribution
$\mathrm{B}(n, p)$	binomial distribution with parameters n and p, where n is the number of trials and p is the probability of success in a trial
q	$q = 1 - p$ for binomial distribution
$\mathrm{N}(\mu, \sigma^2)$	Normal distribution with mean μ and variance σ^2
$Z \sim \mathrm{N}(0,1)$	standard Normal distribution
ϕ	probability density function of the standardised Normal variable with distribution $\mathrm{N}(0, 1)$
Φ	corresponding cumulative distribution function
μ	population mean
σ^2	population variance
σ	population standard deviation
$\bar{x}$	sample mean
s^2	sample variance
s	sample standard deviation
H_0	null hypothesis
H_1	alternative hypothesis
r	product moment correlation coefficient for a sample
ρ	product moment correlation coefficient for a population

Mechanics

kg	kilograms
m	metres
km	kilometres
m/s, m s^{-1}	metres per second (velocity)
m/s^2, m s^{-2}	metres per second per second (acceleration)
F	Force or resultant force
N	Newton
N m	Newton metre (moment of a force)
t	time
s	displacement
u	initial velocity
v	velocity or final velocity
a	acceleration
g	acceleration due to gravity
μ	coefficient of friction

Answers

Full solutions to all of these questions can be found at the link in the page footer.

Chapter 1

Try it 1A

1 a $10x^{10}$ b $6x^7$ c $16x^{24}$ d $\frac{x^6}{9}$

2 a 6 b 9 c $\frac{1}{8}$ d $\frac{1}{16}$

3 a $x^{\frac{2}{5}}$ b $3x^{-\frac{1}{2}}$ c $3x^{\frac{3}{2}}$ d $\frac{1}{3}x^{-\frac{1}{2}}$

4 a $5\sqrt{7}$ b $\frac{4\sqrt{3}}{3}$ c $-3+3\sqrt{2}$ d $5+2\sqrt{5}$

Bridging Exercise 1A

1 a 7 b 3 c $\frac{1}{5}$ d $\frac{1}{4}$
e 27 f 8 g $\frac{1}{25}$ h $\frac{1}{8}$
i 81 j $\frac{2}{3}$ k $\frac{4}{3}$ l $\frac{4}{9}$

2 a $2\sqrt{2}$ b $5\sqrt{3}$ c $4\sqrt{6}$
d $12\sqrt{3}$ e $3\sqrt{5}$ f $\sqrt{3}$
g $14\sqrt{2}$ h $5\sqrt{2}+15\sqrt{5}$ i $5\sqrt{17}$
j $14\sqrt{2}$ k $12\sqrt{2}-2\sqrt{3}$ l $6\sqrt{5}+5\sqrt{2}$

3 a $\frac{\sqrt{7}}{7}$ b $\frac{\sqrt{2}}{2}$ c $4\sqrt{3}$
d $\frac{\sqrt{6}}{3}$ e $\frac{1}{2}(\sqrt{3}-1)$ f $2(\sqrt{2}-1)$
g $-2(1+\sqrt{5})$ h $\frac{1}{2}(\sqrt{5}+1)$ i $2\sqrt{2}-\sqrt{6}$
j $3\sqrt{2}+2\sqrt{3}$ k $-3-2\sqrt{2}$ l $-\frac{3}{2}\sqrt{5}-\frac{7}{2}$

4 a $5+4\sqrt{2}$ b $1+2\sqrt{2}$
c $1-2\sqrt{2}$ d $5-4\sqrt{2}$
e $6\sqrt{3}+11$ f $2\sqrt{3}+5$
g $2\sqrt{3}-5$ h $6\sqrt{3}-11$
i $2\sqrt{3}+3\sqrt{6}+\sqrt{2}+3$ j $2\sqrt{3}-3\sqrt{6}+\sqrt{2}-3$
k $2\sqrt{3}+3\sqrt{6}-\sqrt{2}-3$ l $2\sqrt{3}-3\sqrt{6}-\sqrt{2}+3$

5 a x^{10} b $21x^{11}$ c $40x^{11}$
d x^6 e $4x^{-2}$ f $\frac{1}{4}x$
g x^{35} h x^{-10} i $81x^8$
j $36x^{10}$ k $x^{\frac{3}{2}}$ l $x^{\frac{5}{4}}$
m $5x^{-\frac{1}{2}}$ n $2x^{\frac{3}{2}}$ o $\frac{1}{3}x^{\frac{3}{2}}$
p x^8-x^3 q $x^{\frac{7}{2}}+2x^3$ r $x^{-2}+2x^{-3}$
s $x^{-\frac{1}{2}}+3x^{-1}$ t $3x^{-\frac{1}{2}}-x^{\frac{5}{2}}$ u $x+6\sqrt{x}+9$
v $3x^{-2}+x^{-\frac{3}{2}}$ w $\frac{1}{2}x^{-\frac{1}{2}}-\frac{1}{2}x^{\frac{1}{2}}$ x $\frac{1}{3}x^{-\frac{5}{2}}+\frac{2}{3}x^{-3}$

Try it 1B

1 $x=7$

2 $x>2$

3 $x=\frac{1-3A}{3-B}$

4 $y=3, x=-7$

5 (2, 10)

Bridging Exercise 1B

1 a $x=-\frac{10}{3}$ b $x=-\frac{5}{3}$ c $x=31$
d $x=\frac{13}{3}$ e $x=2.5$ f $x=2.1$
g $x=7$ h $x=\frac{5}{13}$

2 a $x\geq-4$ b $x>-3$ c $x>\frac{17}{4}$
d $x>5$ e $x\leq-\frac{2}{3}$ f $x\geq-1.8$
g $x<\frac{16}{7}$ h $x\leq-\frac{5}{3}$

3 a $x=\frac{3A-6}{2}$ b $x=\frac{3-u}{1-v}$ c $x=\frac{1}{3-2k}$
d $x=\frac{15m-4}{5-2n}$ e $x=\frac{1\pm\sqrt{t}}{3}$ f $x=\frac{pq}{p+q}$
g $x=\pm\sqrt{\frac{1}{10}-k}$ h $x=4B^2-A$

4 a $y=2, x=-6$ b $x=-3, y=7$ c $y=2, x=7$
d $=\frac{3}{5}, y=-\frac{1}{10}$ e $y=12, x=7$ f $y=-\frac{1}{2}, x=\frac{1}{4}$

5 a $(-3, 17)$ b $\left(\frac{1}{2},-\frac{1}{2}\right)$ c $\left(\frac{2}{3},\frac{13}{3}\right)$
d $(1.5, -0.5)$ e $(0.8, 3.4)$ f $(0.2, 6.6)$

Try it 1C

1 a $7x(2x-1)$ b $(x-4)(x-1)$ c $(x+5)(x-5)$

2 a $(5x+1)(x+4)$
b $(3x-1)(2x+3)$
c $(2x-5)(4x-1)$

3 a $x=0$ or $x=2$ b $x=\frac{3}{4}$ or $x=5$

4 a

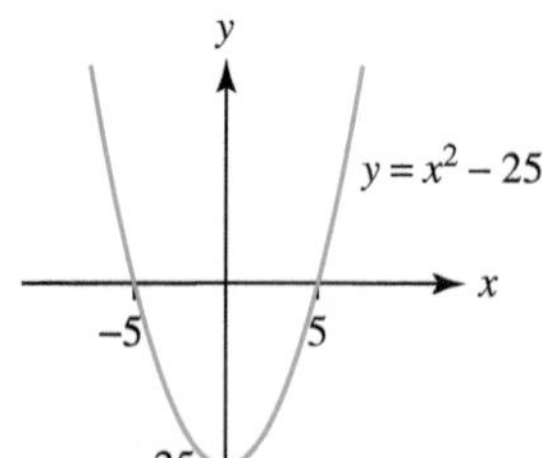

b

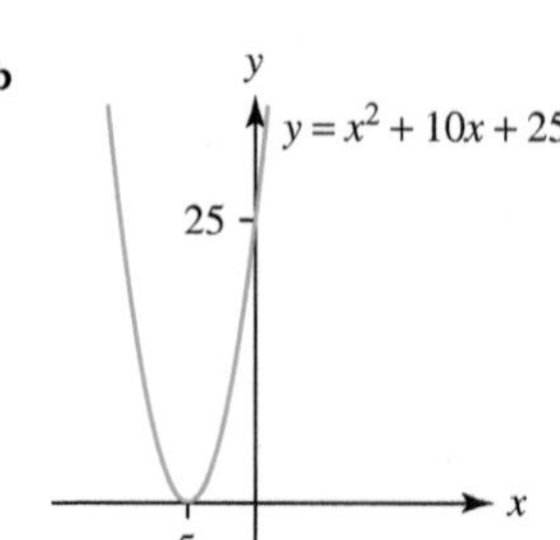

c

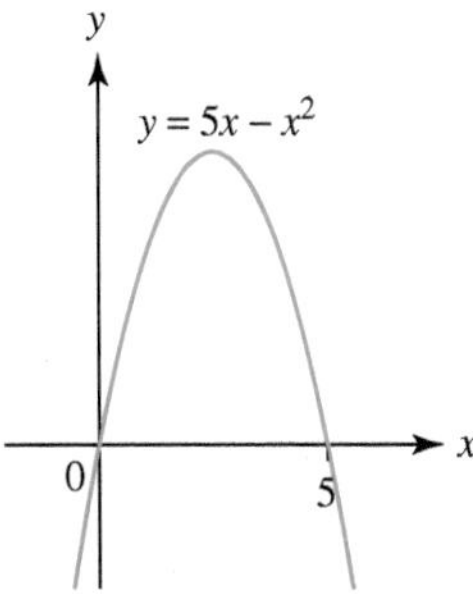

Bridging Exercise 1C

1 **a** $x(3x+5)$ **b** $4x(2x-1)$

c $17x(x+2)$ **d** $6x(3x-4)$

2 **a** $(x+2)(x+3)$ **b** $(x-5)(x-2)$

c $(x-6)(x+1)$ **d** $(x+7)(x-4)$

e $(x-9)(x+8)$ **f** $(x+8)(x-6)$

g $(x-11)(x-1)$ **h** $(x-8)(x+3)$

3 **a** $(x+10)(x-10)$ **b** $(x+9)(x-9)$

c $(2x+3)(2x-3)$ **d** $(8+3x)(8-3x)$

4 **a** $(3x+1)(x+2)$ **b** $(3x+4)(2x+3)$

c $(4x-1)(x-3)$ **d** $(2x+3)(x-5)$

e $(2x+5)(x-1)$ **f** $(7x-3)(x+4)$

g $(4x-5)(2x-3)$ **h** $(4x-1)(3x+5)$

5 **a** $(4x+5)(4x-5)$ **b** $4x(x-4)$

c $(x+12)(x+1)$ **d** $(3x-5)(x+7)$

e $(x+4)(x-3)$ **f** $(10+3x)(10-3x)$

g $2x(x-7)$ **h** $(5x-2)(4x+1)$

6 **a** $x=0$ or $x=\frac{1}{3}$ **b** $x=-6$ or $x=6$

c $x=0$ or $x=-2$ **d** $x=-\frac{1}{2}$ or $x=-\frac{5}{3}$

e $x=-\frac{7}{2}$ or $x=\frac{7}{2}$ **f** $x=9$ or $x=-2$

g $x=6$ or $x=1$ **h** $x=\frac{2}{7}$ or $x=-\frac{1}{3}$

i $x=\frac{2}{5}$ or $x=3$ **j** $x=-\frac{3}{4}$

k $x=\frac{2}{3}$ **l** $x=\frac{3}{8}$ or $x=-\frac{2}{5}$

7 **a**

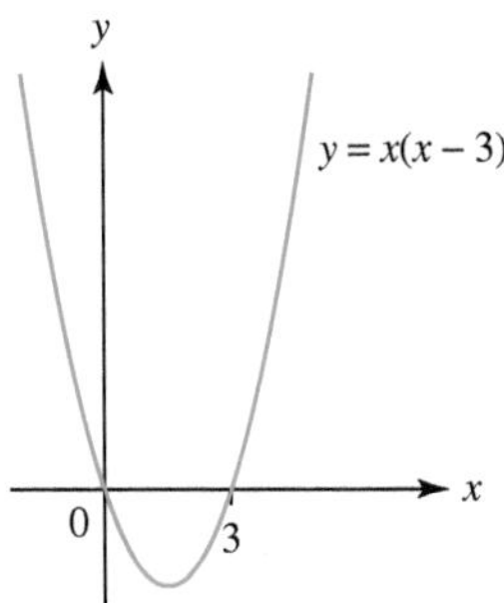

b

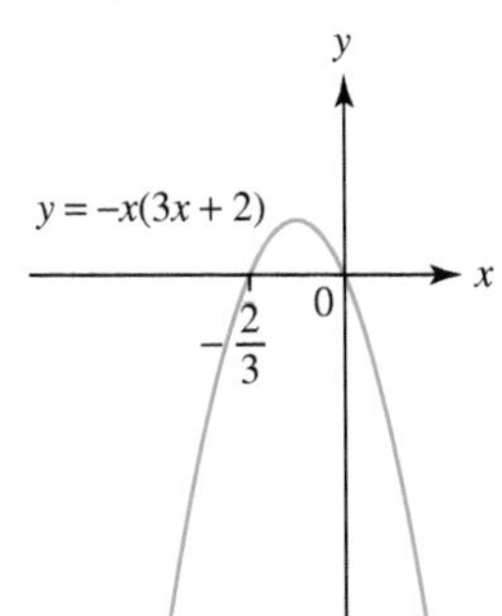

c

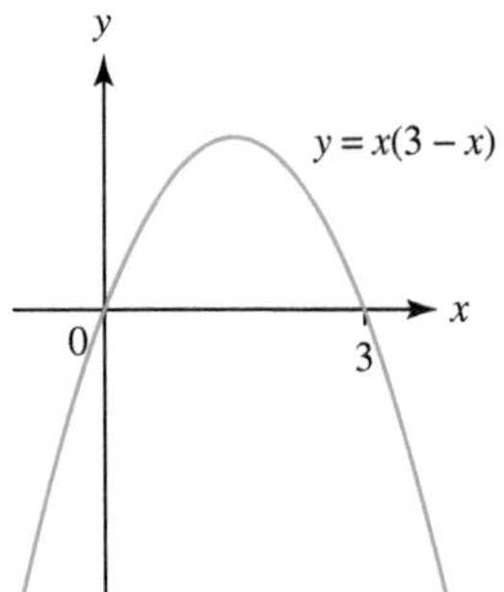

d

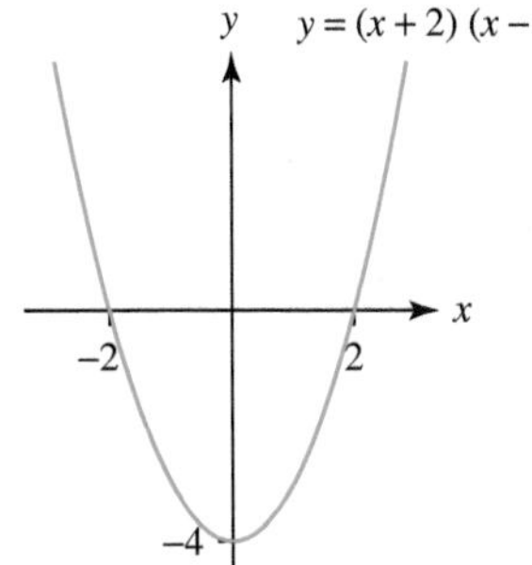

e

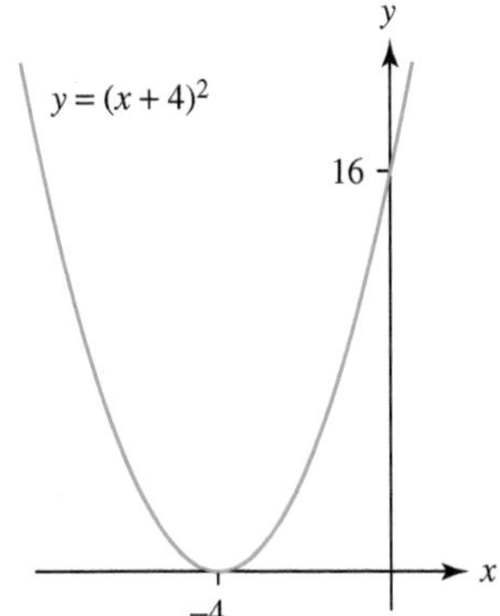

f

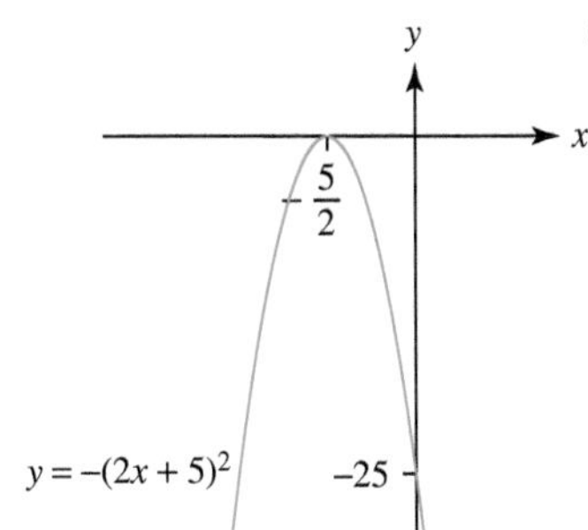

g

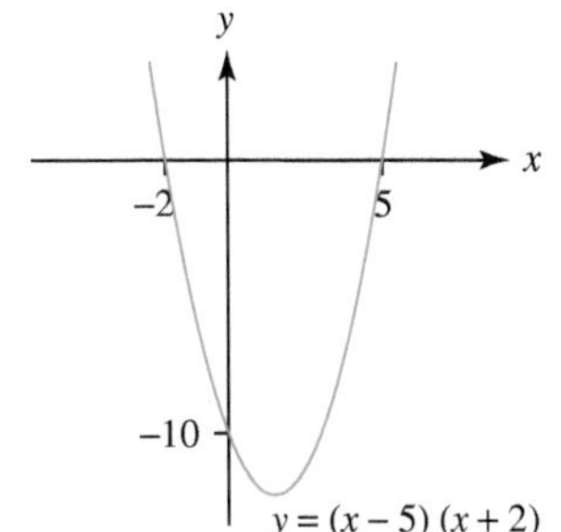

h

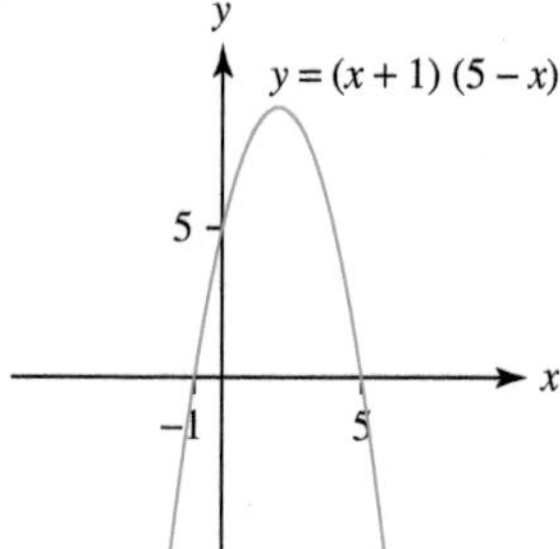

8 a

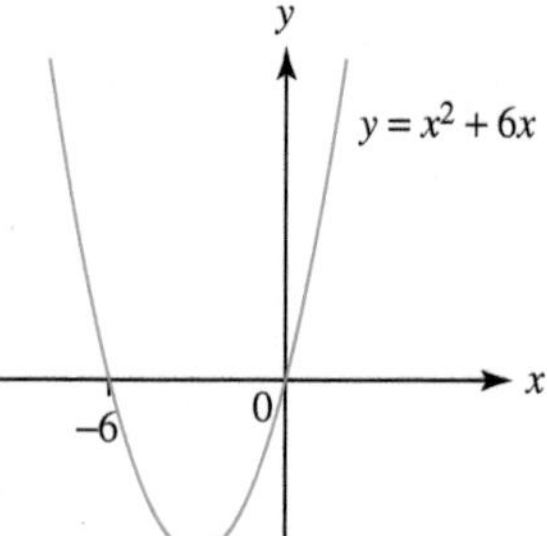

b

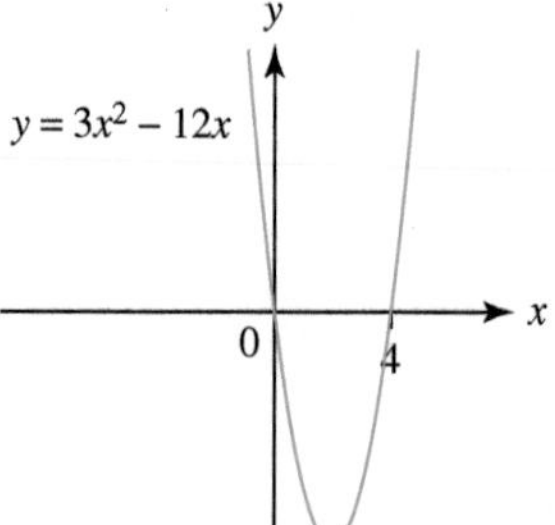

c

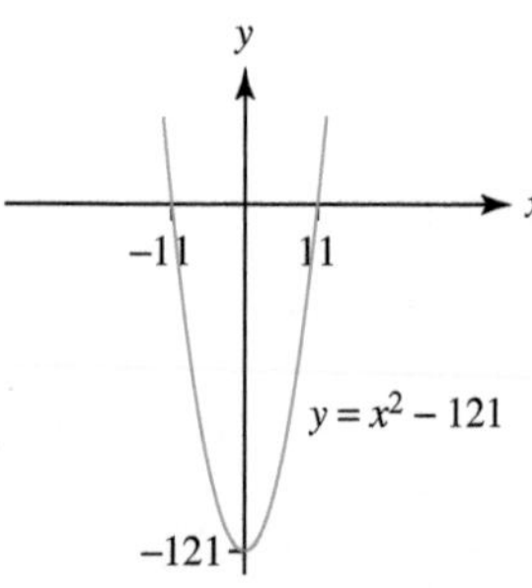

d

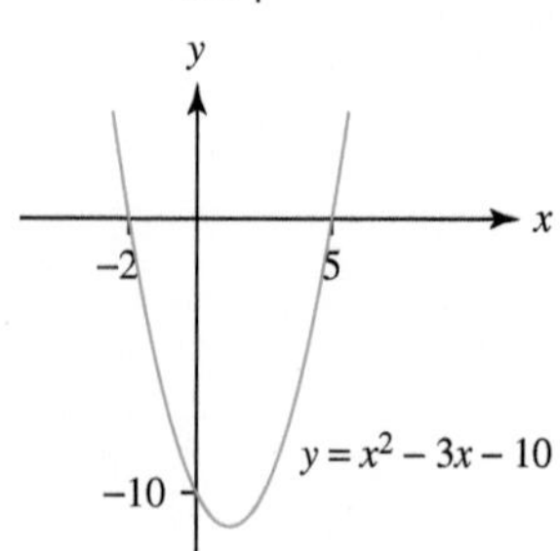

e

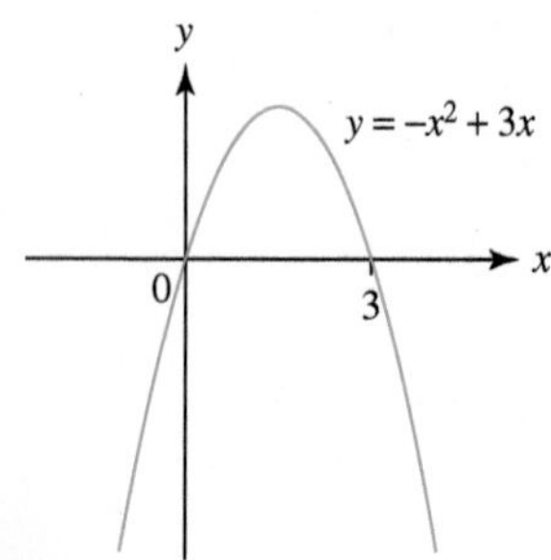

f

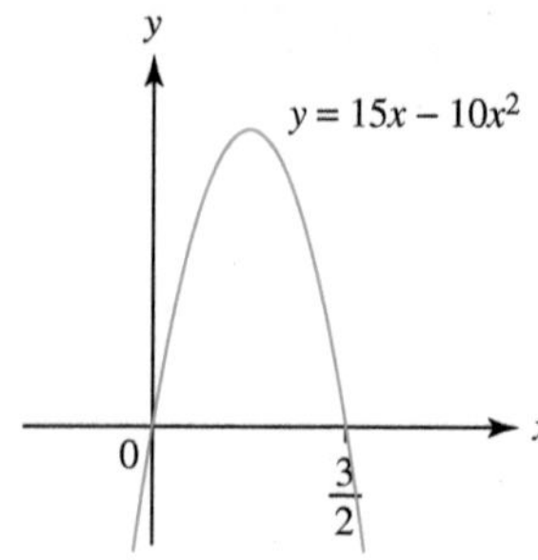

g

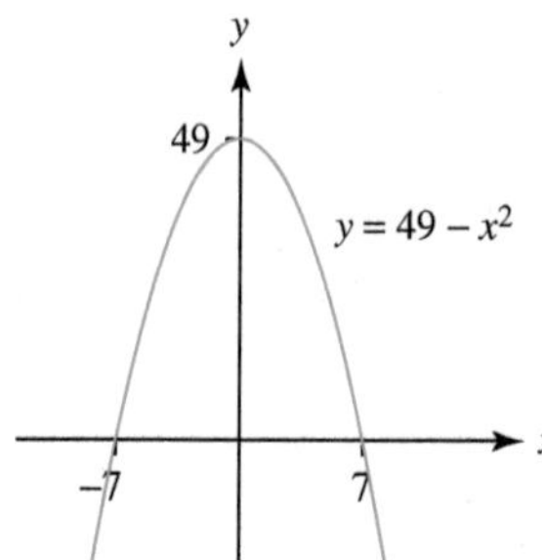

h

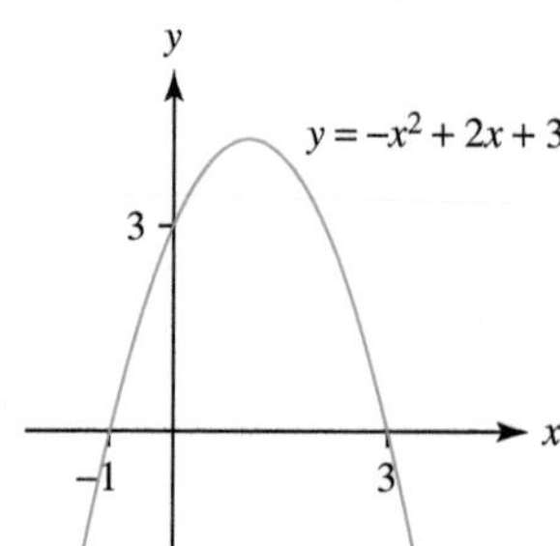

i

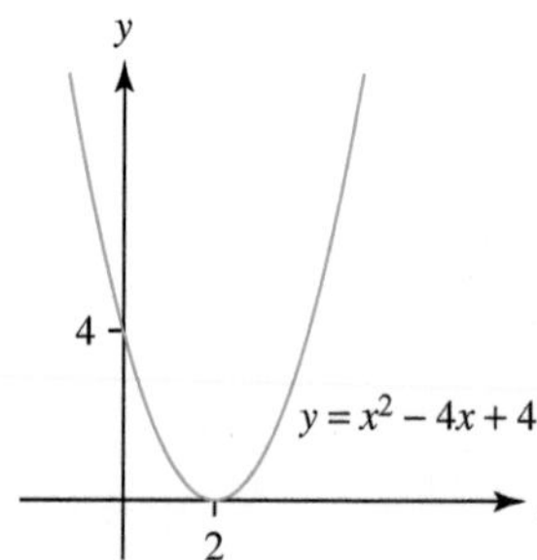

j

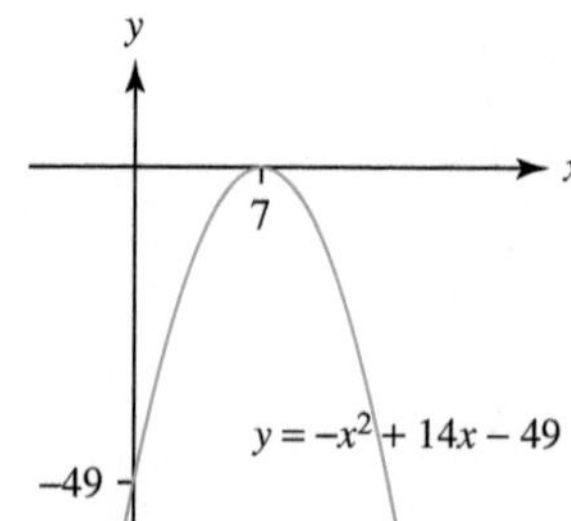

k

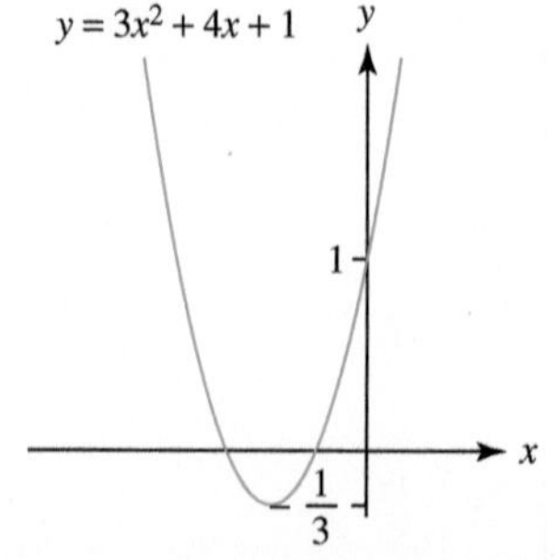

1

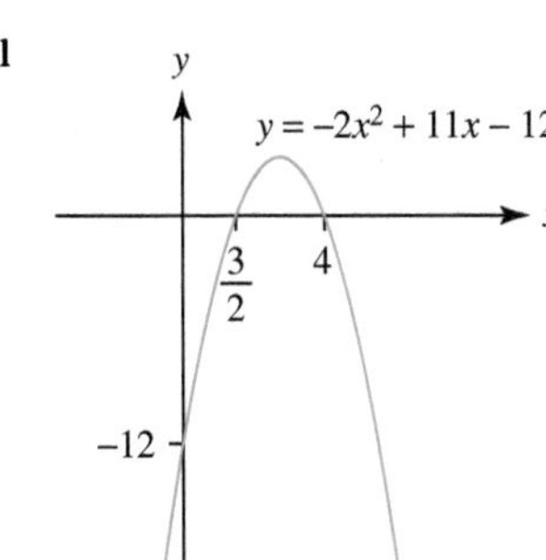

Try it 1D

1 a $(x+11)^2 - 121$ b $2(x-2)^2 - 14$
c $-(x-5)^2 + 25$

2 a $\left(\frac{3}{2}, -\frac{5}{4}\right)$ is a minimum
b $\left(-\frac{7}{2}, \frac{1}{4}\right)$ is a maximum
c $(-1, -3)$ is a minimum

Bridging Exercise 1D

1 a $(x+4)^2 - 16$ b $(x-9)^2 - 81$
c $(x+3)^2 - 6$ d $(x+6)^2 - 41$
e $\left(x-\frac{7}{2}\right)^2 - \frac{9}{4}$ f $\left(x+\frac{5}{2}\right)^2 + \frac{11}{4}$
g $2(x+2)^2 - 4$ h $3(x+2)^2 - 33$
i $2\left(x-\frac{5}{2}\right)^2 - \frac{19}{2}$ j $-(x-6)^2 + 35$
k $-\left(x-\frac{9}{2}\right)^2 + \frac{69}{4}$ l $-2\left(x-\frac{5}{4}\right)^2 + \frac{17}{8}$

2 a $(-7, -49)$ is a minimum point
b $(9, -78)$ is a minimum point
c $\left(\frac{9}{2}, -\frac{81}{4}\right)$ is a minimum point
d $(2, 4)$ is a maximum point
e $\left(-\frac{11}{2}, -\frac{1}{4}\right)$ is a minimum point
f $(3, 2)$ is a maximum point
g $(-4, -37)$ is a minimum point
h $\left(\frac{5}{2}, \frac{67}{4}\right)$ is a maximum point.

Try it 1E

1 $x = 1.25$ or $x = -0.68$

2 $\frac{1}{20}$

3 $k \geq -\frac{9}{4}$

4 $k > \frac{49}{4}$

Bridging Exercise 1E

1 a $x = 0.88$ or $x = -1.30$
b $x = 3.41$ or $x = 0.59$
c $x = 11.66$ or $x = 0.34$

2 a No real solutions.
b Two (distinct) real solutions.
c One real solution (coincidental solutions).

3 a $y = 7x^2 - 5x + 4$ b $y = -4x^2 + 12x - 9$
c $y = 6x^2 - x - 15$ d $y = -x^2 + 2x - 4$

4 a $-\frac{1}{3}$ b $\frac{1}{16}$ c $\frac{65}{8}$

5 a $k \geq -\frac{3}{4}$ b $k \leq \frac{49}{16}$ c $k \leq 7$

6 a $k > \frac{1}{40}$ b $k < -\frac{4}{5}$ c $k < \frac{47}{48}$

Try it 1F

1 a $\frac{1}{3}$ b -2 c $-\frac{7}{2}$

2 a $2\sqrt{2}$ b $3\sqrt{10}$ c $3\sqrt{11}$

3 a $(1.5, 7)$ b $(-3.5, -2)$ c $(1.9, -6.5)$

4 a gradient is -2, y-intercept is 8
b gradient is $-\frac{1}{2}$, y-intercept is $\frac{3}{2}$
c gradient is $\frac{2}{3}$, y-intercept is $-\frac{4}{9}$

5 a $y = -2x + 13$ b $y = 3x - 16$ c $5y = 3x - 11$

6 $3x - 2y - 13 = 0$

7 a neither parallel nor perpendicular
b perpendicular
c parallel

8 $3x - 2y + 13 = 0$

9 $7x - 4y + 39 = 0$

Bridging Exercise 1F

1 a -1 b $\frac{4}{9}$ c -6
d $\frac{4}{3}$ e $\sqrt{3}$ f -2

2 a $5\sqrt{2}$ b $\sqrt{481}$ c 14.2
d $\frac{\sqrt{13}}{5}$ e $\sqrt{41}$ f $k\sqrt{10}$

3 a $(2, 8)$ b $(-0.5, -6.5)$
c $(4.2, -0.1)$ d $(-0.5, -1)$
e $\left(\frac{5}{2}\sqrt{5}, \frac{3}{2}\sqrt{5}\right)$ f $(2m, 0)$

4 a gradient is 7, y-intercept is -4
b gradient is -2, y-intercept is 3
c gradient is 1, y-intercept is -4
d gradient is $-\frac{3}{2}$, y-intercept is $\frac{7}{2}$
e gradient is $\frac{5}{2}$, y-intercept is $-\frac{9}{2}$
f gradient is $\frac{3}{5}$, y-intercept is 0
g gradient is $-\frac{1}{6}$, y-intercept is $-\frac{1}{2}$
h gradient is $\frac{4}{3}$, y-intercept is $\frac{2}{3}$

5 a $y = -\frac{1}{2}x + 6$ b $y = -2x - 1$
c $11x + 3y - 56 = 0$ d $y = \frac{1}{4}x - 4$
e $y = 2x - 1$ f $y = \frac{5}{2}x - \frac{7}{2}\sqrt{2}$

6 a perpendicular
b neither parallel nor perpendicular
c parallel

7 a perpendicular
b neither parallel nor perpendicular
c parallel

8 a parallel
b neither parallel nor perpendicular
c perpendicular

9 a $5x-y-18=0$ **b** $x+5y-1=0$

10 a $x-2y+11=0$ **b** $2x+y-14=0$

11 a $3x+y-22=0$ **b** $x-3y-2=0$

12 a $6x+5y-24=0$ **b** $5x-6y-42=0$

13 a $6x-2y-1=0$ **b** $2x+6y+5=0$

14 a $2x-3y-5=0$

b $5x+7y+14=0$

c $y=-x+6$

d $x-3y-7=0$

e $8x-2y-25=0$

15 a $(1, 1)$ **b** $(-2, 16)$

c $\left(\frac{4}{3},\frac{13}{3}\right)$ **d** $\left(2,\frac{15}{2}\right)$

16 a $(2, -1)$ **b** $(4, 4)$

c $\left(\frac{5}{6},\frac{1}{6}\right)$ **d** $\left(-\frac{1}{3},-\frac{1}{3}\right)$

Try it 1G

1 a centre $(-2, 8)$, radius is 5

b $(x-7)^2+(y+9)^2=64$

2 a centre $(0, 5)$, radius 3

b centre $(-3, 6)$, radius $3\sqrt{5}$

3 $(x-3)^2+(y-1)^2=26$

4 a $(6-1)^2+(1+4)^2=5^2+5^2=50$ so $(6, 1)$ lies on the circle

b $y=-x+7$

5 a $(1.6, 0.2)$, $(-1, 8)$

b $\frac{13}{5}\sqrt{10}$

6 $y=2x+11 \Rightarrow (x-5)^2+(2x+11-1)^2=80$

$\Rightarrow (x-5)^2+(2x+10)^2=80$

$\Rightarrow x^2-10x+25+4x^2+40x+100=80$

$\Rightarrow 5x^2+30x+45=0$

$b^2-4ac=30^2-4\times5\times45=0$ so exactly one solution

Therefore the line and the circle touch once, hence the line is a tangent to the circle.

Bridging Exercise 1G

1 a $(x-2)^2+(y-5)^2=49$

b $(x+1)^2+(y+3)^2=16$

c $(x+3)^2+y^2=2$

d $(x-4)^2+(y+2)^2=5$

2 a centre $(5, 3)$, radius 4

b centre $(-3, 4)$, radius 6

c centre $(9, -2)$, radius 10

d centre $(-3, -1)$, radius $4\sqrt{5}$

e centre $(\sqrt{2}, -2\sqrt{2})$, radius $4\sqrt{2}$

f centre $\left(-\frac{1}{4},-\frac{1}{3}\right)$, radius $\frac{5}{2}$

3 a centre $(-1, 0)$, radius 5

b centre $(0, -6)$, radius 7

c centre $(2, 0)$, radius 1

d centre $(-3, -4)$, radius is $\sqrt{23}$

e centre $(4, 5)$, radius $2\sqrt{11}$

f centre $(-7, 1)$, radius $\sqrt{55}$

g centre $\left(-\frac{5}{2},2\right)$, radius $\frac{1}{2}\sqrt{29}$

h centre $\left(\frac{3}{2},\frac{9}{2}\right)$, radius $\frac{7}{2}\sqrt{2}$

i centre $\left(\frac{1}{2},-\frac{7}{2}\right)$, radius $\frac{\sqrt{2}}{2}$

4 a $(x-2)^2+(y-6)^2=2$

b $(x-3)^2+(y+3)^2=5$

c $(x+4)^2+(y+4.5)^2=27.25$

d $(x-2.5)^2+(y+11.5)^2=50.5$

e $x^2+(y-5)^2=3$

f $(x-\sqrt{3})^2+(y+3\sqrt{3})^2=39$

5 a does not lie on the circle

b does lie on the circle

c does not lie on the circle

d does lie on the circle

6 a lies on this circle

b doesn't lie on this circle

c lies on this circle

7 $x-3y-14=0$

8 $3x+5y+10=0$

9 $y=\frac{1}{4}x-\frac{19}{4}$

10 $y=9x-5$

11 a $(-2, 7)$ and $(7, -2)$

b $(-3, -1)$ and $(5, -1)$

c $(-1.6, 3.8)$ and $(-4, -1)$

d $(-2.2, -3.4)$ and $(-5, -9)$

12 a $(-4, -2)$ and $(-10, -4)$

b $2\sqrt{10}$

13 a $(-7, 6)$ and $(-0.6, 2.8)$

b $\frac{16}{5}\sqrt{5}$

14 $(x-3)^2+(x-3+2)^2=2 \Rightarrow (x-3)^2+(x-1)^2=2$

$\Rightarrow x^2-6x+9+x^2-2x+1=2$

$\Rightarrow 2x^2-8x+8=0$

$b^2-4ac=(-8)^2-4\times2\times8=0$ so only one solution hence a tangent

15 $y=34-4x \Rightarrow (x+1)^2+(34-4x-4)^2=68$

$\Rightarrow (x+1)^2+(30-4x)^2=68$

$\Rightarrow x^2+2x+1+900-240x+16x^2=68$

$\Rightarrow 17x^2-238x+833=0$

$b^2-4ac=(-238)^2-4\times17\times833=0$ so only one solution hence a tangent

16 $x=25-3y \Rightarrow (25-3y)^2+(y-5)^2=10$

$\Rightarrow 625-150y+9y^2+y^2-10y+25=10$

$\Rightarrow 10y^2-160y+640=0$

$b^2-4ac=(-160)^2-4\times10\times640=0$ so only one solution hence a tangent

17 $(x-1)^2+(2x+3+4)^2=1$

$\Rightarrow (x-1)^2+(2x+7)^2=1$

$\Rightarrow x^2-2x+1+4x^2+28x+49=1$

$\Rightarrow 5x^2+26x+49=0$

$b^2-4ac=26^2-4\times5\times49=-304$ negative so no solutions hence they do not intersect

18 $3x=-2-4y \Rightarrow x=-\frac{2}{3}-\frac{4}{3}y$

$\Rightarrow \left(-\frac{2}{3}-\frac{4}{3}y+3\right)^2+(y-6)^2=9$

$\Rightarrow \left(\frac{7}{3}-\frac{4}{3}y\right)^2+(y-6)^2=9$

$\Rightarrow \frac{49}{9}-\frac{56}{9}y+\frac{16}{9}y^2+y^2-12y+36=9$

$\Rightarrow \frac{25}{9}y^2-\frac{164}{9}y+\frac{292}{9}=0$

$b^2-4ac=\left(-\frac{164}{9}\right)^2-4\times\frac{25}{9}\times\frac{292}{9}=-\frac{256}{9}$ negative so no solutions hence they do not intersect

Exercise 1.1A Fluency and skills

1 A prime number, by definition, has exactly two factors: 1 and the number itself. The number 1 has only one factor so is NOT a prime number.

2 Let the numbers be $2m+1$ and $2n+1$
$2m+1+2n+1=2m+2n+2=2(m+n+1)$
$m+n+1=k$, an integer, so the sum is $2k$ which is the definition of an even number.

3 Let the smaller odd number be $2m+1$
The next one is $2m+3$
$(2m+1)(2m+3)=4m^2+8m+3=4(m^2+2m+1)-1$
m^2+2m+1 is an integer so this is one less than a multiple of 4

4 Let the integers be m, $m+1$ and $m+2$
The mean $=\dfrac{m+m+1+m+2}{3}=\dfrac{3m+3}{3}=m+1$, which is the middle number.

5 **a** Let the integers be m and $m+1$
$m^2+(m+1)^2=2m^2+2m+1$
The first two terms are even and the third is odd so the sum must be odd.

b Let the integers be $2m$ and $2m+2$
$(2m)^2+(2m+2)^2=8m^2+8m+4=4(2m^2+2m+1)$
4 is a factor of this expression so the squares of two consecutive even numbers is always a multiple of 4

6 Let the integers be m, $m+1$, $m+2$ and $m+3$
$m+m+1+m+2+m+3=4m+6=2(2m+3)$
2 is even; $2m$ is even so $2m+3$ must be odd. Hence the sum has both odd and even factors.

7 Let the numbers be m and n. $(m+n)^2=m^2+2mn+n^2$ which is $(m^2+n^2)+2mn$, which is $2mn$ more than the sum of the squares, since, if m and n are positive, then mn must also be positive.

8 Let the equal side be a. The hypotenuse $=\sqrt{a^2+a^2}=a\sqrt{2}$
Thus the perimeter is $2a+a\sqrt{2}$
Since $\sqrt{2}>1$, it follows that the perimeter is always greater than three times the length of one of the equal sides.

9 If the sum = the product then $a+b=ab$
Hence $b-2+b=(b-2)b \Rightarrow 2b-2=b^2-2b \Rightarrow b^2-4b+2$
This quadratic does not factorise and hence b cannot be an integer. Consequently, since $a=b-2$, neither can a

10 If $(5y)^2$ is even, then $5y$ is even and, since 5 is odd, y must be even to make $5y$ even.

11 Checking each number, only 25 is a square number and only 27 is a cube.

12 For example: JANUARY, YARN; FEBRUARY, FRAY; AUGUST, STAG; SEPTEMBER, TERM; OCTOBER, BOOT; NOVEMBER, BONE; DECEMBER, BRED

13 There are 5 'cases' and we investigate each one: $(0+1)^3 \geq 3^0 \Rightarrow 1=1$ TRUE; $(1+1)^3 \geq 3^1 \Rightarrow 8>3$ TRUE; $(2+1)^3 \geq 3^2 \Rightarrow 27>9$ TRUE; $(3+1)^3 \geq 3^3 \Rightarrow 64>27$ TRUE; $(4+1)^3 \geq 3^4 \Rightarrow 125>81$ TRUE. Thus the statement is true.

14 Let the square number be m^2
$m=10p+k$ for integers p and k, $0 \leq k \leq p$
If $m=10p$ then $m^2=(10p)^2=100p^2$ which ends in 0, if $m=10p+1$ then $m^2=(10p+1)^2=100p^2+20p+1$ which ends in 1, if $m=10p+2$ then $m^2=(10p+2)^2=100p^2+40p+4$ which ends in 4, … if $m=10p+9$ then $m^2=(10p+9)^2=100p^2+180p+81$ which ends in 1
Therefore all square numbers cannot have a last digit 2, 3, 7 or 8.

15 $2\times3=6$ which disproves the statement.

16 $1+2=3$ which is <6 and disproves the statement.

17 $3-(-4)=3+4=7$ which disproves the statement.

18 $5\times-2=-10$ which disproves the statement.

19 e.g. $a=4$, $b=3$; $4>3$ but $4^3=64<3^4=81$

20 Try any set with two odd numbers, such as $1\times2\times3=6$, which is not divisible by 4

Exercise 1.1B Reasoning and problem-solving

1 **a** Case 1: P is 2 so even. PQ is even × odd which is even.
Case 2: P is odd. PQ is odd × odd which is odd.
Graham is right.

b Case 1: P is 2 so even and $Q+1$ is even.
$P(Q+1)$ is even × even which is even.
Case 2: P is odd. $P(Q+1)$ is odd × even which even.
Sue is right in this case.

2 $99=3\times33$ and 33 is not prime. The statement is false.

3 $9^1-1=9-1=8$ which is 8×1; $9^2-1=81-1=80$ which is 8×10; $9^3-1=729-1=728$ which is 8×91; $9^4-1=6561-1=6560$ which is 8×820; $9^5-1=59\,049-1=59\,048$ which is 8×7381; $9^6-1=531\,441-1=531\,440$ which is $8\times66\,430$
Thus the value of 9^n-1 is divisible by 8 for $1\leq n\leq6$

4 An equilateral triangle is not obtuse. A right-angled triangle is not obtuse.

5 A convex hexagon can be split into 4 triangles. The sum of the interior angles of a triangle is 180°. $4\times180=720$. So the sum of the interior angles of a convex hexagon is 720°

6 False. A Rhombus has equal sides but is not a square.

7 A convex n-sided polygon can be split into $n-2$ triangles. The sum of the interior angles of a triangle is 180°
$(n-2)\times180=180(n-2)$. So the sum of the interior angles of a convex n-sided polygon is $180(n-2)°$

8 $5^2=(-5)^2$ but $5\neq-5$

9 The square of the remaining side $=(2s+a)^2-(2s-a)^2$
$=4as+4as=8as$
Therefore the remaining side is a multiple of 8

10 $\dfrac{1}{1\times2}=\dfrac{1}{1+1}$; $\dfrac{1}{1\times2}+\dfrac{1}{2\times3}=\dfrac{1}{2}+\dfrac{1}{6}=\dfrac{3+1}{6}=\dfrac{4}{6}=\dfrac{2}{3}$;
$\dfrac{1}{1\times2}+\dfrac{1}{2\times3}+\dfrac{1}{3\times4}=\dfrac{1}{2}+\dfrac{1}{6}+\dfrac{1}{12}=\dfrac{6+2+1}{12}=\dfrac{9}{12}=\dfrac{3}{4}$;
$\dfrac{1}{1\times2}+\dfrac{1}{2\times3}+\dfrac{1}{3\times4}+\dfrac{1}{4\times5}=\dfrac{1}{2}+\dfrac{1}{6}+\dfrac{1}{12}+\dfrac{1}{20}=$
$\dfrac{30+10+5+3}{60}=\dfrac{48}{60}=\dfrac{4}{5}$; $\dfrac{1}{1\times2}+\dfrac{1}{2\times3}+\dfrac{1}{3\times4}+\dfrac{1}{4\times5}+$
$\dfrac{1}{5\times6}=\dfrac{1}{2}+\dfrac{1}{6}+\dfrac{1}{12}+\dfrac{1}{20}+\dfrac{1}{30}=\dfrac{30+10+5+3+2}{60}=\dfrac{50}{60}=\dfrac{5}{6}$
Thus $\dfrac{1}{1\times2}+\dfrac{1}{2\times3}+\dfrac{1}{3\times4}+.....+\dfrac{1}{n\times(n+1)}=\dfrac{n}{n+1}$ for $1\leq n\leq5$

11 Let the two-digit number be X with tens digit y and units digit z
Therefore $X=10y+z=9y+(y+z)$. Let the sum of the digits be divisible by 3, i.e. $(y+z)=3P$
Thus $X=9y+3P=3(3y+P)$ and hence X is a multiple of three.

Exercise 1.2A Fluency and skills

1 64
2 -243
3 2401
4 c^{11}
5 p^{12}
6 $-p^{12}$
7 $64c^{-18}$
8 d^{14}
9 $70e^9$
10 $-108f^{12}$
11 $4g^6$
12 $-4k^{55}$
13 $8f^3$
14 $\dfrac{2}{3}e^2$
15 $15ab$
16 $-120wx^2$
17 $24def^2$
18 $-9h^{14}$
19 $5r^8s^{10}$
20 $5r^2s^2$
21 $-g^{10}h^9i^4$
22 $-g^8h^7i^{-4}$
23 $5z^5y^5$
24 $6u^{18}$
25 $6u^{18}$
26 $5t^9$
27 $-5t^9c^4$
28 $\dfrac{1}{5}$
29 5
30 6
31 -50
32 3
33 $\dfrac{1}{16}$
34 $\dfrac{1}{1024}$
35 $\dfrac{1}{9w^2}$

36 $\frac{w^4}{9}$

37 512

38 4

39 256

40 8

41 $\frac{1}{8}$

42 $\frac{6}{7}$

43 $\frac{7}{6}$

44 $\frac{216}{343}$

45 $n=4$

46 $m=5$

47 $t=1$

48 $b=-2$

Exercise 1.2B Reasoning and problem-solving

1 a $4s^4$ inches^2 b $5p^2q^3$ cm

2 a Circumference $=6\pi w^5$ ft; Area $=9\pi w^{10}$ ft^2

b Surface area $=36\pi w^8$ ft^2; Volume $=36\pi w^{12}$ ft^3

3 $6c^{-4}d^{-4}e^{-5}$ 4 $36p^5q^5$

5 a $2y^3z^4$

b Area = 8. Therefore the area is independent of m and n

6 $\frac{4}{3c^{\frac{1}{2}}}$ mph

7 a $2d^3$ cm

b Volume $=225\pi s^6t^0=225\pi s^6$m^3. It is independent of t since $t^0=1$

8 $7\pi v^4z^{-4}$ cm^2

9 a $13n^{\frac{1}{2}}$ b $30n$

10 a $18m^2n^{-7}$ V b $108n^{-10}$ W

11 $\frac{81}{2}mx^{\frac{3}{2}}c^{\frac{3}{2}}$ 12 $\frac{3t^3}{16g}$ 13 $\frac{15}{8}rs^2$ ohms

14 a $A=\sqrt{s(s-a)(s-b)(s-c)}$

b $A=30x^2y^2$

c The sides of the triangle are in the ratio 5 : 12 : 13
Since $13^2=5^2+12^2$ the triangle is right-angled and so the area is $\frac{1}{2}\times 5xy\times 12xy=30xy$

Exercise 1.3A Fluency and skills

1 Parts **a, e** and **f** are rational; parts **b, c, d** and **g** are irrational

2 a $2\sqrt{21}$ b $2\sqrt{14}$ c $5\sqrt{3}$ d $3\sqrt{3}$
e $2\sqrt{2}$ f $16\sqrt{2}$ g $17\sqrt{17}$ h $6\sqrt{6}$
i $210\sqrt{3}$ j $288\sqrt{5}$

3 a 8 b 5

4 a $\frac{4}{10}$ or 0.4 b $\frac{5}{6}$

5 a $3\sqrt{6}$ b $12\sqrt{3}$ c $16\sqrt{5}$ d $22\sqrt{7}$
e $4\sqrt{5}$ f $6\sqrt{7}$ g $3\sqrt{5}$ h $2\sqrt{2}$
i $3\sqrt{6}$ j $7\sqrt{3}$ k $2\sqrt{3}$ l $4\sqrt{5}$
m $7\sqrt{3}$ n $4\sqrt{2}$

6 a $59+6\sqrt{30}$ b $14+7\sqrt{2}$ c $-14+7\sqrt{2}$
d $6-10\sqrt{5}$ e $120-14\sqrt{6}$

7 a $\frac{\sqrt{13}}{13}$ b $\frac{4\sqrt{6}}{3}$
c $\frac{\sqrt{55}}{10}$ d $3(\sqrt{2}+1)$
e 5 f $\frac{13\sqrt{5}}{20}-\frac{\sqrt{30}}{30}$
g $\frac{5(8+\sqrt{5})}{59}$ h $\frac{2(2\sqrt{2}-1)}{7}$
i $11-2\sqrt{30}$ j $\frac{5-\sqrt{77}}{4}$

8 a $\frac{a\sqrt{b}+b}{b}$ b $\frac{a^2+2a\sqrt{b}+b}{a^2-b}$
c $\frac{\sqrt{ac}+bc}{bc}$ d $\frac{a-\sqrt{ab}-b\sqrt{ac}+b\sqrt{bc}}{a-b}$

Exercise 1.3B Reasoning and problem-solving

1 $6\sqrt{6}$ cm^2

2 a $\pi(9\sqrt{3})^2=\pi(81\times 3)=243\pi$ cm^2 b $14\sqrt{5}$

3 a $3\sqrt{5}$ m s^{-1} b $8\sqrt{5}$ m

4 $50+19\sqrt{7}$ m^3 5 A is $\frac{3\sqrt{3}}{\sqrt{7}}$ times bigger than B

6 $5\sqrt{3}$ cm 7 $10\sqrt{3}$ m

8 $\frac{45}{360}\times 2\pi\left(\frac{12}{\sqrt{3}}\right)=\sqrt{3}\pi$m 9 $2\sqrt{3}$

10 $\frac{(6\sqrt{8})^3}{(4\sqrt{2})^3}=\frac{3456\sqrt{2}}{128\sqrt{2}}=27$ 11 $\frac{19\sqrt{a}}{3a}$

12 $\frac{40\pi}{13}$ m^2 13 $40(3-\sqrt{6})$

14 Thus coordinates of the centroid are $\left(\frac{5\sqrt{6}}{2},\frac{5\sqrt{2}}{2}\right)$, so the distance from the origin to the centroid $=5\sqrt{2}$

15 $2s=5+5\sqrt{2}+3+3\sqrt{2}+6+4\sqrt{2}+2+4\sqrt{2}=16+16\sqrt{2}$

$s=8+8\sqrt{2}$

$A^2=(3+3\sqrt{2})(5+5\sqrt{2})(2+4\sqrt{2})(6+4\sqrt{2})$

$=3(1+\sqrt{2})5(1+\sqrt{2})2(1+2\sqrt{2})2(3+2\sqrt{2})$

$=2^2(1+\sqrt{2})^2 15(11+8\sqrt{2})$

$A=2(1+\sqrt{2})\sqrt{15(11+8\sqrt{2})}$

Exercise 1.4A Fluency and skills

1 a $x=\pm 3\sqrt{2}$ b $x=\sqrt{3}$ or $-\sqrt{3}$
c $x=0$ or $\frac{-5}{4}$ d $x=-\sqrt{3}$ twice
e $x=\frac{1}{2}$ or -3 f $x=7$ or $\frac{2}{3}$
g $x=\frac{3}{4}$ twice h $x=\frac{9}{4}$ or -2

2 a i $f(x)=x^2+3x+2=(x+1)(x+2)$

ii

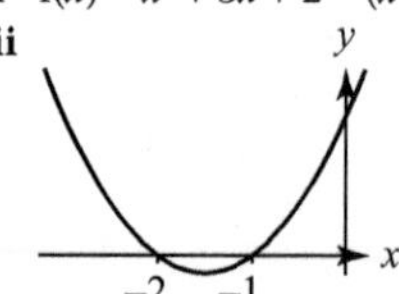

b i $f(x)=x^2+6x-7=(x-1)(x+7)$

ii

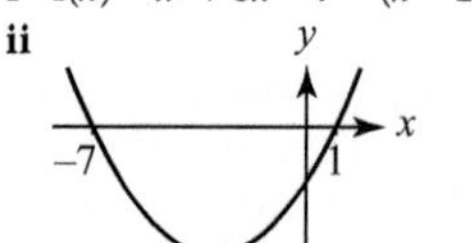

c i $f(x)=-x^2-x+2=(x+2)(1-x)$

ii

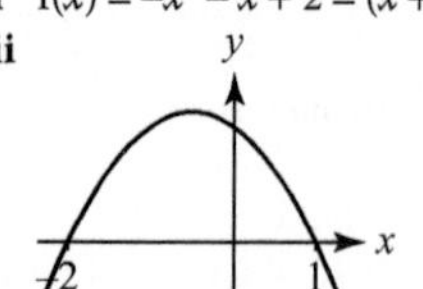

d i $f(x)=-x^2-7x-12=(x+4)(-x-3)$

ii y x −4 −3

e **i** $f(x) = 2x^2 - x - 1 = (2x+1)(x-1)$

ii

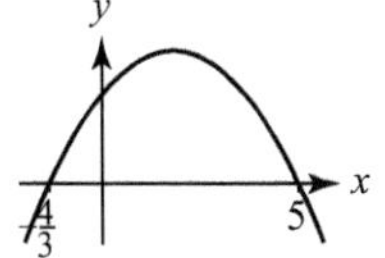

f **i** $f(x) = -3x^2 + 11x + 20 = (x-5)(-3x-4)$

ii

3 **a** $x = \frac{-3}{2} \pm \frac{\sqrt{21}}{6}$ or -2.26 or -0.74

b $x = \frac{-5}{8} \pm \frac{\sqrt{41}}{8}$ or 0.18 or -1.43

c $x = -6 \pm \sqrt{31}$ or -0.43 or -11.57

d $x = -1 \pm \sqrt{29}$ or -6.39 or 4.39

e $x = \frac{-15}{2} \pm \frac{\sqrt{365}}{2}$ or 2.05 or -17.05

f $x = \frac{3}{2} \pm \frac{\sqrt{145}}{2}$ or -4.52 or 7.52

g $x = 4.5$ twice

h $x = \frac{23}{6} \pm \frac{\sqrt{277}}{6}$ or 6.61 or 1.06

i $x = \frac{-8}{5} \pm \frac{\sqrt{19}}{5}$ or -0.73 or -2.47

j $x = \frac{1}{20} \pm \frac{\sqrt{41}}{20}$ or 0.37 or -0.27

4 **a** $f(x) = (x-7)^2$; minimum $(7, 0)$

b $f(x) = (x+1)^2 - 6$; minimum $(-1, -6)$

c $f(x) = -1(x+3)^2 + 4$; maximum $(-3, 4)$

d $f(x) = -1(x-2)^2 + 7$; maximum $(2, 7)$

e $f(x) = 9\left(x - \frac{1}{3}\right)^2 - 6$; minimum $\left(\frac{1}{3}, -6\right)$

f $f(x) = -2(x+7)^2 + 63$; maximum $(-7, 63)$

5 **a** $x = 1 \pm 1 = 2$ or 0 **b** $x = -2 \pm \sqrt{7}$

c $x = 7 \pm 4 = 11$ or 3 **d** $x = -4 \pm \sqrt{6}$

e $x = 3$ twice **f** $x = -5 \pm 1 = -6$ or -4

g $x = -11 \pm \sqrt{3}$ **h** $x = 8 \pm \sqrt{10}$

i $x = \frac{3 \pm \sqrt{7}}{2}$ **j** $x = \frac{-2 \pm \sqrt{6}}{3}$

k $x = \frac{-11 \pm \sqrt{109}}{2}$ **l** $x = \frac{5 \pm \sqrt{57}}{3}$

6 **a** $x = 0$ or 3 **b** $x = -5$ or 3

c $x = -1$ or 6 **d** $x = -2$ or 4

e $x = -2.5$ or 3 **f** $x = -0.5$ or 1.5

7 **a** two real distinct roots **b** two real distinct roots

c two real distinct roots **d** no real roots

e two real coincident roots **f** no real roots

g two real coincident roots **h** no real roots

Exercise 1.4B Reasoning and problem-solving

1 **a** $3x^2 + 25x - 2035 = 0 \Rightarrow x = 22.2$ cm (nearest mm)

b 14.8 litres

2 **a** $x^2 + 3x - 10 = 0 \Rightarrow x = 2$ years

b 28 years

3 The card has side length 39.7 cm and the photo is 29.8 cm by 19.7 cm.

4 **a** Area $= 30x^2 - x^4$ **b** $z^2 - 30z + 85 = 0$

c $x = 5.18$ inches or 1.78 inches

5 **a** $h = -5\left(t - \frac{5}{2}\right)^2 + \frac{325}{4}$

b

c **i** $t = 2.5$ s, $h = 81.25$ m **ii** $t = 5$ s **iii** $t = 6.5$ s

6 **a**

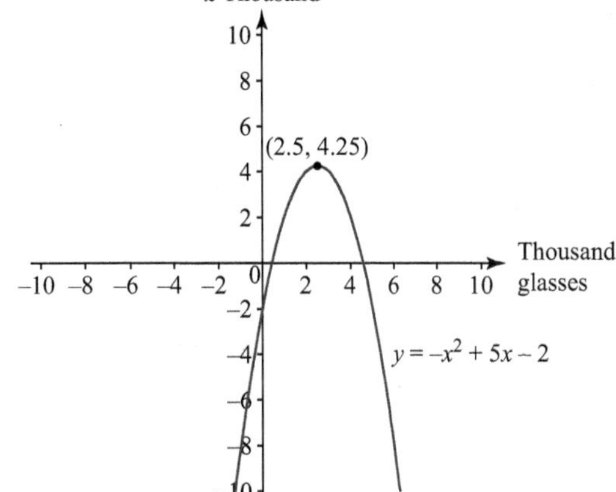

b **i** 2.5 **ii** 4.6 or 0.4 **iii** 1.5 to 3.5

7 **a**

b **i** 10 m **ii** 55 m **iii** 138 m

c 42 mph

8 $k = \sqrt[3]{16}$ or $-\sqrt[3]{5}$

Exercise 1.5A Fluency and skills

1 $x = 4, y = 3$ **2** $a = 6, b = 1$

3 $a = 1, b = 2$ **4** $x = 1, y = -1$

5 $c = 5, d = -9$ **6** $e = 8, f = 3$

7 $e = 6, f = -1$ **8** $g = 0, h = 3$

9 $x = 12, y = -18$ **10** $n = 3, m = -1$

11 $n = 5, m = -\frac{5}{3}$ **12** $a = 5, b = -1$

13 $c = -1, d = 7$ **14** $d = -11, c = -5$

15 $f = \frac{1}{3}, e = 6$ **16** When $x = 0, y = 3$ and when $x = 1, y = 2$

17 When $g = -3, h = 4$ and when $g = 2, h = -1$

18 When $g = 3, h = 2$ and when $g = -\frac{9}{4}, h = \frac{79}{8}$

19 When $n = -1, m = 1$ and when $n = \frac{17}{10}, m = \frac{289}{100}$

20 $(0, 0), (5, 5)$

21 $(-1,1), (3,5)$

22 $\left(\frac{2}{9},\frac{4}{3}\right)$, $(1,-1)$

23 $\left(-\frac{1}{4},\frac{1}{2}\right)$, $(-2,-3)$

24 $y=\frac{20}{x}$ so $\frac{20}{x}=8+x$

$20=8x+x^2$ so $x^2+8x-20=0$ so $(x-2)(x+10)=0$

$x=2$ or -10; $y=10$ or -2; points of intersection are $(2, 10)$ and $(-10, -2)$

Exercise 1.5B Reasoning and problem-solving

1 $x-y=257; x+y=1619$

The candidates polled 938 and 681 respectively.

2 Maggots cost 11 p each and worms cost 12 p each

3 $m=-3, c=-1$

4 $m=1\frac{1}{2}, n=3$

5 $q=25$ cm, $p=40$ cm

6 Florence is 8 years old and Zebedee is 12

7 a Both equations reduce to $y-2x=3$ so they are, in fact, the same line. Therefore there are an infinite number of solutions.

b The equations reduce to $y=2x+3$ and $y=2x+4$ so they are, in fact, parallel lines. Hence they do not meet and so there are no solutions.

8 $x-y+3=0$: $(2, 5)$ or $(8, 11)$; This line intersects the curve in two points.

$y+11x-27=0$: $(2, 5)$ or $(2, 5)$. This is the tangent.

$y+2x+4=0$: the line misses the curve.

9 $2x-9=x^2-x-6 \Rightarrow x^2-3x+3=0 \Rightarrow (-3)^2-4\times1\times3$
$=9-12$ which is negative.

Therefore there are no real roots to the quadratic and the line does not intersect the curve.

10 $n=7$ or -4. However $n=7$ is the only valid solution since n stands for a number of terms and thus cannot be negative so $n=-4$ is not valid.

11 The field is 225 m long and 75 m wide.

12 $3x+4y=25 \Rightarrow y=\frac{25-3x}{4} \Rightarrow x^2+\left(\frac{25-3x}{4}\right)^2=25$

$\Rightarrow 25x^2-150x+225=0 \Rightarrow x^2-6x+9=0$

$\Rightarrow (x-3)(x-3)=0 \Rightarrow x=3, y=4$. The line is a tangent because there are two coincident values of x from the solution of the quadratic. The point of intersection is $(3, 4)$

13 $4(2x+1)^2+9x^2=36 \Rightarrow 25x^2+16x-32=0$

$x=\frac{-16\pm\sqrt{16^2+4\times25\times32}}{50}=\frac{-16\pm\sqrt{3456}}{50}=\frac{-16\pm24\sqrt{6}}{50}$

$=\frac{-8\pm12\sqrt{6}}{25}$

Hence $y=2x+1=2\left(\frac{-8\pm12\sqrt{6}}{25}\right)+1=\frac{9\pm24\sqrt{6}}{25}$

14 a A is 2 units from O and B is 65 units from O

b Gradient is negative, hence x (distance) is decreasing.

c For the first 4 seconds, x increases $(2\to9\to14\to17\to18)$. Hence it is moving away; and then decreases $(17\to14\to9\to2\to-7)$. Hence it is moving back.

d 18

e 7 seconds

f Back towards O

Exercise 1.6A Fluency and skills

1 a $2x+3y-13=0$ b No.

2 The gradient is $\frac{4}{3}$ and the y-intercept is $-\frac{8}{3}$

3 a The gradient of $2x-3y=4$ is $\frac{2}{3}$, the gradient of $6x+4y=7$ is $\frac{-6}{4}=\frac{-3}{2}$, and $\frac{2}{3}\times\frac{-3}{2}=-1$ so the lines are perpendicular.

b The gradient of $2x-3y=4$ is $\frac{2}{3}$ and the gradient of $8x-12y=7$ is $\frac{8}{12}=\frac{2}{3}$

Both lines have the same gradient, so the lines are parallel.

4 Gradient $=\frac{-8}{9}$ and y-intercept $=-\frac{7}{6}$

5 $m=\frac{-1-(-6)}{4--5}=\frac{5}{9}$ and $9y-5x+29=0$

6 $y=-3x-25$ or $y+3x+25=0$

7 a $(4,6)$, length $4\sqrt{5}$ b $\left(\frac{-1}{2},\frac{-7}{2}\right)$ c $\left(\frac{\sqrt{5}}{2},\frac{2\sqrt{3}}{2}\right)$

8 $y=4x+8$ and $3y-12x=7$ are parallel; $2x+3y=4$ and $6x+9y=12$ are parallel; $4x-5y=6$ and $10x+8y=5$ are perpendicular.

9 a $3y+2x+2=0$ b $y-2x-1=0$

10 a $x^2-2x+y^2-16y+40=0$

b $x^2-12x+y^2+14y+76=0$

c $x^2-2\sqrt{5}x+y^2-2\sqrt{2}y-4=0$

11 a centre $=(-9, 7)$; radius $=10$

b centre $=(-6, -5)$; radius $=\sqrt{86}$

c centre $=(\sqrt{3}, -\sqrt{7})$; radius $=\sqrt{11}$

12 $m_{AB}=\frac{18+12}{6+10}=\frac{30}{16}=\frac{15}{8}$; $m_{BC}=\frac{-14-18}{-2-6}=\frac{-32}{-8}=4$;

$m_{CA}=\frac{-12+14}{-10+2}=\frac{2}{-8}=\frac{-1}{4}$

Since $4\times\frac{-1}{4}=-1$, BC is perpendicular to CA and hence ABC is a right-angled triangle.

Since the angle in a semicircle is a right angle it follows that AB is a diameter and the points form a semicircle.

13 $y=3x+5$

14 a $x^2-14x+y^2+2y-274=0$

b $4\sqrt{31}$

c $\left(-3+3\sqrt{62}, 9+3\sqrt{62}\right)$ and $\left(-3-3\sqrt{62}, 9-3\sqrt{62}\right)$

15 a $x^2+y^2-20y=0$

b $x^2-8x+y^2-8y+24=0$

c $x^2-x+y^2-14y-44=0$

d $x^2+(4+\sqrt{2})x+y^2+(5-\sqrt{5})y+(4\sqrt{2}-5\sqrt{5})=0$

16 $x=3, y=4$ or $x=4, y=3$

Coordinates are $(3,4)$ and $(4,3)$.

17 a $y=-\frac{1}{2}x+\frac{3}{2}$

b $(x-2)^2+(y-2)^2=9$

c $(-1,2)$ and $\left(\frac{19}{5},-\frac{2}{5}\right)$

18 a $y=2x-1$

b $(x-1)^2+y^2=\frac{289}{4}$

c $\left(\frac{3}{5}+\frac{\sqrt{1441}}{10},\frac{1}{5}+\frac{\sqrt{1441}}{5}\right)$ and $\left(\frac{3}{5}-\frac{\sqrt{1441}}{10},\frac{1}{5}-\frac{\sqrt{1441}}{5}\right)$

Exercise 1.6B Reasoning and problem-solving

1 PQ: $12y-5x-63=0$, QR: $15y-8x-84=0$, RS: $12y-5x-84=0$, and SP: $15y-8x-105=0$

Since the quadrilateral contains two pairs of parallel sides of different lengths, it must be a parallelogram.

2 a $(-2, -5)$

b Q is $(-5, -1)$; S is $(1, -9)$

c RP: $4y - 3x + 14 = 0$, QS: $3y + 4x + 23 = 0$

d QP: $7y + x + 12 = 0$, PS: $y - 7x + 16 = 0$, SR: $3y - x + 28 = 0$, RQ: $9y - 13x - 56 = 0$

e 100 units^2

f $30 + 10\sqrt{10} + 10\sqrt{2}$ units

3 a AB: $y + 2 = -\frac{4}{3}(x - 6)$ or $3y + 4x - 18 = 0$,

AC: $y + \frac{11}{2} = 7\left(x - \frac{11}{2}\right)$ or $2y - 14x + 88 = 0$

b Centre $= (6, -2)$, equation of circle is $(x - 6)^2 + (y + 2)^2 = 25$ or $x^2 - 12x + y^2 - 4y + 15 = 0$

$m_{BC} = \frac{-6 - 1}{9 - 10} = \frac{-7}{-1} = 7$. $m_{AC} \times m_{BC} = \frac{-1}{7} \times 7 = -1$ so AC and BC are perpendicular. Hence the triangle is right-angled.

4 a $x^2 + y^2 - 4x - 12y + 15 = 0$; $(x - 2)^2 + (y - 6)^2 = 25$ so centre is $(2, 6)$ and radius 5. If radius is 5, lowest possible point is $(2, 1)$ so the circle does not intersect the x-axis.

b $CP = \sqrt{61} = 7.81...$; radius is 5 so P lies outside the circle.

c $k = -15$ or 35

5 160 units^2

6 $(10, -80)$

7 a 4 **b** $y = -2x + 20$

8 a Centre $= \left(\frac{a+b}{2}, \frac{c+d}{2}\right)$, radius $= \frac{1}{2}\sqrt{(a-b)^2 + (c-d)^2}$

Equation of circle:

$$\left(x - \frac{(a+b)}{2}\right)^2 + \left(y - \frac{(c+d)}{2}\right)^2 = \frac{1}{4}((a-b)^2 + (c-d)^2)$$

so $(x - a)(x - b) + (y - c)(y - d) = 0$

b $x^2 - 10x + y^2 - 12y - 39 = 0$

Exercise 1.7A Fluency and skills

1

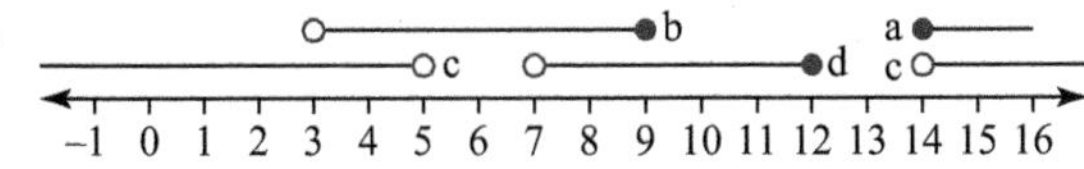

2 a

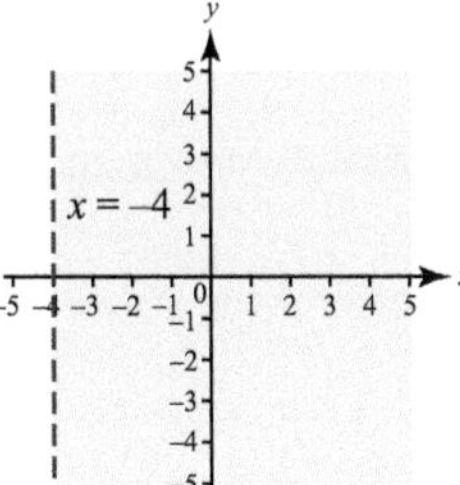

b

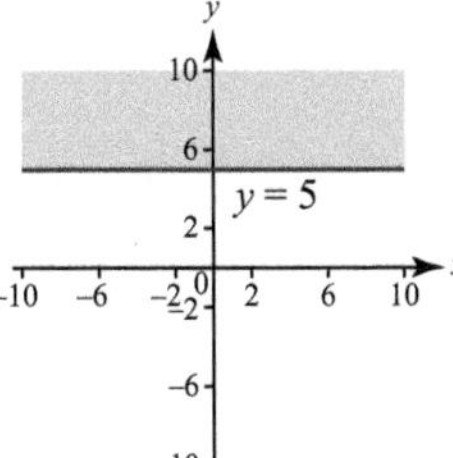

c

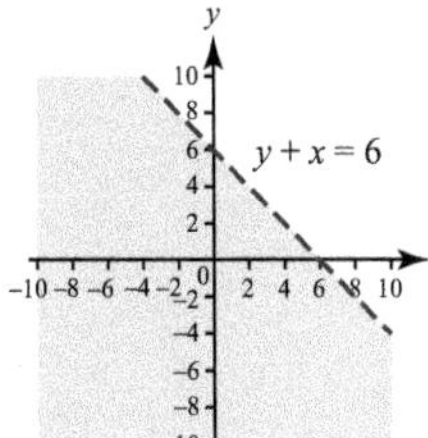

d

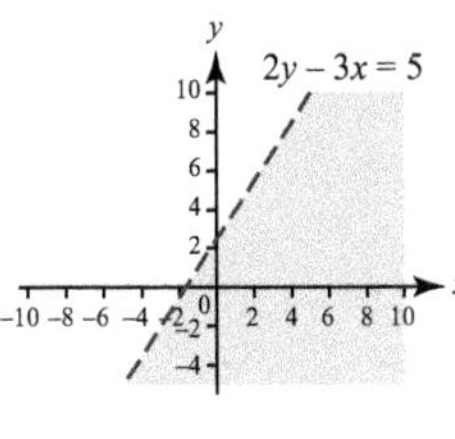

e

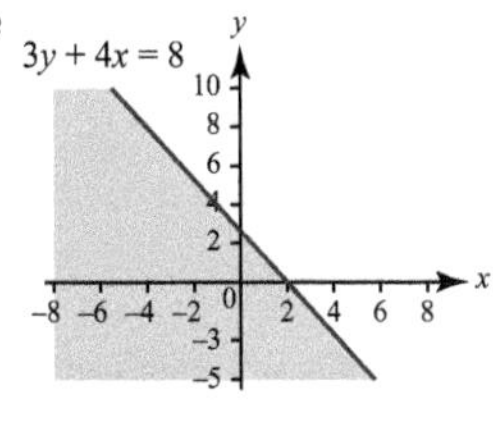

f

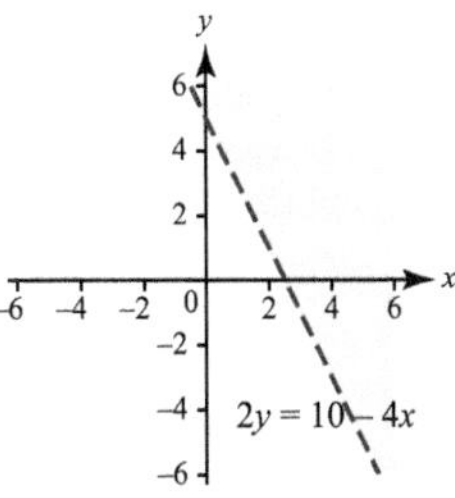

g

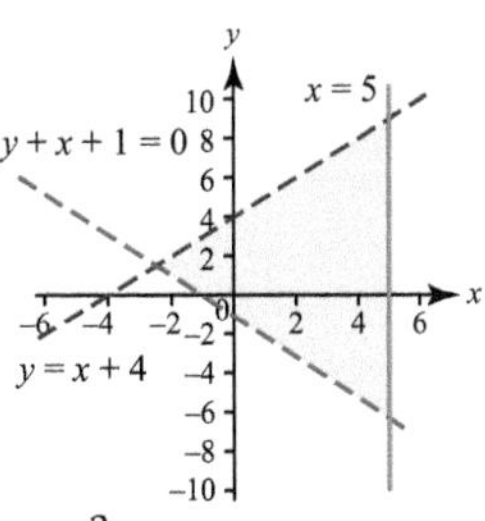

h

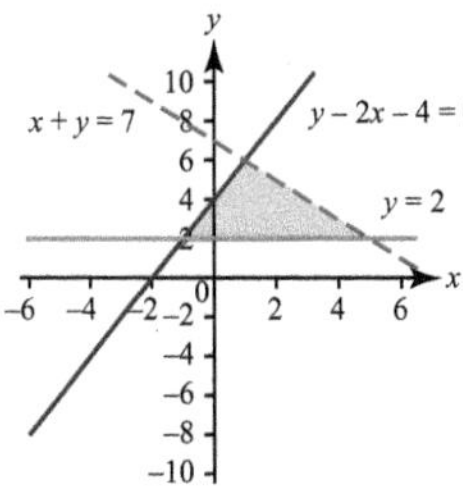

3 a $x > \frac{3}{2}$ **b** $x \le -1.9$ **c** $x < 1$

d $x \ge \frac{-11}{7}$ **e** $x < \frac{-4}{3}$ **f** $x < 0$

4 a $x < -3$ or $x > 2$ **b** $-7 < x < -4$

c $3 \le x \le 8$ **d** $x \le -4$ or $x \ge 6$

e $x < \frac{-1}{2}$ or $x > 2$ **f** $-7 < x < \frac{2}{3}$

g $x \le \frac{1}{4}$ or $x \ge 3$ **h** $x \le \frac{-2}{3}$ or $x \ge -2$

5 a $x < -3.83$ or $x > 1.83$ **b** $-5.56 < x < -1.44$

c $1.76 \le x \le 10.24$ **d** $x \le -3.32$ or $x \ge 6.32$

e $x < -0.91$ or $x > 2.57$ **f** $-4.47 < x < 0.22$

g $1.00 \le x \le 2.40$ **h** $x \le -0.38$ or $x \ge 3.05$

6 a i $y = 2x + 3$; $y = x^2$ **ii** $(-1, 1)$ and $(3, 9)$

iii

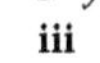

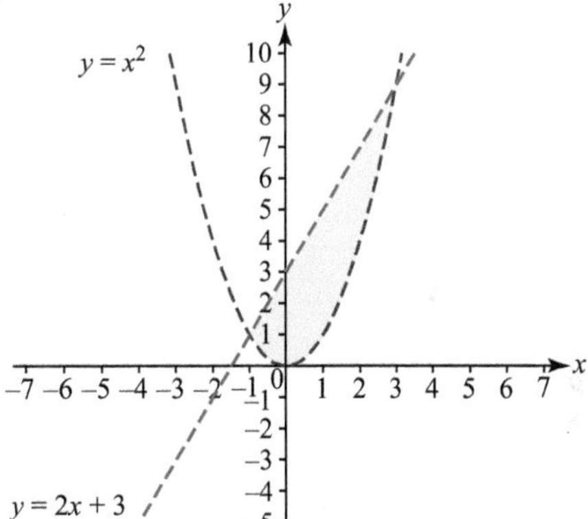

b i $x + y = 4$; $y = x^2 - 5x + 4$

ii $(0, 4)$ and $(4, 0)$

iii

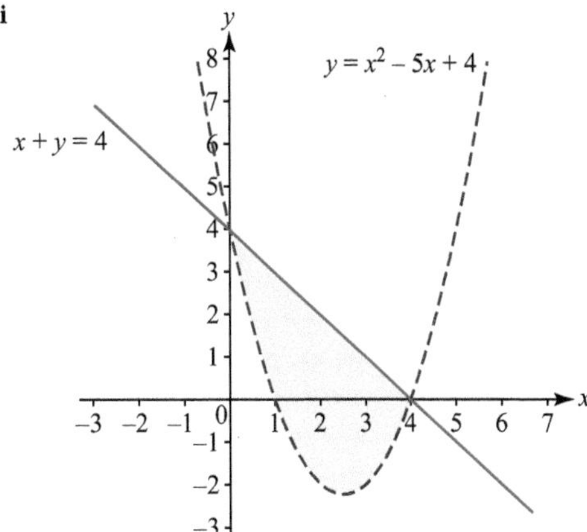

c **i** $y - 4x = 17$; $y = 4x^2 - 4x - 15$; $x = 4$

ii $(-2, 9)$ and $(4, 33)$

iii

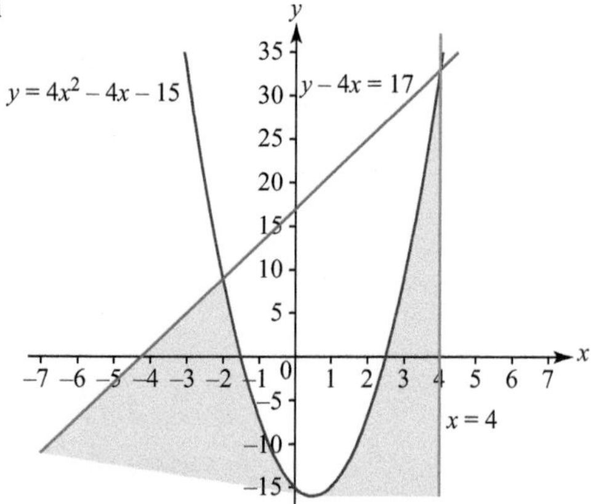

d **i** $y - 2x - 20 = 0$; $y + 4x - 6 = 0$; $y = x^2 - 5x - 24$

ii $(-4, 12)$ and $(11, 42)$, $(-5, 26)$ and $(6, -18)$, $\left(-\frac{7}{3}, \frac{46}{3}\right)$

iii

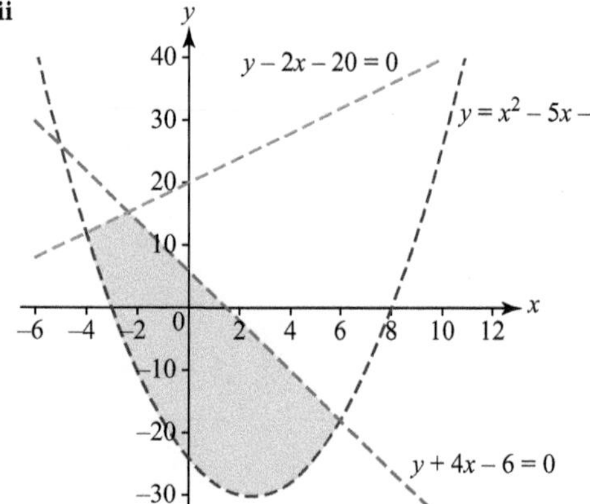

Exercise 1.7B Reasoning and problem-solving

1 **a**

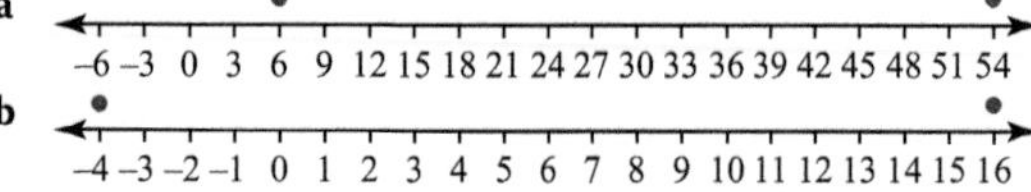

b

2

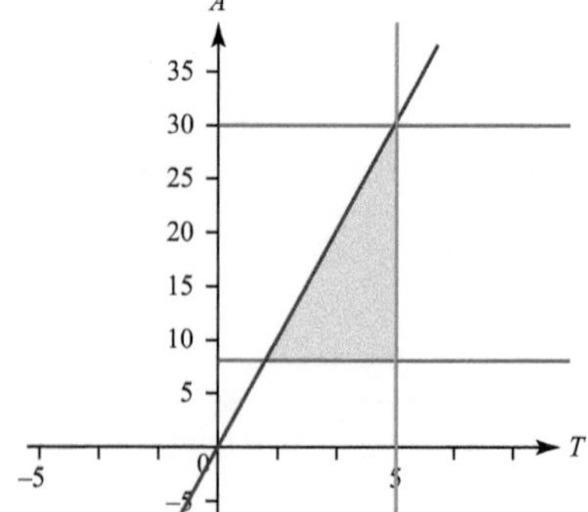

3 **a** $2w + 3p \geq 200$

b **i** $p \geq 17.3...$ i.e. p must be at least 18

ii $w \geq 86.5$ i.e. w must be at least 87

c Yes. She could score 100 in her written paper which would give her a total of 200

4 $m > 3$

5 $99 \leq n \leq 105$

6 $r \leq 8.5$ and $g \geq 4$. Hence we have 8 red, 11 green; 7 red, 10 green; 6 red, 9 green; 5 red, 8 green; 4 red, 7 green; 3 red, 6 green; 2 red, 5 green; 1 red, 4 green.

7 The sister must be more than 10 years old.

8 $\frac{10}{3} < b < 5$

9 $6 \leq n \leq 40$

10 $p = 3, q = 7$ or $p = 4, q = 6$ or $p = 5, q = 5$ or $p = 6, q = 4$ or $p = 7, q = 3$

11 **a** $x = 2.16$ or 57.84

b

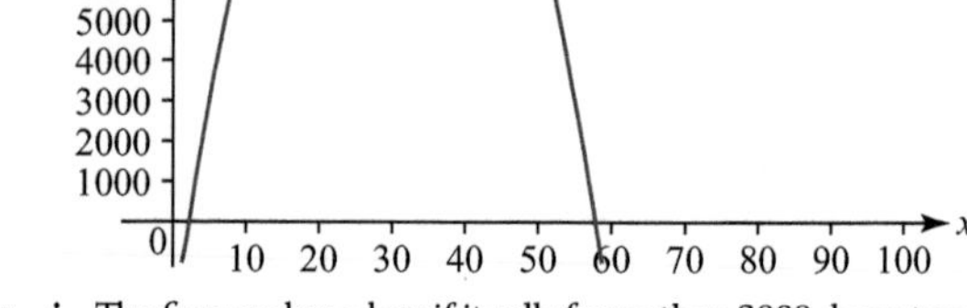

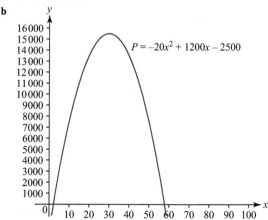

c **i** The firm makes a loss if it sells fewer than 3000 decanters or more than 57 000

ii $13.42 \leq x \leq 46.58$. Thus the firm makes at least £10 000 if they sell between 14 000 and 46 000 decanters.

Review exercise 1

1 Let the numbers be $2m + 1$ and $2n + 1$

Product is $(2m + 1)(2n + 1) = 4mn + 2(m + n) + 1$

$14mn$ and $2(m + n)$ are both even so $4mn + 2(m + n) + 1$ is even + even + odd, which is odd.

2 Find one prime e.g. 41

3 No. $\frac{1}{-3} > -3$ OR $\frac{1}{\left(\frac{1}{4}\right)} > \frac{1}{4}$ i.e. $4 > \frac{1}{4}$

4 **a** $-s^{12}$ **b** $8c^{32}$ **c** $\frac{1}{81}$ **d** $\frac{1}{\sqrt{k^3}}$

5 **a** $5\sqrt{11}$ **b** $\frac{4\sqrt{a} - a - 3}{a - 1}$

6 $6\sqrt{2}$ cm

7 **a** $c = -5$ or $\frac{1}{2}$ **b** $x = -3.43$ or 1.63

8

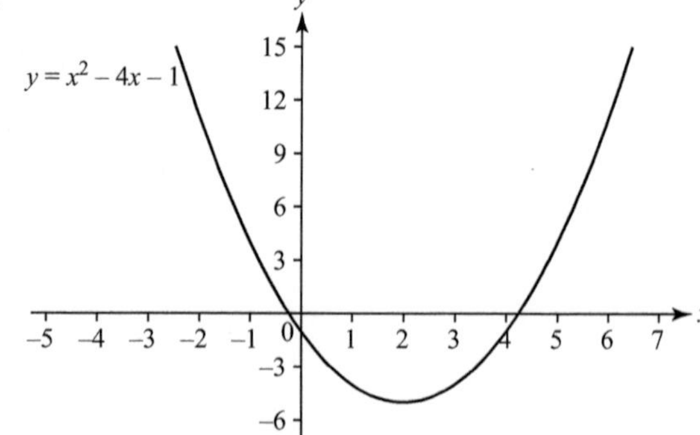

9 **a** **i** $y + 6x = 29$ **ii** $3y = 2x + 18$

b 1st Diagonal: Equation is $14y + 2x = 10$

2nd Diagonal: Equation is $2y - 14x + 20 = 0$

Product of gradients $= -1$ so perpendicular

10 a i $x^2 - 6x + y^2 - 12y - 19 = 0$
ii $x^2 + 6x + y^2 - 18y + 74 = 0$
iii $x^2 + 4x + y^2 + 14y - 68 = 0$

b $12y = 5x + 273$ **c** $(5, -4)$ and $(11, 14)$

11 a $y = -1$ and $x = 3$ **b** $y = -3$ twice and $x = -2$ twice

12 $(-6, -14)$ and $\left(\frac{14}{3}, 18\right)$

13 a i $\{x: x \le 1\}$ ii $\left\{x: x \ge \frac{6}{13}\right\}$

b i $1.26 \le x \le 12.74$ ii $x \le -0.67$ or $x \ge 3.27$

14 a

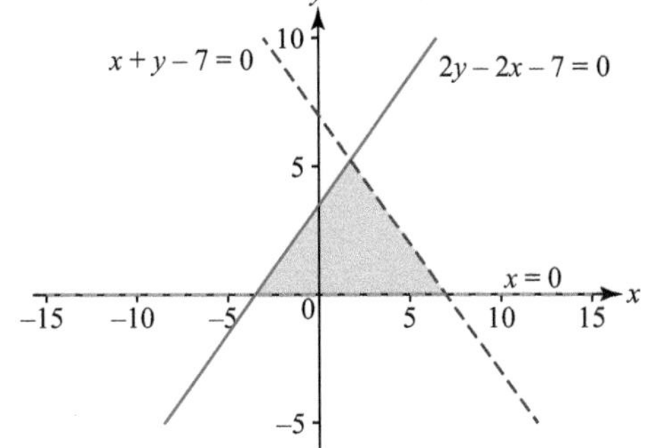

b

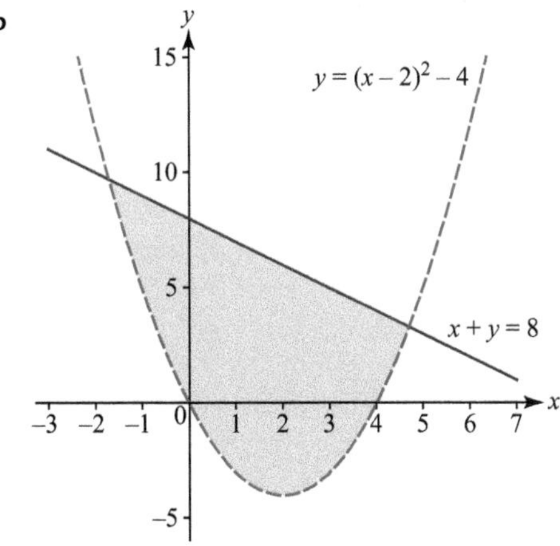

Assessment 1

1 C, $-1 - 2\sqrt{2}$

2 A, $3y - 2x = 3$

3 a i 2^{m+n} ii 5^{m-2n+1} iii $3^{\frac{5m}{2}}$

b 2^{2p+q}

4 a i $4\sqrt{3}$ ii $\frac{-12+5\sqrt{7}}{31}$

b $6 + 2\sqrt{5}$

5 a $(x+3)^2 + 4$

b

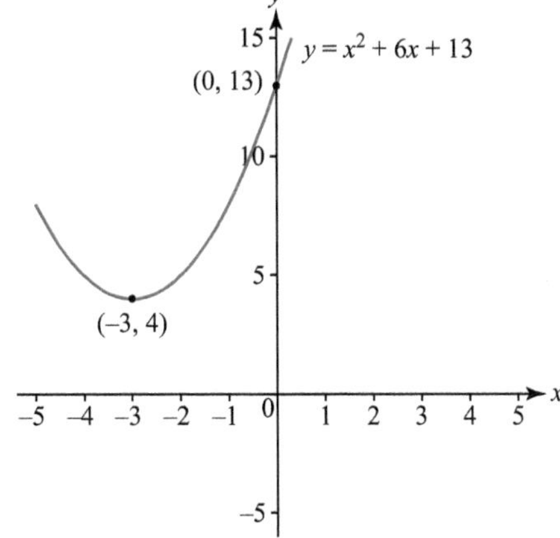

6 $x = 7$, $x = -1$; $y = -11$, $y = 5$

7 $x(x+4) = x + 4 + 2x - 5 \Rightarrow x^2 + x + 1 = 0$
$\Rightarrow \Delta = 1^2 - 4 \times 1 \times 1 = -3 < 0 \Rightarrow$ No real solutions

8 a i $x < 4$
ii $1 \le x \le 5$

b

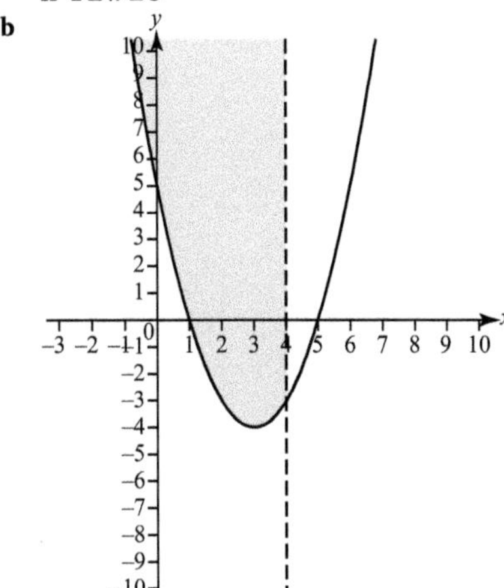

9 $x = \sqrt{5}$

10 a i $(5, -1)$ ii 7

b $\left(1 - \frac{\sqrt{34}}{2}, 3 - \frac{\sqrt{34}}{2}\right), \left(1 + \frac{\sqrt{34}}{2}, 3 + \frac{\sqrt{34}}{2}\right)$

11 $k < -\frac{3}{4}$ or $k > 3$

12 $\left(\sqrt{a} - \sqrt{b}\right)^2 \ge 0 \Rightarrow a - 2\sqrt{a}\sqrt{b} + b \ge 0 \Rightarrow a + b \ge 2\sqrt{ab}$
$\Rightarrow \frac{a+b}{2} \ge \sqrt{ab}$

13 a $(2u-1)(u-8)$ **b** $x = -1$ or $x = 3$

14 a $x^2 + (mx+2)^2 + 4x - 6(mx+2) + 10 = 0$
$\Rightarrow x^2 + m^2x^2 + 4mx + 4 + 4x - 6mx - 12 + 10 = 0$
$\Rightarrow (m^2+1)x^2 + 2(2-m)x + 2 = 0$

b $m = -2 \pm \sqrt{6}$

15 a 4 **b** $x = 4\frac{2}{3}$, $y = -\frac{2}{3}$

16 a False: eg $a = 4$, $b = -5$

b True: $n^2 + n \equiv n(n+1)$; either n or $(n+1)$ must be even

c True: $(b-2a)^2 \ge 0 \Rightarrow b^2 - 4ab + 4a^2 \ge 0 \Rightarrow b^2 \ge 4ab - 4a^2$

d False: eg $n = 4$

17 a $(0, 9)$

b $(2, 1)$

c $C(3-\sqrt{2}, 7-4\sqrt{2})$, $D(3+\sqrt{2}, 7+4\sqrt{2})$

18 $(x+3)^2 - 9 + (y-2)^2 - 4 - 2 = 0$; Centre, $C_1 = (-3, 2)$; Radius, $r_1 = \sqrt{15}$
$(x-1)^2 - 1 + (y-5)^2 - 25 - 55 = 0$; Centre, $C_2\,(1, 5)$; Radius, $r_2 = 9$
Distance $|C_1C_2| = \sqrt{(1--3)^2 + (5-2)^2} = 5$; $|C_1C_2| + r_1 < r_2$

Chapter 2

Try it 2A

1 a $x^2 - 14x + 49$ **b** $25x^2 + 10x + 1$

2 a $x^3 - 3x^2 - x + 3$ **b** $x^3 - 12x - 16$

Bridging Exercise 2A

1 a $x^2 - 8x + 16$ **b** $x^2 + 12x + 36$

c $x^2 - 18x + 81$ **d** $x^2 + 10x + 25$

e $4x^2+4x+1$ f $9x^2-12x+4$
g $16x^2+24x+9$ h $25x^2+20x+4$
i x^2-6x+9 j $4x^2-28x+49$
k $9x^2-48x+64$ l $81x^2-180x+100$

2 a $x^3+11x^2+38x+40$
b $x^3+8x^2+5x-14$
c $-x^3-3x^2+34x-48$
d $2x^3-9x^2-86x+240$
e $6x^3+29x^2-6x-5$
f $-24x^3+98x^2-133x+60$
g $x^3+19x^2+115x+225$
h $x^3-14x^2+57x-72$
i $x^3-11x^2-45x+567$
j $-4x^3+4x^2+39x+36$
k $9x^3-30x^2-287x-392$
l $-8x^3+100x^2-374x+363$

Try it 2B

1 a $\frac{x+2}{x}$ b $\frac{3x-1}{x-1}$

2 $5x-1+\frac{3}{x+4}$

3 $a=-4, b=2, c=8$

4 $2x+7-\frac{14}{3x-5}$

Bridging Exercise 2B

1 a $\frac{x-5}{x^2}$ b $\frac{x+3}{x}$
c $\frac{1}{2x}$ d $\frac{x}{x+5}$

2 a $\frac{x-4}{x+2}$ b $\frac{x-7}{x+2}$
c $\frac{x+2}{x-5}$ d $\frac{x+6}{2}$
e $\frac{x}{x-6}$ f $\frac{3x}{x-7}$
g $\frac{5x^2}{x+3}$ h $\frac{x+8}{3x}$
i $\frac{5+x}{9+x}$ j $\frac{x-4}{x^2}$
k $\frac{3x-1}{2x+1}$ l $\frac{x(x-10)}{6x-4}$
m $6x$ n $\frac{7x+1}{6x-1}$
o $\frac{x(3x-1)}{8x-3}$ p $\frac{9x+2}{2x^3}$

3 a $2x+3+\frac{2}{x-6}$ b $6x+9+\frac{11}{x-1}$
c $5x+6-\frac{1}{x+7}$ d $3x+2$
e $5x+2-\frac{3}{3x+4}$ f $2x-3-\frac{7}{4x+9}$
g $3x^2+3x-6+\frac{49}{x+5}$ h $6x^2-x+7-\frac{3}{2x+1}$

4 a $a=1, b=-1, c=-12$
b $a=1, b=-18, c=81$
c $a=4, b=-4, c=-16$
d $a=18, b=105, c=200$
e $a=5, b=-19, c=-109$
f $a=-2, b=2, c=-13$
g $a=3, b=-12, c=-11$
h $b=3, a=-7, c=84$

5 a $3x+2-\frac{4}{x-7}$
b $4x+3+\frac{2}{x+6}$
c $6x-2+\frac{5}{3x+1}$
d $14x+3-\frac{7}{2x-9}$

6 a $3x^2+x-2+\frac{4}{x-1}$
b $7x^2+3x-1-\frac{1}{2x+3}$
c $3x-2+\frac{x}{x^2+3x-2}$
d $x+\frac{3x-1}{2x^2-4}$

Try it 2C

1 a

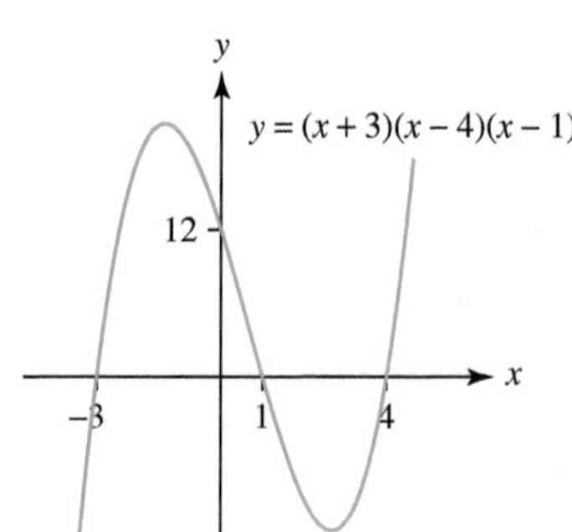

b

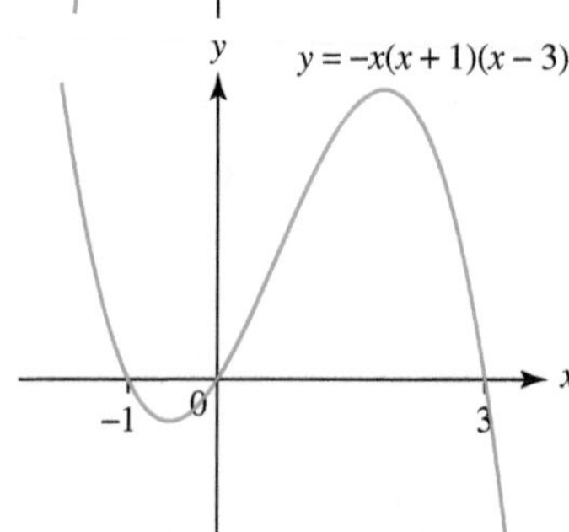

2 a

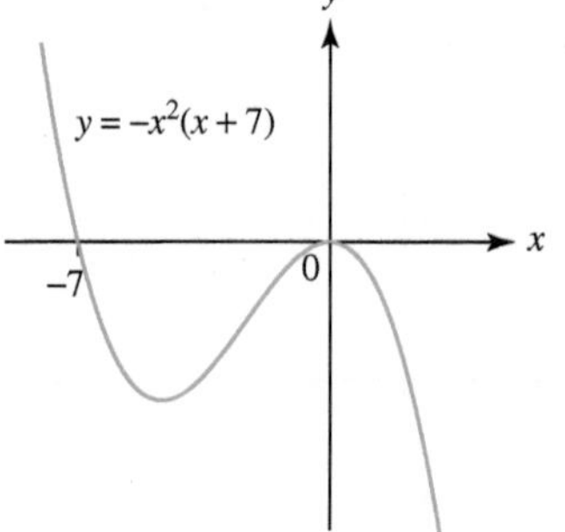

b

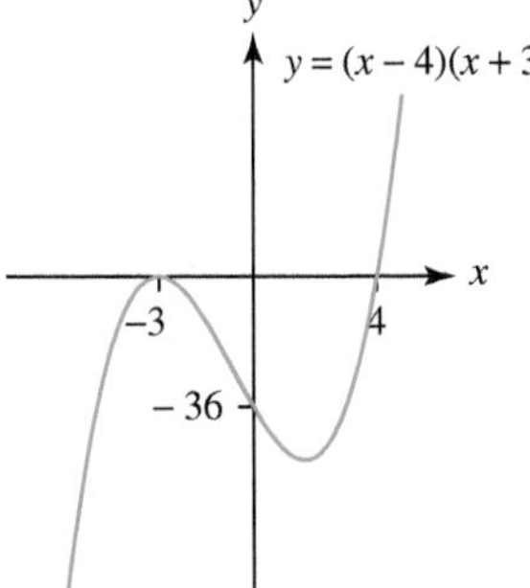

3 a $y = 2x^2(x+3)$

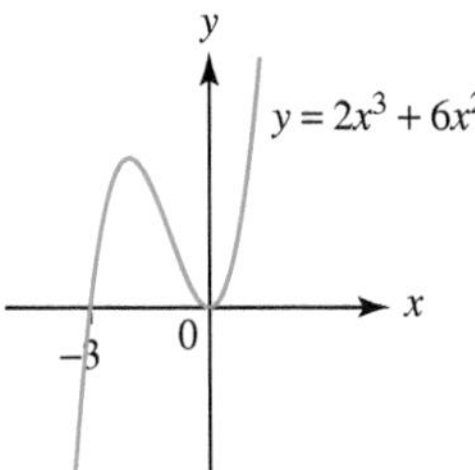

b

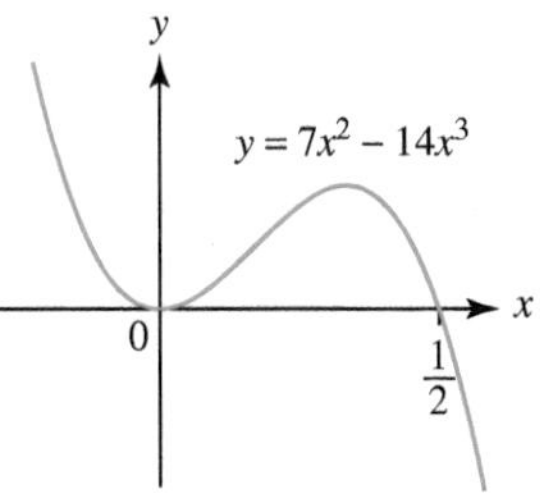

4 a

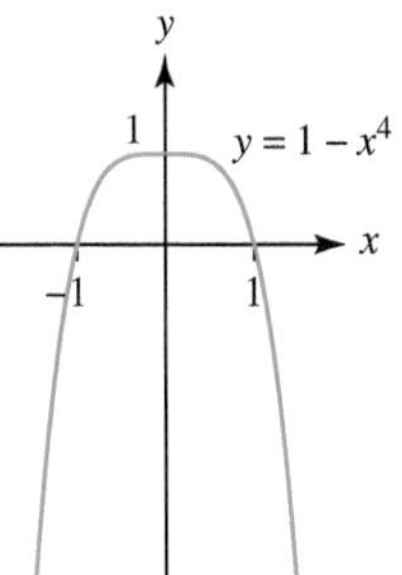

b

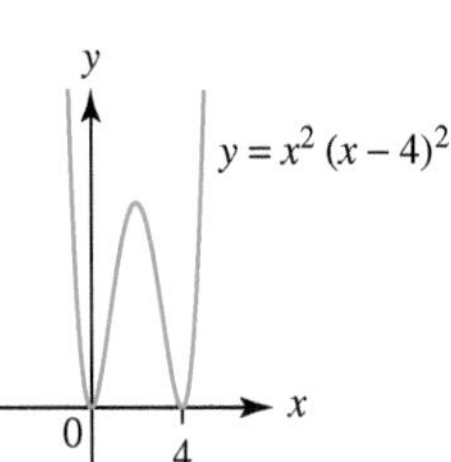

5 a

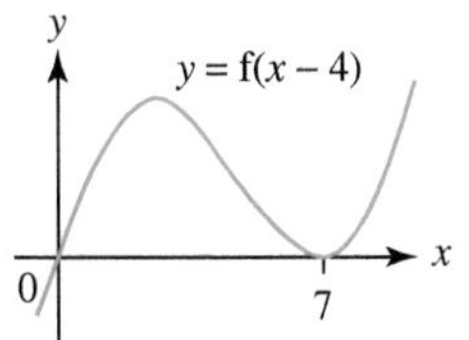

A has coordinates (3, 4)

b

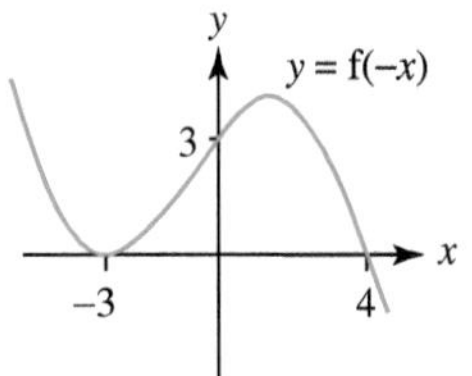

A has coordinates (1, 4)

c

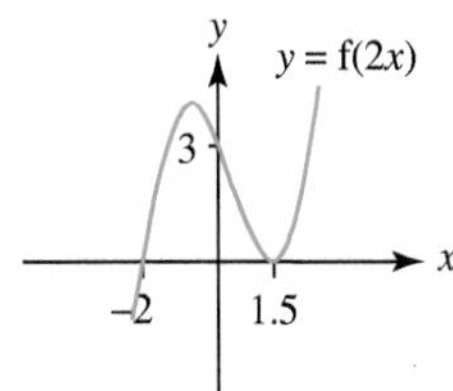

A has coordinates (–0.5, 4)

d

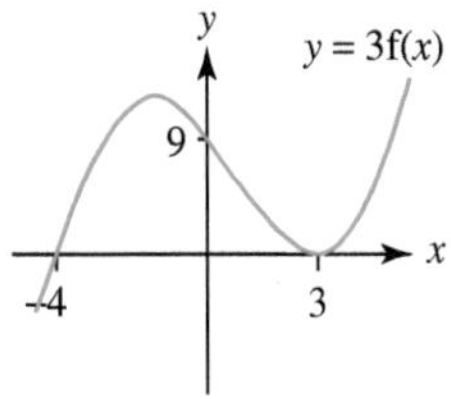

A has coordinates (–1, 12)

6 a Asymptotes at $x = 0$ and $y = 0$

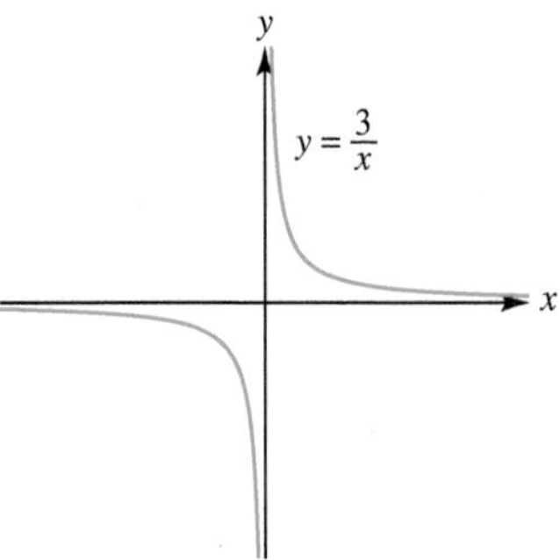

b a Asymptotes at $x = 0$ and $y = 0$

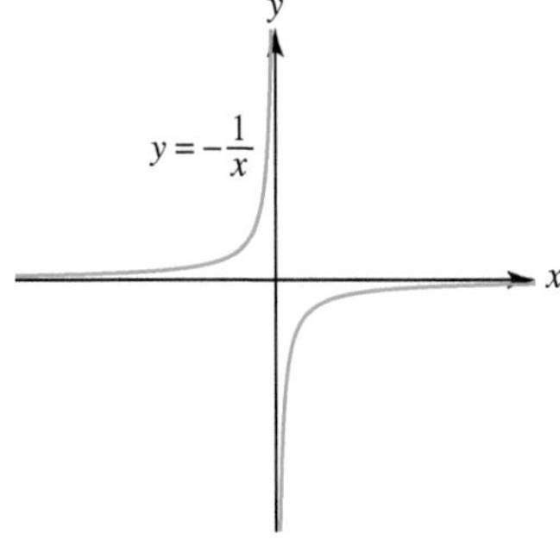

7 Asymptotes at $x = 5$, $y = 0$

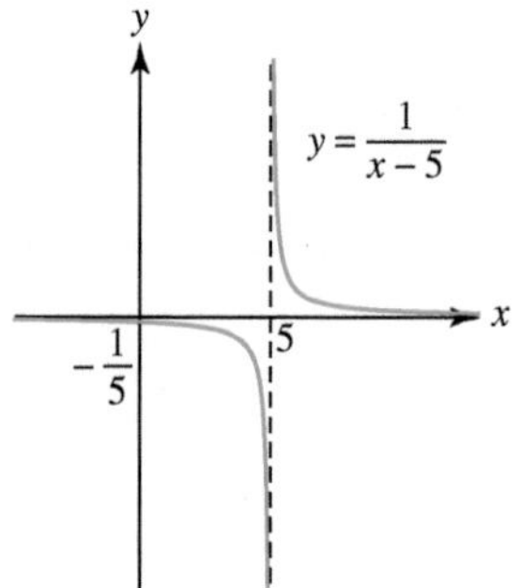

8 Asymptotes at $x = 0$, $y = -3$

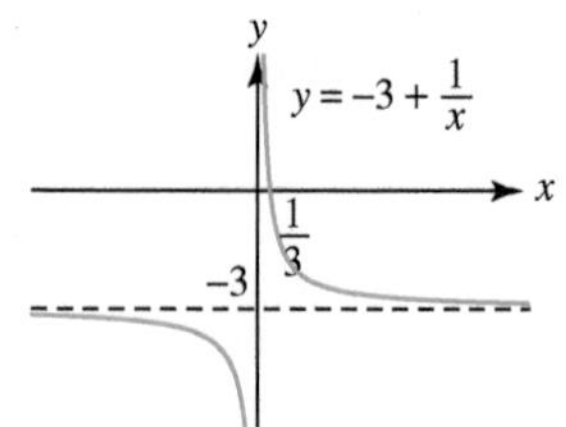

Bridging Exercise 2C

1 a

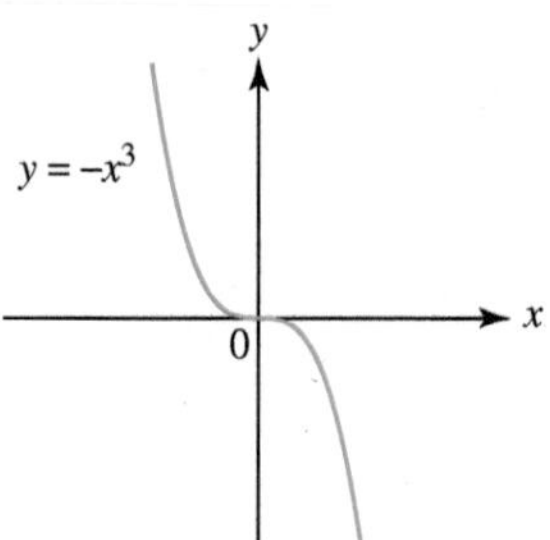

b

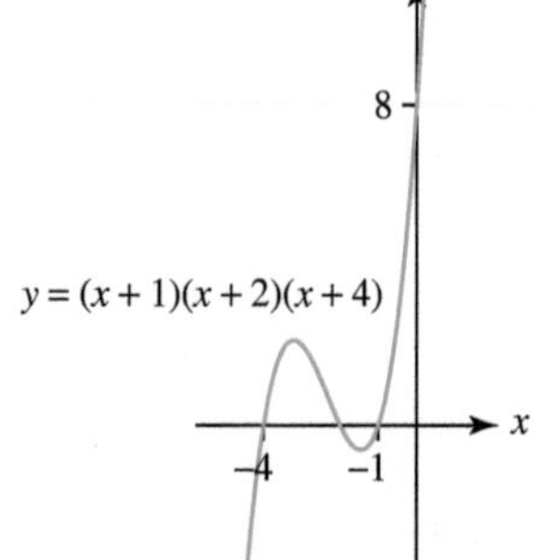

c

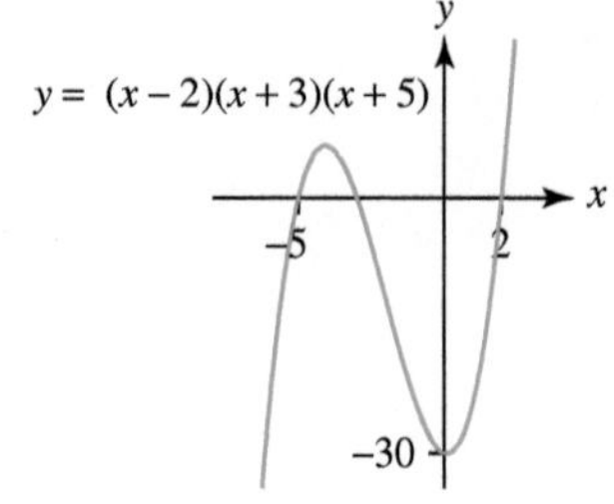

d

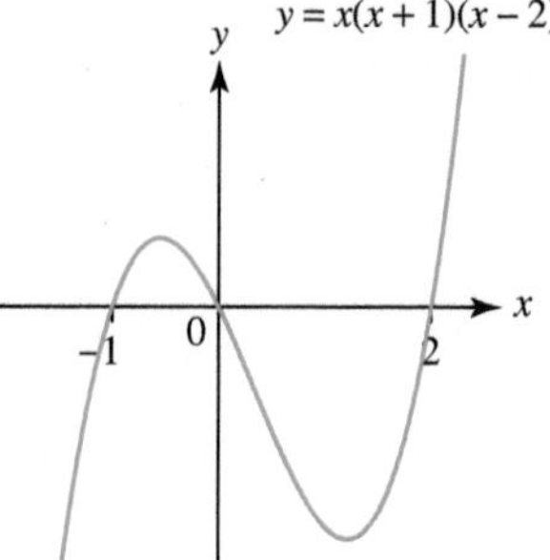

e

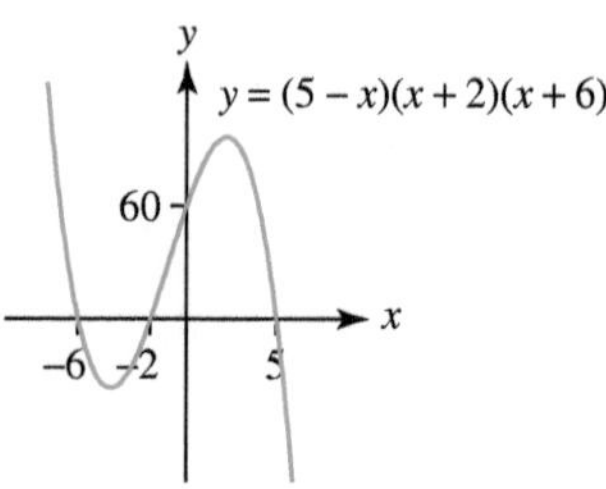

f

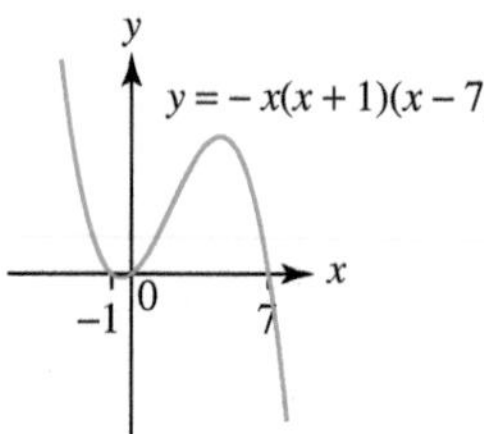

g

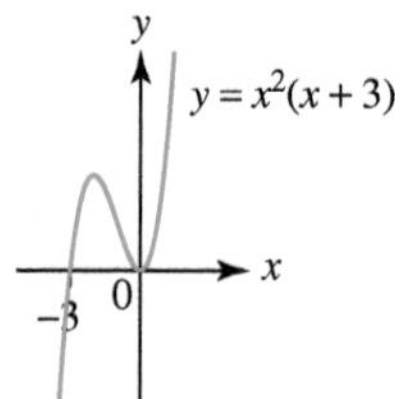

h

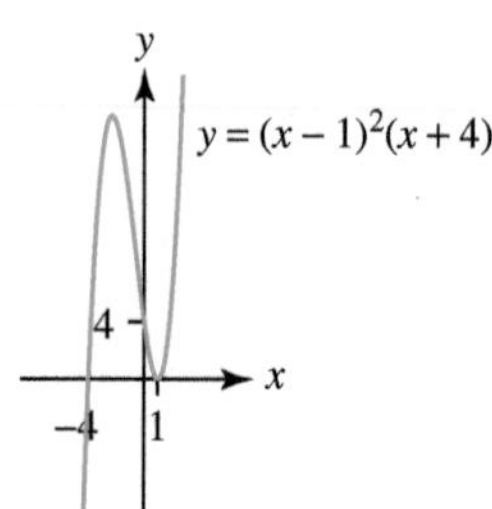

i

y

$y = x(x + 5)^2$

x

−5 0

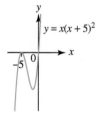

j

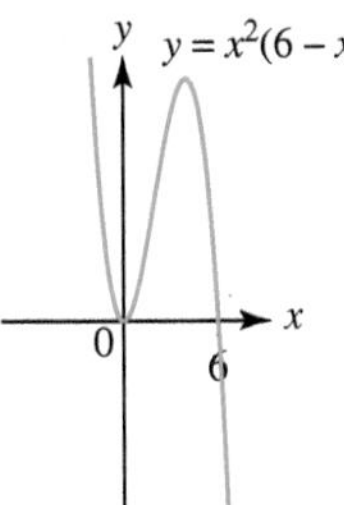

k

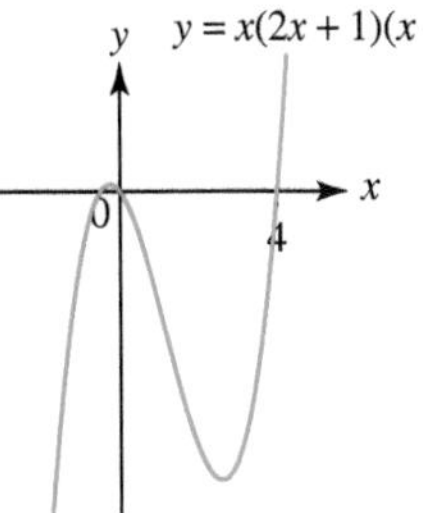

l

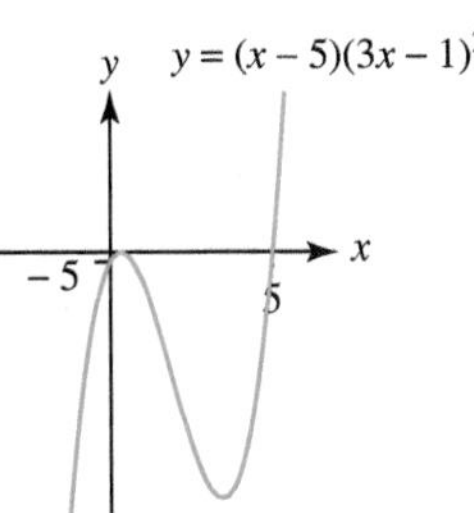

2 a

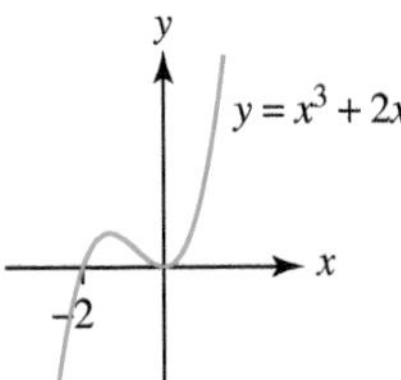

b

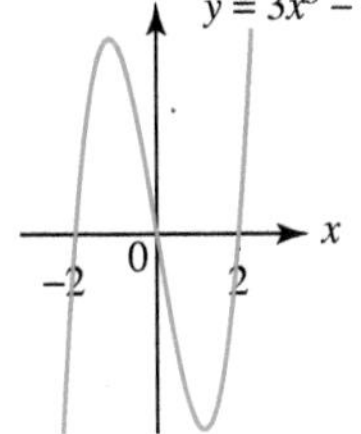

c

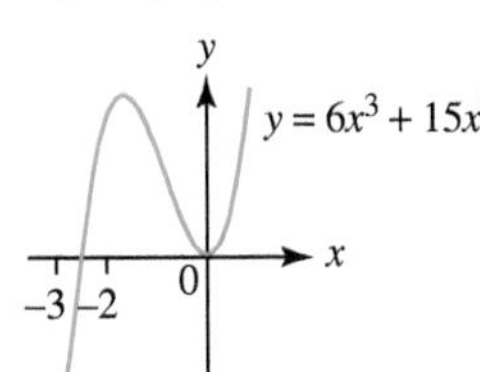

d

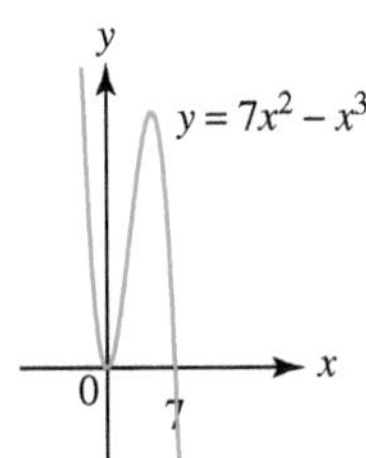

e

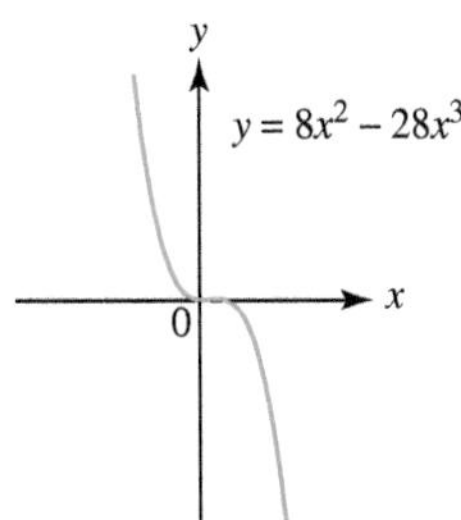

f

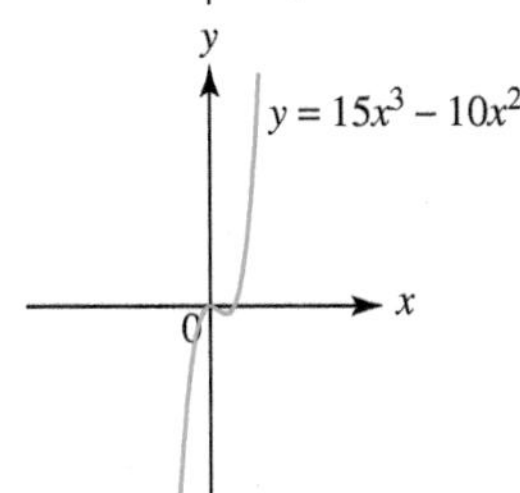

g

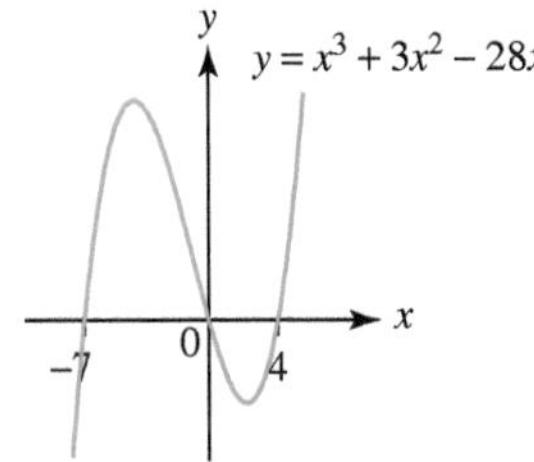

h

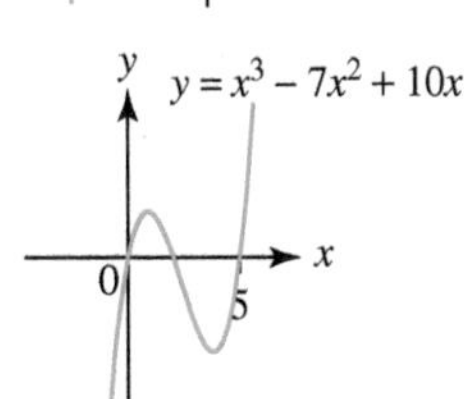

i

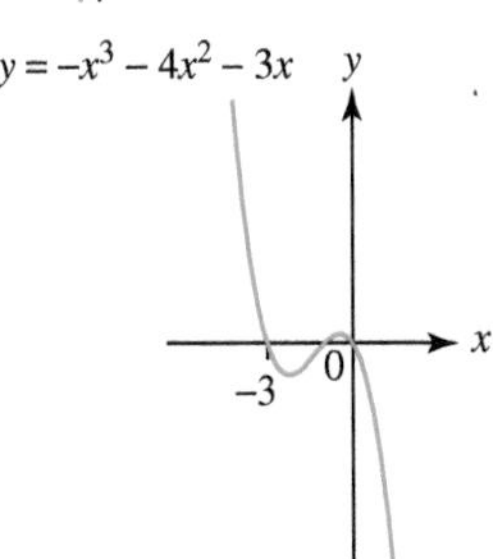

j

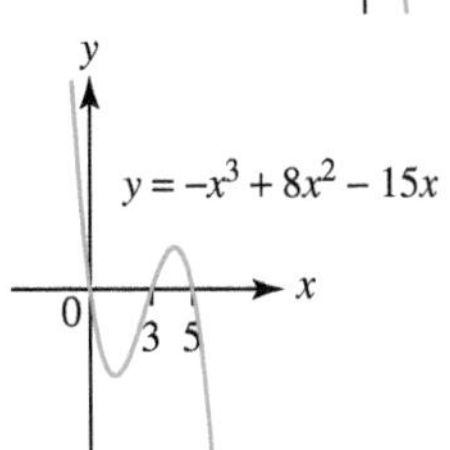

k

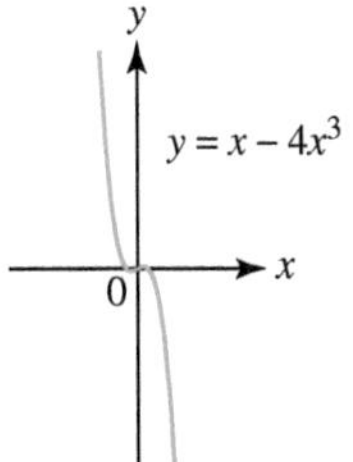

l

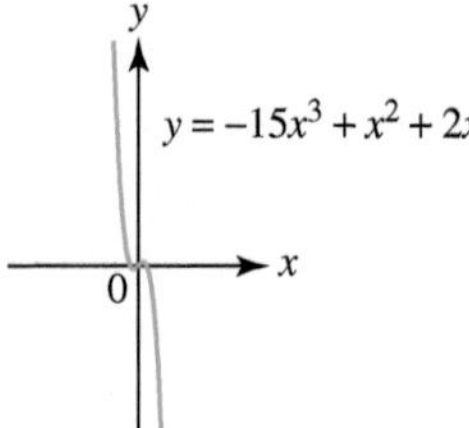

3 a

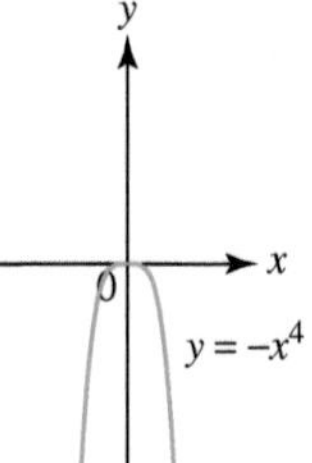

b

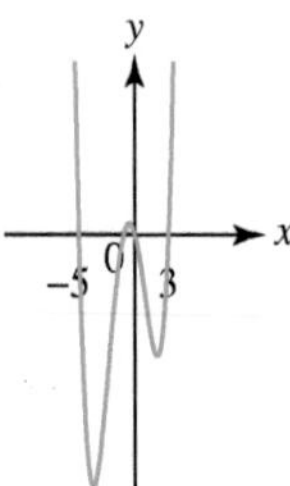

$y = x(x + 5)(x + 1)(x - 3)$

c

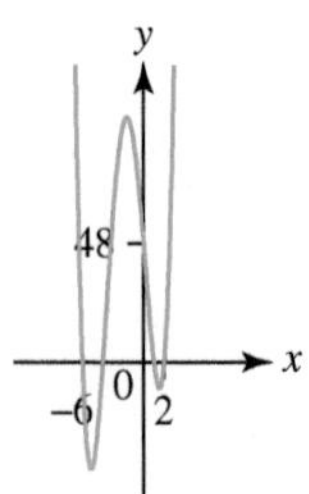

$y = (x + 4)(x + 6)(x - 2)(x - 1)$

d

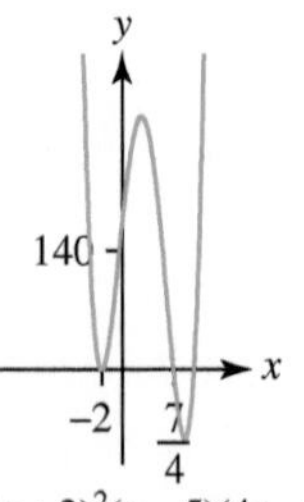

$y = (x + 2)^2(x - 5)(4x - 7)$

e

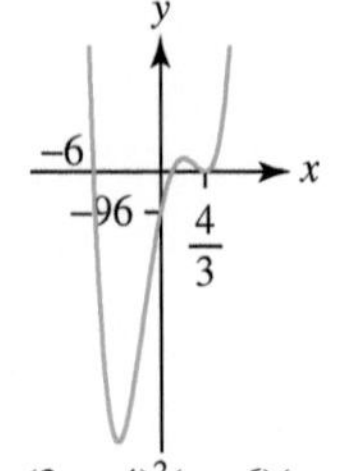

$y = (3x - 4)^2(x + 6)(x - 1)$

f

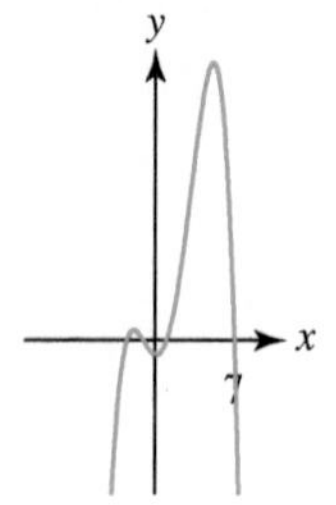

$y = -(x + 1)(2x + 5)(x - 7)(x - 1)$

g

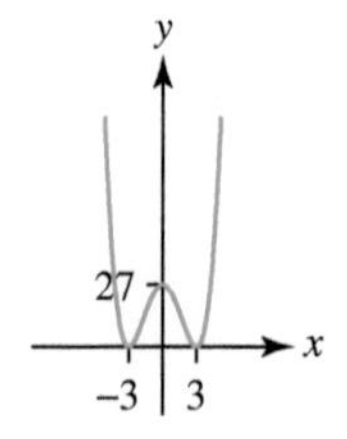

$y = \frac{1}{3}(x + 3)^2(3 - x)^2$

h

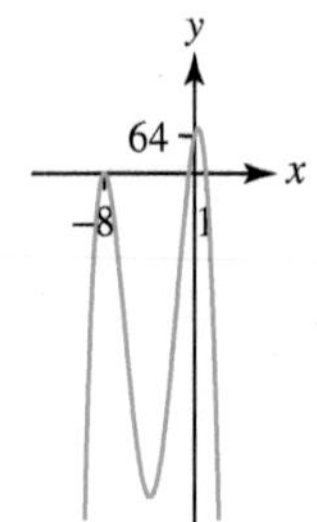

$y = (x + 8)^2(1 - x)(2x + 1)$

4 a

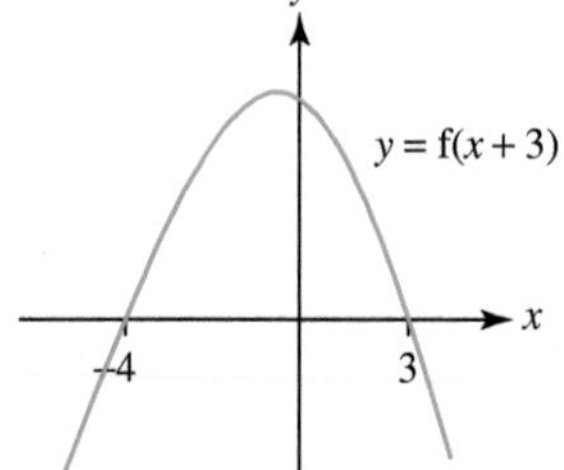

A has coordinates (–1, 5)

b

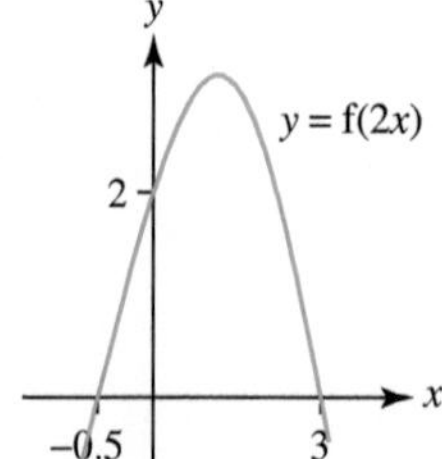

A has coordinates (1, 5)

c

y

$y = 3f(x)$

6

–1 6 x

A has coordinates (2, 15)

d

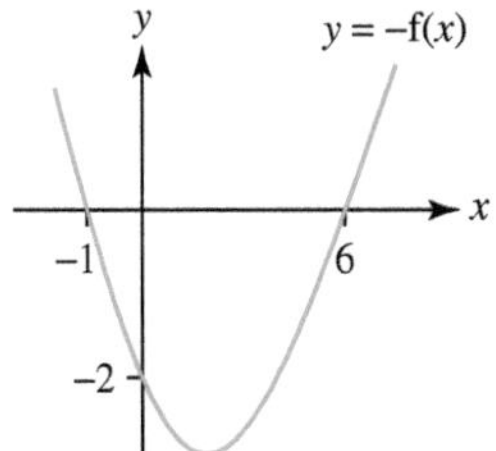

A has coordinates (2, −5)

5 a

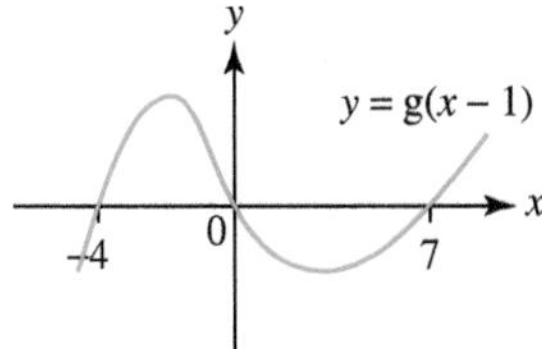

A has coordinates (−3, 4)

b

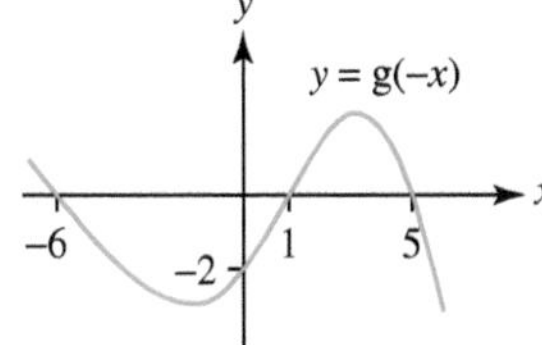

A has coordinates (4, 4)

c

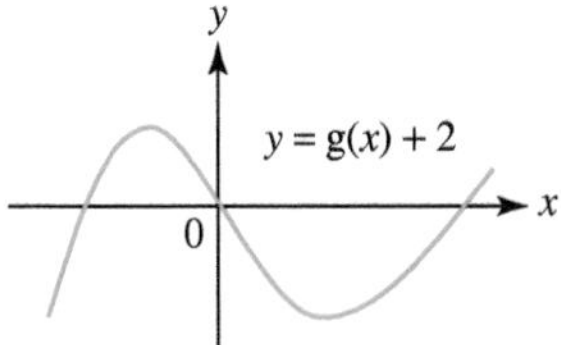

A has coordinates (−4, 6)

d

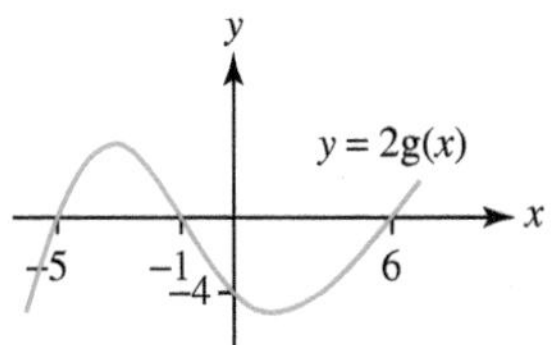

A has coordinates (−4, 8)

6 a

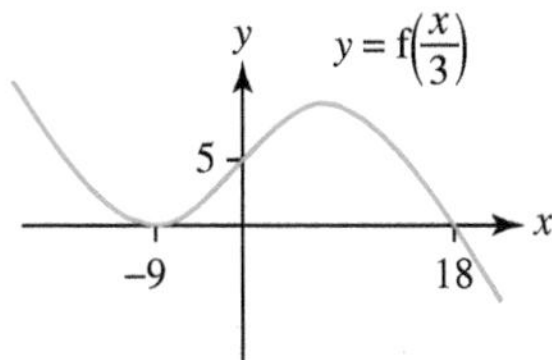

A has coordinates (6, 7)

b

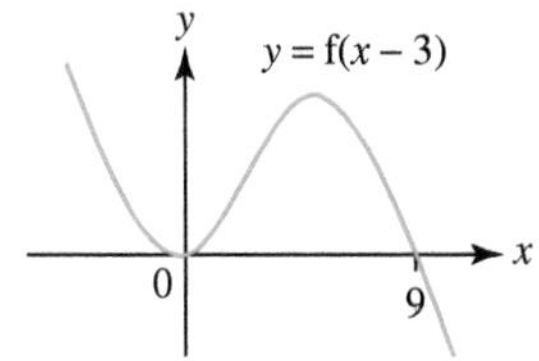

A has coordinates (5, 7)

c

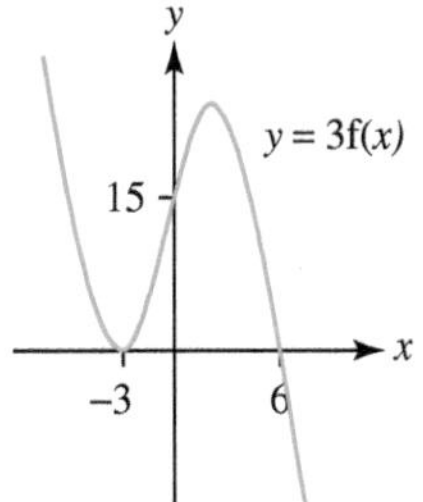

A has coordinates (2, 21)

d

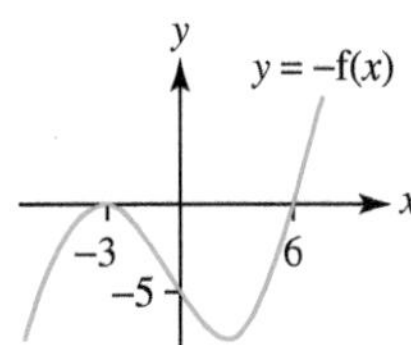

A has coordinates (2, −7)

7

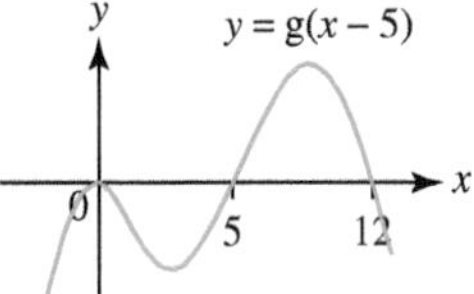

A has coordinates (2, −5)

b

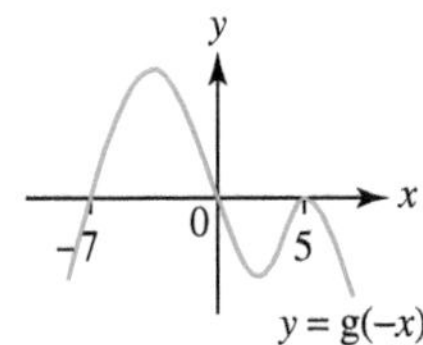

A has coordinates (3, −5)

c

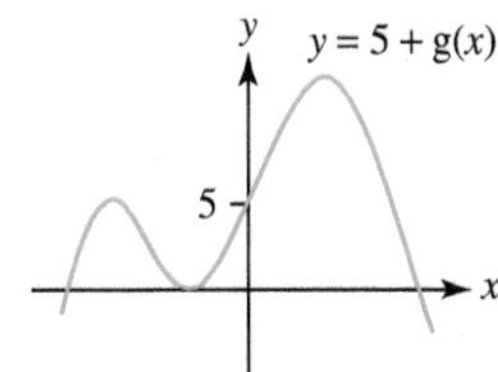

A has coordinates (−3, 0)

d

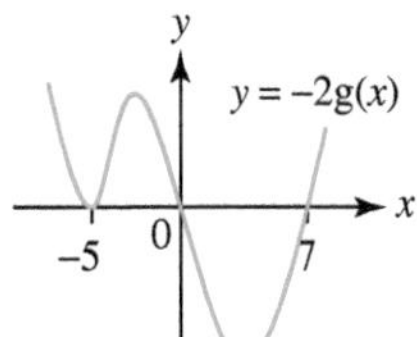

A has coordinates (−3, 10)

8 a

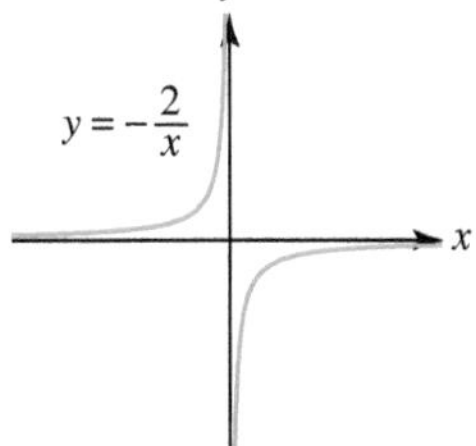

b

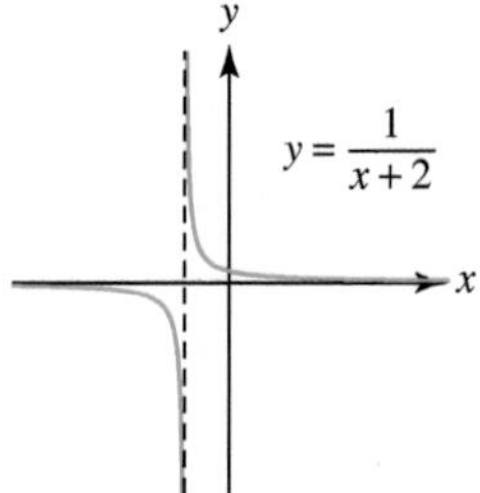

c

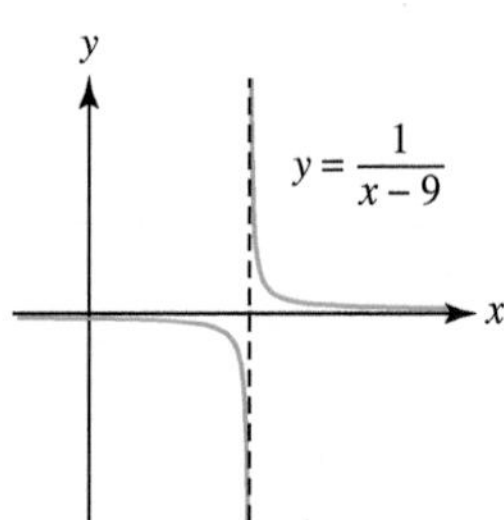

d

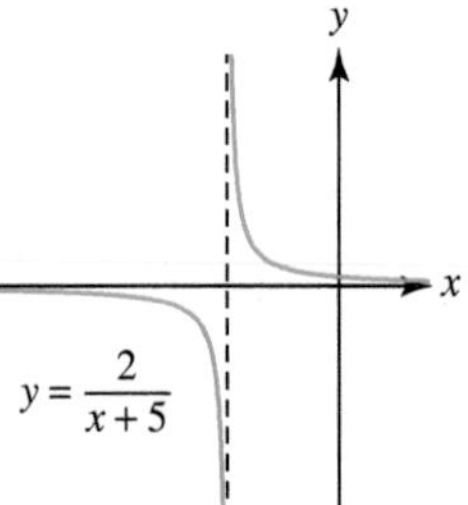

e

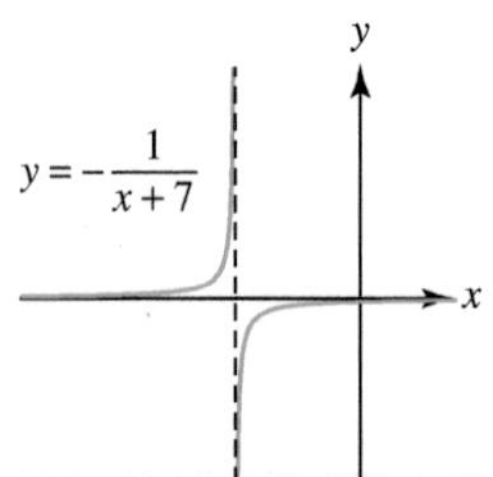

f

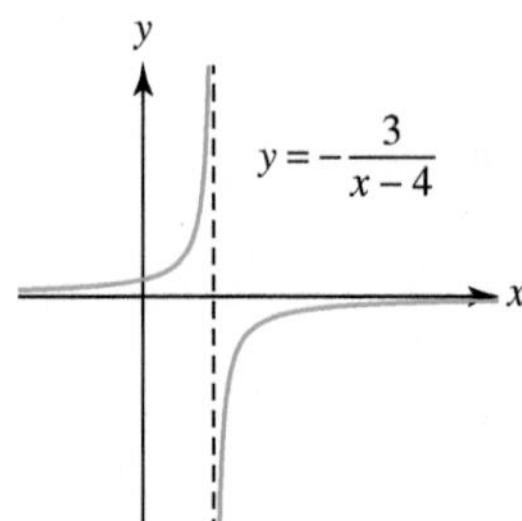

g

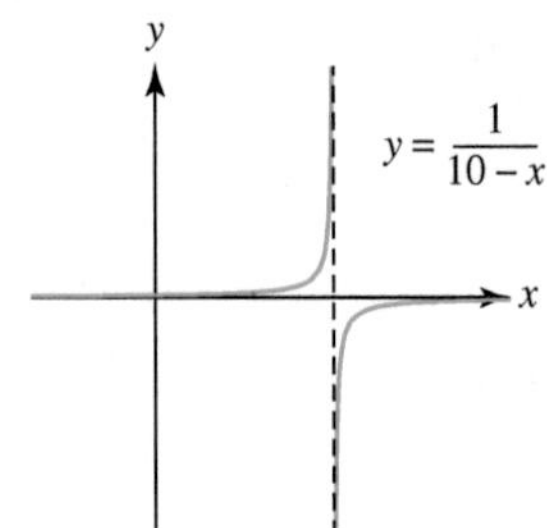

h

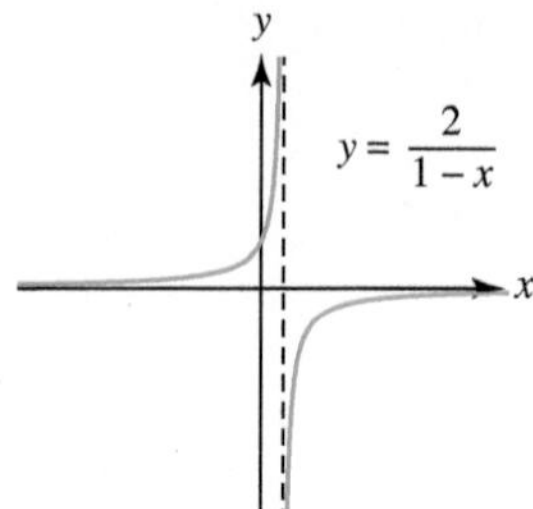

i

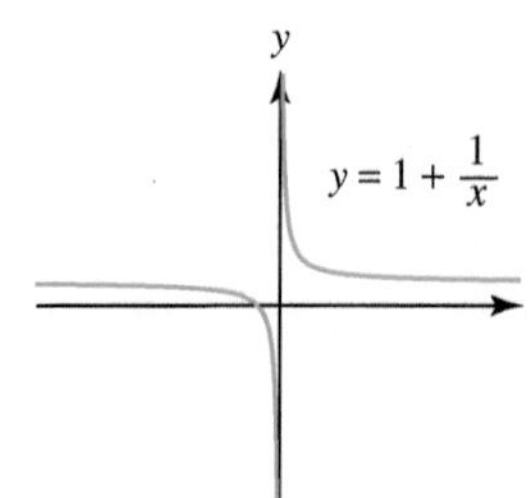

j

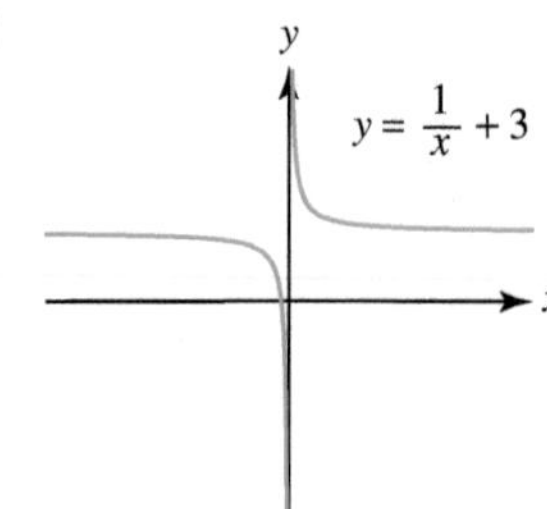

k

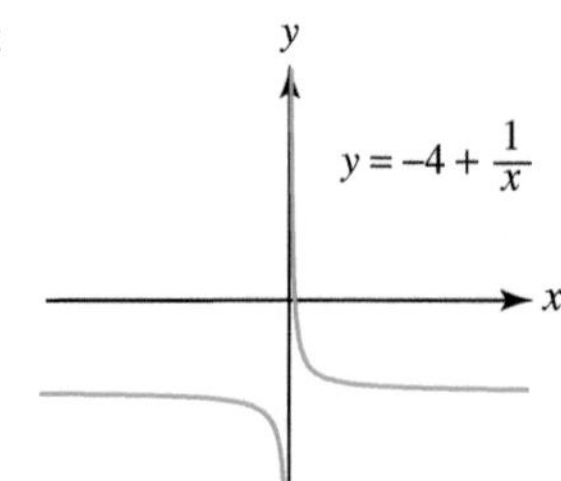

l

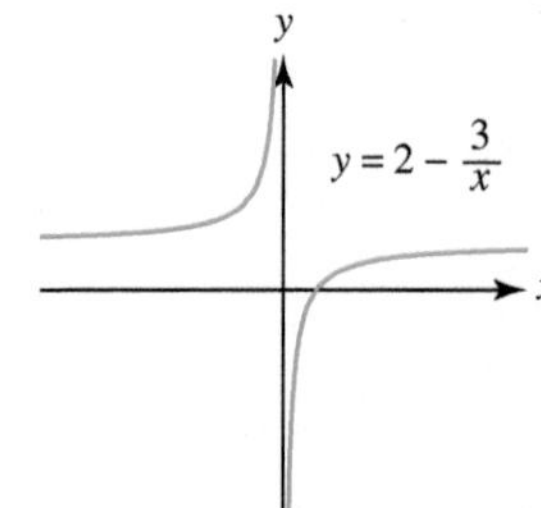

9 **a** e.g. $y = (x+3)(x-1)^2$

b e.g. $y = -x^2(x-9)$

c e.g. $y = 4 - \frac{1}{x}$

d e.g. $y = -x(x+3)(x-1)(x-6)$

e e.g. $y = (x+4)^2(x-3)(x-4)$

Exercise 2.1A Fluency and skills

1 **a** Degree 2 **b** Degree 4 **c** Degree 3

2 **a** $6x^2 + 16x$ **b** $6x^3 + 16x^2 - 18x$
c $12y^2 - 13y - 14$ **d** $12y^3 + 24y^2 - 21y$
e $t^2 - 10t + 25$ **f** $t^3 - 7t^2 - 5t + 75$

3 **a** $2x^2 + 32$ or $2(x^2 + 16)$ **b** $20pq$

4 **a** $2m^2(2m + 3)$ **b** $4n(4n^3 - 3)$
c $p(5p^3 - 2p + 6)$ **d** $3y(3y - 5x)$
e $3x(2x - y + 3)$ **f** $7z(y - 3z^2)$
g $4e(e - 5f)$ **h** $(p - 10)(p + 10)$
i $3q(9 - 4q)$ **j** $\frac{y(24-5y)}{75}$
k $2(d + 1)(d - 1)$ **l** $w(2w + 3)(5w - 2)$

5 **a** $m(2m + 5)(2m - 3)$ **b** $n(7n - 1)(n - 2)$

6 $2(x + 2)(2x - 1)(2x + 3)$

7 **a** $80pq$ **b** $4x(y + z)$
c $6x^2 + 32$ **d** $8x^2 + (8\sqrt{5} - 8\sqrt{3})x + 32$

8 **a** $(x - 1)$ **b** $(2x - 7)$
c $(2y - 1)$ **d** $(z^2 + 3z + 4)$
e $(3a^2 - 4a + 9)$
f $(2x^3 - 5x^2 + 2x + 21) \equiv (x^2 - 4x + 7)(2x + 3)$
g $(k + 7)$ and $(k - 1)$

Exercise 2.1B Reasoning and problem-solving

1 $(2a^2 - 2a + 5) - (a^2 - 2a + 4) = a^2 + 1$, $(a^2 + 1) - (3a - 8)$ $= a^2 - 3a + 9$, $(a^2 - 2a + 4) - (a^2 - 3a + 9) = a - 5$

2 $16b^2 - 56ab + 49a^2\,\text{cm}^2$

3 $(6c^3 - 5c^2 - 27c + 14)\,\text{cm}^3$

4 $(a + 3)(3a + 7)\,\text{cm}^2$

5 $2a^2x - 6ax^2 + 4x^3\,\text{cm}^3$

6 **a** $t = 5$ **b** 31.25 ft

7 **a** $t = 0, \frac{4}{3}$, or 8 seconds
$t = 0$: $v = 32\,\text{ms}^{-1}$, $a = -56\,\text{ms}^{-2}$
$t = \frac{4}{3}$: $v = -\frac{80}{3}\,\text{ms}^{-1}$, $a = -32\,\text{ms}^{-2}$
$t = 8$: $v = 160\,\text{ms}^{-1}$, $a = 88\,\text{ms}^{-2}$
b $x = -\frac{19\,712}{243}$, $v = -\frac{496}{9}\,\text{ms}^{-1}$
c $t = \frac{28}{9} + \frac{4\sqrt{31}}{9}$ s or $\frac{28}{9} - \frac{4\sqrt{31}}{9}$ s
$a = 8\sqrt{31}\,\text{ms}^{-2}$ or $a = -8\sqrt{31}\,\text{ms}^{-2}$

8 **a** $p = -x^2 + 8x - 9$, $q = x^2 + x + 5$
b Perimeter $\equiv 4(x^2 + 4x - 5)$
Area $\equiv x^4 + 8x^3 + 5x^2 - 34x + 16$

9 **a** $66.89^2 - 33.11^2 = (66.89 + 33.11)(66.89 - 33.11) =$ $(100)(33.78) = 3378$
b $\left(2\sqrt{2}\right)^3 - \left(\sqrt{2}\right)^3 = 8 \times 2\sqrt{2} - 2\sqrt{2} = 14\sqrt{2}$

10 $(2h + 1)$ cm

11 $(s - 2)(s - 5)$ cm

12 $b \equiv [2t^2 - 3t + 7]$

13 $V = \pi(20p^3 + 9p^2 - 56p - 45)$

Exercise 2.2A Fluency and skills

1 **a** 120 **b** 5040 **c** 39 916 800

2 **a** 10 **b** 84 **c** 330 **d** 1287

3 **a** 10 **b** 10 **c** 1287 **d** 38 760

4 **a** $1 + 9x + 27x^2 + 27x^3$
b $1 - \frac{5}{2}z + \frac{5}{2}z^2 - \frac{5}{4}z^3 + \frac{5}{16}z^4 - \frac{1}{32}z^5$
c $1 - \frac{4m}{3} + \frac{2m^2}{3} - \frac{4m^3}{27} + \frac{m^4}{81}$
d $1 + \frac{15x}{2} + \frac{45x^2}{2} + \frac{135x^3}{4} + \frac{405x^4}{16} + \frac{243x^5}{32}$

5 **a** $1 + 8x + 28x^2 + 56x^3 + \ldots$
b $1 - 21x + 189x^2 - 945x^3 + \ldots$
c $1 + 18x + 144x^2 + 672x^3 + \ldots$
d $64 - 576x + 2160x^2 - 4320x^3 + \ldots$
e $256 - 1024x + 1792x^2 - 1792x^3 + \ldots$
f $1 - 20x + 180x^2 - 960x^3 + \ldots$

6 **a** $8 - 48y + 96y^2 - 64y^3$
b $81b^4 + 540b^3 + 1350b^2 + 1500b + 625$
c $1024z^5 - \frac{1280z^4y}{3} + \frac{640z^3y^2}{9} - \frac{160z^2y^3}{27} + \frac{20zy^4}{81} - \frac{y^5}{243}$

7 **a** $x^6 + 12x^5 + 60x^4 + \ldots$
b $256x^8 - 1024x^7 + 1792x^6 + \ldots$
c $-x^9 + 27x^8 - 324x^7 + \ldots$
d $x^7 + 28x^6 + 336x^5 + \ldots$
e $1024x^{10} + 15\,360x^9 + 103\,680x^8 + \ldots$
f $\frac{x^{11}}{2048} + \frac{11x^{10}}{256} + \frac{55x^9}{32} + \ldots$

8 **a** $16 + 96t + 216t^2 + 216t^3 + 81t^4$
b $81 - 216p + 216p^2 - 96p^3 + 16p^4$
c $1024p^5 + 3840p^4q + 5760p^3q^2 + 4320p^2q^3 + 1620pq^4 + 243q^5$
d $243p^5 - 1620p^4q + 4320p^3q^2 - 5760p^2q^3 + 3840pq^4 - 1024q^5$
e $81z^4 - 216z^3 + 216z^2 - 96z + 16$
f $64z^6 - 96z^5 + 60z^4 - 20z^3 + \frac{15}{4}z^2 - \frac{3}{8}z + \frac{1}{64}$
g $8 + 8x + \frac{8}{3}x^2 + \frac{8}{27}x^3$
h $\frac{r^8}{6561} + \frac{2r^7s}{2187} + \frac{7r^6s^2}{2916} + \frac{7r^5s^3}{1944} + \frac{35r^4s^4}{10368} + \frac{7r^3s^5}{3456}$
$+ \frac{7r^2s^6}{9216} + \frac{rs^7}{6144} + \frac{s^8}{65536}$
i $\frac{x^3}{8} + \frac{x^2y}{4} + \frac{xy^2}{6} + \frac{y^3}{27}$

9 **a** ${}^5C_3\,p^2 5^3 = 1250p^2$
b ${}^9C_5(4)^4y^5 = 32\,256y^5$
c ${}^{12}C_7(3)^5q^7 = 192\,456q^7$
d ${}^5C_3 4^2(-3m)^3 = -4320m^3$
e ${}^{15}C_{11}(2z)^4(-1)^{11} = -21\,840z^4$
f ${}^8C_2(z)^6\left(\frac{3}{2}\right)^2 = 63z^6$
g ${}^5C_1(3x)^4(4y) = 1620x^4y$
h ${}^{10}C_5(2a)^5(-3b)^5 = -1\,959\,552a^5b^5$ and
${}^{10}C_4\,(2a)^6(-3b)^4 = 1\,088\,640a^6b^4$
i ${}^3C_1(4p)^2\left(\frac{1}{4}\right) = 12p^2$
j ${}^{11}C_6(4a)^5\left(-\frac{3b}{4}\right)^6 = \frac{168399}{2}a^5b^6$ and
${}^{11}C_5(4a)^6\left(-\frac{3b}{4}\right)^5 = -449\,064a^6b^5$
k ${}^{11}C_4\left(\frac{a}{2}\right)^7\left(-\frac{2b}{3}\right)^4 = \frac{55}{108}a^7b^4$ and
${}^{11}C_5\left(\frac{a}{2}\right)^6\left(-\frac{2b}{3}\right)^5 = \frac{-77}{81}a^6b^5$

10 **a** $c^8 + 4c^6d^2 + 6c^4d^4 + 4c^2d^6 + d^8$
b $v^{10} - 5v^8w^2 + 10v^6w^4 - 10v^4w^6 + 5v^2w^8 - w^{10}$
c $8s^6 + 60s^4t^2 + 150s^2t^4 + 125t^6$
d $8s^6 - 60s^4t^2 + 150s^2t^4 - 125t^6$
e $d^3 + 3d + \frac{3}{d} + \frac{1}{d^3}$
f $16w^4 + 96w^2 + 216 + \frac{216}{w^2} + \frac{81}{w^4}$

11 **a** $x^3 + 6x + \frac{12}{x} + \frac{8}{x^3}$

b $x^8 - 8x^6 + 24x^4 - 32x^2 + 16$

c $x^{10} - 5x^7 + 10x^4 - 10x + \frac{5}{x^2} - \frac{1}{x^5}$

d $\frac{1}{x^{12}} + \frac{18}{x^9} + \frac{135}{x^6} + \frac{540}{x^3} + 1215 + 1458x^3 + 729x^6$

12 a $96x^6 - 1200x^5 + 6000x^4 - 15\,000x^3 + 18\,750x^2 - 9375x$

b $16 + 48x + 56x^2 + 32x^3 + 9x^4 + x^5$

13 a $206 + 66x + 276x^2 + 88x^3 + 16x^4$

b $27x + 42x^2 + 334x^3 + 405x^4 + 243x^5$

14 a $125 + 40\sqrt{3}$ **b** $-31 - 174\sqrt{5}$

15 a $1 + 2nx + 2n(n-1)x^2 + \frac{4n(n-1)(n-2)}{3}x^3 + \ldots$

b $1 - 3nx + \frac{9n(n-2)}{2}x^2 - \frac{9n(n-1)(n-2)}{2}x^3 + \ldots$

16 a $1 + 24x + 240x^2 + \ldots$

b 1.264

17 a $1 - 14x + 84x^2 - 280x^3 + \ldots$

b 0.932 07

18 a $7 + 5\sqrt{2}$ **b** $41 - 29\sqrt{2}$

c $577 + 408\sqrt{2}$ **d** $792 - 560\sqrt{2}$

e $\frac{5}{2} - \frac{7}{4}(\sqrt{2})$ **f** $\frac{7537}{81} + \frac{332}{9}(\sqrt{2})$

19 a $28 + 16\sqrt{3}$ **b** $576 - 256\sqrt{5}$

c $13\,100 - 4924\sqrt{7}$ **d** $198\sqrt{6} + 485$

e $112 + 64\sqrt{3}$ **f** $485 - 198\sqrt{6}$

Exercise 2.2B Reasoning and problem-solving

1 120 ways

2 184 756

3 $8s^3 - 36s^2w + 54sw^2 - 27w^3$

4 a 1.340 096 (to 6 dp) **b** 7.5295 (to 4 dp)

5 a 1.0773 (to 4 dp) **b** 973.94 (to 5 sf)

6 362.590 82 (to 5 dp)

7 $56\,700\sqrt{15}$

8 a 480 **b** 270

9 a $-\frac{5}{2}$ **b** $\frac{15}{4}$

10 a 210 **b** 120

11 a 5 **b** 7

12 a 8 **b** −3

13 9

14 14

15 a n **b** $\frac{1}{6}n^3 - \frac{1}{2}n^2 + \frac{1}{3}n$ **c** $-n$

16 a $\frac{1}{n+1}$ **b** $n + 5 + \frac{6}{n}$

17 −2416

18 −168

19 a 15 **b** 0.0198 (to 3 sf)

c 0.297 (to 3 sf) **d** 0.132 (to 3 sf)

Exercise 2.3A Fluency and skills

1 a $x - 10$ **b** $3x + 2$ **c** $4x - 3$

2 a $x^2 + 4x - 7$ **b** $x^2 - x - 1$ **c** $x^2 + 7x + 9$

d $x^2 - 5x + 7$ **e** $x^3 - 4x^2 - 7x + 1$

3 $(4x^2 + 20x + 72)$ rem 293

4

$$\begin{array}{r} 5x^2 + 26x + 5 \\ x-3\,\overline{)\,5x^3 + 11x^2 - 73x - 15} \\ \underline{5x^3 - 15x^2} \qquad\qquad\quad \\ 26x^2 - 73x \qquad\;\; \\ \underline{26x^2 - 78x} \qquad\;\; \\ 5x - 15 \\ \underline{5x - 15} \\ 0 \end{array}$$

5 a $x^2 + x - 1$ **b** $x^2 + x + 1$

c $2x^2 - 7x + 3$ **d** $3x^3 - 5x^2 + 4x - 7$

e $2x^3 + 7x^2 - 10x - 1$

6 a $f(0) = 0$; $f(1) = 9$; $f(-1) = -13$; $f(2) = 20$; $f(-2) = -36$

b $f(0) = -2$; $f(1) = -5$; $f(-1) = -3$; $f(2) = -6$; $f(-2) = -14$

c $f(0) = 2$; $f(1) = 1$; $f(-1) = -3$; $f(2) = 0$; $f(-2) = -20$

d $f(0) = 2$; $f(1) = 0$; $f(-1) = 6$; $f(2) = 12$; $f(-2) = 0$

e $f(0) = 4$; $f(1) = 0$; $f(-1) = 6$; $f(2) = 0$; $f(-2) = 0$

7 a $f(-6) = (-6)^3 + 4(-6)^2 - 9(-6) + 18 =$
$-216 + 144 + 54 + 18 = 0$

b $f(8) = 2(8)^3 - 13(8)^2 - 20(8) - 32$
$= 1024 - 832 - 160 - 32 = 0$

c $f\left(\frac{1}{3}\right) = 3\left(\frac{1}{3}\right)^3 + 11\left(\frac{1}{3}\right)^2 - 25\left(\frac{1}{3}\right) + 7 = \frac{1}{9} + \frac{11}{9} - \frac{25}{3} + 7 = 0$

d $f\left(\frac{-2}{5}\right) = 10\left(\frac{-2}{5}\right)^3 + 19\left(\frac{-2}{5}\right)^2 - 39\left(\frac{-2}{5}\right) - 18$
$= \frac{-16}{25} + \frac{76}{25} + \frac{78}{5} - 18 = 0$

8 $x(4x - 1)(x + 7)$

9 $(x - 1)(x + 1)(2x + 9)$

10 a $(x + 6)(x - 1)(x - 2)$ **b** $(x + 7)(x - 9)(x - 4)$

c $(x + 3)(2x - 1)(3x + 2)$ **d** $(x + 4)(x - 4)(x^2 + 3)$

Exercise 2.3B Reasoning and problem-solving

1 a $p + q + 2 = 0$ **b** $-9p - q + 38 = 0$ **c** $p = 5, q = -7$

2 a $\frac{-9}{4}$ **b** $b = 2$ **c** $p = 5$ and $q = -7$

3 $(x - 2)$

4 $(x + 1)(x - 2)(x + 3)$

5 $(x - 1)(x + 3)$

6 LCM $(x - 3)(x + 4)(2x + 7)$, HCF $(2x + 7)$

7 LCM $(x + 5)(x - 7)(x + 7)(x + 9)$, HCF $(x + 5)(x + 9)$

8 $(x - 1)$. $f(1) - g(1) = 7x^2 + 24x - 31 \equiv 7(1)^2 + 24(1) - 31 = 0$
$\Rightarrow (x - 1)$ is a factor of $f(x) - g(x)$

9 $a = 15$ and $b = -36$

10 a $(2x - 3)$ m **b** $(x - 3)\sqrt{3}$ cm

11 a $t = 2$, 3 and 4.5 secs

b i When $t = 2$, $a = 5$
When $t = 3$, $a = -3$
When $t = 4.5$, $a = 7.5$

ii $t = \frac{19 \pm \sqrt{19}}{6}$

12 a The radius could be $(x - 3)$ m if the height is $(3x + 7)$ m. (Accept alternatives.)

b $x > 3$

13 a $t = 1$, 5 and 6 seconds

b When $5 < t < 6$, $f(t)$ is negative; when $1 < t$, $f(t)$ is negative; hence the ride is above ground level when $1 < t < 5$

14 a The dimensions could be $(x + 1)$, $(x + 2)$ and $(3x + 12)$ (Accept alternatives.)

b $x > -1$

15 $\left(32\pi x^2 + 64\pi x + \frac{608\pi}{3}\right)$ cm^3

Exercise 2.4A Fluency and skills

1 a $(-2, 0)$ and $(3, 0)$ **b** $(\frac{-5}{2}, 0)$ and $(7, 0)$

c $(-2, 0)$ **d** $(3, 0)$

e $(3, 0)$ **f** $\left(\frac{1}{2}(\sqrt[3]{7} - 5), 0\right)$

2 $x = 1$ is a vertical asymptote; $y = 0$ is a horizontal asymptote

3 a $x = 4$ **b** $x = -2$

c $y = 3$ **d** $x = \frac{1}{2}$

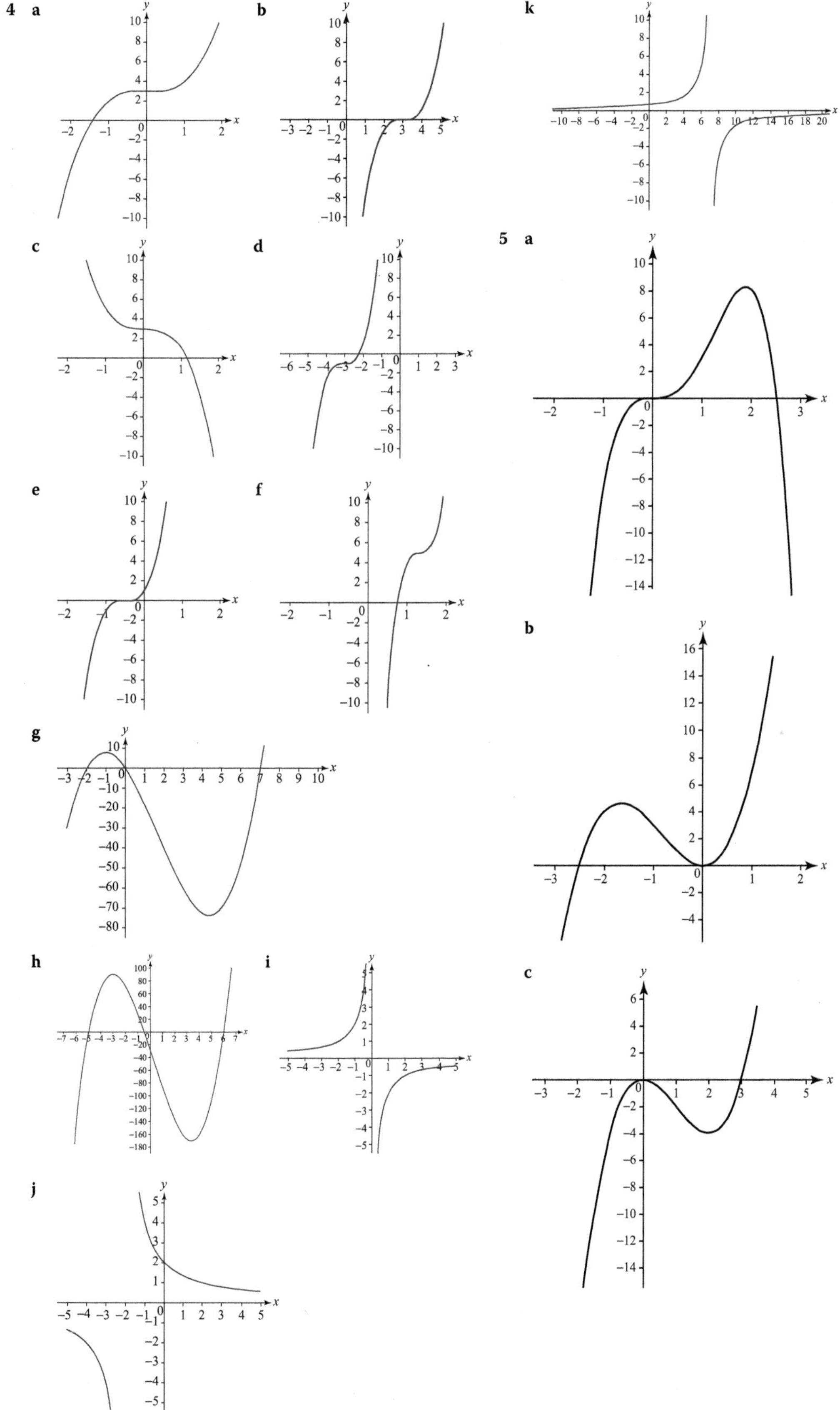

d

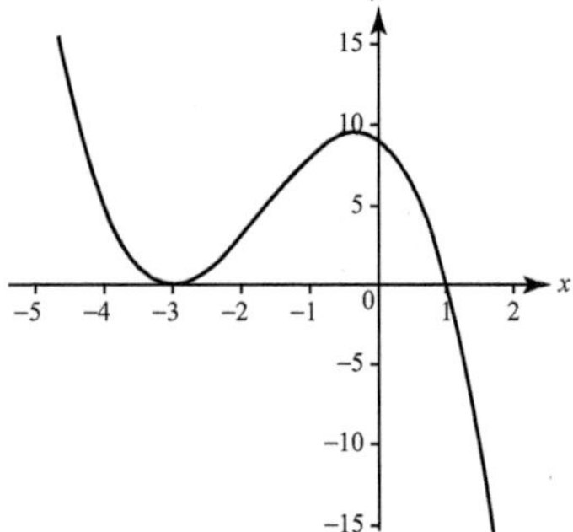

e

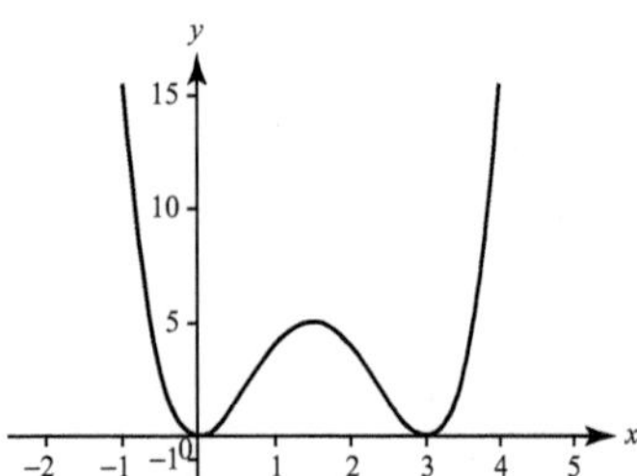

f

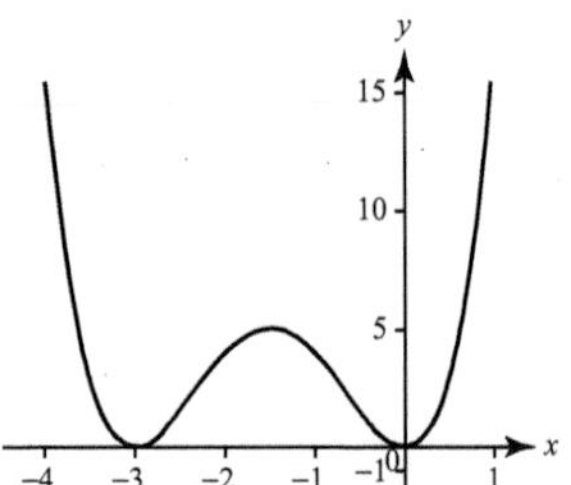

g

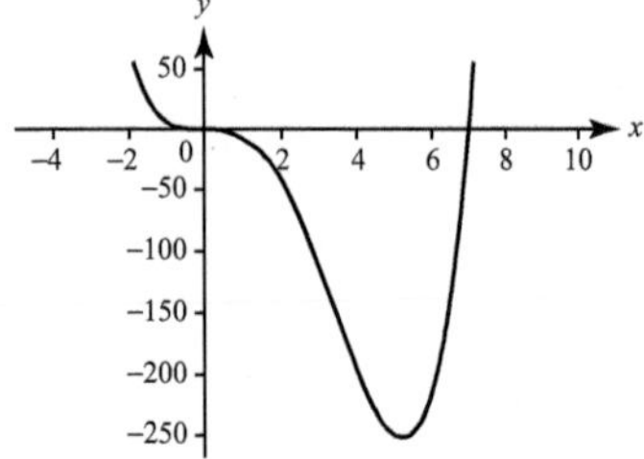

h

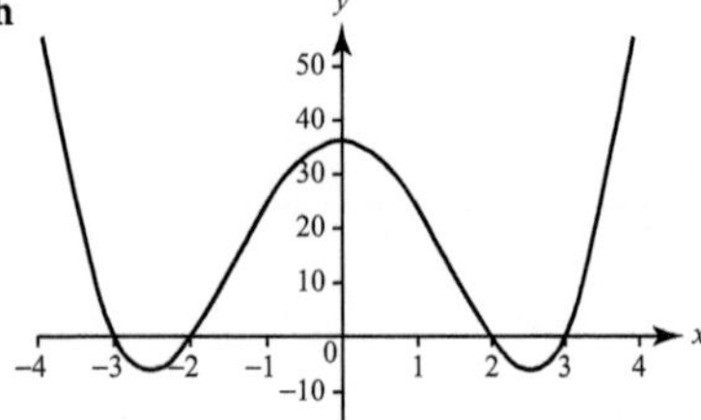

6 a

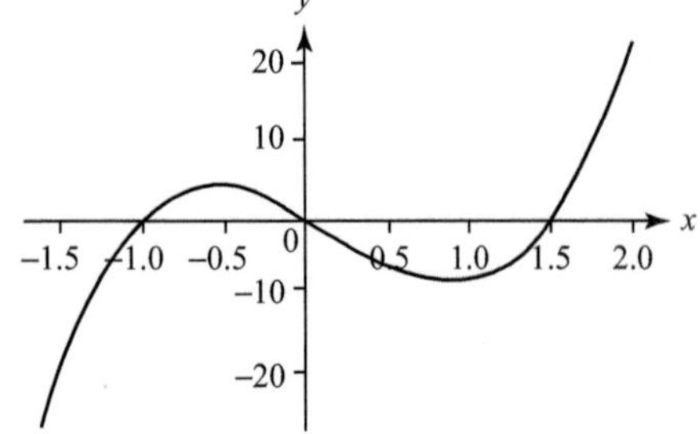

b

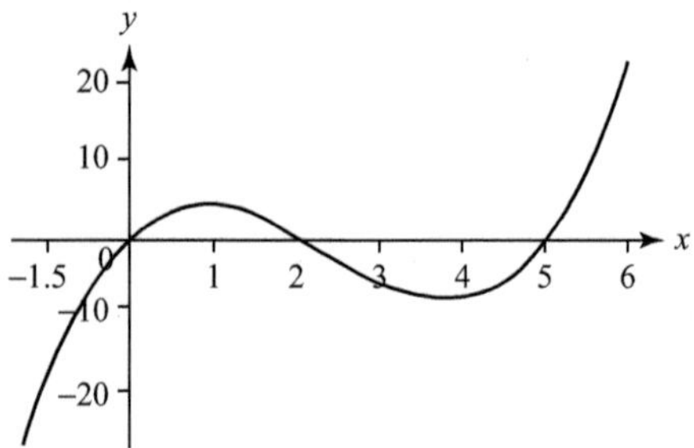

c

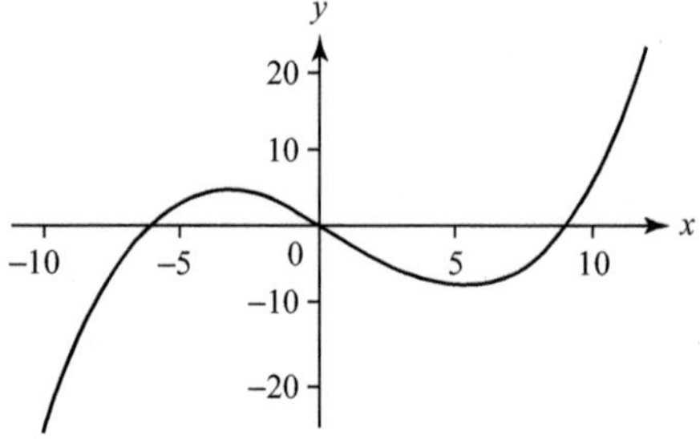

d

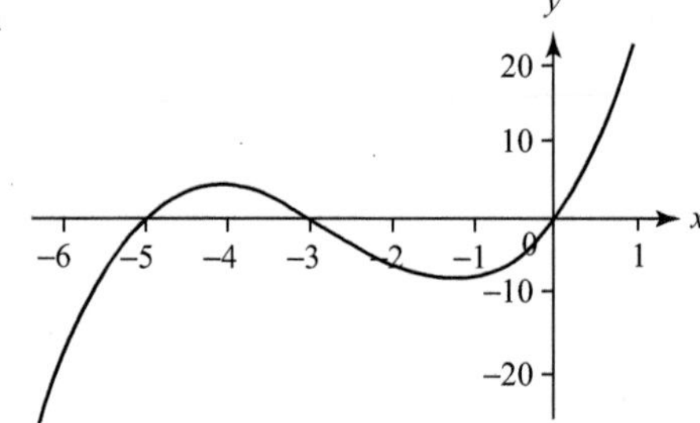

e

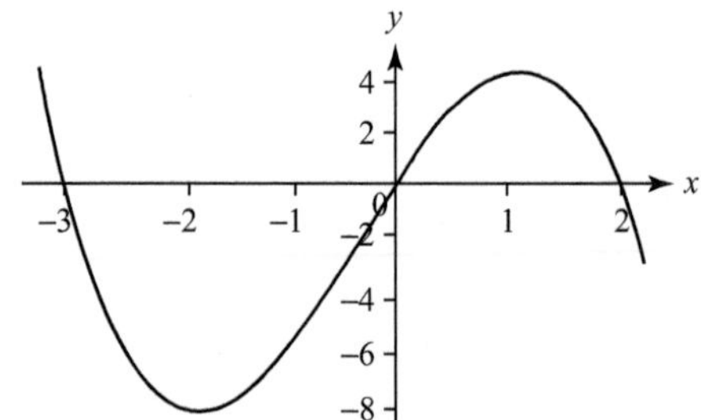

f

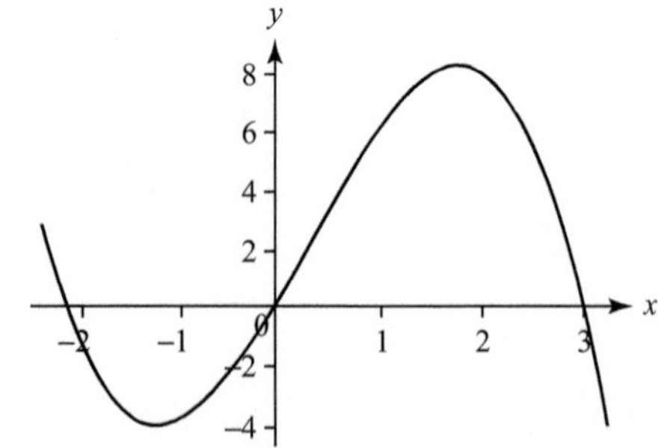

7 a Max (–0.5, 5), Min (2, –4) **b** Max (–2, 15), Min (8, –12)
c Max (–9, 5), Min (1, –4) **d** Max (–2, 9), Min (8, 0)
e Max (–2, 2.5), Min (8, –2) **f** Min (–2, –5), Max (8, 4)
g Max (2, 5), Min (–8, –4) **h** Max (–4, 5), Min (16, –4)

8 a Stretch scale factor ¼ in x-direction
b Stretch scale factor 3 in y-direction
c Translate 7 units left/ by vector $\begin{pmatrix} -7 \\ 0 \end{pmatrix}$

d Translate 4 units up/ by vector $\begin{pmatrix}0\\4\end{pmatrix}$

e Stretch scale factor ½ in y-direction

f Reflect in x-axis (or line $y = 0$)

g Reflect in y-axis (or line $x = 0$)

h Stretch scale factor 2 in x-direction

9 **a** $y=(x+3)^3$ **b** $y=x^3+2$ **c** $y=2x^3$ **d** $y=\left(\frac{x}{3}\right)^3$

10 a

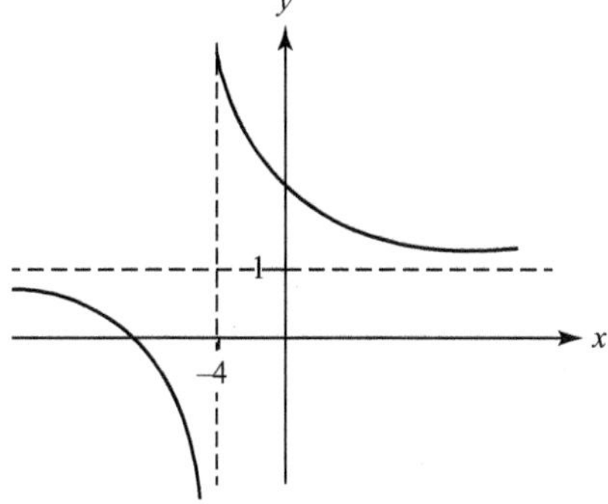

b

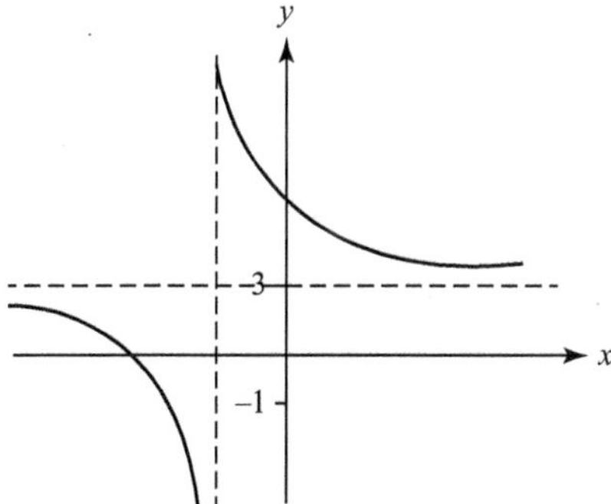

c

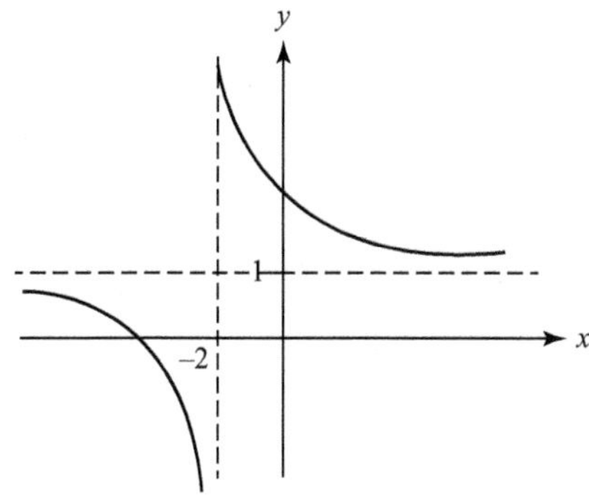

d

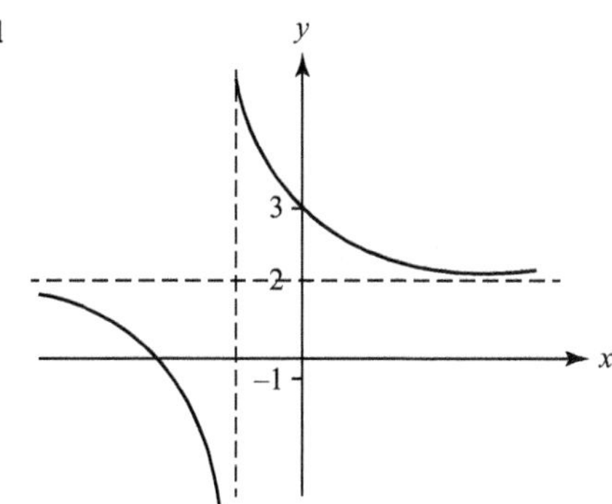

Exercise 2.4B Reasoning and problem-solving

1 **a** $r=\frac{56}{h}$

b

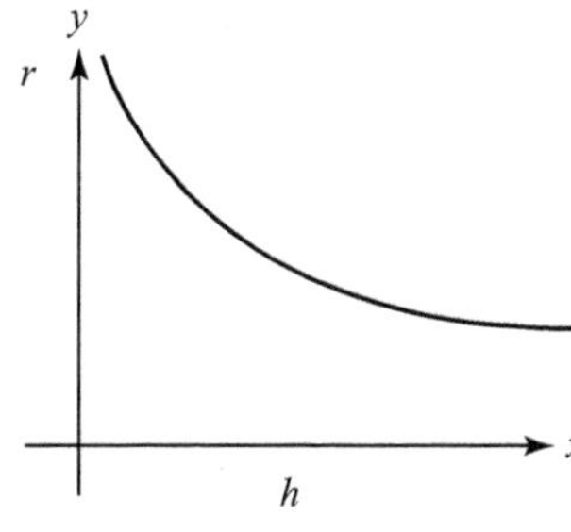

2 **a** $v=60\sqrt{t}$

b

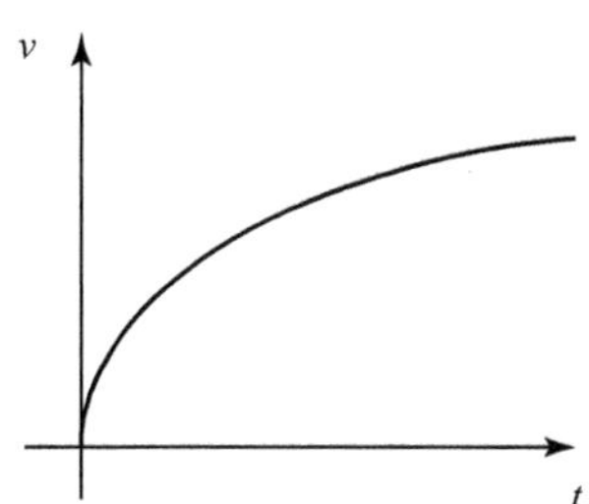

3 **a**

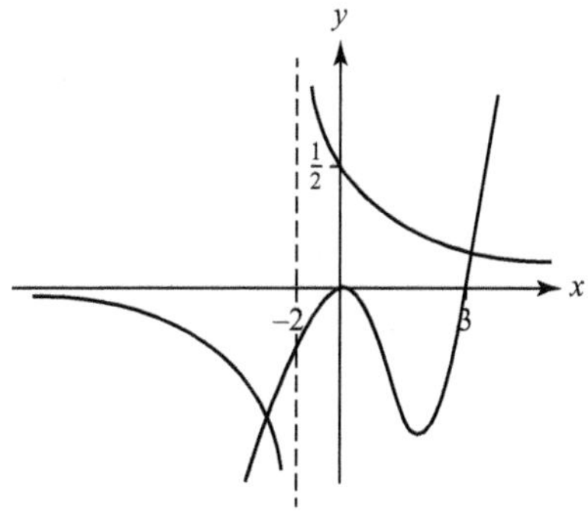

b Two solutions as they intersect twice.

4 **a** $y=\text{f}(x+1)$ **b** $y=\text{f}(2x)$

c $y=-\text{f}(x)$ **d** $y=2\text{f}(x)$

5 $A=-4$, $B=4$ and $C=0$

6 $A=-5$, $B=-3$ and $C=17$

7 $A=4$, $B=-3$ and $C=-3$

8 **a** $a<0 \Rightarrow$ maximum, $a>0 \Rightarrow$ minimum

b $\left(-\frac{b}{2a}, c-\frac{b^2}{4a}\right)$

c $(0, c)$, $\left(\frac{-b+\sqrt{b^2-4ac}}{2a}, 0\right)$ and $\left(\frac{-b-\sqrt{b^2-4ac}}{2a}, 0\right)$

d $x=\frac{-b}{2a}$

Review exercise 2

1 $2x^5-6x^4-3x^3+15x^2+9x-9$

2 $n(2n-3)(2n+5)$

3 **a** $2y^3+9y^2+4y-15$ **b** $2z^3-7z^2+4z+4$

4 **a** $-4(m+4)$ **b** $(d+1)(5-3d)$

5 $a=-5$, $b=-31$, $c=84$

6 $2x^2+3x-13$

7 1.4641

8 $16s^8-128s^6t+384s^44t^2-512s^2t^3+256t^4$

9 a $28+16\sqrt{3}$ b $-1760\sqrt{5}$

10 $\frac{-938223}{16}$

11 a $a=-1, b=4608, c=5376$

b $512+1792x+2304x^2+768x^3+...$

12 $2x^2-9x+1$

13 $(x+1), (x-2)$ and $(x-3)$

14 $4x^2-3x+7$

15 x^2+3 rem 10

16 $f\left(\frac{3}{2}\right)=4\left(\frac{3}{2}\right)^3-8\left(\frac{3}{2}\right)^2+\left(\frac{3}{2}\right)+3=\frac{27}{2}-18+\left(\frac{3}{2}\right)+3=0$

17 $(x-3)(2x+1)(x+3)$

18 a She has moved the graph one unit left, it should be one unit right.

b

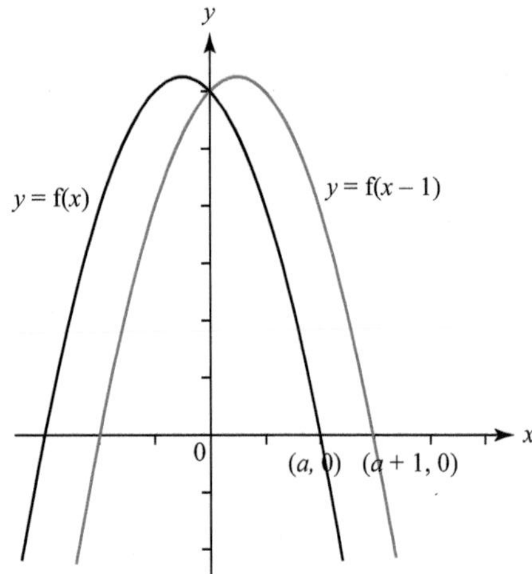

19

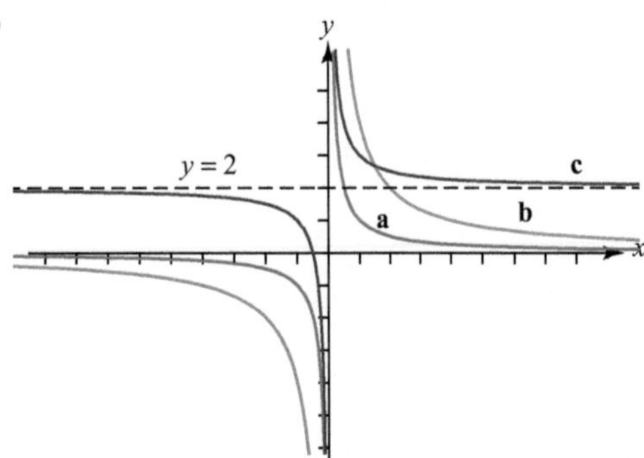

20

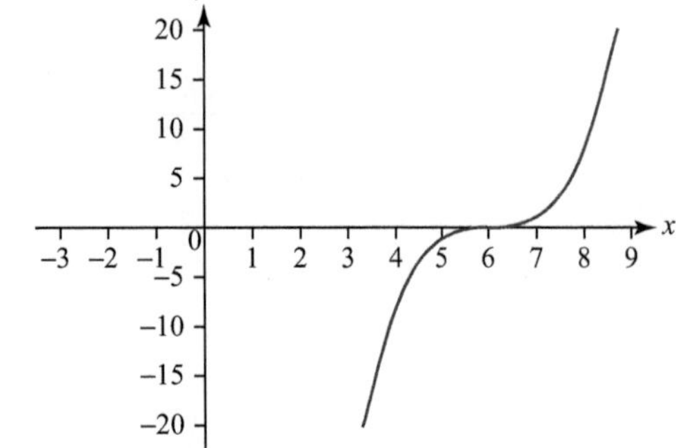

21

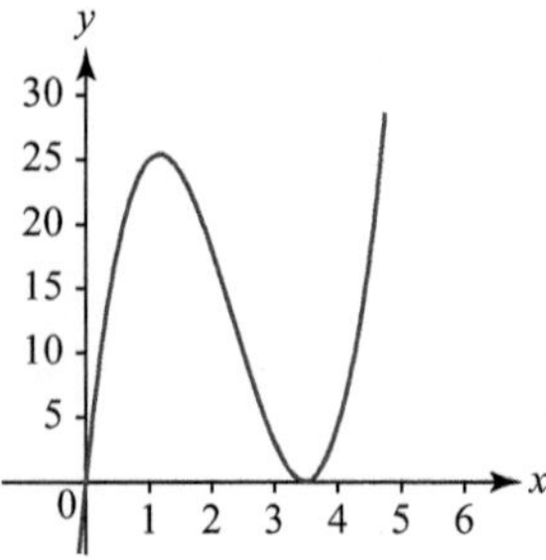

Starts from O at $t=0$ and accelerates to a distance of about 25.5 m in about 1.2 secs; It then returns to O, which it reaches at 3.5 secs. After that it accelerates continuously away from O

22 a Length $=16-2x$, width $=10-2x$

$\Rightarrow$ Volume $=x(16-2x)(10-2x)$

$=160x-52x^2+4x^3$

b

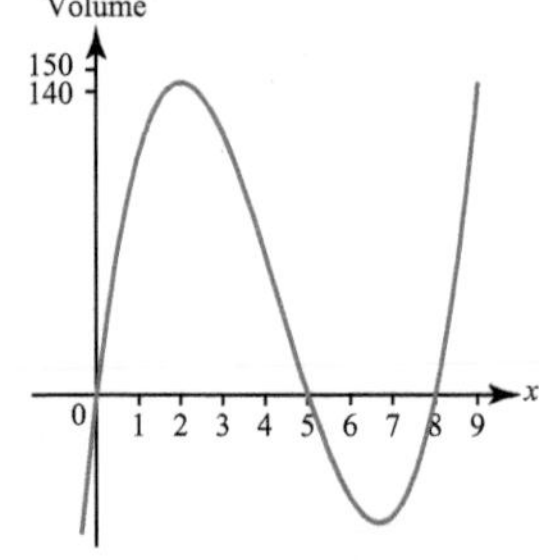

From the graph the maximum value of V is about 145 when x is 2

23 The particle is decelerating away from O for the first $\frac{4}{3}$ seconds and then returns back to O, reaching it when $t=4$ s After that it accelerates away from O indefinitely.

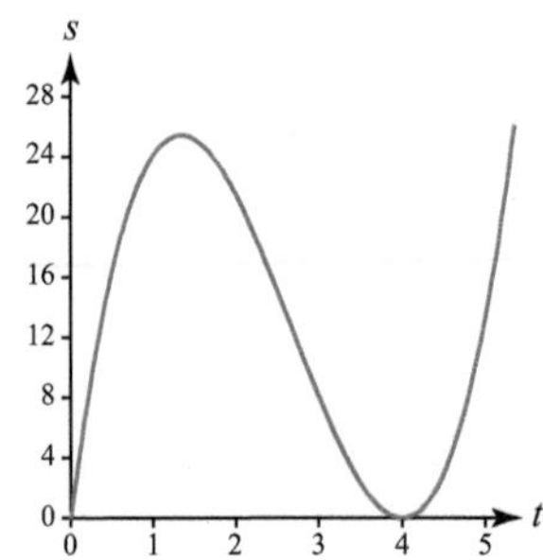

Assessment 2

1 D

2 B

3 a i $12x^2-16x-3$

ii $4a^2-12ab+9b^2$

iii $5x^3-13x^2y-11xy^2-2y^3$

b $a=9, b=0, c=-16$

4 $1+4x+7x^2+7x^3$

5 a $p(p-5)^2$

b $(2x+5)(2x+5-5)^2=4x^2(2x+5)$ so $a=4$

6 $f(3)=20\neq 0$

7 a 18 000 b 469 000 c 15

8 a i $c=3$ ii $A=189$ iii $B=945$

b 2079

9 $(x-2)(x^2-x+3)+7$

10 a i $x^4+4x^3y+6x^2y^2+4xy^3+y^4$

ii $x^4-4x^3y+6x^2y^2-4xy^3+y^4$

b $2x^4+12x^2y^2+2y^4=2(\sqrt{5})^4+12(\sqrt{5})^2(\sqrt{2})^2+2(\sqrt{2})^4$
$=2\times25+12\times5\times2+2\times4=178$

11 5376

12 a **i** $1+12x+60x^2$ **ii** $64-192x+240x^2$

b $64+576x+1776x^2$

13 a $2(2)^3+(2)^2-7(2)-6=16+4-14-6=0$,
hence $(x-2)$is a factor

b $2x^3+x^2-7x-6=(x-2)(2x^2+5x+3)=(x-2)(2x+3)(x+1)$
$x=2,-1\frac{1}{2},-1$

14 a $a=2, b=-1$ **b** $(x+3)(x-1)(2x-5)$

c

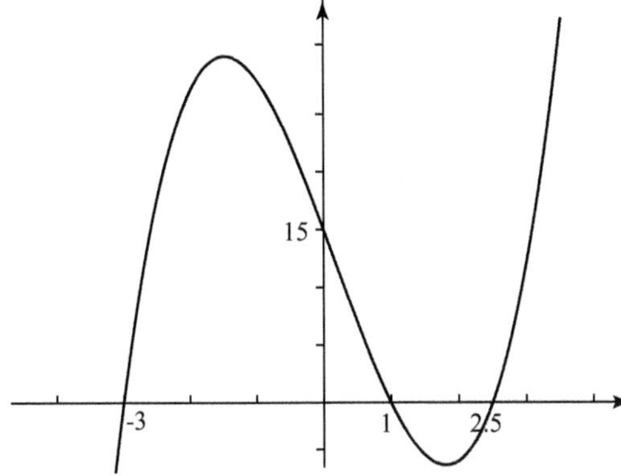

d $-3\le x\le1$ or $x\ge2.5$

15 a $x^6+6x^4+15x^2+20+\frac{15}{x^2}+\frac{6}{x^4}+\frac{1}{x^6}$

b $x^6-6x^4+15x^2-20+\frac{15}{x^2}-\frac{6}{x^4}+\frac{1}{x^6}$

c $12x^4+40+\frac{12}{x^4}=64\Rightarrow x^8-2x^4+1=0\Rightarrow$
$(x^4-1)(x^4-1)=0\Rightarrow(x-1)(x+1)(x^2+1)=0\Rightarrow x=\pm1$

16 a $^nC_r+{}^nC_{r-1}=\frac{n!}{r!\times(n-r)!}+\frac{n!}{(r-1)!\times(n-(r-1))!}$

$=\frac{n!}{r!\times(n-r)!}+\frac{n!}{(r-1)!\times(n+1-r)!}$

$=\frac{n!}{r!\times(n-r)!}\times\frac{(n+1-r)}{(n+1-r)}+\frac{n!}{(r-1)!\times(n+1-r)!}\times\frac{r}{r}$

$=\frac{(n+1)!}{r!\times(n+1-r)!}={}^{n+1}C_r$

b $^{n+2}C_3-{}^nC_3=\frac{(n+2)!}{3!\times(n+2-3)!}-\frac{n!}{3!\times(n-3)!}$

$=\frac{(n+2)\times(n+1)\times n!}{3!\times(n-1)!}-\frac{(n-2)\times(n-1)\times n!}{3!\times(n-1)!}$

$=\frac{n!}{3!\times(n-1)!}\times\left[(n^2+3n+2)-(n^2-3n+2)\right]$

$=\frac{n\times(n-1)!}{6\times(n-1)!}\times[6n]=n^2$

17 a **i** B **ii** D **iii** E **iv** no graph **v** C

b $y=\frac{1}{x+1}$ (or anything of the form $y=\frac{a}{x+b}$ $a>0$ and $b>0$)

c

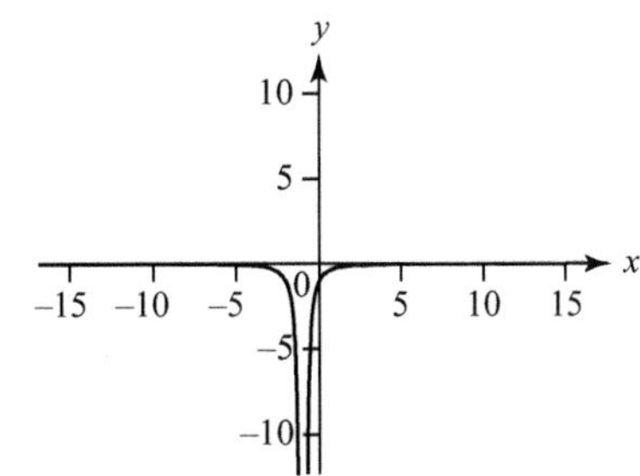

Chapter 3

Try it 3A

1 a 13.9 cm

b 16.9 cm

2 25.4°

3 $\cos(\theta)=\frac{\sqrt{7}}{4}, \tan(\theta)=\frac{3}{\sqrt{7}}$

4 $x=11.5°, 168.5°, -348.5°, -191.5°$

5 $x=84.3°, 275.7°, -84.3°, -275.7°$

6 $x=-16.7°, -196.7°, 163.3°, 343.3°$

7 a $\theta=-63.4°, 116.6°$

b $\theta=97.2°, -97.2°$

c $\theta=16.6°, 163.4°$

8 $x=30°, 60°, 120°, 150°$

9 $x=96.7°, 276.7°$

Bridging Exercise 3A

1 a 3.88 cm **b** 6.50 cm **c** 2.45 cm

d 3.91 cm **e** 11.8 cm **f** 6.42 cm

g 12.0 mm **h** 2.26 mm **i** 1.64 m

2 a 14.5° **b** 48.6° **c** 80.7°

d 60° **e** 30° **f** 45°

3 a 16.5 cm **b** 11.8 cm **c** $7\sqrt{2}$ cm

4 $\sin(\theta)=\frac{4}{5}, \tan(\theta)=\frac{4}{3}$

5 $\sin(\theta)=\frac{1}{\sqrt{37}}, \cos(\theta)=\frac{6}{\sqrt{37}}$

6 $\cos(\theta)=\frac{2}{\sqrt{5}}, \tan(\theta)=\frac{1}{2}$

7 a $x=44.4°, 135.6°$

b $x=210°, 330°$

c $x=20.5°, 159.5°$

d $x=195.7°, 344.3°$

8 a $x=30°, 150°, -330°, -210°$

b $x=-48.6°, 228.6°, -131.4°, 311.4°$

c $x=45°, 135°, -315°, -225°$

d $x=-19.5°, 199.5°, 340.5°, -160.5°$

9 a $x=72.5°, 287.5°$

b $x=0°, 360°$

c $x=154.2°, 205.8°$

d $x=104.5°, 255.5°$

10 a $\theta=60°, 300°, -60°, -300°$

b $\theta=109.5°, 250.5°, -109.5°, -250.5°$

c $\theta=81.4°, 278.6°, -81.4°, -278.6°$

d $\theta=100.4°, 259.6°, -100.4, -259.6$

11 a $\theta = 5.7°, 185.7°, -174.3°, -354.3°$
b $\theta = -45°, 135°, 315°, -225°$
c $\theta = -30°, 150°, 330°, -210°$
d $\theta = 56.3°, 236.3°, -123.7°, -303.7°$

12 a $x = 270°$ b $x = 116.6°, 296.6°$
c $x = 60°, 300°$ d $x = 71.6°, 251.6°$

13 a $x = -48.6°, -131.4°$ b $x = 31.0°, -149.0°$
c $x = 131.8°, -131.8°$ d $x = -45°, -135°$

14 a $\theta = 48.2°, 311.8°$ b $\theta = 23.6°, 156.4°$
c $\theta = 53.1°, 233.1°$ d $\theta = 120°, 240°$

15 a $\theta = 79.7°, -100.3°$ b $\theta = 0°$
c $\theta = -60°, -120°$ d $\theta = 80.4°, -80.4°$

16 a $x = 72.5°, 287.5°, -72.5°, -287.5°$
b $x = 90°, -270°$
c $x = -76.0°, -256.0°, 104.0°, 284.0°$
d $x = 82.3°, 277.7°, -82.3°, -277.7°$

17 a $\theta = 30°, -60°, -150°, 120°$
b $\theta = \pm 22.5°, \pm 157.5°$
c $\theta = -22.5°, 112.5°, 157.5°, -67.5°$
d $\theta = \pm 36.3°, \pm 143.8°$

18 a $\theta = 7.5°, 52.5°$ b $\theta = 5.82°, 54.2°$
c $\theta = 31.7°, 58.3°$ d $\theta = 45°$

19 a $\theta = 34.6°$ b $\theta = -120°, 240°$
c $\theta = \pm 60°$ d $\theta = 27.7°, 297.7°, -242.3°$

20 a $\theta = -15°, 165°$ b $\theta = 125.5°, -25.5°$
c $\theta = -20.6°, 70.6°$ d $\theta = 38.5°, -118.5°$

21 a $\theta = 63.7°, 243.7°$ b $\theta = 181.5°, 338.5°$
c $\theta = 151.4°, 348.6°$ d $\theta = 115°, 295°$

Try it 3B

1 $x = 11.1$ cm, $y = 13.2$ cm
2 36.6°
3 $x = 13.8$ cm
4 $x = 133.1°, y = 10.5°$
5 32.9 cm^2

Bridging Exercise 3B

1 a $x = 11.0$ cm, $y = 6.50$ cm
b $x = 9.35$ cm, $y = 7.31$ cm
c $x = 8.42$ cm, $y = 10.6$ cm

2 a $\theta = 63.3°$
116.7° is not possible since the angle sum would be more than 180°, so the answer is unique.
b $\theta = 66.9°$
113.1° is a possible solution so the answer is not unique.
c $\theta = 49.5°$
130.5° is not possible since the angle sum would be more than 180°, so the answer is unique.

3 a 22.1 cm b 13.2 cm c 19.8 cm

4 a $x = 83.4°, y = 51.9°$
b $x = 82.8°, y = 41.4°$
c $x = 28.3°, y = 108.7°$

5 a 24.2 cm b 11.7 cm c 10.5 cm
d 16.5 mm e 11.8 cm f 5.32 m

6 a 19.7° b 51.0° c 34.8°

7 120°

8 32.3°

9 a 165.4 cm^2 b 36.9 cm^2 c 167.5 cm^2

Exercise 3.1A Fluency and skills

1 a $\sin\theta = \frac{4}{5}$; $\tan\theta = \frac{4}{3}$
b $\sin\theta = 0.6$; $\tan\theta = 0.75$
c $\sin\theta = \frac{5}{13}$; $\tan\theta = \frac{5}{12}$

2 a $-\cos 10°$ b $-\tan 20°$ c $-\sin 20°$
d $-\cos 22°$ e $\tan 35°$ f $-\sin 75°$

3

θ	−90°	0°	90°	180°	270°	360°
$\sin\theta$	−1	0	1	0	−1	0
$\cos\theta$	0	1	0	−1	0	1
$\tan\theta$	–	0	–	0	–	0

4 a i line symmetry about $\theta = 0°, \pm 180°, \pm 360°, \ldots$
Rotational symmetry (order 2) about every point of intersection with θ-axis: $(\pm 90°, 0), (\pm 270°, 0), \ldots$
ii No line symmetry
Rotational symmetry (order 2) about $(0°, 0)$, $(\pm 90°, 0)$, $(\pm 180°, 0) \ldots$

b i

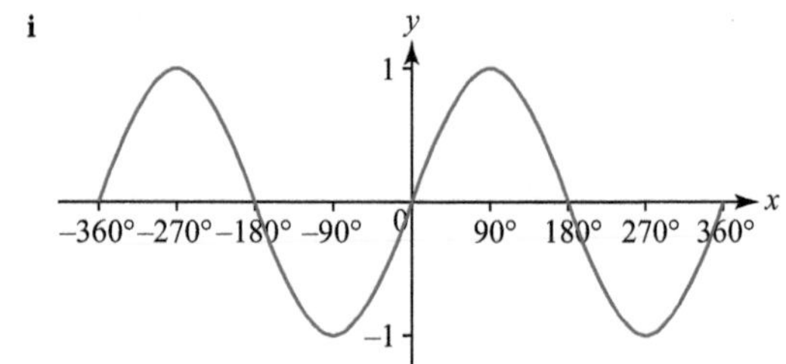

ii

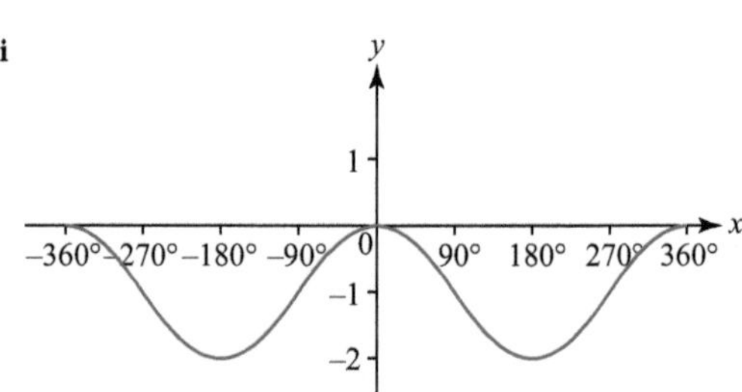

iii

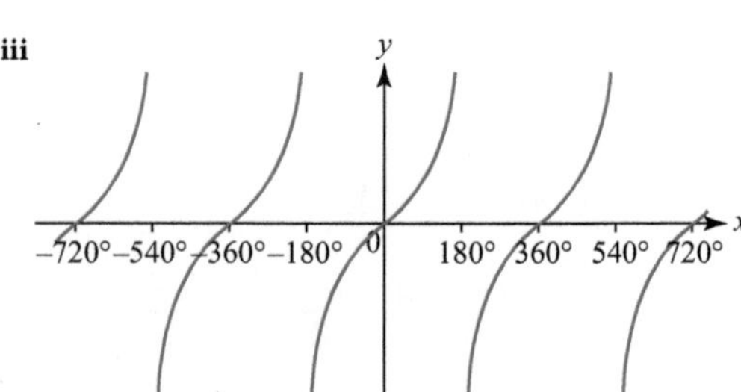

5 a $\tan\theta$ b $\tan\theta$ c 1
d $\sin\theta$ e $\cos\theta$ f $\sin^2\theta$

6 a $\sin 20°$ b $\tan 30°$ c $\cos 20°$
d $-\tan 42°$ e $-\cos 22°$ f $\sin 23°$

7 a $\theta = 45°, -135°$
b $\theta = -45°, 135°$

8 a $\frac{1}{2}$ b $\frac{1}{2}$ c $-\sqrt{3}$
d $-\frac{\sqrt{3}}{2}$ e $\frac{1}{2}$ f $\frac{1}{\sqrt{3}}$

9 a $\theta = 23.6°, 156.4°, -203.6°, -336.4°$
b $\theta = 56.3°, 236.3°, -123.7°, -303.7°$
c $\theta = 120°, 240°, -120°, -240°$

10 a $\theta = 53.1°$ **b** $\theta = 323.1°$

11 a $\theta = 48.6°$ or $131.4°$ **b** $\theta = 53.1°$ or $233.1°$
c $\theta = 210°$ or $330°$ **d** $\theta = 131.8°$ or $228.2°$
e $\theta = 108.4°$ or $288.4°$ **f** $\theta = 224.4°$ or $315.6°$
g $\theta = 138.6°$ or $221.4°$ **h** $\theta = 156.0°$ or $336.0°$

Exercise 3.1B Reasoning and problem-solving

1 a $\theta = 0°, 120°, 360°$ **b** $\theta = 90°, 330°$
c $\theta = 25°, 205°$ **d** $\theta = 180°$ or $300°$

2 a $\theta = 15°$ or $75°$ **b** $\theta = 16.8°$ or $106.8°$
c $\theta = 22.1°, 97.9°$ or $142.1°$ **d** $\theta = 72.3°$ or $107.7°$
e $\theta = 49.7°$ or $139.7°$ **f** $\theta = 97.2°$

3 Any correct transformation, for example:
a A translation of $\begin{pmatrix} -90° \\ 0 \end{pmatrix}$
b A $180°$ rotation about $(0°, 0)$

4 a $(3\cos\theta, 3\sin\theta)$
b $(3\cos\theta)^2 + (3\sin\theta)^2 = 9\cos^2\theta + 9\sin^2\theta$
$= 9(\cos^2\theta + \sin^2\theta)$
$= 9 = 3^2$
$r = 3$
c $\frac{5}{2}$

5 $x^2 + y^2 = 100$

6 a $\theta = 26.6°$ or $-153.4°$
b $\theta = 38.7°$ or $-141.3°$
c $\theta = 9.2°, 99.2°, -80.8°, -170.8°$
d $\theta = 0°, 90°, 180°, -90°$ or $-180°$ and $\theta = 24.1°, 155.9°, -24.1°$ or $-155.9°$
e $\theta = 0°, 180$ or -180 and $\theta = \pm 109.5°$
f $\theta = \pm 90°$

7 a $\theta = 90°$ **b** $\theta = 135°, 315°, 63.4°, 243.4°$
c $\theta = 19.5°, 160.5°, 270°$ **d** $\theta = 0°, 180°, 360°$
e $\theta = 60°, 180°, 300°$ **f** $\theta = 48.6°$ or $131.4°$

8 $a = 5, b = 8$

9 a Max = 5, min = 1 **b** 12 hours **c** 3 am and 3 pm

10 a $\theta = 60°, 300°, 48.2°, 311.8°$ **b** $\theta = 190.1°$ or $349.9°$

11 a

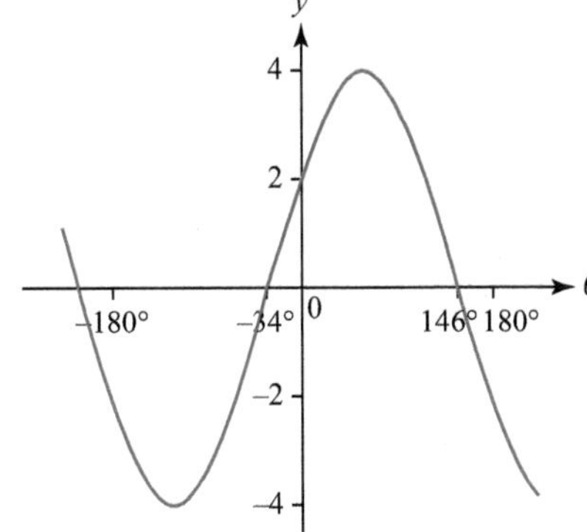

b i From graph, $\theta \approx -34°$ and $146°$
ii $\theta = -33.7°$ or $146.3°$

12 a $\cos^4 x - \sin^4 x = (\cos^2 x - \sin^2 x)(\cos^2 x + \sin^2 x)$
$= (\cos^2 x - \sin^2 x) \times 1 = \cos^2 x - \sin^2 x$

b $\tan x + \frac{1}{\tan x} = \frac{\sin x}{\cos x} + \frac{\cos x}{\sin x}$

$= \frac{\sin^2 x + \cos^2 x}{\cos x \sin x} = \frac{1}{\sin x \cos x}$

13 $\frac{1-\tan^2 x}{1+\tan^2 x} = \frac{1-\frac{\sin^2 x}{\cos^2 x}}{1+\frac{\sin^2 x}{\cos^2 x}} = \frac{\cos^2 x - \sin^2 x}{\cos^2 x + \sin^2 x}$

$= \frac{\cos^2 x - \sin^2 x}{1} = \cos^2 x - \sin^2 x$

$= 1 - \sin^2 x - \sin^2 x = 1 - 2\sin^2 x$

$x = 30°, 150°$ or $210°, 330°$

Exercise 3.2A Fluency and skills

1 $BC = 10.7$ cm, $PR = 19.7$ cm

2 a $62.4°$ **b** $115.7°$ **c** $110.1°$

3 a $BC = 6.97$ cm, $PR = 15.0$ cm
b area of triangle $ABC = 26.8\text{ cm}^2$
area of triangle $PQR = 29.8\text{ cm}^2$

4 a $BC = 7.48$ cm, angle $B = 64.8°$, angle $C = 40.2°$
b $EF = 16.0$ cm, angle $E = 30.8°$, angle $F = 24.2°$
c angle $J = 62°$, $HJ = 10.6$ cm, $HI = 9.82$ cm
d Angle $M = 94°$, $KM = 3.74$ cm, $LM = 5.06$ cm
e Angle $R = 66.8°$ or $113.2°$, angle $Q = 63.2°$ or $16.8°$
Two triangles are possible. $RP = 17.5$ cm or 5.67 cm

5 a Angle $A = 31.5°$, angle $C = 94.0°$, angle $B = 54.5°$
b $DE = 59.2$ cm, angle $D = 81.4°$, angle $E = 47.6°$
c $HI = 40.5$ cm, angle $I = 110.6°$, angle $H = 36.4°$

6 $x = \sqrt{13}$ cm, area $= 3\sqrt{3}\text{ cm}^2$

7 a $\frac{\sin C}{12} = \frac{\sin 55°}{10} \Rightarrow \sin C = \frac{12 \times \sin 55°}{10} = 0.982...$
Angle $C = 79.4°$ or $100.6°$
So angle $B = 180° - 55° - 79.4° = 45.6°$
or $= 180° - 55° - 100.6° = 24.4°$
so both versions of angles B and C are possible.

b $\frac{\sin Z}{12} = \frac{\sin 55°}{13} \Rightarrow \sin Z = \frac{12 \times \sin 55°}{13} = 0.756...$
Angle $Z = 49.1°$ or $130.9°$
So angle $Y = 180° - 55° - 49.1° = 75.9°$
or $= 180° - 55° - 130.9° = -5.9°$
which is impossible.

Exercise 3.2B Reasoning and problem-solving

1 11.4 cm^2

2 a $44.4°$ **b** $111.8°$

3 $BC^2 = 12^2 + (4\sqrt{3})^2 - 2 \times 12 \times 4\sqrt{3} \times \cos 30°$
$= 144 + 48 - 144 \Rightarrow BC = \sqrt{48} = 4\sqrt{3} = AB$
So ΔABC is isosceles.

4 5.44 cm and 12.2 cm

5 a $\frac{9}{4}\sqrt{3}$ **b** $\frac{1}{2}$

6 a $h = b\sin A = a\sin B$
Hence,
$\frac{\sin A}{a} = \frac{\sin B}{b}$
Area of $\Delta ACP = \frac{1}{2}h \times AP$
Area of $\Delta CBP = \frac{1}{2}h \times PB$
So area of $\Delta ACB = \frac{1}{2}h \times (AP + PB) = \frac{1}{2}h \times c$
$h = b\sin A = a\sin B$ so $\frac{1}{2}h \times c = \frac{1}{2}bc\sin A = \frac{1}{2}ca\sin B$

b $CP = b\sin A$
$AP = b\cos A$
$BP = c - b\cos A$
Pythagoras in ΔCBP gives
$a^2 = (b\sin A)^2 + (c - b\cos A)^2$
$a^2 = b^2(1 - \cos^2 A) + c^2 - 2cb\cos A + b^2\cos^2 A$
$a^2 = b^2 + c^2 - 2cb\cos A$

7 $\frac{\sin Z}{2\sqrt{3}} = \frac{\sin 30°}{2} \Rightarrow \sin Z = \frac{2\sqrt{3}}{2} \times \frac{1}{2} = \frac{\sqrt{3}}{2}$
$Z = 60°$ or $120°$
Angle sum of triangle gives $X = 90°$ or $30°$
The angles of the triangle are either 30°, 60°, 90° or 30°, 30°, 120°
The triangle is either isosceles or right-angled.

8 $AC = 14.5$ cm, $AB = 18.3$ cm

9 11.4 cm

10 Angle at centre = 2 × angle at circumference
$B\hat{O}C = 2 \times \hat{A}$
ΔBOC is isosceles and symmetrical
$$B\hat{O}P = \frac{1}{2}B\hat{O}C = A$$
In ΔBOP, $\sin A = \frac{\frac{1}{2}a}{r} = \frac{a}{2r}$
$$2r = \frac{a}{\sin A}$$
Similarly with angles B and C at the apex, giving
$$2r = \frac{a}{\sin A} = \frac{b}{\sin B} = \frac{c}{\sin C}$$

11 In ΔABC, $B = 180° - A - C = 45°$
$$\frac{a}{\sin 22.5°} = \frac{b}{\sin 45°} \Rightarrow a = \frac{b \sin 22.5°}{\sin 45°}$$
In ΔBCP, $h = a \sin 67.5° = \frac{b \sin 67.5° \times \sin 22.5°}{\sin 45°} = 0.5 \times b$
The height is half the length of the base.

12 The cosine rule gives
$(n^2 - n + 1)^2 = (n^2 - 2n)^2 + (n^2 - 1)^2 - 2(n^2 - 2n)(n^2 - 1)\cos Y$
$n^4 - n^3 + n^2 - n^3 + n^2 - n + n^2 - n + 1$
$= n^4 - 4n^3 + 4n^2 + n^4 - 2n^2 + 1 - 2(n^4 - 2n^3 - n^2 + 2n)\cos Y$
$2n^3 + n^2 - 2n = n^4 - 2(n^4 - 2n^3 - n^2 + 2n)\cos Y$
$\cos Y = \frac{n(n^3 - 2n^2 - n + 2)}{2n(n^3 - 2n^2 - n + 2)} = \frac{1}{2}$
Angle $Y = 60°$

Review exercise 3

1 a $\sin\theta = 0.6$, $\tan\theta = 0.75$ b $\cos\theta = \frac{12}{13}$, $\tan\theta = \frac{5}{12}$

2 a $\tan\theta$ b $\sin\theta$

3 a $-\sin 10°$ b $\tan 80°$ c $-\cos 40°$
d $-\tan 42°$ e $\sin 11°$ f $-\cos 60°$
g $\tan 30°$ h $-\cos 20°$ i $\sin 80°$

4 a Max = 4, period = 360° b Max = 5, period = 180°
c Max = 6, period = 72°

5 a $x = 45°$ b $x = 45°$ for $0° \le x \le 180°$

6 a $x^2 + y^2 = 4\cos^2\theta + 4\sin^2\theta = 4(\cos^2\theta + \sin^2\theta) = 4 \times 1 = 4$
$x^2 + y^2 = 4$ is equation of circle of radius $\sqrt{4} = 2$
b At Q, $x = 1$ and $y = \sqrt{3}$
$x^2 + y^2 = 1 + 3 + = 4$
Hence, Q lies on the circle. $\theta = 60°$

7 a $\frac{1}{\sqrt{2}}$ b $-\frac{1}{\sqrt{2}}$ c -1
d $\frac{1}{\sqrt{2}}$ e $-\frac{1}{\sqrt{2}}$ f -1

8 a $\theta = 45.6°, 314.4°, -45.6°$ or $-314.4°$
b $\theta = 68.2°, 248.2°, -111.8°$ or $-291.8°$
c $\theta = 210°, 330°, -30°$ or $-150°$

9 a $\theta = 41.8°$ or $138.2°$ b $\theta = 74.1°$ or $254.1°$
c $\theta = \pm 120°, 240°$ d $\theta = 210°$ or $330°$
e $\theta = 28°$ or $332°$

10 a $\theta = 36.9°$ or $-143.1°$ b $\theta = 0°, \pm 180°$ or $\pm 41.4°$
c $\theta = 0°, \pm 180°, 19.5°$ or $160.5°$ d $\theta = 80°$ or $140°$
e $\theta = 30°$ or $-90°$ f $\theta = 145°$ or $-35°$

11 a $\theta = 9.7°$ or $80.3°$ b $\theta = 10.9°$ or $100.9°$
c $\theta = 3.8°, 56.2°, 123.8°, 176.2°$ d $\theta = 16.1°, 103.9°, 136.1°$
e $\theta = 9.2°, 99.2°$ f $\theta = 53.1°$

12 a $\theta = 210°, 330°, 90°$
b $\theta = 60°, 300°, 180°$
c $\theta = 99.6°, 260.4°, 0°, 360°$
d $\theta = 64.3°, 295.7°, 219.9°, 140.1°$
e $\theta = 90°$
f $\theta = 53.6°, 306.4°, 147.5°, 212.5°$

13 a $BC = 6.46$ cm, area = 25.6 cm^2
b Angle $E = 26.9°$, area = 16.1 cm^2

Assessment 3

1 a C, $x = 78.5°, 281.5°$ b A, $x = 136.3°, 316.3°$

2 a D, $(0°, 1)$ b C, $(360°, -1)$

3 a

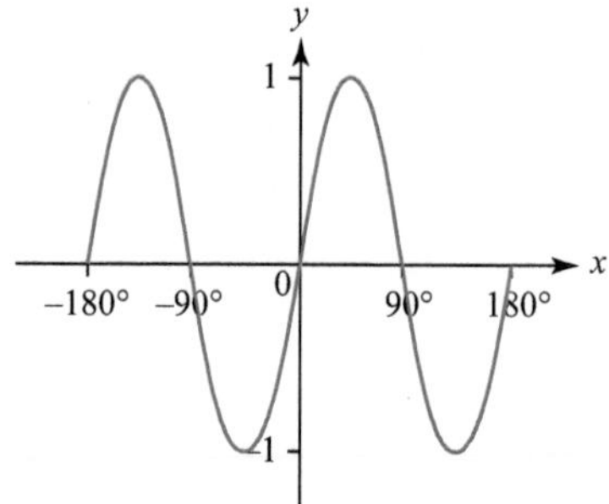

b

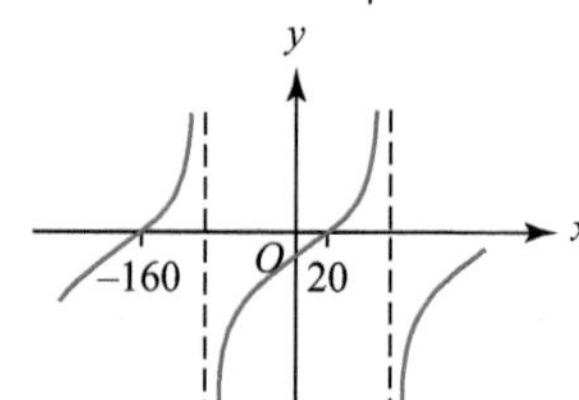

Asymptotes are $x = 110°$, $x = -70°$

4 a

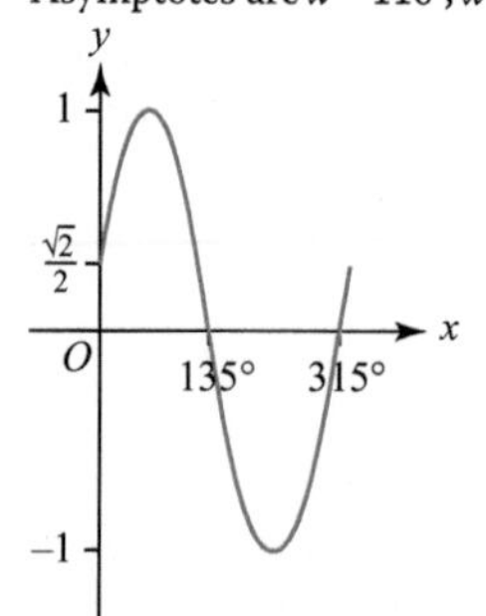

b Maximum at $(45°, 1)$; Minimum at $(225°, -1)$
c $x = 117.5°, 332.5°$

5 a 57.9° b 20.3 m^2

6 15.7 cm 7 2 units

8 a $x = -1.5°, -158.5°$
b $x = 11.7°, 71.7°, 131.7°, -48.3°, -108.3°, -168.3°$

9 a

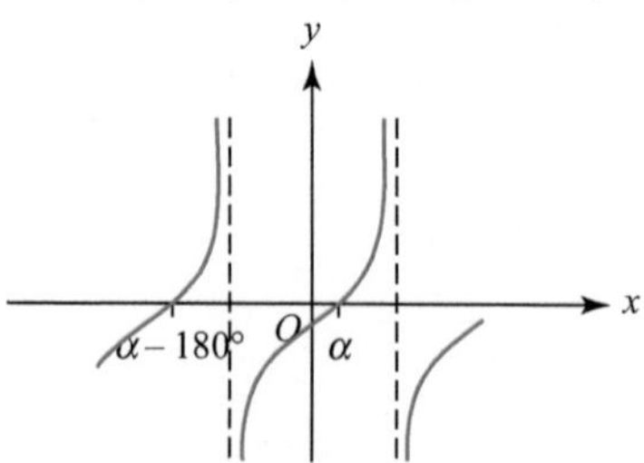

b $x = \alpha + 90°$, $x = \alpha - 90°$ c $x = \alpha + 60°, \alpha - 120°$

10 a

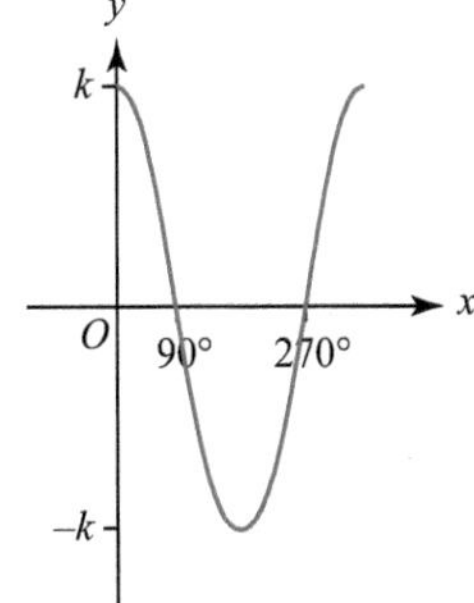

b $x = 71.6°, 251.6°$

11 a

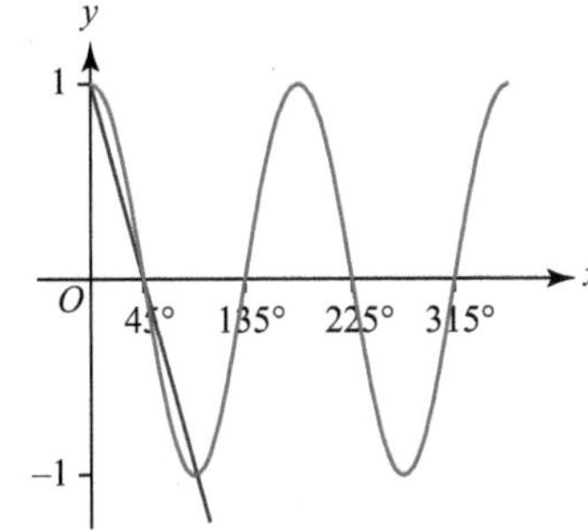

b 3 as the line intersects the curve 3 times

12 a 19.9 or 7.86 cm **b** 31.4 cm²

13 a 31.3°

b 180°–31.3° = 148.7°, but 148.7° + 54° > 180°, which makes the triangle impossible

14 a 10.1 cm **b** 94.1°

15 117.3°

16 $45° < x < 135°$

17 a $\cos\theta + \tan\theta\sin\theta \equiv \cos\theta + \frac{\sin\theta\sin\theta}{\cos\theta}$

$\equiv \frac{\cos^2\theta + \sin^2\theta}{\cos\theta} \equiv \frac{1}{\cos\theta}$

b $\theta = 66.4°, -66.4°$

18 a $x = 53.7°, 126.3°, 233.7°, 306.3°$

b $x = 71.6°, 251.6°, 135°, 315°$

19 $x = 0°, 90°, 110.9°, 159.1°, 180°$

20 $\theta = 0°, 60°, 120°, 180°, 37.9°, 82.1°, 157.9°$

21 a $\frac{2\sqrt{6}}{5}$ **b** $\frac{\sqrt{6}}{12}$

22 a $-8 < k < 0$ **b** $\theta = 30°, 150°, 270°$

23 $\frac{\alpha}{\sqrt{1-\alpha^2}}$

24 $x = 56.9°, 123.1°, 236.9°, 303.1°$

25 157.3 cm

Chapter 4

Try it 4A

1 a i $x < 3$ **ii** $x > 3$

b (3, 16)

c i −5 **ii** −4.5

2 a 0

b 3.9, 1.9, −2.1

c

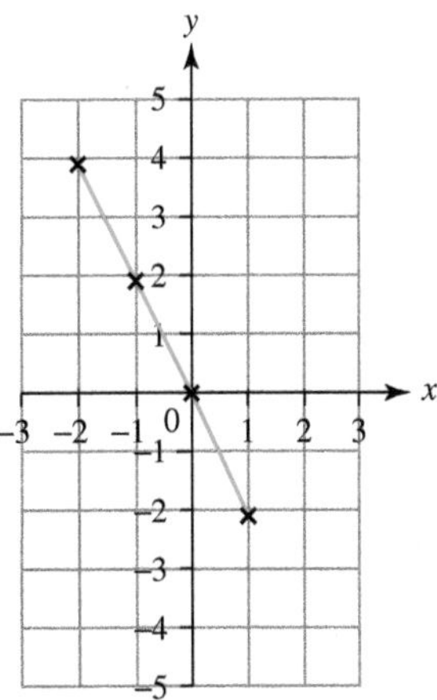

3

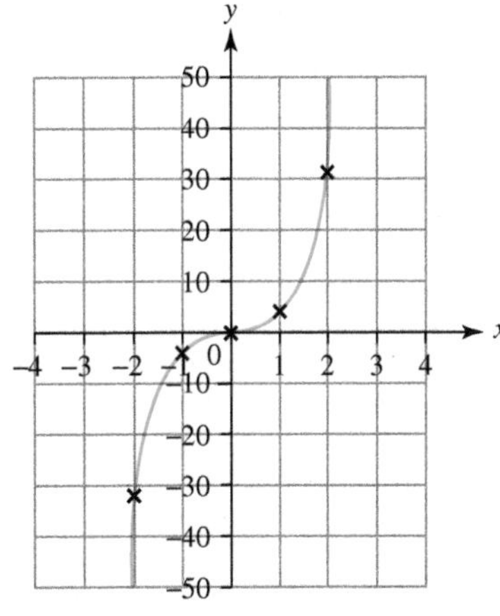

4 $y = x^3 - 2x^2 - 16$

Bridging Exercise 4A

1 a i $x > -1$ **ii** $x < -1$

b $\left(-1, -\frac{25}{4}\right)$

c i 1.75 **ii** 1.55

2 a i $x < -4, x > 4$ **ii** $-4 < x < 4$

b (−4, 26), (4, −6)

c i −5.875 **ii** −5.96875

3 a 0

b −5.9, −3.9, 4.1

c

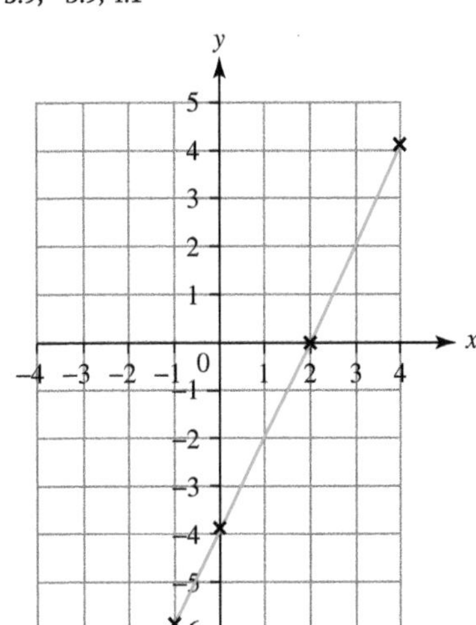

4 **i** C **ii** D **iii** B **iv** A

5 $y=6x^2-8x-18$

6 $y=x^3+2x^2-3x+3$

7 $y=3x^3-7x^2+4x+64$

8 6

9 2

Exercise 4.1A Fluency and skills

1 **a** 6 **b** 8 **c** 8 **d** 10
e 4 **f** 15 **g** 3 **h** 36

2 **a** 2 **b** 12 **c** 27 **d** 4
e 6 **f** 4 **g** 6 **h** 12

3 **a** 24 **b** 18 **c** 4 **d** 2
e –8 **f** 3 **g** 13

4 **a** $4x$ **b** $8x$ **c** $12x$ **d** x
e $2x$ **f** 2 **g** x^2 **h** $6x^2$

5 **a** **i** $1+2x$ **ii** $1+2x$ **iii** $1+2x$
iv $2x+2$ **v** $2x-3$ **vi** $4x+5$
b **i** 0 **ii** 0 **iii** 0 **iv** 0
c **i** 1 **ii** –1 **iii** 2 **iv** –3

6 **a** $6x^2$ **b** $12x^3-4x$ **c** $10x^4$
d $-3x^2$ **e** $2x-6x^2$ **f** $1-2x+3x^2$

Exercise 4.1B Reasoning and problem-solving

1 **a** $x=2$ **b** $x=10$ **c** $x=0.05$
d (1, 5) **e** (1, 5) **f** (4, 80)

2 **a** (1, 1) and (–1, –1) **b** (2, 8) and (–2, –8)
c (3, 27) and (–3, –27) **d** (0.1, 0.001) and (–0.1, –0.001)
e $\left(\frac{1}{3}, \frac{1}{27}\right), \left(-\frac{1}{3}, -\frac{1}{27}\right)$
f $(0.7, 0.343), (-0.7, -0.343)$

3 **a** $x=1$ **b** $x=3$
c $x=2.5$ **d** $x=\frac{b}{2a}$

4 **a** $x=\frac{5}{3}, -1$ **b** $x=3, -1$ **c** $x=\frac{1}{3}, 1$ **d** $x=\frac{2}{3}, 2$
e $x=2$ **f** $x=2.5$ **g** $x=3$ **h** $x=-2, 1$

5 **a** nx^{n-1} **b** anx^{n-1} **c** $\frac{dy}{dx}=6x^2; \frac{dy}{dx}=6x$
d No. No matter how many special cases you show work, you have not shown it works for all n

6 **a**
$$f(x)=\frac{1}{x}\Rightarrow f'\left(\tfrac{1}{2}\right)=\lim_{h\to 0}\frac{\frac{1}{\frac{1}{2}+h}-\frac{1}{\frac{1}{2}}}{h}$$
$$=\lim_{h\to 0}\frac{\frac{1}{2}-\left(\frac{1}{2}+h\right)}{h\frac{1}{2}\left(\frac{1}{2}+h\right)}=\lim_{h\to 0}\frac{-h}{h\frac{1}{2}\left(\frac{1}{2}+h\right)}$$
$$=\lim_{h\to 0}\frac{-1}{\left(\frac{1}{2}\right)^2+\frac{1}{2}h}=-\frac{1}{\left(\frac{1}{2}\right)^2}=-4$$

b
$$f(x)=\frac{1}{x}\Rightarrow f'(x)=\lim_{h\to 0}\frac{\frac{1}{x+h}-\frac{1}{x}}{h}$$
$$=\lim_{h\to 0}\frac{x-(x+h)}{hx(x+h)}=\lim_{h\to 0}\frac{-h}{hx(x+h)}$$
$$=\lim_{h\to 0}\frac{-1}{x^2+xh}=-\frac{1}{x^2}$$

Exercise 4.2A Fluency and skills

1 **a** $21x^6$ **b** $24x^3$ **c** 5
d 0 **e** 0 **f** 0
g $-10x^4$ **h** $-7x^6$ **i** $-2x^{-2}$
j $3x^{-4}$ **k** $\frac{1}{2}x^{-\frac{1}{2}}$ **l** $\frac{2}{3}x^{-\frac{1}{3}}$
m $\frac{25}{3}x^{\frac{2}{3}}$ **n** $\frac{1}{4}x^{-\frac{5}{4}}$ **o** 0

2 **a** $\frac{1}{2}x^{-\frac{1}{2}}$ **b** $\frac{1}{3}x^{-\frac{2}{3}}$
c $\frac{1}{5}x^{-\frac{4}{5}}$ **d** $\frac{3}{2}x^{\frac{1}{2}}$
e $\frac{2}{3}x^{-\frac{1}{3}}$ **f** $\frac{\sqrt[3]{3}}{3}x^{-\frac{2}{3}}$
g $-x^{-2}=-\frac{1}{x^2}$ **h** $-6x^{-3}=-\frac{6}{x^3}$
i $24x^{-5}=\frac{24}{x^5}$ **j** $-\frac{1}{2}x^{-\frac{3}{2}}$
k $-\frac{9}{2}x^{-\frac{5}{2}}$ **l** 0
m 0

3 **a** $2x+2$ **b** $-2-10x$
c $3x^2+4x-3$ **d** $2x-4x^3$
e $1+\frac{1}{x^2}$ **f** $1-\frac{4}{x^3}$
g $3x^2+\frac{3}{x^4}-\frac{3}{x^2}$ **h** $-\frac{10}{x^2}-\frac{1}{10}$
i $\frac{1}{2\sqrt{x}}-\frac{1}{2\sqrt{x^3}}$ **j** $5x^4-\frac{5}{3\sqrt[3]{x^4}}$
k $-x^{-\frac{3}{2}}+\frac{10}{3}x^{-\frac{5}{3}}$ **l** $4x^{-3}-\frac{3}{2}x^{\frac{1}{2}}$

4 **a** **i** $6x+4$ **ii** –8 **b** **i** $6x^2-10x$ **ii** –4
c **i** $10-\frac{8}{x^2}$ **ii** 8 **d** **i** $-\frac{8}{x^2}-\frac{8}{x^3}$ **ii** –3
e **i** $y=x^2-x$ **ii** 7 **iii** 7

5 **a** $\frac{dy}{dx}=4x-5, -1$ **b** $\frac{dy}{dx}=-5-\frac{10}{x^2}, -\frac{30}{4}$
c $\frac{dy}{dx}=4x+1, -11$ **d** $\frac{dy}{dx}=\frac{1}{2}x^{-\frac{1}{2}}-x^{-\frac{3}{2}}, \frac{1}{8}$

6 **a** $y=x^2$ **b** $y=x^2+3x-8$
c $y=2x+3$ **d** $y=x^2-2x+3$
e $y=2x-15$

7 **a** True **b** True **c** True
d False **e** True **f** False

Exercise 4.2B Reasoning and problem-solving

1 **a** $2x+1$ **b** $2x$ **c** $2x-9x^2$
d $\frac{3}{2}x^{\frac{1}{2}}+\frac{3}{2}x^{-\frac{1}{2}}$ **e** $18x^2+6x-9$ **f** $\frac{3}{2}x^{\frac{1}{2}}+1$
g $-4x^{-2}$ **h** $-x^{-2}-2$ **i** $1-x^{-2}$

2 **a** 1 **b** $1+\frac{1}{2}x^{-\frac{1}{2}}$ **c** $-\frac{1}{2}x^{-\frac{3}{2}}$
d $-x^{-2}+x^{-\frac{3}{2}}$ **e** $\frac{1}{2}x^{-\frac{1}{2}}-\frac{3}{2}x^{\frac{1}{2}}$ **f** $-x^{-2}-x^{-\frac{3}{2}}$
g $1+x^{-2}$ **h** $-x^{-2}-2x^{-3}$ **i** $-\frac{1}{3}x^{-\frac{4}{3}}$

3 **a** $2x-2$ **b** $2x-9$ **c** $-2-2x$
d $4x+3$ **e** $3x^2-1$ **f** $\frac{3}{2}x^{\frac{1}{2}}+1+\frac{1}{2}x^{-\frac{1}{2}}$

4 **a** $2x-8$
b $x=1$ or $x=7$
c (4, –9)
d **i** $x<4$ **ii** $x>4$
e **i** falling **ii** rising

5 **a** $10-\frac{40}{x^2}$
b **i** $x=2$ **ii** $y=40$

c i $0.5 \le x < 2$ ii $2 < x \le 5.5$

6 $f(x) = (x+1)^4 = x^4 + 4x^3 + 6x^2 + 4x + 1 \Rightarrow$
$f'(x) = 4x^3 + 12x^2 + 12x + 4$
$\Rightarrow f'(1) = 32$

7 a $2x+1$
b 11
c $x = -\frac{1}{2}, y = -2\frac{1}{4}$
d i $x = \frac{1}{2}, y = -1\frac{1}{4}$
ii $k = 2\frac{1}{4}$

Exercise 4.3A Fluency and skills

1 a 18 b 0 c 5
d $\frac{8}{9}$ e 24 f 0
g $-\frac{27}{2}$ h 9 i 0 j $-\frac{3}{4}$
k 0 l -3 m 2 n -3
o 0 p -2

2 a 18 b $\frac{5}{2}$ c $\frac{3}{4}$ d 46
e 1 f 2 g -2 h $\frac{125}{8}$
i 0 j 16 k 1280 l $-\frac{1}{3}$

3 a i $x = -2$ ii $x > -2$ iii $x < -2$
b i $x = -1.25$ ii $x < -1.25$ iii $x > -1.25$
c i $x = 0$ ii $x > 0$ iii $x < 0$
d i $x = -2, 2$ ii $x < -2$ or $x > 2$ iii $-2 < x < 2$
e i $x = -7$ or $x = 3$
ii $-7 < x < 3$
iii $x < -7$ or $x > 3$
f i $x = 1$ and $x = 2$ ii $x < 1$ or $x > 2$ iii $1 < x < 2$
g i no stationary points
ii increasing for all x in the domain
iii never decreasing
h i no stationary points
ii increasing for all x in the domain
iii never decreasing
i i $x = 0$ ii $x < 0$ or $x > 0$ iii never decreasing
j i $x = 0$ ii $x > 0$ iii $x < 0$

4 a $y = -x^2$ b $y = 5 - 3x$
c $y = x^2 - 2$ d $y = 1 + 10x - x^3$

Exercise 4.3B Reasoning and problem-solving

1 £2.50 per kilometre
2 a -7.5 b It is losing weight.
3 a $4\pi r^2$ b 36π c 24π
4 a i 2 ii -6 iii -12
b $x = -1$
5 a $(3 - 9.8t)\,\text{m s}^{-1}$
b falling at $46\,\text{m s}^{-1}$
c falling at $95\,\text{m s}^{-1}$
d $-9.8\,\text{m s}^{-2}$ or $9.8\,\text{m s}^{-2}$ towards the ground
6 a $(10 - 3.24t)\,\text{m s}^{-1}$
b 3.09 seconds (to 3 sf)
c $3.24\,\text{m s}^{-2}$ towards lunar surface
7 a $360 - 12t$
b i $240\,\text{ml s}^{-1}$ ii $120\,\text{ml s}^{-1}$
c 30 secs
d 5400 ml
8 a i $x^2 + 10x + 30$
ii 30
iii $\frac{dy}{dx} = (x+5)^2 - 5^2 + 30 = (x+5)^2 + 5$, which is positive for all x. So function increasing for all x
b i $x^2 - 6x + 12$
ii 12
iii $\frac{dy}{dx} = (x-3)^2 - (-3)^2 + 12 = (x-3)^2 + 3$, which is positive for all x. So function increasing for all x
c i $x^2 - 5x + 8$
ii 8
iii $\frac{dy}{dx} = \left(x - \frac{5}{2}\right)^2 - \left(-\frac{5}{2}\right)^2 + 8 = \left(x - \frac{5}{2}\right)^2 + \frac{7}{4}$, which is positive for all x. So function increasing for all x

9 $\frac{dy}{dx} = -(x^2 - 8x + 17) = -((x-4)^2 - 4^2 + 17) = -(x-4)^2 - 1$, which is negative for all x. So function decreasing for all x

10 a i $4x^3 - 2x^{-3}$ ii $12x^2 + 6x^{-4} = 12x^2 + \frac{6}{x^4}$
b The second derivative is positive for all $x \ge 1$ so the derivative is an increasing function for all x in the domain.

Exercise 4.4A Fluency and skills

1 a $y - 4 = 7(x-1)$ b $y - 1 = 3(x+2)$
c $y - 5 = 12(x-1)$ d $y - 4 = -12(x-2)$
e $y - 1 = -\frac{1}{3}(x-3)$ f $y - 1 = -\frac{1}{2}(x-4)$
g $y - 3 = \frac{1}{6}(x-9)$ h $y - 5 = -\frac{1}{10}(x-25)$
i $y - 3 = \frac{1}{8}(x-4)$ j $y - 2 = -\frac{3}{2}(x-1)$

2 a $y - 1 = -\frac{1}{6}(x-2)$ b $y - 10 = -(x+3)$
c $y + 5 = \frac{1}{12}(x-2)$ d $y - 4 = -\frac{1}{12}(x-1)$
e $y - 3 = \frac{2}{3}(x-2)$ f $y - 6 = -\frac{4}{5}(x-4)$
g $y - \frac{3}{4} = 8(x-4)$ h $y - 2 = \frac{2}{7}(x-1)$

3 a (3, 27) b 18 c $y - 27 = 18(x-3)$
4 a $k = 3$ b 5 c $y - 3 = 5(x-1)$
5 a (1, −1) b 9 c $-\frac{1}{9}$ d $y + 1 = -\frac{1}{9}(x-1)$
6 a (2, 5) b $\frac{3}{2}$
c $-\frac{2}{3}$ d $y - 5 = -\frac{2}{3}(x-2)$
7 a i 0 ii $y = -4$
b $x = -3$
c i tangent: $y = -25$, normal: $x = -1$
ii tangent: $y = -25$, normal: $x = -5$
iii tangent: $y = 25$, normal: $x = 2$

Exercise 4.4B Reasoning and problem-solving

1 a $x = 2$ b $k = 5$
2 $\left(\frac{3}{2}, \frac{7}{2}\right)$
3 a $y = x - 2$
b (3, 1)
c $m_{\tan B}$ at $(3,1) = 2(3) - 3 = 3 \Rightarrow m_{\text{norm } B} = -\frac{1}{3}$;

$m_{\text{norm }B} = -\frac{1}{3} \neq m_{\text{norm }A}$

4 a Consider the line as tangent to first function and normal to second. In 1st function: $m_{\tan} = \frac{dy}{dx} = 2x$; in second function $m_{\text{norm}} = -\frac{1}{m_{\tan}} = -\frac{1}{\frac{dy}{dx}} = -\frac{1}{-x^{-2}} = x^2$

b $\left(2, 4\frac{1}{2}\right)$ **c** $y - 4\frac{1}{2} = 4(x-2)$

5 a $y+1=1(x-3)$ **b** $y+1=-1(x-3)$ **c** 9 units²

6 a $y-1=2(x-0)$ or $y=2x+1$

b (–1, –1) **c** $\left(-\frac{2}{3}, 0\right)$ **d** $\frac{1}{3}$ units²

7 a $y-3=5(x-2)$ **b** $y-3=-\frac{1}{5}(x-2)$

8 a $p=2, q=9$

b $y-9=12(x-2)$

c $y-9=-\frac{1}{12}(x-2)$

9 a (3, 9) **b** $y-9=-(x-3)$

10 a i $(x+3)^2-10$

ii (–3, –10)

iii tangent: $y=-10$; normal: $x=-3$

b i $(x-5)^2-20$

ii (5, –20)

iii tangent: $y=-20$; normal: $x=5$

c i $\left(x-\frac{3}{2}\right)^2-\frac{37}{4}$ **ii** $\left(\frac{3}{2}, -\frac{37}{4}\right)$

iii tangent: $y=-\frac{37}{4}$; normal: $x=\frac{3}{2}$

d i $20-(x+4)^2$

ii (–4, 20)

iii tangent: $y=20$; normal: $x=-4$

e i $(x+0)^2+4$

ii (0, 4)

iii tangent: $y=4$; normal: $x=0$

f i $2-(x+1)^2$

ii (–1, 2)

iii tangent: $y=2$; normal: $x=-1$

iv In each case at a turning point the tangent was horizontal (gradient = 0) and the normal was vertical (gradient undefined). This might be used to test if a point is a turning point.

11 a i $y-10\sqrt{p}=-\frac{\sqrt{p}}{5}(x-p)$ **ii** $y+10\sqrt{p}=\frac{\sqrt{p}}{5}(x-p)$

b Normal at A cuts x-axis when $y=0$:

$-10\sqrt{p}=-\frac{\sqrt{p}}{5}(x-p) \Rightarrow x=50+p$

Normal at B cuts x-axis when $y=0$:

$10\sqrt{p}=\frac{\sqrt{p}}{5}(x-p) \Rightarrow x=50+p$

Thus both cut x-axis at $(50+p, 0)$

12 a $p=0, q=4$

b 2

c $y-4=-\frac{1}{2}(x-0)$ or $y=4-\frac{1}{2}x$

d i (2, 0) and (–1, 0)

ii $y=\frac{1}{6}(x-2)$ and $y=-\frac{1}{6}(x+1)$

e $x=0$

13 a $y-12=-\frac{2}{5}(x-30)$

b $y-6=10(x-60)$

c i (6, 60) **ii** $y-60=\frac{1}{10}(x-6)$

d i (3, 120) **ii** $k=240$

Exercise 4.5A Fluency and skills

1 a (–2, –9) min **b** (–2, –36) min

c (3, –16) min **d** (0, 1) max

e $\left(-\frac{7}{4}, -\frac{1}{8}\right)$ min **f** $\left(-\frac{1}{2}, 20\frac{1}{4}\right)$ max

g $\left(\frac{1}{12}, -1\frac{1}{24}\right)$ min **h** $\left(-\frac{13}{14}, \frac{225}{28}\right)$ max

i $\left(-\frac{1}{4}, 6\frac{1}{8}\right)$ max **j** (2, 4) min

k (3, 12) min, (–3, –12) max **l** (–1, 12) min, (1, 8) max

m (25, –25) min **n** (4, –48) min

2 a (0, 0), (4, –32) **b** (0, 0) max, (4, –32) min

3 (0, 1) min, (–10, 1001) max

4 a $6x^2-18x+12$

b (1, 12) max, (2, 11) min

c i $3x^2-6x-24$; (–2, 29) max, (4, –79) min

ii $3x^2+6x-45$; (–5, 130) max, (3, –126) min

iii $-6x^2-42x-36$; (–1, 18) max, (–6, –107) min

iv $6x^2-22x-8$; $\left(-\frac{1}{3}, \frac{91}{27}\right)$ max, (4, –78) min

v $-6x^2+10x-4$; (1, 2) max, $\left(\frac{2}{3}, \frac{53}{27}\right)$ min

vi $-12x^2-4x+1$; $\left(\frac{1}{6}, \frac{275}{54}\right)$ max, $\left(-\frac{1}{2}, \frac{9}{2}\right)$ min

5 a $f(x)=3x^4+8x^3-6x^2-24x-1$

$\Rightarrow \frac{dy}{dx}=12x^3+24x^2-12x-24$

At turning points

$\frac{dy}{dx}=0 \Rightarrow 12x^3+24x^2-12x-24$

$=0 \Rightarrow x^3+2x^2-x-2=0$

Checking: at $x=1$, $\frac{dy}{dx}=1^3+2.1^2-1-2=0$

at $x=-1$, $\frac{dy}{dx}=(-1)^3+2.(-1)^2-(-1)-2=0$

at $x=-2$, $\frac{dy}{dx}=(-2)^3+2.(-2)^2-(-2)-2=0$

b $f(1)=(1,-20)$, $f(-1)=(-1, 12)$, $f(-2)=(-2, 7)$

c $(1,-20)$ is a min, $(-1, 12)$ is a max, $(-2, 7)$ is a min

6 a $f'(x)=12x^3-12x^2-72x$. At stationary points $f'(x)=0$

$12x^3-12x^2-72x=0 \Rightarrow x(x^2-x-6)=0$

$\Rightarrow x(x-3)(x+2)=0$

Hence result.

b minima at $x=-2$ and $x=3$; maximum at $x=0$

7 a $f(x)=8x+\frac{72}{x} \Rightarrow f'(x)=8-\frac{72}{x^2}$;

At stationary point $f'(x)=0 \Rightarrow 8-\frac{72}{x^2}=0 \Rightarrow x=\pm 3$

but in the definition $x>0$. So only $x=3$ is a solution.

b A minimum at (3, 48)

c A maximum.

Exercise 4.5B Reasoning and problem-solving

1 Both numbers would be 500
2 Both numbers would be 60
3 $x=8, y=4$
4 a $y=150-3x$ b $V=300x^2-6x^3$ c $x=33\frac{1}{3}, y=50$
5 a $x=4, y=8$ b 20 m
6 a $6x-5x^2$ b $\left(\frac{3}{5},3\right)$
7 2 cm
8 a $h=\frac{1000}{x^2}$ b $2x^2+\frac{4000}{x}$ c $x=10$
9 a $h=\frac{440}{\pi x^2}$ b $x=4.1$ (to 1 dp) c $x=5.2$ (to 1 dp)

Exercise 4.6A Fluency and skills

1 a $10x+c$ b $\frac{3}{2}x^2+c$
c $2x^3+c$ d $3x^4+c$
e $5x^5+c$ f $\frac{x^7}{7}+c$
g $\frac{3x^2}{2}+x+c$ h $5x-2x^2+c$
i x^3+3x^2+2x+c j $4x^3+6x+c$
k $3x-2x^2-2x^3+c$ l $\frac{x^4}{4}+\frac{x^3}{3}+\frac{x^2}{2}+x+c$
m $3x-\frac{x^2}{2}-6x^4+c$ n $\frac{1}{2}x^4+2x^2+x+c$
o $\frac{1}{4}x^4-3x^{-1}+c$ p $-x^{-2}+x^{-1}+c$
q $\frac{2}{3}x^{\frac{3}{2}}-\frac{1}{4}x^4+x+c$ r $\frac{2}{3}x^{\frac{3}{2}}-2x^{\frac{1}{2}}+\frac{1}{2}x^2+6x+c$
s $\frac{9}{4}x^{\frac{4}{3}}-\frac{3}{5}x^{\frac{5}{3}}+c$ t $\frac{4}{5}x^{\frac{5}{4}}-\frac{4}{7}x^{\frac{7}{4}}+2x^{\frac{1}{2}}+c$
u $x+x^{-1}-3x^{\frac{1}{3}}+c$ v $x^{\frac{1}{3}}-\frac{1}{5}x^5+4x+c$
w $\frac{1}{3}\pi x^4-\frac{1}{4}\pi x^{\frac{4}{3}}+\pi x^2+c$ x $\frac{1}{3}x^3-\frac{1}{2}x^2-x+x^{-1}+c$

2 a $x+c$ b $\frac{1}{2}x^2+c$
c $3x^2+7x+c$ d $3x-x^2+c$
e $\frac{1}{3}x^3+\frac{1}{2}x^2+x+c$ f $x-2x^2-x^3+c$
g x^4+x^2-7x+c h $2x+3x^3-3x^4+c$
i $\frac{2}{3}x^{\frac{3}{2}}+c$ j $\frac{3}{4}x^{\frac{4}{3}}+c$
k $2x^{\frac{1}{2}}+c$ l $\frac{3}{5}x^{\frac{5}{3}}+c$

3 a $\pi x+c$ b $3\pi x+\frac{x^2}{2}+c$
c $\frac{x^3}{6}+c$ d $\frac{x^3}{3}+3x^2+c$
e $\frac{4x^3}{3}+2x^2-28x+c$ f $\frac{x^3}{3}-x^{-1}+c$
g $\frac{2x^{\frac{3}{2}}}{3}-4x^{\frac{1}{2}}+c$ h $4x^2+3x^{-1}+c$
i $-\frac{x^{-2}}{2}+\frac{x^{-3}}{3}+c$ j $-x^{-1}-\frac{x^2}{2}+\frac{x^{-2}}{2}+c$
k $\frac{1}{2}x^2+2x^{\frac{1}{2}}+c$ l $\frac{1}{3}x^3-6x^{\frac{1}{2}}+x+c$
m $-1x^{-1}-\frac{3}{2}x^{-2}+c$ n $\frac{1}{6}x^3+\frac{\sqrt{3}}{4}x^2+c$
o $x^3+2x^{-\frac{1}{2}}+c$ p $\frac{1}{2}x^2-x+c$
q $\frac{2}{3}x^{\frac{3}{2}}+c$ r $\frac{2}{3}x^{\frac{3}{2}}+2x^{\frac{1}{2}}+c$

4 a $f(x)=2x^3+3$ b $f(x)=3x^4-30$
c $f(x)=\frac{10}{3}x^{\frac{3}{2}}+10$

5 a $2x^2+3x-10$ b $10x+2$
c x^3+x^2+x+7 d $2x^2+2x^{\frac{3}{2}}+1$

Exercise 4.6B Reasoning and problem-solving

1 a $P(t)=7t-26$ b 9 c 6
2 a $f(-2)=16$ b $x=-3$ or $x=\frac{10}{3}$
3 a $y=r^{\frac{3}{2}}+c$ b $y=r^{\frac{3}{2}}$
c 1.84 (to 3 sf) d 9.53 (to 3 sf)
4 a $h=\frac{2}{3}t^4-2t^2+4$ b $3.54m$ (to 3 sf) c $t=\sqrt{6}$ days
5 a $\frac{dy}{dx}=6x^2+x-3$ b 13
6 19
7 a i $v(t)=t^2+2; s(t)=\frac{1}{3}t^3+2t+2$ ii $3\text{ cm s}^{-1}; \frac{13}{3}\text{ cm}$
b i $v(t)=6t+5; s(t)=3t^2+5t-2$; 11 cm s^{-1}; 6 m
ii $v(t)=\frac{1}{2}t^2+2; s(t)=\frac{1}{6}t^3+2t+1; \frac{5}{2}\text{ cm s}^{-1}; \frac{19}{6}\text{ cm}$
iii $v(t)=\frac{1}{3}t^3-1; s(t)=\frac{1}{12}t^4-t; -\frac{2}{3}\text{ cm s}^{-1}; -\frac{11}{12}\text{ cm}$
8 6 seconds
9 a 0.5 metres to the right of the origin. b 3 m s^{-2}
c 17.5 m d $3n+\frac{5}{2}$

Exercise 4.7A Fluency and skills

1 a 8 b 20 c $\frac{15}{2}$
d −96 e 35 f 6π
g $2\pi+2$ h 48 i 72
j $5\frac{1}{3}$ k $\frac{5}{4}$ l $\frac{23}{24}$
m $\frac{7}{4}$ n 6 o 156π
p 5 q −126 r 2π
2 a 16 b 18 c $15\frac{1}{6}$
d $\frac{1}{6}$ e $\frac{64}{3}$ f 36
g $6\frac{3}{4}$ h $6\frac{2}{3}$ i 378

Exercise 4.7B Reasoning and problem-solving

1 $-\sqrt{67}$
2 $1\frac{1}{3}$ square units
3 $57\frac{1}{6}$ square units
4 a $\frac{2}{3}x^{\frac{3}{2}}+c$ b $\frac{2}{3}$ square units
c $P:Q=7:19$ d $1\frac{2}{3}$ square units
The difference is a square of 1 square unit: 1 unit along the x-axis by 1 unit translation in the y-direction.
5 a $a=10$ b $a=5$ c $a=-3, 2$
6 27:5
7 7
8 In the domain $0\le x\le 360$, the area below the x-axis is equal to the area above the axis. Since each will be of opposite sign, their sum is zero.

9 75 m

10 20.08 km

Review exercise 4

1 a 6 **b** 3 **c** 31
d $-2x$ **e** $1-2x$ **f** $2x\pi$

2 a $3x^2+4x+3$ **b** $\frac{2}{\sqrt{x}}+1$ **c** $1-\frac{1}{x^2}-\frac{2}{x^3}$
d $\frac{1}{4}x^{-\frac{3}{4}}-\frac{1}{3}x^{-\frac{2}{3}}$ **e** $-x^{-2}-6x^{-3}$ **f** $-4x^{-2}-x^{-\frac{3}{2}}$
g $2x^{-3}-\frac{1}{3}x^{-\frac{4}{3}}$ **h** $1-3x^{-2}$ **i** $\frac{1}{2}x^{-\frac{1}{2}}$
j $\frac{1}{6}x^{-\frac{5}{6}}-\frac{1}{3}x^{-\frac{4}{3}}$

3 a 8 **b** $-\frac{1}{3}$ **c** 2 **d** $\frac{1}{6}$

4 a 56 y units per x unit
b 150π cm^3 per cm
c i $1\,\text{cm}\,\text{s}^{-1}$ **ii** $2\,\text{cm}\,\text{s}^{-2}$
d i $0.25\,\text{km}\,\text{m}^{-1}$ (to 2 sf); $0.18\,\text{km}\,\text{m}^{-1}$ (to 2 sf)
ii 20 m (to 2 sf)
iii $0.13\,\text{km}\,\text{m}^{-1}$ (to 2 sf)

5 a tangent: $y-8=6(x-1)$; normal $y-8=-\frac{1}{6}(x-1)$
b tangent: $y-5=\frac{7}{3}(x-3)$; normal $y-5=-\frac{3}{7}(x-3)$
c tangent: $y=9$; normal $x=2$
d Equation of tangent: $y-1000=20(x-25)\Rightarrow y=20x+500$
Equation of normal: $y-1000=-\frac{1}{20}(x-25)$

6 a $(-2, -9)$ min **b** $\left(\frac{2}{3},-\frac{16}{9}\right)$ min, $\left(-\frac{2}{3},\frac{16}{9}\right)$ max
c (0, 0) min **d** $(1, 2a)$ min, $(-1, -2a)$ max
e At the stationary point $x=-\frac{b}{2a}$ the turning point is a minimum.

7 a 2 **b** $6x+2$ **c** $\frac{2}{x^3}$
d $\frac{dy}{dx}=1+x+\frac{x^2}{2}+\frac{x^3}{6}+\frac{x^4}{24}+\frac{x^5}{120}$
$\Rightarrow\frac{d^2y}{dx^2}=1+x+\frac{x^2}{2}+\frac{x^3}{6}+\frac{x^4}{24}$

8 a $x-\frac{x^2}{2}-\frac{x^4}{4}+c$ **b** $-\frac{1}{x}-\frac{4x^3}{3}+c$
c $\frac{2x^{\frac{3}{2}}}{3}-2\sqrt{x}-\frac{3x^{\frac{4}{3}}}{4}+c$ **d** $\frac{2}{5}x^{\frac{5}{2}}-2x^{\frac{1}{2}}+c$

9 a $25\frac{1}{2}$ **b** 18π
c $-33\frac{1}{3}$ **d** $10\frac{2}{3}$

10 a $4\frac{1}{2}$ units^2 **b** $\frac{1}{6}$ units^2 **c** $60\frac{3}{4}$ units^2

Assessment 4

1 a D, $15x^2+6x^{-4}$ **b** B, $\frac{5x^4}{4}+x^{-2}+c$

2 A, 5.5

3 a $\frac{dr}{dt}=-\frac{1}{\sqrt{t}}$ **b** $-\frac{1}{5}\,\text{cm}\,\text{s}^{-1}$

4 a 7 **b** $x+7y-8=0$ or $y=-\frac{1}{7}x+\frac{8}{7}$

5 a x^4-2x **b** 12 **c** $60x-5y-108=0$

6 a $\frac{dy}{dx}=0\Rightarrow 18x^2-6x-12=0\Rightarrow x=-\frac{2}{3}$
b $(1, -4)$
c Minimum

7 $-4<x<\frac{2}{3}$

8 a $x^2+\frac{1}{2}x^6+c$ **b** $-\frac{1}{3}x^{-3}-x^4+c$ **c** $2x^{\frac{3}{2}}+2x^{\frac{1}{2}}+c$

9 a $4\sqrt{3}$ **b** -3.75

10 a $f(x)=\frac{4}{3}x^3+\frac{5}{2}x^2-x(+c)$
b $f(x)=-\frac{7}{2}x^{-2}-\frac{1}{2}x^2+\frac{2}{3}x^{\frac{3}{2}}(+c)$

11 $\frac{4}{3}$

12 a $\frac{dy}{dx}=\lim_{h\to 0}\left(\frac{3(x+h)^2-3x^2}{(x+h)-x}\right)$
$=\lim_{h\to 0}\left(\frac{3x^2+6xh+3h^2-3x^2}{h}\right)$
$=\lim_{h\to 0}(6x+3h)=6x$
b 30

13 a $\frac{dy}{dx}=\lim_{h\to 0}\left(\frac{\left[(x+h)^3-2(x+h)\right]-\left[x^3-2x\right]}{(x+h)-x}\right)$
$=\lim_{h\to 0}\left(\left[3x^2+3xh+h^2-2)\right]\right)=3x^2-2$
b 10

14 a $3x^2+12x+9$ **b** $\frac{1}{2}x^{-\frac{1}{2}}-x^{-\frac{3}{2}}$

15 $\frac{8}{3}$ l per min

16 $x-8y-162=0$ or $y=\frac{1}{8}x-\frac{81}{4}$

17 a $y-2=\frac{11}{4}(x-1)$ **b** $\frac{9}{88}$

18 $f(x)=1+6x+12x^2+8x^3\Rightarrow f'(x)=6+24x+24x^2$
$=6(1+2x)^2\geq 0$

19 $0<x<\frac{3}{5}$

20 $1<x<\frac{5}{3}$

21 a $6x^2-\frac{5}{2}x^{\frac{3}{2}}$ **b** $12x-\frac{15}{4}x^{\frac{1}{2}}$

22 $6+x^{-\frac{3}{2}}$

23 $\frac{dy}{dx}=32-4x^{-3}$; $\frac{dy}{dx}=0\Rightarrow x^{-3}=8\Rightarrow x=\frac{1}{2}\Rightarrow y=9$;
$\frac{d^2y}{dx^2}=12x^{-4}\Rightarrow$ at $x=\frac{1}{2}$, $\frac{d^2y}{dx^2}=192>0$ so a minimum

24 a $h=\frac{200}{\pi x^2}$
b $A=2\times\pi x^2+2\pi xh=2\pi x^2+\frac{400}{x}$
c 3.17 cm (to 3 sf)
d 189.3 cm^2
e $\frac{d^2A}{dx^2}=4\pi+\frac{800}{x^3}\Rightarrow$ at $x=3.17$, $\frac{d^2A}{dx^2}>0$ so a minimum

25 a $x^2l=3000\Rightarrow l=\frac{3000}{x^2}$; $A=4xl+x^2=4x\times\frac{3000}{x^2}+x^2$
$=\frac{12000}{x}+x^2$
b 18.2 cm (to 3 sf)
c 991 cm^2 (to 3 sf)
d $\frac{d^2A}{dx^2}=\frac{24000}{x^3}+2\Rightarrow$ at $x=18.2$, $\frac{d^2A}{dx^2}>0$ so a minimum

26 a $\frac{4}{3}x^3+6x^2+9x+c$

b $2x^{\frac{5}{2}}-\frac{2}{3}x^{\frac{3}{2}}(+c)$

c $2x^{\frac{1}{2}}+\frac{1}{3}x^{\frac{3}{2}}(+c)$

27 a $\frac{5}{4}$ b $\frac{206}{15}$

28 $\frac{343}{6}$

29 $\frac{67+8\sqrt{2}}{24}$ (or 3.26)

30 a $x^5+x^{-2}-1$ b $y-1=-\frac{1}{3}(x-1)$

31 a i $k+kx^{k-1}$ ii $-kx^{-k-1}$

b i $\frac{k}{2}x^2+\frac{x^{k+1}}{k+1}(+c)$ ii $\frac{x^{1-k}}{1-k}-kx(+c)$

32 $P\left(\frac{16}{11}, -\frac{16}{11}\right)$

33 6

34 $Q\left(-\frac{2}{3}, \frac{32}{9}\right)$

35 (0, 1) maximum, (–1, 0) and (1, 0) minimums

36 (0, 20) maximum, (1, 15) and (–2, –12) minimums

37 $x>\frac{3}{8}$ 38 $-3<x<0$ or $x>5$

39 270 cm² (to 3 sf) 40 2010 cm³ (to 3 sf)

41 a $y-2=5(x+1)$ b $\frac{7}{2}$

42 $\frac{4}{3}$ 43 $\frac{243}{2}$

44 $\frac{37}{12}$ 45 $\frac{36-4\sqrt{3}}{9}$ (or 3.23)

46 6 47 $\frac{9}{2}$

Chapter 5

Answers are given to 3 significant figures where appropriate.

Try it 5A

1 a $x=1.49$

b $x=-0.613$

2 $x=1.46$

3 $x=1.43, 0.569$

4 e.g.

x	–2	–1	0	1	2	3
y	0.0625	0.25	1	4	16	64

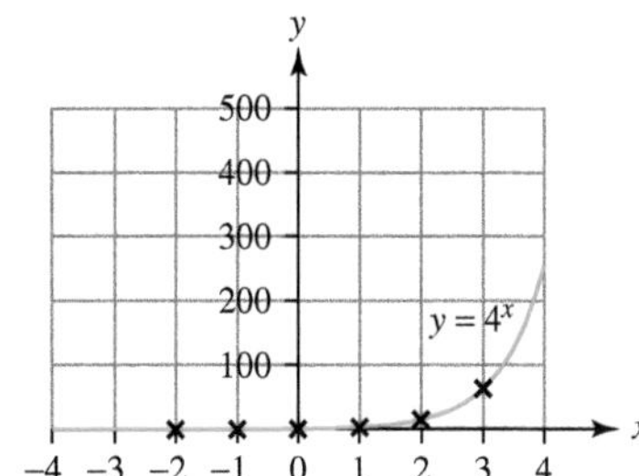

5 e.g.

x	$\frac{1}{9}$	$\frac{1}{3}$	1	2	3	6	9
y	–2	–1	0	0.631	1	1.63	2

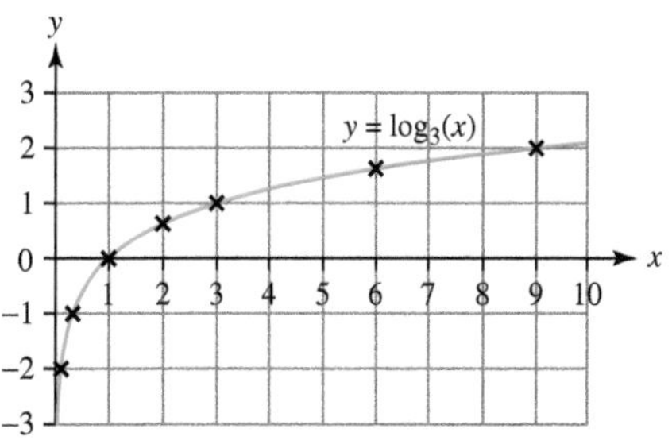

6 a

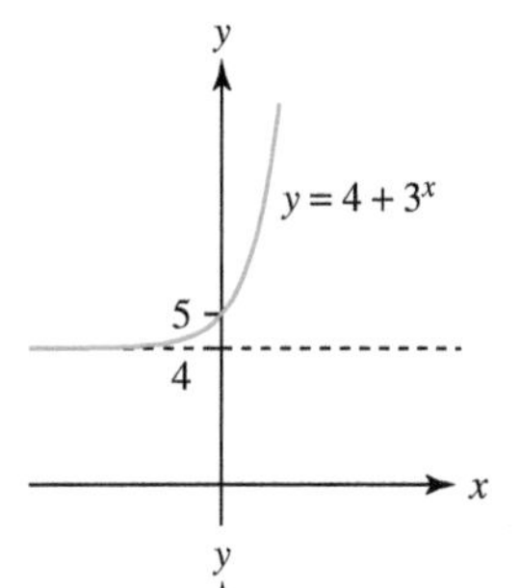

b

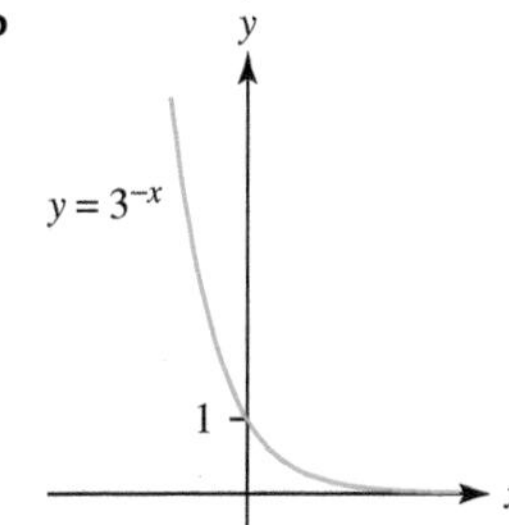

7 a

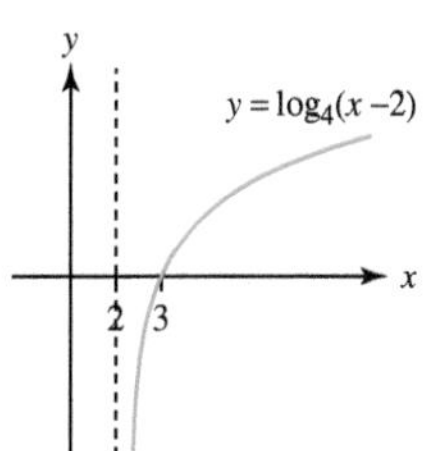

b

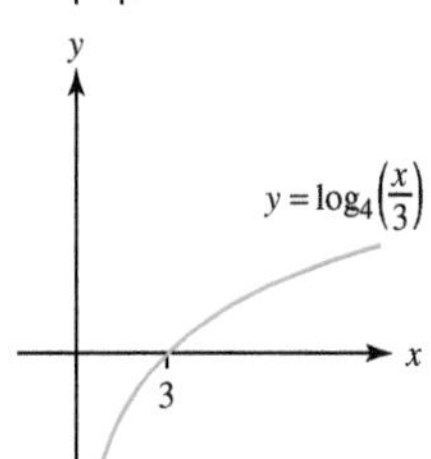

Bridging Exercise 5A

1 a $x=3$

b $x=4$

c $x=-2$

d $x=-5$

e $x=2.33$

f $x=4.86$

g $x=-4$

h $x=-3$

i $x=2.03$

j $x=-0.619$

k $x=4.05$

l $x=-0.898$

m $x=2.70$

n $x=-0.904$

o $x = 0.581$

p $x = -\frac{5}{2}$

2 a $x = 0.631$

b $x = 1, 1.58$

c $x = 1.26$

d $x = 2.32$

e $x = -0.387, 1.51$

f $x = 1, 1.89$

g $x = 3.58$

h $x = 1.07$

i $x = 1.26, 1.46$

3 a

x	−3	−2	−1	0	1	2	3
y	0.008	0.04	0.2	1	5	25	125

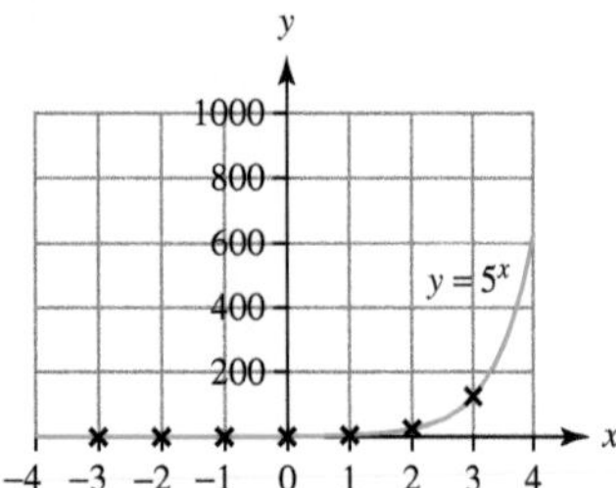

b

x	−3	−2	−1	0	1	2	3
y	8	4	2	1	0.5	0.25	0.125

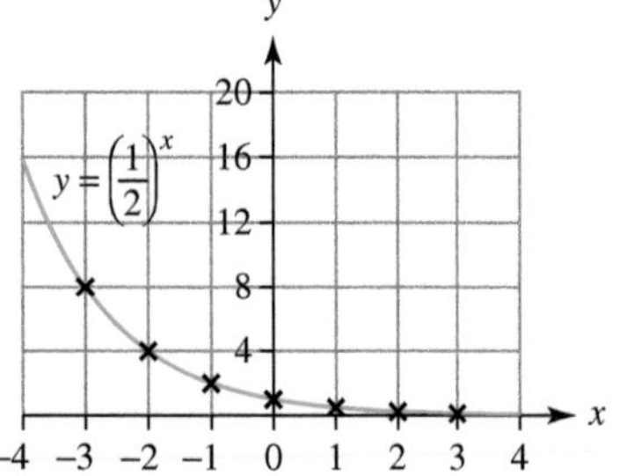

4 a

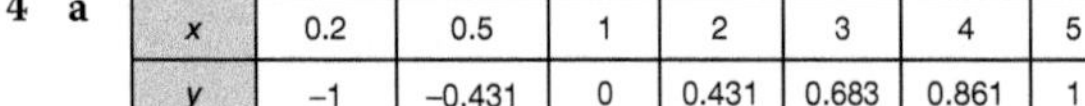

x	0.2	0.5	1	2	3	4	5
y	−1	−0.431	0	0.431	0.683	0.861	1

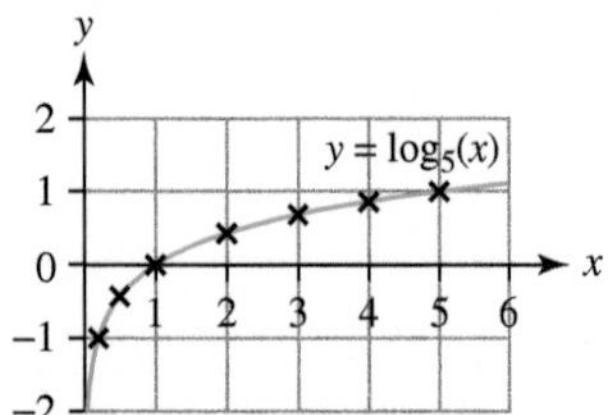

b

x	0.2	0.5	1	2	3	4	5
y	−1.16	−0.5	0	0.5	0.792	1	1.16

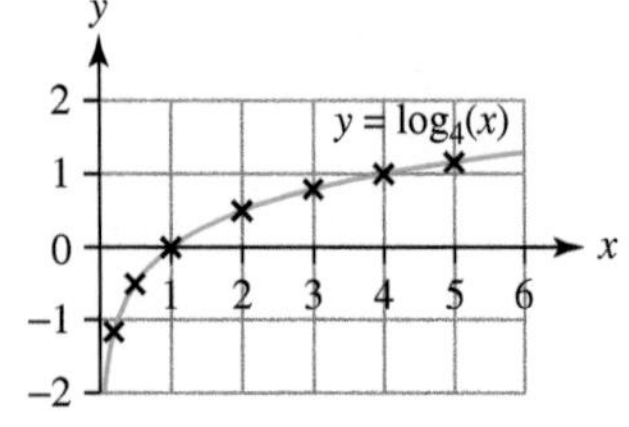

5 a

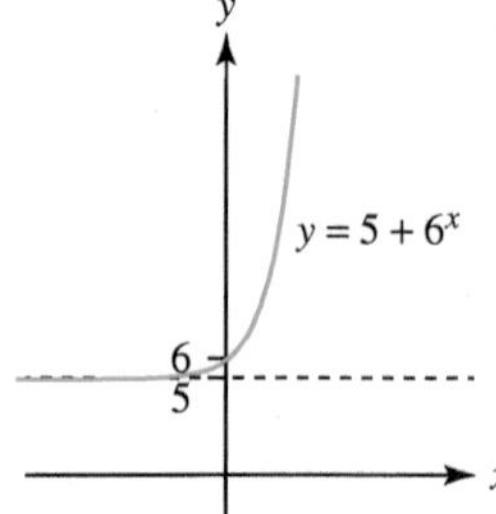

b

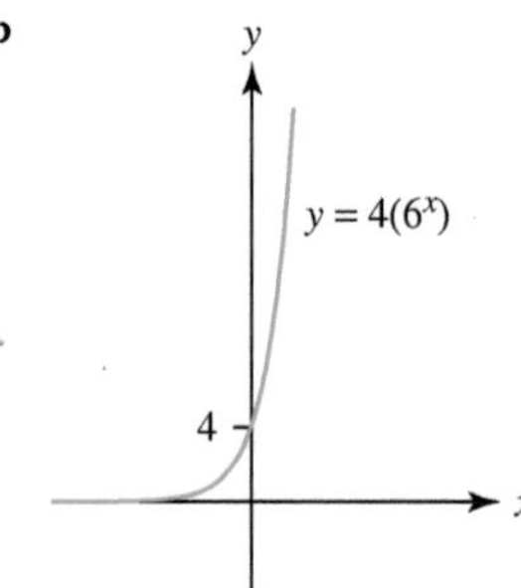

c

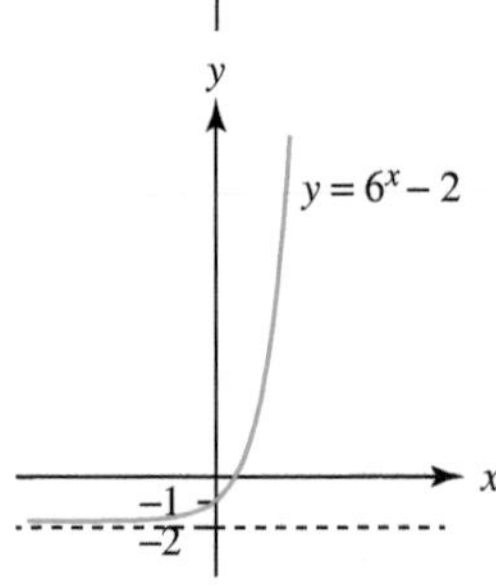

d

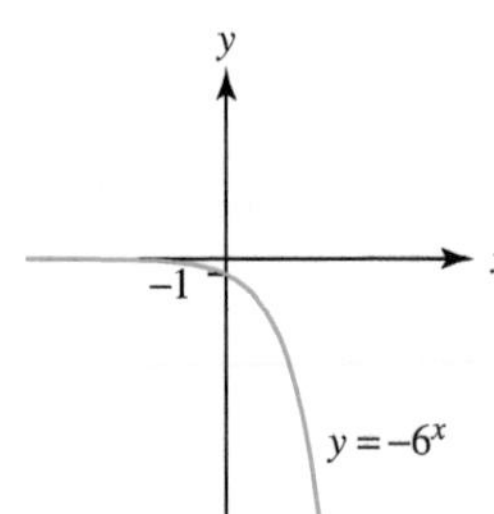

6 a

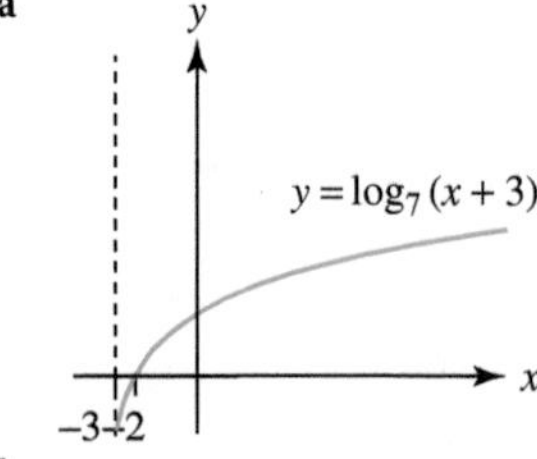

b

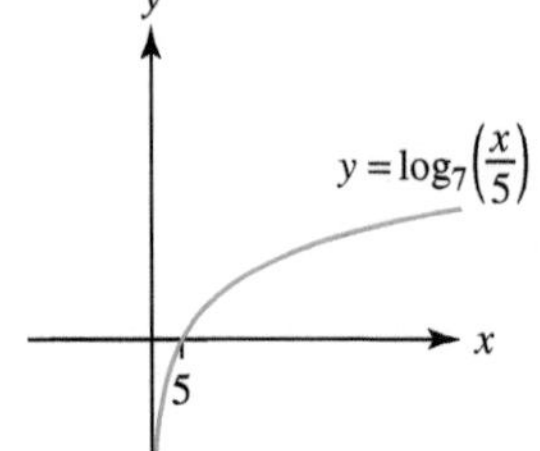

c

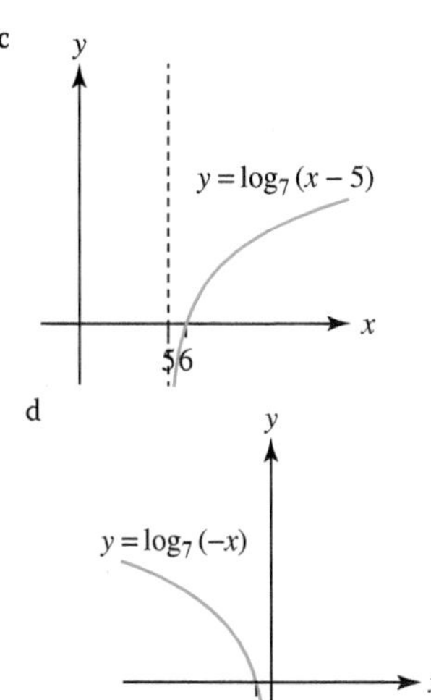

Exercise 5.1A Fluency and skills

1 a $\log_2 8 = 3$ b $\log_3 9 = 2$
c $\log_{10} 1000 = 3$ d $\log_{16} 2 = \frac{1}{4}$
e $\log_{10} 0.001 = -3$ f $\log_4 \frac{1}{4} = -1$
g $\log_2 \frac{1}{8} = -3$ h $\log_8 \frac{1}{4} = -\frac{2}{3}$

2 a $2^5 = 32$ b $2^4 = 16$
c $3^4 = 81$ d $5^0 = 1$
e $3^{-2} = \frac{1}{9}$ f $6^1 = 6$
g $16^{\frac{1}{4}} = 2$ h $2^{-6} = \frac{1}{64}$

3 a 2 b 4 c 2
d 1 e 4 f 1
g 4 h −1

4 a $\log 12$ b $\log 6$ c $\log 4$
d $\log 32$ e $\log 108$ f $\log 72$
g $\log 16$ h $\log 8$ i $\log 6$
j $\log 18$ k $\log(x^4y^2)$ l $\log(x^2y)$

5 a $2\log a + \log b$ b $\log a - \log b$
c $2\log a - 3\log b$ d $\log a + \frac{1}{2}\log b$
e $\frac{1}{2}\log a + \frac{1}{2}\log b - \log c$ f $\frac{1}{2}(\log a + \log b + \log c)$
g $\log a + \frac{1}{2}\log b - \frac{1}{2}\log c$

6 a $\log 3 + 2\log 2 \approx 1.079$ b $\log 2 + 2\log 3 \approx 1.255$
c $2\log 3 - \log 2 \approx 0.653$ d $3\log 3 - \log 2 \approx 1.130$
e $1 - \log 2 \approx 0.699$ f $0 - 3\log 2 \approx -0.903$

7 a $\frac{3\log 5}{2\log 5} = \frac{3}{2} = 1.5$ b $\frac{3\log 3}{5\log 3} = \frac{3}{5} = 0.6$

8 $\log 40 = 3 - 2\log 5 = 1.602060$

Exercise 5.1B Reasoning and problem-solving

1 a 2.32 b 2.10 c 0.861 d 1.81
e 4.99 f 2.66 g −3.17 h −2.41
i 0.461 j 0.892

2 a 520 b 7 c $\frac{2}{7}$ d 1

3 Adding gives $2\log x = 5$
$\log x = \frac{5}{2} = 2.5$
$x = 10^{2.5} = 316$ (3 sf)
$y = 10^{0.5} = 3.16$ (3 sf)

4 a 0, 1 b 1, 2 c 1, 1.58
d 2 e 1 f 2
g 1.32, 2 h 0.631, 1.26 i 0, 0.861
j 1.16

5 a 1 or 0.631 b (1, 5) and $\left(0.631, \frac{10}{3}\right)$

6 (2.32, 20)

7 a i $x > 8.38$ ii $x > 4.29$
b 20 c 7

8 a 4 b 14

9 a $\frac{x}{y} = \frac{a^p}{a^q} = a^{p-q}$
So $\log_a\left(\frac{x}{y}\right) = p - q$
$= \log_a x - \log_a y$
b $x^k = (a^p)^k = a^{pk}$
So $\log_a(x^k) = pk$
$= k \times \log_a x$

10 a Let $y = \log_a b$ so $b = a^y$
Take logs base b on both sides
$\log_b b = \log_b(a^y)$
$1 = y\log_b a$
$1 = \log_a b \times \log_b a$
So $\log_a b = \frac{1}{\log_b a}$
b i 25, $\frac{1}{25}$ ii 9, 81

11 a Let $x = a^y$
Take logs base b on both side
$\log_b x = \log_b(a^y)$
$\log_b x = y\log_b a$
$\frac{\log_b x}{\log_b a} = y$
So $\frac{\log_b x}{\log_b a} = \log_a x$
b $\log_2 12 = 3.32 \times \log_{10} 2$
c i $\frac{\log_{10} 37}{\log_{10} 6} = 2.015$
ii $\frac{\log_{10} 6}{\log_{10} 4} = 1.292$
iii $\frac{\log_{10} 25}{\log_{10} 3} = 2.930$

12 $xyz = \log_y z \times \log_z x \times \log_x y$
$= \log_y z \times \log_z x \times \frac{1}{\log_y x}$ (From Q10a)
$= \frac{\log_y z}{\log_y x} \times \log_z x$ (Rearranging)
$= \log_x z \times \frac{1}{\log_x z}$ (From Q11a and Q10a)
$= 1$

Exercise 5.2A Fluency and skills

1

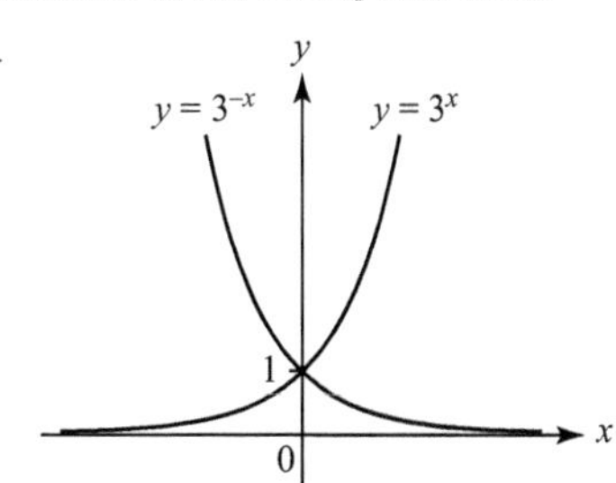

$\left(\frac{1}{3}\right)^x = (3^{-1})^x = 3^{-x}$

2 **a** **i** $y_1 = 2x$
$y_2 = x^2$
$y_3 = 2^x$
ii y_3 is exponential
iii 10, 25, 32

b **i** $y_1 = \frac{12}{x}$
$y_2 = \sqrt{x}$
$y_3 = 2^{-x}$ or $\left(\frac{1}{2}\right)^x$
ii y_3 is exponential
iii 2.4, 2.236, $\frac{1}{32}$

3 **a**

x	−2	−1	1	2	3	4
y	$\frac{1}{9}$	$\frac{1}{3}$	3	9	27	81

$y = 3^x$

b

x	−2	−1	1	2	3	4
y	25	5	$\frac{1}{5}$	$\frac{1}{25}$	$\frac{1}{125}$	$\frac{1}{625}$

$y = \left(\frac{1}{5}\right)^x$ or 5^{-x}

4 **a**

x	y	gradient at (x, y)
0	1	1
1	2.72	2.72
2	7.39	7.39
3	20.1	20.1
4	54.6	54.6

b

x	y	gradient at (x, y)
0	1	3
1	20.1	60.3
2	403	1210
3	8100	24300
4	163000	488000

5 WC, XD, YA, ZB

6 **a** **i** $e^2 = 7.39$ **ii** $-e^{-2} = -0.14$
b **i** $y = e^2x - e^2$ **ii** $y = \frac{3-x}{e^2}$

7 **a** $y = e^3(x-2)$ **b** $y = \sqrt{e}\left(x + \frac{1}{2}\right)$

8 **a** Reflection in y-axis **b** Reflection in x-axis

c Stretch (sf = 2) parallel to y-axis **d** Stretch (sf = $\frac{1}{2}$) parallel to x-axis

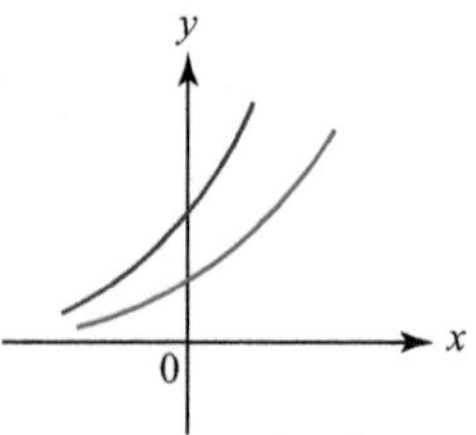

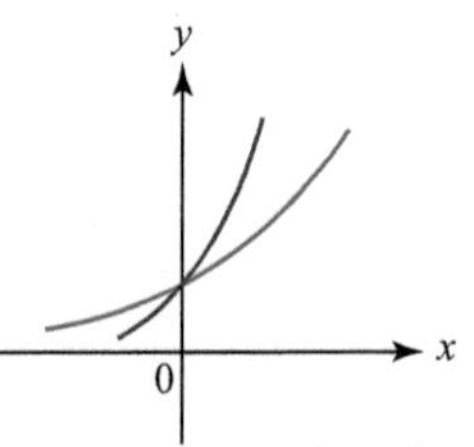

e Translation of $\begin{pmatrix} 0 \\ 1 \end{pmatrix}$ **f** Translation of $\begin{pmatrix} 0 \\ -1 \end{pmatrix}$

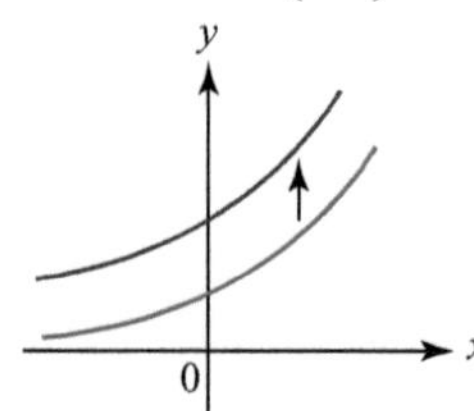

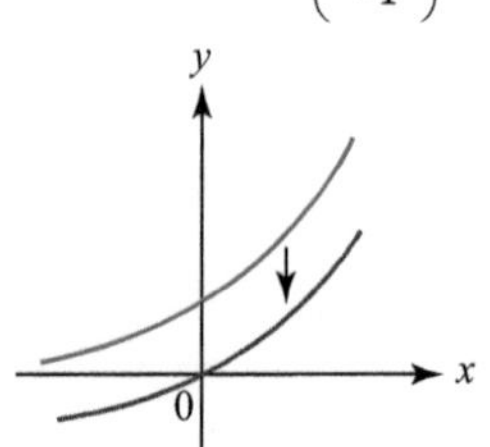

g Translation of $\begin{pmatrix} -1 \\ 0 \end{pmatrix}$ **h** Translation of $\begin{pmatrix} 1 \\ 0 \end{pmatrix}$

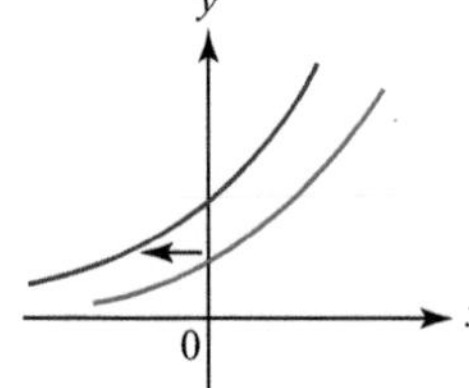

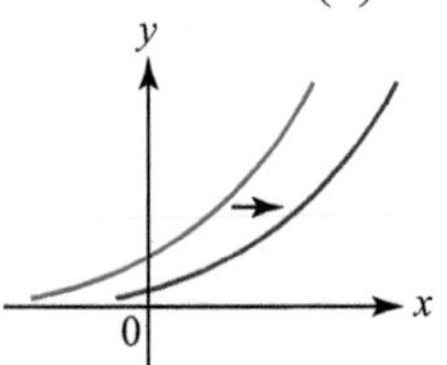

9 **a**

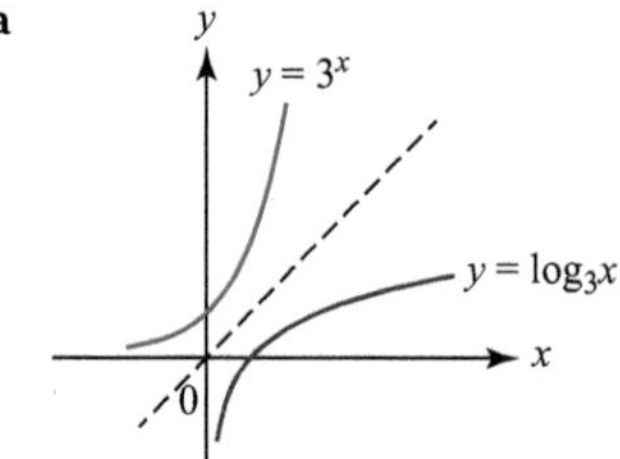

b (3, 1)
c (1.2, 3.74) and (3.74, 1.2)
$\log_3 3.74 = 1.2$
d **i** 5.20 **ii** 1.50 **iii** 1.73
iv 0.50 **v** 1.63 **vi** 1.77

10 **a** $p = e^3, q = 2, r = \ln 9$ **b** $a = e^{-\frac{1}{2}}, b = -3, c = -\ln 9$

Exercise 5.2B Reasoning and problem-solving

1 **a**

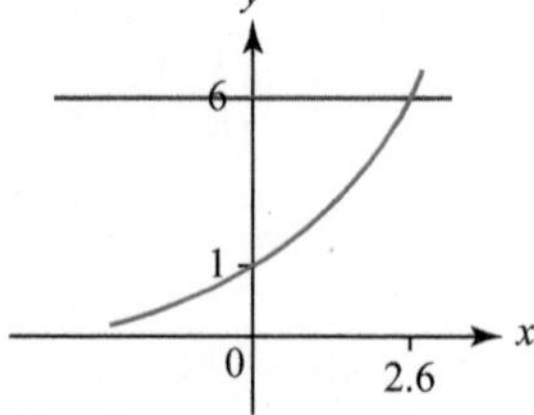

(2.6, 6) so approximately 2.6

b

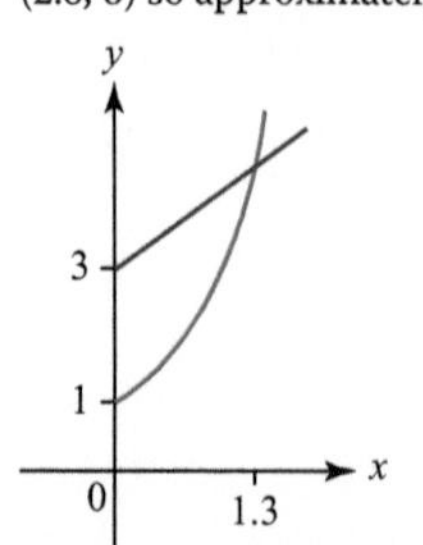

(1.3, 4.3) so approximately 1.3

2 a

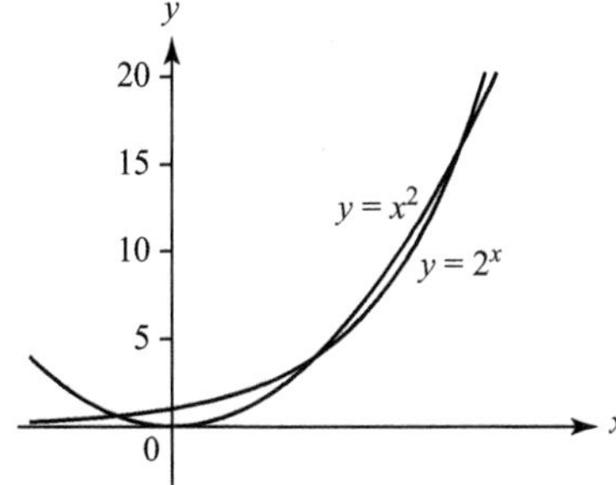

There are 3 points of intersection.

b $x = -0.75, 2, 4$

3 a $(3, 4^9)$ **b** $(1, 4)$ **c** $(4, 81)$ **d** $(-1, 2)$

4 7.25

5 $\frac{y - e^3}{x - 1} = 3e^3$

$y - e^3 = 3e^3x - 3e^3$

$y = 3e^3x - 2e^3$

When $y = 0$, $x = \frac{2}{3}$

Tangent passes through $(\frac{2}{3}, 0)$

6 $(6, 2e^2)$ **7** $y = 3.695x - 3.559$ **8** $(2e^4 + 1, 0)$

9 a $y = \log_a x$

b $y = 2^x$

$\ln y = \ln 2^x$

$\ln y = x \ln 2$

$x = \frac{\ln y}{\ln 2}$

$y = \frac{\ln x}{\ln 2}$

$y = \frac{1}{\ln 2} \times \ln x$

$\frac{1}{\ln 2} \approx 1.44$

So if $2^x = 17$ then $x = 1.44 \times \ln 17 = 4.08$

10 a

a	2.0	2.2	2.4	**2.6**	**2.8**	3
f(a)	0.693	0.788	0.876	**0.956**	**1.030**	1.099

b Between 2.6 and 2.8, because 1 is in between 0.956 and 1.030

c

a	2.6	**2.7**	**2.8**
f(a)	0.956	**0.993**	**1.030**

a	2.7	**2.71**	**2.72**	2.73	2.74	2.75
f(a)	0.993	**0.997**	**1.001**	1.004	1.008	1.012

a	**2.715**
f(a)	**0.9988**

e = 2.72 to 2 dp

d $\frac{a^{\delta x} - 1}{\delta x}$ is the gradient of a chord and only approximates the gradient of the tangent.

11 a

x	1	2	3	e.g. 12
$y = e^x$	2.7183	7.3891	20.086	162755
Gradient	2.7184	7.3894	20.087	162763

b As in Q10, the gradient is of a chord not a tangent.

Exercise 5.3A Fluency and skills

1 a

t	0	1	2	3	4
n	1	2	4	8	16

b 64

2 a $A = 10$

b

t	0	2	4	6	8
Actual units	10.0	9.1	8.2	7.4	6.7
n	10	9.03	8.15	7.35	6.63

Yes, predictions are reasonably accurate.

c Yes, about 44% of the initial insulin is still present.

3 a i 2 **ii** 109

b i 15 **ii** 6050

c i −5 **ii** −4.52

x and y grow over time. z decays over time.

4 a $0.049\,\text{cm h}^{-1}$ **b** $0.0074\,\text{kg h}^{-1}$

5 a i $\frac{1}{5}$°C **ii** 1.48 °C

b $0.639\,°\text{C min}^{-1}$ **c** $1.48\,°\text{C min}^{-1}$ **d** 80.7 °C

6 a 21.5 millions **b** 1.29 millions per year

c 1.08 millions per year

7 a $2\,\text{mg ml}^{-1}$ **b** $0.331\,\text{mg ml}^{-1}$

c i 0.9 (mg/ml)/h **ii** 0.149 (mg/ml)/h

8 a After 1 year, cost $= £P\left(1 + \frac{5}{100}\right)$

$= £P(1.05)$

After 2 years, cost $= £P(1.05) \times 1.05$

$= £P \times (1.05)^2$

b £166.15

9 £92.62 **10** $k = 0.0105$, $t = 65.8$ units

Exercise 5.3B Reasoning and problem-solving

1 a 20

b Model gives $100 - 80e^{-1}$

= 70.6, which is reasonably accurate (roughly 2% error).

Yes, model fits data.

c 87 insects

d As $t \to \infty$, $e^{-\frac{t}{5}} \to 0$

So yes, the model does predict a limiting number and the predicted limiting number = 100

2 a $A = 5 - 2 = 3$

t	0	20	40	60	80	100
n	5	11	25	60	140	265
p	5	10	24	62	166	447

b 29.3 hours

c Accurate for $t < 60$

Fairly accurate for $t = 60$

Not accurate for $t = 100$

No limit is placed on the number of bacteria. It would increase to infinity according to the model.

d 1.1 bacteria per hour

3 a $A = 1.5$, $k = 0.358$

b 5.6 mins

c $0.090\,\text{mg ml}^{-1}\,\text{min}^{-1}$

4 a 180 **b** 20, 9 trees/yr **c** 5 years

d As $t \to \infty$, $e^{-\frac{t}{20}} \to 0$

So $N \to 200$

5 a 393 minutes

b $2.26\,°\text{C min}^{-1}$

c As $t \to \infty$, $e^{-0.005t} \to 0$

So $\theta \to 30$

Minimum temperature = 30 °C

d Any valid reason e.g. the minimum temperature to which the block of steel cools may change depending on environmental factors. The block of steel will cool to the ambient temperature which may be less than 30°C.

6 a $k = 5.0, A = 1500$

b $n = \frac{Ae^{\frac{t}{4}}}{e^{\frac{t}{4}} + k} = \frac{A}{1 + ke^{-\frac{t}{4}}}$

$1 + k, e^{-\frac{t}{4}} > 1$ so $\frac{A}{1 + ke^{-\frac{t}{4}}} < A$ so the population cannot exceed 1500

Exercise 5.4A Fluency and skills

1 $y \approx 2.5x^{1.5}$

2 $y \approx 4 \times 2.5^x$

3 a i

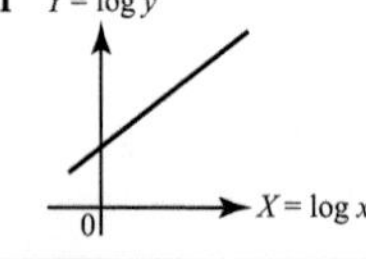

X = logx	0.301	0.477	0.602	0.699	0.778
Y = logy	1.23	1.60	1.87	2.07	2.24

ii $a = 4, n = 2.1, y = 4 \times x^{2.1}$

b i

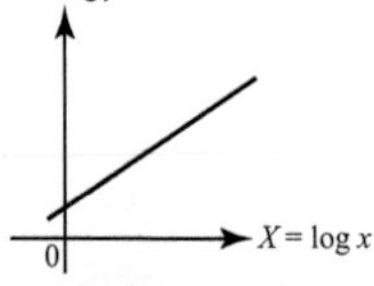

X = logx	0	0.301	0.477	0.602	0.699
Y = logy	0.204	0.362	0.462	0.531	0.580

ii $a = 1.6, n = 0.54, y = 1.6 \times x^{0.54}$

c i

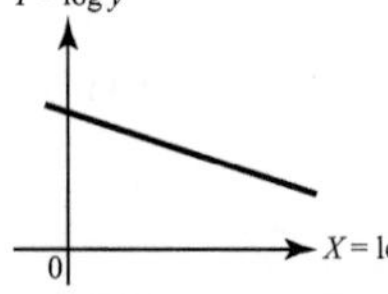

X = logx	1	1.176	1.301	1.398	1.477
Y = logy	2.58	2.52	2.48	2.46	2.43

ii $a = 750, n = -0.3, y = 750 \times x^{-0.3}$

d i

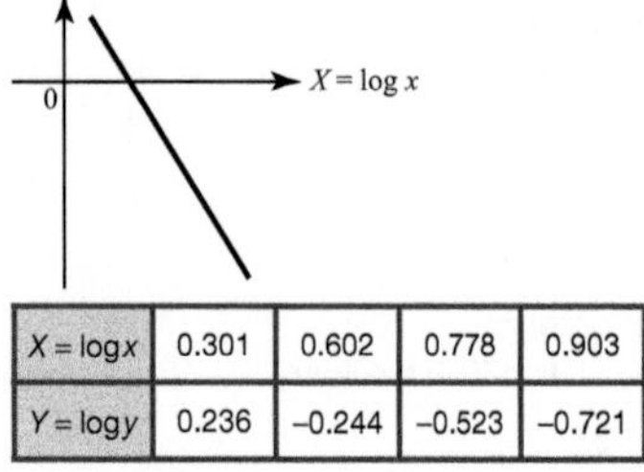

X = logx	0.301	0.602	0.778	0.903
Y = logy	0.236	−0.244	−0.523	−0.721

ii $a = 5.2, n = -1.6, y = 5.2 \times x^{-1.6}$

4 a i

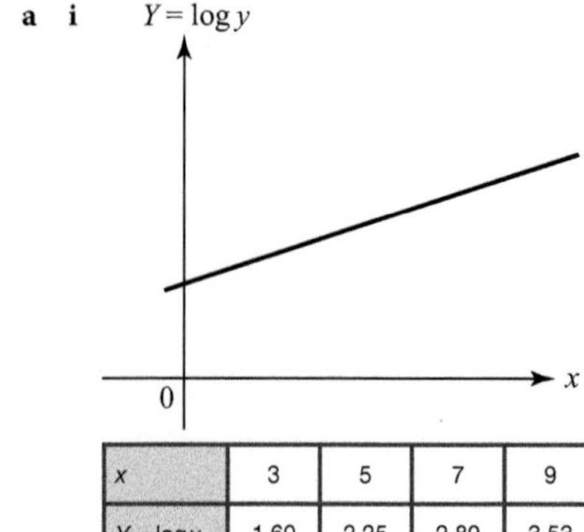

x	3	5	7	9
Y = logy	1.60	2.25	2.89	3.53

ii $b = 2.1, k = 4.3, y = 4.3 \times 2.1^x$

b i

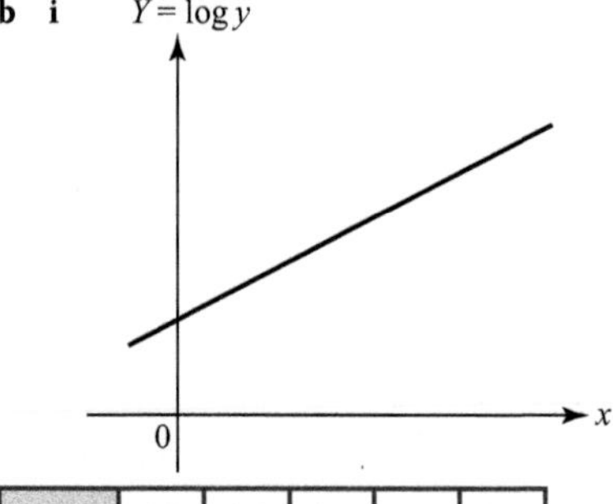

x	0.2	0.3	0.4	0.5	0.6
Y = logy	0.914	0.964	1.017	1.064	1.117

ii $b = 3.2, k = 6.5, y = 6.5 \times 3.2^x$

c i

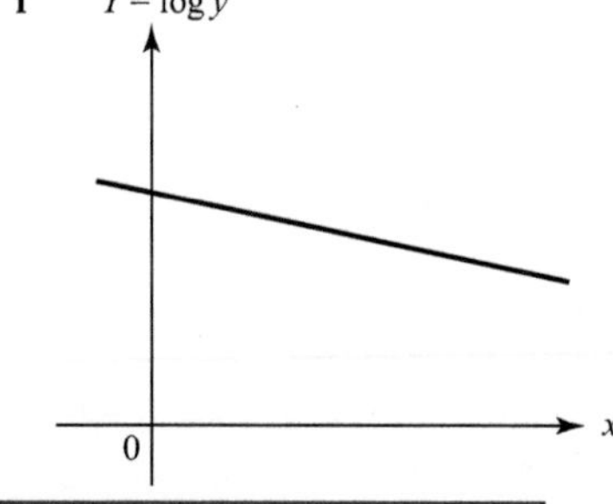

x	2	2.5	3	3.5	4
Y = logy	2.24	2.14	2.04	1.94	1.84

ii $b = 0.625, k = 450, y = 450 \times 0.625^x$

d i

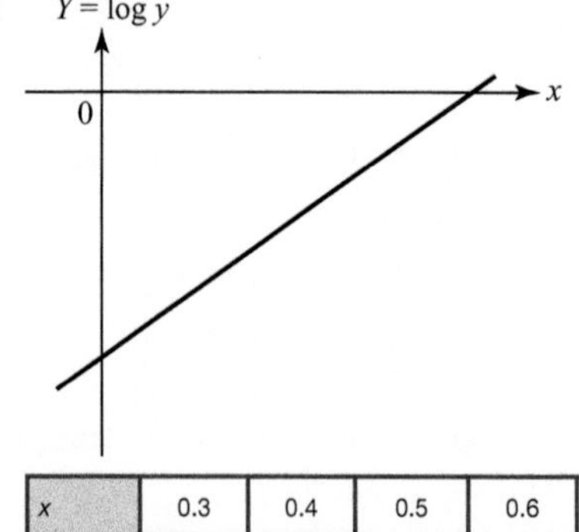

x	0.3	0.4	0.5	0.6
Y = logy	−0.481	−0.409	−0.337	−0.268

ii $b = 5.2, k = 0.2, y = 0.2 \times 5.2^x$

5

x	0.5	1	2	3
y	0.66	1.08	2.92	7.87
$X = \log x$	−0.301	0	0.301	0.477
$Y = \log y$	−0.180	0.033	0.465	0.896

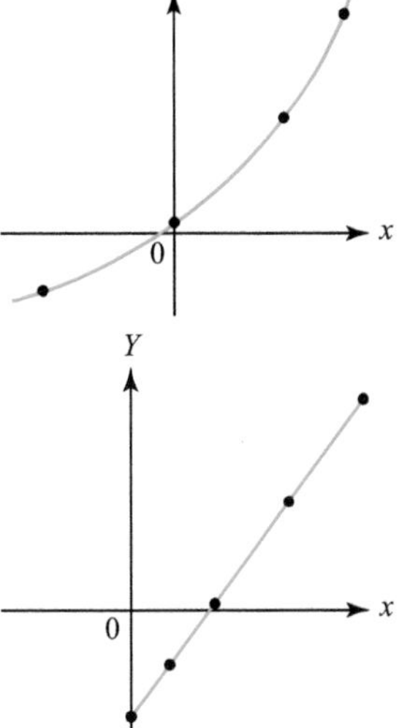

Graph of $Y = \log y$ against x gives straight line so function has the form $y = kb^x$

$y = 0.4 \times 2.7^x$

Exercise 5.4B Reasoning and problem-solving

Some numbers here are subjective so answers might vary.

1 a

$X = \log x$	0.079	0.114	0.146	0.176
$Y = \log y$	0.763	0.881	0.940	1.000

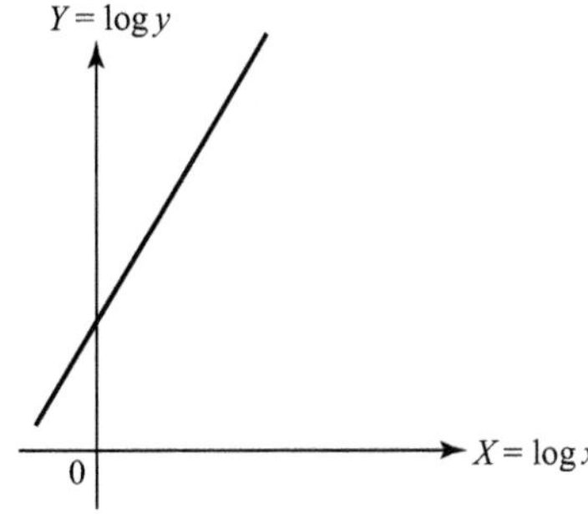

$n = 2.4$, $a = 3.9$, $y = 3.9 \times x^{2.4}$

b i $8\,\text{m}^3$ ii $64\,\text{m}^3$

c (i) is more reliable, because 1.35 is within the range of the task 1.2 to 1.5 (interpolation).
3.2 is well outside 1.2 to 1.5 and liable to inaccuracy (extrapolation).

2

$Y = \log t$	0.301	0.477	0.602	0.699
$Y = \log A$	1.23	1.60	1.86	2.07

a $n = 2.1$, $a = 4.0$ to 2 sf, $A = 4.0 \times t^{2.1}$

b 1 second is only slightly outside the data range 2 to 5 seconds. 1 min (60 sec) is well outside the data range 2 to 5 seconds.

3

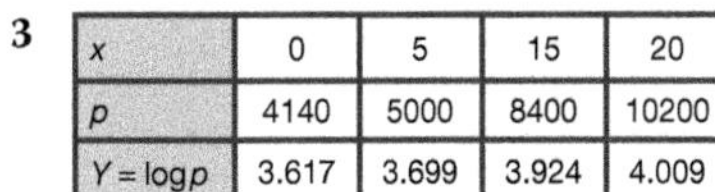

x	0	5	15	20
p	4140	5000	8400	10200
$Y = \log p$	3.617	3.699	3.924	4.009

a

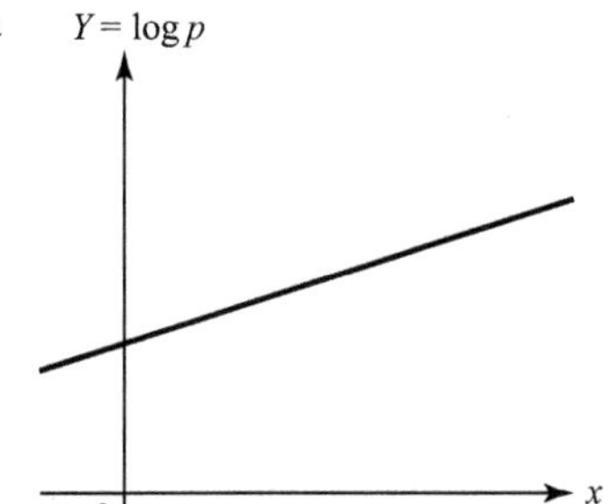

$b = 1.05$, $k = 4140$, $p = 4140 \times 1.05^x$

b $x = 10$, $p = 6740$
$x = 30$, $p = 17\,900$
6740 is better as 10 lies in the range 0 to 20 (interpolation).
17 900 is less good as 30 lies outside the range 0 to 20 (extrapolation).

4

v	200	250	300	350	400
p	78.0	56.5	43.7	36.0	29.7
$X = \log v$	2.301	2.398	2.477	2.544	2.602
$Y = \log p$	1.892	1.752	1.640	1.556	1.473

a $n = -1.40$, $k = 119000$, $p = 119000 \times v^{-1.4}$

b i $v = 308$ ii $v = 973$

c (i) is more reliable as it is found by interpolation, whereas (ii) is found by extrapolation.

5

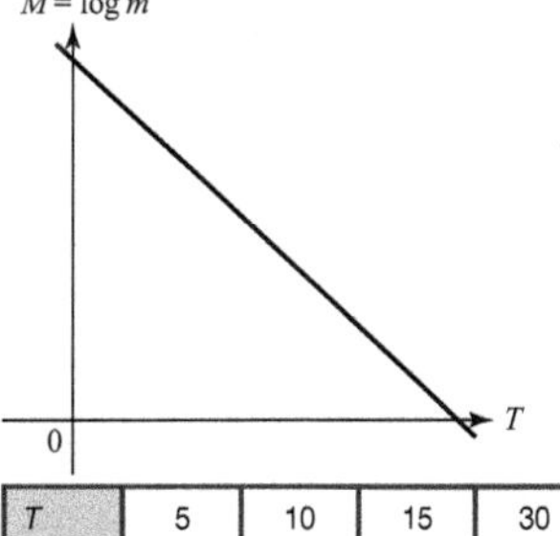

T	5	10	15	30
M	47.6	45.3	43.1	38.0
$\log m$	1.678	1.656	1.634	1.580

a $m = 50 \times 0.991^t$ b $m = 50$ grams c 77 hours

Review exercise 5

1 a $\log_2 32 = 5$
b $\log_{10} 0.0001 = -4$

2 a $2^4 = 16$ b $4^{-2} = \frac{1}{16}$

3 a 5 b 3

4 $\log 27$

5 $2\log a - \frac{1}{2}\log b$

6 a $x = 1.79$ b $x = 3$ or 0 c $x = 57.5$

7 a 19 b 16

8 a $y = e^3x - 2e^3 = e^3(x - 2)$
b $y = \frac{1}{2}ex$, $\left(1, \frac{e}{2}\right)$

9 Stretch sf $\frac{1}{2}$ parallel to x-axis

10 $a = e^3$, $b = \ln 4$ 11 $k = 0.910$ 12 $y = \frac{1}{2} - 2x$

13 $y = 1 - x + e^x$
Equation of normal is $y = -\frac{x-1}{e-1} + e$
When $x = e^2$, $y = -(e+1) + e = -1$
So normal passes through the point $(e^2, -1)$

14

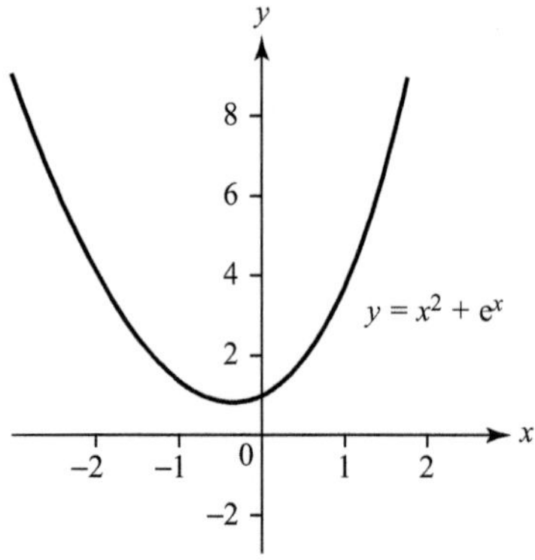

Minimum at (−0.35, 0.83)

15 a After 1 year, $A = (1+\frac{r}{100}) \times 250$

After 2 years,

$$A = 250\left(1+\frac{r}{100}\right) + \frac{r}{100} \times 250\left(1+\frac{r}{100}\right)$$

$$= 250\left(1+\frac{r}{100}\right)\left(1+\frac{r}{100}\right)$$

$$= 250\left(1+\frac{r}{100}\right)^2$$

etc. until after n years $A = 250(1+\frac{r}{100})^n$

b $A = 303.88$ c 12 years

16 a

x	1.0	2.0	3.0	4.0
y	15.0	23.7	43.4	70.2
$Y = \log y$	1.176	1.375	1.637	1.846

$b = 1.67, k = 9.0, y = 9.0 \times 1.67^x$

These numbers are subjective so answer might vary.

b Model gives $9.0 \times 1.67^{7.5} = 421$

This is not sufficiently close to 372, so the model does not predict the results accurately enough.

17 Line of best fit is $y = 3.58 \times x^{2.33}$

x	4	4.8	5.6	6.4
y	90	140	198	270
Model	90.6	138	198	271

$a = 3.58, b = 2.33$

These numbers are subjective so answer might vary.

18 No constraint on the number of deer – it would increase to infinity according to the model.

Continuous model but number of deer is discrete.

Model ignores external factors such as disease, predation, and limited food supply.

Assessment 5

1 a C $\log_5 56$ b D $\log_5 0.8$

2 a B 4 b C −1

3

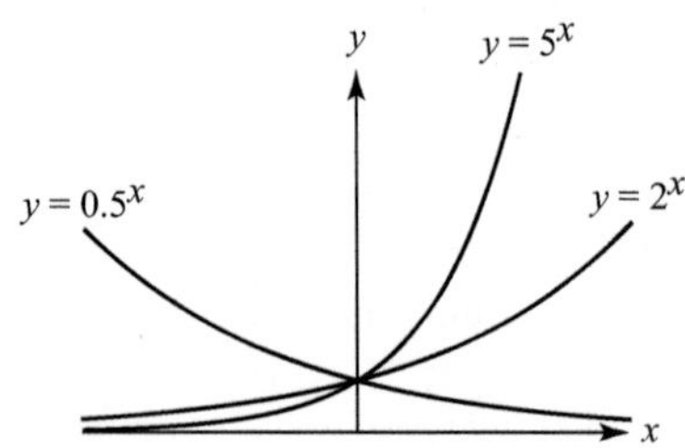

4 a

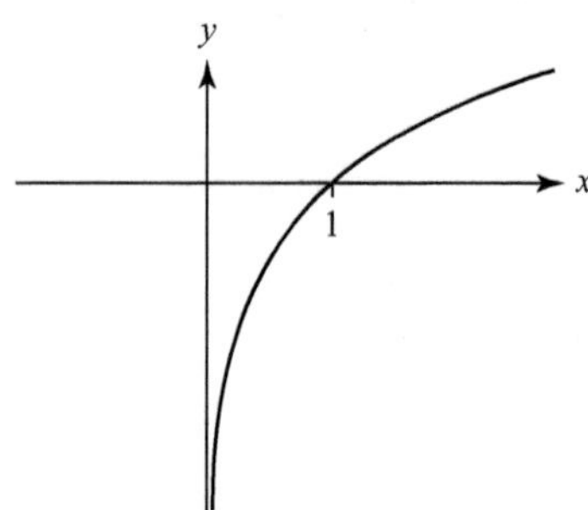

Asymptote is $x = 0$

b i 27 ii $\frac{1}{9}$ iii $\sqrt{3}$

5 a $\log_n 75$ b $\log_n\left(\frac{n}{5}\right)$

6 a $x = 1.43$ b $x = 1.61$

7 a $x = 12$ b $x = 3$

8 a $B = 200$ b $t = 1.3$ hours

c E.g. No constraint on the number of bacteria (i.e. would go off to infinity according to the model).

Continuous model but number of bacteria is discrete.

Ignores extraneous factors such as limited dish size.

9 $t = 9.9$ years 10 $k = \sqrt{15}$ 11 $a = 2, n = 4$

12 a

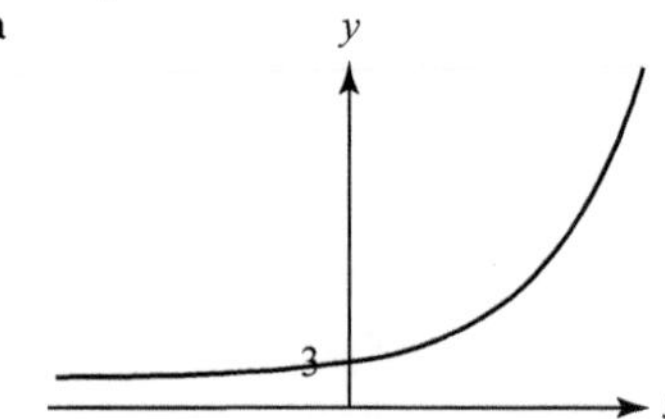

Asymptote is $y = 2$

b $x = \frac{1}{2}\ln 5$ or $\ln\sqrt{5}$

13 a i $2p$ ii $-p$ iii $p+1$

b $n = \frac{1}{3}$

14 a $x = 4$ b $x = \frac{9}{8}$ c $x = \sqrt{2}e^{2.5}$ or 17.2

15 a 500

b i $a = \frac{2}{\sqrt{3}}$ or $\frac{2\sqrt{3}}{3}$ or 1.15 ii $t = 9.64 \approx 10$ years

c $P = \frac{1500a^t}{2+a^t} = \frac{1500}{\frac{2}{a^t}+1}$

But $\frac{2}{a^t}+1 > 1$ so $\frac{1500}{\frac{2}{a^t}+1} < 1500$

So the population can never exceed 1500

d E.g. Model is continuous but population of a species is discrete.

16 a 50 g b $t = 2.31$ days

17 a $a = 5, b = 0.5$

b

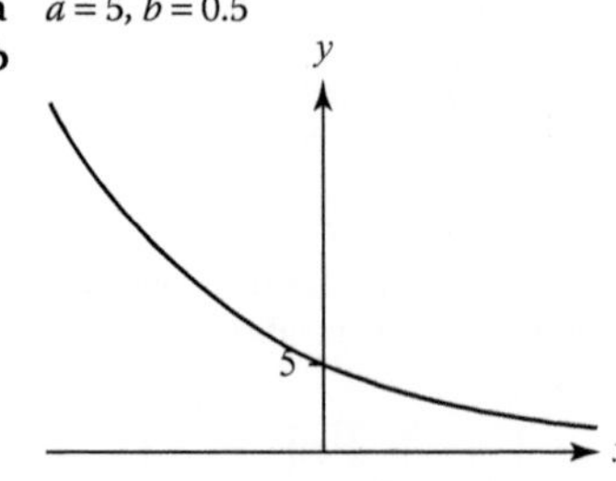

18 40.2 19 $x = 1$

20 a $r = 0.079$ per hr, $C_0 = 10$ mg per litre

b $t = 13$ hours

Assessment answers chapters 1–5: Pure

1 a i C ii D
b B

2 $y = \frac{2}{3}$ or $\frac{1}{4}$, $x = -3$ or 2

3 a $b^2 - 4ac = -15$ so no real solutions
b i $x = -1, -2$ ii $x = \frac{-3 \pm \sqrt{5}}{2}$
c $k = \frac{25}{4}$

4 a $(x-2)^2 + 8$
b $x - 2 = \sqrt{-8}$ so no real solutions
Or use discriminant.
c

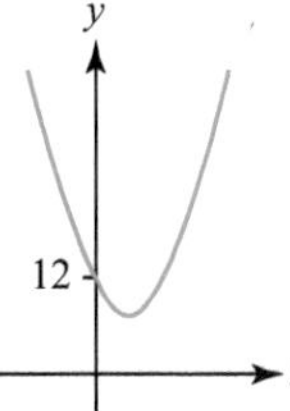

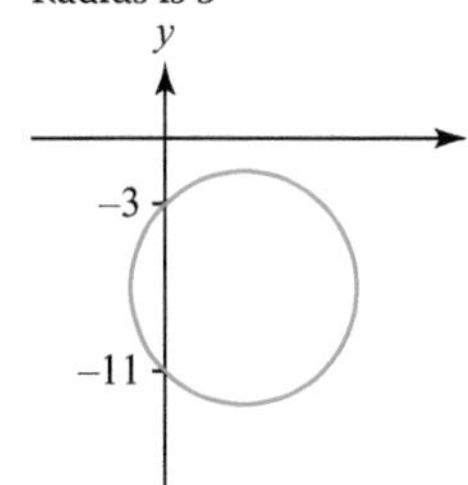

Minimum point is (2, 8)

5 a $-\frac{5}{2} - 2\sqrt{3}$, $p = -\frac{5}{2}$ and $q = -\frac{3}{2}$ b $x = \frac{1}{2}$

6 a Centre is (3, −7)
Radius is 5
b

7 a $-\frac{1}{4}$ b $4x - y - 6 = 0$ c $\sqrt{5}$

8 e.g. $0 \times 1 = 0$ or $-3 \times -2 = 6$
Which is even so this disproves the statement at the product of any two consecutive integers is odd.

9 a 126
b i $1 - 6x + 15x^2 - 20x^3 + 15x^4 - 6x^5 + x^6$
ii $1 + 8x + 24x^2 + 32x^3 + 16x^4$

10 a $p(3) = 3^3 - 2(3^2) - 5(3) + 6$
$= 0$ so $x - 3$ is a factor of $p(x)$
b $x = (3), -2, 1$
c

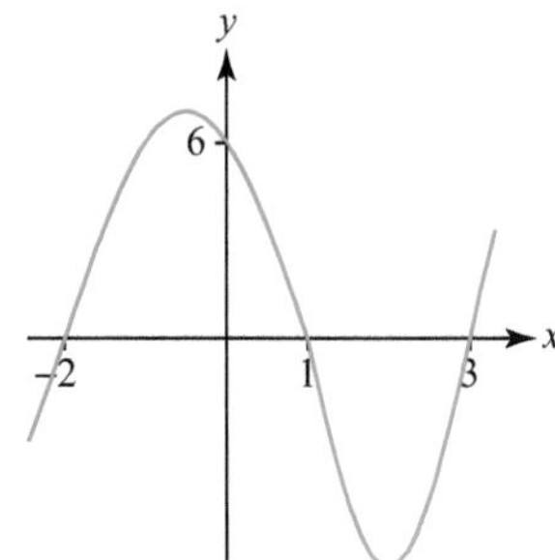

11 a

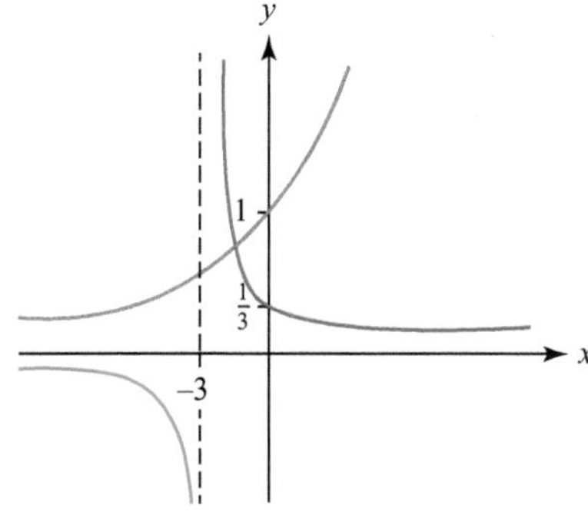

b One as the curves intersect once only.

12 $0 < y \le 45°$, $135° \le y < 180°$

13 a i 30°, 150° ii 20.7°, 159.3°
b

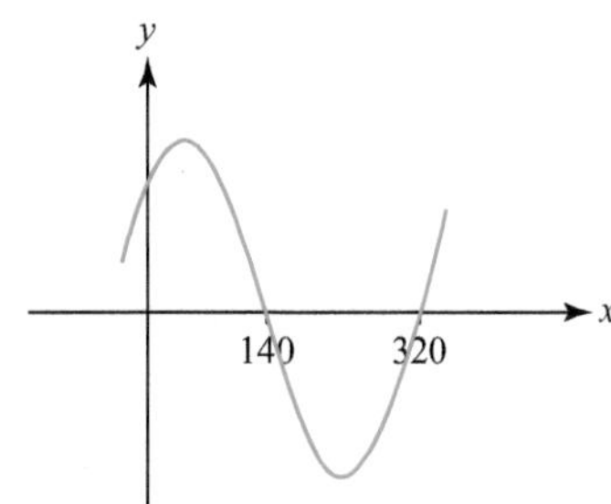

14 a 38.5°, 141.5° b 80.4 cm^2

15 a i $3x^2 - 10x^4$ ii $\frac{1}{2}x^{-\frac{1}{2}} - x^{-2}$
b $x = 4$

16 $x + 10y = 42$

17 a $\frac{4}{3}x^3 + 6x^2 + 9x + c$ b $2x^{\frac{5}{2}} - \frac{2}{3}x^{\frac{3}{2}}$ $(+c)$
c $2x^{\frac{1}{2}} + \frac{1}{3}x^{\frac{3}{2}}$ $(+c)$

18 a (4, 0) b $\frac{10}{3}\sqrt{5} - \frac{22}{3}$

19 a $\frac{dy}{dx} = 3x^2 - 4x - \frac{1}{x^2}$ b $\frac{1}{3}$

20 $y = 0.374x^{3.42}$

21 a i $\log_2 5$ ii $\log_2 54$ iii $\log_2 10$
b $\log_3 17 = 2.58$

22 a

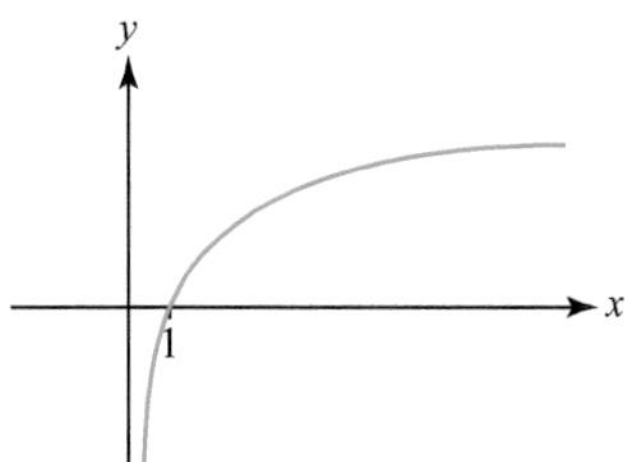

Asymptote at $x = 0$
b i $x = \frac{1}{64}$ ii $x = \frac{1}{4}$

23 a i $x \ge -2$ ii $-3 < x < 2$
b $-2 \le x < 2$

24 a $(x-4)^2 + (y-2)^2 = 26$ b $y - 7 = \frac{1}{5}(x-3)$

25 a (−3, 4) and (3, 1) b $x + 2y - 5 = 0$

26 Consider n^2:
n odd $\Rightarrow n^2$ odd
So m will be even $\Rightarrow m^2$ even

$\therefore n^2 - m^2$ is odd
n even $\Rightarrow n^2$ even
So m will be odd $\Rightarrow m^2$ odd
$\therefore n^2 - m^2$ is odd
Alternatively, Let $m = n - 1$
$n^2 - m^2 = n^2 - (n-1)^2$
$= n^2 - n^2 + 2n - 1$
$= 2n - 1$ which is odd
So $n^2 - m^2$ is odd for any consecutive integers m and n

27 $k \le -8, k \ge 0$

28 a $(x+4)(2x+1)(x-3)$

b

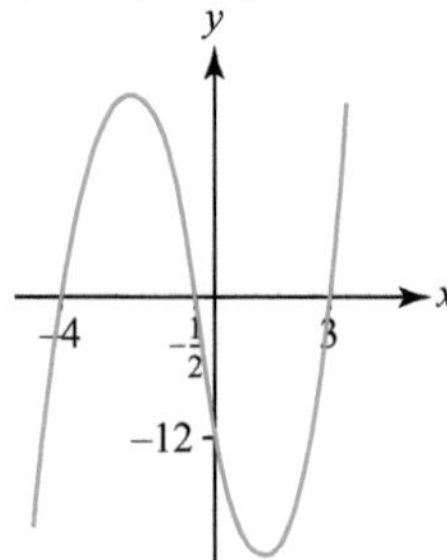

29 a i $176 + 80\sqrt{5}$ **ii** $265 - 153\sqrt{3}$

b $(x+\sqrt{5})^5 - (x-\sqrt{5})^5 = (x^5 + 5\sqrt{5}x^4 + 50x^3 + 50\sqrt{5}x^2 + 125x + 25\sqrt{5}) - (x^5 - 5\sqrt{5}x^4 + 50x^3 - 50\sqrt{5}x^2 + 125x - 25\sqrt{5})$
$= 10\sqrt{5}x^4 + 100\sqrt{5}x^2 + 50\sqrt{5}$
$x^2 \ge 0$ for all values, $x^4 = (x^2)^2 \ge 0$ so
$10\sqrt{5}x^4 + 100\sqrt{5}x^2 + 50\sqrt{5} \ge 50\sqrt{5} > 0$

30 a $64 + 192x + 240x^2 + 160x^3 + 60x^4 + 48x^5 + x^6$ **b** 64.192

31 a $\sin 3x + 2\cos 3x = 0$
$\sin 3x = -2\cos 3x$
$\frac{\sin 3x}{\cos 3x} = -2$
$\tan 3x = -2$

b $x = 39°, 99°, 159°$

32 a i

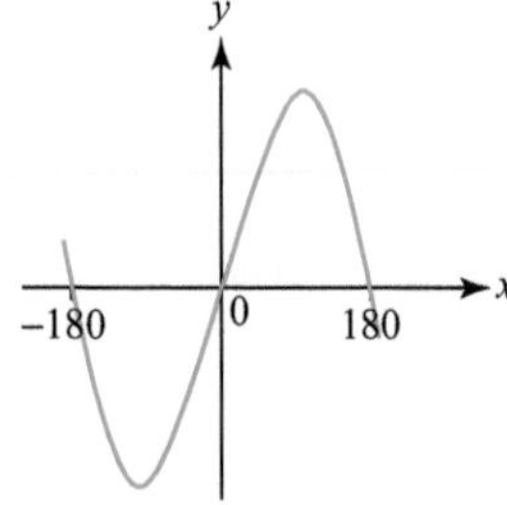

Maximum at (90, 2)
Minimum at (−90, −2)

ii

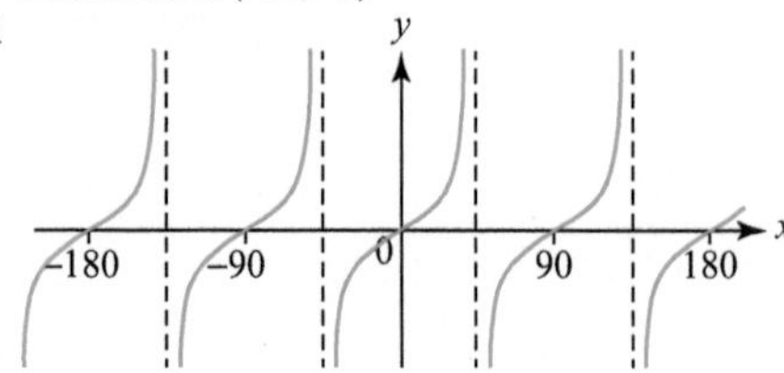

Asymptotes at $x = \pm 45°, x = \pm 135°$

b $x = 30°, 90°$

33 a $\frac{2}{3}\sqrt{2}$ **b** $2\sqrt{2}$

34 a 37.8° **b** 227.9 mm²

35 a 37.9 cm **b** 32.6 cm

36 a $\frac{dy}{dx} = 12x^2$ **b** 48

37 $0 < x < 2.297$

38 (0, 20) is a maximum
(1, 15) is a minimum
(−2, −12) is a minimum

39 a $x^2h = 3000 \Rightarrow h = \frac{3000}{x^2}$
$A = 4xh + x^2$
$= 4 \times \frac{3000}{x^2} + x^2$
$= \frac{12000}{x} + x^2$ as required

b $x = 18.2$ cm

c $A_{min} = 991$ cm²

d $\frac{d^2A}{dx^2} = \frac{24000}{x^3} + 2$
$\left.\frac{d^2A}{dx^2}\right|_{x=18.2} > 0$ so a minimum

40 a $y = 2x^5 - 4x^3 + x - 25$

b $y = 2(-1)^5 - 4(-1)^3 + (-1) - 25$
$= -24$ so passes through (−1, −24)

41 $y = 2.2x^{1.5}$, $a = 2.2$, $n = 1.5$

42 a 5 years and 6 months

b Not suitable as it implies the car will ultimately have a negative value

43 a

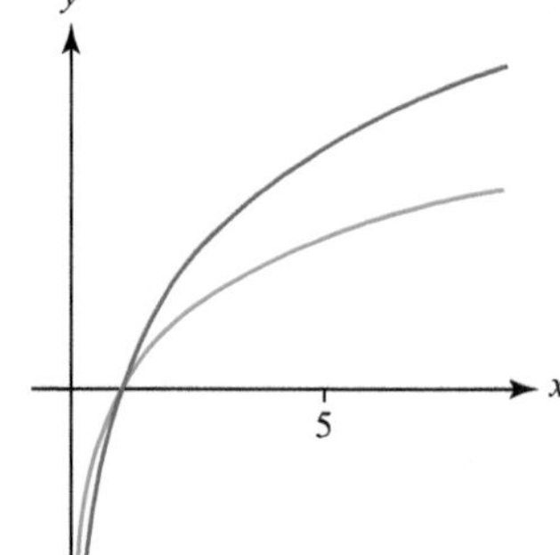

b $x = 9e^5$

44 a i If $n = 3$
Then $2^n + 1 = 8 + 1 = 9$
9 is divisible by 3 so not a prime number.
Hence $2^n + 1$ is not a prime for all positive integers n

ii If $n = 1$
Then $3^n + 1 = 3 + 1 = 4$
$= 2 \times 2$ so even
If $n = 2$
Then $3^2 + 1 = 9 = 10$
$= 2 \times 5$ so even
If $n = 3$
Then $3^3 + 1 = 27 + 1 = 28$
$= 2 \times 14$ so even
If $n = 4$
Then $3^4 + 1 = 81 + 1 = 82$
$= 2 \times 41$ so even
So we have proved that $3^n + 1$ is even for all integers n in the range $1 \le n \le 4$

iii $5^n + 10 = 5(5^{n-1} + 2)$ Which is a multiple of 5 for all positive integers n

b i Proof by counterexample
ii Proof by exhaustion
iii Direct proof

45 $A = -3, B = 16, C = 8$

46 a $n=16$ b $a=-\frac{1}{2}$

47 a 50.8°, 129°, –50.8°, –129° b 180°

48 a $\text{LHS} \equiv 3\sin^2 x - 4(1-\sin^2 x)$
$\equiv 3\sin^2 x - 4 + 4\sin^2 x$
$\equiv 7\sin^2 x - 4$ as required

b 57.7°, 122.3°, –57.7°, –122.3°

49 39.9°, 320.1°, 115.7°, 244.3°

50 a $x=-1, 7, 3$ b 128

51 a $\frac{5}{\sqrt{3}}$ cm

b $V_{max} = \frac{250}{3\sqrt{3}}\pi$ (151.15) cm³

$\frac{d^2V}{dx^2} = -6\pi r < 0$ so a maximum

52 a i $A=11$ ii $k=0.094$

b

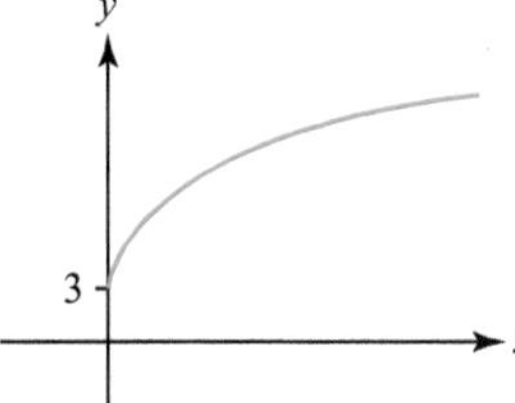

c 4966 d 567 organisms per hour e 11 000

53 $x=\ln\left(\frac{3}{2}\right)$, ln 5

Or 0.405, 1.61

54 a $x=2$ b $y=3$ c $x=0.739, 27.2$

55 $x=\frac{1}{9}, y=216$

Chapter 6

Exercise 6.1A Fluency and skills

1 a **q** b **p** c **–p** d **–q**
e **p** + **q** f **q** – **p** g **p** – **q**

2 a **p** + **q** b 2**p** + **q** c **q** – 2**p** d $\mathbf{p}+\frac{1}{2}\mathbf{q}$

3 a 2**q** b **p** + **q** c **p** + 2**q** d **q** – **p**

4 a Magnitude = 25.6, direction = 038.7°
b Magnitude = 10.3, direction = 078.1°
c Magnitude = 7.88, direction = 231.4°

5 a Magnitude = 12.1, direction = 17.0° to **p**
b Magnitude = 5.69, direction = 38.4° to **p**

6 a Resultant is 7.02 km on a bearing of 098.8°
b Resultant is 23.4 km on a bearing of 217.3°
c Resultant is 23.5 km h⁻¹ on a bearing of 290.6°
d Resultant is 529 N on a bearing of 347.5°

Exercise 6.1B Reasoning and problem-solving

1 a **p** – 2**q**

b $\frac{1}{2}\mathbf{p}$

c *DE* is parallel to *BC* and half as long.

2 $\overrightarrow{BC} = \mathbf{q}-\mathbf{p}$ so $\overrightarrow{BD} = \frac{1}{2}(\mathbf{q}-\mathbf{p})$

$\overrightarrow{AD} = \overrightarrow{AB} + \overrightarrow{BD} = \mathbf{p} + \frac{1}{2}(\mathbf{q}-\mathbf{p}) = \frac{1}{2}(\mathbf{p}+\mathbf{q})$

$\overrightarrow{AG} = \frac{2}{3}\overrightarrow{AD} = \frac{1}{3}(\mathbf{p}+\mathbf{q})$

$\overrightarrow{BG} = \overrightarrow{BA} + \overrightarrow{AG} = -\mathbf{p} + \frac{1}{3}(\mathbf{p}+\mathbf{q}) = \frac{1}{3}(\mathbf{q}-2\mathbf{p})$

But $\overrightarrow{BE} = \frac{1}{2}\mathbf{q}-\mathbf{p} = \frac{1}{2}(\mathbf{q}-2\mathbf{p})$

So $\overrightarrow{BG} = \frac{2}{3}\overrightarrow{BE}$

Hence $\overrightarrow{BG}$ and $\overrightarrow{BE}$ are collinear, and $BG:GE = 2:1$

3 $\overrightarrow{AB} = \mathbf{b}-\mathbf{a}$ so $\overrightarrow{AD} = \frac{1}{2}(\mathbf{b}-\mathbf{a})$

$\overrightarrow{PD} = \overrightarrow{PA} + \overrightarrow{AD} = \mathbf{a} + \frac{1}{2}(\mathbf{b}-\mathbf{a}) = \frac{1}{2}(\mathbf{a}+\mathbf{b})$

Similarly $\overrightarrow{PE} = \frac{1}{2}(\mathbf{b}+\mathbf{c})$ and $\overrightarrow{PF} = \frac{1}{2}(\mathbf{c}+\mathbf{a})$

$\overrightarrow{PD} + \overrightarrow{PE} + \overrightarrow{PF} = \frac{1}{2}(\mathbf{a}+\mathbf{b}) + \frac{1}{2}(\mathbf{b}+\mathbf{c}) + \frac{1}{2}(\mathbf{c}+\mathbf{a}) = \mathbf{a}+\mathbf{b}+\mathbf{c}$

Hence $\overrightarrow{PD} + \overrightarrow{PE} + \overrightarrow{PF} = \overrightarrow{PA} + \overrightarrow{PB} + \overrightarrow{PC}$

4 $\overrightarrow{EF} = \overrightarrow{EB} + \overrightarrow{BF} = \frac{1}{2}\mathbf{p} + \frac{1}{2}\mathbf{q}$

$\overrightarrow{AD} = \mathbf{p}+\mathbf{q}+\mathbf{r}$

$\overrightarrow{HG} = \overrightarrow{HD} + \overrightarrow{DG} = \frac{1}{2}(\mathbf{p}+\mathbf{q}+\mathbf{r}) - \frac{1}{2}\mathbf{r} = \frac{1}{2}\mathbf{p} + \frac{1}{2}\mathbf{q}$

Hence $\overrightarrow{EF} = \overrightarrow{HG}$, so EF and HG are parallel.

$\overrightarrow{FG} = \frac{1}{2}\mathbf{q} + \frac{1}{2}\mathbf{r}$

$\overrightarrow{EH} = \overrightarrow{EA} + \overrightarrow{AH} = -\frac{1}{2}\mathbf{p} + \frac{1}{2}(\mathbf{p}+\mathbf{q}+\mathbf{r}) = \frac{1}{2}\mathbf{q} + \frac{1}{2}\mathbf{r}$

Hence $\overrightarrow{FG} = \overrightarrow{EH}$, so FG and EH are parallel.

Hence $EFGH$ is a parallelogram.

5 a $\overrightarrow{AF} = 3\mathbf{p} + \lambda(-3\mathbf{p}+\mathbf{q})$
b $\overrightarrow{AF} = 2\mathbf{q} + \mu(-2\mathbf{q}+\mathbf{p})$
c $\lambda = 0.8$ and $\mu = 0.6$

6 a It will travels 50 m downstream.
b The boat should steer upstream at 60° to the bank.

7 a 145.9° b 238 km h⁻¹

8 a The velocity is 15 km h⁻¹ on a bearing of 036.9°
b Steer on a bearing of 253.7°

Exercise 6.2A Fluency and skills

1 a 5.80**i** + 3.91**j** b 2.00**i** + 9.39**j**
c 8.92**i** + 8.03**j** d 4**j**
e –3.64**i** + 7.46**j** f –10.1**i** + 21.8**j**
g –3.56**i** – 5.08**j** h 12.5**i** – 6.36**j**

2 a 5.39, 21.8° b 11.4, 52.1° c 5, –90°
d 3.61, 124° e 5.83, –59.0° f 7.81, –140°
g 2, 180°

3 Magnitude is 4.66 and direction is 5.43° to positive x-direction.

4 a 2**j** b –2**i** – 2**j** c –6**i** + 7**j**
d 16**i** + **j** e $\sqrt{5} = 2.24$ f $\sqrt{20} = 4.47$

5 a $u=-10, v=-1$ b $u=-4.5$

6 a –12**i** + 16**j** b –0.6**i** + 0.8**j**

7 a $a=\frac{2}{3}, b=3\frac{1}{3}$ b $a=2, b=1$ or $a=-2, b=-1$

8 17 m (by Pythagoras)

9 $\overrightarrow{AB} = 6\mathbf{i}+4\mathbf{j}$

a $AB = \sqrt{52} = 7.21$ m b $\tan^{-1}\left(\frac{4}{6}\right) = 33.7°$

10 a 9.85 m b 42.8°

11 a –7**i** + **j** b –6**i** + 3**j**

12 $\mathbf{c} = \mathbf{b} + \overrightarrow{BC} = \mathbf{b} + \overrightarrow{AD} = \mathbf{b}+\mathbf{d}-\mathbf{a} = \begin{pmatrix} 21 \\ 22 \end{pmatrix}$

13 $\mathbf{b} = \mathbf{a} + 0.7\overrightarrow{AC} = \mathbf{a} + 0.7(\mathbf{c}-\mathbf{a}) = 3.4\mathbf{i} + 9.6\mathbf{j}$

Exercise 6.2B Reasoning and problem-solving

1 Vectors between each successive checkpoint can be shown to be parallel and share a common point, so they are collinear.

2 Distance = 10.8 m

3 The diagram shows the two possible situations.

C_1 lies between A and B

$\mathbf{c}_1 = \mathbf{a} + \frac{3}{5}(\mathbf{b} - \mathbf{a}) = 7.8\mathbf{i} + 4\mathbf{j}$

C_2 lies on AB produced, with $AC_2 = 3AB$

$\mathbf{c}_2 = \mathbf{a} + 3(\mathbf{b} - \mathbf{a}) = 27\mathbf{i} + 16\mathbf{j}$

4 Vectors $\overrightarrow{BC}$ and $\overrightarrow{CD}$ can be shown to be parallel and share a common point, so they are collinear.

5 $k = 3$ or $-\frac{2}{3}$

6 **a** $((2 + 0.8d)\mathbf{i} + 0.6d\mathbf{j})$ m **b** 8.60 m

7 Journey would reduce by 7.6 km. The tunnel should not be built.

8 **a** $\mathbf{a}' = (1 + 3t)\mathbf{i} + (4 + 3t)\mathbf{j}$

b $\overrightarrow{A'B'} = (4 - t)\mathbf{i} + (0.5t - 2)\mathbf{j}$

c They collide after 4 s at $\mathbf{a}' = 13\mathbf{i} + 16\mathbf{j}$

9 After t s the first particle is at $2t\mathbf{i} + (4 - t)\mathbf{j}$ and the second is at $(6 - t)\mathbf{i} + (8 - 3t)\mathbf{j}$. When $t = 2$, both particles are at $(4\mathbf{i} + 2\mathbf{j})$ m

10 $\mathbf{v} = 7\mathbf{i} + 6\mathbf{j}$

Review exercise 6

1 **a** scalar **b** vector **c** vector
d vector **e** scalar

2 **a** i $3\mathbf{p}$ ii $2\mathbf{q} - 3\mathbf{p}$ iii $3\mathbf{p} - 3\mathbf{q}$

b $\overrightarrow{FE} = \mathbf{p} - \mathbf{q}$, so $\overrightarrow{DF} = 3\overrightarrow{FE}$
They are parallel vectors passing through F, so the points are collinear.

3 **a** $\mathbf{p} + \mathbf{q}$ **b** $2\mathbf{q}$ **c** $2\mathbf{p} + \mathbf{q}$ **d** $\mathbf{q} - \mathbf{p}$

4 **a** $\mathbf{q} - \mathbf{r}$ **b** $\frac{1}{2}(\mathbf{q} - \mathbf{r})$ **c** $\frac{1}{2}(\mathbf{q} + \mathbf{r})$
d $\mathbf{r} - \mathbf{p}$ **e** $\frac{3}{4}(\mathbf{r} - \mathbf{p})$ **f** $\frac{1}{4}(\mathbf{p} + 3\mathbf{r})$

5 **a** $\theta = 74.9°$
b Magnitude = 11.2 m s^{-1}

6 **a** $\mathbf{P} = 5.14\mathbf{i} + 6.13\mathbf{j}$
$\mathbf{Q} = -1.74\mathbf{i} + 9.85\mathbf{j}$
$\mathbf{R} = 5.64\mathbf{i} - 2.05\mathbf{j}$
b Magnitude = 16.6 N
Direction = 57.0°

7 **a** $OA = 5.39$ **b** $\overrightarrow{AB} = 4\mathbf{i} - 7\mathbf{j}$ **c** $AB = 8.06$

Assessment 6

1 **a** C **b** A
2 **a** B **b** C
3 **a** 1346 m **b** 330°
4 **a** $x = 7, y = -4$ **b** $x = \frac{4}{3}$

5 **a** i $\frac{2}{3}\mathbf{a}$ ii $\frac{3}{4}\mathbf{b}$ iii $\frac{1}{3}\mathbf{a} + \frac{1}{2}\mathbf{b}$
iv $-\frac{2}{3}\mathbf{a} + \frac{1}{2}\mathbf{b}$ v $-\frac{1}{3}\mathbf{a} + \frac{1}{4}\mathbf{b}$
b 2:1

6 $3\sqrt{5}$, –26.6°, or on a bearing of 116.6°

7 **a** i $3\mathbf{i} - 2\mathbf{j}$ ii $9\mathbf{i} - 6\mathbf{j}$
b $\overrightarrow{BC} = 3(3\mathbf{i} - 2\mathbf{j})$
Since $\overrightarrow{BC}$ is a multiple of $\overrightarrow{AB}$, $\overrightarrow{AB}$ and $\overrightarrow{BC}$ are parallel, and since they have a point in common, A, B and C are collinear.

8 $\lambda = -4.8$ or 4

9 **a** i $2\mathbf{a} + \mathbf{c}$ ii $\mathbf{a} + \mathbf{c}$ iii $\mathbf{a} + \mathbf{c}$
iv $\frac{3}{2}\mathbf{a} + \frac{1}{2}\mathbf{c}$ v $-\frac{1}{2}\mathbf{a} + \frac{1}{2}\mathbf{c}$
b $\frac{4}{3}\mathbf{a} + \frac{2}{3}\mathbf{c}$

10 $\overrightarrow{PQ} = \mathbf{q} - \mathbf{p} = \begin{pmatrix} -3 \\ -5 \end{pmatrix} - \begin{pmatrix} 6 \\ 3 \end{pmatrix} = \begin{pmatrix} -9 \\ -8 \end{pmatrix}$

$\overrightarrow{SR} = \begin{pmatrix} -9 \\ -8 \end{pmatrix}$

$\overrightarrow{QR} = \mathbf{r} - \mathbf{q} = \begin{pmatrix} 1 \\ -3 \end{pmatrix} - \begin{pmatrix} -3 \\ -5 \end{pmatrix} = \begin{pmatrix} 4 \\ 2 \end{pmatrix}$

$\overrightarrow{PS} = \begin{pmatrix} 4 \\ 2 \end{pmatrix}$

Since $\overrightarrow{PQ} = \overrightarrow{SR}$ and $\overrightarrow{QR} = \overrightarrow{PS}$, $PQRS$ is a parallelogram

11 **a** $\mathbf{x} + 3\mathbf{y} = \begin{pmatrix} 1 \\ -1 \end{pmatrix} + \begin{pmatrix} 15 \\ 9 \end{pmatrix} = \begin{pmatrix} 16 \\ 8 \end{pmatrix}$

$= 8 \times \begin{pmatrix} 2 \\ 1 \end{pmatrix}$

Since $\mathbf{x} + 3\mathbf{y}$ is a multiple of $\mathbf{z}$, they are parallel

b –8

12 **a** 343.4° **b** 5.1

Chapter 7

Try it 7A

1 80 kmh^{-1}
2 **a** –1.5 ms^{-1}
b 1.5 ms^{-1}
3 **a** 7.5 ms^{-2}
b 45 m

Bridging Exercise 7A

1 **a** 4.5 kmh^{-1}
b 0.625 ms^{-1}
c 22.5 kmh^{-1}
2 **a** 32.5 km
b 690 m
c 231.2 km
3 **a** 0.9 hours (54 minutes)
b 16.25 s
c 1.2 s
4 **a** 25 ms^{-1}
b 1 s
c –12 ms^{-1}, 12 ms^{-1}
d 110 m
e 13.75 ms^{-1}
5 **a** 2.5 ms^{-2}
b –1 ms^{-2}
c the particle is travelling at a constant velocity of 3 ms^{-1}
d 3 ms^{-2}
e 10.5 m

Exercise 7.1A Fluency and skills

1 **a** Force **b** Mass
c Speed or velocity **d** Acceleration
2 **a** 8500 m **b** 2300 mm
c 4.82 m **d** 1.65 km
e 20 m s^{-1} **f** 50.4 km h^{-1}
g 0.9 km h^{-1} **h** 24000 cm^2
i 1400 g **j** 1600 kg
3 **a** 120 km h^{-1} **b** 33.3 m s^{-1} (to 3 sf)
4 18 km
5 0.0154 m s^{-2}
6 45 kg

Exercise 7.1B Reasoning and problem-solving

1 29 min (to 2 sf) **2** 31.5 km (to 3 sf)
3 36 s **4** 236 kg (to 3 sf)

Exercise 7.2A Fluency and skills

1 **a** Between 5 and 15 s
b Between 0 and 5 s
c Between 15 and 20 s; the gradient is steepest

2 **a** It travels forward 12 m in 5 s, then backwards 18 m in the next 15 s
b 2.4 m s^{-1} in first stage, then -1.2 m s^{-1} in second stage
c **i** 1.5 m s^{-1} **ii** -0.3 m s^{-1}

3 **a** 37.5 m **b** -1 m s^{-2}

4 **a** Between 8 and 11 s **b** 0.73 m s^{-2} (to 2 sf) **c** -2 m s^{-2}

5 **a**

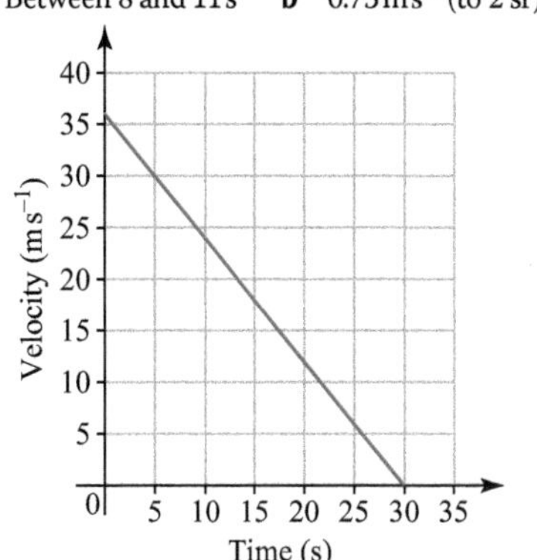

b 540 m

6 **a**

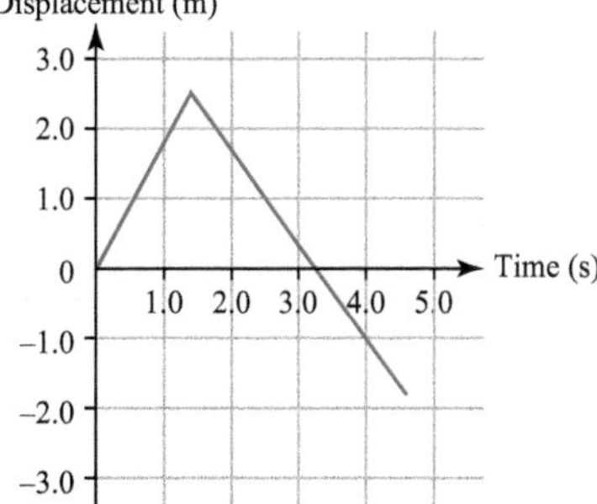

b **i** 1.8 m s^{-1} (to 2 sf)
ii -1.3 m s^{-1} (to 2 sf)
iii 1.5 m s^{-1} (to 2 sf)

7 **a** 2.5 m s^{-2} **b** 360 m

Exercise 7.2B Reasoning and problem-solving

1 **a**

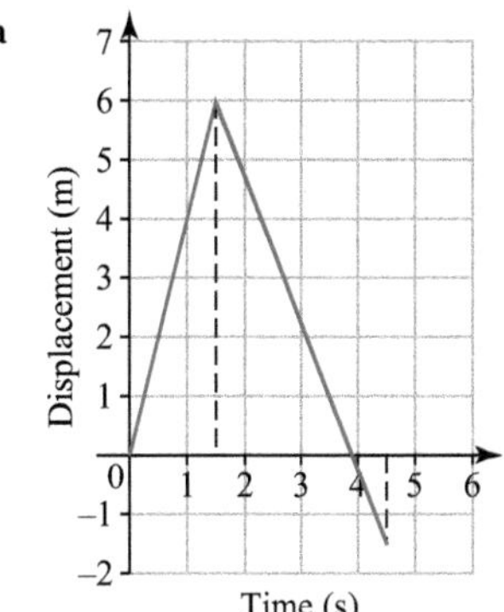

b **i** 3 m s^{-1} **ii** $-\frac{1}{3}$ m s^{-1}

2 **a** The cat goes forward 9 m in 6 s at constant speed, is still for 4 s, then (turns around and) goes backwards 15 m in 10 s at constant speed.
b From 0 to 6 s, $v = 1.5$ m s^{-1}
From 6 to 10 s, $v = 0$ m s^{-1}
From 10 to 20 s, $v = -1.5$ m s^{-1}
c **i** 1.2 m s^{-1} **ii** -0.3 m s^{-1}
d The cat's velocity would not change instantaneously, so there would be curved sections of the graph where it's accelerating or decelerating.

3 **a**

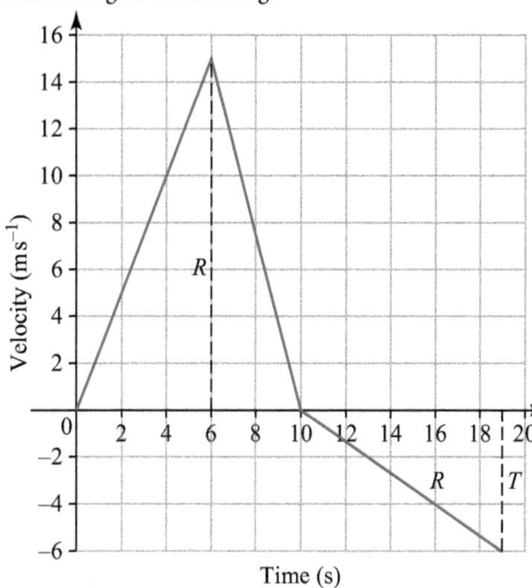

b 75 m **c** 35 s **d** -0.24 m s^{-2}

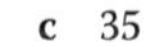

4 **a**

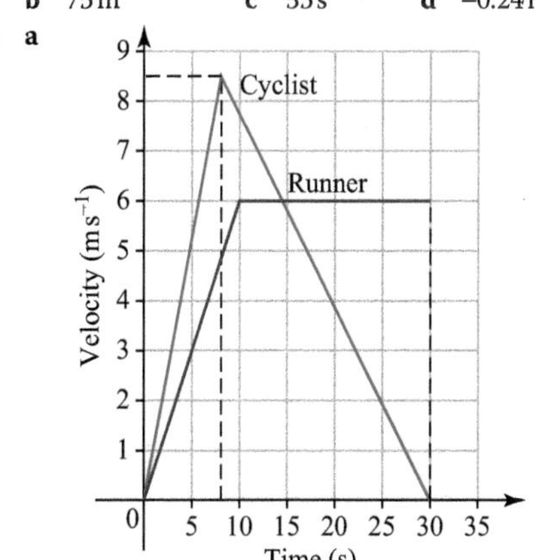

b **i** 150 m **ii** 10 m s^{-1}

5 **a**

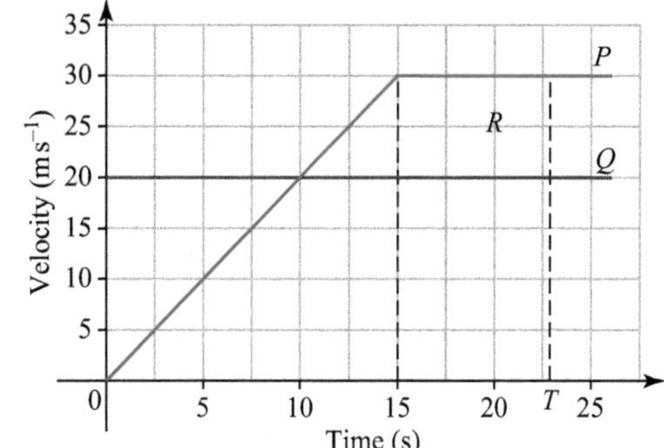

b 75 m **c** 7.5 s

d E.g. It would be impossible for the cars to move with uniform acceleration due to factors like air resistance and the fact that human drivers aren't perfect.

6 Max is 7 min 33.6 s when the roadworks start more than 230 m from the start of the journey and end more than 230 m from the end of the journey.
Min is 7 min 28.8 s when the roadworks start at the beginning of the journey or end at the end of the journey.

Exercise 7.3A Fluency and skills

1 $s = ut + \frac{1}{2}at^2$

2 **a** 18 m s^{-1} **b** 24 m s^{-1} **c** 9 s

3 **a** 36 m **b** 3 m s^{-2} **c** 3 s

4 **a** 2 m **b** 5 s **c** 15 m

5 **a** 18.0 m s^{-1} (to 3 sf) **b** 4.89 s (to 3 sf)

6 **a** 3 m s^{-2} **b** 30 m s^{-1}

7 **a** 0.01 m s^{-2} **b** 0.058 m s^{-1}

8 **a** -4 m s^{-2} **b** 10 s

9 **a** 70 m **b** 4 m s^{-1}

10 **a** 102 km h^{-1} (to 3 sf) **b** 1.16 s (to 3 sf)

Exercise 7.3B Reasoning and problem-solving

1 750 m

2 In first stage it travels $\frac{1}{2} \times 0.2 \times 120^2 = 1440\text{ m}$
$v = 0 + 0.2 \times 120 = 24\text{ m s}^{-1}$
In second stage $24 \times 240 = 5760$ m
In third stage $0^2 = 24^2 - 2 \times 1.5s \Rightarrow s = 192$ m
Distance AB = 1440 + 5760 + 192 = 7392 m

3 **a** 2.4 m s^{-1} **b** 14.4 m
c $(-)0.51\text{ m s}^{-2}$ (to 2 sf)
d The ferry is of negligible size. It would need to decelerate quicker than this to prevent the front of the ferry crashing into the opposite bank (the values calculated are for the back of the ferry).

4 first part: $\frac{5}{36}\text{ m s}^{-2}$, last part: $-\frac{5}{72}\text{ m s}^{-2}$

5 **a** -25 m **b** 65 m

6 5 m s^{-2}

7 **a** B comes to a safe stop 19.375 m behind A (they don't collide).
b The lorries have negligible size. If the lorries are longer than 19.375 m then they will actually collide.

8 See Ch7.3 Fluency and skills for full derivations.

9 **a** $23\frac{5}{6}$ m **b** $3\frac{2}{3}$ s
c The actual distance between humps would be greater than the distance calculated by an amount equal to the length of the car. The time would be the same, as the distance between humps is irrelevant in its calculation.

Exercise 7.4A Fluency and skills

1 **a** $v = 8t - 3t^2$ **b** 4 m s^{-1}

2 **a** $v = 2t - 4$ **b** 16 m s^{-1} **c** 196 m s^{-1}

3 **a** $a = 12t - 8$ **b** 28 m s^{-2}

4 **a** $a = 3t^2 + 7$ **b** 7 m s^{-2}

5 **a** 14 m s^{-1} **b** 13 m s^{-2}

6 **a** $v = 6t - 12$ **b** 2 s

7 **a** $a = 14 - 2t$ **b** 4 s

8 **a** $s = 24t - 3t^2$ **b** 36 m

9 **a** $v = 2t^3 + 3t + 4$ **b** 26 m s^{-1}

10 **a** $12\frac{2}{3}$ m **b** 22 m

11 **a** -7 m s^{-1} **b** -10 m s^{-1}

12 **a** 3 m s^{-1} **b** 5 m

Exercise 7.4B Reasoning and problem-solving

1 **a** $v = 30t - 3t^2$ and $s = 15t^2 - t^3$
b 15 s **c** 500 m

2 **a** 6 s **b** 72 m

3 45 m

4 **a** 4 m **b** 2 s
c 12 m s^{-1} **d** 7.2 m

5 **a** **i** -48 m **ii** 48 m s^{-2}
b 96 m s^{-1} **c** 81 m

6 **a** 4 s

7 $s = 7\frac{1}{3}$ m and $s = 7\frac{1}{6}$ m

8 **a** -9 m s^{-1} **b** 6 m

9 **a** $v = \frac{dx}{dt} = \frac{1}{2700}t^2(t-30)^2$ **b** 30 s
c 300 m **d** 18.75 m s^{-1}

10 **a** Integrate: $[at]_0^t = [v]_u^v \Rightarrow v = u + at$
b $\int_0^s ds = \int_0^t (u + at)\,dt$
Integrate: $[s]_0^s = \left[ut + \frac{1}{2}at^2\right]_0^t \Rightarrow s = ut + \frac{1}{2}at^2$

Review exercise 7

1 0.473 km

2 12 m s^{-1}

3 **a**

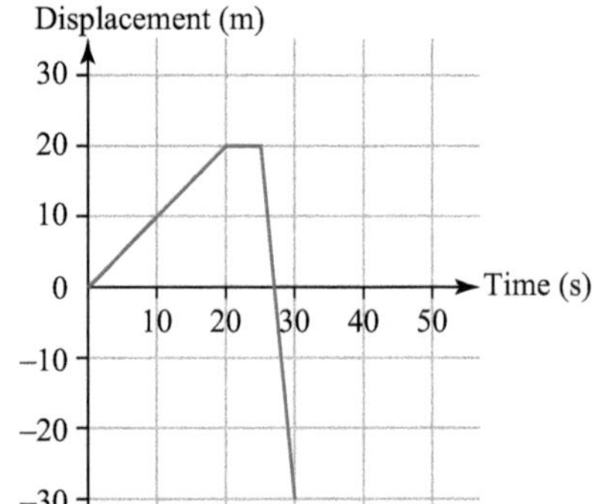

b In the graph, the changes in speed are instantaneous (i.e. there is no acceleration). In reality, this would not be possible and the graph would curve slightly in places.

4 **a** First stage: 2.4 m s^{-2}
Second stage: 0 m s^{-2}
Third stage: -2 m s^{-2}
b **i** 27.5 m **ii** 35.5 m
c 2.75 m s^{-1}

5 **a** 0.4 m s^{-2} **b** 75 m

6 **a** $v = 3t^2 - 16$ **b** 30 m s^{-2}

7 **a** $v = 2t^2 - 3t + 5$ **b** $38\frac{2}{3}$ m

Assessment 7

1 C

2 **a** A **b** D

3 **a**

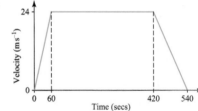

24
Velocity (ms^{-1})
0
0 60 420 540
Time (secs)

b 10.8 km

4 a $6\,\text{m s}^{-1}$ b 3 s

5 a $v = 3t^2 - 18t + 24$ b 2 s or 4 s c 28 m

6 a 8 s b $39\,\text{m s}^{-1}$

7 a

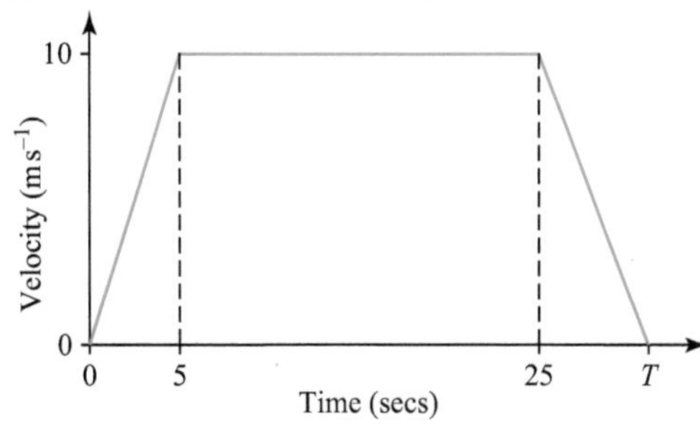

b $T = 33$ c $-1.25\,\text{m s}^{-2}$

8 $29.2\,\text{m s}^{-1}$

9 a $20\,\text{m s}^{-1}$ b 15 s

10 a $2\,\text{m s}^{-1}$ b $\frac{5}{3}\,\text{m s}^{-2}$

c constant velocity of $7\,\text{m s}^{-1}$ d $T = 12\frac{6}{7}$

11 a $v = t^2 - 6t + 10$ b 2 s or 4 s

c $s = \frac{1}{3}t^3 - 3t^2 + 10t$ d 24 m

12 a

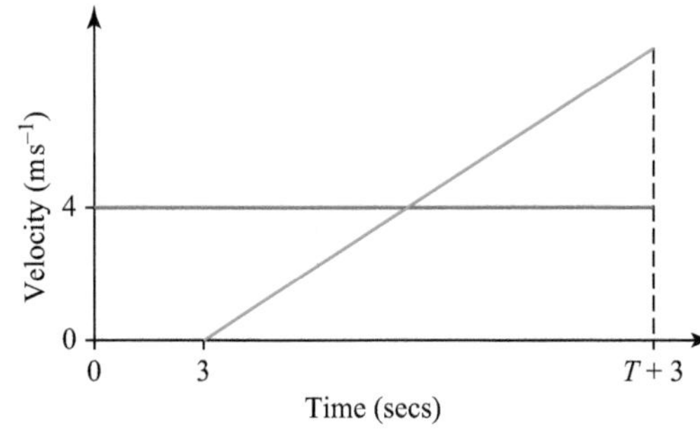

b $T^2 - 4T - 12 = 0$ c $T = 6$ d $12\,\text{m s}^{-1}$

Chapter 8

Try it 8A

1 $T = 5$ N, $P = 17$ N

2 $2.5\,\text{ms}^{-2}$ in the negative horizontal direction

3 $Y = 7$ N, $X = 140$ N (2 sf)

4 $a = 7\,\text{ms}^{-2}$, $T = 34$ N (2 sf)

Bridging Exercise 8A

1 a $X = 6$ N, $Y = 8$ N

b $X = 5$ N, $Y = 11$ N

c $X = 6$ N, $Y = 12$ N

2 $0.857\,\text{ms}^{-2}$ to the right

3 $5.9\,\text{ms}^{-2}$ (2 sf) vertically downwards

4 a 1.33 kg

b 4.5 kg

c 0.75 kg

5 $X = 17$N (2 sf), $Y = 17$ N (2 sf)

6 $X = 15$ N (2 sf), $Y = 15$ N (2 sf)

7 $a = 5.9\,\text{ms}^{-2}$(2 sf), $T = 63$ N (2 sf)

8 $a = 5.4\,\text{ms}^{-2}$(2 sf), $T = 3.0$ N (2 sf)

Exercise 8.1A Fluency and skills

1 a

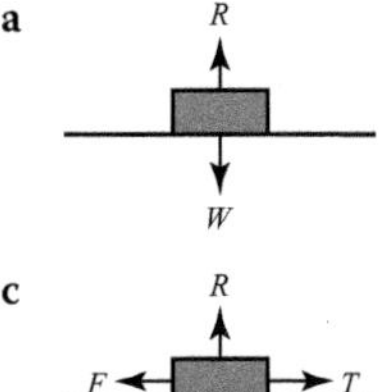

b

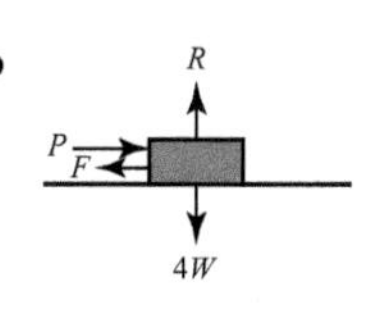

c R F T W

d R T 40° F W

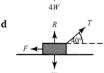

2 a Resolve vertically $R - W = 0$

b Resolve horizontally $P - F = 0$

Resolve vertically $R - 4W = 0$

c Resolve horizontally $T - F = 0$

Resolve vertically $R - W = 0$

3 a 40 N b 40 N

4 a Resolve horizontally $90 - T = 0$

Resolve vertically $Y - 50 = 0$

b Resolve horizontally $Y + 20 - 30 = 0$

Resolve vertically $40 - T = 0$

c Resolve horizontally $50 - Y = 0$

Resolve vertically $20 - X = 0$

d Resolve horizontally $P + Z - X = 0$

Resolve vertically $X - Q = 0$

5 a 900 N b The speed decreases.

Exercise 8.1B Reasoning and problem-solving

1 a $(48\mathbf{i} + 18\mathbf{j})$ N

b $R = 51.3$ N (to 3 sf), $\alpha = 20.6°$ (to 1 dp) above **i**

2 a $(125\mathbf{i} + 125\mathbf{j})$N

b $R = 177$ N (to 3 sf)

$\alpha = 45°$ above **i**

3 a $a = -65$, $b = 85$ b $R = 130$ N, $\alpha = 67.4°$ above $-\mathbf{i}$

4 a $A = 20$N, $B = 30$N b $C = 18$N, $D = 6$N

c $E = 20$N, $F = 20$N d $G = 10$N, $H = 20$N

5 $R = 24.2$ N on a bearing of 119.7°

6 a $R = 5$ N, $\alpha = 36.9°$ above the 12N force.

b $R = 26$ N, $\alpha = 22.6°$ (to 1 dp) below the 94N force.

7 a If $Y = 50$ then $Y \neq 20$

b $4X = 3Y \Rightarrow Y = \frac{4}{3}X$

$5X = 4Y \Rightarrow 5X = 4\left(\frac{4}{3}\right)X \Rightarrow X = 0$

This contradicts the condition $X > 0$

8 a 300 N

b The resistance to motion might decrease due to a change in wind strength or direction.

9 a 50 N b 5 N

10 a Tension. The bar is pulled forwards by the engine and is pulled backwards by the resistance force on the carriage.

b Yes, the bar is now in thrust. The bar is pushed forwards by the engine and pushed backwards by the resistance force on the carriage.

11 There is an overall upward force on the block. Therefore the rod must exert a downward force and so the rod is in tension.

12 $Y = 33.2$N (to 3 sf), $X = 56.2$ N (to 3 sf)

Exercise 8.2A Fluency and skills

1 15 N   2 1.5 m s^{-2}

3 1.2 m s^{-2} 4 712.5 N

5 a $a=(1.8\mathbf{i}+3.6\mathbf{j})\text{ms}^{-2}$ b $a=\begin{pmatrix}1.5\\2.5\end{pmatrix}\text{m s}^{-2}$

c $a=0.28\text{ms}^{-2}$ d $a=(2\mathbf{i}+1.25\mathbf{j})\text{ m s}^{-2}$

e $a=(60\mathbf{i}+84\mathbf{j})\text{ms}^{-2}$

6 3900 N

7 a $R=98\text{N}, a=9\text{ms}^{-2}$ b $R=196\text{N}, X=30\text{N}$

8 0.7 m s^{-2}

9 $m=14$ kg

10 a 3.75 m s^{-2} b 8 s   c 120 m

Exercise 8.2B Reasoning and problem-solving

1 $m=2800$ kg   2 202.5 N

3 30 m   4 53000 N

5 a 25 N   b 550 N (to 2 sf)

c The resistance to motion (wind etc.) will decrease as she slows down, so the actual braking force will be higher than the answer in **b**.)

6 $m=\frac{32}{7}$ and $p=\frac{42}{25}$

7 a 2400 N

b The acceleration will not be constant over this time because the resistance to motion will decrease as the parachutist slows down.

8 2000 kg

9 a 49 N   b 0.95 s (to 2 sf)   c 1.9 m s^{-1} (to 2 sf)

10 $a=6$ and $b=8$

3.75 m s^{-2}

11 $x=-13$ or $x=4$ 12 $m=20$ kg, $R=200$ N

Exercise 8.3A Fluency and skills

1 a 0.098 N   b 98000 N   c 10 kg (to 2 sf)

2 a $t=0.41$ s (to 2 sf)

b The model assumes that there is no air resistance or wind and it models the stone as a particle that has no size and does not spin.

c $v=6.3\text{ m s}^{-1}$ (to 2 sf)

3 $a=6.2\text{ m s}^{-2}$ 4 14.7 m s^{-1} (to 3 sf)

5 20 m s^{-1} (to 1 sf)   6 $m=25$ kg, $a=6\text{ ms}^{-2}$

7 $m=4$ kg, $a=5\text{ ms}^{-2}$ 8 3 m s^{-2}

9 8 kg

10 a 3900 N (to 2 sf)   b 4300 N (to 2 sf)   c 3100 N (to 2 sf)

11 120 N (to 2 sf)   12 $R=15.6$ N

13 20 kg   14 2060 N (to 3 sf)

15 583 N (to 3 sf)

Exercise 8.3B Reasoning and problem-solving

1 $a=1\text{ m s}^{-2}, T=13\,000$ N (to 2 sf)   2 $R=50\text{N}, a=5\text{m s}^{-2}$

3 1 : 6   4 2.5 m

5 4.0 s (to 2 sf)

6 a i $8.5g$N   ii $(8.5g+17)$N

b The rope is light and inextensible.

7 $a=4\text{ m s}^{-2}, s=50$ m   8 1.2 s (to 2 sf)

9 a 27 N   b 17 m s^{-2} (to 2 sf)

10 a 3 m s^{-2} b 5 m s^{-2}

11 $T=3g$

Exercise 8.4A Fluency and skills

1 a 2000 N (to 2 sf)   b 400 N (to 2 sf)

2 $a=2.9\text{ m s}^{-2}$ (to 2 sf), $R=250$ N (to 2 sf)

3 a 1.875 m s^{-2} b 1600 N

4 $a=2\text{ m s}^{-2}, F=9$ N

5 a 3790 N   b 840 N   c 130 m (to 2 sf)

6 a $T=5350$ N (to 3 sf), $R=753$ N (to 3 sf)

b 18.4 N (to 3 sf)

7 T_1 (both crates) = 3800 N (to 2 sf)

T_2 (bottom crate) = 1600 N (to 2 sf)

Exercise 8.4B Reasoning and problem-solving

1 $T=23.5$ N (to 3 sf), $a=1.96\text{ m s}^{-2}$ (to 3 sf)

2 a $a=4.6\text{ m s}^{-2}$ (to 2 sf), $T=26$ N (to 2 sf)

b $v=1.4\text{ m s}^{-1}$ (to 2 sf)

c 0.43 m (to 2 sf)

d 37 N (to 2 sf)

3 a 7 m s^{-2} b 33 N   c 20 m s^{-1} d 30 m

4 a 1.40 m s^{-2} b 33.6 N   c 7.01 m s^{-1} d 20.0 m

5 1.9 s (to 2 sf)

6 $T=\frac{2xyg}{x+y}, a=\left(\frac{y-x}{x+y}\right)g$

Review exercise 8

1 $(\mathbf{i}-\mathbf{j})$ N

2 20.2 N (to 3 sf), $\alpha=8.5°$ (to 2 sf) above **i**

3 $(-20\mathbf{i}+29.4\mathbf{j})$ N

4 $B=10000$N

5 a 7.2 N, 5.2 N (to 2 sf)

b 1.3 N, 0.95 N (to 2 sf)

6 $T=4400$ N (to 2 sf), $R=360$ N (to 2 sf)

7 a $T=46$ N, $a=5.2\text{ m s}^{-2}$

b 1.5 s (to 2 sf)

c 0.36 m (to 2 sf)

d 16 N (to 2 sf)

Assessment 8

1 A   2 C

3 a 3000 N   b 4 ms^{-2} c 62.5 m

4 a 2.0 ms^{-2} (to 2 sf)   b 235.2 N

5 a 4 ms^{-2} b 32 ms^{-1} c 130 m (to 2 sf)

6 a 0.35 m s^{-2} b 580 N

7 a

Velocity (ms^{-1})
6
0
0 3 8 10
Time (secs)

b $R_1=1200$ N (2 sf), $R_2=980$ N, $R_3=680$ N

8 a 470 N (2sf)   b 390 N (2sf)   c 330 N (2 sf)

9 a $2900-700-T=1900a, T-400=800a$

b i $\frac{2}{3}\text{ms}^{-2}$ ii $933\frac{1}{3}$ N

c 144 m

10 a $a=\frac{3g}{10}\text{ms}^{-2}$ b $v=\sqrt{\frac{9g}{5}}$ $(=4.2\text{ m s}^{-1})$

c $\frac{3g}{4}\text{ms}^{-2}$ d $v^2=\left(\frac{9g}{5}\right)-2\times\left(\frac{3g}{4}\right)\times1\Rightarrow$

$v=\sqrt{\frac{3g}{10}}\text{ m s}^{-1}$

11 a 1.7 m s^{-1} (to 2 sf)   b 0.86 s (to 2 sf)   c 160 cm (to 2 sf)

12 a 2.9 m s^{-2} (to 2 sf)   b 25 N (to 2 sf)   c 34 N (to 2 sf)

13 a 12420 N   b 1104 N

c -0.2 m s^{-2} d 768 N

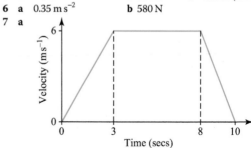

14 a 1800 N b 4600 N c 5300 N
15 a 1400 N b 200 N
16 a 2 m s^{-2} b 240 N (to 2 sf)
17 a 3500 N b 9500 N
 c 333 N (to 3 sf) d $t = 60$ s

Assesment Chapters 6–8: Mechanics

1 a D
 b B
2 a $\mathbf{c} = -22\mathbf{i} - 6\mathbf{j}$
 b 22.8, 195.3° from positive **i** direction
 c They are parallel
 Since $\mathbf{c} = -2(11\mathbf{i} + 3\mathbf{j})$
3 a 4.1 m s^{-1} b 346°
4 a 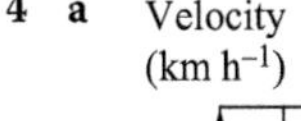

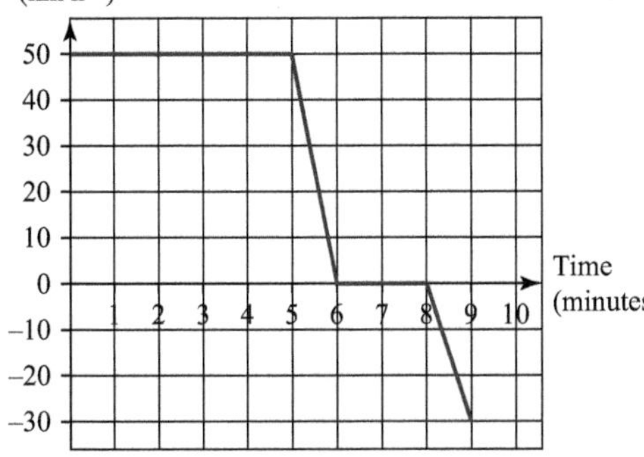

 b 8.83 km c 4.33 km
5 a i 4 m s^{-2} ii 18 m
 b 1.25 kg
6 i 48.3 m s^{-1} ii 34 m s^{-2}
7 $\mathbf{a} = (2\mathbf{i} + 0.75\mathbf{j})\text{ m s}^{-2}$
8 a i 1.76 m ii 5.88 m s^{-1}
 b Book is modelled as a particle, no air resistance, no spin, gravity constant etc.
9 a i $T = 5g$ N ii $T = (5g + 10)$ N
 b The rope is light/has no mass, the rope is inextensible/cannot be stretched.
10 6.5 square units 11 $\left(\frac{15}{2}\sqrt{3}\mathbf{i} + 7.5\mathbf{j}\right)\text{m s}^{-1}$
12 $-3\mathbf{i} + 0\mathbf{j}$
13 465 m, 256°
14 a i 5 s ii 5 cm s^{-1} iii 76 cm
 b 9.33 cm s^{-1}
15 a 112 m b $v = 3t^2 - 3$ c 30 m s^{-2}
 d Velocity (m s^{-1})

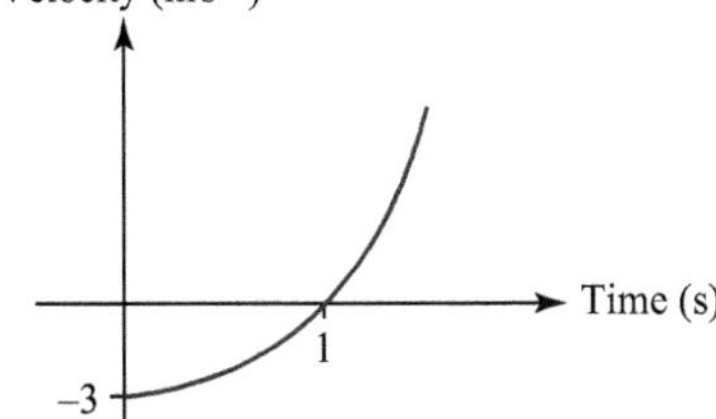

16 a 240 m b 12 m s^{-1}
17 a 8300 N b 268 N
 c It will be larger the heavier the person.
18 a 1.2 m s^{-2} b 4.6 s
19 a 6 m b 3 s
20 a i 1.5 m s^{-1} ii 5100 N
 b 422.5 m
21 a

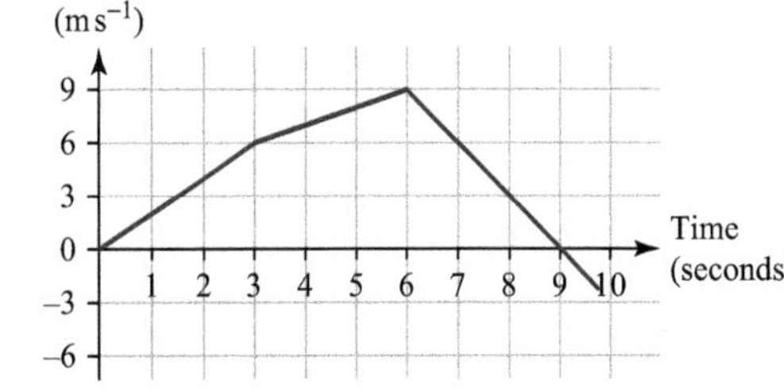

 b 43.5 m c 4.65 m s^{-1}
22 51.3 m
23 a 1.2 s b 8.4 m s^{-1}
24 a $t = 1$ or 1.5 b 1.33 m c 1.42 m
25 a 4.74 s b 20.2 m
26 1.46 m above the ground at time $t = \frac{1}{3}$ seconds
27 a i 0.892 m s^{-2} ii 26.8 N
 b 4.18 m
 c The only force acting horizontally on the 2.5 kg box is now the tension so the resultant force for the whole system will be greater.
 So the acceleration will be greater.

Chapter 9

Try it 9A

1 mean = 5.67
 median = 6
 mode = 6
2 mean = 8.03
 median = 8
 mode = 8
3 mean = 5.56
 modal class is $2 \le x < 5$
4 3.97
5 50

Bridging Exercise 9A

1 a mean = 4
 mode = 1
 median = 3.5
 b mean = 7
 mode = 6
 median = 6
 c mean = 1.1
 mode = 1.3
 median = 1.3
 d mean = 20
 mode = 19 and 23
 median = 19
 e mean = 3.41
 mode = 3
 median = 3
 f mean = 5.3125
 mode = 6
 median = 6
 g mean = 14
 mode = 15
 median = 15
 h mean = 1.16
 mode = 0.5
 median = 1
2 a mean ≈ 20.75
 modal class is $10 \le x < 20$
 median ≈ 19.4
 b mean ≈ 5.47
 modal class is $6 \le x < 8$
 median ≈ 6.17
 c mean ≈ 22.9
 modal class is $20 \le x < 25$
 median ≈ 23.0
 d mean ≈ 10.2
 modal class is $5 \le x < 10$
 median ≈ 8.41
3 78%
4 68.7%
5 1.4
6 5.4
7 64
8 1.8×10^9

Try it 9B

1 range = 8
IQR = 7
2 4.36
3 mean = 6.5
Standard deviation = 3.45
4 0.897

Bridging Exercise 9B

1 a range = 7
IQR = 6.5
b range = 10
IQR = 4
c range = 2.3
IQR = 1.05
d range = 12
IQR = 4.5
e range = 4
IQR = 2
f range = 6
IQR = 2
g range = 20
IQR = 5
h range = 1.5
IQR = 1
2 a 2.45
b 9.92
c 18.3
d 2.48
3 mean = 27.3
standard deviation = 3.29
4 mean = 2.78
standard deviation = 3.22
5 mean = 9.69
standard deviation = 7.07
6 mean = 3.31
standard deviation = 2.91
7 a $\sum x = 76$
$\sum x^2 = 196$
$n = 37$
mean = 2.05
standard deviation = 1.04
b $\sum x = 230$
$\sum x^2 = 3150$
$n = 20$
mean = 11.5
standard deviation = 5.02

Try it 9C

1 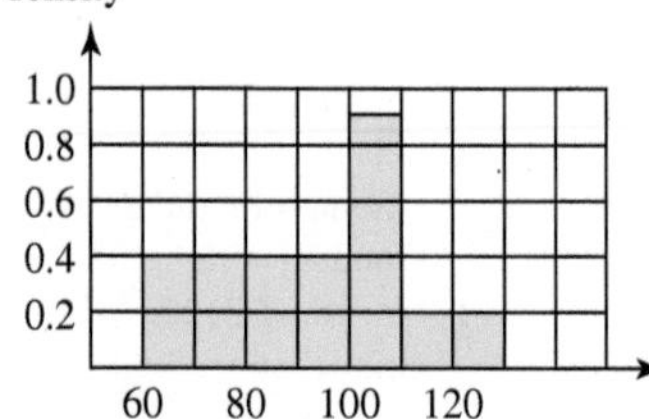

2 a width = 2
Height = 4
b 25 books

Bridging Exercise 9C

1 a 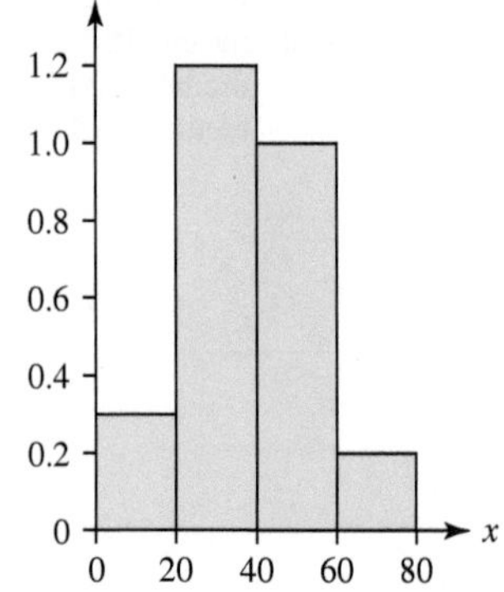

b 0.233

2 a width of 4
Height = 0.25
b 1 person (to nearest whole person)
3 a i 8
ii 17
b 16 or 17 shots

Exercise 9.1A Fluency and skills

1 Population is the adults in their extended family. Parameter is the average height. Sample is the adults in their extended family who live in their city. Statistics are the heights of the sampled adults.
2 Population is the trees of different species in the country's forest. The parameter for each species is the maximum height. The samples are the trees measured in the chosen forest. The statistics are the maximum heights of each species within that forest.
3 a Simple random sampling – Take the list of students in the school and order them from 1 upwards. Generate 40 different random numbers and choose those students.
b Cluster Sampling – Take one class from each year group and sample from the students in those classes using the method described above.
4 291 people
5 a Simple random sampling
b Systematic sampling
c Stratified sampling
6 13 red bags, 17 blue bags, 6 green bags.

Exercise 9.1B Reasoning and problem-solving

1 a The average over three months will be biased depending on the season it was taken in. You would be best to conclude that the average temperature is 21 °C.
b Systematic sampling will give you only data values about a single month's average temperature and not the yearly average.
2 The newspaper might conclude that the average UK house price is £1.7 million. This is unlikely to be reliable, as the newspaper has only sampled from a very small area of the country and it is one that is not likely to be representative of the whole population.
3 Each book in the living room has a $\frac{1}{10}$ chance of being selected but each book in the bedroom has a $\frac{1}{15}$ chance.
These are not equal so it is not simple random sampling. It might not matter if we expect the books to be distributed randomly across the rooms.
4 a True for Idea 1, but not for Idea 2.
b Idea 1 is simple random sampling. Idea 2 is cluster sampling.
c Idea 1 is fairest as each student has an equal probability of being chosen. Idea 2 is a bad idea as each school is unlikely to have students whose views on location are similar to those of each other school.
d The teacher should use stratified sampling to make sure that the number of students chosen from each school is proportional to the number of students at that school. The teacher should then use the list of students at each school and use random numbers to randomly select the appropriate number of students from each school.
5 a Some people give their title as Dr, Rev, etc. and it is not possible to determine their gender. The LDS classifies such people as 'unknown'.
b Men, 400; Women, 236; Companies, 364
6 a A census means that every British car owner is included in the investigation.
Advantage: the results will have no uncertainties.
Disadvantage: a census is expensive in terms of monetary cost, time, organisational effort, etc.

b i A stratified sample is one in which the population is divided into different groups and the number of car owners in the sample taken from each group is in direct proportion to the size of the group in the population.

ii Any two of

By the type of car owned: MPV, estate, hatchback, etc.

By the type of area where the owner lives: city, town, countryside, etc.

By the owner's income group: high income, middle income, low income.

By the owner's occupation: medical professions, transport industry, office worker, etc.

By the number of children in the owner's household: none, one, two, three or more.

Any other factor which could affect an owner's attitudes to pollution.

c A biased sample is one which is not representative of the whole population.

d London is a very large city, with good public transport, and people may have more extreme views about pollution than, for example, a country dweller.

If Londoners are less (more) tolerant of pollution than the British population as a whole then the researcher's results will be biased towards being less (more) tolerant than the true level of tolerance.

Exercise 9.2A Fluency and skills

1 a i 8 ii 6.85 iii 7

iv $11-2=9$ v $\sigma=2.41$

b i 68, 71 ii 68.8 iii 69

iv $75-62=13$ v $\sigma=3.40$

2 a $16-21$ b Mean 19.6

3 a $0-4$ b Mean 5.44

4 150.1 mg (1dp)

5 a $Q_1=14.5$, $Q_3=30.5$, IQR $=16$ b $Q_1=2.1$, $Q_3=3.8$, IQR $=1.7$

6 $\bar{x}=7.41$

$s^2=0.425$ (3 dp)

7 a 3.76 b 3.49

8 a 6 b 1.08

9 a 0.816

b i s^2 because the researcher is estimating a parameter for a larger population using a sample.

ii $s^2=2.17$

10 a The researcher will use this information to estimate the spread of the larger population.

b $s=2.26$

Exercise 9.2B Reasoning and problem-solving

1 a $M=16.5$

Range $=4$

IQR $=1$

b The range is affected by the outlier 19.5 so does not represent the spread of most of the data.

c 14, 19

d $14<14.5<19$, do *not* test the batch.

2 a $\mu=15.9$, $\sigma=4.37$

b $15.9-2\times4.37=7.16$

$15.9+2\times4.37=24.64$

$27>24.64$

No other values in the list are above 24.64, no values in the list are below 7.16

c i The outlier might be have been recorded incorrectly or might represent a very unusual occurrence.

ii The data set is small at 16 items so by chance might give a disproportionate number of small values. In this case, removing 27 could lead to an underestimate for the standard deviation.

d This makes μ and σ smaller.

3 a Mean = 1405 kg, Variance = 31 530 kg^2

b No

Different car makers have different shares of the market. The average should be weighted by the car makers' market shares.

4 a Median = 156 g km^{-1}, IQR = 21 g km^{-1}

b Typical Vauxhall and Ford cars emit the same amount of CO_2: median $156\approx152.5$

Compared to Ford cars, Vauxhall cars show a larger variability in the level of their CO_2 emissions: IQR $45>21$

5 a £2.43 b £31

c The median would decrease.

6 a No mode

Median $=107$

Mean $=109$

b The median. There is no mode and the mean is distorted by the outlier 145. The median is not distorted by outliers.

c $a=102$

The first team has a lower median score of 107 compared with 110 so are, in general, better golfers. The first team has a larger interquartile range of 9 compared with 7 so exhibit more diversity in achievement. The median and interquartile range are not affected by the outlier unlike the mean, standard deviation and range.

Exercise 9.3A Fluency and skills

1 Class 2 did better on average, the median is $82>73$ for Class 1. The Class 2 scores include an outlier, 50, but even with this included are generally more consistent with an interquartile range of $86-74=12<89-65=24$ for Class 1.

2 a 220 houses b 208 houses

3 a 12 b $Q_1\approx90$, $Q_3\approx110$

c Interquartile range $\approx110-90=20$

$90-1.5\times20=60$

Reading up from 60 on the graph the cumulative frequency is 1 so 1 value is below 60.

$90+1.5\times20=120$

Reading up from 120 on the graph the cumulative frequency is 28 so 2 values are above 120.

Therefore there are three outliers.

4 a i 23 ii 32 iii 9

b This group drinks more milk on average with median 250 ml > 230 ml for fruit juice. There more variety in fruit juice consumption with interquartile range $380-130=250>320-190=130$ for milk.

5 a $0.5-0.7$ frequency $=18$

$0.8-1.0$ frequency $=6$

b Width $=0.4$ (= 4 squares)

Height = 3.75 squares = 37.5 units

c i 39% ii 31%

Exercise 9.3B Reasoning and problem-solving

1 a i 0.75 ii 0.625

b The end of the upper whisker will move to 165 kg

2 A box-and-whisker plot would show the five-number summary but the histogram would show the full distribution and would give the art student a better idea of how much money they could make. So a histogram may be the better choice.

3 a Volkswagen estates may be made with only a small number of engine sizes. Thus many cars may have the same engine size causing quartiles to coincide.

b 0.5

c 0.25

d It is possible that a quarter of BMW estates have engines of size 2000 cm³ and a quarter of Vauxhall estates have engines of size 2175 cm³
Therefore it is possible that only 75% of BMW estates have larger engines than Vauxhall estates

4 a $0.15 \le x < 0.20$: frequency 14
$0.20 \le x < 0.30$: frequency 4
Bar 0.075 to 0.10, height = 11 squares / 440
Bar 0.10 to 0.15, height = 2 squares / 160
Bar 0.15 to 0.20, height = 7 squares / 280

b i 59%
ii 95%

Exercise 9.4A Fluency and skills

1 a Moderate negative correlation.
b Moderate positive correlation.
c Strong positive correlation.
d Zero correlation.

2 a Moderate negative correlation. **b** –0.7

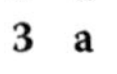

3 a i

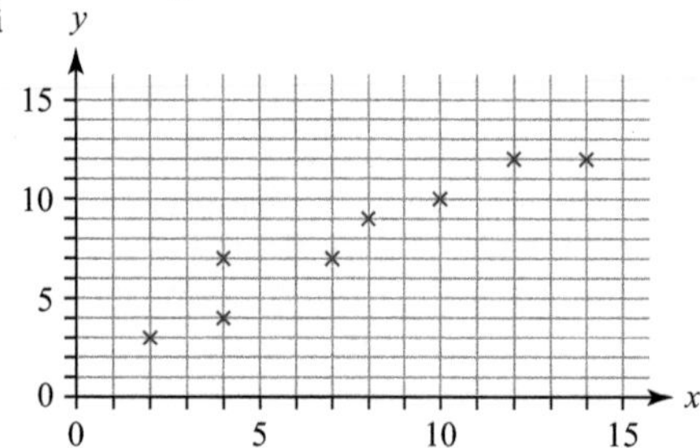

Strong positive correlation

ii

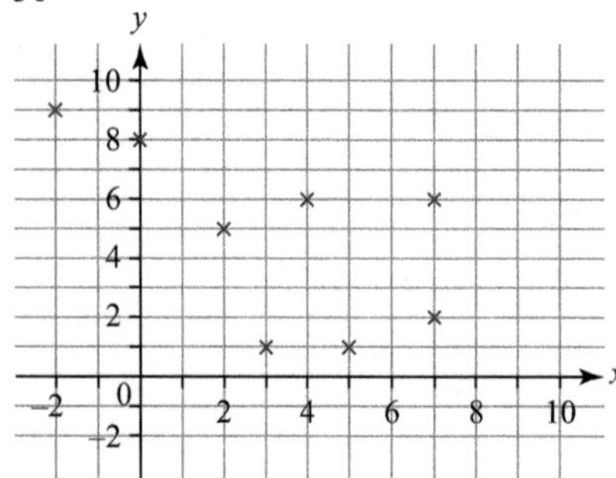

No correlation

b These correlation coefficients indicate perfect straight lines but it is impossible to draw a straight line that passes through every data point in this set. The straight line that passes through (4, 4) and (7, 7) is $y = x$ but the point (2, 3) does not lie on this line.

4 a

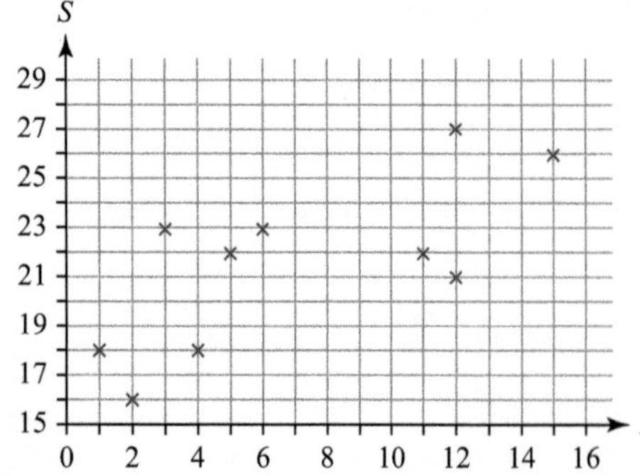

b i +0.73
ii As L increases, S increases so the value of the correlation coefficient must be positive. The correlation coefficient cannot take a value greater than 1.

5 a

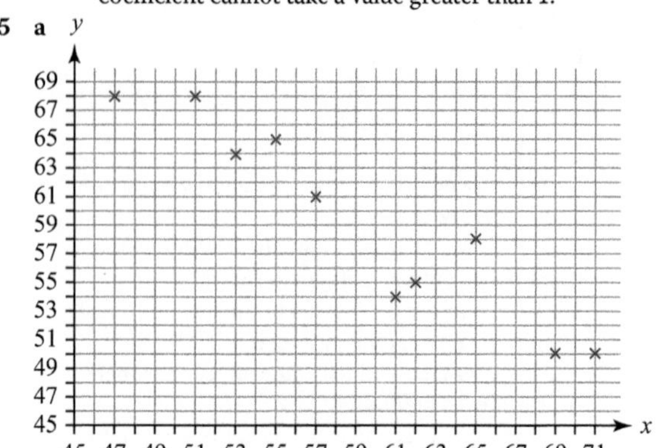

b Moderate/strong negative correlation. Value in the range –0.7 to –1.0

Exercise 9.4B Reasoning and problem-solving

1 a Positive correlation, causal.
b Positive correlation, causal.
c Zero correlation.
d Negative correlation, not causal.

2 a

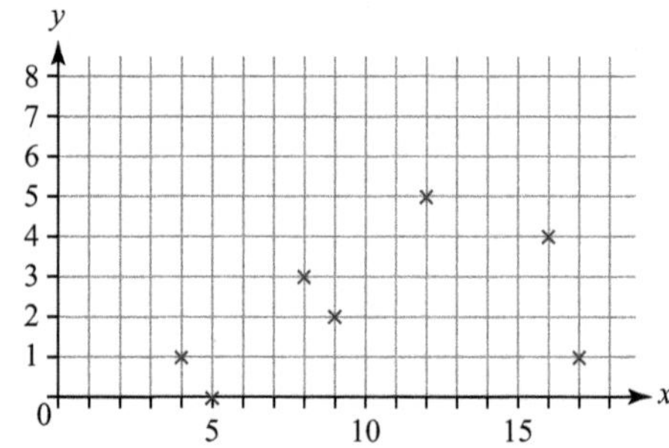

(17, 1)

b 5

3 a, c

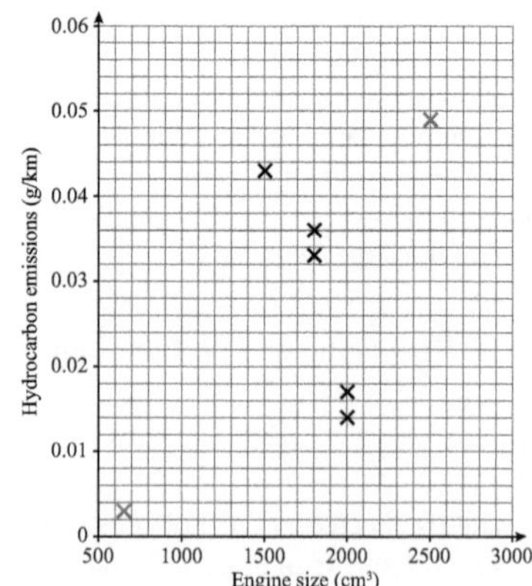

b Strong, negative correlation: cars with larger engine sizes usually have smaller hydrocarbon emissions.

c Cars B and C now appear to be outliers.
Moderate, positive correlation: cars with larger engine sizes are likely to have larger hydrocarbon emissions.

4 a i Positive
ii Negative
iii Positive

b Spurious means that there is no direct causal link. Increasing the number of cars does not cause the number of vans to increase. Another factor is likely to cause both to rise, for example, the growth in the population.

5 a

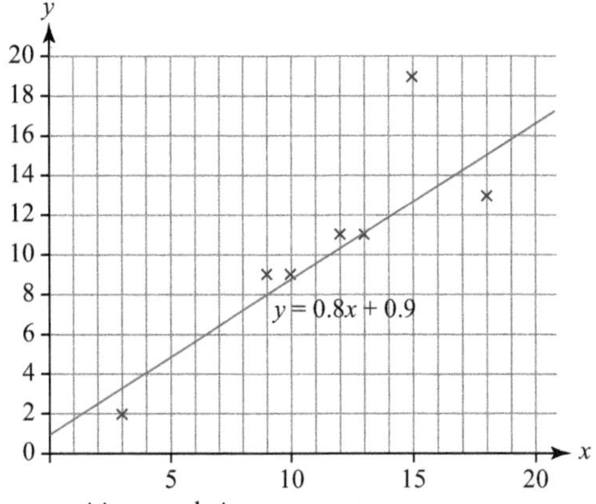

Strong positive correlation.

b (15, 19) **c** 6.5

Review exercise 9

1 a Population: all LED light bulbs.
Sample: the 23 bulbs measured.
Parameter: the temperature of a lit LED light bulb.
Statistics: the temperatures of the 23 light bulbs lit.

b Population: all casks of that type of whiskey.
Sample: the nine casks measured.
Parameter: the alcohol level of a cask of whiskey.
Statistics: the alcohol content of the nine casks measured.

2 a Simple random sampling.
b Opportunity sampling.
c Cluster sampling.
d Opportunity sampling and cluster sampling.

3

Class	Frequency (any multiple of this column)	Frequency Density
0–15	21	1.4
15–50	7	0.2
50–70	10	0.5
70–90	16	0.8
90–105	33	2.2
105–130	20	0.8
130–160	6	0.2
160+	0	0

(The frequency density column is not required.)

4 a

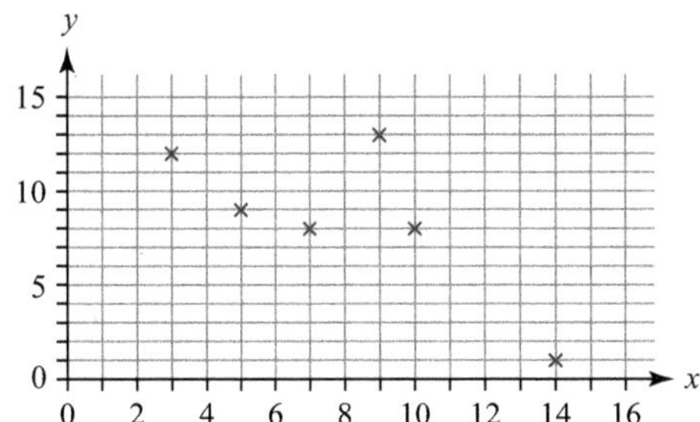

Moderate negative correlation

b Incorrect point: (9, 13)
More plausible point: (9, 6)
Accept $y = 5$ to 7 also

5 Median value = 10.75, mean = 10.8, modal interval: $8 \le x < 12$, variance $s^2 = 18.56$

6 a IQR = 6 **b** Outlier: 26

Assessment 9

1 B

2 C

3 a All the batteries produced.
b You may be testing to destruction.
c The mean/median length of life of the batteries in the sample.

4 a Moderate, negative correlation: hatchback cars with higher CO emissions tend to have smaller NOX emissions.
b −0.6

5 a 5 **b** 9% **c** £49.11

6 a i 0.0698 g km^{-1}
ii 0.0429 g km^{-1}
b Non-outlier range [0, 0.156]
Outliers: 0.172
c 0.048 g km^{-1}
d Median is the best representative as it uses all the data but is largely unaffected by the outlier, unlike the mean. The mode is less reliable due to the small number of data points.
e Mean = 0.0632 g km^{-1}
Sample, standard deviation = 0.0360 g km^{-1}
Non-outlier range [0, 0.135]
Outliers: 0.172
Median = 0.050 g km^{-1}
Median is the best representative as it uses all the data but is largely unaffected by the outlier; unlike the mean. The mode is the assumed value for the missing data.

7 a 3 **b** 2.3, 1.49
c Median: 2, Q1: 1, Q3: 3
d Mode/median
e.g. any one of
Mode is the most likely number of assignment to be set.
Mode doesn't take into account fact that most students set 3 or fewer assignments.
Median and mode are unaffected by outliers.
Mean is affected by the outlier who claims to have been set 8 homework assignments.

8 a £1.60 **b** £2.60 **c** 20%

9 a e.g. sample will not be representative of the population as there are more girls than boys in each year / more year 12s than 13s.
b 20 year 12 girls
17 year 13 girls
13 year 12 boys
10 year 13 boys
c e.g. students in the common room may use the library less often.

10 A, C, D

11 a Temperature
b Moderate positive correlation.
A higher temperature implies a higher germination rate.
c An answer between 0.6 and 0.85
d Not necessarily – we do not have data for temperatures this high/high temperature may kills the seeds.

12 a i 0.122 g km^{-1} **ii** 0.154 g km^{-1}
b i 0.0332 g km^{-1} **ii** 0.0215 g km^{-1}
c There has been a significant, factor 4, decrease in the average level of NOX emissions: mean $0.0332 < 0.122$
There has been a significant, factor 7, decrease in the spread of NOX emissions: standard deviation $0.0215 < 0.154$

13 a Method A – stratified random sampling
Method B – quota sampling
b Method A is preferable as each member is equally likely to be chosen, as opposed to Method B where the person chosing the sample will possibly just choose the most convenient people.

14 a Men: 1475 kg
Women: 1225 kg
b Men: 400 kg
Women: 350 kg
c On average men drive heavier cars: median 1475 > 1225. There is more variation in the mass of cars driven by men: IQR 400 > 350 and range 1350 > 925.
d If the numbers of men's cars and women's cars in the sample were equal Eric's claim would be true as the cars above 1500 kg, 50% of men's cars and 25% of women's cars, would represent a 2:1 ratio. In fact, there are more men's cars than women's cars in the sample so a car with mass greater than 1500 kg is *more than* twice as likely to be registered to a man as to a woman.

16 a Width = 80 mm, Height = 12 mm
b i 133 g km^{-1}
ii 39 g km^{-1}

Chapter 10

Exercise 10.1A Fluency and skills

1 (1, 2), (1, 3), (1, 4), (2, 1), (2, 3), (2, 4), (3, 1), (3, 2), (3, 4), (4, 1), (4, 2), (4, 3)

2 a $a = 12, b = 8, c = 7, d = 25, e = 45$
b i $\frac{4}{9}$ ii $\frac{7}{45}$ iii $\frac{4}{9}$

3 a $\frac{3}{31}$ b $\frac{8}{31}$ c $\frac{20}{31}$

4 a 0.1 b 0.9

5 a 0.5 b 0.5 c 0.7 d 0

6 P(square number or prime) = P(1, 2, 3, 4 or 5) $= \frac{10}{12} = \frac{5}{6}$

7 a $\frac{1}{3}$
b

x	0	1	2	3	4	5
$P(X = x)$	$\frac{1}{6}$	$\frac{5}{18}$	$\frac{2}{9}$	$\frac{1}{6}$	$\frac{1}{9}$	$\frac{1}{18}$

8 a $0.2+0.3+0.4+b=1; b=0.1$ b 0.8

Exercise 10.1B Reasoning and problem-solving

1 a

Score	1	2	3
Probability	0.3	0.6	0.1

b i 0.9 ii 0.4

2 a i 0.457
ii 0.549
b P(Ford ∩ SW) = 0.247
P(Ford) × P(SW) = 0.251
Yes, the difference in the two expressions for the probability is consistent with sampling errors.
c 0.138

3 a $a = 0.32, b = 0.20$
b i 0.43
ii 0.57
iii 0.73

4 a A and B are mutually exclusive.
b i 0.2 ii 0.2

5

	1	2	3	4	5	6
1	2	3	4	5	6	7
2	3	4	5	6	7	8
3	4	5	6	7	8	9
4	5	6	7	8	9	10
5	6	7	8	9	10	11
6	7	8	9	10	11	12

No. For fair dice P(sum of scores > 10) $= \frac{1}{12}$ so at least one dice is not fair.

6 a 0.3 b 0.14 c 0.56

7 a $P(X = 2) = (1 - p)p = 0.21$(given) so $p^2 - p + 0.21 = 0$
$(p - 0.3)(p - 0.7) = 0$ so $p = 0.3$ or 0.7
Since $p < 0.5, p = 0.3$
b 0.657

8 Yes. $\frac{3}{4} \times \frac{1}{2} = \frac{3}{8}$

9 12

10 a i $P(E) = 0.5, P(F) = 0.4, P(E \text{ or } F) = 0.8$ ii 0.2
b $P(E) + P(F) - P(E \text{ and } F) = 0.5 + 0.4 - 0.1 = 0.8 = P(E \text{ or } F)$

Exercise 10.2A Fluency and skills

1 a 0.132 b 0.837 c 0.832

2 a 0.00066 b 0.17 c 0.41 d 0.95

3 a 0.087 b 0.00038 c 0.29

4 a $P(X = x) = {}^5C_x \times 0.4^x \times 0.6^{5-x}$
b

x	0	1	2	3	4	5
$P(X = x)$	0.078	0.259	0.346	0.230	0.077	0.010

5 a $X \sim B\left(4, \frac{1}{6}\right)$
b i 0.001 ii 0.016 iii 0.502

6 0.016 (to 3 dp)

Exercise 10.2B Reasoning and problem-solving

1 a A fixed number of 'yes/no' trials that are independent of each other and with fixed probability of success.
b i 0.089 ii 0.673 iii 0.661

2 a Yes. $n = 4, p = \frac{3}{7}$
b Yes. If the population is sufficiently large when compared to the sample, as the patients are chosen at random the probability of getting a patient who will be prescribed antibiotics remains constant. $n = 8, p = 0.12$
c Yes. $n = 5, p = 0.6$
d No. Number of trials not fixed.

3 No. If it was, the probability of two zeros would be 0.0486

4 a i 0.083 ii 0.209
b i 0.689 ii 0.211

5 a 0.247
b Lower
The smaller probability means that it is less likely that a large number of cars would be diesel fuelled.
[Probability = 0.0001]

6 0.939 (to 3 sf)

7 0.5

8 Let X be a random variable for the number of girls born in five children.
P(girl) = 0.5 so $X \sim B(5,0.5)$
$P(A) = P(X \text{ is not 0 or 5}) = 0.9375$
$P(B) = P(X \geq 3) = 0.5$
$P(A \text{ and } B) = P(X = 3 \text{ or } 4) = 0.46875$
$P(A) \times P(B) = 0.46875 = P(A \text{ and } B)$ so events are independent.

Review exercise 10

1

x	1	2	3
$P(X=x)$	p	$(1-p)p$	$(1-p)^2p+(1-p)^3$

$p+(1-p)p+(1-p)^2p+(1-p)^3 = p+p-p^2+p-2p^2+p^3$
$+1-3p+3p^2-p^3 = 1$

2 a $\frac{1}{84}$ b $\frac{83}{84}$

3 a 0.004 (to 3 dp) b 0.473 (to 3 dp)

4 a $X \sim B(5,0.5)$

b i 0.31 (to 2 dp) ii 0.97 (to 2 dp) iii 0.47 (to 2 dp)

5 0.788 (to 3 sf)

6 0.993 (to 3 sf)

7 0.100 (to 3 sf)

8 Your friend. Let Y be the number of double 6s in 24 throws.

$P(Y \geq 1) = 1 - P(Y=0) = 1-\left(\frac{35}{36}\right)^{24} = 0.491$ (to 3 sf)

9 0.613 (to 3 sf)

10 a $X \sim B(12,0.68)$ b 0.432 (to 3 sf)

Assessment 10

1 B

2 D

3 a 56212

b i $\frac{53}{94}$ ii $\frac{17}{46}$ iii $\frac{901}{4324}$

c Yes

4 a The probability that one event occurs is not affected by the occurrence or non-occurrence of any other event. (Or $P(A \cap B) = P(A) \times P(B)$)

b $\frac{1}{84}$ c $\frac{1}{140}$

5 a $\frac{1}{20}$ b 0.35 c The second bag

6 a 0.0773 (to 3 sf) b 0.997 (to 3 sf) c 0.992 (to 3 sf)

7 a Each discarded sweet has an equal probability of being misshapen and each sweet's probability of being misshapen is not affected by any other sweet being misshapen.

b 0.0806 (to 3 sf)

c 0.403 (to 3 sf)

d 0.668 (to 3 sf)

8 0.991 (to 3 sf)

9 a 184

b i $\frac{43}{92}$ ii $\frac{65}{184}$ iii $\frac{25}{184}$

c No, since $\frac{43}{92} \times \frac{65}{184} \neq \frac{25}{184}$

10 a 0.547

b 0.833

c 0.994

d 28.4 weeks

11 0.335

Chapter 11

Try it 11A

1 a There isn't strong evidence that the coin is biased towards tails since there is an 18.08% chance that a fair coin would give this result. This is greater than 5% so assume the coin is fair.

b 10 or fewer

2 2 heads from 12 flips implied that the first coin was biased towards tails, therefore 20 heads from 120 flips will imply that the second coin is biased since a larger number of trials requires a less 'extreme' result.

Bridging Exercise 11A

1 a 32 or fewer

2 a i There isn't strong evidence that the coin is biased towards tails since there is an 10.13% chance that a fair coin would give this result. This is greater than 5% so assume the coin is fair.

ii The probability of 18 or fewer heads is 3.25%, this is less than 5% so supports the claim that the coin is biased towards tails.

iii 18 or fewer heads implied that the coin is biased towards tails so 15 heads will provide even stronger evidence for this.

iv 20 or fewer heads was not strong enough evidence to support a claim that the coin is biased towards tails so 22 heads will not be strong enough evidence either. Therefore, assume the coin is fair.

b 36 out of 100 is the same proportion as 18 out of 50. 18 heads out of 50 implied a coin was biased towards tails. Therefore 36 heads out of 100 will imply that a coin is biased since a larger number of trials requires a less 'extreme' result.

c 10 out of 25 is the same proportion as 20 out of 50. 20 heads out of 50 was not good enough evidence that a coin was biased towards tails. Therefore, 10 heads out of 25 will not be enough evidence to imply a coin is biased towards tails since a smaller number of trials requires a more 'extreme' result.

Exercise 11.1A Fluency and skills

1 a Critical region: $X \leq 5$, acceptance region: $X \geq 6$

b Critical region: $X \geq 5$, acceptance region: $X \leq 4$

c Critical region: $X \leq 5$ or $X \geq 12$, acceptance region: $6 \leq X \leq 11$

2 a Accept H_0 b Reject H_0 c Accept H_0

3 a Reject H_0 b Accept H_0

4 a p is the probability that a game is lost. H_0: $p = 0.15$ and H_1: $p \neq 0.15$

b Critical region: $X \leq 2$ or $X \geq 14$, acceptance region: $3 \leq X \leq 13$

c i Accept H_0 ii Reject H_0 iii Reject H_0

5 a Critical region: $X \leq 33$, acceptance region: $X \geq 34$

b i Reject H_0 ii Accept H_0

6 a Critical region: $X \geq 9$, acceptance region: $X \leq 8$

b No.

7 a Critical region: $X \geq 5$, acceptance region: $X \leq 4$

b $p < 10$

Exercise 11.1B Reasoning and problem-solving

1 a H_0: $p = 0.6$ b H_0: $p = 0.7$
H_1: $p > 0.6$ H_1: $p < 0.7$

c H_0: $p = 0.4$
H_1: $p \neq 0.4$

2 a Let X = number of cars with an engine size above 1975 cm^3

$X \sim B(20, p)$

p = Probability that a sold car has an engine size above 1975 cm^3

H_0: $p = 0.25$, H_1: $p \neq 0.25$

b $2 \leq n \leq 9$

3 a Let X = number of cars in the sample with carbon monoxide emissions above 0.5 g km^{-1}
$X \sim B(150, p)$
p = Probability that a car has high CO emissions
$H_0: p = 0.250$, $H_1: p > 0.250$
b Accept H_0, there is insufficient evidence, at the 15% significance level, to say that the fraction of Vauxhall cars with high CO emissions is greater than 25.0%.
c 44 cars
4 a p is the probability that James hits the bullseye.
$H_0: p = 0.25$
$H_1: p < 0.25$
b The critical value at the 10% significance level is 6 so the critical region is $X \le 6$. 5 is in the critical region so if James hit the bullseye with 5 of his darts you reject the null hypothesis. You conclude that there is evidence, at the10% significance level, to say that his rate of hitting the bullseye is less than 25%.
c $n \ge 7$
d $x < 10$
5 a $a \ge 8$ and $b \le 24$ b $x < 5$

Exercise 11.2A Fluency and skills

1 a $X \le 1$ b Reject H_0
2 $X \le 3$ or $X \ge 13$
3 a $P(X \le 1) = 0.0480$ (to 4 dp). Reject H_0
b $P(X \le 2) = 0.1514$ (to 4 dp). Accept H_0
4 a p-value = 0.0274 (to 4 dp), accept the null hypothesis.
b p-value = 0.0468 (to 4 dp), accept the null hypothesis.
5 a $P(X \ge 34) = 0.0462$. Reject H_0
b $P(X \le 21) = 0.0758$. Accept H_0
6 $X \ge 8$
7 a Reject the null hypothesis.
b Accept the null hypothesis.
8 a i $X \ge 26$ ii $X \ge 23$ iii $X \ge 19$
b i Accept the null hypothesis.
ii Accept the null hypothesis.
iii Reject the null hypothesis.
c $2 < p < 20$
9 a i $X \ge 66$ ii $X \ge 64$. iii $X \ge 62$.
b i Accept H_0 ii Reject H_0 iii Reject H_0
c 4
10 $y = 1$
11 a $X = 0$ b 28
12 $a = 5$ or 6

Exercise 11.2B Reasoning and problem-solving

1 a Let X = number of cars owned by men.
$X \sim B(40, p)$
p = Probability that a BMW is owned by a man
$H_0: p = 0.85$, $H_1: p < 0.85$
Critical region: $n \le 29$
b Accept H_0, there is insufficient evidence, at the 5% significance level, to say that the fraction of BMW cars owned by men is less than 85%.
2 a Let X = number of company-owned cars.
$X \sim B(70, p)$
p = Probability that a new car is owned by a company
$H_0: p = 0.65$, $H_1: p \ne 0.65$
p-value = 0.01997
b Accept H_0, there is insufficient evidence, at the 1% significance level, to say that the fraction of new cars owned by companies is not 65%.
3 a $H_0: p = 0.15$, $H_1: p \ne 0.15$
Reject H_0
b Yes: accept H_0

4 a Under H_0, $X \sim B(n, 0.5)$
i Accept H_0 ii Reject H_0
b 18
5 a 0.1 b 10%
6 a $H_0: p = 0.1$, $H_1: p < 0.1$ b 1
7 a $n = 40$ b $X \le 11$ or $X \ge 29$

Review exercise 11

1 $X \sim B(n, p)$
2 a $H_0: p = 0.2$, $H_1: p \ne 0.2$ b $X = 0$ or $X \ge 10$
c 0.0211 (to 4 dp) d The distribution is discrete.
3 a $H_0: p = 0.5$, $H_1: p > 0.5$ b 13
4 a $H_0: p = 0.2$, $H_1: p > 0.2$
b Conclude that there is not enough evidence to suggest that the student was not guessing the answers.
5 a Fixed n and p; independent and identical trials; each trial results in exactly one of two possible outcomes, "success", "failure"; interested in the number of successes (or failures) only.

$$P(X = x) = \binom{n}{x} p^x (1-p)^{n-x}; x = 0, 1, 2..., n$$

b

x	0	1	2	3	4	5	6
$P(X = x)$	0.0827	0.2555	0.3290	**0.2260**	**0.0873**	**0.0180**	0.0015

c x-values 5 and 6. Significance level 1.95%
6 a $H_0: p = 0.1$, $H_1: p \ne 0.1$ b No. c No.

Assessment 11

1 D
2 B
3 a $H_0: p = \frac{1}{2}$ $H_1: p \ne \frac{1}{2}$
b X-values in the critical region are 0, 6
Significance level = $P(X \le 0$ or $X \ge 6) = 0.0312 = 3.12\%$
4 a $H_0: p = 0.5$, $H_1: p > 0.5$
b $n \le 8$
c It has increased
5 a Accept H_0 b Yes
6 a Ready-meals chosen at random.
b $H_0: p = 0.4$, $H_1: p < 0.4$. No reason to reject H_0
7 Critical region: $n \le 2$
Yes, there is evidence that a smaller proportion of diesel cars have high particulate emissions than petrol cars.
8 a Let a successful trial be the event a car has mass ≤ 1600 kg
There are only two outcomes to a trial.
There are a fixed number of trials, 14
The probability of success, 0.81, is fixed and the same for all trials
Trials are independent; for randomly chosen cars in a large sample the mass of one car should not influence the mass of another car.
b ≤ 7, 0.00871; ≤ 8, 0.0349; ≤ 9, 0.109
c $P(X \le 8) = 0.0349 < 10\%$. The journalist's belief is supported by the evidence.
9 a $H_0: p = 0.6$, $H_1: p < 0.6$ b 4
10 a Random sample. So that a binomial distribution is a valid model.
b $H_0: p = 0.35$, $H_1: p > 0.35$, $X \sim B(30, 0.35)$
c Accept H_0
11 a $k^n + nk^{n-1}(1-k)$ b $0 < k \le 0.1$

Assessment chapters 9–11: Statistics

1 a A 20 **b** B 12

2 a i Simple random
ii Opportunity
iii Stratified
b i The sample is likely to be unbiased
Many of the car owners contacted will not respond to the survey.
ii The sample will be straightforward to collect.
The sample is likely to be biased to car owners in a small area of Britain who move their cars at a certain point in the day.
iii The sample should avoid one potential source of bias, income.
It may be difficult to find out a car owner's income group.

3 a 1.5
b e.g. The range/IQR is wider 2 hours after the meal than before the meal.
e.g. The average blood glucose level is higher 2 hours after the meal than before the meal.
c 75% **d** 0.5625

4 a Moderate/strong negative correlation. The longer spent training, the quicker the trainees can complete the task.
b –0.8
c Increase/be closer to zero

5 a i $\frac{5}{12}$
ii $\frac{4}{12} = \frac{1}{3}$
b No
$\text{P(BMW)} \times \text{P(Man)} = \frac{7}{18} \neq \frac{1}{3} = \text{P(BMW} \cap \text{Man)}$

6 a 0.0146 **b** 0.1256 **c** 0.1275

7 a i $\frac{1}{7}$
ii $\frac{3}{7}$
b i 0.0100
ii 0.110
iii 0.803

8 a 22 is the error. 77 is the correct rate.
b 103
c 99 is an outlier.
d Median is 68; Median will not be affected by the outlier.

9 a Medium
b Mean = 59.3 g, standard deviation = 9.7
c Median = 58, IQR = 14.1
d Mean will increase, standard deviation will decrease.

10 a i 0.9 **ii** 0.6
b 0.1

11 a i 0.107
ii 0.678
iii 0.00637
b The binomial distribution should provide a good model. There are only two event outcomes. The events are independent and with identical probabilities, 0.2. There is a fixed number of events, 10.

12 a Critical region $X \geq 6$
5 is not in the critical region so accept H_0. There is insufficient evidence to support the rival's claim that their cars have significantly lower NOX emissions.
b 7.32%

13 a i Opportunity
The survey will be expensive.
The people who visit car parks at the time of the survey may not be representative of the whole population, potentially biasing results.
ii Simple random.
The people who appear in the phone book may not be representative of the location's population. The location's population may not be representative of the population in the whole country.
b BMW, 1282; Ford, 3589; Vauxhall, 2821; VW,2308

14 a Width 2.5cm; Height 2.4 cm **b** 15 **c** 0.0486

15 9

16 a MPVs are typically above average size cars and likely to have bigger engines producing more pollutants, therefore it is not a fair comparison.
b Reject H_0, there is evidence that the level of CO_2 emissions for Joe's sample of MPVs is lower.
c Significance level > 0.000116

Index

A

acceleration
- constant 248–51
- definition 243
- differentiation 137, 252
- forces 262–5, 272, 274, 276
- integration 168, 253
- kinematics 235, 248–51, 252
- rates of change 150
- units 240–1
- *v-t* graphs 238, 243–4, 246, 252
- variable 252–5

acceptance regions 361–2
acute angles 123–4
adding vectors 220, 225
addition rule, probability 340, 343, 348
air resistance 281
algebra 5–70
- circles 55–9
- division 73–6, 94–7
- index laws 6–9, 36–9
- inequalities 60–5
- proof 32–5
- quadratic functions 13–15, 44–9
- simultaneous equations 11–12, 29, 50–3
- surds 6–9, 40–3

algebraic division 73–6, 94–7
- factor theorem 95, 96
- methods 73–6
- remainders 94–5

alternative hypotheses 356–7, 360–9
ambiguous case, sine rule 128
angles
- in quadrilateral 34
- in semi-circle 58
- trigonometry 110–11, 122–31

approximations, gradient of curves 138
areas
- area of triangle 120, 129, 130
- histogram bars 306–7, 322–3
- normal/tangent/axis 159
- under curves 172–5
- under graphs 243, 246

argument and proof 32–5
asymptotes 80–2, 98, 102, 187–8, 195
average speed/velocity 235, 242, 243
averages *see* mean; median; mode
axioms 32

B

balance method, solving linear equations 10–12
base of logarithms 186, 190
base of terms 36
bearings 221
best fit lines 204, 206, 330
biased sampling 310
binomial distribution 346–9
- conditions for 346, 348
- cumulative probabilities 347
- distribution function 347
- hypothesis testing 361, 368
- ^nC_r notation 347
- ^nC_x notation 346
- shape of distribution 347

binomial theorem 71, 88–93
- constructing an expansion 92
- general expansion 88–93
- Pascal's triangle 71, 88–9, 92

bisection of chords 56
bivariate data 328–31
box-and-whisker plots 320, 325, 326
brackets, expanding 72, 84–7, 88

C

calculus 137
- *see also* differentiation; integration

CAST (cos/all/sin/tan) mnemonic 124
causation/causal connections 330
central tendency measures 312–19
- *see also* mean; median; mode

choose function 89–90
chords of circles 29–30, 56
chords of curves 138–9, 142
circles
- equations of 27–31, 55–9
- theorems 56, 58
- trigonometric ratios 123–4

cluster sampling 309
coded data 300
coefficients
- algebraic division 94–5
- binomial theorem 88–9
- comparing 74–5
- correlation 328
- index laws 36
- Pascal's triangle 88–9
- polynomials 74–5, 84–5, 88–9, 94–5

collecting data 297–338
collinear points 220, 222
column vectors 224
combinations 89, 92
comparing coefficients of polynomials 74–5
complementary events 340
completing the square 16–17, 45–6, 48
constant acceleration 248–51
constant of integration 168, 173
constant velocity 266–7
constants in differentiation 143
continuity correction 322
continuous data 312–22
coordinate geometry 138–41
correlation 328–31
cosine (cos) 122–31
- CAST mnemonic 124
- cosine ratio 110, 122–7
- cosine rule 118, 119–20, 128–31, 221
- graphs 112, 114, 115, 123
- quadrant diagrams 124
- use of unit circle 123–4

counter examples, disproof 33, 34
critical regions 361–2, 364, 366–9
critical values 361–2, 364, 366
cubic graphs 77–8, 98
cumulative frequency 321, 322, 325
cumulative probabilities 346
cumulative probability 346–7
curve sketching 98–103, 165
curves
- area under 172–5
- fitting 144–7, 204–7
- gradient at given point 138–43
- intersecting straight lines 51
- normals 156–61
- tangents 156–61

D

data
- bivariate 328–31
- central tendency/averages 298–301, 312–19
- coded 300
- collection 297–337
- continuous 312–22
- discrete 312–22
- distribution 322
- grouped 312, 322
- interpretation 297–337
- representation 297–337
- sampling 308–11
- single-variable 320–7
- spread 302–5, 312–19

deceleration 244
deductive proof 32
definite integrals 172–3
degree of polynomial 84, 94
denominators, rationalising 8, 40–1
dependent variables 328
derived functions (derivatives) 143, 147
- first derivatives 137, 150
- second derivatives 137, 150, 163, 165

difference of two squares 13
differentiation 137–84
- ax^n 146–9
- from first principles 142–5
- kinematics 252
- Leibniz notation 146–9, 172
- rates of change 137, 143, 150–5
- rules 143
- tangents to curves 156–61
- turning points 162–7
- *see also* integration

dimensions 240–1
direct proof 32–4
direction
- forces 266, 272
- vectors 219, 220–1, 224–5

discrete data 312–22
discrete random variables 339–53, 366
discriminants, quadratic formula 18–19, 30, 46–7, 51
dispersion (spread) 302–5, 312, 313–14
- *see also* interquartile range; range; standard deviation

displacement
- definition 236, 242
- differentiation 137
- integration 168
- motion equations 248–51

resultant 242, 243
units 240–1
v-t graphs 238, 243–4
displacement-time (s-t) graphs 237, 242–3, 246, 252
disproof, by counter example 33, 34
distance
between two points 54
definition 242
distance-time graphs 150
see also displacement
distribution of data 322
distribution function 347
division, algebraic 73–6, 94–7
divisors 75
dots above variables, symbol 252
dy/dx notation 146, 150–1
dynamics 272–5

E

e (irrational number) 194
elimination method, simultaneous equations 11, 50
equal and opposite forces 280
equal vectors 220, 226
equally likely outcomes 340–1
equations
circles 27–31, 55–6
exponentials 194–9
motion 248–51
simultaneous 11–12, 29, 50–3
straight lines 54, 156, 198
equilibrium 261, 262, 266–8
equivalent angles 124
equivalent fractions 40
equivalent statements 190–1
events, definition 340
exhaustion, proof by 32–3
exhaustive events 340
expanding brackets 72, 84–7, 88
binomial expansion 88–93
experimental data, lines of best fit 206
the exponential function (e^x) 194
exponential relationship 204–7
exponentials 185–218
conversion to logarithms 186
curve fitting 204–7
functions 187–9, 194–9
general equation 194–9
graphs 187–9, 194–5
logarithms 185–93
processes 200–3
exponents *see* index laws; powers

F

factor theorem 95, 96
factorials 89
factors/factorising 13–15, 84–7, 96
first derivatives 137, 150
five-number summary 318–20
flipping coins, hypothesis testing 356–9
force = mass × acceleration ($F = ma$) 240, 272–4, 276–8, 282
forces 261–92
acceleration 262–5
diagrams 262–5
dynamics 272–5
frictional 267–8, 273, 277
gravity 240, 266, 276–9
man in lift 280–1
notation 266
systems 280–5
fractions, equivalent 40
frequency density 306, 322
frequency distributions 312
frequency of outcomes 322
frictional forces 267–8, 273, 277
functions
derived 137, 143, 147, 150, 163, 165
exponential/logarithmic 187–9, 194–9
increasing/decreasing 151
periodic 109, 115, 123
quadratic 13–15, 44–9
sketching 165
stationary 151
fundamental quantities 240–1

G

gradient functions 138–40, 151
gradients
curves 138–40, 142–3, 145
differentiation 142–5, 150–1
increasing/decreasing 162–3
normals 156
perpendicular 156
rates of change 150–1
s-t graphs 237, 242–3
straight lines 54–5, 58, 59
tangents 146, 156
turning points 162
v-t graphs 238, 243–4, 246
velocity 150, 242–4, 246
$y = Ae^{kt}$ 200
$y = e^x$ 194
graphs
areas under 172–5, 243, 245–9
cubic/quartic/reciprocal 77–83
curve sketching 98–103
displacement-time 237, 242–3, 246, 252
distance-time 150
exponential and logarithmic functions 187–9, 195
intercepting axes 98
quadratic equations 44–7
s-t 242–3, 246, 252
simultaneous equations 50, 51
straight lines 20–6, 54–5
trigonometric functions 111–15, 123, 126
v-t 238, 243–4, 246, 252
$y = a^x$ 194
$y = e^x$ 194
gravity 276–9
forces 266
motion under 276–9
units 240
greater than/greater than or equal to 60
grouped data 312–22
growth, exponential 185

H

highest common factor 73
histograms 306–7, 322–3
hypothesis testing 355–73
alternative hypotheses 356, 360–9
conclusions 361, 364
critical regions 361–2, 364, 366–9
formulating tests 360–5
null hypotheses 356–7, 360–9

I

identical trials 346–7, 348
identities 84, 86, 122
independent events 341, 343
independent trials 346–7, 348
independent variables 328
index form 6, 36
index laws 6–9, 36–9, 186, 190
see also powers
index notation, conversion to logarithmic 192
inequalities 60–5
linear 60–1
quadratic 62
signs 60, 62
infinity 98, 165
integration 137, 168–75
area under curve 172–5
ax^n 169
kinematics 252–4
see also differentiation
intercepts 98
interpolation, linear 300, 302
interpreting data 297–337
interquartile range (IQR) 302, 313
intersections of circles 29–30
intervals, data 312
inverse functions, exponentials and logarithms 186, 188, 195–6, 198
IQR *see* interquartile range
irrational numbers 8, 40, 194

K

kinematics 236–9, 240, 242–55
constant acceleration 248–51
motion equations 236, 248–51
straight line motion 242–7
variable acceleration 252–5

L

laws of indices 6–9, 36–9, 186, 190
laws of logarithms 190–3
Leibniz, Gottfried 146
Leibniz notation 146–9, 172
less than/less than or equal to 60
like terms, simplifying expressions 8, 74, 84–5
limiting values (limits) 143, 172, 202
line graphs 20–6
linear equations 10–12, 54, 156, 198
see also simultaneous equations
linear inequalities 60–1
linear interpolation 300, 302
lines 54–9
of best fit 204, 206, 330
distance between two points 54
midpoints 54, 55, 58
number 60
parallel 55
perpendicular 55, 56, 58, 59
see also straight lines
logarithmic functions 187, 195
logarithmic notation, conversion to index notations 192
logarithms 185–93

graphs 187–9, 195
laws 190–3
lines of best fit 204, 206
natural/Naperian 195, 202
three cases 190
$y = ax^n$ 204
$y = e^x$ 195
$y = kb^x$ 204, 206
long division, algebraic 73–4, 94–7
lower limits 172
lower quartiles 302, 313, 320–1

M

magnitude 219, 220–1, 224–5
magnitude of force 266
mass 240, 276
mathematical models 200, 202
maximum turning points 44, 138, 162–3
maximum values 320
mean 298–301, 312–19
measures of central tendency (averages) 312–19
measures of dispersion (spread) 302–5, 312–19
median 298–301, 312–19, 320–1
midpoints of lines 54, 55, 58
minimum turning points 44, 138, 162–3
minimum values 320
modal interval/group 312
mode 298–9, 312
motion
equations 248–51
s-t graphs 242–3, 246, 252
straight line 242–7
v-t graphs 238, 243–4, 246, 252
see also kinematics
multiplication
probability rule 341, 343
vectors by scalars 220, 225
mutually exclusive events 340, 343–4, 348

N

N (newtons) 240, 266
natural (Naperian) logarithms 195, 202
negative acceleration (deceleration) 244
negative correlation 328
Newton, Isaac 146
Newton's laws 261–92
first law 266
second law 272, 276
third law 280–2
newtons (N) 240, 266
normal reaction force 266
normals to a curve 156–61
null hypotheses 356, 360–9
number lines 60

O

one-tailed hypothesis tests 360–1, 366
opportunity sampling 309
optimisation, differentiation 166
outliers 317, 325, 330

P

p-values 366–7
parabolas 14, 44, 173
parallel lines 23–4, 55
parallel vectors 225, 228
parameters 308
Pascal's triangle 71, 88–9, 92
perfect correlation 328
perfect square quadratics 16, 45
periodic functions 109, 115, 123
perpendicular bisectors 25, 55
perpendicular gradients 156
perpendicular lines 23–4, 55, 56, 58, 59
points
collinear 220, 222
distance between two 54
gradient at point on curve 138–40, 142–3
midpoints of lines 54, 58
see also stationary points; turning points
polynomial relationship 204–7
polynomials 71–108
algebraic division 94–7
binomial theorem 71, 88–93
curve sketching 98–103
differentiation 148
division 74–5
expanding brackets 72, 84–7, 88
factorising 84–7, 96
gradient functions 138–40
graphs 77–80
simplifying 84–5
populations 308
position vectors 226, 242
positive correlation 328–9
powers
logarithms 190–1
polynomials 84
see also index laws
prime numbers 33, 35
probability 339–53
addition rule 340, 343, 348
binomial distribution 346–9
discrete random variables 339–53
distributions 341, 346–9
equally likely outcomes 340–1
hypothesis testing 364, 366–7
multiplication rule 341, 343
probability spaces 340
probability distribution function 342
proof 32–5
by exhaustion 32–3, 35
direct 32–4
disproof by counter example 33, 34
statements 34–5
types of 32–3
proportional relationship graphs 102
Pythagoras' theorem 54, 111, 122, 270, 284

Q

quadrant diagrams 124
quadratic formula 17–19, 44, 46
quadratic functions 13–15, 44–9
completing the square 16–17, 45–6, 48
curves 14–15, 44
expanding and factorising 84–5
factorisation 13–15, 45
inequalities 62
quadrilaterals 34
quartics, graphs 79–80
quartiles 302, 313, 320–1
quota sampling 309, 310
quotients, algebraic division 75, 94, 96

R

random experiments 340
random variables
binomial probability distribution 346
discrete 339–53, 366
range 302, 313
rates of change 137, 143, 150–5
exponential processes 200–2
integration 170
$y = e^{kt}$ 200, 202
rational numbers 40
rationalising the denominator 8, 40–1
ratios, trigonometric 110, 122–7
raw data 312
rearranging formulae 10–12
reciprocal graphs 80–2, 98
regions 61, 62
relationships, exponential and polynomial 204–7
remainders 94–5
repeated roots, polynomials 78
representing data 297–337
resolving forces 266–7, 270, 278
resolving vectors 224
resultant displacement 242, 243
resultant forces 262, 269–70
resultant vectors 220–1
right-angled triangles 110–11, 122–7
roots of numbers 36, 40–3
see also surds
roots of polynomials 45, 78
rounding up/down 319

S

s-t (displacement-time) graphs 237, 242–3, 246, 252
sample spaces 340
sampling 308–11
scalars 220, 225, 240, 242
scatter diagrams 328, 330
second derivatives 137, 150, 163, 165
semi-circles 58
set notation 60
SI units (Système International d'Unités) 240
significance levels 356–7, 361, 366
significant figures 276
simple random sampling 308
simplifying fractions 73
simultaneous equations 11–12, 29, 50–3
sine rule 118–19, 128–31
ambiguous case 128
area of triangle 120, 129, 130
vectors 221
sine (sin) 122–31
CAST mnemonic 124
graphs 111, 112, 114, 115, 123
quadrant diagrams 124
sine ratio 110, 122–7
sine rule 118–19, 128–31
use of unit circle 123–4
single-variable data 320–7
continuous data 312–22
discrete data 312–22
five-number summary 318–20
speed
average 242, 243
definition 242

kinematics 235, 236
see also velocity
spread 302–5, 312–17
spurious correlation 330
standard deviation 302–5, 313, 314
standard units 240–1
statements, proving 34–5
stationary functions 151
stationary points 162
see also turning points
statistics
definition 308
summary 298–301, 312–19, 320
test 361
straight lines
equations 54, 156, 198
gradients 54–5, 58, 59
graphs 20–6
intersecting curves 51
motion in 242–7
stratified sampling 308
stretching graphs 99, 194
substitution method, simultaneous equations 11–12, 29, 50–1, 52
subtracting vectors 220, 225
sum of, symbol 312
sum rule 168–9
summary statistics 298–305, 312–19
surds 6–9, 40–3
suvat equations 248
symmetry 123
systematic sampling 308
Système International d'Unités (SI units) 240
systems of forces 280–5

T

tangent (tan) 122–7
CAST mnemonic 124
graphs 112, 113, 115, 123
quadrant diagrams 124
tangent ratio 122–7
use of unit circle 123–4
tangents to curves 156–61
area calculations 159
circles 28–9, 30, 55, 56
gradients 146, 156
where tangent meets curve 158
$y = e^x$ 194, 198
tension 267, 277, 284
test statistics 361
testing hypotheses 355–73
thrust 267
time 64
see also displacement-time graphs; distance-time graphs; velocity-time graphs
transformations 99–100, 194
tree diagrams 346
triangles
areas 120, 129, 130
non-right-angled 118–21
Pascal's 71, 88–9, 92
right-angled 110–11, 122–7
see also trigonometry
triangulation 109
trigonometry 109–36
cosine rule 128–31, 221
equation solving using graphs 112–14
function graphs 111–16
non-right-angled triangles 118–21
sin/tan/cos ratios 110, 122–7
sine rule 128–31, 221
turning points 162–7
maximum/minimum 44, 138, 162–3
optimising a given situation 166
quadratic equations 44
sketching a function 165
stationary points distinction 162
two-tailed hypothesis tests 361, 368

U

unbiased estimate 314
unit vectors 220
units 235–41
dimensions 240–1
kinematics 240–1
standard 240–1
upper limits 172
upper quartiles 302, 313, 320

V

v-t (velocity-time) graphs 238, 243–4, 246, 252
variable acceleration 252–5
variables
binomial distribution 346–8
correlated 328
dependent 328
discrete random 339–53, 366
independent 328
variance 313–14
vectors 219–34
definitions 220–3
kinematics 240, 242
properties 220–3
x/y components 224–9
velocity
average 242, 243
constant 266–7
definition 237, 242
differentiation 137, 150, 252
forces 266–7
integration 168, 170, 252–4
motion equations 248–51
rates of change 150
s-t graphs 243
units 240–1
velocity-time (*v-t*) graphs 238, 243–4, 246, 252
vertex of a parabola 44, 45

W

weight
forces 266, 276, 278
units 240
see also gravity

X

x-components of vectors 224–9

Y

y-components of vectors 224–9

Z

zero correlation 328
zero vectors 220, 221